American Romanian Academy of Arts and Sciences

MIRCEA IFRIM
LIANA MOȘ
CRIS PRECUP

CSONGOR TOTH
CĂLIN POPA

ZSOLT GYORI
DAN GOLDIȘ
CASIANA STĂNESCU

HUMAN ATLAS OF TOPOGRAPHICAL, FUNCTIONAL AND CLINICAL ANATOMY VISCERA

ISBN: 978-1-935924-20-3

ARA Publisher: http://www.AmericanRomanianAcademy.org

Printed in the United States

PREFACE

This work can be considered a work of art, because it is treating the most complex creature in the universe, the human being.

Analyzing this fine worked atlas, I found that it is structured in six chapters, the first four refers to the topographic anatomy of the cephalic extremity (nose area, mouth, throat area, parotid and submandibular region, laryngeal region, tyro-trachea-esophageal region), chest (wall chest, thoracic cavity and mediastinum), abdomen (range and wall, abdominal cavity, supramezocolic floor, inframezocolic floor, retroperitoneal space) and pelvis (pelvic peritoneal, internal male genital organs, internal female genital organs, female urethra, external genital organs of men and women), and the last two chapters refers to a synthesis of medical and surgical applications in the regions and structures described, live anatomy and endoscopy notions where it can be systematize basic concepts, useful and absolutely necessary for medical practice work of students and specialists.

Anatomical regions and organs described clearly and fluently in English, are presented in topographic and functional aspect, together with great accuracy charts, which are made with high scientific rigor, highlights the importance of anatomical elements, useful for training medical thinking students with certain application in medical practice.

Different topographic regions are dealt differently with their specific features, giving larger clinical anatomical and functional aspects. Text is clear and concise, well structured, focused on what needs to be known by medical doctor and practitioner, and the images have exceptional quality accompanying anatomical descriptions completing descriptive terms.

The entire atlas by its content is well organized, that demonstrates successful restructuring of classical notions of anatomy in relation to the massive increased volume of scientific information, through judicious selection of the essential elements that must be learned by students and physicians.

Correlated and very well knitted together in the last two chapters of the atlas are the concepts of clinical anatomy, endoscopy, radiological anatomy, surgical and ·medical applications, through systematic and anatomical representation of useful concepts and procedures for clinical functional and surgical exploration, exemplified gradually by clear and suggestive plates, considerably enhancing their value with practical application. This aspect gives a clearly certain character of originality and competitiveness.

In conclusion, I express my conviction that the "Human Atlas of Topographical, Functional and Clinical Anatomy - Viscera" fully meets all the criteria for publication of scientific works.

Professor Doctor Alexandru T. Ispas,
Head of the Anatomy and Embryology Department from
University of Medicine and Pharmacy "Carol Davila", Bucharest, Romania

SUMMARY

Chapter 1
The Cephalic Extremity

Chapter 2
The Thorax

Chapter 3
The Abdomen

Chapter 4
The Pelvis

Chapter 5
Some Medicosurgical Applications

Chapter 6
In Vivo Anatomy and Endoscopy Notions

The Cephalic Extremity

Chapter 1

Preliminaries

The cephalic extremity consists of two components: the head (cephalaeum) and the neck (cervix), the boundary between them being formed by a line which starts from the external occipital protuberance (inion), advances laterally, on either side, along the superior occipital lines, up to the mastoid processes, surrounds them, arrives below the external acoustic meatus and passes to the posterior border of the mandibular ramus, along which it descends up to its angle (gonion) and, advancing then on its lower border, reaches the mental protuberance (gnathion).

The inferior boundary of the neck, which separates it from the trunk, is formed by the line which starts from the spinous process (prominence) of the seventh cervical vertebra, advances laterally, reaches on either side the respective acromion, then passes along the anterior border of the clavicles, anteriorly and medially, up to the upper border of the manubrium of sternum, where it forms the inferior boundary of the jugular fossa.

The organs forming the cephalic portion of the "matter import apparatus"(Rainer), i.e. of the digestive and respiratory systems, were improperly also called "viscera". The thoracic and abdominal components of the gastropulmonary system were called "viscera"by the old anatomists, because they are glossy (viscous), moist and situated in serous cavities (pleura, pericardium, peritoneum), from which they can be rather easily removed (eviscerated) by opening them through a simple incision, sufficiently deep to pierce the whole depth of the walls covering them. The term "viscera" used for the cephalic organs of the gastropulmonary apparatus is improper, since they are not situated in a serous cavity, hence they are not covered by a glossy, moist, serous membrane, but they are arranged so that, to be reached, they should be dissected, carved out of the regions harbouring them.

These regions are median, both in the cephalic extremity proper, i.e. in the head, and in the anterior region of the neck, conferring their names to the respective areas.

Thus, at the level of the head, in the middle of the face, lies the nasal region, formed by the external bone and the nasal fossae; below it the buccal or oral region is situated, with the upper lip and the lower lip, and the mental region (the chin) (fig.1). In the anterior region of the neck lie in continuity the submental, the hyoid, the thyroid and the cricoid regions (making up together the laryngeal region), as well as the tracheal region up to the jugular notch (incisura jugularis), in the depth of the jugular fossa (see vol. I). The lateral boundaries of the organs of the digestive-respiratory tract, harboured in the anterior region of the neck, consist of two conventional vertical lines, traced from the sternoclavicular joints to the lower border of the mandible.

However, the cephalic segment of the digestive system also contains paired organs, which are situated on either side of the median line. These are the two pairs of large salivary glands: the submandibular glands and the parotid glands, which confer their name to the respective region. The submandibular glands lie below the mandible, on either side of the submental region, and the parotid glands, below the external acoustic meatuses, between the sternocleidomastoid muscle and the posterior edge of the mandibular ramus. Between the parotid region and the oral region lies the masseteric region, above which the parotid sends its anterior projection with the Stensen's duct.

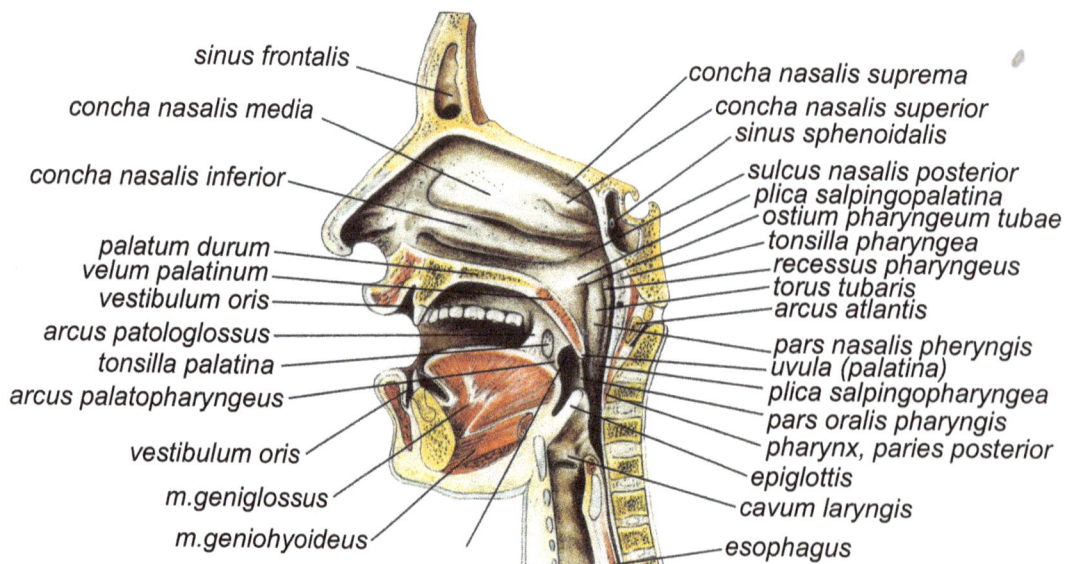

Fig. 1. The cephalic extremity (sagittal section).

The Nasal Region
(Regio nasalis)

The nasal region has the following boundaries: above, a horizontal line which passes through the nasion (the point where the nasofrontal suture and the mediosagittal plane meet); below, a horizontal line which passes through the subnasal point (a point situated in the angle between the inferior border of the nasal septum and the upper lip); laterally, the nasogenian and nasopalpebral grooves, on either side .It has the shape of a triangular pyramid, with an anterior edge - the back of the nose (dorsum nasi) - which extends from the root of the nose, at the point called nasion, to the tip of the nose termed apex, and which separates the two anterolateral surfaces of the nasal pyramid, extending up to the nasogenian and nasopalpebral grooves, which make up the other two edges of the pyramid. The third side of the nasal pyramid is deep and continuous with the nasal fossae. The base of the pyramid is occupied by the two nostrils (nares).

It has an external, cutaneous surface and an internal, cutaneomucous surface, continuous with the respiratory mucosa of the nasal fossae. Between these surfaces lie the other planes of the region:
- the subcutaneous connective-fatty tissue;
- the muscular plane and
- the osteocartilaginous skeleton.

The nasal region is prominent and exhibits a great variability in shape and size. The lower part of the anterolateral surfaces is mobile, forming the wings of the nose (alae nasi). From the apex of the nose courses posteriorly a cutaneous fold. The nasal septum and the free borders of the alae nasi delineate the two orifices: the nostrils; the most lateral parts of the nostrils form the alar points.

The height of the nose is measured between the nasion and the punctum subnasale (subnasal point) and the width, between the two alar points. The ratio between the height and the width of the nose gives the nasal index, the distribution of which on the globe has an adaptive character and, according to Thomson - Buxton's law, increases from the cold towards the warm areas (fig .1-6).
Structure. The skin, thin and mobile in the upper part, becomes thick and adherent in the lower part of the region. It is well supplied by blood vessels and furnished with numerous sebaceous glands. In the upper part it may be easily detached; in the lower half it adheres firmly to the subjacent fibrocartilaginous plane. The subcutaneous connective-fatty tissue is poorly represented, being present especially towards the superior part of the region and lacking at the level of the wings and of the apex of the nose.

The muscular plane is formed by four platysma muscles:
- the procerus muscle (or pyramidal muscle of the nose) is inserted into the lateral cartilages of the wing of the nose and into the nasal bones and ends on the integument of the region between the eyebrows;
- the nasal muscle (musculus nasalis) consists of two regions: pars transversa or transverse muscle of the nose and pars alaris or dilator muscle of the nostrils. The transverse part is inserted into the dorsal aponeurosis of the nose and ends on the deep surface of the nasogenian groove. The alar part has its origin on the deep surface of the nasogenian groove and ends in the integument of the lower border of the nostrils;
- the levator muscle of the upper lip and of the wing of the nose (m. levator labii superioris alaeque nasi) is inserted into the lateral surface of the frontal process of the maxilla and on the nasal process of the frontal bone (the internal orbital apophysis) and is distributed to the integument of the wing of the nose and of the upper lip;
- the depressor muscle of the septum (m. myrtiformis) arises from the myrtiform fossa of the maxilla and from the canine fossa and is inserted into the integument of the mobile part of the nasal septum (fig. 5). The osteocartilaginous skeleton. The nasal region is supported by an osteocartilaginous skeleton. The bony portion of this skeleton is formed by the frontal processes of the maxillary bones, the nasal bones proper, the anterior nasal spine and the palatine processes of the maxillae which circumscribe the pyriform aperture. Alongside the frontal processes of the maxillae, the mastication pressure of the anterior teeth is transmitted towards the skull cap (calvaria), forming an area of bone condensation: the nasofrontal resistance pillars. Towards these pillars converge also the infraorbital lines of the zygomatic pillars, alongside which the pressure is transmitted from the first molars.

The sum of these pressures is transmitted, through the nasofrontal pillars, to the frontal bone and to the nasal bones, determining on the skull the anterior, fronto-ethmoid resistance pillar, formed by the ethmoid and the metopic (interfrontal) suture. Thus, the anterior part of the nasal fossae is enclosed by two resistance pillars, which protect it and serve at the same time as a support to the nasal region.

The nasal bones, two quadrilateral plates of compact bony tissue, join with each other on the median line, forming a vault based on the anterior edge of the frontal processes of the maxillary bones and above on the nasal spine of the frontal bone.

Medially and posteriorly, this vault is articulated with the perpendicular plate of the ethmoid, reinforcing the rather fragile,median supporting system of the nasal fossae (the perpendicular plate and the vomer). The anterior edge of the palatine processes supports the posterior border of the orifices of the nostrils and on the nasal spine is attached the nasal septum.

The skeleton of the lower half of the nose is cartilaginous, formed by the lateral nasal cartilages (cartilago nasi lateralis), two triangular plates situated on each side of the median line on the lateral surfaces of the nose, between the bones proper and the wings of the nose (alae nasi) (fig. 2).

The alar cartilages (cartilago alaris major), two in number, situated below the lateral cartilages, enclose like a horseshoe the orifices of the nostrils, the medial branch participating in the formation of the mobile septum and the lateral branch, in the formation of the skeleton of alae nasi (fig. 3, 4).

The space left between these cartilages is filled by a fibrous membrane, in the depth of which may be found a variable number of cartilaginous nodes: the accessory nasal, lesser alar, sesamoid and vomerine cartilages. This membrane is continuous with the periosteum and the regional perichondrium.

To the shape of the nose contribute the contact angle between the bony and the fibrocartilaginous portion as well.

The cutaneomucous layer. At the level of the orifices of the nostrils, the skin rolls up inwards, towards the nasal fossae, and at the level of the lower border of the lateral cartilages it is continuous with the respiratory mucosa; here the skin adheres intimately to the perichondrium, which explains the intense pain caused by furuncles in this region.

Nasal cavity. Posteriorly, the external nasal region, on either side of the nasal septum, is continuous with the most anterior part of the nasal fossae. The inferior segment, lined with skin, has the shape of a channel 15 mm in length (the nasal vestibule). Its lateral wall corresponds to the ala nasi, and the medial wall, to the nasal septum.

The inferior aperture makes up the orifices of the nostrils. The superior orifice is situated at the level of the nasal threshold (limen nasi), i.e. at the passage between the cutaneous portion of the vestibule and the nasal mucosa; it has the shape of a triangular cleft. In the anterior part, the inferior border of the ala nasi is curved posteriorly, bounding a small cavity termed the recess of the apex. In the inferior part of the nasal vestibule, the skin is furnished with numerous sebaceous glands and hairs (vibrissae), which are longer and rougher in the aged.

The wings of the nose convey the air to the olfactory mucosa; their lack brings about a lowering of the sense of smell.

For the exploration of the anterior part of the nasal fossae are used the narinoscopy and the anterior rhinoscopy.

Vessel and nerve supply

The arteries of the nasal region are:

- the dorsal artery of the nose, a terminal branch of the ophthalmic artery, which is the terminal branch of the internal carotid artery; it is distributed to the root and the crest of the back of the nose and anastomoses with the angular artery, a terminal branch of the facial artery, from which may sometimes detach also the dorsal artery of the nose, of the wing of the nose and the artery of the lacrimal dome.

- the alar artery, a collateral branch of the facial artery.

The veins drain the blood to the facial vein through the alar vein and the angular vein. The venous facial system anastomoses at the level of the inner angle of the eye with the superior ophthalmic vein, afferent to the cavernous sinus, which explains the danger of a thrombophlebitis of this sinus by squeezing a furuncle in the nasal region.

The lymph vessels of the nasal region are grouped in:

- the superior group, afferent to the superior parotid and preauricular lymph nodes;
- the middle group, afferent to the inferior parotid lymph nodes;
- the inferior group, afferent to the submandibular lymph nodes.

The motor nerve supply is assured by the facial nerve.

10

The sensory nerve supply is assured by the trigeminal nerve through:

- the infratrochlear (external nasal) nerve, a terminal branch of the nasociliary nerve, which is the terminal branch of the ophthalmic nerve; it supplies the root of the nose;

- the infra-orbital (suborbital) nerve, a terminal branch of the maxillary nerve; it supplies the lateral parts of the nose;

- the external nasal branch (the nasolobar nerve), a branch derived from the anterior ethmoidal nerve (the internal nasal nerve); which is a terminal branch of the nasociliary nerve, which, in turn, is a terminal branch of the ophthalmic nerve. After its detachment, the external nasal branch courses on the dorsal surface of the bones proper of the nose up to the articulation of the bony portion with the cartilaginous portion of the nose, beyond which it passes and continues under the skin up to the apex of the nose; it supplies the antero- inferior part of the nasal region.

Behind the nasal pyramid (the external nose) are the nasal fossae.

The Nasal Fossae
(Cavum nasi)

The nasal fossae make up a system of anfractuous cavities, occupying the centre of the facial massif, divided into two halves by a medial septum (see vol. I).

They communicate with the pneumatic cavities or paranasal sinuses (sinus paranasales): frontal, maxillary, sphenoidal and ethmoidal. Anteriorly, they open outwards through two orifices: the nostrils or nares. Posteriorly they communicate with the nasopharynx through two large orifices termed choanae. The nasal fossae are the site of smell; by their ceiling they form the superior portion of the airways. The nasal fossae have four walls: the lateral wall, which is irregular, and a medial, a superior and an inferior wall, with a more simple structure.

The medial wall (septum nasi) consists of an osteocartilagi-nous skeleton, composed of the perpendicular plate of the ethmoid (above), the vomer (behind and below) and the septal cartilage (anteriorly). The septum is covered by a detachable mucous membrane. At 1.5 cm away from the posterior border of the nostril there may be sometimes found the small orifice of the Jacobson's vomeronasal organ (organum vomeronasale), a small vestigial canal lined with a mucosa for a length of a few millimeters. In the inferior part of the septum, the mucosa exhibits the "vascular area", a zone very richly supplied with blood vessels, which is the source of epistaxis. The superior wall, the arch of the nasal fossae, is an anteroposterior gutter, concave inferiorly and of only a few millimeters in width. It is formed, from before backwards, by the root of the nasal bones, the nasal spine of the frontal bone, the cribriform plate of the ethmoid bone (lamina cribrosa) and the body of the sphenoid bone. This wall separates the nasal fossae from the cranial cavity and represents a brittle point, exposed to breaking, owing to the thinness of the cribriform plate. The inferior wall, the floor of the nasal fossae, has also the shape of a gutter, but wider and shorter than that of the ceiling; it separates the nasal cavity from the oral cavity. It is formed by the palatine process of the maxilla in the anterior 3/4 and by the horizontal plate of the palatine bone in the posterior 1/4. The mucosa which covers it closes the incisive duct or anterior palatine duct (ductus incisivus).

The lateral wall of the nasal fossae is formed by the medial surface of the maxilla and the medial plate of the pterygoid process, on which articulate successively:

- the lacrimal bone (os lacrimale);
- the perpendicular plate of the palatine bone;
- the ethmoidal labyrinth (or latetal mass of the ethmoid);
- the inferior concha (concha nasalis inferior).

The ethmoid is the chief element in the formation of the lateral wall of the nasal fossae, as it is "the nasal bone by excellence" (Guerran). By applying on each other, these bones delineate two canals and an orifice:

- the nasolacrimal canal (canalis naso-lacrimalis), enclosed by the maxilla laterally, the lacrimal bone and the lacrimal process of the inferior concha medially; through it the orbita communicates with the nasal fossae and the tears are conveyed into the inferior meatus of the nose;

-the greater palatine canal or posterior palatine canal, between the maxilla and the palatine bone; it is isolated from the nasal fossa and opens into the palatine vault;

-the sphenopalatine foramen (foramen sphenopalatinumj is delineated by a notch bearing the same name and by the sphenoid bone; through it the pterygopalatine fossa communicates with the nasal cavity and it contains the vasculonervous hilum of the nasal fossae.

The conchae. The lateral wall of the nasal fossae has an irregular shape owing to the presence of the conchae. These are bony laminae, convex towards the lumen of the fossae and elongated.

The inferior concha (concha nasalis inferior) is an independent bone. It is the longest of the conchae; it crosses the orifice of the maxillary sinus, obturating it at the inferior part, and articulates with the turbinated crests of the maxillary and palatine bones and with the lacrimal bone,

The superior and the middle conchae (concha nasalis superior et media) are ethmoidal. They are fastened through their upper border to the inner wall of the ethmoidal labyrinth (lateral mass).

The middle concha forms a medial shell-valve-shaped projection, in proximity of the septum, and divides the nasal fossa into two areas: the upper, olfactory region (regio olfactoria) and the lower, respiratory region (regio respiratoria). It passes beyond the anterior and posterior ethmoid and fastens itself on the ethmoidal crests (or superior turbinals) of the maxilla and of the palatine bone. The superior concha is smaller, but has the same shape as the former; a fourth concha or supreme concha (Santorini) may also exist. Above lies the supraturbinal zone (or spheno- ethmoidal recess), in front of which the orifice of the sphenoidal sinus opens. Each of the conchae borders, with the corresponding part of the lateral wall of the respective nasal fossa, a space termed meatus (meatus nasi).

The meatuses. The superior meatus (meatus nasi superior), developed to a small extent and situated in the posterior part of the respective nasal fossa, exhibits the two or three orifices of the posterior ethmoidal cells (cellulae posteriores), which open into it.

The middle meatus (meatus nasi medius) contains in its lateral wall numerous formations:

-the ethmoidal bulla (bulla ethmoidalis) consisting of a swallow-nest-shaped ethmoidal cell, the superior wall of which is confounded with the lamina of origin of the middle concha;

-one or several orifices of the anterior and middle ethmoidal cells, below and in front of the bulla (in the groove behind the bulla);

-the nasal atrium or the orifice between the meatuses;

-the unciform process, a bony lamina in the shape of a yataghan blade, which detaches itself in the agger nasi region -another bulging formed by an ethmoidal cell -, crosses the orifice of the maxillary sinus and articulates downwards with the inferior concha and behind with the palatine bone;

-the semilunar hiatus (hiatus semilunaris) or the unci - bullar groove, bounded by the ethmoidal bulla and the unciform process;

-the ethmoidal infundibulum or the frontonasal canal is continuous with the semilunar hiatus, crossing the ethmoidal labyrinth; this is the high "chimney" on which the middle concha is formed.

The inferior meatus (meatus nasi inferior) exhibits anteriorly, alongside its upper border, the lower orifice of the nasolacrimal duct.

The hiatus of the maxillary sinus is obliterated downwards by the maxillary process of the inferior concha; the maxillary sinus opens into the middle meatus.

The nasal mucosa. The mucosa of the lateral wall of the nasal fossae is intimately adherent to the periosteum: it is a periosteal mucosa. It lines the bony bulgings and invaginates in the paranasal sinuses and the ethmoidal cells. The maxillary sinus is an exception to this rule, since the mucosa closes its two orifices and leaves only a communication posteriorly and superiorly, situated above the unciform process. The opening of all the orifices of the sinuses into the middle meatus is accounted for by the fact that these sinuses have as origin a unique embryonic pattern, in the respective nasal fossa, a pattern which secondarily is diverging. Topographically, we distinguish in the lateral wall of the nasal fossa an anterior smooth region, a posterior irregular region of conchae and a superior olfactory region.

Vessel and nerve supply (fig. 6).

The blood supply of the nasal fossae has as the main source the terminal branch of the maxillary artery, the sphenopalatine artery and, as accessory sources, branches of the ophthalmic and facial arteries. The sphenopalatine artery (a. sphaenopalatina) leaves the infratemporal fossa through the sphenopalatine foramen (the hilum of the nasal fossae) and divides into the posterior, lateral and medial nasal arteries (aa. nasales posteriores, laterales et septi).

The medial nasal artery (or artery of the septum) gives off a superior and a lateral nasal branches, for the region of the superior concha, then crosses obliquely the septum and anastomoses, at the level of the incisive duct, with the descending palatine artery, which has traversed from before backwards the velum palatinum (soft palate).

The lateral nasal arteries, sometimes united in a common trunk, are distributed to the conchae and to the inferior and middle meatuses.

12

The accessory arteries are:

-the anterior and posterior ethmoidal arteries (a. ethmoidales anterior et posterior), branches of the ophthalmic artery; they supply the olfactory region and the mucosa of the anterior preturbinal region, as well as the frontal sinus;

-the subseptal artery, a branch of the facial artery; it supplies the antero - inferior part of the septum.

The vascular zone is a mucous area situated in the antero-inferior part of the septum, where the mentioned arteries anastomose. Severe epistaxes are nearly always located behind the vascular area and arise from the internal nasal artery.

The veins are satellites of the arteries. They form two networks: a superficial mucous network and a deep, periosteal one. The intraosseous veins run directly towards the hilum and are sources of haemorrhages in resections of conchae. The venous blood drains through the posterior nasal veins to the pterygoid plexus, through the superior nasal veins to the cavernous sinus and through small venules to the facial vein.

The lymph vessels form rich networks in the nasal mucosa and drain, like the veins, in three directions: to the retropharyngeal nodes, to the deep cervical lymph nodes and, to a lesser extent, to the submandibular lymph nodes.

The sensory nerve supply is assured by the olfactory nerve, consisting of filaments which arise from the olfactory part of the mucosa and assemble on a surface of 2 square centimeters, on the upper surface of the superior concha and on the part of the septum situated at its level, it unites in fascicles the axons of the cells of the olfactory mucosa, disseminated in this mucosa, between the supporting cells. The bipolar olfactory cells enter in a synaptic contact with the mitral cells of the olfactory bulb after the passage of their axons through the foramina of the cribriform plate. From here the nerve influx reaches directly the rhinencephalon, without any thymic relay, hence through a chain of only two neurons.

The sensitive nerve supply is achieved through the terminal arborization of the sphenopalatine nerve, a branch of the maxillary nerve, the second branch of the trigeminal nerve. One distinguishes the superior nasal nerves (nn. nasales superiores) supplying the mucosa of the inferior and middle conchae, the nasopalatine nerve (n. nasopalatinus), which supplies the nasal septum, and filaments of the anterior and middle palatine nerves which supply the floor of the nasal fossae. Moreover, a group of vegetative fibres is included in the sphenopalatine nerve, arising from the pterygopalatine or sphenopalatine ganglion. The ganglion receives them from the nerve of the pterygoid canal or vidian nerve, through the sympathetic root of the internal pericarotid plexus and the petrosal nerves via the facial and the glossopharyngeal nerves.

The internal nasal or anterior ethmoidal nerve, a branch of the ophthalmic nerve, supplies the nostrils and the anterior part of the nose.

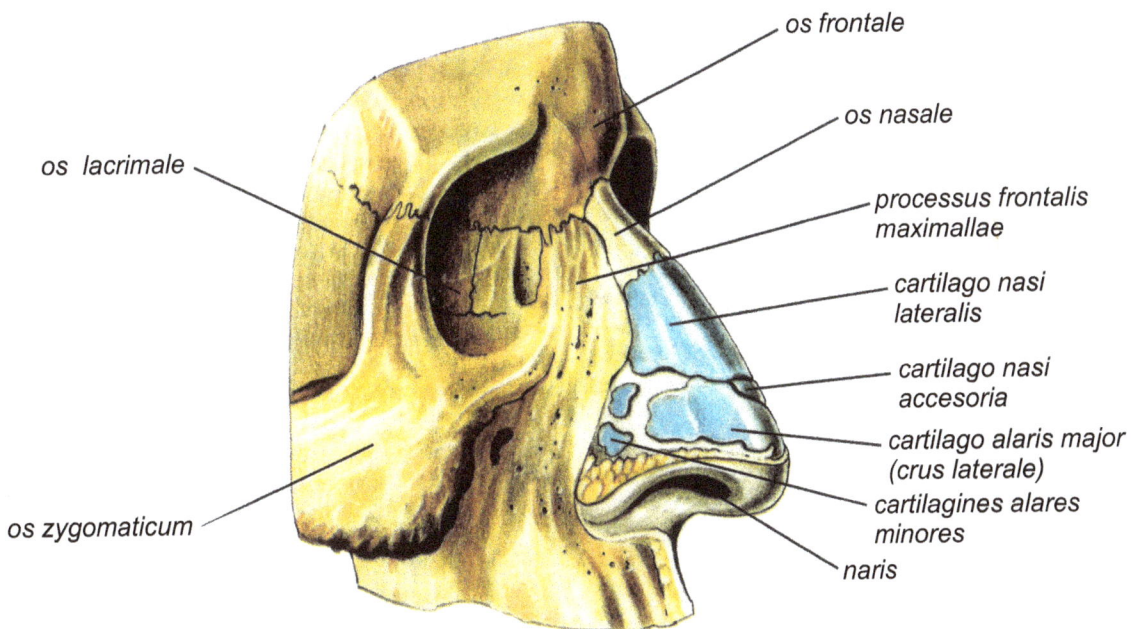

Fig. 2. The nasal cartilages (lateral view).

13

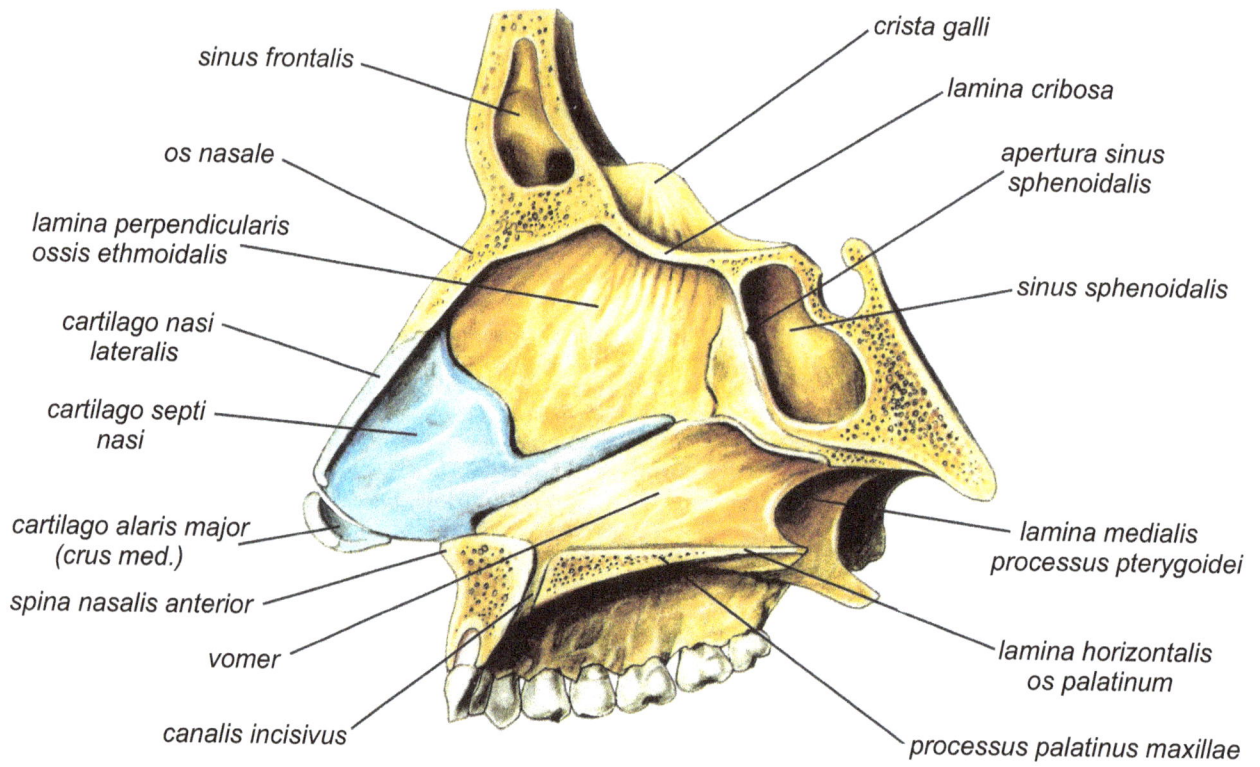

Fig. 3. The nasal cartilages (medial view).

sinus frontalis

crista galli

os nasale

lamina cribosa

lamina perpendicularis ossis ethmoidalis

apertura sinus sphenoidalis

cartilago nasi lateralis

sinus sphenoidalis

cartilago septi nasi

cartilago alaris major (crus med.)

lamina medialis processus pterygoidei

spina nasalis anterior

vomer

lamina horizontalis os palatinum

canalis incisivus

processus palatinus maxillae

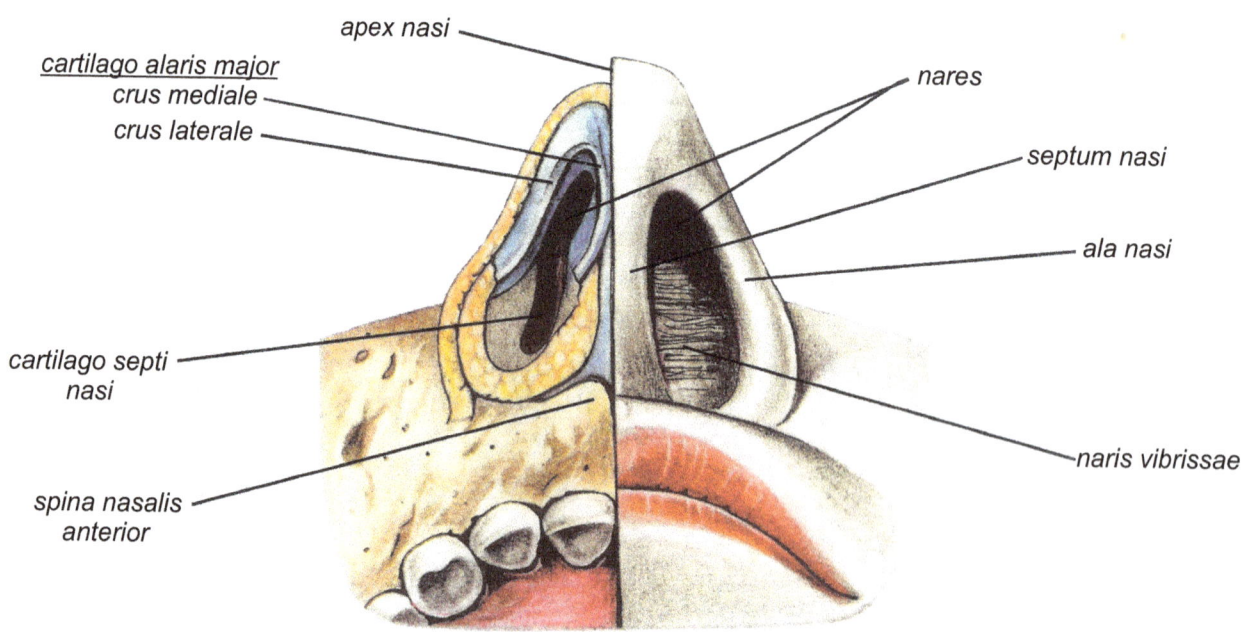

Fig. 4. The nasal cartilages (inferior view).

apex nasi

cartilago alaris major
crus mediale
crus laterale

nares

septum nasi

ala nasi

cartilago septi nasi

spina nasalis anterior

naris vibrissae

14

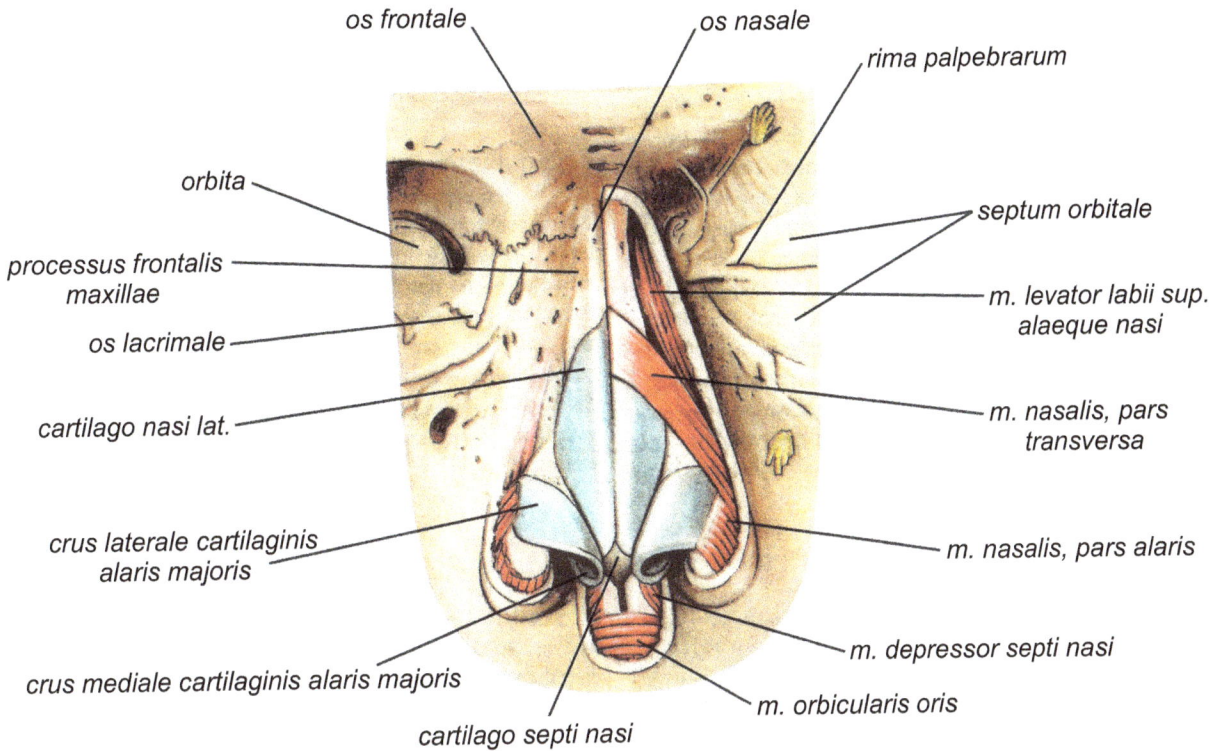

os frontale   **os nasale**

rima palpebrarum

orbita

septum orbitale

processus frontalis maxillae

os lacrimale

m. levator labii sup. alaeque nasi

cartilago nasi lat.

m. nasalis, pars transversa

crus laterale cartilaginis alaris majoris

m. nasalis, pars alaris

m. depressor septi nasi

crus mediale cartilaginis alaris majoris

m. orbicularis oris

cartilago septi nasi

Fig. 5. The nasal pyramid (skeleton and musculature).

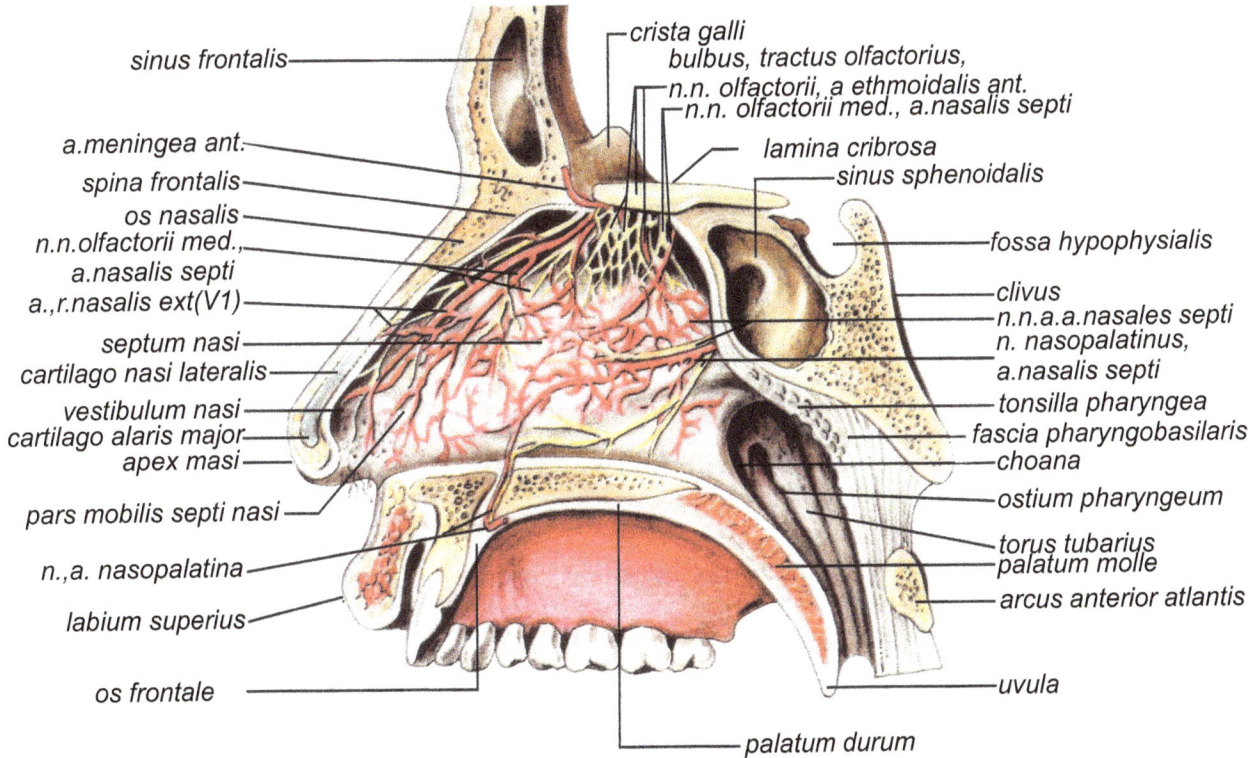

sinus frontalis

crista galli
bulbus, tractus olfactorius,
n.n. olfactorii, a ethmoidalis ant.
n.n. olfactorii med., a.nasalis septi

a.meningea ant.

lamina cribrosa

spina frontalis

sinus sphenoidalis

os nasalis

n.n.olfactorii med.,
a.nasalis septi

fossa hypophysialis

a.,r.nasalis ext(V1)

clivus

n.n.a.a.nasales septi

septum nasi

n. nasopalatinus,

cartilago nasi lateralis

a.nasalis septi

vestibulum nasi

tonsilla pharyngea

cartilago alaris major

fascia pharyngobasilaris

apex masi

choana

pars mobilis septi nasi

ostium pharyngeum

torus tubarius

n.,a. nasopalatina

palatum molle

labium superius

arcus anterior atlantis

os frontale

uvula

palatum durum

Fig. 6. Vesel and nerve suppy of the mucosa.

15

The Paranasal Sinuses
(Sinus paranasales)

The paranasal sinuses are pneumatic cavities annexed to the nasal fossae from which they arise and from which they receive air. They are distributed into four groups: ethmoidal, frontal, sphenoidal and maxillary and are often invaded by infections of nasal origin, causing sinusitides.

The Ethmoidal Cells
(Sinus ethmoidalis seu labyrinthus ethmoidalis)

The ethmoidal cells, 8-10 in number, make up an anfractuous system contained in the depth of the lateral masses of the ethmoidal bone. These cells open medially into the superior and middle meatuses of the nasal fossae and are clearly bounded laterally by the orbital plate of the ethmoid. They are in relation: superiorly with the dura mater and the brain (entailing numerous possibilities of complications), laterally with the orbit, posteriorly with the sphenoidal sinus, inferiorly with the maxillary sinus, anteriorly and superiorly with the frontal sinus, which may be considered as a large ethmoidofrontal cell.

The Frontal Sinus
(Sinus frontalis)

In the shape of a triangular pyramid, this sinus appears in childhood and has a variable development. Its anterior, thickened wall corresponds to the eyebrow region; its posterior, thinner wall corresponds to the meninges and the frontal lobe (the anterior pole of the brain). The medial wall makes up the intersinus boundary which separates the two frontal sinuses, always unequal; its inferior wall, or the base of the sinus, is in relation with the orbit and the ethmoid bone. The frontal sinus communicates with the middle meatus through a frontonasal duct, which opens into the infundibu-lum situated at the upper extremity of the semilunar hiatus. The frontal sinus may be catheterized through the middle meatus.

The Sphenoidal Sinus
(Sinus sphenoidalis)

Situated in the lateral half of the body of the sphenoid bone, the sphenoidal sinus is separated from its fellow sinus by a thin wall. It is in relation: superiorly with the sella turcica (Turkish saddle), which harbours the pituitary gland (hypophysis), posteriorly with the clivus, laterally with the cavernous sinus, which contains the internal carotid artery and the cranial nerves III, IV, V1, VI, and inferiorly with the pharynx. Its anterior wall corresponds to the nasopharynx and exhibits a small orifice, diaphragmated by a mucosa and concealed from view by the mass of the middle concha. The sinus can be small, middle-sized or large; in the latter case, if may send projections to the wings of the sphenoid, into the base of the pterygoid process, to the optic canal, to the maxillary sinus, as well as into the basilar part of the occipital bone.

The Maxillary Sinus
(Sinus maxilliaris seu antrum Highmori)

It is a cavity dug in the body of the maxillary bone, its walls being reduced to a mere bony plate. The shape of the sinus is that of the body of the maxilla, of a pyramid with the base medially situated.

The anterior wall corresponds to the canine fossa and is the "surgical" wall, the sinus being approached trough the gingivolabial groove; it is crossed in its inferior part by the superior alveolar plexus. The infratemporal (posterior and lateral) wall is in relation with the pterygopalatine fossa, in the thickness of which passes the posterior alveolar nerve (or posterior dental nerve). The orbital or superior wall makes up the floor of the orbit; it is traversed by the infraorbital groove, then by the infraorbital canal, which bulges info the sinus.

The nasal wall or the base of the pyramid presents a wide orifice, the maxillary hiatus, and corresponds to the middle and inferior meatuses; at the level of the middle meatus, the hiatus is crossed by the unciform process and is to a great extent, obliterated by the mucosa.

At the level of the inferior meatus, the orifice is obturated by the maxillary process of the inferior concha; this is the site of choice for the sinus puncture. The anterior border corresponds to the nasolacrimal duct. The inferior border is involved in pathology, it corresponds to the roots of the first two molar teeth and of the second premolar. The sinus cavity contains, normally, air. It is variable in shape and in size. A small sinus, of a capacity of less than 8 cube centimeters, can be either natural or the consequence of a chronic infection. A large sinus, of a capacity exceeding 15 cube centimeters, can send projections: above into the ascending branch of the maxilla, laterally up to the zygomatic bone, below into the alveolar border of the palatine vault, above and behind to the superior part of the palatine bone.

The Oral Cavity
(Cavum oris)

The oral cavity or the mouth (gr. stoma ="mouth") is the initial or facial segment of the digestive tract. Owing to its structures, it accomplishes the first phases of digestion and the quality control of foodstuffs: mastication, insalivation and deglutition of the food bolus; likewise, owing to its walls and organs, the sounds emitted by the larynx acquire the articulated and intelligible form of speech (fig. 7-27). The oral cavity consists of a skeleton formed by the two maxillaries, with regard to which the soft planes are disposed superficially or in depth, forming its walls.

The superficial walls are confounded with the superficial regions of the face: the lips, the chin and the masseteric region. The deep walls make up: above and inside, the palatine region; below, the floor, containing on the median line the tongue, and laterally the regions of the submandibular and sublingual glands. Posterolaterally, the oral cavity is in relation with two adjacent regions: the infratemporal region and the pterygomaxillary region which, through the vasculonervous elements composing it, assure to the oral cavity the greatest part of the blood vessel and nerve supply.

The oral cavity opens anteriorly through the oral fissure (rima oris) which, in the rest state of the lips, represents a transverse slit of 4-5 cm, bounded by the free margins of the lips. It communicates behind with the pharynx through a permanently open orifice, the pharyngeal vestibule, the isthmus (pharyngo-oral isthmus), in the shape of an arch open below and behind, bounded: above by the velum palatinum, laterally by the palatoglossal folds (anterior pillars) and below by the tongue, at the level of the junction of its base with its body. When the lower and upper dental arches are in an occlusal contact, the oral cavity is capillary, the mass of the tongue occupying its central part; it becomes true only by the introduction or presence of a solid, liquid or gaseous element, by retraction of the tongue or by removing the jaws away from each other; the distance between the jaws may attain values as high as 4-5 cm, measured between the edges of the opposing incisors.

The dental arches and the oral segment of the jaws divide the oral cavity into two parts:
- one outside the osteodental relief: the vestibule;
- the other inside this relief: the oral cavity proper.

These two parts communicate with each other through the interdental spaces and through a vertical space, situated behind the last molars, called retromolar space, which is present also in the position of occlusion of the dental arches.

The Oral Vestibule
(Vestibulum oris)

In the rest state of the oral cavity, the oral vestibule is a capillary space in the shape of a posteriorly open horseshoe, enclosed between the lips and the cheeks, outwards, and the dental arches covered by the gingival mucosa, inwards.

Of an average height of 4-5 cm, it ends in a cul-de-sac, forming the grooves of the superior and inferior infundibula (or the upper and lower fornices) of the vestibule. Each of the grooves of the vestibule is divided into two half - grooves, right and left, by a median vertical falciform fold, termed the bridle of the lip (frenulum labii superioris et inferioris).

The jugal portion of the vestibule exhibits, at the level of the upper second molar, the oral opening of Stensen's duct, the excretory duct of the parotid gland (papilla parotidea).

The vestibule communicates, behind the molars, with the oral cavity proper, its posterior boundary being made up in the depth by the pterygomaxillary ligament vertically situated, in the open mouth, medially to the coronoid fossa.

The Oral Cavity Proper
(Cavum oris proprium)

It is the posterior segment of the oral cavity, situated posteromedially to the dental arches; it has an ovoid shape, slightly flattened from above downwards, wider at the posterior extremity, its dimensions, like those of the adjacent osteodental relief, being the following: the anteroposterior diameter 7.5 cm, the vertical diameter 2.5 cm and the transverse diameter 4-5 cm between the lingual surfaces of the last molars and about 2.5 cm between the premolars.

The oral cavity is bounded above by the osseous palatine vault and lined by the mucosa which projects posteriorly on the musculomembranous wall of the velum palatinum (fig. 7-9).

Anterolaterally, it is bounded by the lingual part of the dental arches and by their osteogingival skeleton. Posteriorly and laterally it narrows towards the pharyngeal vestibule and lies in the depth between the anterior border of the ascending branch of the mandible, situated laterally, which bulges in the open oral cavity, and the vertical wall of the anterior projection of the pharynx, situated medially to the pterygopalatine region, which in this place may be approached through the mouth.

The inferior boundary is represented by the floor of the mouth. It is occupied, in the median part, by the tongue, the bulging mass of which conceals, in the lateral parts, the alveololingual grooves.

The alveololingual groove is an elongated triangular space, the posterior summit of which is bounded by the gingivo-alveolar wall, situated laterally to the implantation base of the tongue. It is occupied, in its anterior part, by the relief of the sublingual gland, which meets posteriorly the anterior projection of the submandibular gland, accompanied, on its turn, by the Wharton's duct. The glandular relief forms a sublingual fold, doubled sometimes by a crest-shaped fold, the sublingual crest.

In the medial part, at 1 cm behind its insertion, the lingual bridle (frenulum linguae) exhibits a small elevation, the salivary caruncula, into the tip of which opens the orifice of Wharton's duct (ductus submandibularis); slightly outside it, opens the orifice of the main duct of the sublingual gland, the Rivinus' duct. The secondary excretory orifices (in number of 10-12) of the sublingual gland open directly at the level of the mucosa, into the sublingual groove and are visible only by means of a magnifying glass.

The oral mucosa. The oral cavity is uniformly lined by a mucous membrane, which lacks only on the coronary portion of the dental arches.

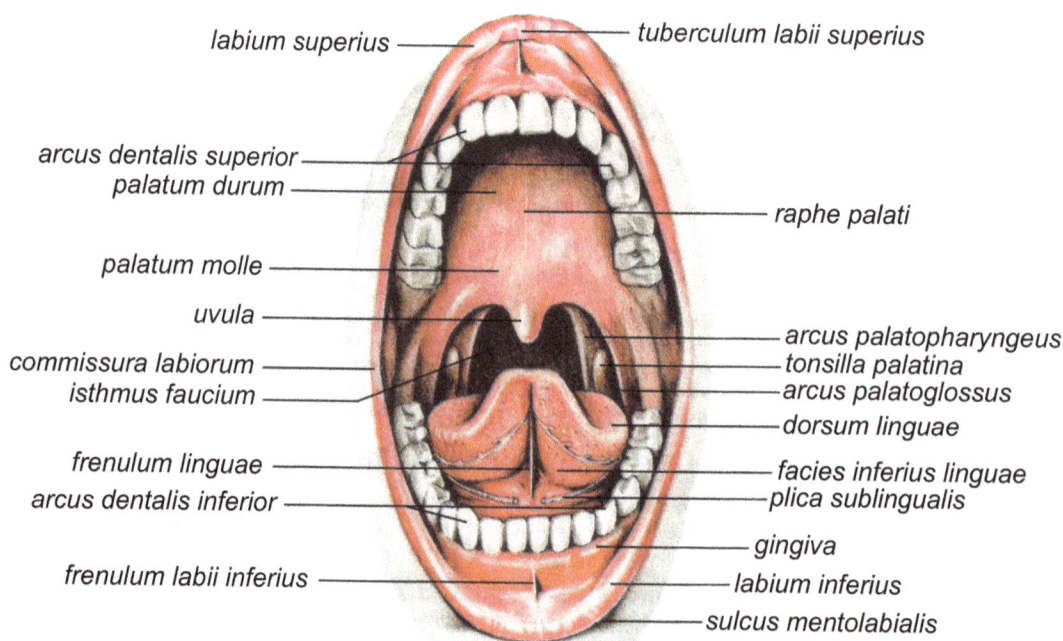

labium superius — tuberculum labii superius
arcus dentalis superior
palatum durum
raphe palati
palatum molle
uvula
arcus palatopharyngeus
commissura labiorum
tonsilla palatina
isthmus faucium
arcus palatoglossus
dorsum linguae
frenulum linguae
facies inferius linguae
arcus dentalis inferior
plica sublingualis
gingiva
frenulum labii inferius
labium inferius
sulcus mentolabialis

Fig. 7. The oral cavity - anterior view.

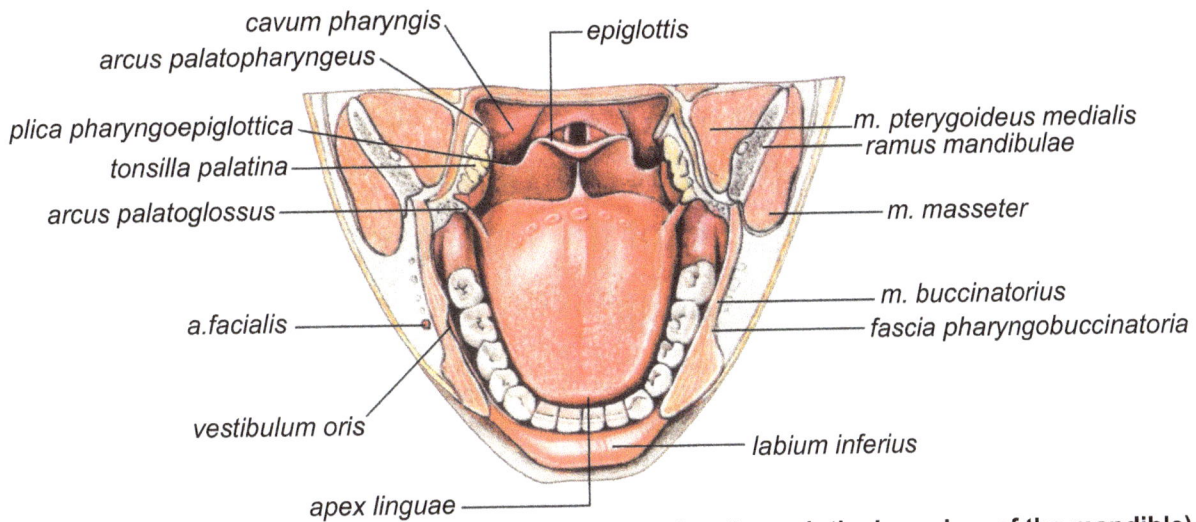

Fig. 8. The oral cavity (superior view with a horizontal section through the branches of the mandible).

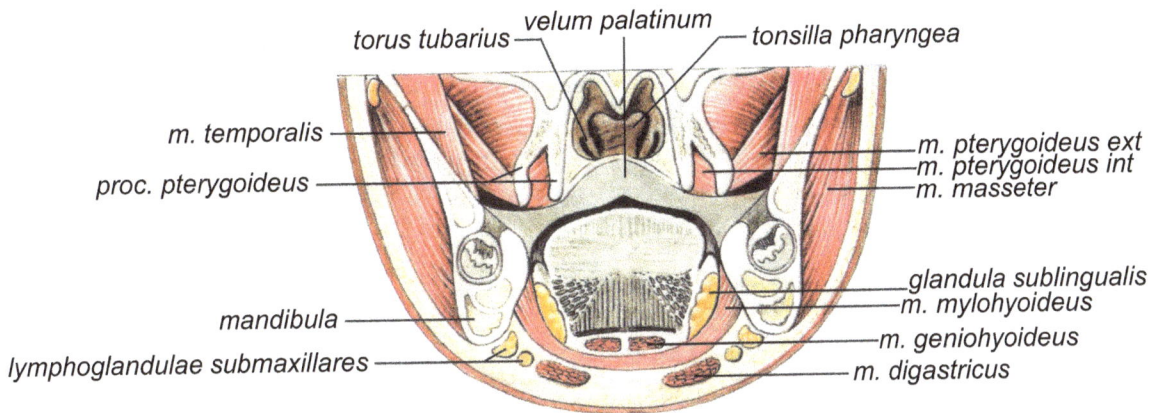

Fig. 9. The oral cavity - muscles of mastication (frontal section).

According to the level at which the examination is performed, this oral mucosa appears very diversified and functionally adapted to the various phases of predigestion of the food bolus.

The oral mucosa begins at the oral orifice, at the level of the cutaneomucous junction of the lips; it continues on the inner surface of the lips and cheeks and contains the seromucous salivary glands disseminated in its epithelial layer. It should be mentioned that this dissemination occurs with a specific concentration at the level of the zones where it is more functional, for example in the molar glands of the dental vestibule.

By its reflection on the alveolar wall, the oral mucosa adapts itself also to a new function, that of bearing pressures. Hence, it becomes denser, more resistant, less loose and richer in fibrous tissue, transforming itself into a fibromucosa or an alveolar gingiva (gum). The gum reflects itself, in a perfect continuity, from the vestibular part of the alveolar wall on the lingual portion, passing through the interdental spaces and forming a resistant sheath around the neck of the teeth, which overlaps slightly the anatomical dental neck. Here, it contributes to the formation of a small circumdental cul-de-sac, the depth of which is represented by the circular segment of the alveolodental ligament; consequently, normally no solution of continuity exists between the mucosa and the dental surface.

The alveolar gingiva of the lingual portion does not display the same characters at the level of the maxilla and of the mandible.

At the level of the maxilla, the mucosa which lines the palatine vault maintains, on all its surface, its characters as a functional fibromucosa, which has been mentioned in relation with the gingiva. At its level is collected and formed the food bolus, before being swallowed; moreover, the deep layer of the mucosa is furnished with a glandular tissue in the median and posterior areas.

Gradually, the mucosa looses its fibrous character, beginning from the posterior boundary of the bony palatine vault, to become a resistant, but suppler mucosa, furnished itself with a glandular substrate.

This mucosa covers the pillars of the soft palate (velum palatinum) with a disseminated glandular apparatus. The alveolar gingiva of the lower lingual part looses rapidly its fibrous character, becoming a sublingual and lingual mucosa, which lines the upper part of the oral floor.

19

This sublingual mucosa, with its reflection on the lower surface of the tongue, is supple and very thin, but relatively resistant. Hence, it is comparable with the mucosa of the cheeks and lips. Without dwelling on the special character of the mucosa of the dorsal surface of the tongue, we underline only its special adaptation not only to the movements of the tongue, but also to the gustative function.

The Superficial Regions of the Oral Cavity and of the Face

The Lips and the Labial Region

The lips are a kind of mobile walls, which close anteriorly the oral cavity as much necesary. Each forms a mobile musculomembranous fold, respectively a superior fold or the upper lip and an inferior fold or the lower lip, which get into contact through their free, horizontal margin, forming the slit of the oral orifice: the oral fissure (rima oris). The lips join at the lateral extremities of this orifice, forming the right and left commissures of the mouth (fig. 7).

External configuration of the lips. Each lip consists of a superficial, cutaneous part and a deep, mucous, internal part; the two parts unite at the level of the free margin of the lip.

The upper lip lies beneath the nasal pyramid, below the level of the subnasal point, situated at the base of the nasal septum. It is bounded laterally, on either side, by a groove, situated behind the wings of the nose, the nasogenian groove, which becomes nasolabiogenian at this level.

Superficially, in the middle of the upper lip is a vertical groove which descends below the nasal septum, named the philtrum; it ends in the margin of the mucous portion of the lip.

The free margin is marked by the red colour of the mucosa, which prooves the abundance of submucous blood capillaries in this region. In the median portion of the upper lip, beneath the philtrum and at its level, lies a median tubercle.

The mucous portion of the lip, which corresponds to the gingivodental segment of the upper incisors and canines, reflects on the gingiva, forming a cul-de-sac or the superior groove of the oral vestibule (fornix vestibuli superior).

This groove is divided into two halves, right and left, by a median, vertical, falciform fold, the frenum or frenulum of the lip (frenulum labii superioris); it unites the deep part of the lip with the gingiva.

The lower lip is situated, in most cases, slightly behind the upper lip; it is bounded superficially by the mental region of its cutaneous portion, by the labiomental groove, and laterally by the projection of the nasolabiogenian grooves. The cutaneomucous zone or seam (pars intermedia) describes a regular curve, slightly concave upwards. The deep surface of the lower lip (pars mucosa) forms, by reflection of the mucosa on the lower incisivo-canine gingiva, the inferior groove of the oral vestibule (fornix vestibuli inferior).

Like at the upper lip, in the deep surface of the lower lip lies a median vertical fold; the frenum or frenulum of the lower lip (frenulum labii inferioris). The frenula of the two lips represent the main landmarks of the median sagittal plane of the oral cavity.

Structure off the lips. From the surface to the depth, the lips are made up of five tissue layers disposed successively: the skin, the subcutaneous connective-fatty tissue, the muscular layer, the submucous layer and the mucous epithelial and glandular layer.

•	The skin is thickened, rich in hair follicles and sebaceous glands. In the deep portion lies a significant network of lymph vessels.

•	The subcutaneous connective-fatty tissue lacks in the zone where the musculature adheres to the skin, in the median zone and at the level of the commissures where, however, true connective fibrous nodes are present.

The musculature of the lips is represented by the orbicular muscle or the constrictor of lips and by a complex of facial, perioral platysma muscles which end in the deep part of the integument of the lips, forming on either side by crossing of their fibres, behind the commissure, a retrocommissural connective node. These muscles are disposed on two planes, superficial and deep: the greater and lesser zygomatic muscles, the canine muscle, the superficial and deep common levator muscles of the upper lip and of the wings of the nose, the triangular muscle of the lower lip and the quadrate muscle of the chin. The buccinator muscle makes up the deepest plane. In the new-born, the lips are crossed anteroposteriorly by a few thin muscle bundles, which form the compressor muscle of the lips (Klein's muscle). In the depth of the muscle layer lies a stratum of very densely disposed mucous glands: the labial glands.

20

The mucosa has a stratified pavement epithelium and is very closely adherent to the muscle layer in the region of the free margin of the lips. On its deep surface is situated a network of deep lymph vessels, apparently independent of the superficial lymphatic network.

Vessel and nerve supply

The arterial supply of the lips is assured, in the first place, by the superior and inferior coronary (labial) arteries, branches of the two facial arteries. They cross the muscular plane of the lips, spread out on the deep surface, in the proximity to the free margin, and anastomose with those of the opposite side. Secondarily, the lips receive labial branches of the suborbital artery and of the mental artery, respectively for the upper and the lower lip.

The motor nerves are derived from the branches of the facial nerve, whereas the sensitive nerves are branches of the trigeminal nerve and especially of the maxillary nerve (V_2) through its suborbital branch for the upper lips and of the mandibular **nerve** through the mental nerve for the lower lip.

The Chin and the Mental Region

The chin forms the median and inferior prominences of the anterior and inferior parts of the face (vol. I). The mental region is separated superiorly from the lower lip by the labiomental groove and inferiorly, from the regions of the neck, by the lower margin of the anterior part of the body of the mandible. The **chin** appears as a tuberosity *(gnathion)* with the anterior convexity directed vertically and transversally. It is covered by a thick and adherent skin, rich in hair follicles and sebaceous glands on the deep surface in the median portion of the region, to which the muscle fibres of the regional mimicry muscles *(platysma, levator of the chin, triangular muscle)* adhere directly.

The mental region corresponds, along its whole surface, to a "dumb" zone of the face (the "dumb" zones of the face are spaces in which, as a matter of fact, no significant vasculonervous elements are present). These median and inferior "dumb" zones have the shape of a triangle, the summit of which is situated at the level of the cutaneomucous junction of the lower lip.

The Cheek and the Genian Region
(Regio buccalis)

The cheek forms the lateral wall of the oral cavity. It represents generally the anterolateral part of the face or the genian region.

This region has the following boundaries: superiorly, the lower border of the orbital cavity; inferiorly, the lower border of the mandible; anteriorly, the nasogenian groove and a vertical line projecting it downwards and passing through the commissure of the lips; posteriorly, the anterior border of the masseter muscle, which separates it from the masseteric region. The genian region is composed of two zones: a superior zone or the infraorbital region, which protects the infraorbital vasculonervous bundle, and an inferior zone or the buccal region, which makes up actually the lateral wall of the oral cavity and which is currently named the genian region, a term entered in usage.

Configuration. The genian region has the shape of a vertically elongated quadrilater, diagonally traversed by the facial vessels and divided by Stensen's duct into two parts: an upper and a lower part.

The cheek has two surfaces: an external, cutaneous surface and an internal, buccal surface, the first being wider than the second.

The external surface, free and mobile in its middle portion, adheres to the bony planes made up of the external surfaces of the maxilla, superiorly, and of the mandible, inferiorly.

The internal, mucous surface forms the external or lateral wall of the oral vestibule; it exhibits at the level of the second upper molar the opening of the Stensen's duct, around which are disseminated the jugal or molar glands.

Structure. The cheek is composed, from the surface to the depth, by five layers, with the exception of the plane formed by the osteoperiosteal skeleton, to which it is attached both superiorly and inferiorly.

• *The skin,* very thin and glabrous in the upper part, is thickened in the lower part and bears at this level a dense pileous apparatus; on its deep surface is present a rich lymphatic network.

• *The subcutaneous connective-fatty tissue,* more abundant in the posterior portion of the region, lines the skin in the depth. It is susceptible to charge itself with fat in stout individuals and in children. This panniculus adiposus has the shape of a fatty bulla -Bichat's fat-pad or Bichat's bulla -in the posterosuperior part of the region.

21

• The adipose mass insinuates itself laterally to the muscular plane of the buccinator and anteriorly to the temporal tendon and the anterior border of the masseter muscle, projecting towards the connective-fatty spaces of the infratemporal and temporal fossae, which it fills. Consequently, owing to this mass, these spaces communicate with each other and with the genian region.

The muscles of the face, which converge towards the retrocommissural muscle node, are disposed in two discontinuous planes: superficial and deep. The superficial plane is composed from above downwards of the following muscles: the greater and lesser zygomatic muscles, the superficial common levator muscle of the upper lip and of ala nasi, the risorius muscle and, inferiorly, the platysma of the neck and the triangular muscle of the lower lip.

The deep plane consists mainly of the buccinator, the canine, the deep common levator of ala nasi and of the upper lip and the quadrate muscle of the chin.

The intermuscular space represents the passage site of the vessels, the nerves and the Stensen's duct. The facial vessels cross diagonally the region lying between the antero-inferior border of the masseter and the interior angle of the eye. Alongside the vessels are situated the genian and buccinator lymph nodes, when they exist.

The Stensen's duct crosses nearly horizontally the posterior half of the region, coursing on a line which unites the lobe of the ear with the nasal wing and alongside which it may be palpated. Then it surrounds the anterior border of the masseter and, medially, the tendon of the temporal muscle, enters in relation with Bichat's bulla, penetrates the buccinator muscle and opens into the oral vestibule at the level of the second upper molar.

A layer interrupted by mucous glands lines on the deep surface the jugal mucosa, which adheres to the deep surface of the buccinator' muscle. This mucosa contains a rich network of lymph vessels.

The upper bony plane is made up of the genian wall of the maxilla (forming the infraorbital region), where the orifice of the infraorbital duct opens, in which the infraorbital vasculonervous bundle is located. The inferior bony wall is formed by the external portion of the body of the mandible, where the mental orifice and the mental vasculonervous bundle are located.

Vessel supply

The region is mainly supplied by the facial artery and vein and secondarily by branches of the superficial temporal artery, of the internal maxillary artery and of the ophthalmic artery. These branches form anastomoses between the external and the internal carotid system.

The Masseteric Region

The masseteric region is continuous posteriorly with the genian region: with it end laterally the superficial regions of the face.

It is bounded by the borders of the masseter muscle, from which it borrows its name: the anterior boundary of the region corresponds to the anterior border of the muscle, directed obliquely upwards and forwards, and the inferior boundary corresponds to the inferior margin of the mandibular insertion of the masseter muscle. The superior boundary is represented by the zygomatic arch and the posterior boundary, by the posterior margin of the ascending branch of the mandible, which makes up the anterior border of the parotid region.

Bounded in this way, the masseteric region has the shape of a vertically elongated quadrilater, like the genian region, which it actually continues backwards.

It is formed, from the surface to the depth, by integument, subcutaneous tissue, aponeurosis and muscle. The skin is resistant, furnished with a dense hairy system and a network of lymphatics which drain into the parotid lymph nodes.

It is lined with a layer of subcutaneous connective-adipose tissue, more or less rich in fat, which is crossed horizontally by the following elements:

- the transverse facial artery, parallel and 1 cm beneath the zygomatic arch;
- Stensen's duct, directed from behind-forwards, accompanied by the masseteric projection of the parotid gland, sometimes isolated, as an accessory parotid. The duct courses on a line which joins the tragus to the nasal wing, on which it is palpable;
- the diverging ramifications of the temporofacial and cervicofacial branches of the facial nerve, which pass through the region and are distributed to the afferent muscles.

The masseteric aponeurosis covers the muscles and adheres intimately to them, transforming the area into a nearly closed space.

By its anterior border it gives insertion to the fascia of the buccinator muscle, the only muscle of mimicry which has a fascia, and to the posterior fibres of the risorius muscles which are located on the middle region of the face. By its posterior border it offers insertion to the superficial cervical aponeurosis, which is nothing else than the superficial leaflet of the fascia of the parotid region. Superiorly it is continuous with the temporal fascia.

The excretory duct and the projection of the parotid gland pass anteriorly, being situated inside the masseteric fascia, which has two leaflets at this level.

The masseter muscle fills completely its space and adheres intimately to the osteoperiosteal plane of the mandible into which it is inserted; it recedes only superiorly, to attach itself on the zygomatic process (fig. 9).

The Floor off the Oral Cavity

It is the space situated between the concavity of the body of the mandible and the convexity of the hyoid bone, bounded superiorly by the oral mucosa and inferiorly by the subhyoid cutaneous plane. By the plane of the mylohyoid muscles, united by a median raphe, forming the *diaphragma oris,* the floor is divided into two parts: a superior, supramylohyoid part, and an inferior, submylohyoid part (fig. 9).

The supramylohyoid part of the median region of the floor makes up the sublingual region, which will be studied concomitantly with the tongue.

The muscular plane of the floor of the oral cavity is formed by three muscles:

1.The mylohyoid muscle is attached to the whole length of the mylohyoid line *(linea mylohyoidea)* of the mandible. From here it descends towards the anterior surface of the body of the hyoid bone and towards the median raphe which unites the two muscles from the mental symphysis to the hyoid. In this way is formed a muscle girth, which supports, in the middle, the geniohyoid muscles and the tongue and, on the borders, the sublingual glands. It is supplied by the trigeminal nerve through its mylohyoid branch from the mandibular nerve (V_3). Above lies the geniohyoid muscle.

2.The geniohyoid muscle (paired muscle) extends from the inferior genian process of the mandibular mental spine to the anterior surface of the hyoid bone and is supplied by the hypoglossal nerve, its motor axons deriving from the medullary segments C1/C2.

3.The anterior belly of the digastric muscle is inserted into the digastric groove of the lower border of the mandible, from where it runs backwards, giving rise to the intermediate tendon, which is fastened to the hyoid bone by loop-shaped tendinous expansions. Like the preceding muscle, it is supplied by the mylohyoid nerve. It is covered by the platysma muscle and by integument. Thesse three suprahyoid muscles are depressors of the mandible or levators of the hyoid bone.

The submylohyoid part corresponds to the suprahyoid region and belongs to the anterior trigone of the neck. The median or submental region of this part is situated between the two anterior bellies of the two digastric muscles. The lateral regions situated between the two bellies - anterior and posterior - of the digastric muscle form, on either side of the submylohyoid part, the triangular submandibular regions, which are studied in the chapter regarding the submandibular glands.

The Tongue-the Lingual Region

It is a musculomembranous organ, covered by a thick mucosa and attached by its base to the floor of the mouth, forming the median region of the supramylohyoid part. The mucosa of the tongue is the site of the gustatory organ, which represents the receptor for the sense of taste and is the trigger of the secretory reflex of the adnexal salivary glands of the oral cavity (fig. 9-12).

The tongue is free in its anterior part and endowed with a great mobility. It occupies the whole oral space situated beneath the palatine vault and bounded by the inferior and superior dental arches.

It fills the oral cavity in the state of occlusion and converts it into a capillary cavity. In edentate individuals, through the disappearance of the teeth, it tends to occupy also the new space formed by their absence.

It is an organ of the digestive tract by its gustatory papillae, which allow, by the perception of the basic taste, the qualitative appreciation of foodstuffs. It intervenes in suction, contributes to the prehension of foodstuffs and distributes them between the triturating surfaces of the teeth. It aids in the formation of the food bolus, which it pushes posteriorly, to the pharynx, at the moment of deglutition. It is an organ of speech and perfects the sounds emitted by the larynx.

External configuration. The tongue has the shape of an anteroposteriorly elongated ovoid, flattened from above downwards, with the tip directed forwards. It has two surfaces, two borders, a tip and a vertical root *(radix linguae)*.

The superior (or dorsal) surface *(dorsum linguae)* is flattened transversally and convex in the sagittal plane. At the junction of the anterior two thirds with the posterior third lies a furrow, termed the lingual V *(sulcus terminalis),* open anteriorly, the tip of which contains a small pit, the *foramen caecum,* marking the point of junction of the three embryonic buds of the tongue and, at the same time, the site of the embryonic emergence of the thyroid gland (fig. 10).

The foramen caecum may appear in the adult under the form of a more or less deep invagination. When it preserves its embryonic character, it has the shape of a duct which connects the *foramen caecum* to the thyroid isthmus.

The anterior part of the lingual V, or the oral part of the tongue, is horizontal and marked by a longitudinal anteroposterior lingual groove, termed median groove *(sulcus medianus).* On either side of this groove, the lingual mucosa is furnished with small prominences, named papillae, which in the order of their increasing size and decreasing number are filiform, foliate, fungiform and caliciform. The caliciform or circumvallate papillae, 9-11 in number, make up the lingual v, situated behind the sulcus terminalis; at the level of these papillae and of the foliate and fungiform papillae are located gustatory corpuscles, taste receptors, whereas the filiform papillae have a mechanical function.

The posterior part of the tongue, behind the lingual V, or the pharyngeal, vertical part is characterized by the presence of collections of irregularly disposed lymphoid follicles, forming the lingual amygdala or tonsil. At the postero-inferior boundary of this part of the tongue, three glosso-epigloittic folds, of which a median fold *(plica glossoepiglottica media)* and two lateral folds *(plicae glossoepiglotticae laterales),* unite the tongue with the epiglottis, separating from each other the two valleculae *(valleculae epiglotticae).*

The inferior surface of the tongue is less smooth then the posterior surface and presents on the median line a fold, its frenulum *(frenulum linguae),* which unites the tongue with the gingivo-alveolar groove and limits the movements of its tip.

At the base of the frenulum, on either side of its relief, lies a small elevation, the salivary sublingual carunculae, into the tip of which opens (in each of them) the Wharton's duct; laterally to it open the ducts of the sublingual gland (Bartholin's duct), laterally to which are situated the small sublingual glands (5-10 in number), which open on a fold of the mucosa by the Rivinus' duct.

Parallel to the lingual frenulum, on the inferior surface, appears a bluish relief formed by the canine vein, bounded laterally by a small crest or indented fold: the fimbriated fold or *plica fimbriata,* parallel to the border of the tongue.

The borders of the tongue are rounded and smooth; they correspond to the lingual surface of the teeth. Alongside the inner part of these borders open the ducts of Weber's mucous glands.

The tip of the tongue *(apex linguae)* is flattened from above downwards; it is the most mobile part of the organ. Towards its apex, the tongue is also furnished with a salivary gland of mixed secretion, the Biandin-Nuhn's paired gland.

The root of the tongue is the portion of its insertion on the skeleton and the zone of confluence of the muscles which enter in its structure. On the lateral and inferior parts of its base, the tongue receives the nutrient vessels and the nerves.

The skeleton off the tongue. The tongue as such has no skeleton proper; however, as it is structured of various, extrinsic and intrinsic, types of muscles, owing to which it has an extraordinary mobility, an adjacent bony skeleton may be described for the extrinsic muscles and a fibrous skeleton, for the intrinsic muscles.

The bony skeleton is formed by the hyoid bone and the styloid process and the fibrous skeleton, by the lingual or hyoglossal aponeurosis, attached to the upper border of the hyoid bone, between the lesser horns, situated under the mucosa of the lingual dorsum, and to the lingual septum, a vertical, median, falciform, fibrous lamina, the base of which is implanted on the hyoglossal membrane and on the hyoid bone and the tip of which is directed towards the apex of the tongue. On these skeletons are attached the muscles of the tongue (fig. 11).

The muscles of the tongue. The tongue is a fleshy mass, the mobility of which is due to the concentric action of 17 muscles, 16 of which are constant and lateral, grouped in 7 pairs, while two muscles are unpaired; the amygdaloglossus is inconstant (fig. 12, 16, 17).

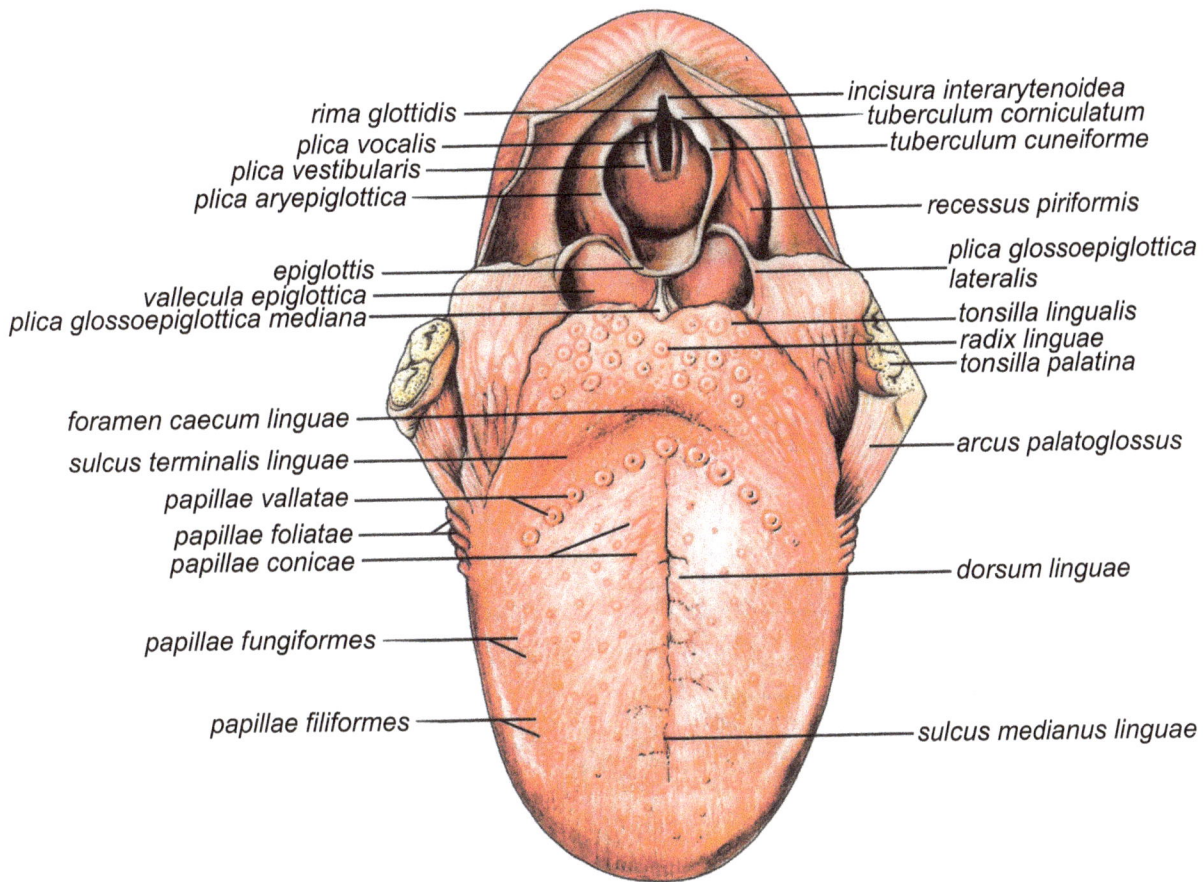

Fig. 10. The tongue (horizontal and vertical portions) and the inlet of the larynx.

Labels (Fig. 10):
- rima glottidis
- plica vocalis
- plica vestibularis
- plica aryepiglottica
- incisura interarytenoidea
- tuberculum corniculatum
- tuberculum cuneiforme
- recessus piriformis
- epiglottis
- vallecula epiglottica
- plica glossoepiglottica mediana
- plica glossoepiglottica lateralis
- tonsilla lingualis
- radix linguae
- tonsilla palatina
- foramen caecum linguae
- sulcus terminalis linguae
- papillae vallatae
- papillae foliatae
- papillae conicae
- arcus palatoglossus
- papillae fungiformes
- dorsum linguae
- papillae filiformes
- sulcus medianus linguae

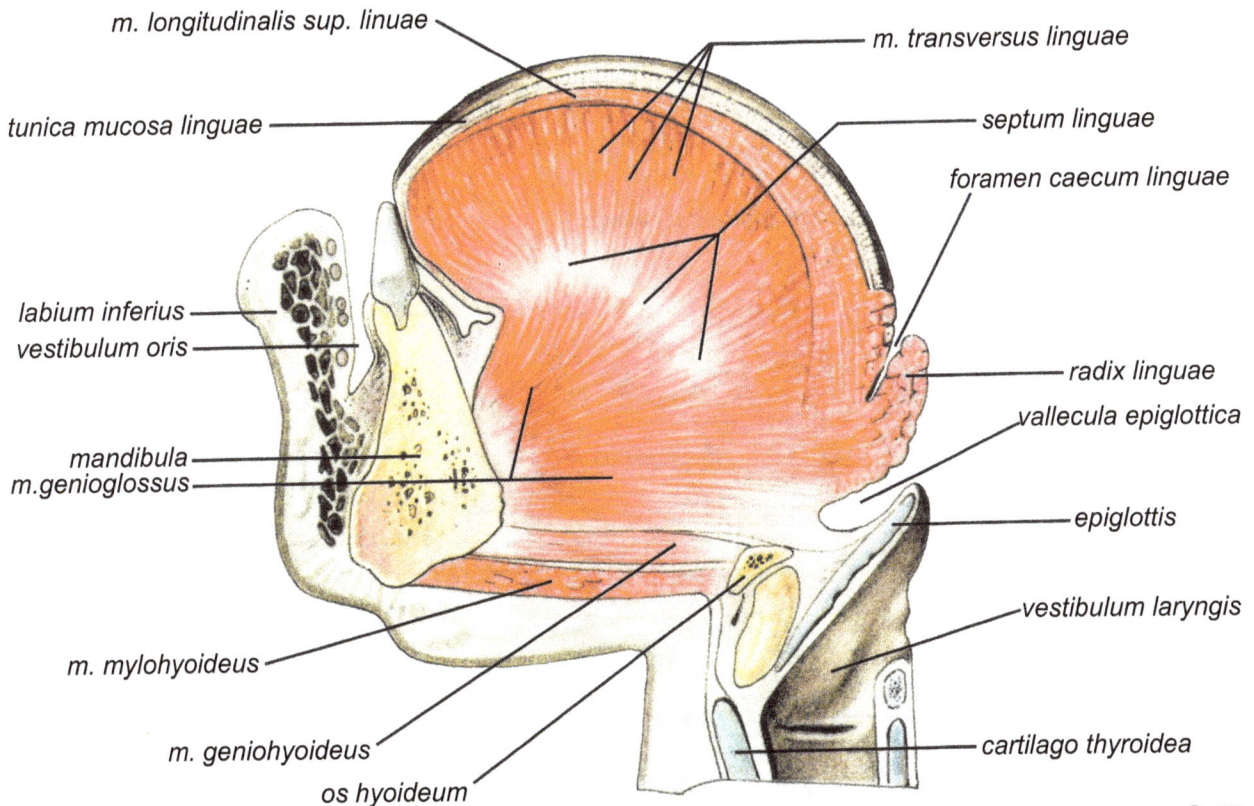

Fig. 11. The tongue and the plate of the oral cavity - mediosagittal section.

Labels (Fig. 11):
- m. longitudinalis sup. linuae
- m. transversus linguae
- tunica mucosa linguae
- septum linguae
- foramen caecum linguae
- labium inferius
- vestibulum oris
- mandibula
- m.genioglossus
- radix linguae
- vallecula epiglottica
- epiglottis
- vestibulum laryngis
- m. mylohyoideus
- m. geniohyoideus
- os hyoideum
- cartilago thyroidea

25

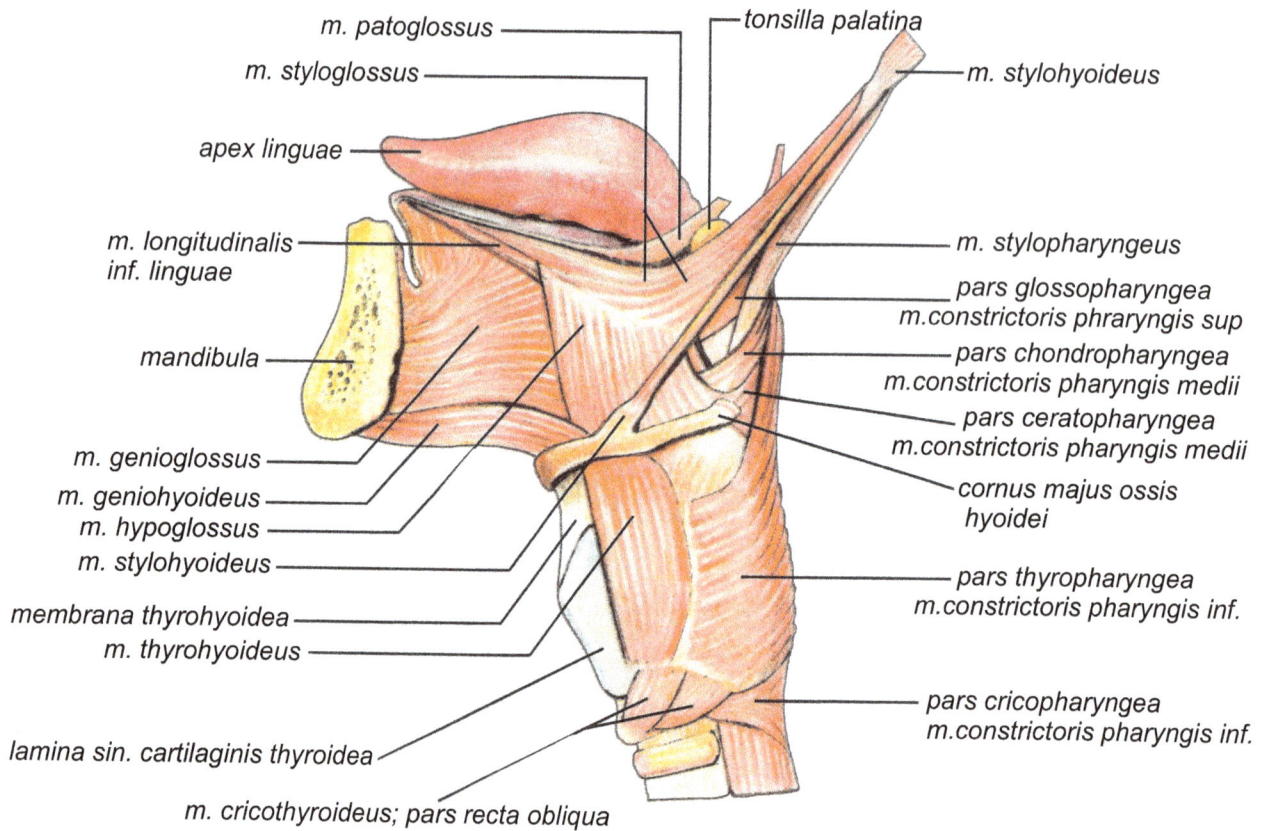

m. patoglossus
m. styloglossus
apex linguae
m. longitudinalis inf. linguae
mandibula
m. genioglossus
m. geniohyoideus
m. hypoglossus
m. stylohyoideus
membrana thyrohyoidea
m. thyrohyoideus
lamina sin. cartilaginis thyroidea
m. cricothyroideus; pars recta obliqua

tonsilla palatina
m. stylohyoideus
m. stylopharyngeus
pars glossopharyngea m.constrictoris phraryngis sup
pars chondropharyngea m.constrictoris pharyngis medii
pars ceratopharyngea m.constrictoris pharyngis medii
cornus majus ossis hyoidei
pars thyropharyngea m.constrictoris pharyngis inf.
pars cricopharyngea m.constrictoris pharyngis inf.

Fig. 12. The muscle of th tongue

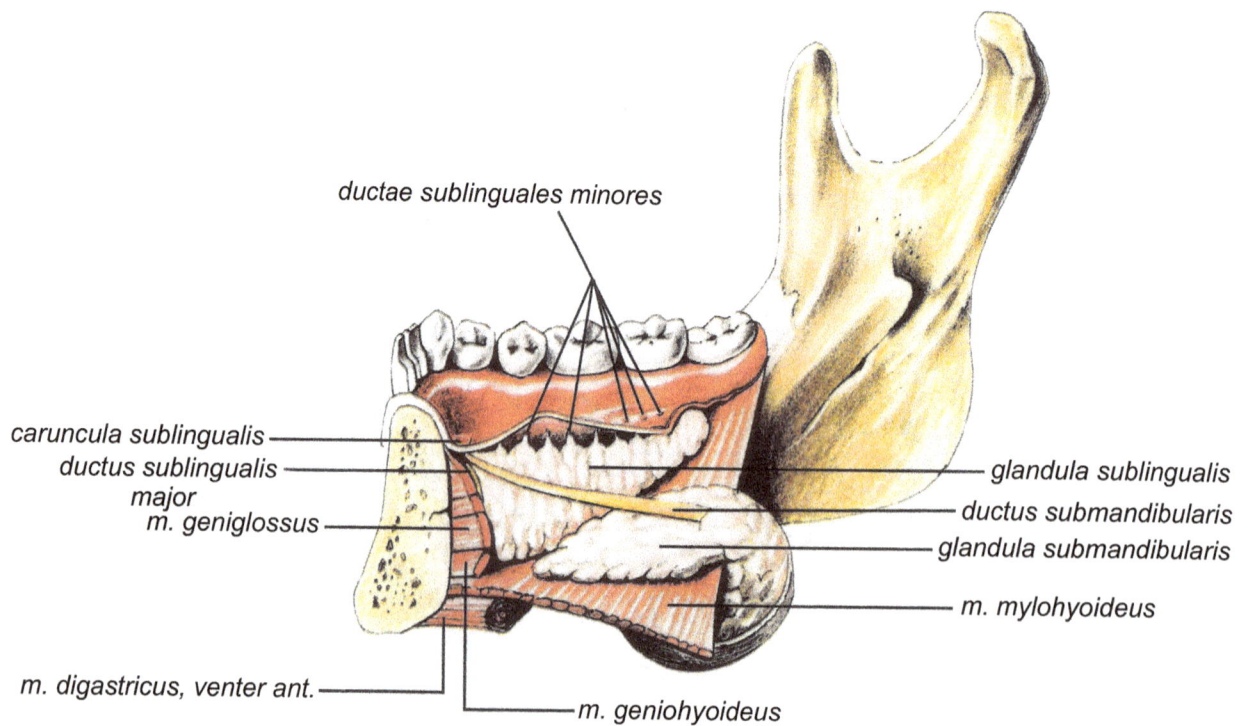

ductae sublinguales minores
caruncula sublingualis
ductus sublingualis major
m. geniglossus
m. digastricus, venter ant.
m. geniohyoideus

glandula sublingualis
ductus submandibularis
glandula submandibularis
m. mylohyoideus

Fig. 13. The sublingual gland

26

A) *The extrinsic muscles:*

The genioglossus muscle *(m. genioglossus),* a paired, symmetrical muscle, makes up the whole antero-inferior region of the mobile part of the tongue. It arises anteriorly by a tendon attached to the superior genian process of the mental spine of the mandible; from here it spreads out in a broad fan, in which the fibres have various orientations, imparting to the tongue various movements. The superior fibres describe a curve with the concavity anterior and run towards the apex, attaching to the deep surface of the lingual aponeurosis. The inferior fibres, nearly horizontal, run towards and attach to the superior border of the hyoid bone and of the hyoglossal lingual aponeurosis, even to the epiglottis. The dense, oblique, middle fibres run towards the posterior part of the tongue.

Action: the contraction of the superior fibres leads the apex of the tongue behind the mental symphyseal region. The contraction of the middle fibres presses the tongue against the floor of the mouth and the inferior fibres draw the hyoid bone upwards or project the apex of the tongue outside the mouth.

The styloglossus muscle *(m. styloglossus)* is a paired muscle, thin at its posterior origin on the styloid process, where it spreads out in a fan-like shape up to the region of the root and the posterior region of the tongue. It extends from the styloid process to the lateral portion of the body of the tongue.

The superior fibres form the border of the tongue, coursing from the root to the apex; a group of fibres runs medially towards the lingual septum and another one forms the lateral portion of the posterior surface of the tongue.

The inferior fibres decussate anteriorly with the main fibres of the keratoglossus muscle, then with the main, basioglossal part of the hyoglossus muscle, medially to the latter with the inferior lingual muscle and, at the apex of the tongue, with its homonym.

Action: it draws the tongue backwards, upwards and at the same side; the simultaneous contraction of the two styloglossus muscles draws the tongue upwards and backwards to the velum palatinum. The hyoglossus muscle *(m. hyoglossus)* is a rectangular, paired muscle, situated on the lateral part of the root of the tongue, above the hyoid bone. It consists of two portions, which differ according to their attachment level on the hyoid bone. The keratoglossal part is attached to the whole length of the superior border of the greater horn and to the adjacent part of the lesser horn of the hyoid bone. The basioglossal part is inserted into the body of the bone, into the concavity of the insertion of the genioglossus muscle and above the mylohyoid muscle. From their inferior origin, the two portions, separated by an interstice, ascend obliquely forwards and medially, towards the lateral part of the tongue. They are united by the superior fascicle of the styloglossus muscle, with the fibres of which they decussate, and course towards the apex of the tongue and the lingual septum.

Action: The hyoglossus muscle draws the tongue downwards. Its action is synergistic with that of the genioglossus muscles and, to some extent, with that of the styloglossus muscles. If both muscles contract concomitantly, they draw the tongue outside the mouth.

The palatoglossus muscle *(m. palatoglossus),* a paired muscle, makes up the anterior pillar of the velum palatinum (the oropharyngeal isthmus). It runs downwards and medially, reaching the lateral part of the root of the tongue, where its fibres blend with those of the styloglossus, pharyngoglossus and amygdaloglossus muscles.

Action: it contributes to the elevation of the root of the tongue, diminishing the calibre of the pharyngeal isthmus.

The amygdaloglossus muscle *(m. amygdaloglossus),* mentioned by the French authors as an inconstant muscle, is a small muscle bundle inserted into the pharyngeal aponeurosis, which is directed towards the deep part of the tonsil. From here the muscle courses downwards to the root of the tongue and medially, where it unites with its homologue of the opposite side.

Action: The two muscles elevate the root of the tongue towards the palate.

The pharyngoglossus muscle *(m. pharyngoglossus)* is actually an individualized fascicle of the superior constrictor of the pharynx, which runs downwards and forwards below the hyoglossus muscle, blending its fibres with those of other muscles of the tongue.

Action: It draws the tongue backwards and slightly upwards.

B) *The intrinsic muscles*

The inferior lingual muscle *(m. longitudinalis inferior)* makes up the inferior portion of the tongue; it is situated posteriorly between the genioglossus and the hyoglossus muscles and anteriorly between the styloglossus and the genioglossus muscles.

Action: it draws the apex of the tongue backwards and downwards, shortening the tongue and vaulting it with the convexity towards its dorsum.

27

The superior lingual muscle *(m. longitudinalis superior)* is situated beneath the mucosa and the aponeurosis of the dorsal surface of the tongue; it is elongated longitudinally, from the root to the apex, on which if is attached.

Action: it elevates the apex of the tongue and draws the tongue laterally.

The transverse muscle *(m. transversus linguae)* consists of transverse fibres - hence it is not paired - which lie beneath the superior longitudinal muscle. It is inserted into the lateral surface of the median septum and into the deep surface of the corium of the mucosa of the lingual border.

Action: the contraction of the transverse muscle narrows the tongue and, consequently, elongates it or converts it into a median longitudinal groove.

The vertical muscle *(m. verticalis linguae)* arises from the deep surface of the lingual aponeurosis and is attached on the submucosa of the inferior surface of the tongue.

Action: by its contraction it flattens the tongue.

Vessel and *nerve* supply (fig. 16 and 17)

Arteries. The tongue is richly supplied by arteries, mainly by the lingual artery and secondarily by the ascending palatine and ascending pharyngeal (branch of the external carotid) arteries.

There are two triangular zones of approach to the lingual artery, which is concealed only by the hyoglossus muscle and by the submandibular gland.

The Beclard's triangle is bounded by the greater horn of the hyoid bone, the posterior margin of the hyoglossus muscle and the posterior belly of the digastric muscle; it is crossed by the hypoglossal nerve.

The other triangle, Pirogoff's triangle, is bounded by the intermediate tendon of the digastric muscle, the posterior margin of the mylohyoid muscle and the hypoglossal nerve. In practice, in the case of a severe haemorrhage of the tongue, the ligation of the trunk of the external carotid artery is preferred.

Veins. In its segment of origin, the main lingual vein is a satellite of the lingual artery; then it separates from it and runs laterally, appearing with the hypoglossal nerve between the hyoglossus muscle and the posterior margin of the mylohyoid muscle.

It receives the dorsal veins of the tongue, the deep lingual veins and the sublingual vein, passes medially to the stylohyoid muscle and empties into the internal jugular vein, either directly or through the thyrolinguofacial trunk.

Lymph vessels. The lymph vessels of the lingual region may be distributed topographically into three main groups:

1.The lymphatic group of the base and that of the borders of the tongue drain into the subdigastric lymph node; towards this lymph node run also the deep lymph vessels of the tongue.

2. The group of lymph vessels which arise from the dorsal part of the tongue, situated anteriorly to the lingual V *(sulcus terminalis linguae),* drains the lymph into the lymph nodes situated at the level of the thyrolinguofacial trunk. The superficial portion of the borders of these regions sends its lymph most frequently into the submaxillary lymph nodes.

3. The group of lymph vessels of the apex of the tongue converges towards the submental lymph nodes, which course towards the anterior jugular chain of lymph nodes, in the group of omohyoid lymph nodes.

This scheme can be considered as variable, especially when the lymph vessels derive from a region situated near the middle border of the tongue.

The motor nerves of the tongue derive almost in totality from the hypoglossal nerve, which divides into a great number of branches for each of the muscle bundles.

This disposition, as well as the great number of muscles, account for the extreme variety of the movements of the tongue. Of the other cranial nerves, only the glossopharyngeal nerve supplies the palatoglossus muscle.

The sensitive nerve supply of the tongue is assured by three nerves: the trigeminal nerve, through the lingual nerve, supplies the anterior segment of the lingual V; the glossopharyngeal nerve supplies the posterior part of the tongue, leaving a small juxtaepiglottic median part innervated by the vagus nerve, through the superior laryngeal nerve.

The sensory nerve supply is assured for the anterior twothirds of the tongue by the chorda tympani nerve (Wrisberg's intermediate nerve -VII bis), which anastomoses with the lingual nerve.

In the posterior third, the sensory nerve supply is assured by the glossopharyngeal and the vagus nerves.

The Glandular System of the Oral Cavity

Numerous salivary glands empty their secretion into the oral cavity. They are divided into large and small salivary glands.

The large salivary glands are paired and situated, in topographical order, from behind-forwards: the parotid gland, the submandibular gland and the sublingual gland (fig. 13).

The submandibular and sublingual glands belong to the floor of the mouth, whereas the parotid gland ranges among the constituent elements of the lateropharyngeal spaces.

The small salivary glands of the oral mucosa. The oral mucosa is constantly moistened by its small glands, which secrete a serous and mucous saliva. They are topographically distributed into three groups:

1. The labiojugal glands are situated in a thin connective premuscular layer or immediately beneath the mucosa, causing its bulging and conferring it a mamelonated appearance. They are more numerous around the oral orifice and in the middle part of the vestibule, where they make up the group of molar glands, in the region around the orifice of Stensen's duct.

2. The palatine glands are situated in the thin layer of the submucosa and of the mucosa of the velum palatinum, both paramedian areas. At the level of the bony palatine vault, the glandular layer adheres intimately to the mucosa and the periosteum, being denser in the middle part, on either side of the median raphe. At the level of the velum palatinum, this glandular layer diminishes in thickness, remaining thin at the level of the aponeurosis; it covers the whole palatine region and the uvula.

3. The lingual glands, already studied, are situated in the deep layer, beneath the mucosa, of the region of the tongue, behind the lingual V, under the follicular layer. A group of glands situated in the mucosa of the lingual border extends from the lingual V to the apex; in the inferior part of the apex are the Blandin-Nuhn's glands, which form a small, 10-15 mm sized glandular mass, located in the depth of the muscles and open through several ducts lateral to the frenulum of the tongue (they secrete a mixed, seromucous saliva). At the level of the root of the tongue lie Ebner's serous glands.

The Sublingual Region and Gland

In the space situated between the root of the tongue, the mandible, the alveolodental mucosa and the floor of the mouth lies a prominence which makes up the sublingual space. This space contains the sublingual gland with its excretory ducts, the sublingual projection of the submandibular gland and the excretory duct of the latter (Wharton's duct), the lingual vasculonervous bundle (the lingual nerve with the annexed vegetative ganglia, as well as the sublingual vessels) and the hypoglossal nerve. The region makes up the supramylohyoid part (the superior part) of the floor of the mouth; it is prismatic in shape and is bounded: medially, by the external surface of the root of the tongue; laterally, by the internal surface of the mandible, on which it leaves a pit, the sublingual fossa, above the mylohyoid crest; inferolaterally, by the mylohyoid muscle, which separates the sublingual region, in the anterior third, from the submental trigone, in the middle third from the anterior portion of the submandibular fossa, while in the posterior third the two regions communicate with each other behind the posterior border of this muscle. Above, the upper part of the floor of the mouth is bounded by the oral mucosa which unites the alveolar margin of the mandible with the base of the tongue. This wall is continuous with the superior wall of the submandibular region.

The sublingual region contains the sublingual gland, the projection of the submandibular gland, the Wharton's duct, the lingual and hypoglossal nerves, the lingual artery and the lymph vessels.

The sublingual gland occupies the greatest part of the sublingual space. It is an agglomeration of glandules, which all together have the size of an almond weighing 6-8 g, of 25-30 mm in length, 11-15 mm in height and 5-6 mm in thickness. It is enveloped in a thin fibroconnective lamina, covered partially (inferomedially) by a thin lamina of adipose tissue. The gland has two surfaces, two borders and two extremities (fig.13).

The medial, plane surface is in relation with the genioglossus and the geniohyoid muscles, from which it is separated by the interstice which contains the lingual nerve, the sublingual vessels, the ending of the hypoglossal nerve and the submandibular (Wharton's) duct.

The inferolateral convex surface is accommodated in the sublingual fossa of the mandible and lies above the mylohyoid muscle.

The inferior, thin, convex border penetrates into the angle bounded by the mylohyoid, the geniohyoid and the genioglossus muscles; in this interstice may develop the phlegmons of the floor of the mouth.

Sometimes, the sublingual region can communicate with the submandibular region through the dissociated fibres of the mylohyoid muscle, covered only by a connective fenestrated lamina.

The ducts of the gland are a main duct *(ductus lingualis major* - Bartholin's duct), which is accompanied by the submandibular duct and ends on the sublingual caruncula, and a series of accessory ducts *(ductus sublinguales minores),* which open by a series of orifices diposed linearly alongside the sublingual fold.

The projection of the submandibular gland is variably developed: it may attain the anterior pole of the sublingual duct or it does not pass beyond the posterior third of the gland. In any topographical situation it may be observed that the anterior extremity of the projection of the submandibular gland is closely united to the inferior border of the sublingual gland through small connective fibrous lamellae, through vessels and through nerves.

• *Wharton's duct* extends in the sublingual region on a distance of 30-40 mm and is situated in the interstice between the gland and the root of the tongue, at a few mm beneath the mucosa of the floor, above the projection of the submandibular gland.

• *The lingual nerve* is a sensitive and sensory nerve, a branch of the mandibular nerve; it penetrates the sublingual region, surrounding from outside- inwards and from below- upwards the Wharton's duct. In the interstice between the sublingual gland and the projection of the submaxillary gland, the nerve gives off an important branch for the sublingual gland.

Then it occupies the interstice between the base of the tongue and the sublingual gland and runs towards the apex of the tongue, where it ends.

• *The hypoglossal nerve* surrounds the anterior border of the hyoglossus muscle, penetrates the inferior part of the sublingual fossa, then reaches the internal wall of the fossa, divides into a bundle of branches and penetrates inside the tongue through the interstices of the genioglossus muscle.

Between the two nerves, the lingual and the hypoglossal nerves, several anastomoses in the form of an arch are formed.

• *The lingual artery* runs alongside the hyoglossus muscle, then laterally, in the genioglossus muscle, up to the apex of the tongue.

• *The lymph vessels.* In the region are numerous lymph ducts, which collect the lymph from the alveolar border of the mandible, from the mucosa of the floor of the oral cavity, from the apex and the respective border of the tongue, and empty into the submandibular lymph nodes. In neoplasms of the submandibular gland, 1-2 hypertrophied lymph nodes could be detected in the sublingual region.

The floor of the mouth can be approached by the following routes:

- the inferior route, by section of the cutaneous wall between the inferior border of the mandible and the hyoid bone;

- the superior route, through the mouth, by incision of the mucosa of the floor of the mouth, either between the sublingual gland and the mandible, or between the sublingual gland and the inferior surface of the tongue (route of approach to Wharton's duct);

- the anterior route, passing through the submental trigone. The incision may be made, either on the median line or more laterally, on the anterior border of the anterior belly of the digastric muscle. In case of need, this muscle can be disinserted from the mandible;

- a wide route of approach to the floor, unfrequently used, consists in the section of the mandible, either on the median line, or at the junction of the horizontal with the vertical branch.

The Roof of the Mouth

The oral cavity is bounded posterosuperiorly by a concave wall, which makes up the palatine region.

This wall is contiguous anteriorly and superiorly with the nasal fossae, laterally with the gingivodental (maxillary) region and with the maxillary sinuses, superiorly and posteriorly with the nasal pharynx and with **the** auditory tubes, laterally and posteriorly with the lateral walls of the pharynx and inferiorly with the oral cavity and the tonsillar region.

The superoposterior surface of the palatine region corresponds to the nasal fossae, to the maxillary sinuses (the floor of which it forms) and to the nasal pharynx; the inferior surface is oriented towards the oral cavity, the roof of which it forms.

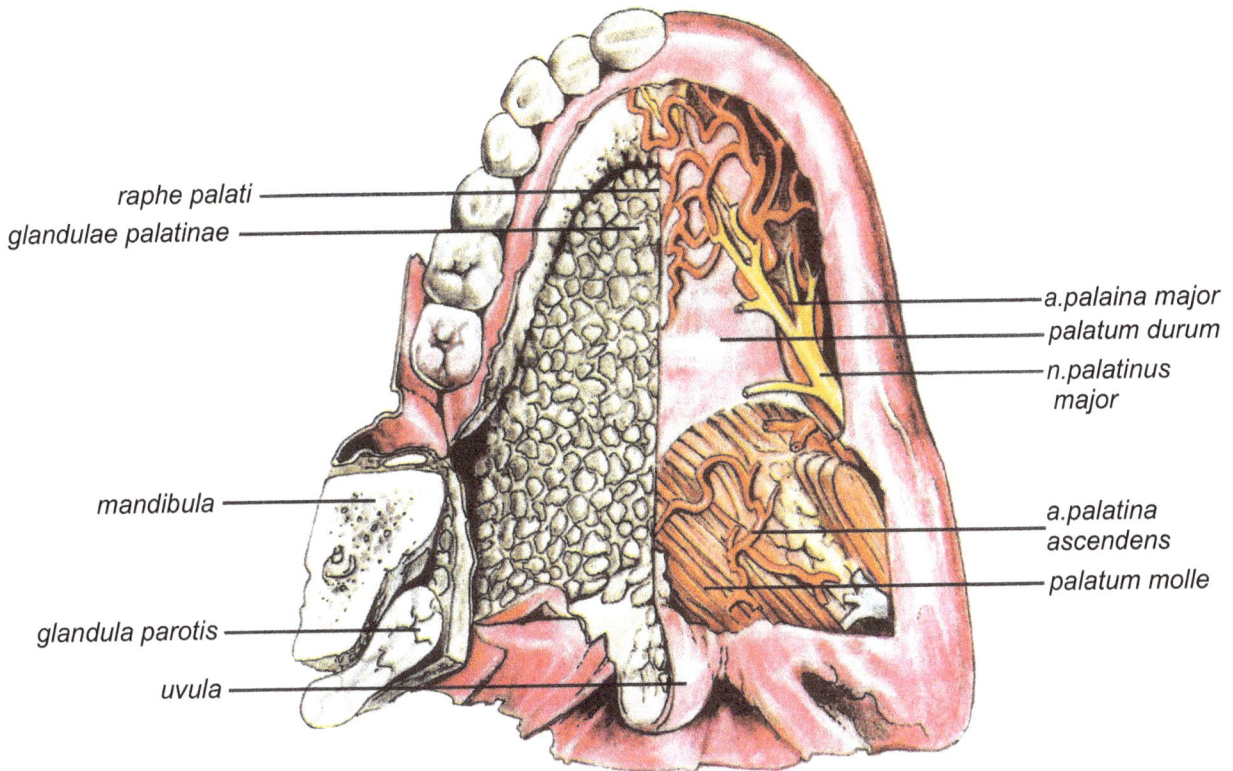

raphe palati
glandulae palatinae
a.palaina major
palatum durum
n.palatinus major
mandibula
a.palatina ascendens
palatum molle
glandula parotis
uvula

Fig. 14. The hard palate and the soft palate (superficial plane on the left - deep on the right)

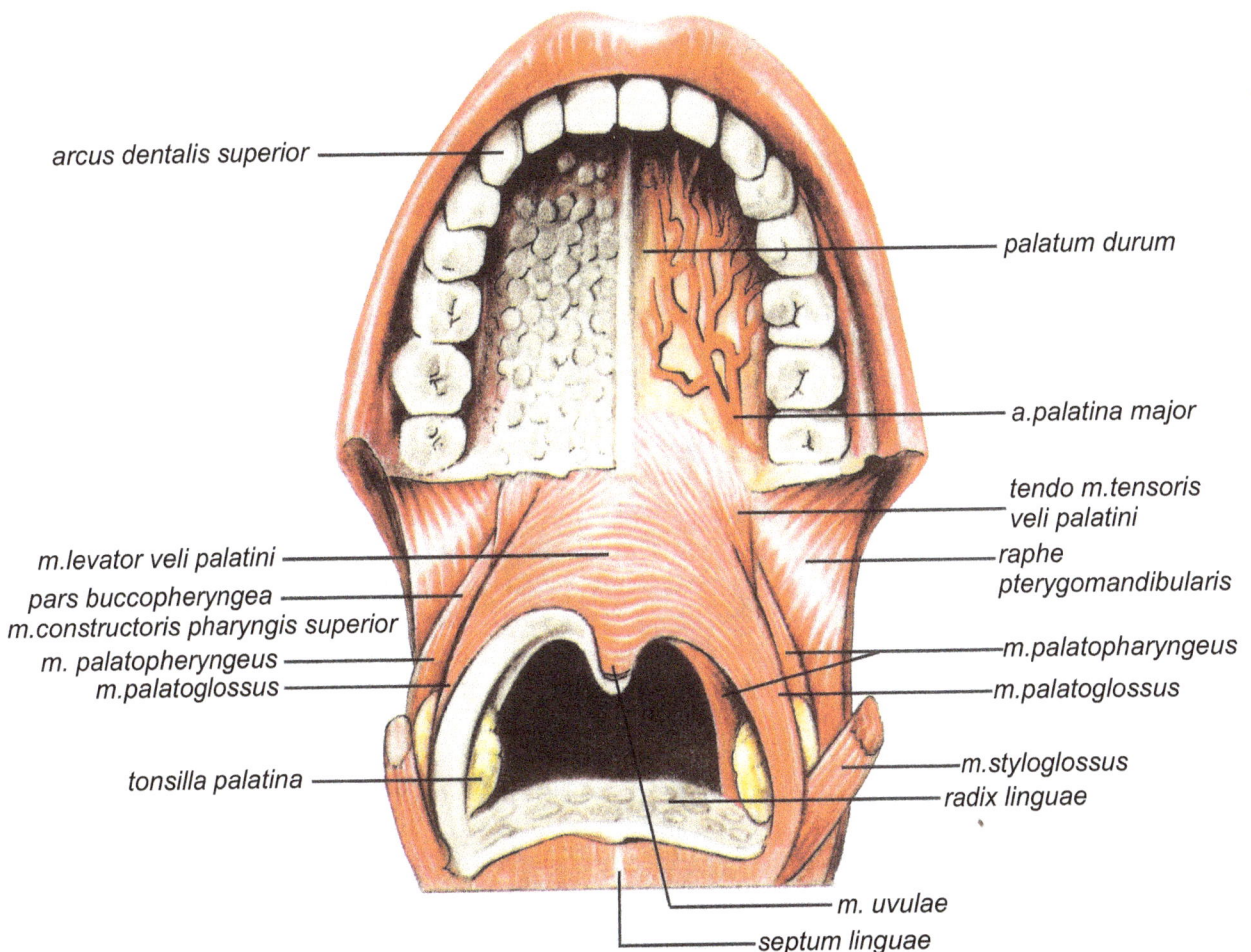

arcus dentalis superior
palatum durum
a.palatina major
tendo m.tensoris veli palatini
m.levator veli palatini
raphe pterygomandibularis
pars buccopheryngea
m.constructoris pharyngis superior
m. palatopheryngeus
m.palatoglossus
m.palatopharyngeus
m.palatoglossus
m.styloglossus
radix linguae
tonsilla palatina
m. uvulae
septum linguae

Fig. 15. The hard and the soft palate, the tonsillar region (muscles of the region).

31

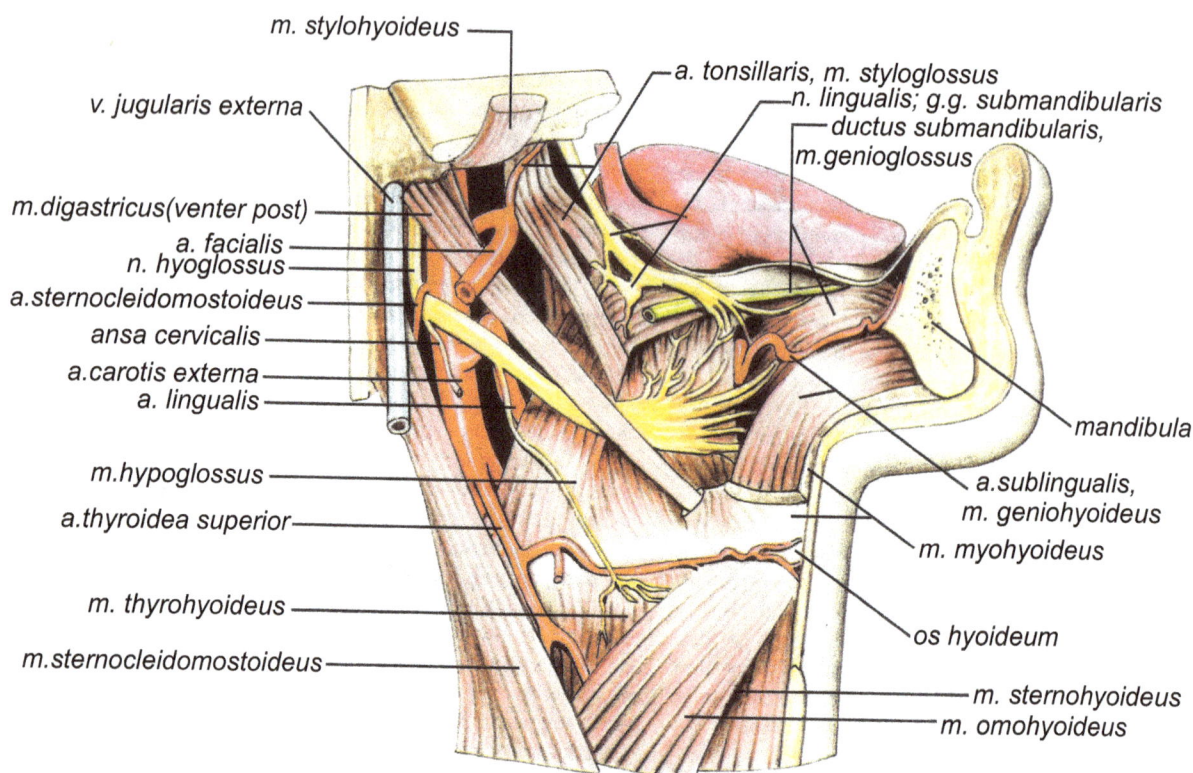

Fig. 16. The sublingual and the suprahyoid regions
(hypoglossal nerve, lingual nerve, hypoglossal muscle, sublingual artery - relations).

Labels for Fig. 16:
- m. stylohyoideus
- v. jugularis externa
- m.digastricus(venter post)
- a. facialis
- n. hyoglossus
- a.sternocleidomostoideus
- ansa cervicalis
- a.carotis externa
- a. lingualis
- m.hypoglossus
- a.thyroidea superior
- m. thyrohyoideus
- m.sternocleidomostoideus
- a. tonsillaris, m. styloglossus
- n. lingualis; g.g. submandibularis
- ductus submandibularis, m.genioglossus
- mandibula
- a.sublingualis, m. geniohyoideus
- m. myohyoideus
- os hyoideum
- m. sternohyoideus
- m. omohyoideus

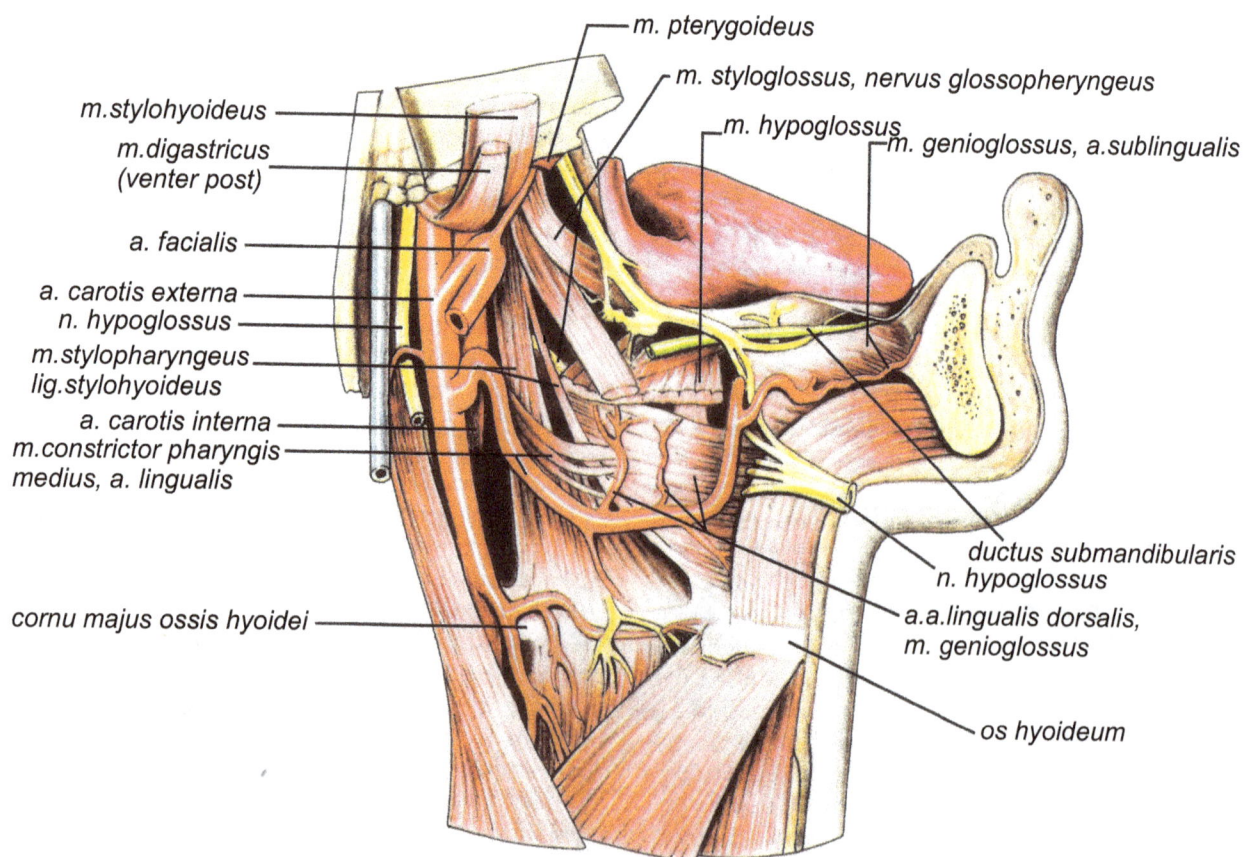

Fig. 17. The sublingual region
(the lingual artery and the lingual nerve after the section of the posterior belly
of the digastric muscle - relations)

Labels for Fig. 17:
- m.stylohyoideus
- m.digastricus (venter post)
- a. facialis
- a. carotis externa
- n. hypoglossus
- m.stylopharyngeus
- lig.stylohyoideus
- a. carotis interna
- m.constrictor pharyngis medius, a. lingualis
- cornu majus ossis hyoidei
- m. pterygoideus
- m. styloglossus, nervus glossopheryngeus
- m. hypoglossus
- m. genioglossus, a.sublingualis
- ductus submandibularis
- n. hypoglossus
- a.a.lingualis dorsalis, m. genioglossus
- os hyoideum

32

The palatine region consists of two distinct parts: an anterior part, nearly horizontal, bony, fixed, the palatine vault or the hard palate *(palatum durum)*, and a posterior, oblique, nearly vertical, soft, membranous, mobile part, the velum palatinum or soft palate *(palatum molle)*.

The palatine region can be examined and explored through the oral cavity, through the nasal fossae and through the nasal pharynx. The exploration the most at hand is through the oral cavity. The exploration is aimed at establishing a change in the shape of the vault, the existence of a solution of continuity, the presence of fissures and perforations; the digital palpation of its surface permits the detection of a thickening, of a tumour or of a collection. The posterosuperior surface of the velum can be explored with the finger passed through the mouth into the pharynx.

The palatine region, concave in both directions and circumscribed by the maxillary dental arch, is oriented downwards and forwards; it is 7-9 cm long. The length of the palate varies with the length of the posterior part of the velum, which may be short, middle-sized or long. The mean height of the vault is 1.5 cm. There are deep and narrow vaults, others are lower, stretched up to flattening (especially in partially or totally edentate subjects).

In children, the palatine velum is longer and may sometimes persist in this manner, producing a nasalizing voice. The palatine velum vibrates during a deep sleep; in the state of deep sleep the jaws relax and we inspire both by the nostrils and by the half-open mouth and the palatine velum flutters between the two air currents formed, producing the characteristic snoring (fig. 14, 15).

The palatine region is made up of four tissue layers:
- the mucosa which covers the inferior surface of the palate with an epithelium of a digestive type (stratified pavement epithelium), of which develop:
- the glandular layer of the palate;
- the skeletal, bony layer of the hard palate, which is continuous with a fibromuscular plane, for the soft palate;
- the mucosa which covers the superior surface of the palate (the floor of the nasal fossae) with a ciliated columnar epithelium of a respiratory type.

The Hard Palate
(Palatum durum)

The oral mucosa lines the palatine vault; in the dental arch it is thick, very closely adherent to the osteoperiosteal plane. In the anterior two-thirds and on the margins it presents 3-7 transverse palatine red-gray folds. In the posterior third, the mucosa is smooth, slightly vaulted and furnished with numerous glandular excretion orifices.

In the anterior two/thirds of the medium line, a whitish streak may be distinguished; in the posterior third, this streak forms the palatine raphe, which appears either elevated as a small crest or depressed as a groove. This groove ends anteriorly, behind the median inter-incisive space, on a small piriform fossa or, sometimes, on an elevation, the palatine tubercle, which corresponds to the incisive foramen.

The depression or the tubercle indicates the place of the Jacobson's canal, a vestige of the primitive bucconasal communication. At this level a connective fascicle penetrates the anterior palatine canal and fastens firmly the raphe. Posteriorly, the palatine raphe ends on the uvula.
The glandular layer extends on either side of the midline in the form of two parallel, 2-3 mm -thick bands, which extend up to the soft palate. This layer cannot be dissociated from the submucosa, as it is fixed to the deep plane of the latter.

The skeleton (the hard palate), reinforced at the periphery by the upper alveolar arch, is anteroposteriorly and fransversally concave. It is formed, in the anterior two-thirds, by the palatine processes of the maxillary bones, articulated with the horizontal processes of the palatine bones, which make up the posterior third of the palatine vault.

Smooth in its posterior part, the palatine vault is rugous in its anterior part, presenting parallel and concentric prominences. It is marked by two sutures perpendicular on each other: one median anteroposterior, the median palatine suture, and another one which crosses it transversally. The cruciate suture represents the articulation line of the four bony processes which form the hard palate.

Slightly posterior to the transverse suture, the posterior and accessory palatine canals open into the lesser palatine foramen *(foramen palatinum minor)*, at the level of the last molar, in proximity of the dental border; this foramen opens upwards, widening funnel-shaped, into the pterygopalatine fossae.

The posterior border of the vault is concave anteriorly and presents in the median plane a bony prominence oriented posteriorly, the posterior nasal spine *(spina nasalis posterior)*. On either side, this border ends on the extremity of the pterygoid process. The medial wing of the pterygoid process presents a projection oriented transversally and outwards, the hamulus of the pterygoid process. Around the main foramen may be seen also one or several secondary foramina. On either side, from the anterior border of each palatine foramen begins a small gutter which follows the border of the vault, where the latter unites with the alveolar arch; through this gutter pass the palatine vessels and nerves. On the sides of the gutter may be seen small fossae separated from each other, some of them through pointed crests containing glandular lobules. The periosteum which covers the bony plane is well developed and adheres at the level of the raphe. It is united with the mucosa and detaches with difficulty from the bony plane, especially because of the rugosity of the bony surface. This periosteum has a practical significance in the surgery of the palatine vault, either to create a provisional route to the nasal fossae or in plastic operations in children with cleft palate *(palatoschisis)*. The latter plane corresponds to the nasal fossae and is formed by the nasal mucosa.

Vessel and nerve supply

The buccal plane of the hard palate is mainly supplied by the *descending palatine artery,* which is the artery proper, and secondarily by the branch of the nasal septum (septal branch) of the sphenopalatine artery.

The descending palatine artery (posterior palatine artery) is the artery of the hard palate and arises in the infratemporal fossa from the internal maxillary artery. It passes through the greater palatine canal, over the anterior prominence of the palatine foramen and is continuous with the vascular groove of the palatine vault, giving off medial and lateral branches; the latter ramify and anastomose with each other in arches on the gingival border of the palatine vault. |

As the course of the main arteries of the vault is known, to avoid their lesion during surgery, the incision for the mucosa flap intended to the obliteration of a defect should be started anteriorly to the last molar. The secondary pedicle of the vault is formed by the internal nasal artery, a branch of the sphenopalatine artery, which penetrates through the incisive (anterior palatine) foramen, is distributed in the anterior portion of the vault and anastomoses with the descending palatine artery.

The veins accompany the arteries and reach the pterygoid plexus and the veins of the nasal mucosa; absolutely secondarily they merge into the veins of the pharynx or of the tonsil.

The lymph vessels run towards the deep cervical lymph nodes and the retropharyngeal lymph nodes. *The nerve supply* is assured by the branches of the maxillary nerve. The anterior and middle palatine nerves (from the greater palatine foramen) pass beside the artery and divide into a bunch of branches which are distributed to the mucosa, coursing among the glandular lobules; some of the filaments pass posteriorly towards the velum palatinum. The innervation territory corresponds to the posterior two-thirds of the hard palate, at the level of the molars and premolars. The nasopalatine nerve penetrates through the incisive foramen and is distributed in the anterior part of the mucosa of the palatine vault up to a transverse line which joins the two canines. The incisive foramen, covered by the incisive papilla, lies on the median line, at 1 cm behind the medial incisors; small in the lower portion, it is divided in "V" or "Y" by a partition wall, courses in an oblique posterosuperior direction and opens on the floor of the nasal fossae, on either side of the nasal septum. In order to perform a correct anaesthesia, one should penetrate the canal for approximately 5 mm, in a parallel direction with the axis of the central incisor. The greater palatine foramen is situated at 5 mm anteriorly to the posterior border of the hard palate, at 1 cm behind the hamulus of the internal wing of the pterygoid process and at 1 cm medially to the gingival border. In the frontal plane, it is situated between the second molar and the wisdom tooth. The palatine canal is oriented downwards and anteriorly. To perform a good anaesthesia it is not necessary to penetrate the canal, it is sufficient to infiltrate the zone of the foramen, especially its anterior zone.

The Soft Palate or Velum Palatinum
(Palatum molle)

The soft palate is a musculo-aponeurotic bridge between the rhinopharynx and the oropharynx. Obturator during deglutition and regurgitation, it modifies the emission of sounds. It is attached to the posterior part of the hard palate and is posteriorly continuous with the lueta or uvula.

Inferiorly it is attached to the root of the tongue and to the pharynx by two muscular arches, the palatoglossal arch *(arcus palatoglossus)* or anterior pillar and the palatopharyngeal arch *(arcus palatopharyngeus)* or posterior pillar of the velum, which enclose the palatine tonsil. It is made up of aponeurosis, muscles and mucosa.

The palatine aponeurosis .It is a fibrous lamella, attached behind to the bony palate, on which it continues, and laterally to the hamulus of the medial wing of the pterygoid process. It is a triangular aponeurosis, inserted on the mentioned bony zones; on the superior and inferior surface of this aponeurosis are attached a series of muscles which ascend from its superior surface towards the base of the skull (the levator veli palatini muscle and the tensor veli palatini muscle) or which descend from its inferior surface (the palatoglossus muscle and the palatopharyngeal muscle). In addition, in the medial part, short muscle fibres start from the posterior nasal spine towards the tip of the uvula (muscle of lueta or uvula).

The muscles of the soft palate

The levator veli palatini muscle *(m. levator veli palatine)*(internal petrostaphylinus muscle) is situated on the posterolateral part of the posterior nasal aperture. It has its origin:

-through a small tendon, on the rugous area of the inferior surface of the temporal bone, anterolaterally to the lower opening of the carotid canal;

- on the vaginal process of the tympanic part of the temporal bone, which forms the superior part of the carotid sheath;

- through a few fleshy fibres starting from the inferior surface of the cartilaginous part of the auditory tube.

At its origin, the muscle is situated inferiorly to the auditory tube, since it crosses only the medial part of the tube at the level of the lateral lamina of the pterygoid process and inserts into the superior surface of the palatine aponeurosis, crossing at the level of the median line the homonymous muscle on the opposite side. It forms in the submucosa of the rhinopharynx the fold of the levator.

Action: it elevates the soft palate and, accessorily, dilates the opening of the auditory tube.

The tensor muscle of the soft palate *(m. tensor veli palatini,* external petrostaphylinus muscle) is situated laterally to the medial lamina of the pterygoid process, to the auditory tube and to the levator muscle of the soft palate. It arises from the scaphoid fossa of the pterygoid process, the lateral lamina of the cartilage of the auditory tube and the bony part of the latter, between the oval foramen and the foramen spinosum, as well as from the medial surface of the spine of the sphenoid bone. Its fibres descend between the medial pterygoid muscle and the levator veli palatini, flexing, united in a tendon, round the pterygoid hamulus; then it passes through the origin of the buccinator muscle and inserts into the palatine aponeurosis, into the surface situated behind the palatine crest and into the horizontal plate of the palatine bone. Between the tendon and the pterygoid hamulus *(hamulus pterygoideus)* lies a small serous bursa.

Action: it tenses the anterior part of the soft palate during deglutition, opening at the same time the auditory tube.

The pharyngopalatine muscle *(m. pharyngostaphylinus, m. palatopharyngeus)* arises from the inferior surface of the palatine aponeurosis, from the medial extremity of the cartilage of the tube and from the pterygoid hamulus, together with the fibres of the superior constrictor muscle of the pharynx. It makes up the posterior pillar (the palatopharyngeal arch) *(arcus palatopharyngeus)* of the soft palate; in the inferior part, most muscle fibres are continuous with the submucous plate of the lower part of the pharynx.

Action: it elevates the pharynx or depresses the soft palate; it narrows the isthmus of the pharynx *(isthmus faucium).* The palatoglossus muscle *(m. palatoglossus)* arises from the infero-medial surface of the palatine aponeurosis and passes to the lateral part of the base of the tongue, forming the palatoglossal arch *(arcus palatoglossus),* the anterior pillar of the soft palate.

Action: the same as that of the pharyngopalatine muscle.

The muscle of uvula (m. of lueta, *m. uvulae)* detaches from the posterior nasal spine and ends in the uvula.

Summing up the action of the tensor and levator muscles of the soft palate, we retain that the first tenses the anterior part of the soft palate, during deglutition, up to the level of the plane formed by the pterygoid hamulus, while the second also tenses the soft palate, but it can elevate it beyond the plane which passes through the two pterygoid muscles, applying the soft palate on the posterior surface of the pharynx. Both muscles are dilators of the tube. By their topography they surround the pharyngotympanic tube: the tensor inserts anteriorly to the tube and the levator behind it. Both muscles open the tube and allow the ventilation of the tympanic cavity.

The mucous membrane of the soft palate resembles, in part, that of the hard palate. However, it is thinner, less adherent and separated from the glandular layer by a submucosa of connective tissue.

The tonsilllar region *(regio tonsillaris)* is situated at the boundary belween the oral cavity and the pharynx and participates in the formation of the pharyngeal isthmus. It consists of the tonsillar fossae *(fossae tonsillares)* and the palatine tonsils *(tonsillae palatinae)*. The tonsillar fossa is ogival, owing to the distal divergence of the pillars of the soft palate.

The anterior pillar. The anterior palatoglossal arch *(arcus palatoglossus)* is formed by the lateral prominence of the palatoglossus muscle covered with mucosa. It arises in proximity of the base of the uvula, passes on the pharyngeal wall and runs towards the lateral border of the tongue. The palatopharyngeal arch or posterior pillar *(arcus palatopharyngeus)* is a musculomembranous fold situated more medially than the anterior pillar and inserts into the pharyngeal wall. It is triangular, has an oblique direction inferiorly, laterally and posteriorly and is covered by a fold of the pharyngeal mucosa. The lateral wall of the tonsillar fossa is formed by a muscular part, made up by the superior pharyngeal constrictor, by the pharyngeal aponeurosis and, very rarely, by the amygdaloglossus muscle, when the latter exists. More laterally lie the styloglossus and the stylopharyngeal muscles, situated in the paratonsillar space. The base, situated inferiorly, corresponds to the lateral glosso-epiglottic fold; the lateral wall is formed by the intrapharyngeal aponeurosis. In the superior half of this fossa is lodged the palatine tonsil.

The palatine tonsil *(palatine amygdal)* consists of masses of lymph follicles, situated on the pharyngeal wall in the tonsillar region. It is oval in shape, elongated supero-infersorly, flattened lateromedially. The medial surface is plane or convex, covered by the smooth pharyngeal mucosa, presenting orifices (18-20) which correspond to the tonsillar crypts. The lateral surface is bounded by a layer of connective tissue, forming the tonsillar capsule, detachable with difficulty from the parenchyma. Between the superior two-thirds of the capsule and the pharynx lies a cleavage plane, consisting of loose connective tissue - the peritonsillar space used in tonsillectomies; in this space collect the tonsillar abscesses, improperly termed phlegmons of the tonsil. The inferior third of the capsule Is more adherent to the adjacent muscles and presents zones of access to the main tonsillar pedicle (fig. 15).

The superior pole is separated from the hard palate by the supratonsillar fossa, a remnant of the second branchial groove.

The posterior border of the tonsilla is separated from the posterior pillar by a groove of a variable depth.

The inferior pole remains generally at 2 cm above the lateral glosso-epiglottic fold, separated from the surrounding lymphoid tissue.

The palatine tonsil can be easily inspected through the mouth. When the mouth is wide open and the tongue is drawn out and pressed downwards and forwards with a spatula, the tonsil, situated belween the two palatine pillars, appears more or less bulky; it is of a greater,size in childhood and atrophic in the aged. If may be approached through the oral cavity, digitally or instrumentally.

The exploration by facial route is difficult, since the tonsil is covered by the ramus of the mandible. If we vigorously press with the tip of the finger the soft planes beneath the angle of the mandible, a considerably enlarged tonsil may be felt; the examination is much facilitated if a finger is introduced into the tonsillar region, but this is a very delicate procedure.

The tonsillar region is contiguous: superiorly, with the softpalate, towards which spreads usually a purulent collection of the tonsil; antero-inferiorly, with the base of the tongue, towards which spreads usually a neoplasm; inferiorly, with the submaxillary fossa, towards which propagate the lateral phlegmons of the tonsil.

The external surface of the region is in relation, through the phasyngeal wall, with the parapharyngeal space (or "maxilopharyngeal space"), filled with connective and fatty tissue, in which are buried: the styloglossus and stylohyoid muscles, the ascending palatine, the facial, the external and internal carotid arteries, the internal jugular vein and the glossopharyngeal, the vagus, the accessory and the hypoglossal nerves. Most of them are situated posterolaterally to the tonsil projection area on the pharyngeal wall. The internal carotid artery lies at a distance of 1.5 cm laterally and posteriorly to the tonsilla (sometimes in its immediate proximity), obliging the surgeon to be very cautious in performing operations in the tonsillar region. The facial artery may sometimes bend in proximity of the inferior pole of the tonsil, requiring the same measures of precaution in order to prevent haemorrhages. Of the mentioned nerves, the glossopharyngeal nerve, situated very close to the pharyngeal wall, is the most liable to be damaged.

Vessel and *nerve* supply of the tonsil

The arteries derive from the lingual, the descending pharyngeal and the facial arteries, which all reach the tonsil on its lateral surface (termed also *"tonsillar hilum")*. The most important of these is the ascending palatine artery *(a. palatina ascendensj* (derived from the facial artery), which gives off a branch *(ramus tonsillaris),* sometimes thick, generating massive haemorrhages during the tonsillectomy.

The veins form on the surface of the tonsil a fine network, drained by the veins of the pharynx.

The lymphatics converge on the submandibular lymph nodes, situated around the angle of the mandible, and on the deep cervical lymph nodes *(nodi lymphatici jugulodigastrici).* In the case of inflammatory processes, they tumefy and become sensible on palpation, forming sometimes a painful block beneath the angle of the mandible.

The nerves form on the surface of the tonsil a fine network, made up of fibres detached from the lingual (trigeminal) and glossopharyngeal nerves.

Vessel and nerve supply of the palatine vault and of the velum palatinum

The arteries of the palatine vault and of the velum palatinum derive from the facial artery *(a. palatina ascendens)* and from the last segment of the internal maxillary artery *(a. palatina descendens),* through the greater palatine artery *(a. palatina major),* respectively the lesser palatine arteries *(aa. palatinae minores),* which ramify in the pterygomaxillary fossa.

The veins form plexuses which make up small trunks, draining the blood into the pterygoid plexus and the veins of the pharynx.

The lymph vessels drain the lymph towards the superior group of the deep lymph nodes of the neck. *The sensitive nerves* are branches of the nasopalatine nerve *(nn. palatini major, medius et minores),* which reaches the region through the anterior palatine canal, and is a ramus of the maxillary nerve. In addition, the region is supplied also by sensitive fibres of the glossopharyngeal and vagus nerves. The motor nerves derive from the mandibular *(n. tensoris veli palatini),* glossopharyngeal and vagus nerves, respectively from the facial and hypoglossal nerves (the latter supplies only the glossostaphyline muscle).

Anexa of the Oral Cavity
The Gingiva
(Gingivae)

If is the portion of the oral mucosa adherent to the periosteum of the maxilla and mandible, which covers the alveolar processes on their vestibular and buccal (palatine or lingual) surfaces. Between the teeth, the vestibular gingiva is continuous with the oral gingiva; around the teeth if forms a gingival ring, which contributes to the fixation of the teeth to the alveola, forming here, with the alveolodental periosteum, a firm, resilient connection, which should be undone before the extraction of the respective tooth. The gingival mucosa adheres firmly to the periosteum, therefore the novocain infiltration of the gingiva is very painful

The Teeth
(Denies)

They are hard structures implanted in the alveolae of the jaws, serving to mastication and to articulation of words.

Arranged on two rows, they make up the dental arches. In the closed mouth, the maxillary arch is usually superimposed on the mandibular arch, achieving the so-called Individual bite" or dental occlusion. In humans, the teeth are replaced only once (diphyodont dentition), a process which starts at the age of about 6 1/2 years and ends at 14-15 years, even somewhat later for the third molar ("wisdom tooth" - *dens sapientiae).*

The dentitions. The first dentition comprises 20 deciduous teeth or milk teeth *(denies decidui lactei):* 8 incisors, 4 canines and 8 milk molars, the last molars occupying the place of the premolars of the permanent dentition. The period of their appearance extends from the 6th to the 36th month after birth. Initially appear the central incisors (months 6-9), then the lateral ones (months 9-18), the first molars (months 22-26), the canines (months 28-34) and, finally, the molars of the second row (months 32-36) (fig. 18, 25).

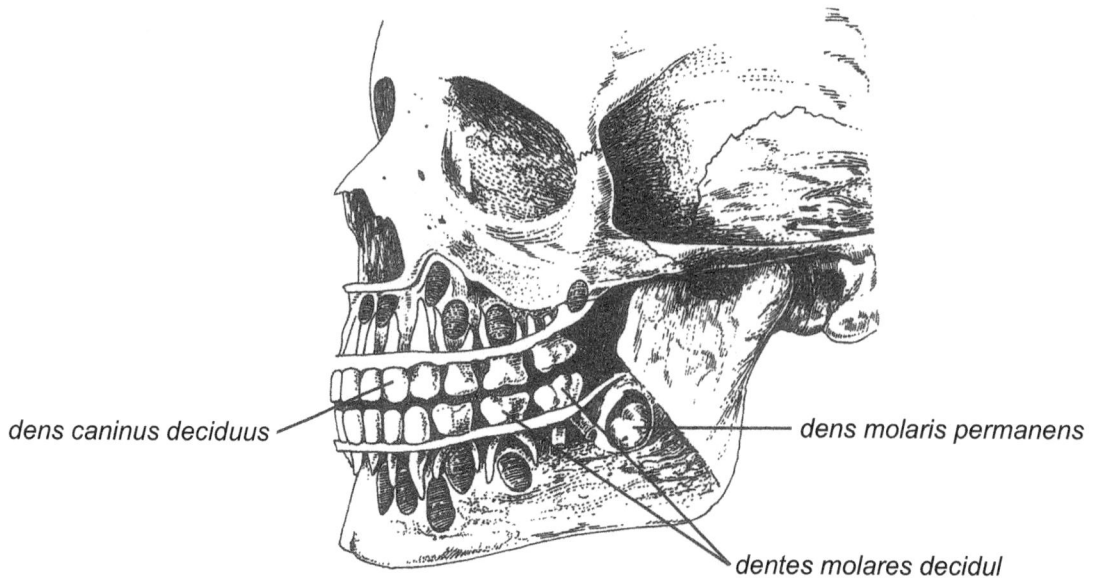

Fig. 18. The dentition (set) in a 6 year old child

dens caninus deciduus

dens molaris permanens

dentes molares decidul

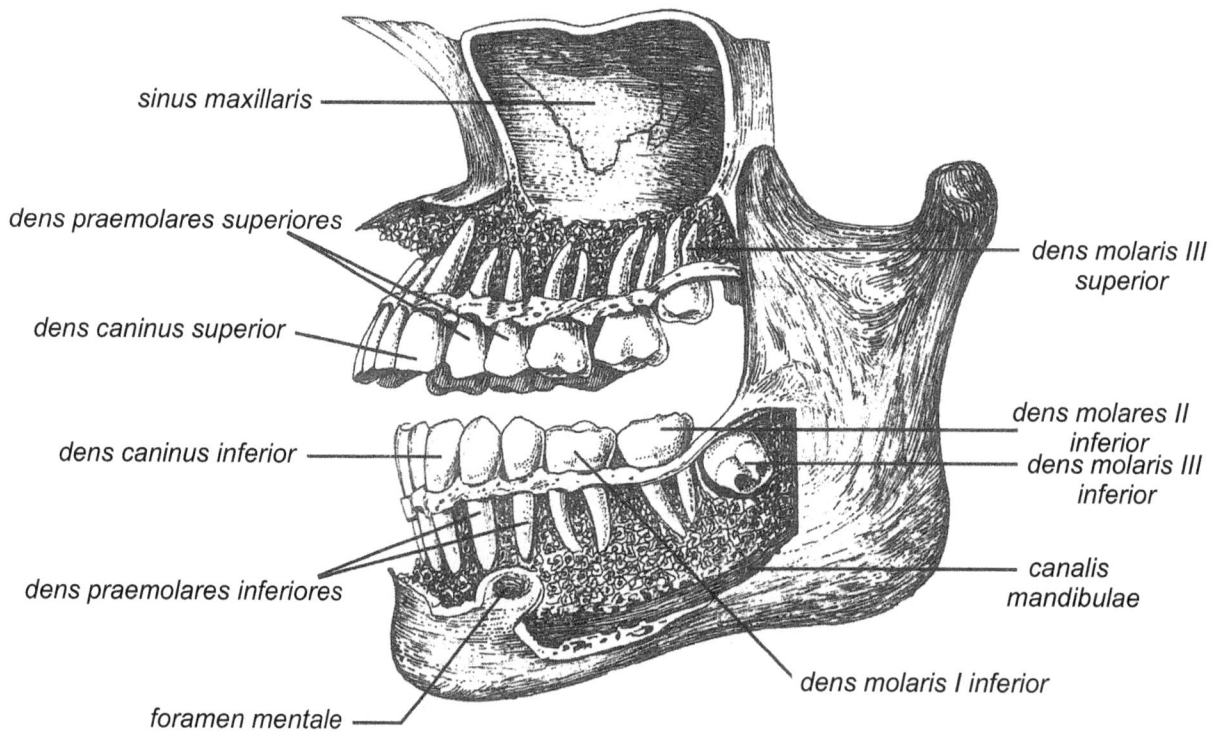

Fig 19. The dentition (set) in the adult

sinus maxillaris

dens praemolares superiores

dens caninus superior

dens caninus inferior

dens praemolares inferiores

foramen mentale

dens molaris III superior

dens molares II inferior
dens molaris III inferior

canalis mandibulae

dens molaris I inferior

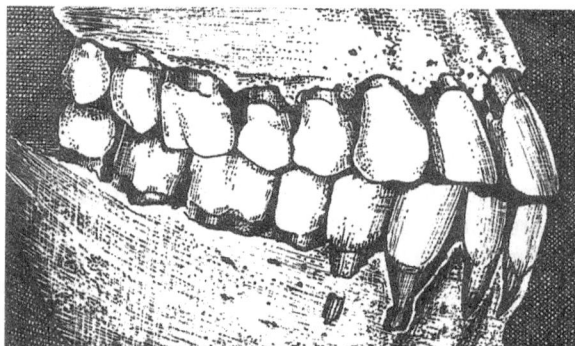

Fig. 20. The dentition - set in occlusion

The permanent teeth *(denies permaneniesj* appear after the expulsion of the deciduous teeth, the root of which is resorbed under the pressure of the permanent teeth (with the exception of the permanent molars, which are not represented in the first dentition). Here the eruption order is the following: the first molars (between 5 and 7 years), the central incisors (between 6 and 8 years), the lateral incisors (between 8 and 9 years), the canines (between 10 and 12 years), the premolars (between 11 and 12 years) and the second molar (between 12 and 14 years). The third molar (called also "wisdom tooth") is the last to appear, between 19 and 30 years. Generally, the mandibular teeth erupt before the maxillary ones, at 2-4 month - intervals.

The permanent teeth, 32 in number, are: 8 incisors, 4 canines, 8 premolars and 12 molars, uniformly distributed on the two dental arches (fig. 19, 20).

In the stomatological practice, for the numbering and recognition of teeth is used the "dental formula ", which expresses the running number of the respective tooth in relation with the median line, on each of the two arches:

$$\begin{array}{c|c} 87654321 & 12345678 \\ 87654321 & 12345678 \end{array}$$

$$\text{or simpler} \begin{array}{c|c} & 2123 \\ & 2123 \end{array}$$

General aspect. Usually, each tooth consists of three parts: the crown, the neck and the, root (fig. 21, 22).

The crown *(corona dentis)* is the visible, white-nacreous part of the tooth. One may observe an external surface, directed towards the oral, vestibule *(fades labialis sive buccalis),* an internal surface oriented towards the tongue or the palatine vault *(fades lingualis sive palatina),* surfaces in contact with the neighbouring teeth on the same arch *(fades contactus)* - according to the orientation, a medial and lateral contact surface at the incisors and canines, respectively an anterior and posterior contact surface at the premolars and molars - as well as a grinding surface *(fades masticatoria),* well individualized in the premolars and molars. On this last surface may be observed elevations termed dental tubercles *(tubercula dentis),* which have the role of contributing to the grinding of the foodstuffs.

The neck *(collum dentis)* joins the crown with the root and is situated at the level of the dental boundary of the gingiva.

The root *(radix dentis)* is the zone fixed in alveoli, separated from their wall through the alveolodental periosteum. The neck and the root of the tooth are yellowish. There are teeth with one or several roots, each of them ending with a tip *(apex radios dentis).*

Particular aspect (fig. 23, 24). The incisors *(denies incisivi)* are chisel-shaped, flattened antero-posferiorly. The lower incisors have a vertical position, while the upper ones are slightly bent forwards

The canines *(denies canini)* have a stronger crown and bulge on the labial surface. The conic, longer, strong root is oriented laterally.

The premolars *(denies premolares)* are characterized by the presence of a grinding surface with two tubercles (vestibular and lingual or palatine), partially separated by a transverse groove. The root is short, usually simple, although sometimes the first premolar can have a bifid root.

The molars *(denies molares)* are characterized by a masticatory surface of the crown presenting 3-5 tubercles, which have a significant role in mastication.

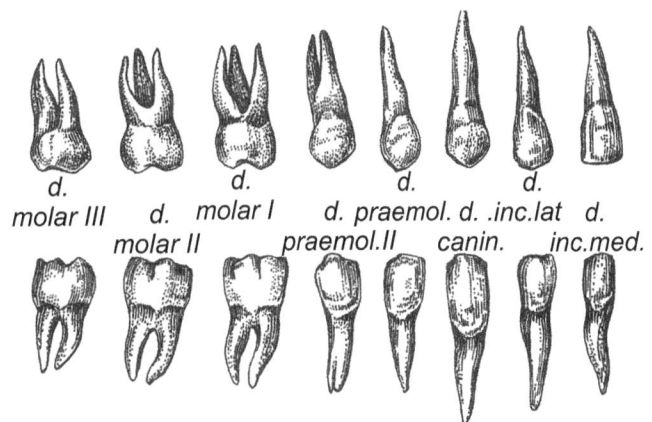

d. molar III d. molar I d. praemol. d. .inc.lat d.
d. molar II praemol.II canin. inc.med.

Fig. 21. The teeth (incisors 1-2, canines 3, premolars 4-5, molars 6-7-8).

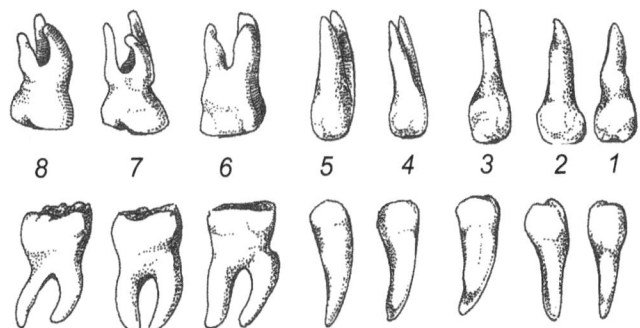

8   7   6   5   4   3   2   1

Fig. 22. The permanent dentition.

39

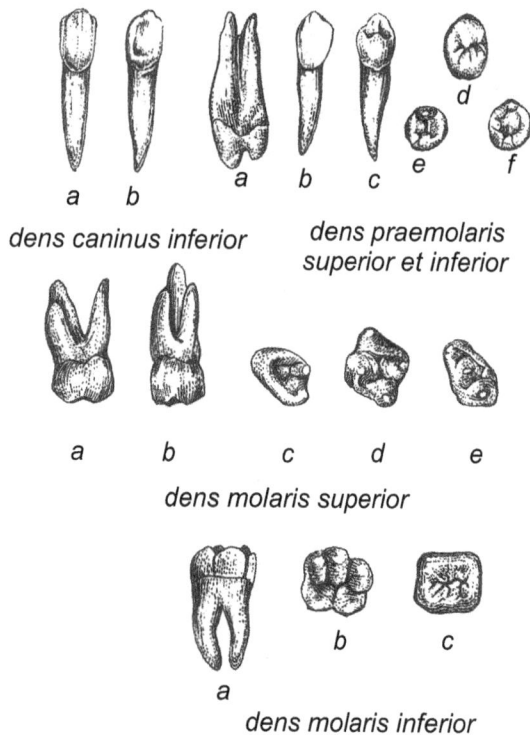

dens caninus inferior

dens praemolaris
superior et inferior

dens molaris superior

dens molaris inferior

Fig. 23. Dental surfaces

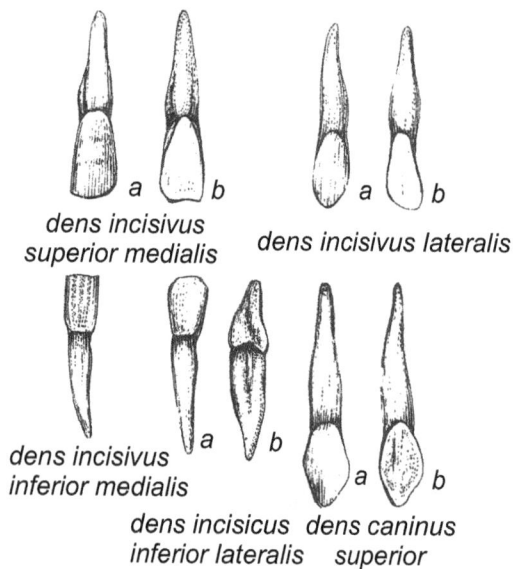

dens incisivus
superior medialis

dens incisivus lateralis

dens incisivus
inferior medialis

dens incisicus
inferior lateralis

dens caninus
superior

Fig. 24. Dental surfaces.

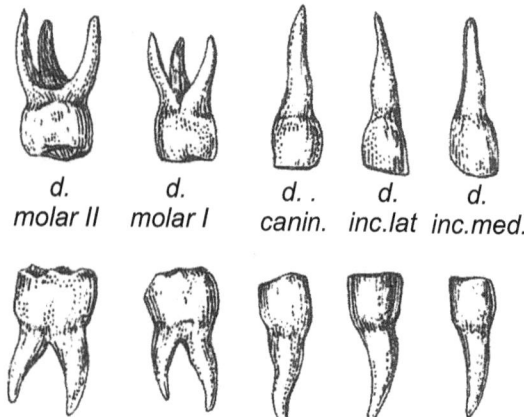

d.
molar II

d.
molar I

d.
canin.

d.
inc.lat

d.
inc.med

40 **Fig. 25. The deciduous dentition.**

The crown of the upper molars is rhomboid or ellipsoid and has on its grinding surface four tubercles (two buccal and two lingual or palatine tubercles), separated by a groove. The crown of the lower molars is usually cubical, larger and has generally a supernumerary tubercle (the Carabelli's tubercle) in comparison with the corresponding upper molar.

The third molar (the "wisdom tooth"- *dens serotinus sive sapientiae)* is the most poorly developed and may sometimes lack.

The upper molars have three roofs (two buccal and one palatine), while the lower molars have two roots (anterior and posterior), their axis being oriented obliquely backwards.The deciduous teeth have a smaller, bluish crown. The cavity of the tooth is larger.

The tooth are arranged on the dental arches in a row, the contact surfaces of the crowns getting in touch reciprocally. At the level of the neck of the neighbouring teeth are situated the interdental spaces *(spatia interdentaria).* The upper premolars and molars are oriented vertically, while the crown of the lower molars is directed medially, towards the oral cavity.

The upper dental arch *(arcus dentalis superior)* is ellipsoid and the lower arch *(arcus dentalis inferior)* is parabolical, with the possible existence of numerous variants.

The contact type of the dental arches is named "occlusion" or "articulation". In occlusion, the upper incisors pass beyond the lower ones, overlapping them partially.

Structure of the teeth (fig.26, 27). Inside each tooth is a cavity surrounded by dentine *(cavum dentis),* which is continuous with one or several axially arranged dental canals *(canalis radicis dentis),* furnished with an opening at the apex of the roots *(foramen apicis dentis).*

The cavity of the tooth and the canals of the roots are filled with a soft, reddish connective tissue, rich in blood vessels and sensitive nerves: the dental pulp *(pulpa dentis).*

Around the pulp is a bone-like substance, the ivory or dentine *(substantia eburnea),* which is yellowish, harder than bone and richer in mineral salts.

The dentine of the root of the tooth is covered by a thin, brown-yellowish layer of bony tissue, named cement *(cementum sive substantia ossea),* connected to the alveolodental periosteum by a system of collagen fibres (Sharpeys fibres). The dentine of the crown is enveloped in enamel *(substantia adamantina),* a white glistening substance, which represents the hardest tissue of the organism, consisting almost only of mineral salts and impregnated with apatite of the category of phosphates. It is made up of enamel prisms, elements of an elongated hexagonal shape, situated perpendicularly on the surface of the crown and united by an amorphous binding substance.

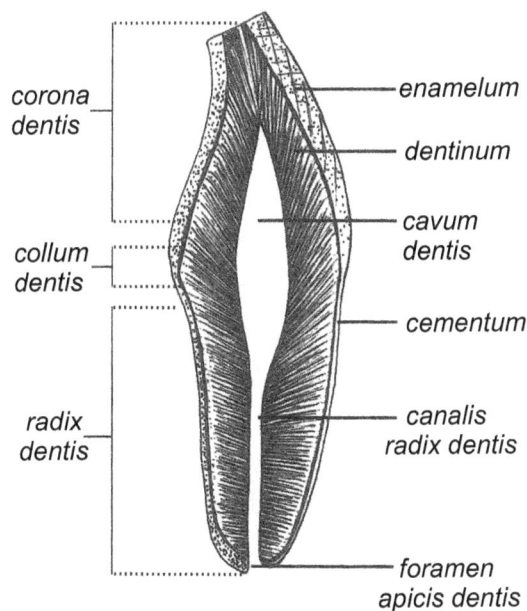

Fig. 26. Structure of a canine (longitudinal section)

corona dentis — enamelum, dentinum, cavum dentis, cementum, canalis radix dentis, foramen apicis dentis
collum dentis
radix dentis

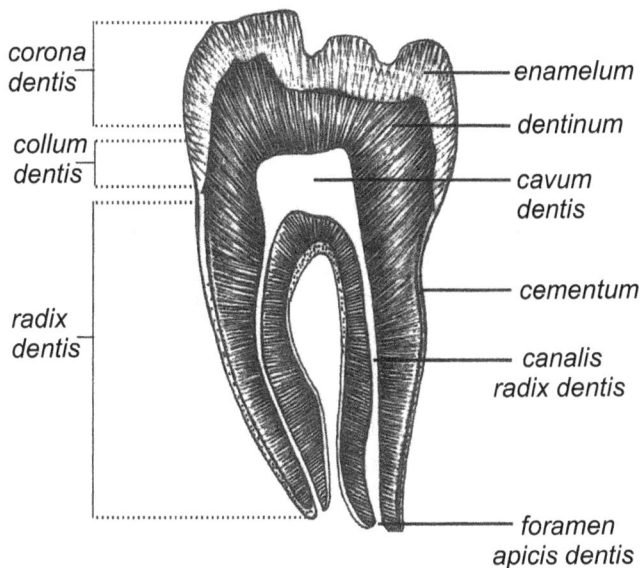

Fig. 27. Structure of a premolar.

corona dentis — enamelum, dentinum, cavum dentis, cementum, canalis radix dentis, foramen apicis dentis
collum dentis
radix dentis

The supporting and attachment apparatus of the tooth (parodontium) consists of four elements:

1. The periodontium (periodontium). It is formed of the connective tissue situated around the root of the tooth, between the root and the alveolar wall. The collagen fibres attach the tooth elastically and damp the pressure exerted by the tooth,During mastication, on the hard foodstuffs. At the inlet in the alveola, the fibres are oriented radially, slightly bent towards the surface of the tooth (ligamentum circulare dentis), contributing to the fastening of the gingiva to the tooth. Inferiorly, the fibres are directed more obliquely, forming at the level of the apex a complex of longitudinal and radial fibres - the apical ligament (ligamentum apicale).

2. *The gingiva* (previously described).

3. *The alveolar processes and dental alveoli.* The dental alveoli or tooth sockets are cavities situated at the level of the alveolar processes of the maxillary bone and of the mandible, in which the teeth are attached. The alveolar bone undergoes permanently changes in dependence on the functional mechanical factors acting at this level.

4. *The cement,* bound to the alveolodental periosteum by Sharpey's fibres.

Vessel and nerve supply of the teeth

The arteries supplying the alveoli and the teeth derive from the inferior alveolar artery *(a. alveolaris inferior)* and from the superior alveolar arteries *(aa. alveolares superiores anferiores et a. alveolaris superior posterior);* all are primary or secondary branches of the internal maxillary artery. Traversing the inferior dental canal of the mandible and the narrow canaliculi of the maxillary, these arteries form multiple anastomoses, sending to each tooth one or several dental branches *(rr dentales).* Before entering the orifice of the dental apex, they give off small rami meant to the gingiva and to the alveolodental periosteum, then they penetrate into the canal of the roof, where they capillarize.

The veins are satellites of the arteries.

The lymph vessels convey the lymph from the upper teeth to the middle and lateral groups of submandibular lymph nodes, while the lymph vessels which drain the lymph of the lower teeth run towards the superior group of deep cervical lymph nodes.

The nerves are sensitive and belong to the innervation territory of the trigeminal The superior dental arch sends branches of the maxillary nerve; they arise from the dental plexus, of which are given off here and there bony, gingival and dental branches *(rr. gingivales et dentales superiors).* The inferior arch is supplied by the inferior dental nerve (a branch of the mandibular nerve), which forms in the canal of the mandible an inferior dental plexus, of which ramify filaments to the gingiva and the teeth *(rr. gingivales et dentales inferiores).* All the mentioned branches emerge from the canals of the roots lying in the pulp, in which they form a very rich network.

The Pharyngeal Region

The Pharynx
[Pharynx]

The pharynx is the organ in which the respiratory tract (from the nasal fossae to the larynx) and the digestive tract (from the mouth to the oesophagus) cross each other; hence the name "pharyngeal crossroads". It has the shape of a musculomembranous funnel, situated anteriorly to the cervical column and extending on a length of about 15 cm from the base of the skull to the level of the inferior border of the cricoid cartilage (corresponding to the vertebrae C_6 in the adult, C_4 in the newborn and C_7 in the aged). These limits vary with the movements of the pharynx, by approximately the height of a vertebra.

The pharynx belongs to the head by the superior two-thirds contained in the concavity of the mandible; through the inferior third, where it is in relation with the larynx and continuous with the oesophagus, it belongs to the region of the neck.

Anteriorly, the pharynx communicates: in the superior third, with the nasal fossae through the choanae *(aditus nasopharyngeus);* in the middle third, with the oral cavity through the isthmus faucium *(aditus buccopharyngeus);* in the inferior third, with the larynx through the ary-epiglottic orifice *(aditus laryngis).* On these communications depends also its topographical division in regions: the superior region-the epipharynx or rhinopharynx *(pars nasalis pharyngis),* the middle region-the mesopharynx *(pars oralis pharyngis)* and the lower region - hypopharynx *(pars laryngea pharyngis)* (fig. 28-34).

Exterior configuration and relations.

The pharynx has a posterior surface, two lateral surfaces, a superior extremity and an inferior extremity. The posterior surface is nearly plane; it is continuous on either side with the lateral surfaces, with which it forms two nearly rounded angles. This surface corresponds to the cervical vertebral column, with the body and the transverse processes of the first 6 cervical vertebrae, covered by a muscle layer formed of the prevertebral muscles *(mm. rectus colli et rectus capitis)* and of the prevertebral fascia. In addition, it is in relation with the cervical portion of the sympathetic ganglionic chain. Between the pharynx, situated anteriorly, and the prevertebral fascia with the prevertebral muscles, situated posteriorly, lies the retropharyngeal space *(spatium retropharyngeum).*

This space contains loose connective tissue and two lymph nodes (the retropharyngeal lymph nodes), which receive lymph vessels from the hypophysis, the pharyngeal tonsil, the auditory tube and the nasal fossae. These lymph nodes are very susceptible to infections (retropharyngeal adenophlegmonsj.

The retropharyngeal space assures the mobility of the pharynx in relation with the vertebral bodies. Through the posterior wall of the pharynx it is possible to explore the superior vetebrae, as well as to incise adenophlegmons and to puncture cold abscesses appeared subsequently to a cervical Pott's disease. The cephalic segment is laterally in relation with the parotid gland.

The cervical segment is in relation with the sternocleidomastoid region and can be divided into two regions by a plane passing through the superior border of the thyroid cartilage: an inferior and a superior region.

In the superior region, the lateral wall is in relation with the origin of the two carotid arteries, internal and external, as well as with the superior thyroid artery, the ascending pharyngeal artery, the lingual artery, the internal jugular artery and inferiorly with the fhyrolinguo-facial venous trunk. Medially and superiorly to the confluence of these two veins courses the hypoglossal nerve. It crosses the common carotid artery at 5=20 mm above the bifurcation, more frequently above the origin of the lingual artery.

On the lateral surfaces of the pharynx descend the vagus nerves with the superior cardiac nerves. In the inferior region, the lateral wall is in relation with the common carotid artery,the Internal jugular vein, the vagus nerve, with the posterior part of the lobes of the thyroid gland and with its vascular pedicles, with the ansa of the hypoglossus, which accompanies the internal jugular vein.

The connective tissue, situated laterally to the pharynx *(spatium parapharyngeum),* contains the stylian muscles (with the insertion into the styloid process), which form with their fasciae a frontal septum (pharyngeal wings - Thoma Ionescu), that divides the lateropharyngeal space Into two zones: pre- and retrostylian.

The presfylian zone communicates directly with the parotid region; the retrostylian zone Is traversed by the Internal carotid artery, the jugular vein and the vagus nerve (X), as well as by the accessory nerve (XI), the glossopharyngeal nerve (IX), and the sympathetic trunk, formations Initially located posteriorly to the internal carotid artery; then the accessory spinal nerve runs laterally to the sternocleidomastoid and trapezius muscles; the glossopharyngeal nerve descends obliquely, coming towards the pharynx and the base of the tongue; the hypoglossal nerve, which has a course similar to that of the glossopharyngeal nerve, but situated lower, lies between the interna! jugular vein and the carotid artery, entering in the sublingual region. At this level are located also the deep cervical lymph nodes.

The cavity of the pharynx (endopharynx) *(cavum phafyngis)* is divided into three regions, called according to their communication: nasopharynx, oropharynx and laryngopharynx (fig. 31-34).

The nasopharynx *(pars nasalis pharyngis)* extends between the base of the skull, which participates in the formation of its superior wall (the vault of the pharynx) *(fornix pharyngis),* and the soft palate. It has a vertical diameter of 5 cm, a transverse diameter of 4 cm and an anteroposterior diameter of 2 cm. It communicates infersorly with the oral part of the pharynx *(pars oralis pharyngis)* and anteriorly with the nasal fossae, through the two choanae. The choanae are two oval openings with the great axis vertical; they are bounded laterally by the nasopharyngeal groove, medially by the posterior edge of the nasal septum, superiorly by the body of the sphenoid bone and inferiorly by the soft palate.

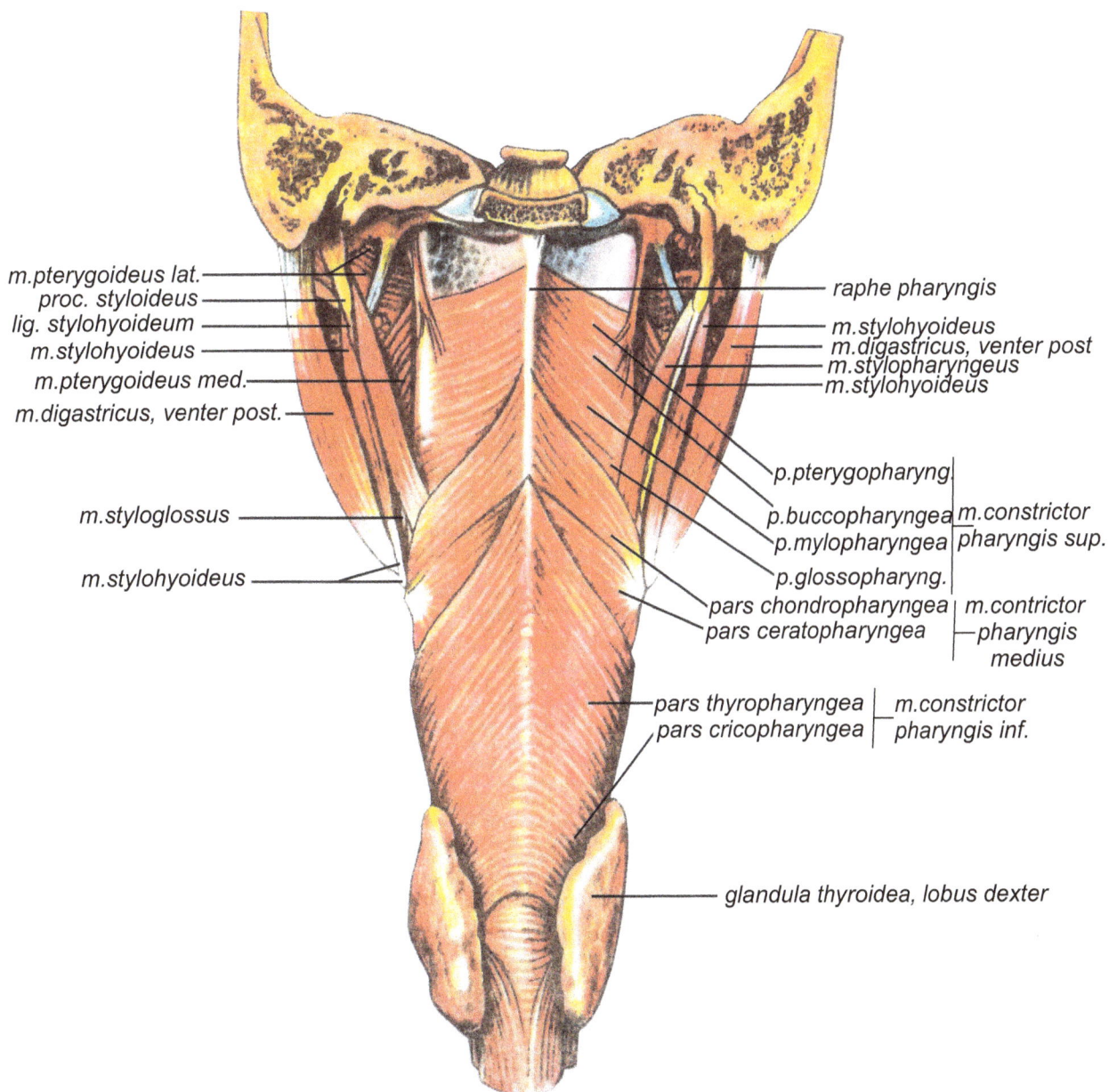

Fig. 28. Posterior surface of the pharynx – musculature

43

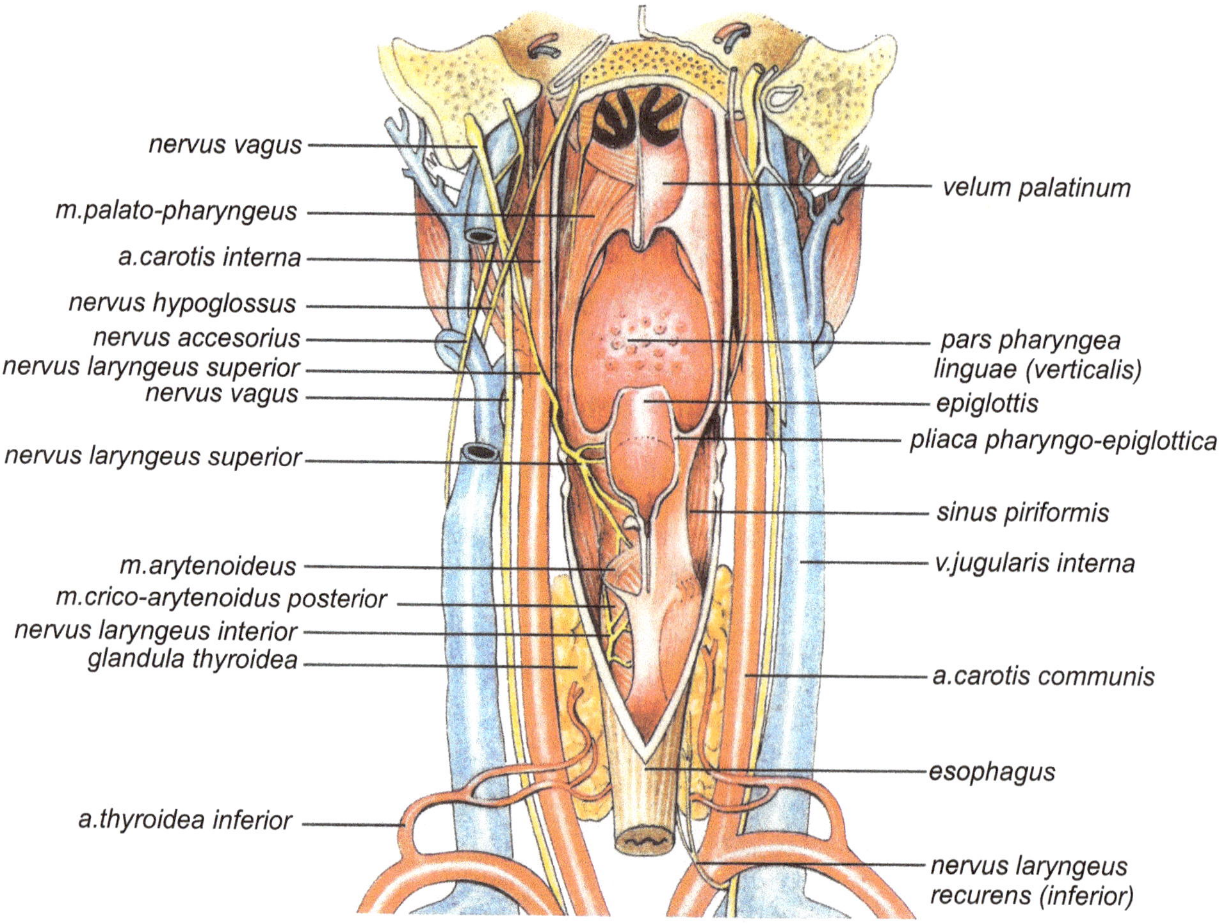
nervus vagus
m.palato-pharyngeus
a.carotis interna
nervus hypoglossus
nervus accesorius
nervus laryngeus superior
nervus vagus
nervus laryngeus superior
m.arytenoideus
m.crico-arytenoidus posterior
nervus laryngeus interior
glandula thyroidea
a.thyroidea inferior
velum palatinum
pars pharyngea
linguae (verticalis)
epiglottis
pliaca pharyngo-epiglottica
sinus piriformis
v.jugularis interna
a.carotis communis
esophagus
nervus laryngeus
recurens (inferior)

Fig. 29. Lateral relations of the pharyngolarynx (posterior view).

The superior wail or the roof *(fornix pharyngis),* formed by the body of the sphenoid bone and the basilar process of the occipital bone, is oriented obliquely downwards and backwards; it is continuous with the posterior wall formed by the prevertebral plane, situated anteriorly to the occipito-atlantoid ligament, to the anterior arc of the atlas and to the body of the axis vertebra. The superior wall presents an elongated lymphoid mass, which makes up Luschka's pharyngeal tonsil *(fonsiila pharyngea).*

The tonsil is a yellow lymphoid organ, of soft consistency, made up of 7-8 irregular lobules, separated by grooves - tonsillar crypts *(cryptae fonsillares).* One of them is median, deeper and ends at its posterior extremity by a blind diverticulum: the pharyn-geal bursa *(bursa pharyngea).* Towards this bursa converge also the other grooves. The pharyngeal tonsil is about 3 cm in length and increases in size up to the age of 16 years, after which begins its involution. The hypertrophy of this tonsil gives rise to the adenoid vegetations. Owing to its rich vessel supply, the palpation of vegetations may cause small haemorrhages. If is a part of Waldeyer's lymphoid ring.

Anteriorly to the pharyngeal tonsil may be sometimes found epithelial cell agglomerations, remnants of the primitive hypophysial duct of which the adenohypophysis (Rathke's pouch} has developed; they represent the pharyngeal hypophysis.

At the level of the pharyngeal vault, the body of the sphenoid and the basilar process are covered by a thick periosteum, in which nasopharyngeal fibromas may appear.The lateral wall, at 1 cm behind the posterior extremity of the inferior concha and of 1 cm above the soft palate, opens the pharyngeal orifice of the auditory tube *(ostium pharyngaeum tubae auditivae),* through which the pharynx communicates with the middle ear, making up a route by which the infections of the pharynx can spread to the middle ear. In the posterior lip of the tubal orifice lies a prominence of the mucosa, due to the cartilage of the auditory tube *(torus tubarius),* situated below, from which a fold runs downwards and laterally, more prominent towards the soft palate *(plica salpingopharyngeal;* in this fold is located the salpingopharyngeal muscle. Behind this fold and the tubal opening lies a depression, termed Rosenmuller's fossa or pharyngeal recess *(recessus pharyngis).*

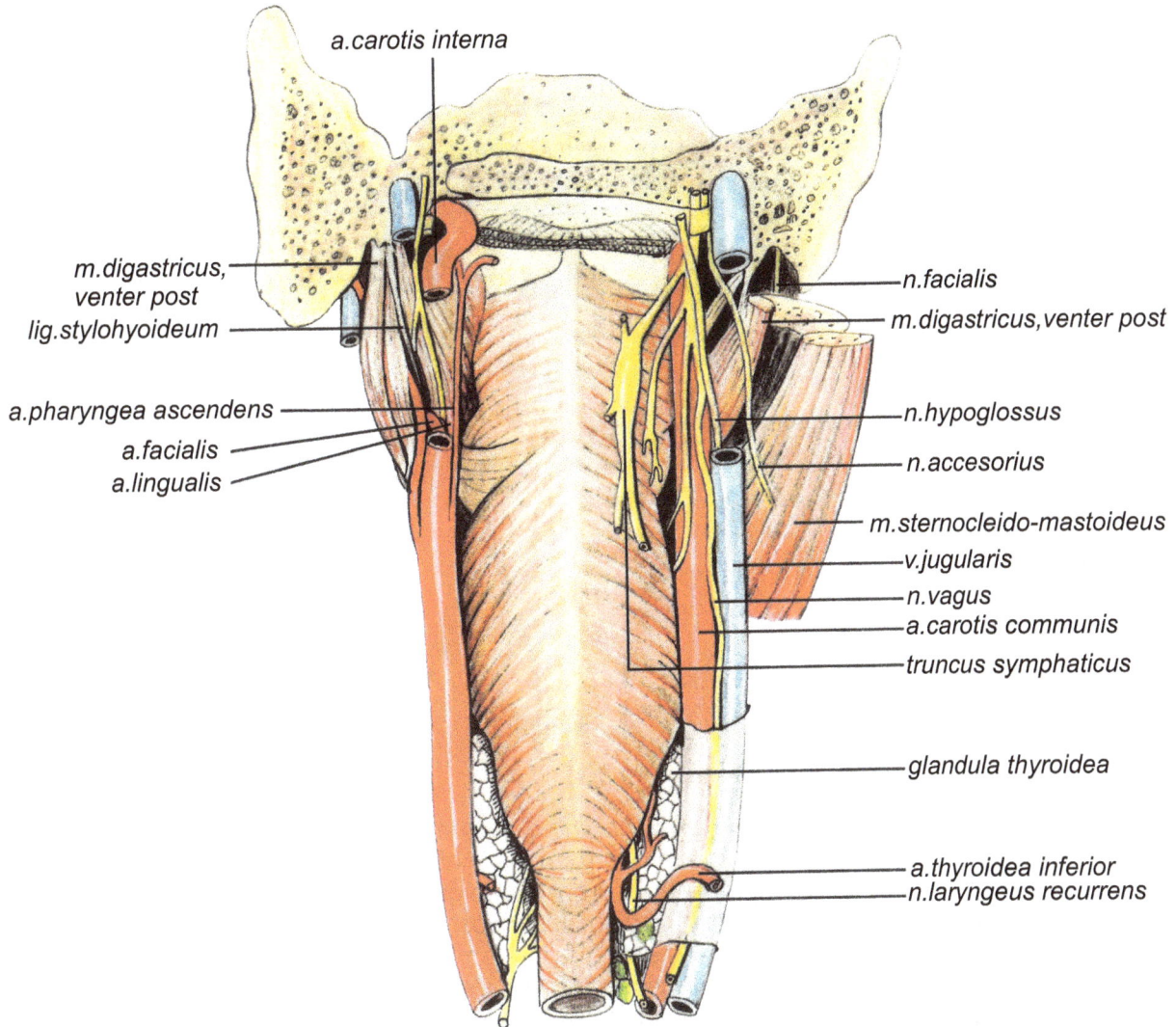

Labels on figure:

- a.carotis interna
- m.digastricus, venter post
- lig.stylohyoideum
- a.pharyngea ascendens
- a.facialis
- a.lingualis
- n.facialis
- m.digastricus, venter post
- n.hypoglossus
- n.accesorius
- m.sternocleido-mastoideus
- v.jugularis
- n.vagus
- a.carotis communis
- truncus symphaticus
- glandula thyroidea
- a.thyroidea inferior
- n.laryngeus recurrens

Fig. 30. The pharynx, the last three pairs of cranial nerves and the symphalhetic trunk (posterior view).

The pharyngeal opening of the auditory tube is so oriented as to allow its direct exploration by means of a probe introduced through the inferior part of the nasal fossae, respectively through the inferior nasal meatus. The mucosa surrounding the tubal ostium presents a lymphoid mass, which makes up Gerlach's tubal tonsil *(tonsilla tubaria).*

The inferior wall appears only at the moment of deglutition, when the soft palate gets in contact with the posterior wall of the pharynx, through Passavant's torus or cushion (caused by the contraction of the superior constrictor muscle). The palatine, pharyngeal, tubal and lingual tonsils form together Waldeyer's lymphoid circle (ring).

The oropharynx (pars oralis pharyngis) is bounded above by the soft palate and below by a plane passing through the body of the hyoid bone.

The three diameters are of about 4 cm each. At this level the pharynx communicates anteriorly, widely, with the oral cavity, through the buccopharyngeal isthmus *(isthmus faucium,* bounded above by the glossopalatine arch, consisting of the two anterior pillars of the soft palate, and below by the tongue. On the lateral wall descend inferiorly and posteriorly the posterior pillars of the soft palate - the palatopharyngeal arches *(arci palatopharyngei).*

Between the anterior and the posterior pillars, the lateral wall is occupied by the tonsillar fossa, in which lies the palatine tonsil The tonsil is in relation through the pharyngeal wall with the contents of the lateropharyngeal space.

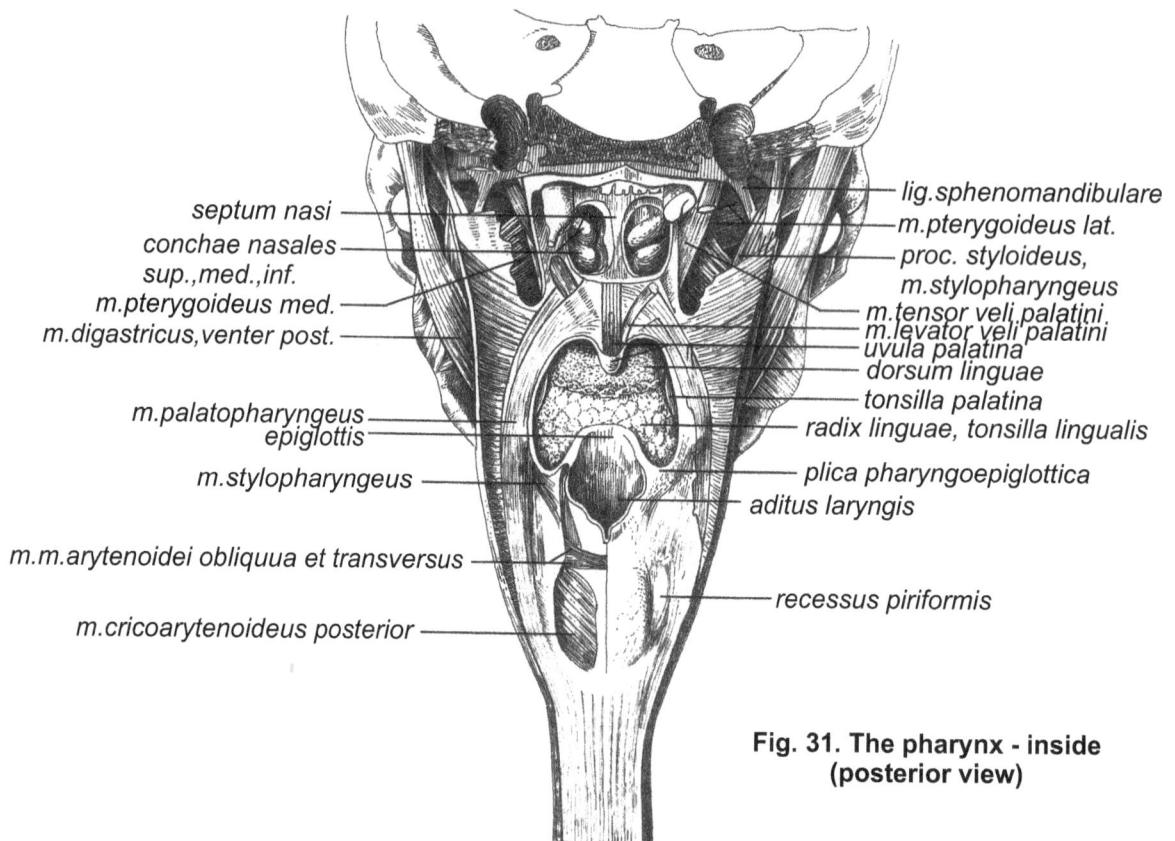

septum nasi

conchae nasales sup.,med.,inf.

m.pterygoideus med.

m.digastricus,venter post.

m.palatopharyngeus
epiglottis

m.stylopharyngeus

m.m.arytenoidei obliquua et transversus

m.cricoarytenoideus posterior

lig.sphenomandibulare

m.pterygoideus lat.

proc. styloideus,
m.stylopharyngeus
m.tensor veli palatini
m.levator veli palatini
uvula palatina
dorsum linguae
tonsilla palatina
radix linguae, tonsilla lingualis
plica pharyngoepiglottica
aditus laryngis

recessus piriformis

Fig. 31. The pharynx - inside (posterior view)

The *laryngopharynx (pars laryngea pharyngis)* is comprised in a horizontal plane passing through the hyoid bone and through the inferior boundary of the larynx, at the level of the cricoid cartilage. This segment is 5 cm in height Anteriorly are the epiglottis, the base of the tongue, the lingual tonsilla and the glossoepiglottic folds with the valleculae; posteriorly are located the inlet of the larynx *(aditus laryngis),* the afferent anatomic elements (the inferarylenoid incisure, the arytenoid cartilages, the aryepeglottic folds) and the piriform recesses on either side of the larynx, in the wall of which the mucosa presents a fold produced by the superior laryngeal nerve (Hyrtl's fold), the only fold formed by a nerve in the human organism. The posterior wall corresponds to the bodies of the C$_3$-C$_6$ vertebrae.

Structure of the walls of the pharynx. The pharynx is lined by a mucosa *(tunica mucosa)* of pavement epithelial type, with the exception of the rhinophorynx in which the epithelium is of ciliated prismatic (respiratory) type, but islets with pavement epithelium may be present If has in its structure the pharyngeal glands *(glandulae oharyngeae),* of mixed type, superficial, in the rhinopharyngeal zone, and of mucous type in the area of the oro- and laryngopharynx, as well as lymph follicles. Beneath the mucosa lies a connective tissue *(tela submucosa pharyngis),* which becomes dense, resistant, assuming the aspect of a compact fibrous lamina, termed the pharyngeal aponeurosis. This lamina is stronger developed in the inferior part of the pharynx, at the level of its cranial insertion into the exobase, making up a supporting means for the organ, named pharyngobasilar fascia *(fascia pharyngohasilaris);* at this level it represents excluively the pharyngeal wall, as there are no muscles in its structure. Musculature.The muscular coat *(tunica muscularis)* is divided into two groups: constrictor muscles and levator muscles (fig. 28,30,31).

A. The constrictor muscles. They are three flat, even, medially arched muscles, posteriorly united to those of the opposite side by a raphe, the median pharyngeal raphe *(raphe pharyngis);* they overlap in a reverse manner than the files on the roof: the inferior constrictor covers partially the middle constrictor which, on turn, covers partially the superior constrictor. The superior constrictor muscle *(m. constrictor pharyngis superior)* has its origin on: the inferior half of the posterior edge of the medial plate of the pterygoid process (the pterygopharyngeal part) *(pars pterygopharyngea),* on the pterygomandibular ligament *(raphe pterygomandibularis),* situated between the pterygoid process and the alveolar process of the mandible (the buccopharyngeal part) *(pars buccopharyngeal),* on the posterior extremity of the mylohyoid line (the mylopharyngeal part) *(pars mylopharyngea)* **and** on the roof of the tongue, forming the pharyngoglossal muscle (the glossopharyngeal part) *(pars glossopharyngeal.*

46

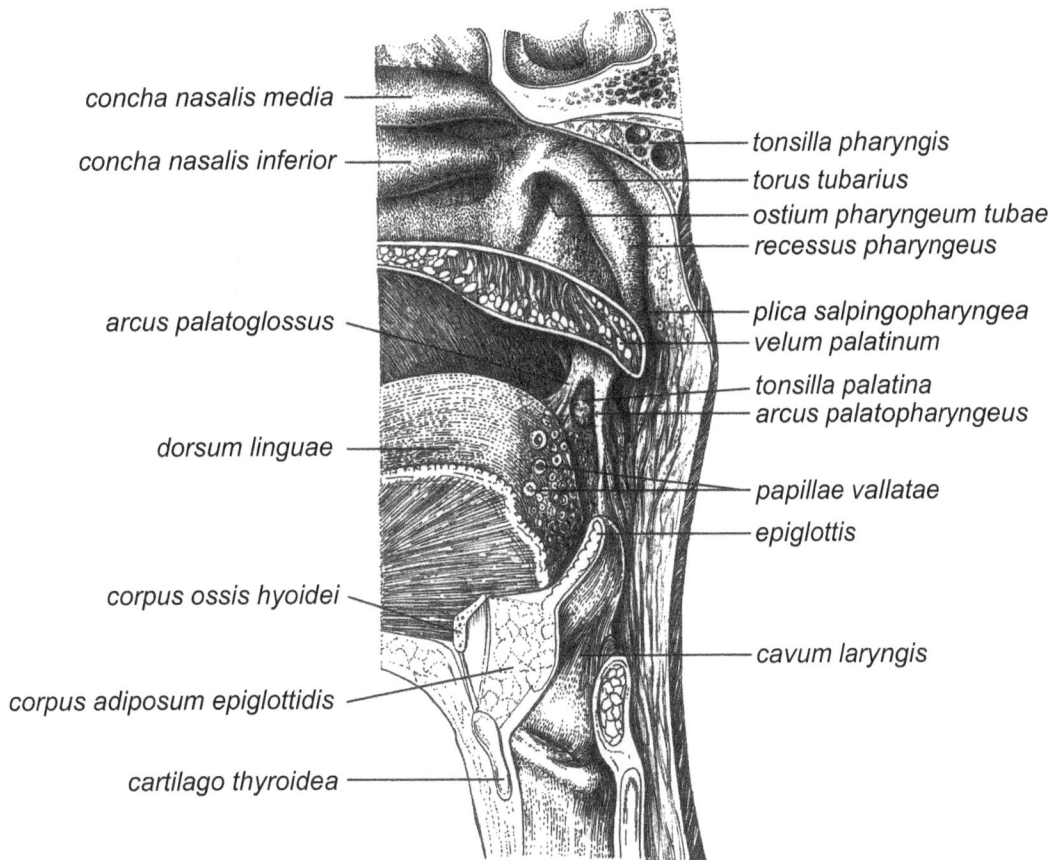

concha nasalis media

concha nasalis inferior

arcus palatoglossus

dorsum linguae

corpus ossis hyoidei

corpus adiposum epiglottidis

cartilago thyroidea

tonsilla pharyngis

torus tubarius

ostium pharyngeum tubae

recessus pharyngeus

plica salpingopharyngea

velum palatinum

tonsilla palatina

arcus palatopharyngeus

papillae vallatae

epiglottis

cavum laryngis

Fig. 32. Inside of the pharynx
(the lateral wall viewed after a sagittal section through the cephalic extremity)

It inserts into the median pharyngeal raphe: the superior fibres arch upwards and backwards, ending on the pharyngeal tubercle of the basilar pro cess of the occipital bone and leaving a portion deprived of musculature immediately beneath the exobase of the skull. Through its contraction, it closes the rhinopharynx during deglutition, by forming a prominence at the level of the posterior wall of the pharynx (Passavant's pad or cushion), which supports the contracted soft palate, so that the communication between the oro- and rhinopharynx is closed.

The middle constrictor muscle *(m. constrictor pharyngis medius)* has its origin on the posterior border of the lesser horn of the hyoid bone (the chondropharyngeal part) *(pars chondropfiaryngea)* and on the upper border of the greater horn (the ceratopharyngeal part) *(pars ceratopharyngea)*. It is inserted in a fan - like shape on the pharyngeal raphe into the middle two quarters of its total length.

The inferior constrictor muscle *(m.constrictor pharyngis inferior)* (laryngopharyngeal muscle) has its origin on the external surface of the thyroid cartilage, at the level of the oblique line, and on its inferior horn (thyropharyngeal part) *(pars thyropharyngea);* on the lateral surface of the ring of the cricoid cartilage (cricopharyngeal part) *(pars cricopharyngea)* and on the lateral surface of the first tracheal ring (fracheopharyngeal part) *(pars tracheopharyngea)*. It inserts into the pharyngeal raphe almost in all its length.

The tonsilloglossal muscle *(m. amygdaloglossus)* is inconstant and extends, when it exists, from the capsule of the tonsil to the lateral surface of the tongue.

Action: By their contraction, the constrictor muscles assure the progression of the food bolus.

B. The levator (longitudinal) muscles of the pharynx arise from the exobase, from which they descend and often cross the fascicles of the constrictors, inserting into the laryngeal cartilages and the pharyngeal aponeurosis.

The stylopharyngeal muscle *(m. stylopharyngeus)* has its origin on the styloid process and inserts into the superior and posterior borders of the thyroid cartilage, after penetrating the pharyngeal wall, between the superior and the middle constrictors.

The palatopharyngeal muscle *(m. palatopharyngeus)* arises from the dorsal wall of the pharynx and from the dorsal border of the thyroid cartilage and inserts into the palatine aponeurosis, on the median line, where it blends with the fibres of the homonymous muscle.

47

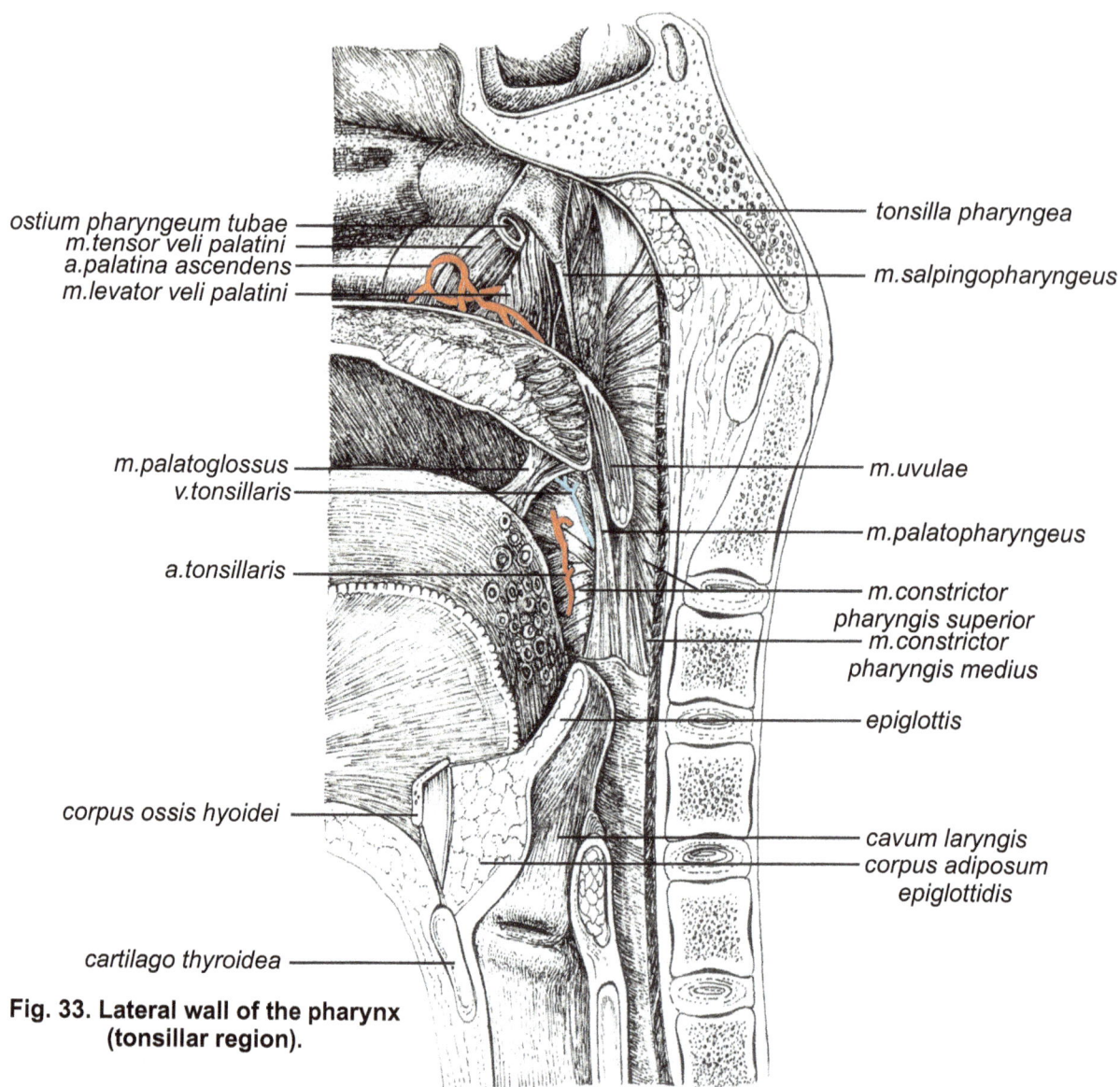

ostium pharyngeum tubae
m.tensor veli palatini
a.palatina ascendens
m.levator veli palatini

m.palatoglossus
v.tonsillaris

a.tonsillaris

corpus ossis hyoidei

cartilago thyroidea

tonsilla pharyngea

m.salpingopharyngeus

m.uvulae

m.palatopharyngeus

m.constrictor
pharyngis superior
m.constrictor
pharyngis medius

epiglottis

cavum laryngis
corpus adiposum
epiglottidis

**Fig. 33. Lateral wall of the pharynx
(tonsillar region).**

The salpingopharyngeal muscle *(m. salpingopharyngeus)* arises from the inferior surface of the cartilage of the auditory tube and is arranged in the same way as the palatopharyngeal muscle, penetrating info the lateral wall of the pharynx.

Action: The levator muscles have the role of raising the pharynx at the moment of deglutition, facilitating the progression of the food bolus, achieved by the constrictors.

The musculature of the pharynx has as external covering a connective tissue ("adventitia of the pharynx"), interposed between the pharynx and the adjacent organs, facilitating its movements.

Vessel and nerve supply

Arteries. The pharynx is mainly supplied by the ascending pharyngeal artery *(a. pharyngea ascendens),* a branch of the external carotid, and secondarily by branches of the ascending palatine artery (or facial artery), of the sphenopalatine artery and of the vidian artery (branches of the external maxillary artery).

The veins form in the submucosa and the adventitia a venous plexus which courses towards the internal jugular vein and, to a lesser extent, towards the pterygoid venous plexus.

The lymph is drained into the retropharyngeal and deep cervical lymph nodes (the inferior group).

The nerve supply is assured, for the musculature of the larynx, by a nerve plexus (the pharyngeal plexus) *(plexus pharyngeus)* located at the level of the lateral zone of the middle constrictor and consisting of branches of the glossopharyngeal and vagus nerves.

This plexus contains also vegetative fibres of the superior cervical sympathetic ganglion, for the mucosa, arising, too, from this level, to which are associated fibres of the trigeminal nerve.

48

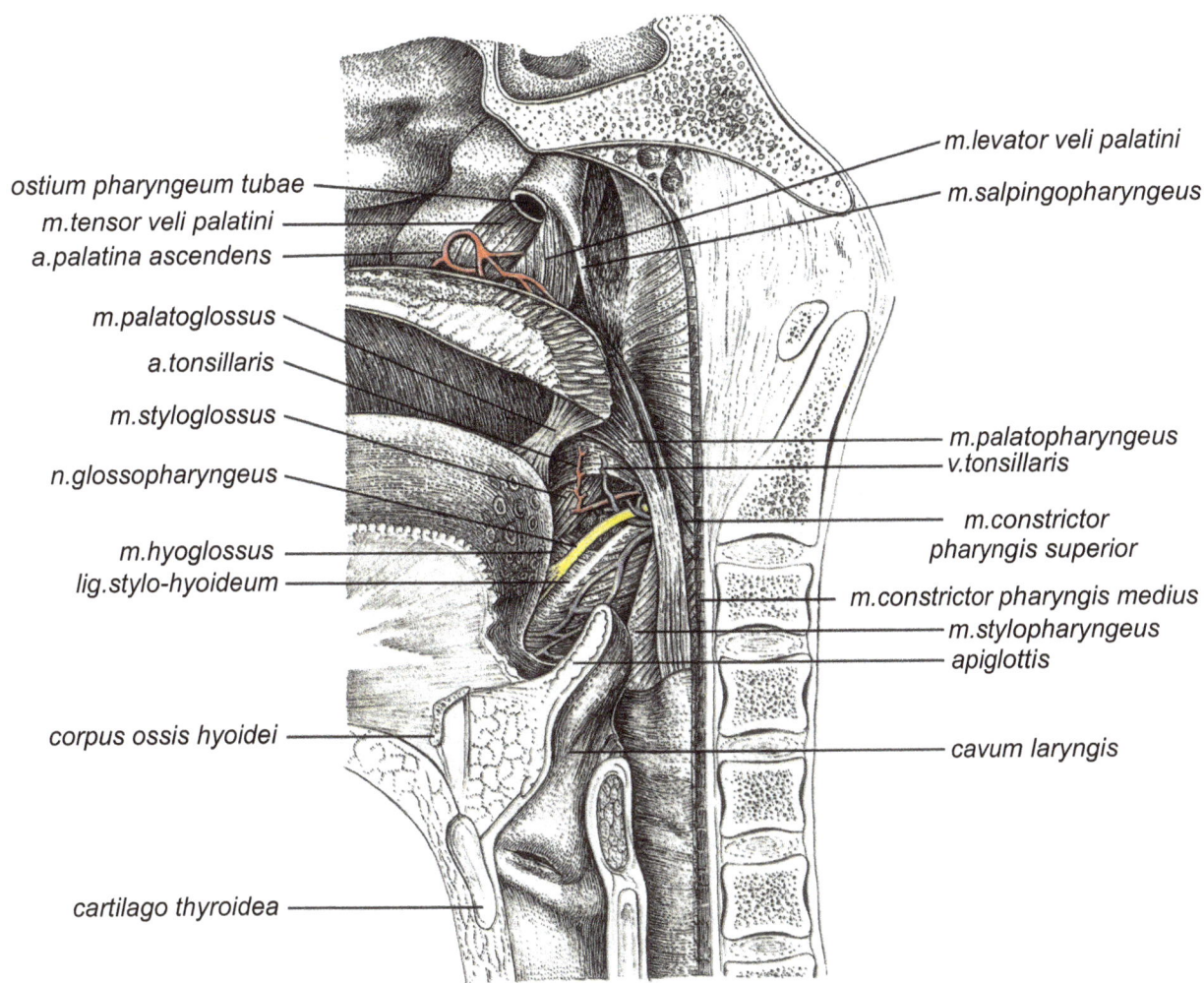

**Fig. 34. Inside of the pharynx - musculature of the naso -
and oropharyngeal parts, after the removal of the palatine tonsil (lateral view).**

Labels on the figure:
- ostium pharyngeum tubae
- m.tensor veli palatini
- a.palatina ascendens
- m.palatoglossus
- a.tonsillaris
- m.styloglossus
- n.glossopharyngeus
- m.hyoglossus
- lig.stylo-hyoideum
- corpus ossis hyoidei
- cartilago thyroidea
- m.levator veli palatini
- m.salpingopharyngeus
- m.palatopharyngeus
- v.tonsillaris
- m.constrictor pharyngis superior
- m.constrictor pharyngis medius
- m.stylopharyngeus
- apiglottis
- cavum laryngis

The Submandibular and Parotid Regions

The Submandibulae Region
(Glandula submandibularis)

Superficially, the region is bounded above by the inferior border of the mandible and below by the anterior and posterior bellies of the digastric muscle, forming a trigone. The contents of the region are mainly represented by the submandibular gland, which is covered by a connective fascia, deriving from the superficial cervical fascia, which delineates the region of the gland.

The region has three walls:

- laterosuperior wall, represented by the medial surface of the mandible, beneath the insertion of the mylohyoid muscle;

- lateroinferior wall, formed by the superficial cervical fascia, the plafysma and the skin;

- medial wall, formed by the mylohyoid, hyoglossus, styloglossus and stylohyoid muscles.

The submandibular gland does not remain limited within the boundaries of the trigone. At the posterior border of the mylohyoid muscle, between the latter and the hyoglossus muscle, lies a hiatus, through which the submandibular region communicates with the sublingual region, this being the place where an anterior projection of the submandibular gland, as well as the submandibular duct (Wharton's duct) *(ductus submandibularis),* penetrate into the sublingual region. In this zone the gland is in relation with the posterior extremity of the sublingual gland, with which it forms a common glandular mass. The submandibular duct is located on the medial surface of the sublingual gland and opens into the sublingual caruncula *(papilla sublingualis)* (fig. 35).

49

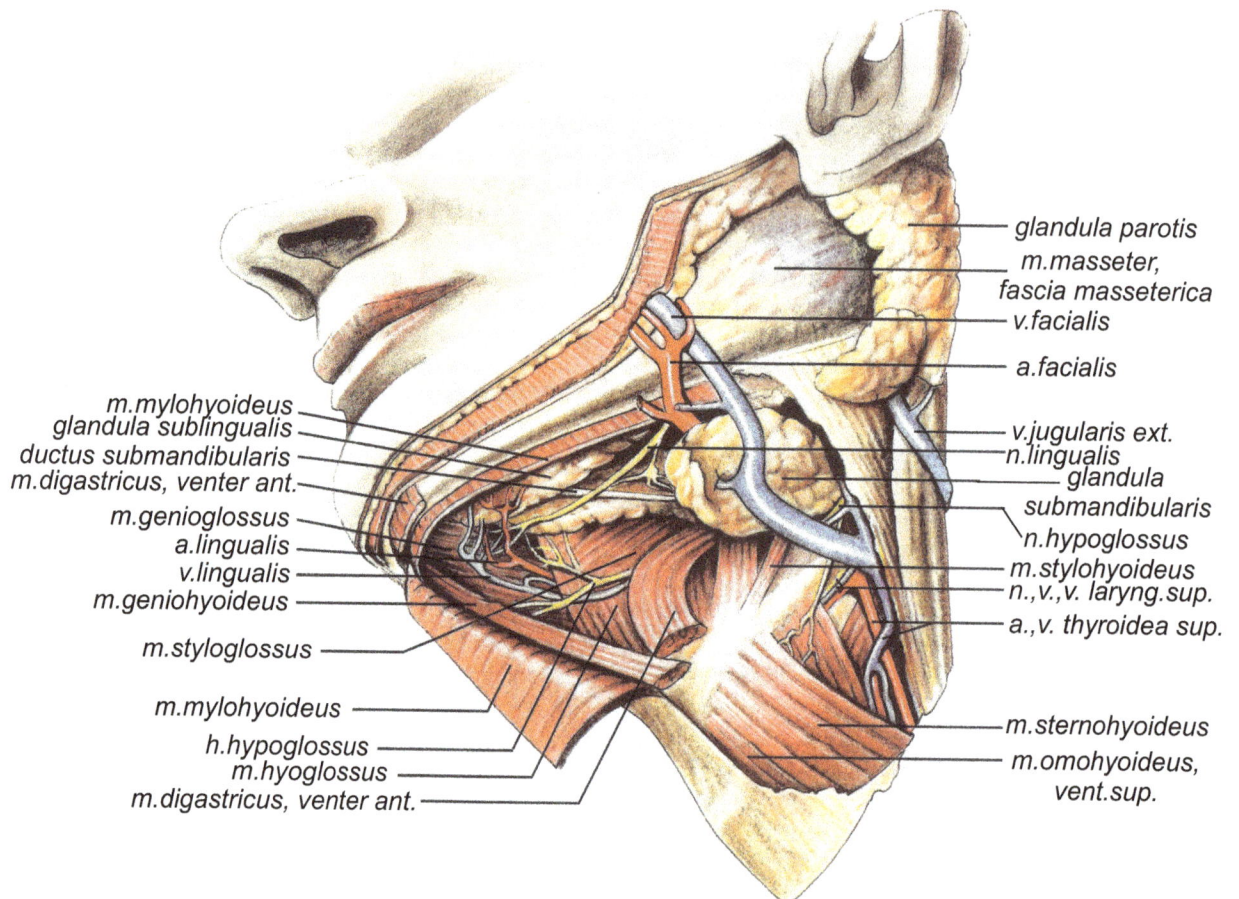

Labels on figure:

- glandula parotis
- m.masseter, fascia masseterica
- v.facialis
- a.facialis
- v.jugularis ext.
- n.lingualis
- glandula submandibularis
- n.hypoglossus
- m.stylohyoideus
- n.,v.,v. laryng.sup.
- a.,v. thyroidea sup.
- m.sternohyoideus
- m.omohyoideus, vent.sup.

- m.mylohyoideus
- glandula sublingualis
- ductus submandibularis
- m.digastricus, venter ant.
- m.genioglossus
- a.lingualis
- v.lingualis
- m.geniohyoideus
- m.styloglossus
- m.mylohyoideus
- h.hypoglossus
- m.hyoglossus
- m.digastricus, venter ant.

Fig. 35. Submandibular, parotid and sublingual glands (relations).

Vessel and nerve supply

Arteries. The submandibular gland is supplied by branches of the facial and lingual arteries.

The veins are satellites of the arteries and tributary of the internal jugular vein.

The lymph drains info the submandibular lymph nodes.

The parasympathetic nerve supply of the submandibular gland is assured by fibres of the nerve of the chorda tympani (Wrisberg's intermediate nerve - VII bis nerve), arrived here with the ingual nerve, which gives off branches to the submandibular ganglion, where actually they form a synapse with the preganglionic fibres of the superior salivatory nucleus.

The sympathetic nerve fibres reach the gland through the perifacial and perilingual plexuses, via the glandular branches of the facial and lingual arteries.

The region is crossed by the facial artery, the facial vein, the lingual artery, the hypoglossal nerve and the lingual vein.

50

The Parotid Region
(Glandula parotis)

The boundaries of the region are: anteriorly, the posterior border of the mandible; superiorly, the external acoustic meatus; posteriorly, the mastoid and the sternocleidomastoid muscle; inferiorly, a conventional line continuous with the inferior border of the mandible up to the place where it joins the sternocleidomastoid muscle.

The parotid region is characterized by the presence of the gland bearing the same name, the largest salivary gland. The parotid gland is enclosed within a fibrous capsule *(parotid fascia),* which closes the parotid region.

The parotid region is parallelepiped-shaped, with the great axis vertical, and has four walls and two bases:

- the external cutaneous wall corresponds to the superficial fascia of the parotid, which is continuous, at the face, with the parotidomasseteric fascia and at the neck, with the fascia of the sternocleidomastoid muscle, deriving from the superficial cervical fascia;

- the supero-anterior base is formed by the external (bony and cartilaginous}meatus;

- the posterior muscular wall is formed by the sternocleidomastoid muscle and the posterior belly of the digastric muscle;

- the inferior base is made up by a connective plate, represented by the interglandular fibrous septum, which separates the parotid region from the region of the submandibular gland;

- the anterior osteomuscular wall is formed by the dorsal border of the ramus of the mandible and by the posterior borders of the masseter and medial pterygoid muscles; the internal medial pharyngeal wail is formed by the sfylian aponeurosis, which lies round the styloid musculo-ligamentous bunch (Riolan's bouquet: the styloglossal, stylopharyngeal and stylohyoid muscles and the stylomandibular and stylohyoid ligaments). Through the aponeurosis, the pharyngeal projection of the gland enters the lafero(para)pharyngeal space, where it is in relation with the lateral wall of the pharynx and the retrostylian space.

The parotid gland. Well covered by the fascia through which it has the same extrinsic relations as the parotid region, the parotid gland sends two projections (fig. 35):

- the facial, anterior projection, which passes above the surface of the masseter muscle and is continuous with the extraglandular portion of Sfensen's excretory duct;

- the pharyngeal projection.

The parotid duct passes above the masseter, crosses the buccinator muscle and opens into the vestibule of the oral cavity, info the superior gingivo-alveolar groove, at the level of the root of the second molar.

Vessel and nerve sypply of the parotid gland

Arteries. The arterial supply of the gland is assured by parotid branches deriving from the external carotid artery and from its bifurcation branches in the depth of the gland, as well as by the superficial temporal and maxillary arteries.

The satellite veins empty into the internal jugular vein.

Lymph vessels.

The lymph is collected by the superficial and deep (intraglandular) parotid lymph nodes and conveyed to the cervical lymph nodes of the cranial group.

Nerves. The gland is supplied by the facial nerve, which spreads out, forming its temporo-and cervicofacial trunks that anastomose with branches of the superficial cervical plexus and with branches of the auriculotemporal nerve. Its parasympathetic nerve supply is assured by the inferior salivatory nucleus located in the medulla oblongata. Through Jacobson's nerve *(n. tympanicus)* and through the lesser superficial petrosal nerve, the fibres reach the otic ganglion (Arnold's ganglion), attached to the mandibular nerve, from which, through the auriculotemporal nerve and its anastomosis with the facial nerve, they run to the parotid gland. The preganglionic neuron is situated in the salivatory nucleus and the postganglionic neuron, in the otic ganglion. The orthosympdthetic innervation derives from the cervical sympathetic (the intermediate lateral tract), in which is situated the preganglionic neuron, that forms a synapse with the postganglionic neuron in the superior cervical ganglion; from here, through the pericarotid plexus and the superficial temporal and maxillary arteries, which give off parotid branches, the fibres reach the gland. Moreover, in the parotid region lie a series of vessels and nerves of which we mention: the external carotid artery, the external jugular vein, the facial, auriculotemporal and greater auricular nerves, as well as superficial and deep lymph nodes.

51

The Laryngeal Region
(Regio laryngica)

It is bounded above by the superior border of the thyroid cartilage and below by the inferior border of the cricoid cartilage; laterally if is in relation with the thyroid gland.

The Larynx
[Larynx]

It is the organ of phonation, situated below the hyoid bone, to which it is suspended by the hyothyroid membrane, and above the first tracheal ring, to which it is connected by the cricotracheal membrane. Posteriorly lies the pharynx (pars laryngica pharyngis) and anteriorly the larynx is covered by the sternothyroid and thyrohyoid muscles, which belong to the system of rectus muscles, extending between the pubis and the inferior border of the mandible, as well as by the platysma and by skin. The superior border of the thyroid cartilage, in thenormal position of the head, projects in the adult on the superior edge of the C_5 vertebra and the inferior border of the cricoid cartilage proiects on the inferior edge of.the C_6 vertebra. Hence, in height, the larynx projected on the vertebral column extends up to the level of C_5-C_6 vertebrae. This situation varies according to the age. In the erect posture, the larynx descends and its inferior border may attain the superior edge of the C_7 vertebra, especially in the aged.

The position of the larynx varies according to age and to sex. In the foetus and newborn, the larynx projects on the body of the C_3-C_4 vertebrae, while in the aged it corresponds to the C_6-C7 vertebrae. Owing to the anatomical connections, any movement occurred between the oral floor and the diaphragm influences the shift of the laryngotracheal canal. The movements of the vertebral column (flexion and extension of the head) modify the length of this duct by 10-30%. In dorsal decubitus, the larynx and the trachea lie higher by a half of vertebra or even by an entire vertebra. The larynx moves vertically: it ascends during the extension of the head, during deglutition and during the emission of high-pitched sounds; it descends during the emission of low-pitched sounds and in flexion of the head. The size of the larynx varies with the sex, the age and from one individual to the other. It is poorly developed at birth, increases up to the age of about 3 years, when it begins to develop; in boys at puberty it develops intensely. The mean dimensions are: 35-50 mm vertically, 35-43 mm transversally and 25-36 mm sagitfally (anteroposteriorly).

The larynx is a hollow organ; it consists of a cartilaginous skeleton, joints, ligaments and fibre-elastic membranes which connect these cartilages, as well as of motor muscles, vessels and nerves (fig. 36-59).

The skeleton of the larynx. It is formed of three unpaired and four paired cartilages.

The thyroid cartilage (cartilago thyroidea), the largest cartilage of the larynx, is visible, palpable and consists of two lateral laminae (lamina dextra et sinistra), united anteriorly in an angle (of about 90° in the male) which, through the upper extremity, is palpable subcutaneously; this projection is termed the laryngeal prominence (prominentia laryngica) or Adam's apple (fig. 38-41). Above the junction of the lamina dextra with the lamina sinistra may be observed the superior thyroid notch (incisura thyreoidea cranialis). From the posterior border of the laminae start, above, the hyoid horn of the thyroid cartilage (cornu hyoideum cartilagii thyreoidei) and below, the cricoid horn (cornu cricoideum), which has on its infernal surface an articular facet, facies articularis cricoidea, for the articulation with the cricoid cartilage.

The posteromedial surfaces of the thyroid cartilage enclose not only the larynx, but also the lateral surfaces of the pharynx, so that between the laminae of the thyroid cartilage and the prominence of the posterior surface of the larynx are formed two vertical fossae, the laryngopharyngeal grooves or piriform recesses (recessus piriformis), situated in the laryngopharynx.

The cricoid cartilage (cartilago cricoidea) has the shape of a signet ring with the signet posteriorly; the signet has approximately the shape of a quadrilateral lamina with the great axis vertical (lamina cartilaginis cricoidae). On the upper part of the lamina are the arytenoid articular surfaces - facies articularis arytenoidea - for the arytenoid cartilages and laterally, near the inferior border, the thyroid articular surfaces - facies articularis thyroidea -, for the articulation with the thyroid cartilage (fig. 45-47).

The epiglottis (cartilago epiglottidis) is leaf-like shaped, with the broad side upwards and the stalk (petiolus epiglottidis) directed downwards, which at the laryngoscopic examination is viewed in the form of a tubercle - tuberculus epiglottidis. It inserts by the stalk into the thyroid dihedral angle, above the insertion of the vocal folds (fig. 36,37,44).

The lateral borders of the epiglottis form the anterior segment of the inlet of the larynx - aditus laryngis. The epiglottis contributes to the closure of the inlet of the larynx, directing thus the liquids and the food bolus towards the pharynx and the oesophagus. The epiglottis is furnished with a number of orifices, which are the site of mucous glands.

The arytenoid cartilage (cartilago arytenoidea) is paired and has the shape of a triangular pyramid. The medial and posterior surfaces are smooth, unlike the lateral surface, which exhibits irregularities. The base (basis) presents articulation facets with the cricoid cartilage (facies articularis cricoidea) and two processes: an anterior process (processus vocalis), for the insertion of the vocal ligaments, and a lateroposterior process (processus muscularis), for the insertion of the internal muscles of the larynx. The apex is bent dorsally and articulates with the corniculate cartilage (fig. 42-47).

By their articulation with the cricoid cartilage, the arytenoid cartilages coordinate, under the action of the motor muscles, the mobility of the vocal folds and the whole mobility of the larynx.

The corniculate cartilage (cartilago corniculata) (Sanforini's cartilage) is paired, symmetrical, small, round, situated in the aryepiglottic fold and forms a tubercle (tuberculum corniculatum).

The cuneiform cartilage (cartilago cuneiformis) (Wrisberg's cartilage) is also paired, small, situated laterally to the corniculate cartilages, in the aryepiglottic fold, and forms also a tubercle (tuberculum cuneiforme), visible at laryngoscopy.

In the depth of the lateral hyothyroid ligament are also the two triticeal cartilages (of the size of grains of wheat) (fig. 36, 48, 50, 51).

The ligamentous apparatus of the larynx. The anterior parts of the hyoid bone and of the thyroid and cricoid cartilages are united by membranes, which attach the skeleton of the larynx to the hyoid bone. The latter, through the stylohyoid ligament and the muscles of the tongue, is fixed on the base of the skull The membranes, reinforced by ligaments, represent at the same time the routes of surgical access to the larynx, to its different regions (fig. 36, 37, 48, 55, 57).

The ligamentous apparatus which connects the larynx to the adjacent organs is formed by: the hyothyroid membrane, the hyoepiglottic membrane, the glosso-epiglottic, pharyngo-epiglottic and thyro - epiglottic ligaments and the cricotracheal ligament or membrane.

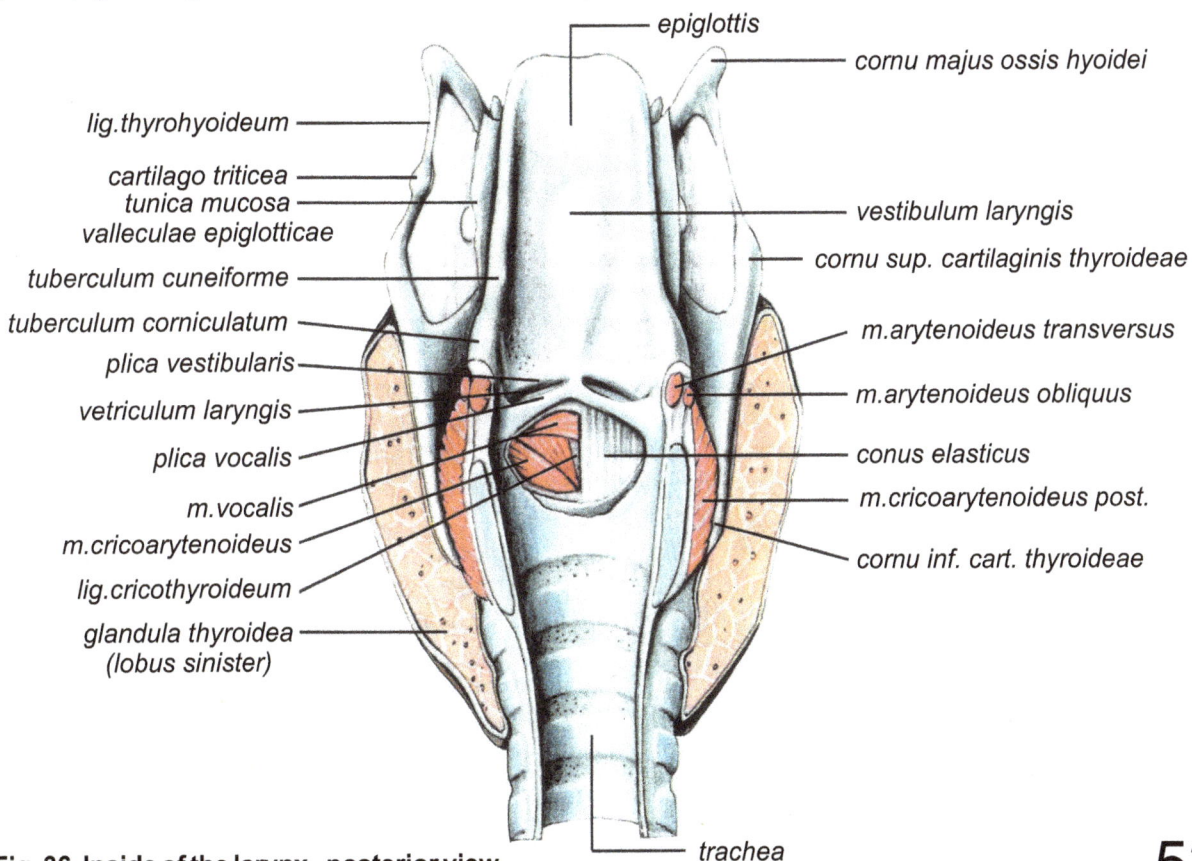

Fig. 36. Inside of the larynx - posterior view.

epiglottis
cornu majus ossis hyoidei
lig.thyrohyoideum
cartilago triticea
tunica mucosa
valleculae epiglotticae
vestibulum laryngis
cornu sup. cartilaginis thyroideae
tuberculum cuneiforme
tuberculum corniculatum
m.arytenoideus transversus
plica vestibularis
m.arytenoideus obliquus
vetriculum laryngis
conus elasticus
plica vocalis
m.vocalis
m.cricoarytenoideus post.
m.cricoarytenoideus
cornu inf. cart. thyroideae
lig.cricothyroideum
glandula thyroidea
(lobus sinister)
trachea

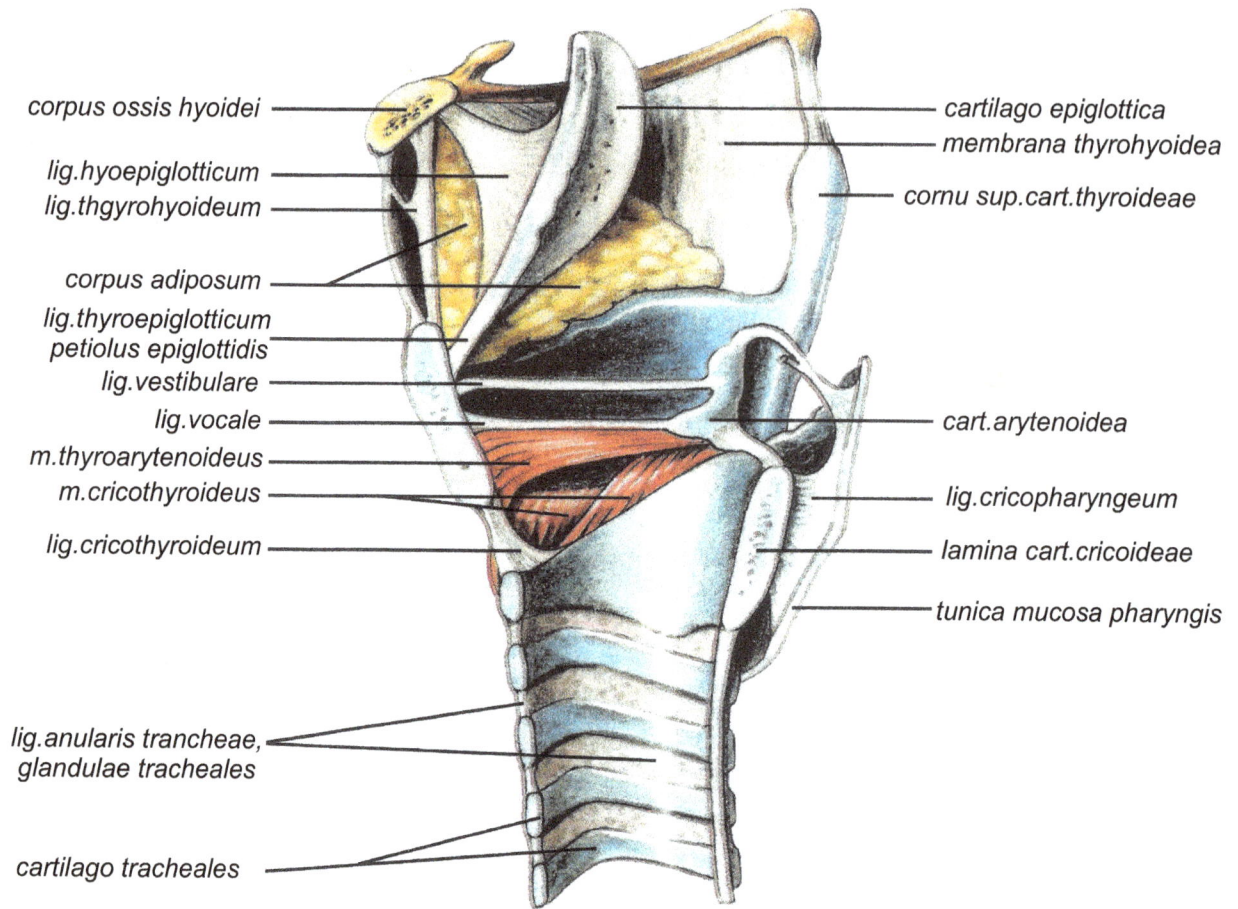

corpus ossis hyoidei

lig.hyoepiglotticum
lig.thgyrohyoideum

corpus adiposum
lig.thyroepiglotticum
petiolus epiglottidis
lig.vestibulare
lig.vocale
m.thyroarytenoideus
m.cricothyroideus
lig.cricothyroideum

cartilago epiglottica
membrana thyrohyoidea
cornu sup.cart.thyroideae

cart.arytenoidea

lig.cricopharyngeum
lamina cart.cricoideae
tunica mucosa pharyngis

lig.anularis trancheae,
glandulae tracheales

cartilago tracheales

Fig. 37. Inside of the larynx - lateral view .

The hyothyroid membrane (membrana hyothyreoidea) extends from the hyoid bone to the superior border of the thyroid cartilage and on the greater hyoid horn. The median portion is reinforced by a ligament (ligamentum hyothyreoideum medium) and its free dorsal borders form another ligament (ligamentum hyothyreoideum laterale), in the depth of which are located the triticeal cartilages (cartilago triticea).

By its close connection to the hyoid bone, the larynx participates in the movements of this bone.

The lateral side of the hyothyroid membrane is furnished with orifices for the superior laryngeal artery and vein and for the internal branch (ramus internus) of the superior laryngeal nerve (n. laryngeus superior).

The hyoepigloftic membrane (membrana hyoepiglottica) connects the anterior surface of the epiglottis to the upper border of the hyoid bone. Beneath it, between the epiglottis and the hypothyroid membrane, lies the fat-pad of the larynx. The hyoepigloftic membrane is discontinuous; it delineates, with the hyothyroid membrane (anteriorly), with the epiglottis (posteriorly) and with the thyroid cartilage superiorly, the hyothyro - epiglottic space.

The glosso - epiglottic ligaments (lig. glossoepiglotticum) raise the mucosa and extend between the root of the tongue and the epiglottis; they are three in number and form the folds bearing the same name: the median glosso-epiglottic fold and the lateral glosso-epiglottic folds. Between them are the valleculae.

The thyro - epiglottic ligament (lig. thyreoepiglotticum) (which forms the articulation of the epiglottis with the thyroid cartilage), situated beneath the preceding ligaments, connects the stalk of the epiglottis to the posterior surface of the thyroid notch.

The cricotracheal ligament (lig. cricotracheale) extends between the inferior border of the cricoid cartilage and the first tracheal ring.

-The vestibular ligaments *(lig. vestibulare)* enter in the structure of the vestibular folds.

-The vocal ligaments *(lig. vocale)* consist of an elastic tissue and are inserted into the angle of the thyroid cartilage, into an elastic formation *(macula flava)* and into the vocal process of the arytenoid cartilages.

-The fibro-elastic membrane of the larynx (membrana fibroelastica laryngis) is situated beneath the mucosa of the larynx and presents significant thickenings. In the fibro-elastsc membrane may be distinguished two parts: superior and inferior.

The superior part, the quadrangular membrane *(membrana quadrangularis)*, is double, symmetrical, a right one and a left one, which start at the level of the aryepiglottic folds and end below at the level of the ventricular ligament, with which they fuse. Anteriorly, the membranes insert into the epiglottic cartilage and posteriorly, into the anterolateral surface of the arytenoid cartilage. They confer to the laryngeal vestibule the shape of a frustum of cone with the base upwards. The inferior part, the elastic cone *(conus elasticus)*, corresponds to the infraglottic cavity and is an important connective-elastic component of the larynx. It is attached downwards on the arch and the lateral borders of the lamina of the cricoid cartilage and from here it passes upwards and attaches itself on the internal surface of the angle of the thyroid cartilage, at 0.5 cm beneath the superior thyroid notch *(incisura thyreoidea cranialis)* and backwards, on the vocal process of the arytenoid cartilages. Its superior free border makes up the vocal ligament. The anterior segment of the elastic cone, situated between the inferior border of the thyroid cartilage and the arch of the cricoid cartilage, is termed pars libera coni elastici or cricothyroid ligament *(lig. cricothyreoideum)*. The shape of the elastic cone changes at each opening of the glottis.

The laryngeal joints (fig. 45, 47).

The crico-arytenoid joint *(articulus crycoarytenoideus)* is adiarthrosis; by the union of the base of the arytenoid cartilage (fades articularis cricoidea) with the upper border of the cricoid cartilage (fades articularis arytenoidea), a complete joint is formed, with a synovial membrane and a capsule reinforced by a medial ligament In this joint are performed transverse gliding movements and rotation movements about a vertical axis, by which the arytenoid cartilage approaches or recedes the vocal processes and the vocal folds. These adduction - abduction movements are indispersable to the laryngeal functions (phonation and respiration). All the intrinsic muscles of the larynx act directly or indirectly upon this joint.

The cricothyroid joint (articulus cricothyreoideus) is of the plane diarthrosis type; if is located between the inferior horns of the thyroid cartilage and the cricoid cartilage *(fades articularis thyreoidea)*, which are kept in contact by an articular capsule, reinforced by three ligaments: lateral, posterior and anterior.

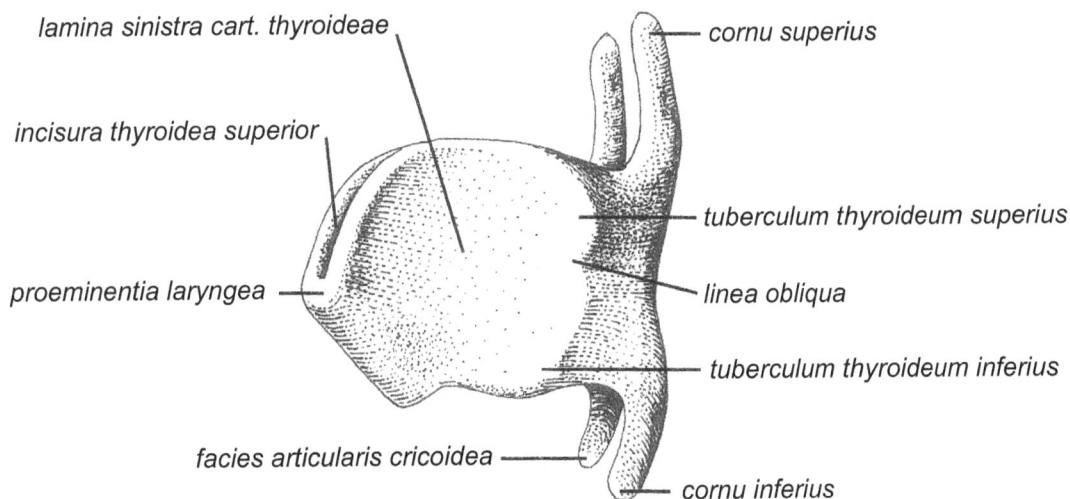

Fig. 38. The thyroid cartilage - lateral view.

incisura thyroidea superior

cornu superius

lamina dextra cart. thyroideae

cornu inferius

incisura thyroidea inferior

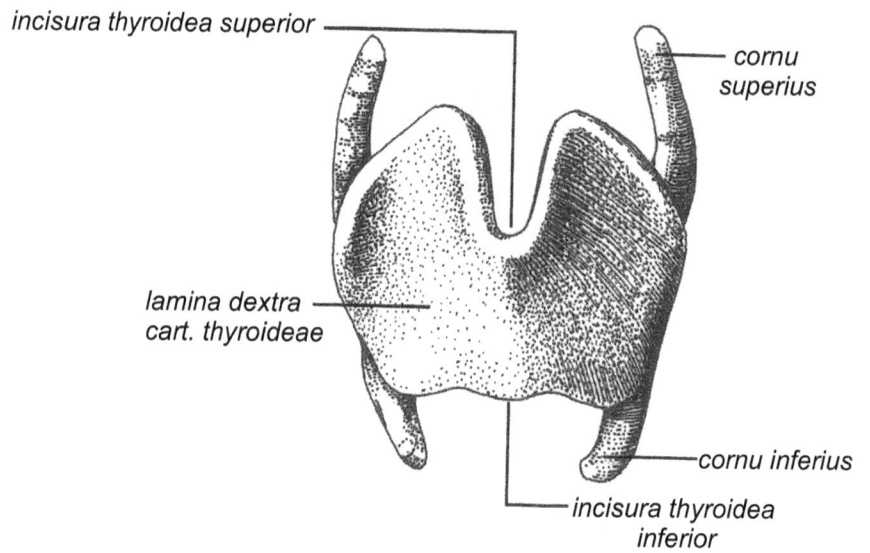

Fig. 39. The thyroid cartilage - anterior view.

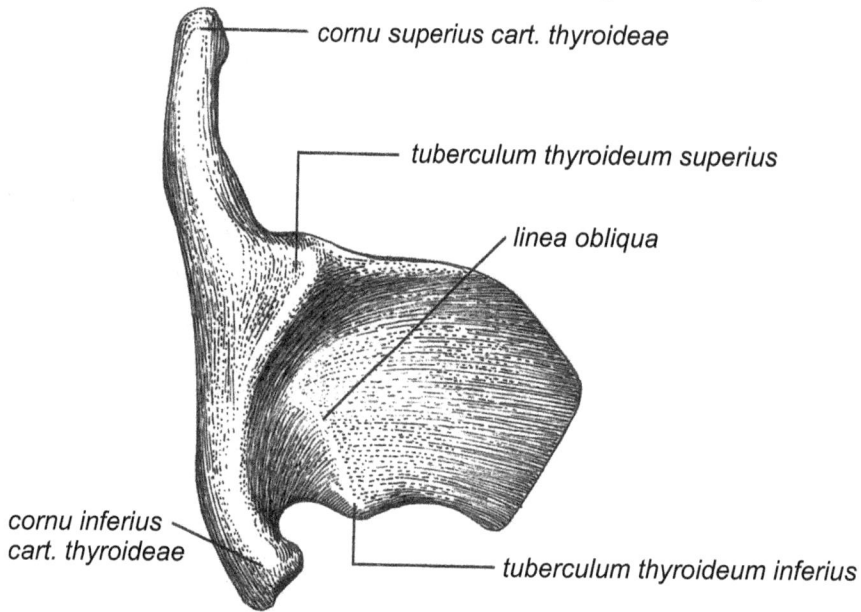

cornu superius cart. thyroideae

tuberculum thyroideum superius

linea obliqua

cornu inferius cart. thyroideae

tuberculum thyroideum inferius

Fig. 40. The thyroid cartilage in the male.

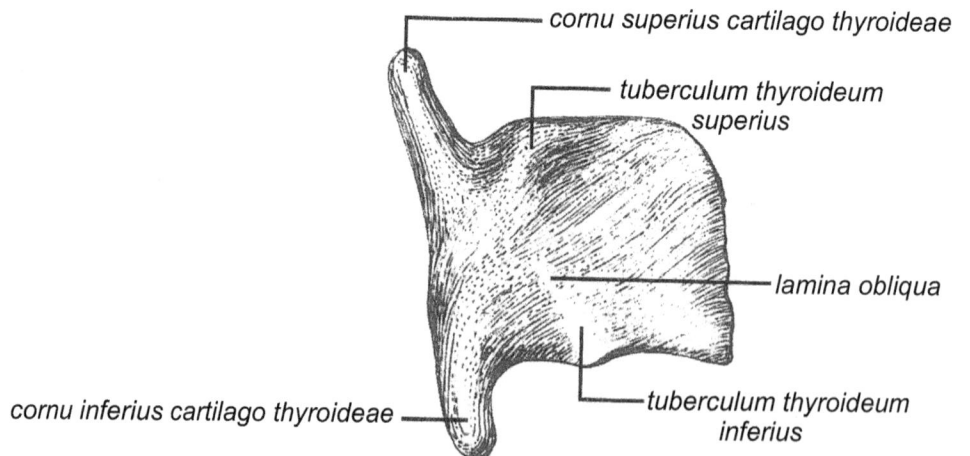

cornu superius cartilago thyroideae

tuberculum thyroideum superius

lamina obliqua

cornu inferius cartilago thyroideae

tuberculum thyroideum inferius

Fig. 41. The thyroid cartilage in the female.

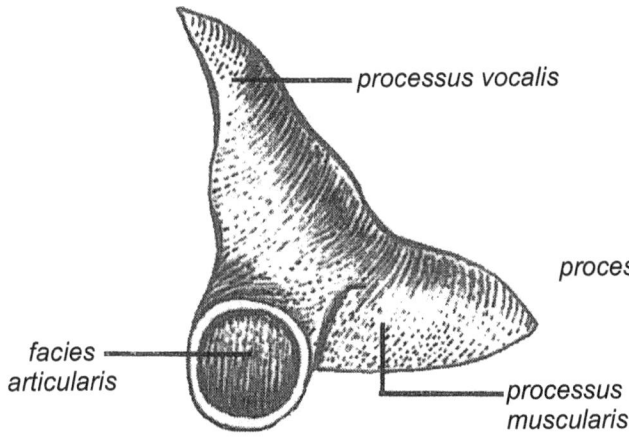

Fig. 42. The arytenoid cartilage (articular surface).

processus vocalis

facies articularis

processus muscularis

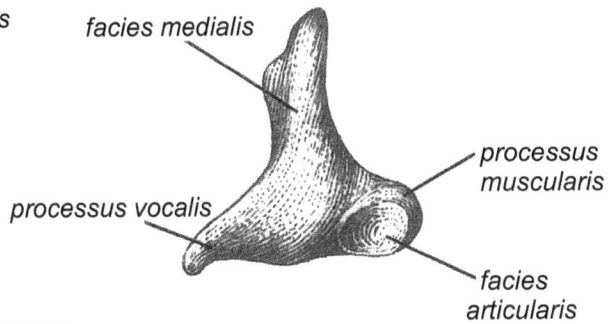

Fig. 43. The arytenoid cartilage (medial surface).

facies medialis

processus muscularis

processus vocalis

facies articularis

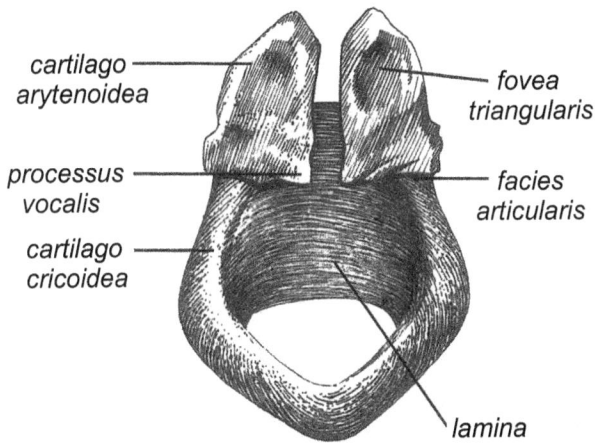

Fig. 45. The arytenoid cartilage (the cricoarytenoid joints - anterior view).

cartilago arytenoidea

processus vocalis

cartilago cricoidea

fovea triangularis

facies articularis

lamina

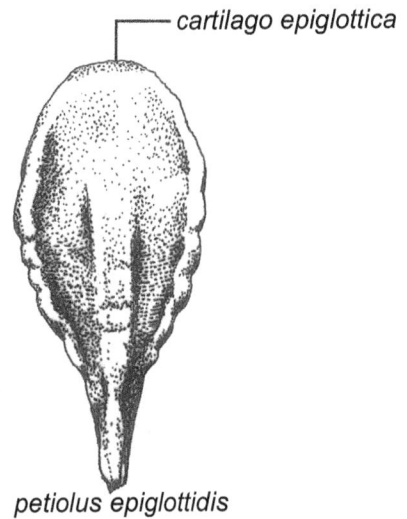

Fig. 44. The epiglottis

cartilago epiglottica

petiolus epiglottidis

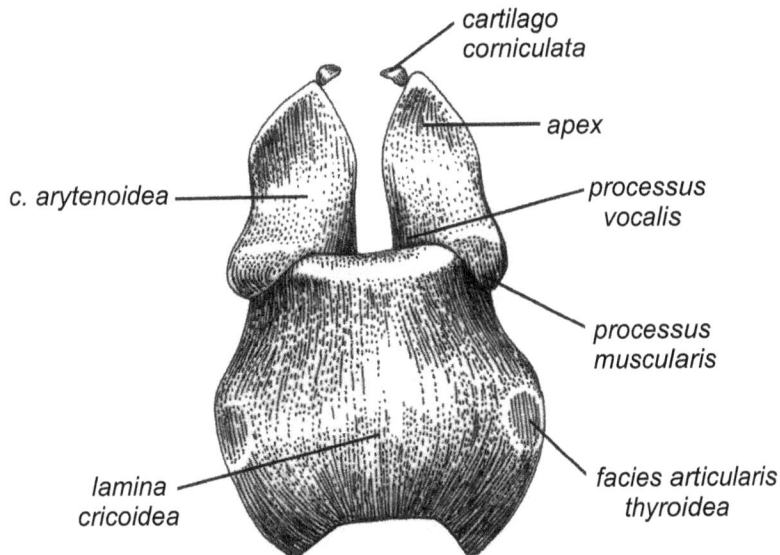

Fig. 46. The cricoarytenoid joints (posterior view).

cartilago corniculata

apex

c. arytenoidea

processus vocalis

processus muscularis

lamina cricoidea

facies articularis thyroidea

57

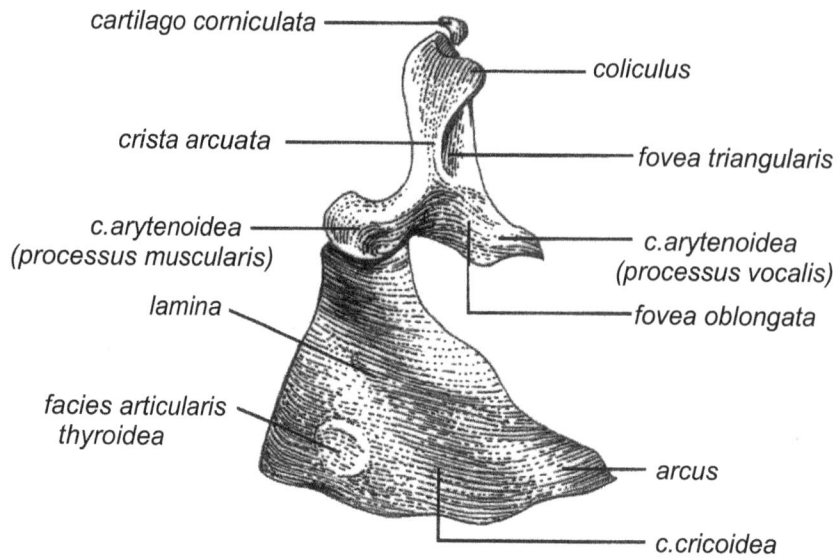

cartilago corniculata

coliculus

crista arcuata

fovea triangularis

c.arytenoidea
(processus muscularis)

c.arytenoidea
(processus vocalis)

fovea oblongata

lamina

facies articularis
thyroidea

arcus

c.cricoidea

Fig. 47.The cricoarytenoid joints (lateral view).

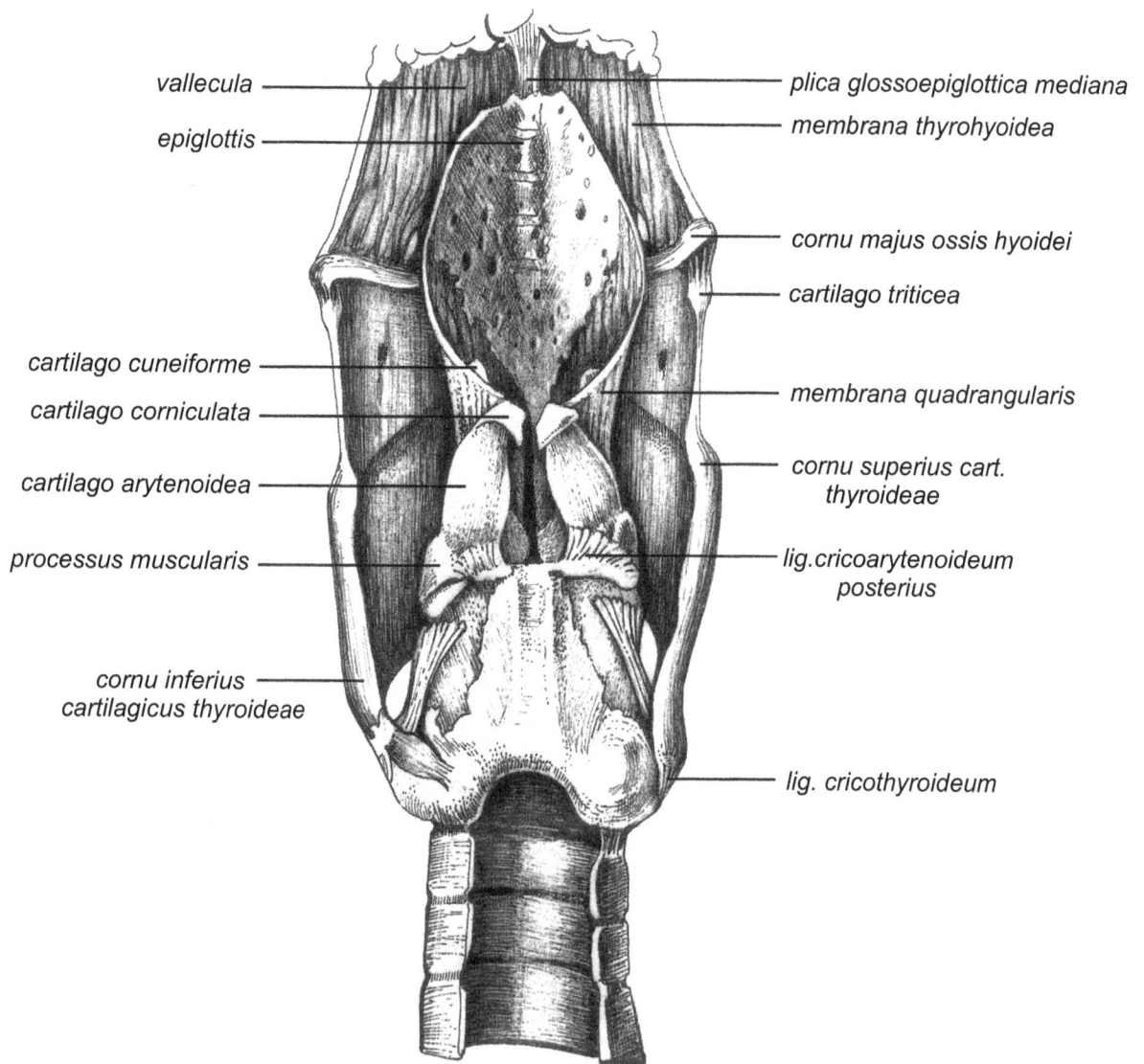

vallecula

plica glossoepiglottica mediana

epiglottis

membrana thyrohyoidea

cornu majus ossis hyoidei

cartilago triticea

cartilago cuneiforme

membrana quadrangularis

cartilago corniculata

cartilago arytenoidea

cornu superius cart.
thyroideae

processus muscularis

lig.cricoarytenoideum
posterius

cornu inferius
cartilagicus thyroideae

lig. cricothyroideum

**Fig. 48. The skeleton of the larynx - posterior view
(cartilages, membranes, ligaments).**

58

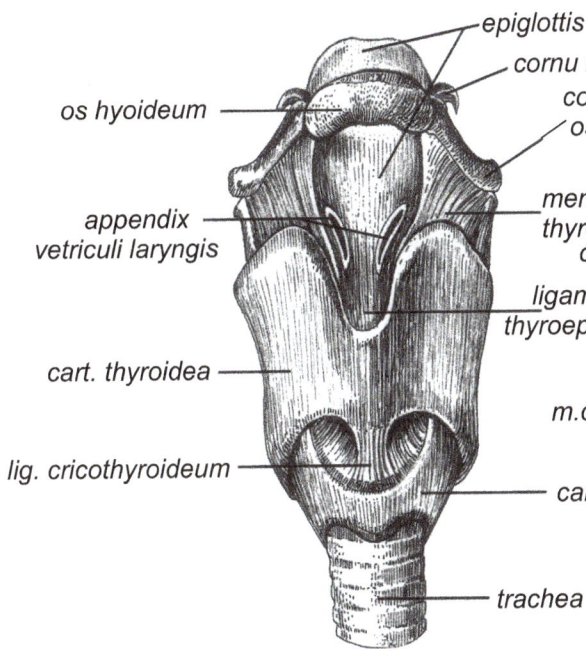

Fig. 49. The cricothyroid ligament and the hyothyroid membrane (anterior view).

epiglottis

cornu minus ossis hyoidei

os hyoideum

cornu majus ossis hyoidei

tuberculum cuneiforme

membrana thyrohyoidea

appendix vetriculi laryngis

cart. corniculata

ligamentum thyroepiglotticum

cart. thyroidea

mm. arytenoideus obliqui

m. cricoarytenoideus posterius

lig. cricothyroideum

cart. cricoidea

trachea

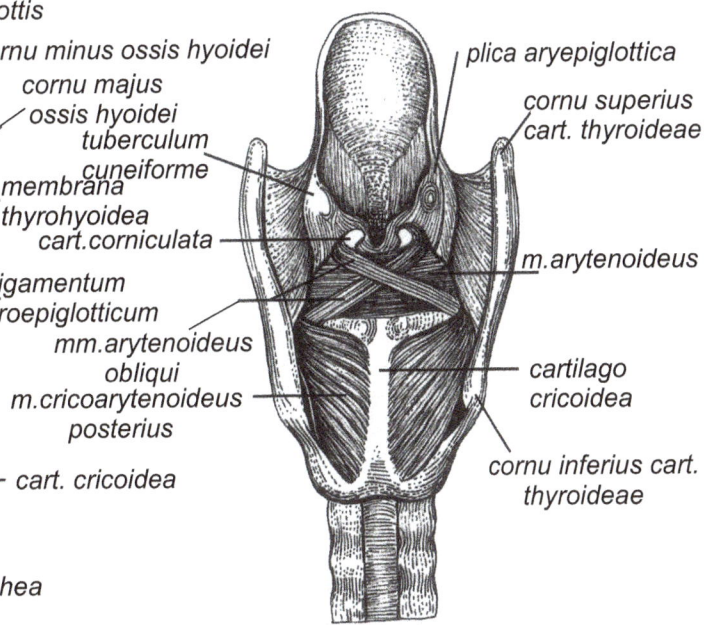

Fig. 50. Musculature of the larynx (posterior view)

plica aryepiglottica

cornu superius cart. thyroideae

m. arytenoideus

cartilago cricoidea

cornu inferius cart. thyroideae

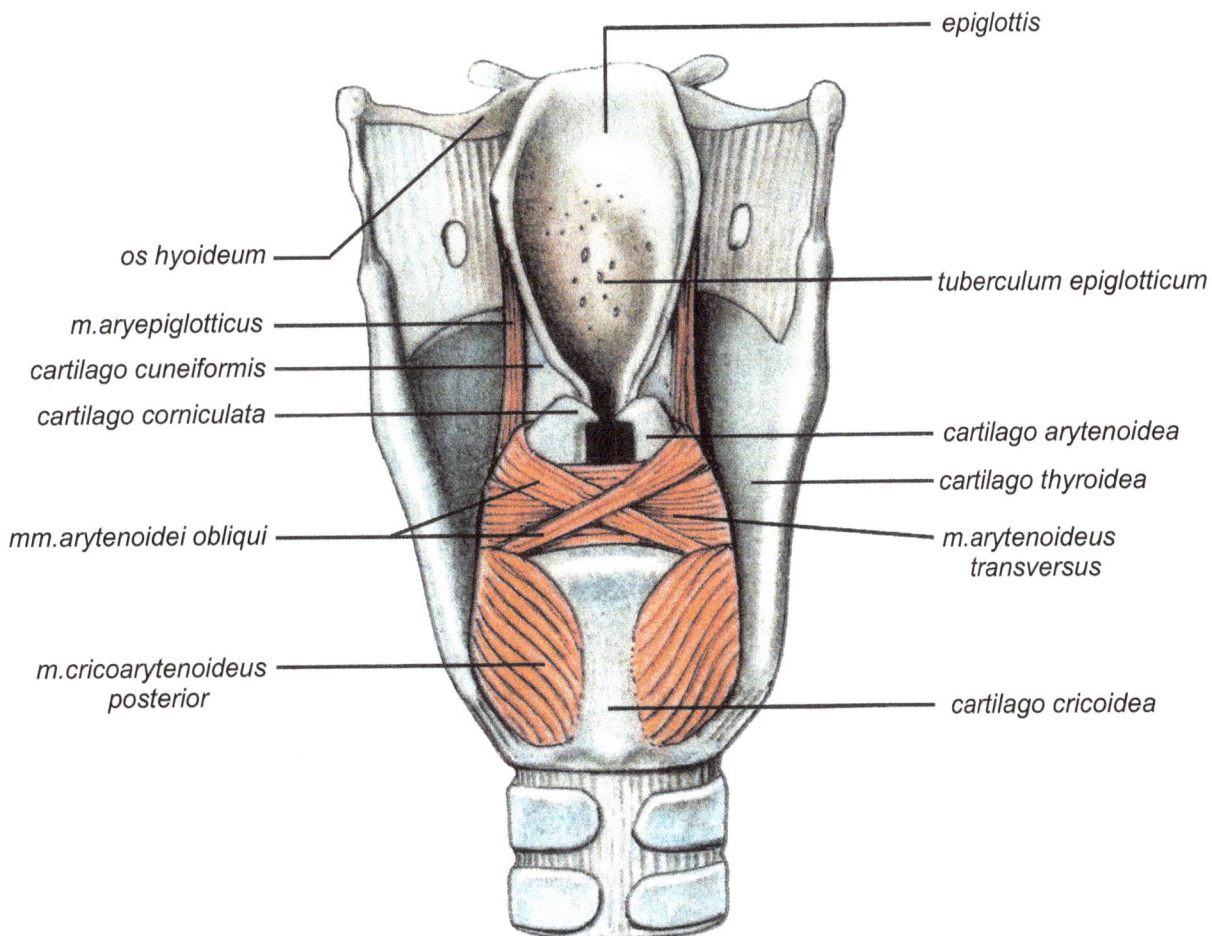

epiglottis

os hyoideum

tuberculum epiglotticum

m. aryepiglotticus

cartilago cuneiformis

cartilago corniculata

cartilago arytenoidea

cartilago thyroidea

mm. arytenoidei obliqui

m. arytenoideus transversus

m. cricoarytenoideus posterior

cartilago cricoidea

Fig. 51. The larynx (posterior view).

59

At this level are produced gliding movements of small amplitude and tilting movements about the transverse axis which passes through the two joints, rendering the vocal folds tense.

By raising the arch of the cricoid cartilage, the upper part of the cricoid lamina, with the arytenoid cartilages (which articulate with it), is dorsally overturned, so that the vocal cords (the vocal ligaments) are made tense - they are stretched. By descent of the cricoid arch, the insertion points of the vocal ligaments approach each other, relaxing the ligaments.

Musculature of *the larynx* (fig. 50, 51, 52, 54, 58).

The cricothyroid muscle (m. cricothyreoideus) extends from the lower border and the cricoid horn of the thyroid cartilage to the upper border of the arch of the cricoid cartilage. It raises the arch of the cricoid cartilage, bends backwards the lamina of the cricoid cartilage or bends forwards the lamina of the thyroid cartilage, making thus the vocal folds tense. It is made up of two parts: the anterior part, called the straight part or pars recta, and the posterior part, termed the oblique part or pars obliqua.

The posterior crico - arytenoid muscle (m. cricoarytenoideus posterior) extends from the dorsal surface of the lamina of the cricoid cartilage to the muscular process of the arytenoid cartilage. It draws dorsally the muscular process, so that the latter is shifted laterally, and opens the slit of the glottis *(rima glottidis)*, the vocal folds being stretched laterally; it is the single abductor of the vocal folds. In the case of unilateral paralysis, the vocal fold of the respective part is not shifted laterally.

The lateral crico - arytenoid muscle (m. cricoarytenoideus lateralis) extends from the upper border of the lateral surface of the arch of the cricoid cartilage to the muscular process of the arytenoid cartilage. It draws forwards the muscular process and closes the orifice of the glottis *(rima glottidis)*, relaxing the vocal folds. It is an adductor of the vocal folds. By its contraction, the two arytenoid cartilages recede from each other.

The thyro - arytenoid muscle (m. thyreoarytenoideus) extends from the internal surface of the inferior segment of the thyroid cartilage to the lateral surface of the arytenoid cartilage. It is made up of two fascicles: one lateral *(pars lateralis)* and the other medial *(pars medialis)*. It narrows the rima glottidis, approaching the arytenoid cartilages to each other *(pars lateralis)*, and regulates the tension of the vocal folds *(pars medialis)*, contracting or relaxing the vocal folds; therefore it is also called vocal muscle (especially its internal part).

The arytenoid muscle (m. arytenoideus) consists of a transverse fasciculus *(pars transversa)*, between the two posterior surfaces and the two muscular processes of the arytenoid cartilage, and an oblique fasciculus *(pars obliqua)*, extending from the muscular process of an arytenoid cartilage to the apex of the opposite arytenoid cartilage. It narrows the orifice of the glottis *(rima glottidis)* and partially the inlet of the larynx *(aditus laryngis)*.

The thyro - epiglottic muscle (m. thyreoepiglotticus) extends from the internal surface of the thyroid cartilage to the lateral borders of the epiglottis and to the quadrangular membrane. If closes the aditus laryngis, drawing the epiglottis downwards.

The ary-epiglottic muscle (m, aryepiglotficus) extends from the apex of the arytenoid cartilage to the lateral borders of the epiglottis. It closes the adifus laryngis and draws the epiglottis backwards and downwards. Both muscles protect reflexly the larynx from the penetration of foreign bodies info it.

The musculature of the larynx, on the whole, accomplishes the following functions: if protects in a reflex way from the penetration of foreign bodies into the respiratory apparatus; it diminishes or widens the rima glottidis and contracts or relaxes the vocal folds. In this way an entire gamut of modulations in the emission of sounds is achieved, the larynx being at the same time the organ of phonafion. The cavity of the larynx *(cavum laryngis)* has the shape of a clepsydra, which may be divided info three portions. The superior part, the epilosynx or epiglottic space or laryngeal vestibule *(vestibulum laryngis)*, extends from the epiglottis above to the ventricular folds below. Between the ventricular folds and the vocal folds lies the middle part, the mesolarynx or the laryngeal ventricle *(venfriculus laryngis)*. From the vocal folds to the lower border of the cricoid cartilage extends the inferior part, the hypolarynx or infragloftic cavity *(cavum infraglotticum laryngis)*.

The superior part contains the laryngeal vestibule *(epilarynx)*, which presents above the inlet of larynx - *aditus laryngis* -, oriented backwards and upwards. in the formation of this superior aperture participate anteriorly the epiglottis, laterally the two ary-epiglottic folds *(plicae aryepigiotticae)*, directed from before backwards and downwards and approaching each other medially, where the interarytenoid notch *(incisura interarytenoidea)* is formed. In the depth of the ary-epiglottic folds lie the cuneiform (Wrisberg's) cartilages and the corniculate (Santorini's) cartilages. The inlet of the larynx is wide open during normal respiratory movements; during deglutition it is covered by the epiglottis and the roof of the tongue, beneath which the larynx is drawn.

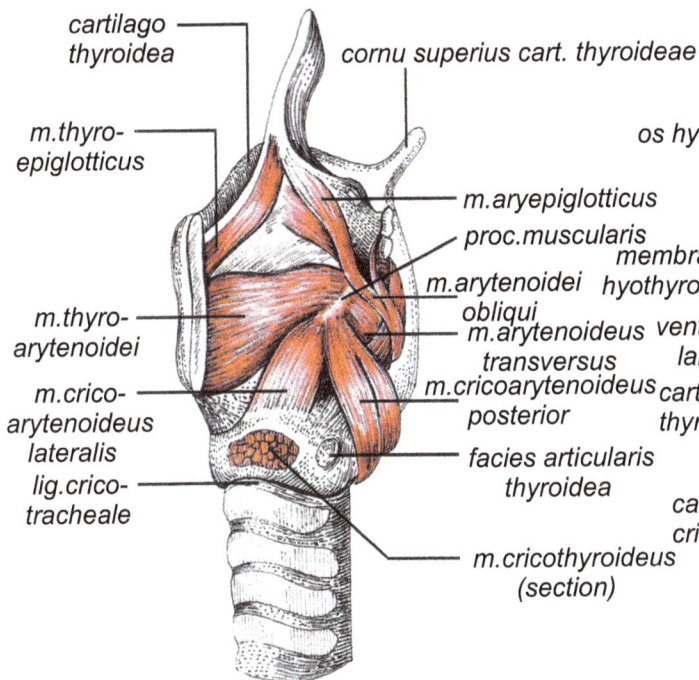

Fig. 52. Muscles of the larynx (lateral view)

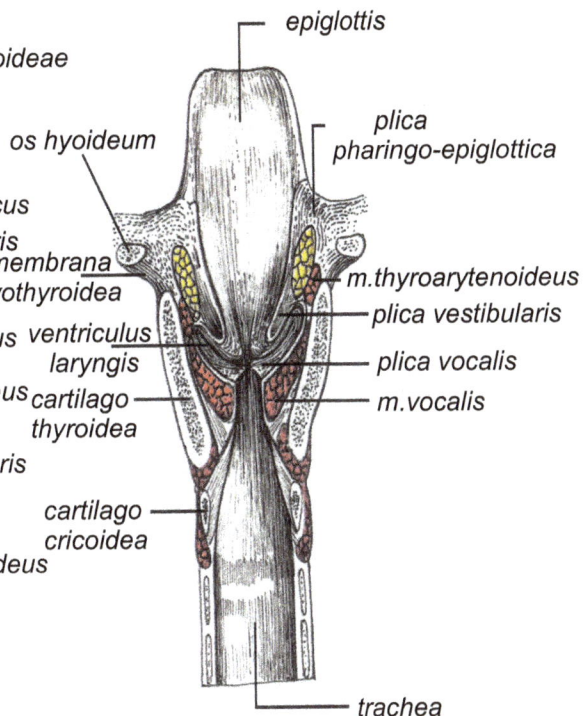

Labels for Fig. 52:
cartilago thyroidea
cornu superius cart. thyroideae
m.thyro-epiglotticus
m.aryepiglotticus
proc.muscularis
m.arytenoidei obliqui
m.arytenoideus transversus
m.cricoarytenoideus posterior
m.thyro-arytenoidei
m.crico-arytenoideus lateralis
lig.crico-tracheale
facies articularis thyroidea
m.cricothyroideus (section)

Fig. 53. Frontal section through the larynx.

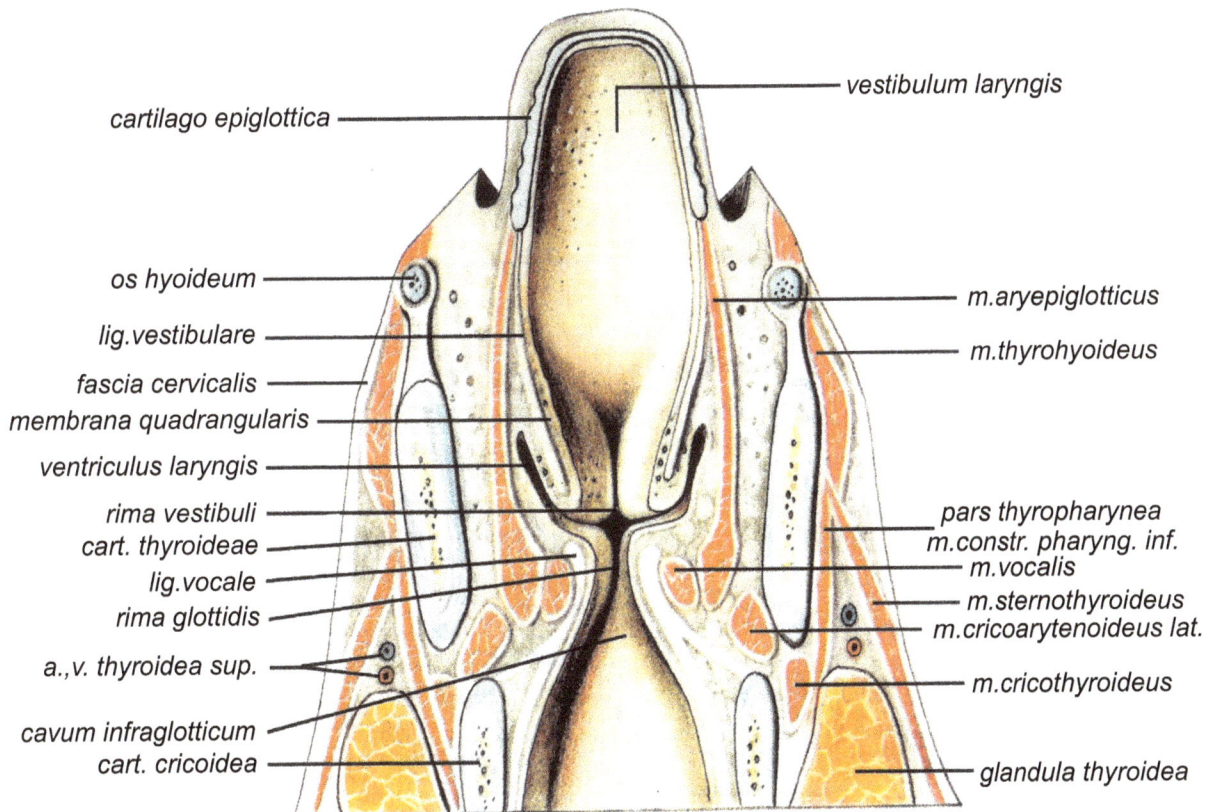

Labels for Fig. 53:
epiglottis
plica pharingo-epiglottica
os hyoideum
membrana hyothyroidea
m.thyroarytenoideus
plica vestibularis
ventriculus laryngis
plica vocalis
m.vocalis
cartilago thyroidea
cartilago cricoidea
trachea

Fig. 54. Frontal section through the larynx.

Labels for Fig. 54:
cartilago epiglottica
vestibulum laryngis
os hyoideum
m.aryepiglotticus
lig.vestibulare
m.thyrohyoideus
fascia cervicalis
membrana quadrangularis
ventriculus laryngis
rima vestibuli
cart. thyroideae
lig.vocale
rima glottidis
a.,v. thyroidea sup.
cavum infraglotticum
cart. cricoidea
pars thyropharynea m.constr. pharyng. inf.
m.vocalis
m.sternothyroideus
m.cricoarytenoideus lat.
m.cricothyroideus
glandula thyroidea

61

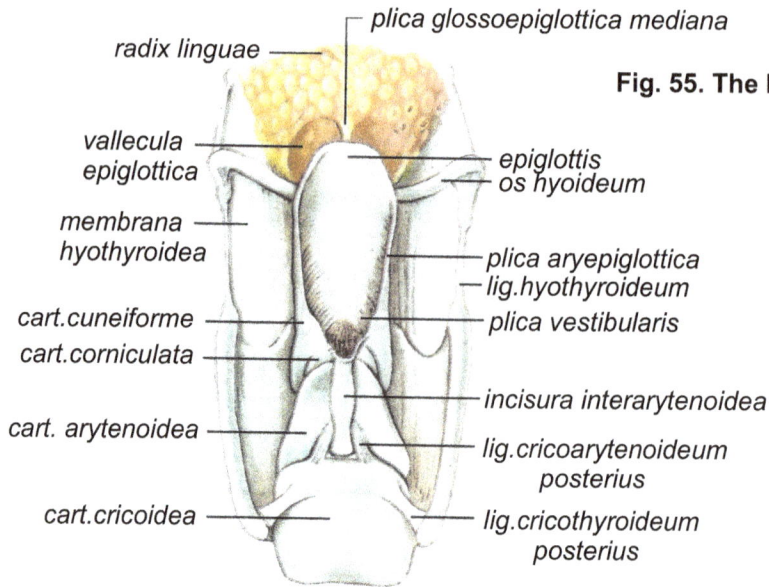

Fig. 55. The larynx: the hyothyroid membrane (posterior view).

- plica glossoepiglottica mediana
- radix linguae
- vallecula epiglottica
- epiglottis
- os hyoideum
- membrana hyothyroidea
- plica aryepiglottica
- lig.hyothyroideum
- cart.cuneiforme
- plica vestibularis
- cart.corniculata
- incisura interarytenoidea
- cart. arytenoidea
- lig.cricoarytenoideum posterius
- cart.cricoidea
- lig.cricothyroideum posterius

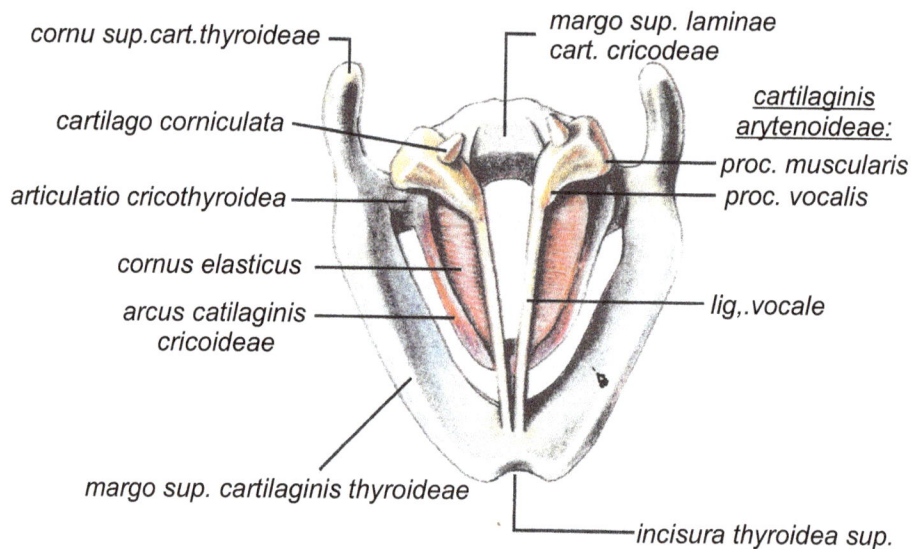

Fig. 56. The vocal cords.

- cornu sup.cart.thyroideae
- margo sup. laminae cart. cricodeae
- cartilaginis arytenoideae:
- cartilago corniculata
- proc. muscularis
- proc. vocalis
- articulatio cricothyroidea
- cornus elasticus
- arcus catilaginis cricoideae
- lig,.vocale
- margo sup. cartilaginis thyroideae
- incisura thyroidea sup.

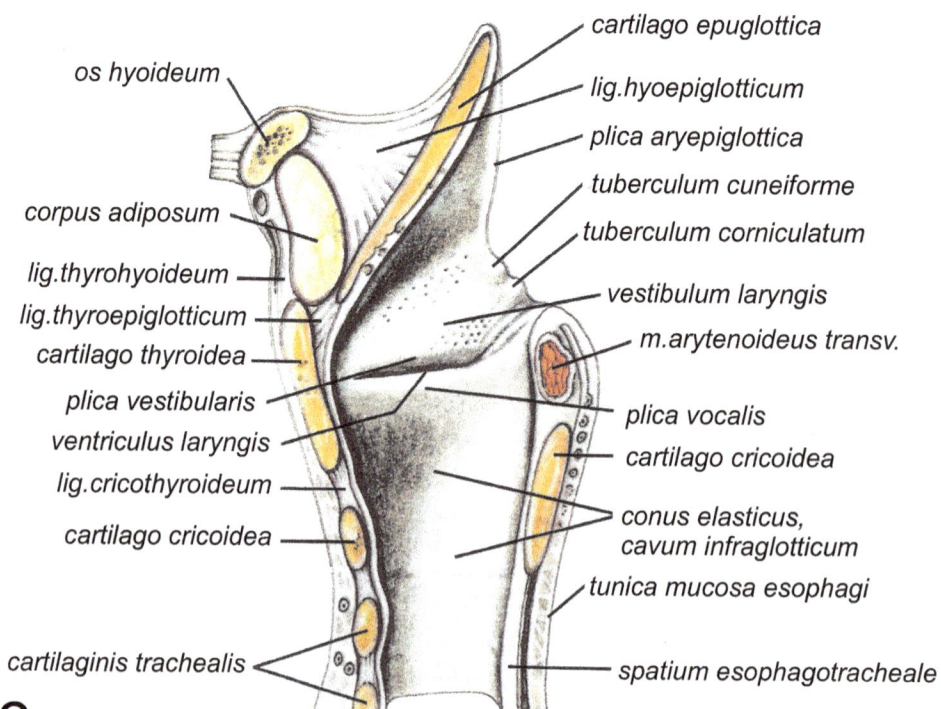

Fig. 57. Sagittal section through the larynx.

- os hyoideum
- cartilago epuglottica
- lig.hyoepiglotticum
- plica aryepiglottica
- corpus adiposum
- tuberculum cuneiforme
- tuberculum corniculatum
- lig.thyrohyoideum
- vestibulum laryngis
- lig.thyroepiglotticum
- m.arytenoideus transv.
- cartilago thyroidea
- plica vestibularis
- plica vocalis
- ventriculus laryngis
- cartilago cricoidea
- lig.cricothyroideum
- cartilago cricoidea
- conus elasticus, cavum infraglotticum
- tunica mucosa esophagi
- cartilaginis trachealis
- spatium esophagotracheale

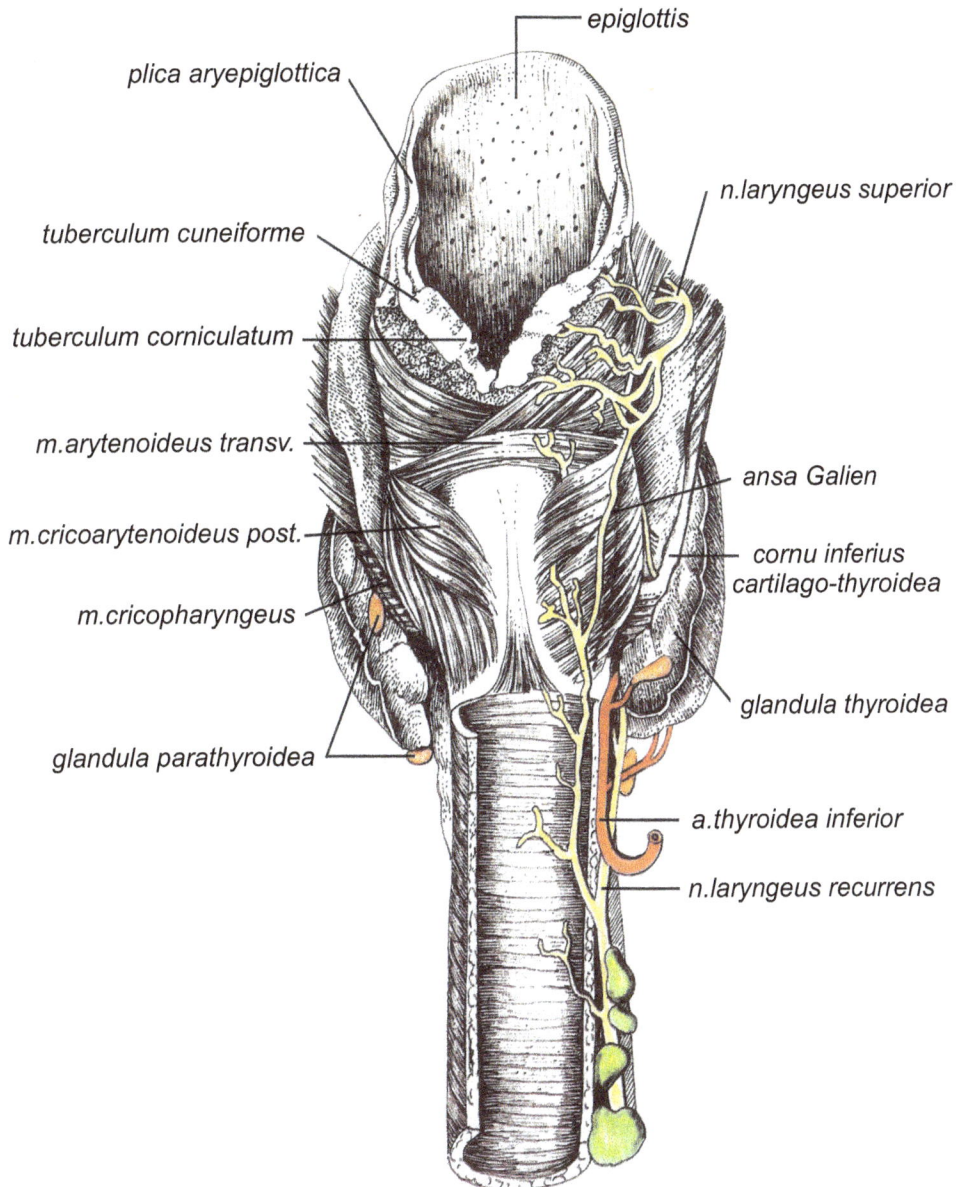

Fig. 58. Nerve supply of the larynx.

Labels (Fig. 58):
- epiglottis
- plica aryepiglottica
- tuberculum cuneiforme
- tuberculum corniculatum
- m.arytenoideus transv.
- m.cricoarytenoideus post.
- m.cricopharyngeus
- glandula parathyroidea
- n.laryngeus superior
- ansa Galien
- cornu inferius cartilago-thyroidea
- glandula thyroidea
- a.thyroidea inferior
- n.laryngeus recurrens

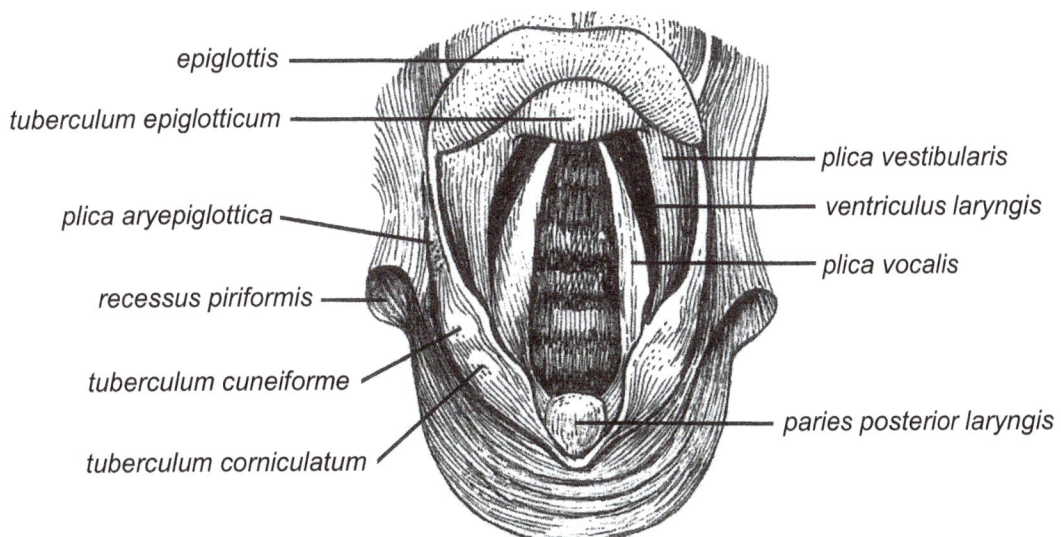

Fig. 59. The inlet of the larynx (aditus laryngis).

Labels (Fig. 59):
- epiglottis
- tuberculum epiglotticum
- plica aryepiglottica
- recessus piriformis
- tuberculum cuneiforme
- tuberculum corniculatum
- plica vestibularis
- ventriculus laryngis
- plica vocalis
- paries posterior laryngis

63

On the lateral sides of the ary-epiglottic folds, the laryngeal mucosa is continuous with the pharyngeal mucosa, forming on each side a groove towards the mouth of the oesophagus, the piriform recess *(recessus piriformis)*. The epiglottis is connected to the roof of the tongue, situated anteriorly and above if, by the median glosso-epiglottic folds and laterally, on either side, by the lateral glosso-epiglottic folds (fig. 36, 37, 53, 54, 57, 59).

The mucosa of the laryngeal vestibule consists of a stratified pavement epithelium and of a rich submucous loose connective tissue which, in case of inflammations, brings about the glottic oedema, named so in spite of its situation above the glottis.

When, for various reasons (serum sickness, croup etc.), the glottic oedema is produced, anafomopathoiogically the folds are tumefied, the inlet of the larynx is considerably narrowed and the respiration becomes difficult, sometimes impossible, threatening with the patients suffocation. Without an emergency intervention by intubation or tracheotomy, the patient's life is jeopardized by asphyxia. The laryngeal ventricle *(mesolarynx)* corresponds to the laryngeal segment situated between the ventricular fold above and the vocal fold below. Between these two folds lies a lateral space, termed laryngeal recess *(recessus laryngis)* (Morgagni's ventricle). The two ventricular folds delineate a slit, named *rima vestibuli,* and between the two vocal folds lies the *rima glottidis;* thus, the middle part communicates with the superior part through the *rima vestibuli* and with the inferior part through the *rima glottidis* (fig. 54, 57).

The ventricular folds are represented by a pair of mucous folds which begin anteriorly, at the insertion of the epiglottis info the thyroid cartilage, and are -disposed horizontally. In their thickness lie a few muscle fibres attached on a ligament *(ligamentum ventriculare).* They are also called false vocal cords. At this level ends the quadrangular membrane.

Beneath the ventricular folds is located the vocal apparatus proper, which makes up the glottis. If is formed by the two folds, the vocal cords or *labia vocales - plicae vocales -*, which delimit the rima glottidis. In the thickness of the vocal fold lies the vocal muscle *(m. vocalis),* attached on the vocal ligament. The free border of each labium is the vocal fold. In this fold ends the elastic cone, the inferior border of which makes up the vocal ligament (fig. 56).

The vocal folds are anteriorly attached to the thyroid cartilage, immediately below the insertion of the epiglottis, and are inserted info the vocal process of the arytenoid cartilage.

Thus, the rima glottidis is divided into an anterior part, the infermembranous part *(pars intermembranacea),* and a posterior part, the intercartilaginous part *(pars intercartiiaginea).*

The infragloftic portion *(cavum infraglotticum) (hypolarynx)* is located in the cricoid segment and is bounded anteriorly and superiorly by the arch of the cricoid cartilage and the elastic cone and posteriorly and laterally by the lamina of the cricoid cartilage and its lateral parts; downwards if is continuous with the trachea.

On the superior border of the cricoid cartilage is attached the elastic cone, which ascends towards the cricoid cartilage, in the median segment, forming the cricothyroid ligament (the free part of the elastic cone - *pars libera coni elastici);* the rest reaches the vocal cords, ending in the vocal ligament. In the space between the elastic cone and the lamina of the thyroid cartilage lies the thyroarytenoid muscle.

The mucosa of the larynx is firmly attached to the infernal surface of the elastic cone. If is of respiratory type *(ciliated columnar epithelium),* with the exception of the vocal folds, on which are exerted the mechanical strains during phonafion and respiration and which are covered by a stratified pavement epithelium.

Vessel and nerve supply of the larynx

The laryngeal arteries are branches of the superior and inferior thyroid arteries.

The superior laryngeal artery *(a. laryngea cranialis)* arises from the superior thyroid artery, above the superior border of the thyroid cartilage. It crosses, with the internal branch of the superior laryngeal nerve, the hyothyroid membrane, supplying the walls of the laryngeal vestibule. If is accompanied by the superior laryngeal veins.

If receives branches from the cricothyroid artery, too, which is also a branch of the superior thyroid artery, that supplies the wall of the larynx in the lower segment. The inferior laryngeal artery *(a. laryngea caudalis),* a branch of the inferior thyroid artery, is a small ramus which ascends behind and above the cricothyroid joint and gives off branches to the posterior musculature and to the mucosa of the larynx.

The veins of the larynx empty as follows: the superior veins into the superior thyroid vein and the inferior ones into the thyroid plexus.

64

The lymph vessels of the larynx are divided into two groups: superior and inferior. In the superior segment, the lymphatics follow the course of the superior laryngeal artery and drain into the deep cervical lymph nodes, approximately at the level of the carotid bifurcation. The lymph vessels of the inferior part drain info the deep cervical lymph nodes, after passing through the prelaryngeal lymph nodes, especially the lymphatics of the supracricoid segment, while those of the subcricoid segment run along the inferior laryngeal nerve, to the pretracheal lymph nodes. All the lymph vessels of the larynx course finally to the inferior deep cervical lymph nodes.

The laryngeal nerves (fig. 58) are the superior laryngeal nerve *(n. laryngeus cranialis)* and the inferior laryngeal nerve *(n. laryngeus caudalis),* which contain both motor and sensitive fibres.

The superior laryngeal nerve is a branch of the vagus nerve. If divides, at the level of the greater horn of the hyoid bone, into two branches: infernal and external. The external branch supplies the laryngo-pharyngeal muscle and the cricothyroid muscle. The internal branch penetrates through the hyothyroid membrane, courses beneath the mucosa of the piriform recess and supplies the mucosa of the epiglottis and of the larynx up to beneath the vocal cord.

The inferior laryngeal nerve is the terminal branch of the recurrent nerve. If reaches, with the inferior laryngeal artery, the posterior wall of the larynx, at the level of the cricothyroid joint. Its motor branches supply all the muscles of the larynx, except the cricothyroid muscle (which is supplied by the external branch of the superior laryngeal nerve). Its sensitive fibres supply the laryngeal mucosa, passing above beyond the rima glottidis and below up to the level of the second and third tracheal rings.

In the superior segment, the territories of the two laryngeal nerves interpenetrate each other.

Examination of the larynx. The palpation of the larynx is performed with the pulp of the fingers, for the median line, and between the thumb and the index, for the lateral surfaces of the pharynx. If should be carried out patiently and systematically, in order to identify the various anatomical elements accessible by this method.

In the superior part of the subhyoid region, a poorly marked prominence, represented by the hyoid bone, is palpated. Follow a depression of about 2 cm, which is the hyothyroid membrane, then the notch of the thyroid cartilage, the thyroid angle, the laryngeal prominence or Adam's apple, a second, smaller depression which is the cricothyroid membrane and, finally, the arch of the cricoid cartilage. Laterally are felt the greater horns of the hyoid bone and, a little lower, those of the thyroid cartilage and the lateral laminae of the thyroid cartilage, covered partially by the two lobes of the thyroid (G. Portman).

The exploration of the lymph nodes is of a particular interest, especially in the case of the cancer of larynx, for the detection of the eventual metastases at this leve.The prelaryngeal lymph nodes, the jugulocarotid chain and the spinal chain can be palpated. Practically we are especially interested in the jugulocarofid chain, which is most frequently affected. The palpation of the lymph nodes of this chain is made with the pulp of the fingers, which are insinuated info the carotid region, anteriorly and below the sternocleidomastoid muscle. A technique which facilitates this exploration is the following: the patient is seated on a stool and the physician stands behind him; the patient's head being flexed, the palpation is performed concomitantly for both chains, with the fingers considerably flexed.

Laryngoscopic examination. In the framework of this inspection we mention only the examination of the motility of the vocal folds and of the arytenoid cartilages. The larynx should be examined during phonafion and respiration. For this purpose, we ask the patient to emit the vowels "e" or "i" and then to inspire deeply. In this way we obtain data regarding the action of the three muscle groups which condition the physiology of the larynx (the tensors and constrictors - phonation, the dilators - respiration), as well as the functionality of the two crico-arytenoid joints.

By examining the larynx under these conditions, we obtain the three characteristic images of the glottic apparatus:

-during quiet respiration, the vocal cords are in a slight abduction and, as a consequence, the glottis has the shape of an isosceles triangle with the base posteriorly (inferiorly in the mirror);

-during forced inspiration, the arytenoid cartilages recede at the most from each other, drawing with them also the vocal folds. The base of the triangle increases in size and the triangle becomes a pentagon;

-during phonafion, the cords become parallel and the arytenoid cartilages approach each other; the glottic orifice has the smallest size. As a matter of fact, during phonafion, the vocal folds vibrate, receding from and approaching each other at a high speed; the naked eye cannot perceive these movements dissociated, but only their resultant, which is exactly the mentioned image.

Because of the physiological phenomenon of fusion of images, the oscillations exceeding the frequency of 10 cycles/s cannot be observed by the naked eye, but they can be perceived by means of a pulsatile light source, of an adequate frequency, owing to the stroboscopic effect. By means of the stroboscopic examination, frequency and amplitude changes, as well as the manner in which the occlusion of the glottic fissure is achieved may be observed, permitting in this way to detect the vocal diseases at the functional stage.

The Thyro-Tracheo-Oesophageal Region

It is a median region of the neck, which includes the thyroid gland, the trachea and the cervical oesophagus. If belongs to the subhyoid region and has the fibrous skeleton made up of the superficial cervical fascia with the platysma muscle and of the middle cervical fascia.

Didactically, the thyroid region may be distinctly delimited from the tracheo- oesophageal region. The thyroid region appears in the form of two triangles, which are in relation medially with the trachea, superoiaterally with the superior belly of the omohyoid muscle and inferolaferally with the sternocleidomastoid muscle. The thyroid gland extends above up to the middle third of the thyroid cartilage and below, up to the fifth-sixth tracheal ring, corresponding to the C_5-C_7 vertebrae.

The tracheo-oesophageal region is rectangular, bounded above by the cricoid cartilage, below by the jugular notch and laterally by two lines uniting the sternoclavicular joint to the lateral borders of the chin.

The Thyroid Gland

It is an unpaired gland, more developed in women, and confers the round and wide form to the neck. If is H-shaped and consists of two lateral, almost equal, pyramidal triangular lobes (with a superior, thin pole and an inferior pole - the base), united by an intermediate isthmus. It presents also a pyramidal process (Lalouette's pyramid) which, when if exists, arises from the isthmus or from the left lobe and ascends to the hyoid bone (fig. 60).

The isthmus. The thyroid isthmus lies anteriorly to the two first tracheal rings, sometimes even above the cricoid cartilage. Its relation with the trachea divides the tracheotomies into supra-,trans- or subisthmic. The isthmus is situated at 2.5-3 cm above the jugular notch of the sternum, a landmark important to be known for the performance of tracheotomy. Anteriorly, the isthmus is covered by the sternohyoid muscles, the middle and superficial cervical fasciae, the platysrna muscle and the skin.

The lobes. The lateral lobes appear as two vertically elongated glandular masses, with three surfaces: muscular (the anterolateral surface), laryngotracheal (the medial surface) and vascular (posterior); a vertebral border, a superior pole and an inferior pole - the base.

The anterolateral surface is in relation, through the connective adipose tissue, with the sternothyroid, sternohyoid and omohyoid muscles, enclosed in the middle fascia of the neck (fascia colli media).

The posteromedial surface is in relation with the first tracheal rings, the cricoid and thyroid cartilages. Behind, this surface passes over the trachea, reaches the oesophagus and gets into contact, in the groove between these two structures, with the inferior laryngeal nerve and the tracheal lymph nodes.

The posterior border is in relation with the vasculonervous bundle of the neck, especially with the common carotid artery.

The superior, pointed pole is the place where the superior thyroid artery, a branch of the external carotid artery, penetrates through the thyroid capsule and spreads, through its branches, predominantly on the anterolateral surface of the gland.

The artery is accompanied by the satellite vein and by the branches of the superior laryngeal nerve. The external branch of this nerve (the cricothyroid nerve) has an oblique descending course above the lateral lobe, towards the cricothyroid muscle.

The inferior pole of the lateral lobe is normally situated at 1.5-2 cm above the sternum (jugular notch) (incisura jugularis).

The connective coverings of the thyroid are the following:
-the internal capsule proper, a fibrous layer intimately adhering to the gland and
-the external perithyroid capsule, which derives from the fascia colli media (cervical fascia).

The perithyroid capsule (the fibrous or external capsule) differs from the internal capsule proper of the gland, which covers the thyroid body, adhering to it. It is formed of thin fibroconnective laminae, more resistant posteromedially and very loose in the rest; they represent the condensation of the connective lamellar tissue of various origins. The perithyroid capsule is, as a matter of fact, a joining of separate connective membranes:

-anteriorly, the fascia covering the sternothyroid muscles, which is itself dependent on the middle lamina of the cervical fascia;

-laterally and posteriorly, the vascular sheath of the posterior jugulocarotid bundle;

-the carotid sheath (vagina carotica);

-medially, it is formed of the connective lamellae around the oesophagus and the trachea, which are part of the visceral sheath. The external capsule is also reinforced by the perivascular tissue of the 4-5 glandular pedicles. To the peritracheo-oesophageal portion of the capsule adjoin the connective lamellae, condensed along the recurrent laryngeal nerves, which pass outside the capsule, between the trachea and the oesophagus. The approach to the recurrent laryngeal nerve in the middle and superior portion of its cervical course brings about the dilaceration of the capsule and opens the interfascial space of the gland. Several fibrous prevertebral lamellae accompany both the inferior thyroid artery and the thyroid loop of the sympathetic chain. At the level of the crossing of these lamellae is formed, by condensation, a unique membrane, which reinforces the fibrous capsule. Along these formations, the collections proper or those absceded into the interfacial space of the thyroid may spread into the superior mediastinum, into the scalenovertebral region and into the prevertebral fascia and space;

- superiorly, the perithyroid capsule continues up to the hyoid bone, into which it inserts;

- beneath the isthmus and the lower extremity of the thyroid lobes, the capsule reflects itself and surrounds the lobes and the isthmus, under the form of a cul-de-sac, which is posteriorly attached on the trachea, fusing with the rest of the pretracheal lamina.

The subisthmic connective tissue appears as a unique vertical lamina, which descends up to the mediastinum, mentioned by some authors under the name of thyropericardial lamina. Between the perithyroid capsule and the capsule proper of the gland lies the perithyroid or thyroid space, a true operative cleavage plane, crossed by vessels, well detachable anteriorly, inferiorly and postero-laterally. The clue to the access to the cleavage space consists in the localization of the sternothyroid muscle, which adheres more intimately to the perithyroid capsule than to the rest of the middle cervical fascia. The localization of the two muscles and the approach to their deep surface, with the annexed connective tissue, lead us into the perithyroid space. Moreover, the perithyroid capsule is firmly connected to the adventitia of the trachea and of the oesophagus and to the vascular sheath. On a transverse section, the whole connective tissue mentioned appears as a transverse connective lamina, owing to which some authors describe these formations as a separate entity - the transverse aponeurosis of the neck - independent from the middle lamina of the cervical fascia.

Posteriorly and medially, the periglandular space is diminished up to disappearance, by fusion of the two capsules, the proper and the periglandular capsules, which form a unique lamina, including also the adventitia of the larynx and of the trachea. This lamina becomes a true supporting apparatus, through the laryngoand tracheoglandular ligaments, which attach the whole formed by the gland and the capsules to the cricoid cartilage and to the trachea. Therefore, at this level, the strictly intracapsular detachment of the gland is actually impossible and the attempts at interfascial dissociation are dangerous. The parathyroid glands are in an intimate relation with the thyroid gland. They are small glandular structures, usually four in number, two on each side, one superior and one inferior. The parathyroid glands are the site of numerous anatomofopographical variations.

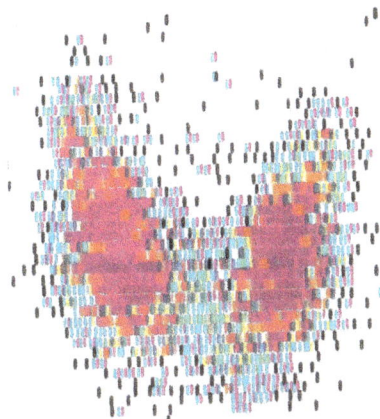

Fig. 60. Thyroid scintiscan.

The inferior parathyroid glands are located either near the inferior poles of the thyroid gland, in the angle between the gland and the oesophagus, or in a median position, within the capsule, at the level of the inferior thyroid artery penetration. The superior parathyroids lie at the level of the anterolateral surface of the cricoid cartilage - high position (Silveirs); they are often situated lower, at the level of the second tracheal ring - median position (Crisoli). Unfrequently, they can also have an extracapsular or iniraglandular position. In order to preserve them in the case of a subtotal thyroidectomy, the posterior branch of the superior thyroid artery is spared and the selective ligation only of the anterolateral branch of the artery is performed.

The parathyroids are most often supplied by branches of the inferior thyroid artery; each arteriole is terminal, without any replacement possibility. Thus, the ligation of the thyroid arteries brings about a physiological parathyroidectomy through ischemia; the same effect is produced by the complete enucleation of the lateral thyroid lobe outside the perithyroid capsule, hence the surgical necessity to spare a posterior parenchyma "wall", in order to protect the glandular branches of the inferior thyroid artery, the inferior parathyroids and the inferior laryngeal nerve.

Their lesion, which leads to postoperative tetany, is avoided by ligating only the anterolateral branch of the superior thyroid artery and sparing the posterior branch.

Vessel and nerve supply of the thyroid gland

The thyroid gland is a highly vascular organ. The size of the thyroid varies according to the blood amount which reaches the gland, regulating in this way the circulation. In the case of a blood pressure increase in the cephalic region, the thyroid gland may take over a significant amount of blood, which it conveys to the heart, the carotid sinus being involved in this mechanism of blood output regulation. Each of the four thyroid arteries has a calibre equal to that of the radial artery. The thyroid arteries convey to the thyroid a blood amount equal to that carried by the cerebral arteries to the brain.

The thyroid gland has five vascular pedicles: two lateral pedicles on each side and one inferior.

Arteries. The superior thyroid artery arises from the anterior surface of the external carotid artery, at 5 cm above the carotid bifurcation. Its course is initially horizontal in the region of the carotid trigone, then it becomes juxtaglandular and ends by trifurcation, at the superior pole of the lateral lobe of the thyroid gland. In the region of the carotid trigone, it corresponds laterally to the covering planes and medially it is applied on the larynx, on which descends also the superior laryngeal nerve.

In the juxtaglandular portion, the artery is accompanied by the homonymous vein and by the external branch of the superior laryngeal nerve, forming the superior thyroid pedicle. It corresponds laterally to the sternothyroid muscle and to the anterior belly of the omohyoid muscle; medially, to the thyrohyoid membrane and to the thyroid cartilage; anteriorly, to the posterior border of the thyrohyoid muscle and posteriorly to the jugulocarotid bundle.

The collateral branches of this artery are: the infrahyoid branch to the infrahyoid muscles, the middle sternocleidomastoid branch, the superior laryngeal artery, which crosses with the homonymous nerve the thyrohyoid membrane and is distributed to the laryngeal wall, and the cricothyroid branch, anastomosed with its fellow on the opposite side.

The terminal branches are:

- the medial branch, which is continuous with the course of the trunk of the artery and descends vertically, along the thyroid lobe; it forms with the opposite homologous artery the supraisthmic anastomosis;

- the posterior branch, which runs along the posterior border of the thyroid gland and anastomoses with the homologous branch of the inferior thyroid artery;

- the lateral branch, which ramifies on the anterolateral surface of the thyroid lobe.

The superior thyroid artery supplies the two superior two-thirds of the thyroid gland and partially the isthmus and the laryngeal wall.

The inferior thyroid artery arises in most cases as a medial branch of the thyrocervical trunk, unfrequently directly from the subclavian artery. The thyroid artery may form a common trunk with anyone of the collaterals of the subclavian artery. Of a surgical interest is the trunk common with the vertebral artery. The presence of an arterial cone, formed by the portion of origin of the inferior thyroid artery and of the vertebral artery, is mentioned in 6% of cases.

The inferior thyroid artery is made up of three portions. In the first, vertical portion, the artery, covered by the prevertebral lamina and fatty tissue, passes between the vasculonenervous bundle of the neck and the vertebral artery (anterolaterallly)and describes a first medio-inferior curve beneath the carotid tubercle which it never touches, as it courses at a distance of 1-3 cm, on the average 1.5 cm, from the latter.

The second, horizontal portion describes a curve with the concavity anterior and crosses the posterior surface of the common carotid artery and the anterior surface of the vertebral artery. In 12% of cases on the left and in 3,5% on the right, it passes behind the vertebral artery, owing to the variation of the origin of the thyroid artery and to the high penetration level (C5-C$_4$) into the transverse canal of the vertebral artery.

In this portion, the artery is fastened by the prevertebral lamina and by the thyroid loop of the sympathetic trunk and has neighbourhood relations with the middle cervical ganglion. At this level, the artery receives branches from the sympathetic trunk. These relations explain the lesions of the sympathetic trunk in the ligations of the inferior thyroid artery.

At the level of the medial border of the common carotid artery, it forms a curve - the third portion - with the concavity superior, after which it either ramifies, in 70% of cases, or courses for 3-4 cm as an arterial trunk. This portion gives off the artery to the parathyroid glands and has relations with the recurrent laryngeal nerve. The artery crosses always the nerve at the level of the posterior border of the thyroid gland. More frequently, the right recurrent laryngeal nerve crosses anteriorly the arterial trunk and not the terminal branches of the inferior thyroid artery. The left recurrent laryngeal nerve is more medial and crosses the artery posteriorly in its internal third. The second and third portions of the inferior thyroid artery are always in relation with the postero - inferior surface of the lobes of the thyroid gland, the third portion being even intracapsular. The collateral branches are meant to the trachea, the pharynx and the oesophagus, when they do not arise from the terminal branches of the artery.

The terminal branches of the inferior thyroid artery are:

- the posterior branch, which ascends behind the lateral lobe of the thyroid gland, forming exceptionally the posterior longitudinal anastomosis;

-the medial branch, which surrounds the inferior pole of the lateral lobe and ascends its internal border, forming the subisihmic arch, rather unfrequently encountered. When it exists, it detaches from an anteromedial branch, which runs along the inferior border of the isthmus;

- a lateral branch, distributed to the posterolateral surface of the lobe;

- a deep branch, which is continuous with the direction of the trunk and runs along the medial border of the lobe, in the space between the gland and the trachea, giving off in its course glandular branches. These terminal arteries supply the inferior half of the lateral lobes and anastomose in the parenchyma, both with each other and with the superior thyroid arteries. The isthmus lies on the "crossroads" of four arteries, but presents only a few trans-verse anastomoses, hence the possibility of sectioning it with a scanty haemorrhage.

The middle thyroid artery or *arferia thyreoidea ima* (Neubauer's artery) is an inconstant arteriole (present in 10% of cases); it arises from the brachio-cephalic trunk or directly from the aortic arch and ascends up to the isthmus the pericardial lamina. When this artery is present, one of the inferior thyroid arteries is usually absent. It should be identified in the sub- and transisthmic tracheotomies.

Veins. The middle thyroid veins may lack or may be single or multiple; always short, they start from the lateral surface of the thyroid lobes, pass in front of the common carotid artery and empty directly into the internal jugular vein at another level than that of their origin, without following the course of a thyroid artery.

The inferior thyroid vein arises from several roots, descends into the thyropericardial lamina and empties into the venous angle or into the left brachiocephalic vein.

Lymph vessels and lymph nodes of the thyroid and parathyroid glands. From the lateral and superoposterior portions of the thyroid lobes, the lymph collectors run along the superior thyroid pedicle and empty into the deep cervical (medial jugular) lymph nodes.

The lymph vessels of the anterior and superior portions, as well as of the thyroid isthmus form a medial trunk which, passing on the anterior surface of the thyroid, may intercept some of the superficial (medial prelaryngeal) cervical lymph nodes. Between the thyroid and the cricoid cartilages they unite with the lymph collectors of the superior portion of the larynx and, together, empty into the deep jugular lymph nodes. From the inferior portion of the lobes and from the thyroid isthmus the efferent lymphatics have a lateral course and a medial course. The lymphatics with a lateral course accompany the middle and inferior thyroid veins and empty into the inferior deep cervical (medial jugular) lymph nodes. The efferent lymphatics with a medial course have as intermediaries 5-6 deep (pre- and paratracheal) lymph nodes, situated along the recurrent laryngeal and the vagus nerves; they unite with the efferent lymphatics from the dorsal and inferior portions of the larynx and from the superior portion of the trachea and empty into the deep cervical (medial jugular) lymph nodes.

The medial efferent lymphatics may have another course (avoiding the jugular nodes), through the posterior mediastinal lymph nodes and from here in the bronchomediastinal lymph trunk.

In a fifth of cases, the efferent lymph vessels of the thyroid empty into the retropharyngeal lymph nodes.

Nerve supply of the gland. The thyroid is supplied by perivascular sympathetic (cervical) amyelinic fibres and by the superior and inferior laryngeal nerves. The latter nerve has relations with the inferior thyroid artery (which we have mentioned) and should be attentively spared by the surgeon, since its lesion leads to respiration and phonation disturbances. Therefore, in the case of ligation of some of the branches which are in contact with the thyroid gland, the lesion of the vascular bundle of the inferior thyroid artery should be avoided.

The Cervical Trachea

The trachea is continuous with the larynx, as it is part of the inferior respiratory tract. In the adult, it begins at the level of the space between the sixth and the seventh cervical vertebrae, this position varying with the age, and ends at its bifurcation, which corresponds to the fourth and fifth lumbar vertebrae.

Topographically, in the trachea may be distinguished a cervical segment *(pars cervicalis tracheae)* and a thoracic segment *(pars thoracalis tracheae)*.

In the adult, the cervical part of the trachea is 5-7 cm in length, about half of the total length. In children it is shorter, of 4-6 cm. The first tracheal ring corresponds to the sixth and seventh cervical vertebrae or even to the first thoracic vertebra (according to Corning-Hafferl). It begins beneath the inferior border of the cricoid cartilage, into which it inserts, and ends at the level of the horizontal plane that passes through the upper border of the manubrium of sternum. In this region the trachea has a median position, it is slightly deviated to the right and descends towards the thorax, following the curve of the vertebral column, obliquely downwards and backwards.

The most important surgical relation of the trachea is with the pretracheal fascia. This is a dense connective lamina which covers the anterolateral surface of the trachea. The deep surface of the lamina may detach from the anterior surface of the trachea, permitting the access to a pretracheal space which descends up to the mediastinum, an ideal surgical cleavage plane, along which the mediastinal endoscopy can be performed.

The identification of the cleavage plane during the tracheotomy transforms this operation into a bloodless intervention. Through the anterior surface, the pretracheal fascia is related to all the visceral and neurovascular elements of the subhyoid plane, which it protects during the manoeuvre of freeing the trachea.

Anteriorly, through the pretracheal lamina, the trachea is in relation at the level of the first tracheal rings (second, third, fourth) with the isthmus of the thyroid gland, with the peritracheal connective tissue, that forms a connective layer - the thyropericardial lamina -, in the depth of which lie the lymph nodes (the pretracheal group, towards wich course the laryngeal lymph nodes and the lymphatics of the isthmus of the thyroid gland), with the thyroid veins, with the brachiocephalic venous trunk, if it passes beyond the manubrium sterni and, eventually, with the middle thyroid artery, when it exists.

Laterally it is related to the large vessels of the neck, on each side; the common carotid artery recedes in an acute angle, being pushed laterally by the thyroid lobes which it surrounds.

In the space between the carotid artery, the oesophagus and the trachea penetrates the inferior thyroid artery, which divides, beneath the thyroid lobes, into its branches.

The right recurrent laryngeal nerve is situated more posteriorly in the space between the trachea, the oesophagus and the vertebral column, whereas the left nerve is located more anteriorly in the angle between the lateral surface of the trachea and the oesophagus; both recurrent laryngeal nerves cross the branches of the inferior thyroid artery.

All these elements are covered by the subhyoid musculofascial plane and by skin. At the level of the cricoid cartilage (C6, the trachea is in relation with the orifice of the oesophagus.

Posteriorly, through the oesophagus, the trachea is related to the vertebral column; it is joined to the oesophagus by elastic tissue and muscle fibres (the tracheo - oesophageal ligaments). Both organs are covered with a connective lamina (Monteiro's visceral fascia), which arises from the base of the skull, covers the pharynx, forming the pharyngeal adventitia, and courses up to the mediastinum, including also the trachea.

70

Structure. The trachea has the shape of a cylindrical tube, the posterior fourth of wich is flattened. It is structured of:

- an internal mucous coat, continuous with that of the larynx and made up of a ciliated stratified epithelium and a connective submucosa;

- a middle, musculocartilaginous coat, formed of 16-20 incomplete cartilaginous rings (the posterior part lacks, hence they are horse-shoe shaped), slightly flattened laterally in the cervical region and flattened sagittaly in the thorax; the first ring is higher than the others; they are linked to each other by elastic ligaments which are continuous with the perichondrium of the rings. The posterior ends of the cartilaginous segments are connected by a muscular membrane, made up of fransversally disposed smooth muscles;

- an external coat, consisting of a fibro - elastic membrane which covers the cartilages, connects them to each other (annular ligaments) and covers the posterior part of the trachea (transverse lamina), making up the adventitial coat.

Vessel and nerve supply

The arteries of the cervical portion of the trachea arise from the inferior thyroid artery, as well as from the middle thyroid artery (when it exists). The veins, disposed horizontally in the spaces between the rings, empty into the thyroid veins and into the perioesophageal network. The lymphatics cross the fibrous membrane and open on each side into the recurrent lymph nodes; they are inferiorly continuous with the tracheobronchial lymph nodes.

Nerves. The trachea shares with the larynx its nerve supply, which is assured by the two pairs of laryngeal nerves: superior and inferior.

The superior laryngeal nerve, predominantly sensitive, arises from the inferior (plexiform) ganglion of the vagus nerve, runs obliquely downwards and forwards alongside the larynx, crossing the carotid bifurcation at the level of the greater horn of the hyoid bone (approach for anaesthesia) and gives off its two terminal branches.

The medial (or superior) branch, lying under the superior laryngeal artery, crosses the thyrohyoid membrane and supplies the mucosa of the superior portion of the larynx, that of the adjacent part of the pharynx and of the root of the tongue. It anastomoses frequently with the inferior (recurrent) laryngeal nerve, forming the inconstant Galien's loop.

The lateral (or inferior) branch runs laterally, along the insertion of the inferior constrictor muscle of the pharynx, supplies the cricothyroid muscle, then pierces the cricothyroid membrane and supplies the mucosae of the ventricle, of the vocal fold and of the subglottic region of the larynx. Moreover, it assures the tone of all the muscles of the larynx.

The recurrent (inferior) laryngeal nerve is predominantly motor. It derives from the vagus nerve: on the right beneath the origin of the subclavian artery, on the left beneath the aortic arch. From here, it ascends into the tracheo-oesophageal angle, slightly obliquely on the right, vertically on the left. It passes on the posterior surface of the lobe of the thyroid gland, in a variable relation with the inferior thyroid artery, then it runs beneath the inferior constrictor muscle of the pharynx, penetrates the larynx and supplies all the muscles of the larynx (with the exception of the cricothyroid muscle), which are phonatory or respiratory, constrictor or dilator muscles. It supplies also the posterior mucosa and anastomoses with the medial branch of the superior laryngeal nerve. In its course, the recurrent nerve gives off the middle cardiac nerve and tracheal, oesophageal and pharyngeal branches.

The Cervical Oesophagus

The oesophagus is a musculomembranous tube, which connects the pharynx to the stomach. It divides, according to the regions which it crosses, into three segments: cervical, thoracic and abdominal. (At the study of the thoracic oesophagus we shall give more precise details about the division of the oesophagus).

The cervical segment of the oesophagus is situated on the deepest plane of the subhyoid region, behind the cervical portion of the trachea. It is 5-6 cm long, begins at the level of the lower border of the cricoid cartilage, at 15-16 cm beneath the dental arch (anteriorly to the sixth cervical vertebra), and ends at the level of the horizontal plane passing through the superior border of the manubrium of sternum.

Its course corresponds to the median line and is slightly deviated to the left. By its left border it passes beyond the left lateral surface of the trachea; its right border lies in the dihedral angle between the trachea and the vertebral column.

With the trachea it is enveloped in a coat consisting of connective tissue, forming a fibrous sheath which is continuous with that of the pharynx and projects into the mediastinum up to the diaphragm muscle. This sheath contains, in the tracheo - oesophageal angle, the recurrent nerves (branches of the vagus nerve).

It has relations, anteriorly, with the posterior wall of the trachea. A small portion passes beyond the trachea on the left and remains uncovered by it. This part is in relation with the left recurrent nerve, the branches of the inferior thyroid artery, the satellite veins, the recurrent lymph nodes and the posterior border of the left thyroid lobe. Behind, the oesophagus is related to the anterior surface of the cervical column, covered by the prevertebral muscles and fascia.

The left border, uncovered by the trachea, is more accessible and it is through it that the external oesophagotomy is performed, this being the surgical border of the oesophagus.

The right border, hidden at a depth of 2-3 mm between the vertebral column and the trachea, is above in relation with the recurrent nerve, which gradually approaches this border, and with the inferior thyroid artery and its branches, which spread out in the shape of a fan between the trachea and the vertebral column.

The two borders of the oesophagus have remote lateral relations with the cervical sympathetic chain (the sympathetic middle ganglion), the common carotid artery, the internal jugular vein and the vagus nerves.

The left, inferior border of the oesophagus is related to the thoracic duct.

Vessel and nerve supply

The arterial supply is mainly assured by the inferior thyroid artery through its superior oesophageal branches; *the venous supply* is represented by the superior oesophageal veins, which empty into the superior thyroid veins.

The lymph drains into the deep cervical lymph nodes.

The nerve supply is assured by branches of the vagus nerve (X), of the glossopharyngeal nerve (IX) and by filaments of the cervical sympathetic trunk.

The Thorax

Chapter 2

The Thoracic Wall

The skeletal thorax *(osteothorax) is* formed by the 12 pairs of ribs, the 12 thoracic vertebrae and the sternum, which delineate a space, a "cage", in totality occupied by the thoracic viscera.

The upper boundary of the thorax is formed by the first thoracic vertebra, the two ribs of the first pair and the suprasternal notch *(incisura jugularis sterna/is),* which delineate the superior thoracic aperture *(apertura thoracis cranialis).*

The lower boundary, called inferior thoracic aperture *(apertura thoracis caudalis),* is made up of the xiphoid appendix, the lower border of the ribs, the upper extremity of the last three ribs and the transverse processes of the twelfth thoracic vertebra.

The thoracic visceral cavity *(cavum viscerum thoracis)* extends beyond these upper bony boundaries. Thus, the pleura and the lung, at certain respiratory moments, exceed the first rib and the pleural cupula attains even the half of the clavicle. In the lower part, the boundary of the cavity is determined by the position of the diaphragm, which extends sometimes up to the fourth-fifth intercostal space.

The thoracic wall proper is made up of the following elements: the osteothorax, the external intercostal muscles, the extrathoracic fascia, the continuation of the superficial fascia of the neck *(fascia colli superficialis),* the internal intercostal muscles covered by the intrathoracic fascia. Between the intrathoracic fascia and the pleura lies the tela subserosa *(endothoracic fascia),* which adheres intimately to the parietal pleura and, together with the pleura when it is inflamed, generates the process of pachypleuritis.

At the level of the thoracic wall, the mammary region is of a particular clinical interest (fig. 61-63).

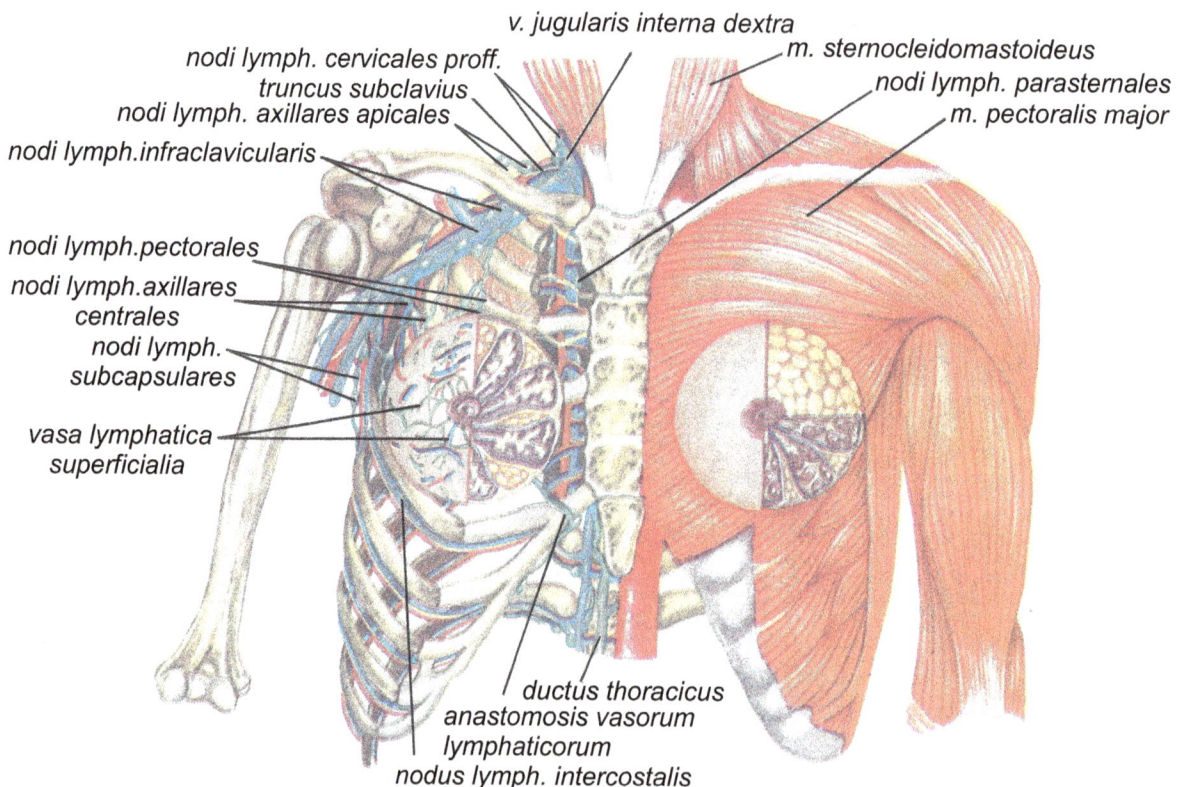

Fig. 61. Lymph vassals of the mammary gland.

The Mammary Region
(Regio mammae)

Topographically this region extends supero-inferiorly on the midclavicular line, from the fourth to the seventh rib, and midlaterally from the parasternal to the anterior axillary line. Laterally the gland exhibits a constant axillary projection as well.

The mammary region extends in depth up to the fascia of the pectoralis major muscle, being thus separated from the ribs and the intercostal spaces by the anatomic formations which make up the anterior wall of the axilla.

There are two mammary glands. The presence of supranumerary glands (polymastia) or the uni- or bilateral absence of the breasts (amastia) is sometimes observed.

The form of the mammary gland is variable, usually hemispherical, sometimes pyriform, discoid or cylindrical.

The size varies according to the age, the physiological state (pregnancy, postpartum) and the respective individual. Till puberty the mammary gland is poorly developed. With the appearance of puberty the gland begins to develop more completely and after menopause if undergoes involution (fig. 62).

The skin which covers the mammary gland presents in the center the areola and the nipple (papilla mammae). The areola is a circular zone of 15-25 mm in diameter, which surrounds the base of the nipple. The integument of this zone is thin and pink-coloured. On the areola, several small eminences, which represent the opening beneath the skin of some slightly hypertrophied sebaceous glands, known under the name of tubercles, may be observed.

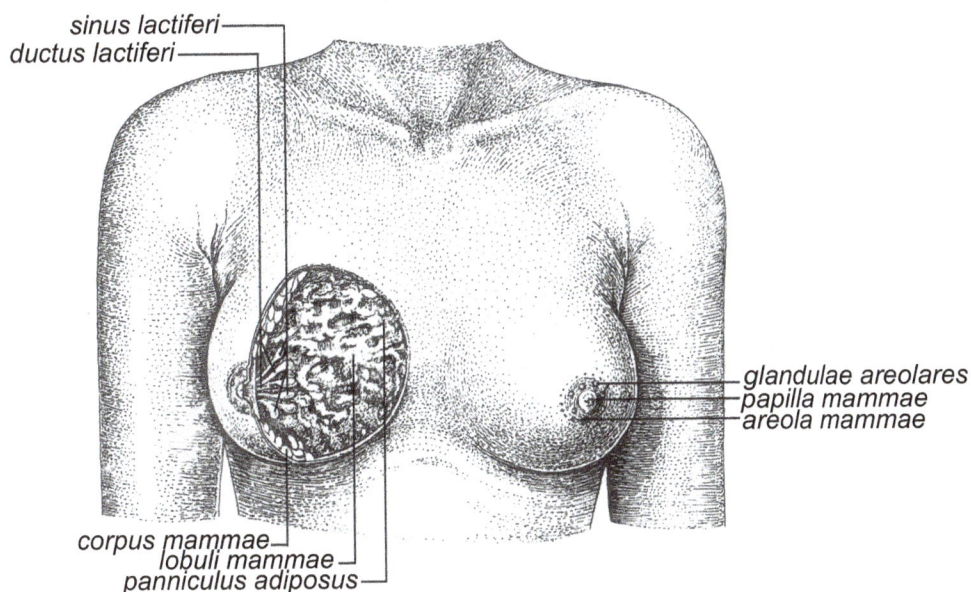

Fig. 62. Mammary glands

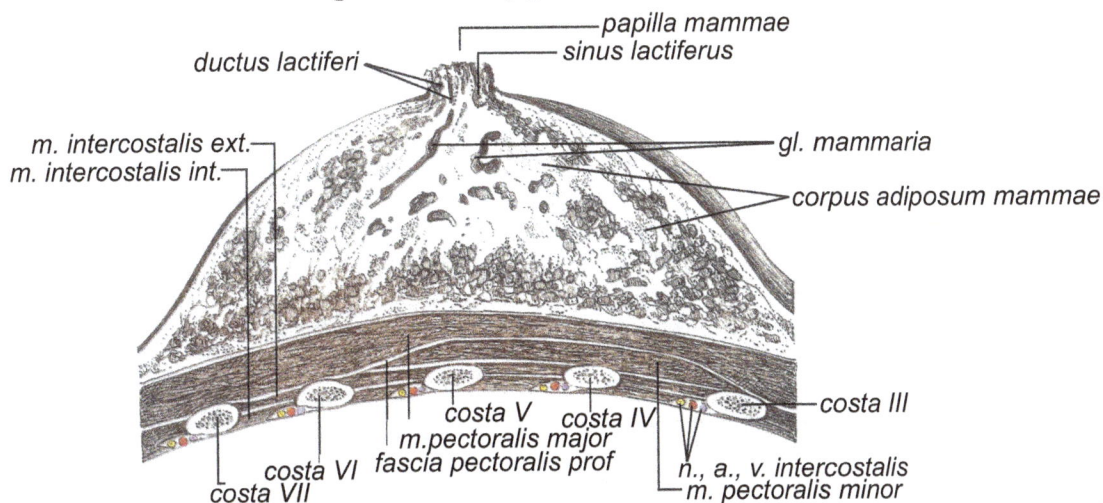

Fig. 63. The mammary gland - sagittal section (structure).

75

During pregnancy, small tubercles considered by some authors as accessory mammary glands, called Montgomery's tubercles, appear at the level of the areola.

In the centre of the areola is situated the nipple, a cylindrical eminence of 10-12 mm in length and 9-10 mm in width. Sometimes, instead of being an eminence, the nipple is deep-set (umbilicated nipple). The surface of the nipple is irregular, wrinkled and its colour is darker than that of the areola, turning into brown. At the level of the apex, the nipple is perforated by 15-20 orifices, which represent the apertures of the galactophorous ducts *(ducti lactiferi)*. At this level, the skin is thin, adhesive, provided on the deep surface with a series of smooth muscle fibres, which form the subareolar muscles, by the contraction of which the nipple is brought into erection. The fatty connective tissue of the mammary region is divided into two layers: one sited anteriorly to the gland, the premammary fatty tissue, more developed, which confers to the gland its shape, and the other, posterior to the gland, the retromammary fatty tissue, less developed. These two layers communicate with each other through the parenchyma of the gland. The glandular tissue is of the tubulo-acinous type. A group of acini makes up a lobule and a group of lobules forms a glandular lobe. The acini consist of a thin basal membrane, distinguishable with difficulty, and of a layer of stellate cells, called myoepithelial, since they are contractile and have an active role in milk expulsion. The glandular canaliculi are made up of a layer of myoepithelial cells and one or several layers of cubic or cylindrical cells. The lobules are arranged radially, each of them opening at the level of the nipple by a lactiferous duct, which presents before the opening a dilation called lactiferous sinus. The glandular acini are surrounded by a mass of fat, situated in front of and behind the mammary gland (fig. 63).

The mammary gland has several projections, some inconstant - subclavicular, sternal and inferior - and one constant, the axillary projection, which runs towards the axilla, avoiding the lower border of the greater pectoral muscle.

The mammary gland is invested in a fascia resulted from the differentiation of the hypoderm, which topographically is divided into a premammary fascia and a retromammary fascia.

The retromammary fascia is separated from the fascia of the greater pectoral muscle by a loose connective tissue, which confers mobility to the gland (on the muscular plane, tissue described under the name of retromammary bursa. (Chassaignac).

The retromammary fascia is fixed above on the fascia of the greater pectoral and even on the periosteum of the clavicle; it unites with the premammary fascia, forming a true suspensory ligament of the mammary gland. The relaxation of this ligament leads to mammary ptosis, a state in which the nipples are sunken down and which is accompanied by a painful hypertrophy (mastodynia).

Vessel and nerve supply

The arteries derive from the internal mammary artery, a branch of the subclavian artery (especially for the medial and upper segment of the gland), from the external mammary artery (the lateral thoracic artery), a branch of the axillary artery (especially for the lateral segment of the gland), and from the intercostal arteries, by the perforating branches (for the deep segment of the gland).

The veins have in a reverse sense the same course as the arteries.

The lymphatics (fig. 61) are divided into three groups: external, internal and inferior. The external group runs towards the thoracic group of axillary lymph nodes; the internal group reaches the lymph nodes situated on the course of the mammary vessels; the inferior group runs towards the axillary and subclavian lymph nodes.

The lymphatics form a superficial network and a deep one. The superficial network makes up a dermal and a subdermal plexus, which are denser in the areolar region; the deep network lies between and around the lobules. The afferent lymphatics at this level can be systematized in main and secondary pathways.

The main pathways are axillary and *internal thoracic* The efferent vessels of the axillary pathway at the lateral border of the gland surround the lower border of the greater pectoral muscle (sometimes they even pass through it) and reach the axillary lymph nodes, especially the group of the external mammary lymph nodes, but also the central, scapular and subclavicular group. They collect especially the lymph of the external quadrant. The axillary lymph nodes drain the lymph into the supraclavicular lymph nodes.

The internal mammary or parasternal lymph nodes are situated along the course of the internal mammary vessels. They collect the lymph from the depth of the gland, mainly from the medial quadrant, and drain either into the innermost nodes of the supraclavicular region, or directly into the thoracic duct (on the left) or into the great lymphatic vein (on the right).

Accessory pathways. The direct supraclavicular pathway (Mornard) is represented by efferent vessels which start from the upper segment of the mammary region (regarding the breast divided into four quadrants by two perpendicular lines drawn across the nipple) and drain the lymph into the supraclavicular lymph nodes, either above the clavicular region or below it.

The centrolateral axillary pathway (Gerota) passes from one mammary region to that on the opposite side, reaching the contralateral axillary lymph nodes.

The inferior accessory pathway, less frequent, starts from the lower segment of the mammary region and runs towards the upper part of the greater rectus abdominis muscle.

It is obvious that this distribution of the lymphatic territories of the mammary region entails for the surgeon a great difficulty -if not even the impossibility- when it is necessary to lift en bloc, concomitantly with the mammary gland invaded by a neoplastic process, the totality of tributary lymphatic territories.

Hence the imperative to establish the earliest possible the diagnosis of mammary neoplasmic before the invasion of the groups of lymph nodes, especially of the internal mammary lymph nodes.

The nerve supply of the mammary gland derives from the intercostal nerves, the pairs II - VI, from the supraclavian branch of the cervical plexus and from the thoracil branches of the brachial plexus. The autonomic innervation consists of fibres running along the vessels.

The Thoracic Cavity

The thoracic cavity, bounded above and below by the two thoracic apertures, anferolaterally by the sternum,fhe ribs and the intercostal musculature and posteriorly by the vertebral column, contains a visceral space in which lie various viscera (fig. 64, 65, see also fig. 120).

Some of the organs (the lung and the heart) are covered with serosae, while others lie in the subserous connective tissue.

The serosae are two in number: a pleural serosa, which invests the two lungs, and a pericardial serosa, which covers the heart, so that in the thorax there are three large serous cavities. The anatomic structures contained in the subserous connective tissue have generally a longitudinal direction and the connective tissue forms around them adventitiae and ligaments or is arranged under the form of mediastinal connective tissue *(lamina fibrosa mediastini)* and subserous connective tissue *(tela subserosa),* in a close relation with the serosae, forming the endothoracic fascia or the fibrous pericardium.

The Pleura

The pleura is the serosa which envelops the lungs.

The pleural serosa is made up of a visceral layer, which invests the lung, and of a parietal layer, which lines the inner surface of the thoracic wall at the level of the endothoracic fascia. The two layers meet at the level of the hilum of the lung, forming an uninterrupted serosa and enclosing between them a virtual, closed cavity, called pleural cavity.

Topographically, the parietal pleura consists of the following regions: the portion comprised between the lateral part of the vertebral column, the ribs, the intercostal spaces and a part of the sternum, which is called the costovertebral pleura *(pars costovertebralis pleurae parietalis);* the pleural portion which covers the upper surface of the diaphragm and is called the diaphragmatic pleura *(pars diaphragmatica pleurae parietalis);* the pleura which invests on either side the organs of the median (mediastinal) region of the thoracic cage, hence the organs lying in its middle *(quod in medio stay),* and which is called the mediastinal pleura *(pars mediastinalis pleurae parietalis);* it has a pericardial portion which covers the fibrous pericardium, separating laterally this mediastinum from the two pleural cavities (fig. 66-68).

The mediastinal pleura, by its reflection at the level of the hilum in the pulmonary pleura, forms the pulmonary ligament below the root and the hilum of the lung, up to the diaphragmatic pleura; this ligament has a triangular form and is called by Waldeyer mesopneumonsum. At the level of the passage from one wall to the other, the parietal pleura gives risc to several culs-de-sac, called pleural sinuses. The costomediastinal sinus is formed at the level of the passage of the costal pleura on the mediastinal pleura, the phrenicomediastinal sinus at the passage of the diaphragmatic pleura on the mediastinum and the costovertebral sinus at the level where the pleura passes from the ribs on the body of vertebrae.

77

In continuation, the mediastinal pleura will give rise to a series of sinuses in relation with the organs of the posterior mediastinum (aorta, oesophagus and azygos vein), such as the interaortico-oesophageal sinus on the left, at the insinuation of the pleura between the aorta and the oesophagus - in the left hemifhorax - and the interazygo-oesophageal sinus on the right, at the passage of the serous membrane from the azygos vein on the oesophagus. The two culs-de-sac are connected with each other by an interpleural ligament (Morozov), consisting of a thin plate of dense fibrous connective tissue, arranged frontally.

At the level where the costal pleura passes on the diaphragm, becoming the diaphragmatic pleura, is formed the phrenseocostal sinus, in which the lung penetrates only partially in forced inspiration, leaving a pleural cul-de-sac.

On the anterior wall of the thorax, behind the sternum, may be seen a space uncovered by the pleura, which is shaped like two triangles situated with the bases upwards and downwards, united by their summits (thymic trigone and cardiac trigone).

The superior triangle, termed area interpleuralis cranialis, contains, in the child, the thymus and in the adult, fatty connective tissue and remnants of the thymic gland.

In the inferior triangle, called area interpleural caudalis, lies the pericardium, in contact with the anterior wall of the thorax, uncovered by the pleura.

The pericardial tap is performed in this triangle, without penetrating in the pleural cavity.

In the area of the apex of the lung (apex pulmonis), above the first rib, the mediastinal pleura is continuous with the costovertebral pleura, bringing about the formation of a vault, the cupula of the pleura (cupula pleurae), in which the apex of the lung finds its shelter.

Anteriorly, the cupula of the pleura passes beyond the first rib and sometimes also beyond the clavicle; posteriorly it is situated at the level of the first thoracic vertebra.

The cupula of the pleura is fastened by the connective tissue of the endothoracic fascia which, according to its topography, is termed costopleural, vertebropleural, vertebropleurocostal, broncho- and oesophagopleural ligament, forming Sebileau's pleural suspensory apparatus. It is fastened also by the adjacent vascular and nervous connective sheaths, as well as by the fascia of the scaleni muscles.

The cupula of the pleura has the following relations: posteriorly with the head and the neck of the first rib, the longus colli muscle, covered by the prevertebral fascia, and the orthosympathetic trunk, wich presents here the inferior stellatum ganglion; anteriorly and laterally the cupula is in relation with the scaleni muscles, the subclavian artery in the retroscalenic portion, the internal thoracic artery, the supreme intercostal artery, wich is a branch of the aorta, and with the phrenic nerve.

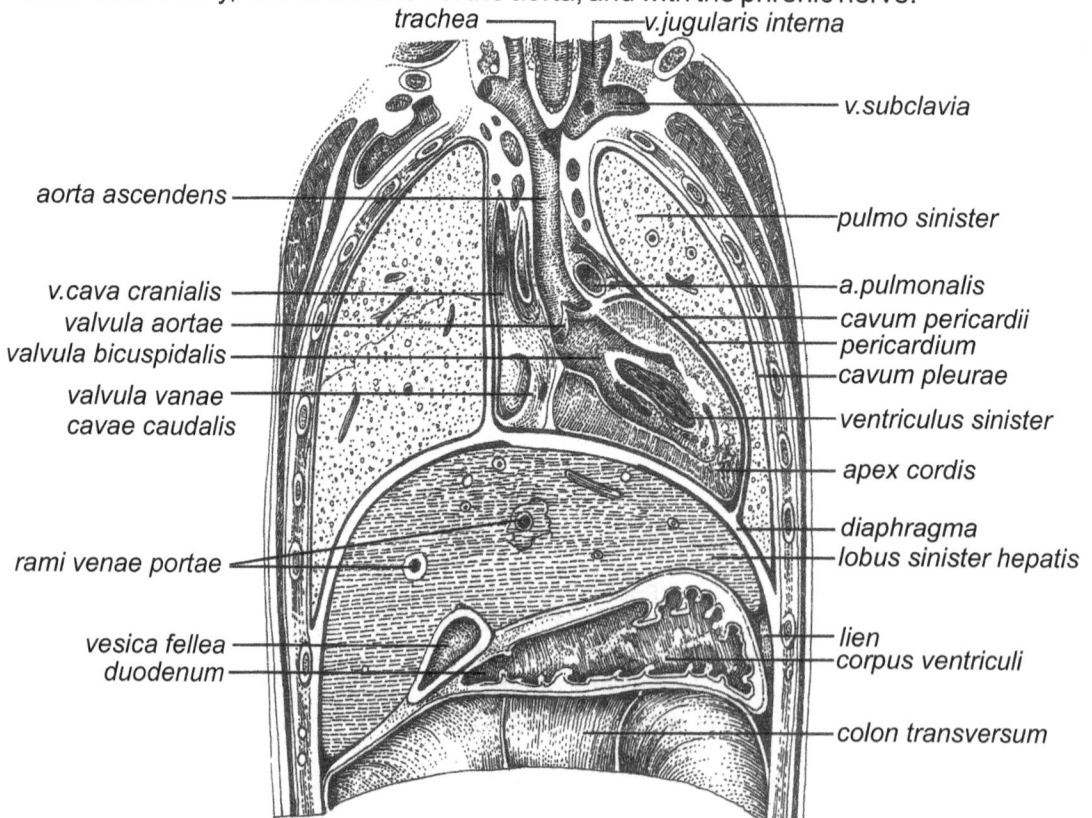

Fig. 64. Frontal section through the thoracic cavity and the abdominal supramesocolic region.

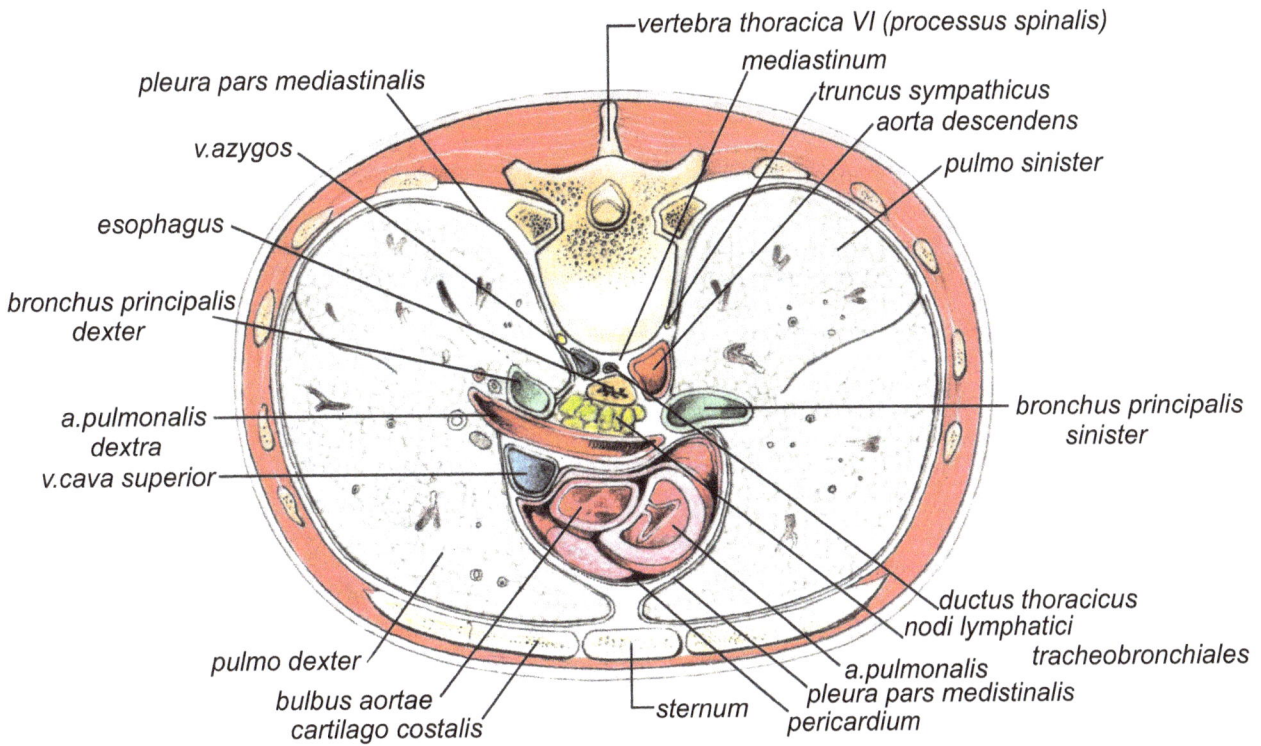

vertebra thoracica VI (processus spinalis)
mediastinum
truncus sympathicus
aorta descendens
pulmo sinister

pleura pars mediastinalis

v.azygos

esophagus

bronchus principalis
dexter

bronchus principalis
sinister

a.pulmonalis
dextra
v.cava superior

ductus thoracicus
nodi lymphatici
tracheobronchiales

pulmo dexter

a.pulmonalis
pleura pars medistinalis
pericardium

bulbus aortae
cartilago costalis

sternum

Fig. 65. Transverse section of the level of the sixth thoracic vertebra.

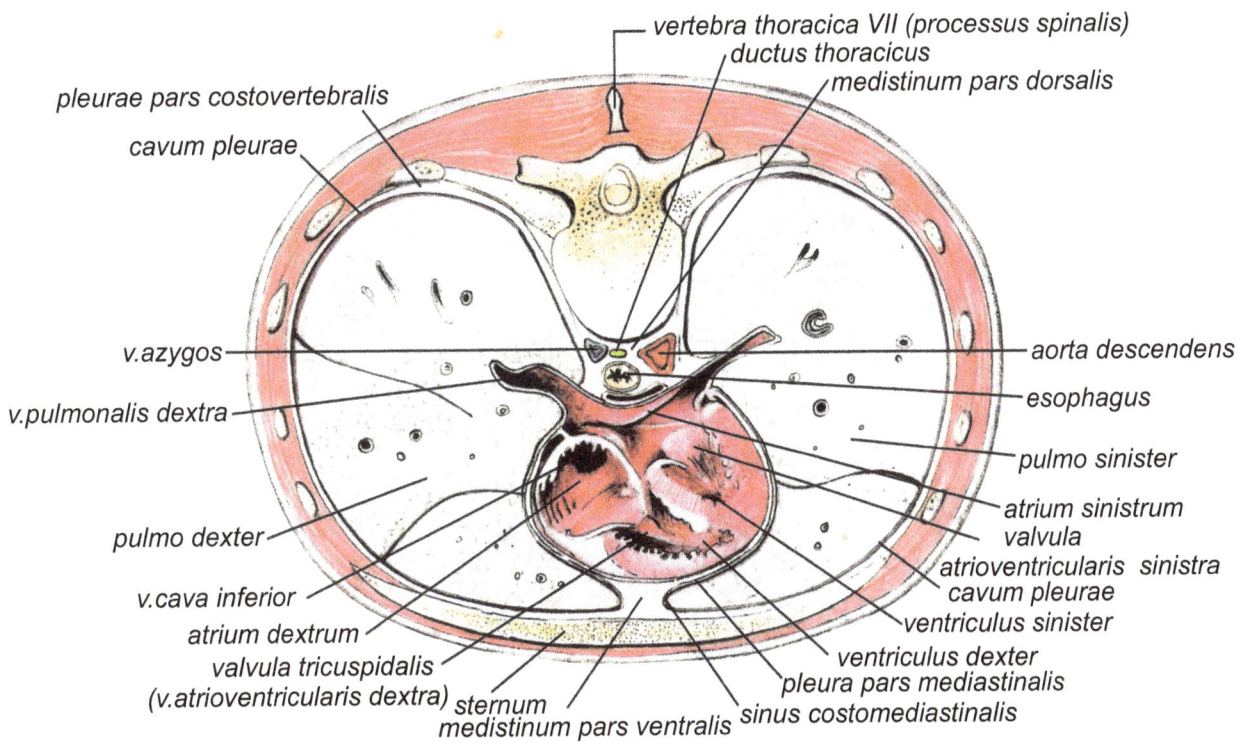

vertebra thoracica VII (processus spinalis)
ductus thoracicus
medistinum pars dorsalis

pleurae pars costovertebralis

cavum pleurae

v.azygos

aorta descendens

v.pulmonalis dextra

esophagus

pulmo sinister

atrium sinistrum
valvula
atrioventricularis sinistra
cavum pleurae
ventriculus sinister

pulmo dexter

v.cava inferior

atrium dextrum

valvula tricuspidalis
(v.atrioventricularis dextra)

ventriculus dexter
pleura pars mediastinalis
sinus costomediastinalis

sternum
medistinum pars ventralis

Fig. 66. Transverse section at the level of the seventh thoracic vertebra.

79

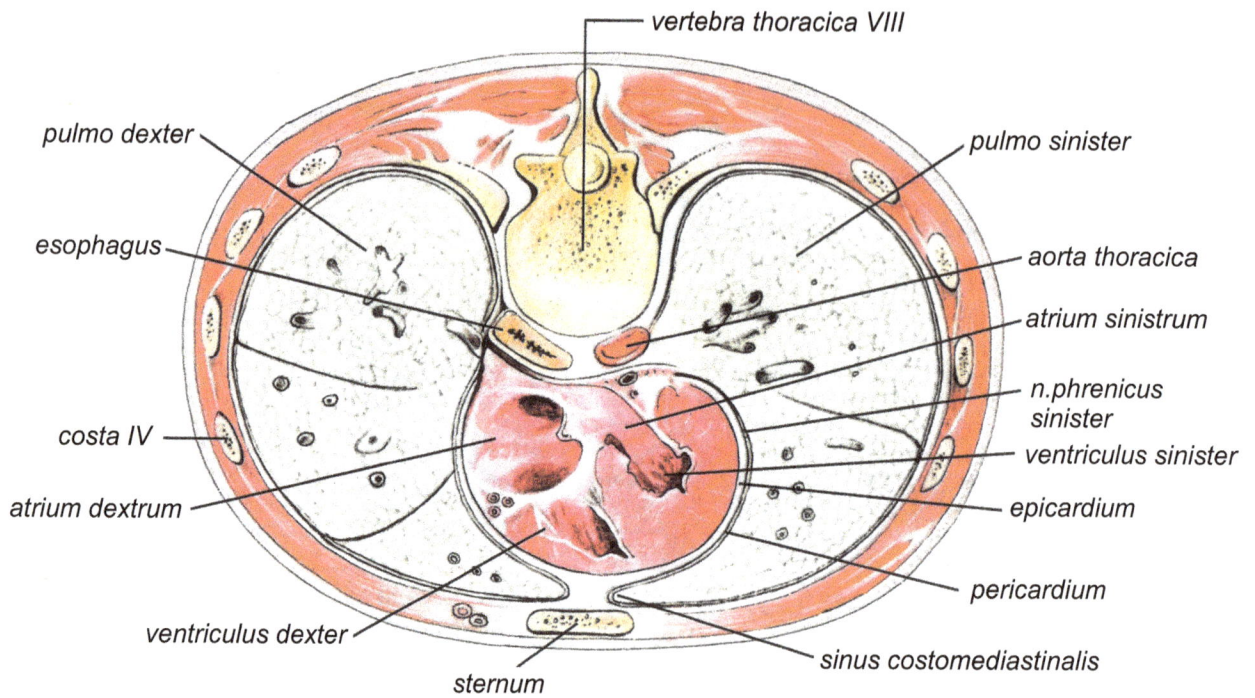

Fig. 67. Transversa section at the level of the eighth thoracic vertebra.

Labels for Fig. 67:
vertebra thoracica VIII
pulmo dexter
pulmo sinister
esophagus
aorta thoracica
atrium sinistrum
n.phrenicus sinister
costa IV
ventriculus sinister
atrium dextrum
epicardium
pericardium
ventriculus dexter
sinus costomediastinalis
sternum

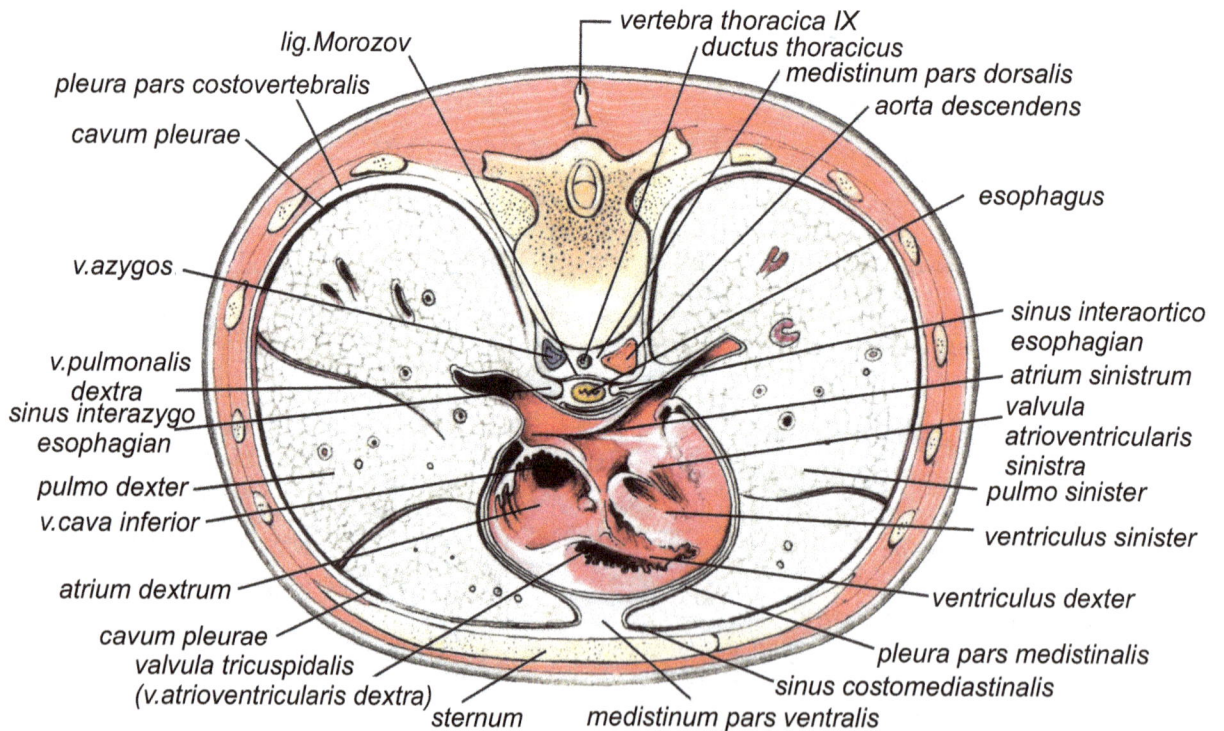

Fig. 68. Transverse section at the level of the ninth thoracic vertebra.

Labels for Fig. 68:
lig.Morozov
vertebra thoracica IX
ductus thoracicus
medistinum pars dorsalis
pleura pars costovertebralis
aorta descendens
cavum pleurae
esophagus
v.azygos
sinus interaortico esophagian
v.pulmonalis dextra
atrium sinistrum
sinus interazygo esophagian
valvula atrioventricularis sinistra
pulmo dexter
pulmo sinister
v.cava inferior
ventriculus sinister
atrium dextrum
ventriculus dexter
cavum pleurae
valvula tricuspidalis (v.atrioventricularis dextra)
pleura pars medistinalis
sinus costomediastinalis
sternum
medistinum pars ventralis

80

The Main Bronchi
(Bronchi principaies)

The main bronchi, right and left *(bronchi principales dexteret sinister),* for the two lungs, arise by bifurcation of the trachea, at the level of the disk between the fourth and sixth thoracic vertebrae. They are directed towards the hilum of the corresponding lung and then ramify into lobar and segmental bronchi (fig. 69).

The main bronchi have an oblique, downwards and lateral direction, forming between them an angle of about 70°. The right bronchus is more vertical, shorter, heavier and slightly directed forwards, while the left bronchus is nearly horizontal, slightly sinuous, italic S-shaped, longer and thinner, initially with a curvature laterosuperiorly concave, related to the presence of the arch of aorta, and then with a second curvature, medially and inferiorly concave, adapted to the convexity of the heart (see also fig. 75). The main bronchi have an extrapulmonary and an intrapulmonary portion (see also fig. 82).

The roots of the lungs *(radix pulmonis)* are formed by the totality of anatomical elements which enter and leave the lungs at the level of the hili of lungs. Each root is composed of: the main bronchus, the branches of the pulmonary artery and vein, the bronchial arteries, nerves and lymphatics (see also fig. 71). The relations of the bronchi with the other elements of the root of the lung are different on the right as compared to the left (see also fig. 113 and 114).

Each bronchus is in relation: anteriorly with the respective pulmonary artery which, at the level of the hilum of the lung-on the right -, is strictly anterior to the bronchus, whereas on the left it is above and anterior to the left bronchus; anteriorly and inferiorly to the bronchus lie the two pulmonary veins, which are situated one above and somewhat in front of the other and below the level of the arteries; posteriorly, the bronchi are in relation with the bronchial vessels. Likewise, at the level of the root, the bronchus is in relation with the lymph nodes and with the anterior and posterior autonomic nervous pulmonary plexuses.

In addition, the bronchi are in relation with the organs adiacent to the root

Thus, the left bronchus is crossed superiorly by the arch of the aorta and is posteriorly in relation with the descending thoracic aorta and the left vagus nerve. Above lies the left recurrent nerve, which passes beneath the arch of the aorta, and posteriorly is the oesophagus, which passes beyond the left border of the trachea. The right bronchus is crossed postero - superiorly by the arch of the azygos vein and posteriorly to it lies the vagus nerve. Anteriorly it is in relation with the superior vena cava and the right phrenic nerve, which descends laterally to this vein.

Vessel and nerve supply

The bronchi ore supplied with *arterial* blood by the bronchial arteries, branches of the thoracic aorta. The *venous* blood is drained by the two bronchial veins tributary on the right to the azygos vein and on the left to the hemiazygos vein. The *lymphatics* drain into the peritracheobronchial lymph nodes and the *nerves* derive from the pulmonary plexuses.

Fig. 69. Bifurcation of the trachea.

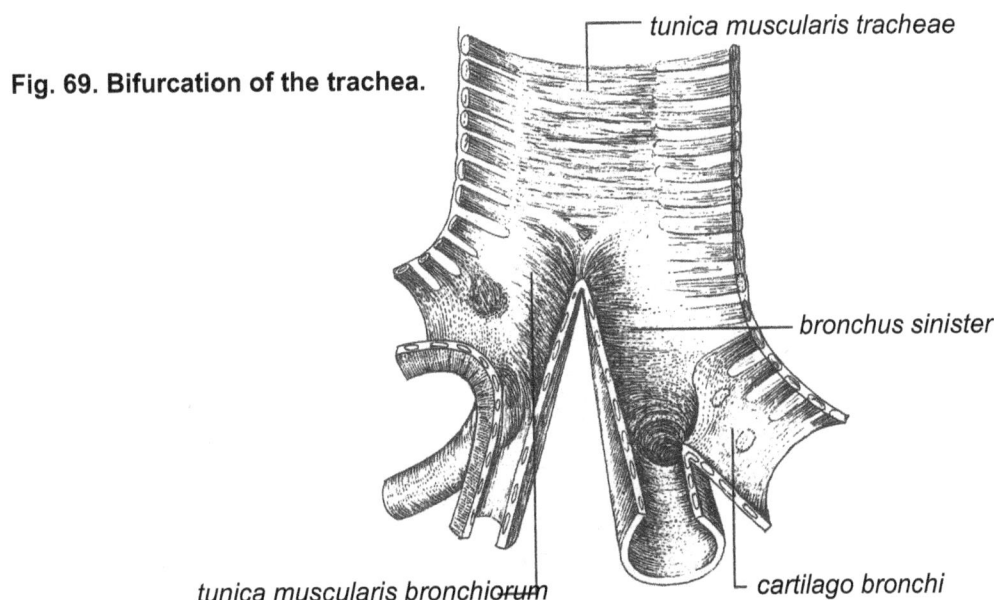

tunica muscularis tracheae

bronchus sinister

tunica muscularis bronchiorum

cartilago bronchi

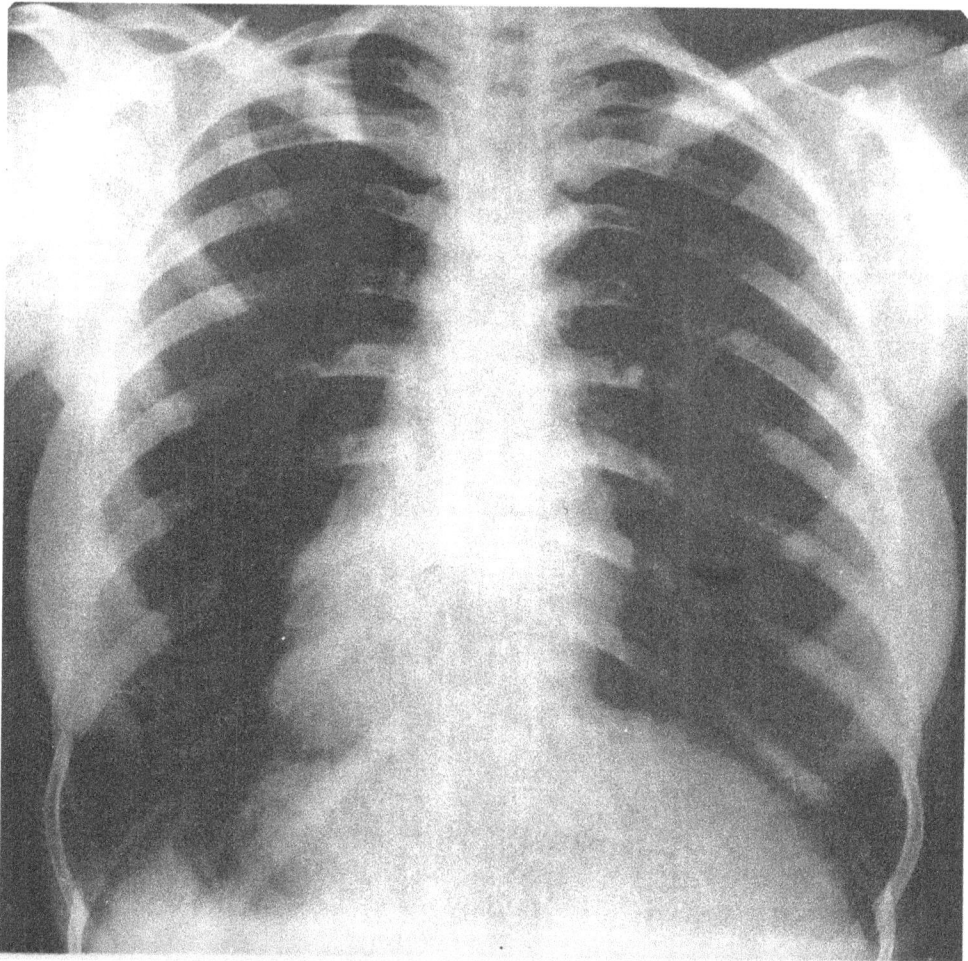

Fig. 70. Radiological aspect of the lungs
(collection of the Department of Radiology of the Central Military Hospital).

The Lungs
(Pulmones)

The lungs represent the organs of respiration at the level of which the venous blood becomes arterial blood. They are two in number, separated from each other by the mediastinum. They show individual variations in shape and size, according to the thoracic capacity and to the phases of inspiration and expiration. The right lung is larger than the left, both are hemicone shaped, with a convex surface, a more or less plane surface, a base and an apex (fig. 70, 82).

The lungs present a costal surface *(facies costalis),* a medial surface *(facies mediastinalis),* with an anterior portion in relation with the pericardium and a posterior portion in a wider relation with the vertebral column, a diaphragmatic base *(facies diaphragmatica),* an apex in relation with vascular and venous structures at the base of the neck, a sharp anterior sternal border *(margo sternalis),* a posterior vertebral border rounded like a roll *(margo vertebralis)* and a diaphragmatic semicircular border *(margo diaphragmatica)* (fig. 71-74).

The relations of the pulmonary surfaces are marked by the imprints left on them by the neighbourhood of the thoracic organs.

On the costal surface may be seen the traces left by the ribs and the intercostal spaces (fig. 73,74).

On the right mediastinal surface, ventrally to the mesopneusnonsum (pulmonary ligament), the right atrium leaves an impression called *impressio cardiaco,* which extends forwards up to the anterior border *(margo sternalis)* and above up to the impression left by the superior vena cava *(sulcus venae cavae cranialis).* The large azygos vein leaves an impression, too, which lies above the hilum over which this vein passes, to empty into the superior vena cava, forming the arch of the azygos vein. Posteriorly to the hilum of the lung is the impression left by the oesophagus *(impressio esophagea).*

82

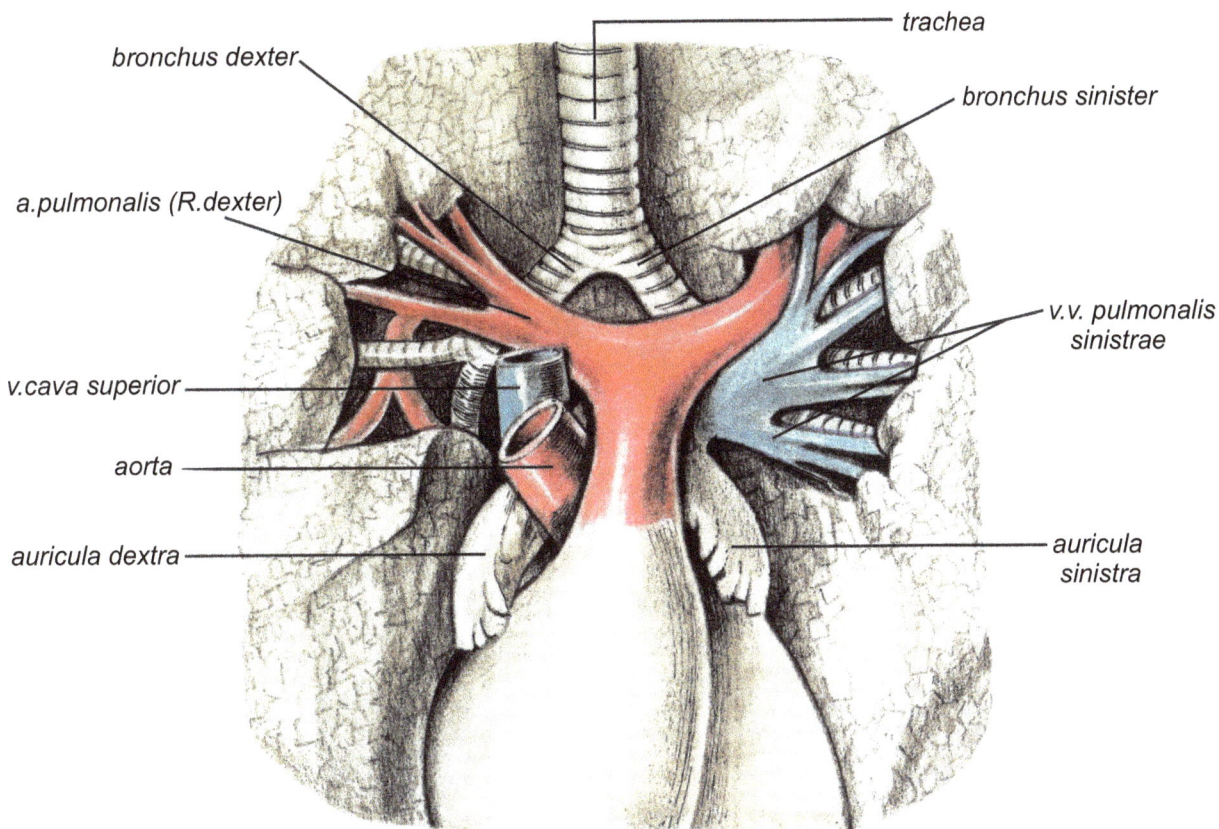

Fig. 71. The pulmonary artery.

bronchus dexter

trachea

a.pulmonalis (R.dexter)

bronchus sinister

v.v. pulmonalis sinistrae

v.cava superior

aorta

auricula dextra

auricula sinistra

Fig. 72. Normal pulmonary scintiscan
(collection of the Laboratory of Nuclear Medicine of the Central Military Hospital).

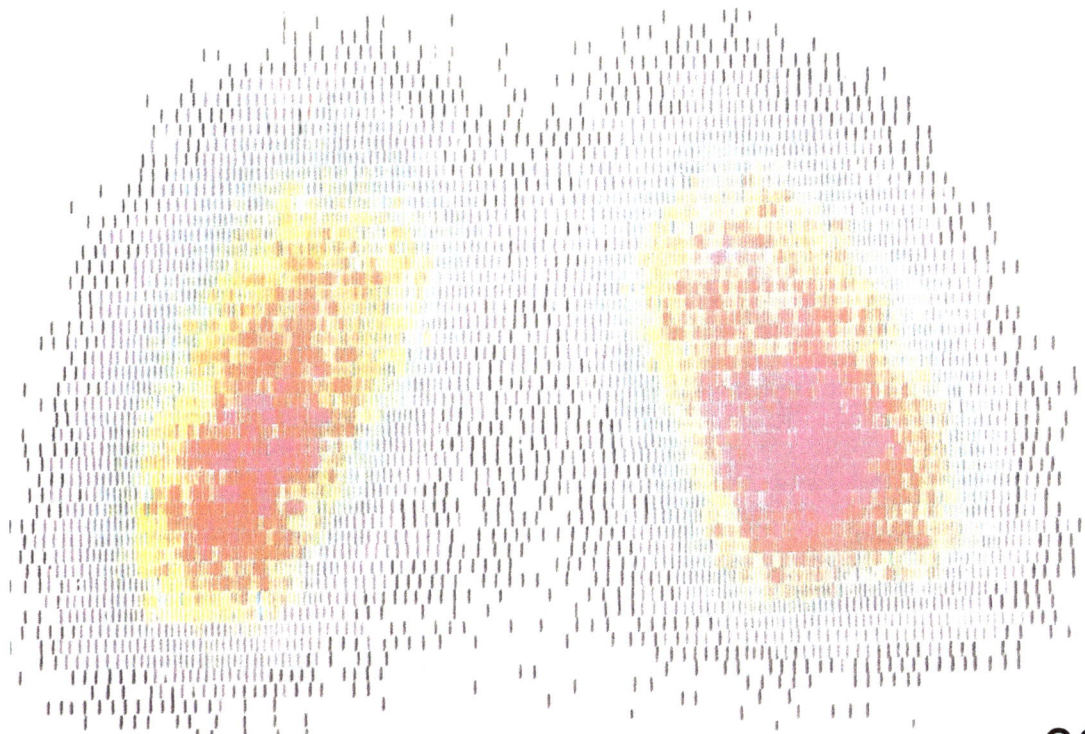

83

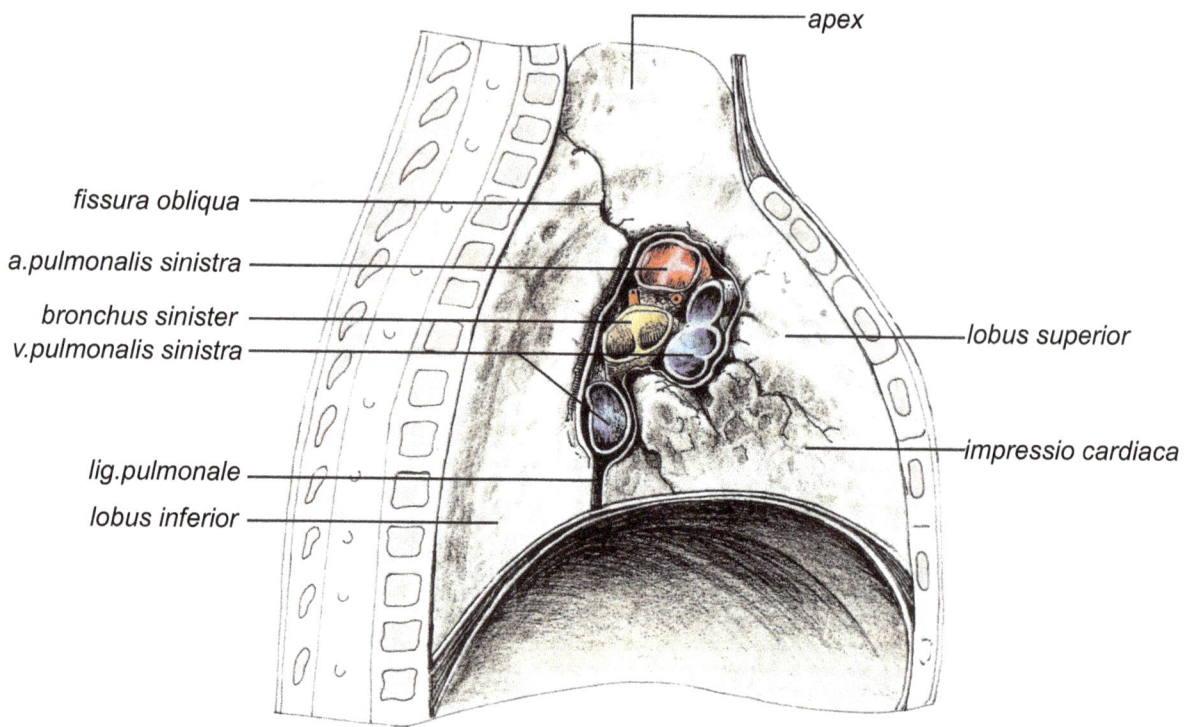

apex

fissura obliqua

a.pulmonalis sinistra

bronchus sinister

v.pulmonalis sinistra

lig.pulmonale

lobus inferior

lobus superior

impressio cardiaca

Fig. 73. Medial surface of the left lung.

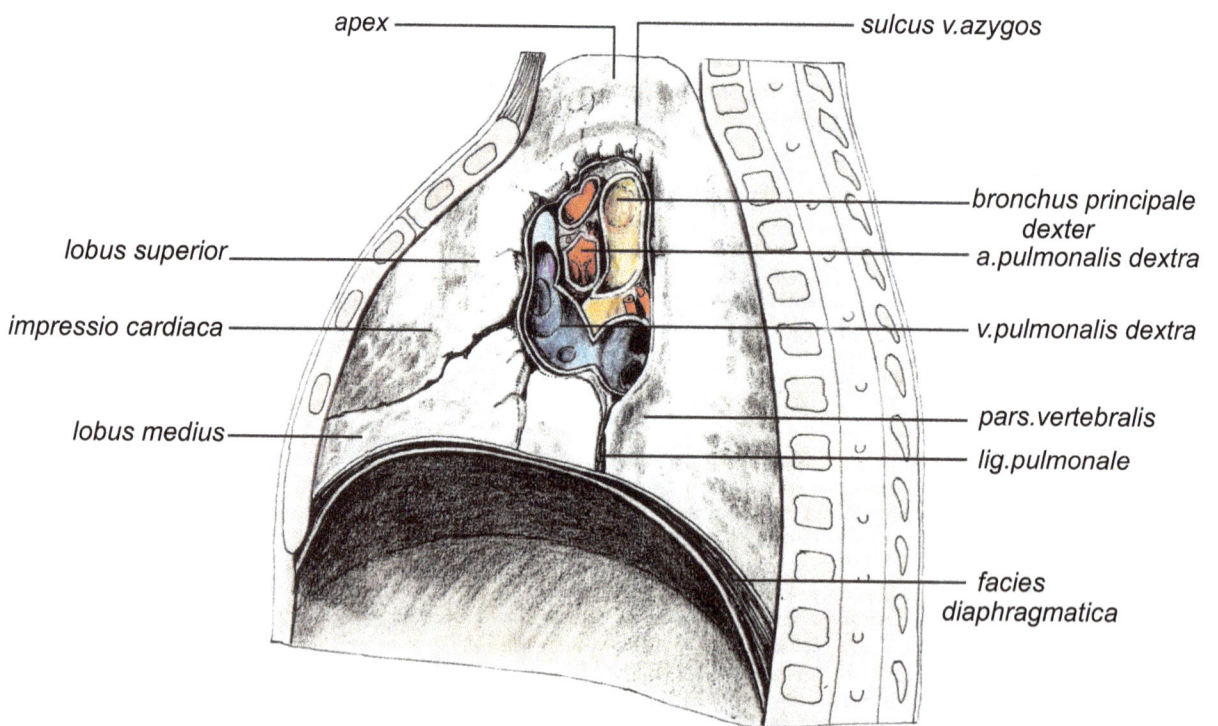

apex

sulcus v.azygos

lobus superior

impressio cardiaca

lobus medius

bronchus principale dexter

a.pulmonalis dextra

v.pulmonalis dextra

pars.vertebralis

lig.pulmonale

facies diaphragmatica

Fig. 74. Mediastinal surface of the right lung.

84

On the left mediastinal surface lies the cardiac impression, much more evident than on the right Superiorly it is continuous with the aortic groove *(sulcus aorticus)* left by the thoracic aorta. At the highest point of this arched groove is formed the groove of the left subclavian artery *(sulcus arterae subclaviae)*, which continues also on the apex of the lung *(apex pulmonis)*.

The mediastinal surfaces have relations, in addition to those mentioned, also with the oesophagus, the trachea, the vagus and phrenic nerves and the thymus.

The posterior portions of the medial surfaces *(pars vertebralis)* have relations with the thoracic vertebral bodies, the posterior extremities of the ribs, the thoracic sympathetic ganglia, the intercostal nerves, arteries and veins. The diaphragmatic surface *(fades diaphragmatica)*, called also base of the lung *(basis pulmonis)*, has different relations at the level of the two lungs. Thus, on the left, it is in contact, through the diaphragm, with the spleen, the fundus of the stomach and the left lobe of the liver, and on the right, with the diaphragmatic surface of the liver, which explains the possibility that sometimes the subphrenic abscesses pierce the diaphragm and affect the pleural cavity and the right lung. The costal surface *(facies costalis)* has relations with the ribs. The apex of the lung *(apex pulmonis)* has the same relations as the pleural cupula.

The three borders of the lung are: an anterior border *(margo anterior)*, which separates the costal surface from the mediastinal one and which presents at the left lung a notch in relation with the heart *(incisura cardiaca pulmonis sinistri)*, prolonged downwards by an area of parenchyma called lingula *(lingula pulmonis sinistri)*; a rounded posterior border *(margo posterior)*, which separates the costal surface from the vertebral one, and an inferior border *(margo inferior)*, which circumscribes the base of the lung. The lungs consist of two lobes, separated from each other by deep grooves extending from the surface to the hilum of the lung *(hilus pulmonis)*, termed fissures or scissures - two for the right lung and one for the left lung.

At the left lung, the oblique fissure *(fissura obliqua)*, which runs from above downwards and forwards, divides it into two lobes: the superior lobe *(lobus superior)* and the inferior lobe *(lobus inferior)*. The fissure begins on the mediastinal surface, above the hilum, runs upwards and backwards up to the costal surface, from where it reaches the diaphragmatic border and continues on the diaphragmatic surface and then on the mediastinal surface up to the hilum.

The right lung has two scissures, which divide it into three lobes: the oblique fissure, which has almost the same course as on the left, delineating the superior lobe *(lobus superior)* and the inferior lobe *(lobus inferior)*, and the horizontal fissure *(fissura horizontalis pulmonis dextri)*, which begins posteriorly, at the level of the oblique fissure on the costal surface of the lung and has an almost horizontal course towards the sternal border of the lung and then on the mediastinal surface, up to the hilum, delineating the middle lobe *(lobus medius)*, comprised between the horizontal fissure and the oblique fissure. Supernumerary lobes may be present, too, such as the azygos lobe of the lung, the accessory inferior lobe *(lobus cardiacus)* and the posterior lung.

The supernumerary lobes are more frequently present in the right lung.

The azygos lobe of the lung is incompletely detached from the apical lobe, owing to a supplementary fissure, formed by an anomaly in the course of the azygos vein. The cardiac lobe (retrocardiac inferior lobe) is more frequent than the azygos lobe. Radiologically it may be seen in sagittal projection, in the angle formed by the heart and the diaphragm, on the right, and has the aspect of a rectangular triangle; on the left, it is hidden by the shadow of the heart. The posterior lobe is formed by the apical segment of the posterobasal lobe, which is transformed into an independent lobe.

The hilum of the lung *(hilus pulmonis)* represents the place where the pulmonary root penetrates into the lung and is situated on the mediastinal surface of the two lungs (fig. 71).

The hilum of the lung contains: the bronchus, a branch of the pulmonary artery, the pulmonary veins, the bronchial arteries and veins, lymphatics and the pulmonary plexus. Their arrangement is different on the right side in comparison with the left side.

In the right hilum, the main bronchus lies most often superiorly *(eparterial)*; the right branch of the pulmonary artery bifurcates here and lies below and anteriorly to the bronchus. More inferiorly and anteriorly are the two pulmonary veins.

On the left, the branch of the pulmonary artery is situated above the bronchus, which lies inferiorly and more posteriorly *(hyparterial)*; the pulmonary veins lie inferiorly and anteriorly.

Anatomical structure of the lungs.

The lungs are made up of:
-the intrapulmonary portion of the bronchial tree;
-the vascular arterial, venous and lymphatic network;
-nerve branches;
-connective elastic tissue lying between the broncho - vasculo-nervous elements.

85

The Bronchial Tree - The Right Main Bronchus
(Bronchus principalis dexter)

It divides into three lobar bronchi: the bronchus of the right superior lobe (bronchus lobaris superior dexter), the bronchus of the right middle lobe (bronchus lobaris medius dexter) and the bronchus of the right inferior lobe (bronchus lobaris inferior dexter).

Each of the lobar bronchi divides into several segmental bronchi, which penetrate into certain territories of the lobe. These pyramid-shaped territories make up the pulmonary segments. The superior lobar bronchus gives off three segmental bronchi: apical, posterior and anterior; the middle lobar bronchus gives off two segmental bronchi: medial and lateral, and the inferior lobar bronchus divides successively into: the apical segmental bronchus of the inferior lobe, the mediobasal (paracardiac) bronchus, the anterobasal bronchus, the laterobasal bronchus and ends by a bronchial trunk afferent to the posterior basal segment, on the whole five segmental bronchi (fig. 75).

The Left Main Bronchus
(Bronchus principalis sinister)

It divides into two lobar bronchi: the left superior lobar bronchus (bronchus lobaris superior sinister) and the left inferior lobar bronchus (bronchus lobaris inferior sinister).

The left superior lobar bronchus does not directly give off the segmental bronchi, but if divides into two stems, one superior and another inferior. The inferior stem penetrates into the anterior and inferior parts of the left superior lobe (termed lingula), giving rise to two segmental bronchi: superior and inferior. The upper stem ramifies into the superior and posterior parts of the superior lobe, dividing into three segmental bronchi: anterior, apical and posterior (fig. 75).

The left inferior lobar bronchus displays the same arrangement as the homologous bronchus of the right lung, the only difference consisting in the fact that the anterobasal and paracardiac segmental bronchi arise from a common stem.

The segmental bronchi (bronchi segmentales) are the last air ducts situated in the pulmonary stroma and endowed with a discontinuous cartilaginous skeleton. From these segmental bronchi arise the lobular bronchioli (bronchioli), then the intralobular broochioli which correspond to the pulmonary lobules. Each lobular bronchiolus ramifies into several terminal bronchioli, which in the structure of their walls do no more contain cartilaginous pieces, but only smooth musculature. The terminal bronchioli give rise to the respiratory bronchioli (bronchioli respiratores), which are furnished with pulmonary alveoli and participate in respiratory processes. The respiratory bronchioli, in turn, ramify info alveolar ducts (ductuli alveolares), ending by alveolar sacs (air sacs) (sacculi alveolares), into which open the pulmonary alveoli (alveoli pulmonis). In the structural organization of the lung, the respiratory bronchioli hold a particular position, since they belong both to the conduction segment, representing an air duct, and to the respiratory segment, as they are endowed with pulmonary alveoli (fig. 75, 87). In this way, afferent to the segmental bronchi are delineated the pulmonary segments.

The pulmonary segment represents a surgical anatomical unit with a segmental bronchus; as a matter of fact, it is a small lung possessing: an anatomical hilum, an artery, a vein, a bronchus, lymphatics and well individualized nerves, permitting a separate exeresis and exhibiting a radioclinically characteristic pathology (fig. 76-81, 83-86).

The summit of the segment is directed towards the pulmonary hilum. The segment, in addition to the morphological and functional differentiations which characterize it, is also a clinical unit, since the pathological changes (bronchopneumonia, tuberculosis, bronchiectasis) are usually located, initially, at the level of a single segment. Under normal conditions, the differentiation of the segments from each other in the radiological picture may be impossible, but this differentiation is clear in certain diseases (infiltrate, atelectasis etc.), when the absorption possibilities of the X rays are modified. Usually, between the segments there is no regular, visible limit, neither on the surface of the lung nor in its depth. The limits of the segments exhibit numerous individual variations, in dependence on the shape of the thorax and on the situation of the lung in the thoracic cavity; they are marked only by the interlobar scissures, the intersegmental limit forming seldom a true septum. The intersegmental septa have a zigzag course and contain the lymphatics and vascular branches, their knowledge being of an utmost importance for the surgeon in: the carrying out of segmentectomies, which represent a functional and conservative surgery of the lung.

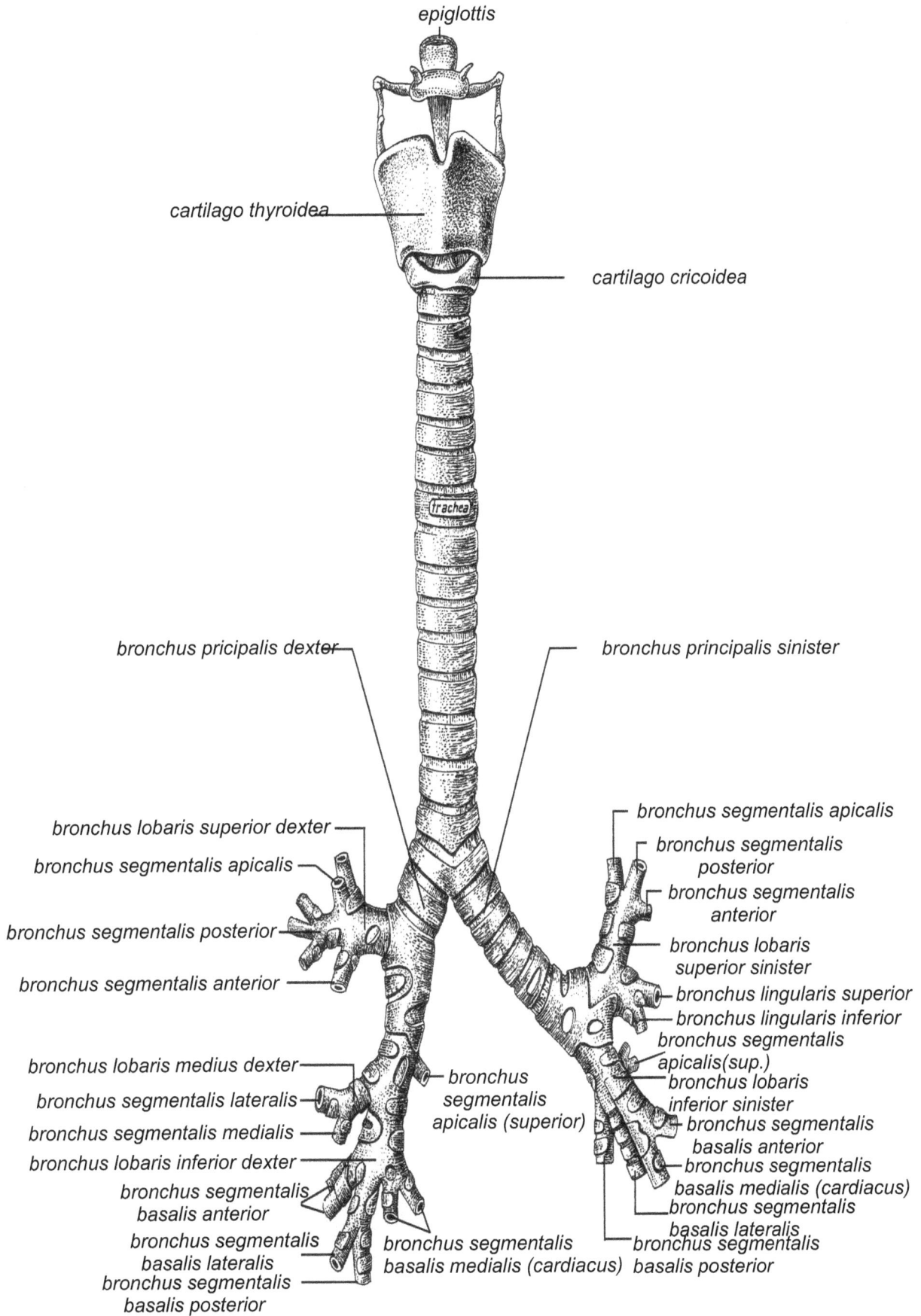

epiglottis

cartilago thyroidea

cartilago cricoidea

trachea

bronchus pricipalis dexter

bronchus principalis sinister

bronchus segmentalis apicalis

bronchus lobaris superior dexter

bronchus segmentalis apicalis

bronchus segmentalis posterior

bronchus segmentalis anterior

bronchus segmentalis posterior

bronchus segmentalis anterior

bronchus lobaris superior sinister

bronchus lingularis superior

bronchus lingularis inferior

bronchus segmentalis apicalis(sup.)

bronchus lobaris medius dexter

bronchus segmentalis lateralis

bronchus segmentalis medialis

bronchus lobaris inferior dexter

bronchus segmentalis basalis anterior

bronchus segmentalis basalis lateralis

bronchus segmentalis basalis posterior

bronchus segmentalis apicalis (superior)

bronchus segmentalis basalis medialis (cardiacus)

bronchus lobaris inferior sinister

bronchus segmentalis basalis anterior

bronchus segmentalis basalis medialis (cardiacus)

bronchus segmentalis basalis lateralis

bronchus segmentalis basalis posterior

Fig. 75. The tracheobronchial tree.

87

Structure of the pulmonary alveoli. The pulmonary alveoli are small hemispheric cavities, of a vesiculous aspect, which open into the alveolar ducts and the respiratory bronchioli; they communicate with each other by alveolar pores or stomata, which are in number of 1-6 per alveolus. These pores, considered in the past as wear-induced orifices, appear subsequently to the desquamation of the alveolar cells and favour the extension of the pathological processes from one alveolus to another.

The pulmonary alveoli are actually small cavities included in the reticulinic elastic ground substance of the lung. Each alveolus has its wall made up of a monostratified pavement epithelium proper, the alveolar epithelium; the alveoli are separated from eachother by interalveolar septa (alveolar walls). In the adult, their mean diameter is 100-300 microns, size which varies in expiration and inspiration (fig. 87).

The structural "alveolocapillary" complex. The pulmonary alveoli are delineated by an epithelio-mesenchymal complex, which has the significance of a blood-air barrier (the pathological change of which represents the lesional substrate of the so-called alveolocapillary block). This complex, running from the alveolar cavity to the blood capillary, consists of the following structures: (1) the surfactant pellicle situated between the atmospheric air and the alveolar epithelium; (2) the alveolar epithelium; (3) the epithelial basement membrane; (4) the ground substance (*substantia fundamentalis*) located between the epithelial and capillary basement membranes; (5) the capillary basement membrane and (6) the cytoplasm of the endothelial cells of interseptal (pulmonary) capillaries.

The morphofunctional units of the lung are represented by the pulmonary acini. A pulmonary acinus consists of the whole parenchyma tributary to a respiratory bronchiolus, which is formed by alveolar ducts and sacs and pulmonary alveoli.

Vessel and nerve supply of the lungs

The arterial and venous circulation of the lungs is of two categories: functional, consisting of the pulmonary arteries and veins, and nutrient, consisting of the bronchial arteries and veins (fig. 71-74, 88; see also fig. 92, 94-96, 98).

Pulmonary arteries. The pulmonary artery, arising from the right ventricle, divides info two branches, right and left, each running to the corresponding lung. They cross the anterior surface of the main bronchus, below the superior lobar bronchus, on the right, and above the superior lobar bronchus, on the left, after which, generally, they follow the course of the lobar and segmental bronchi.

The right pulmonary artery gives off a branch to the superior lobe, which divides into 5 segmental rami, a branch to the middle lobe, which divides into two segmental rami, and two branches to the inferior lobe, of which one to the apical segment and one that divides into four segmental rami.

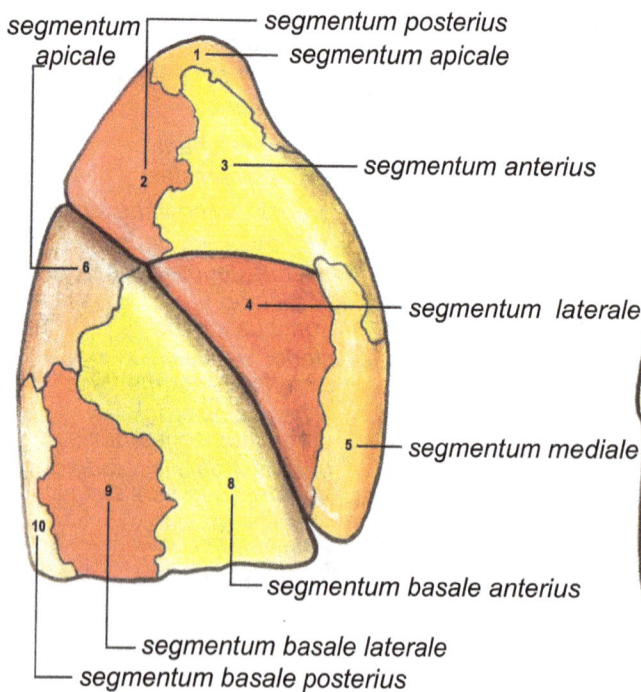

Fig. 76. Pulmonary segments (lateral view, right lung).

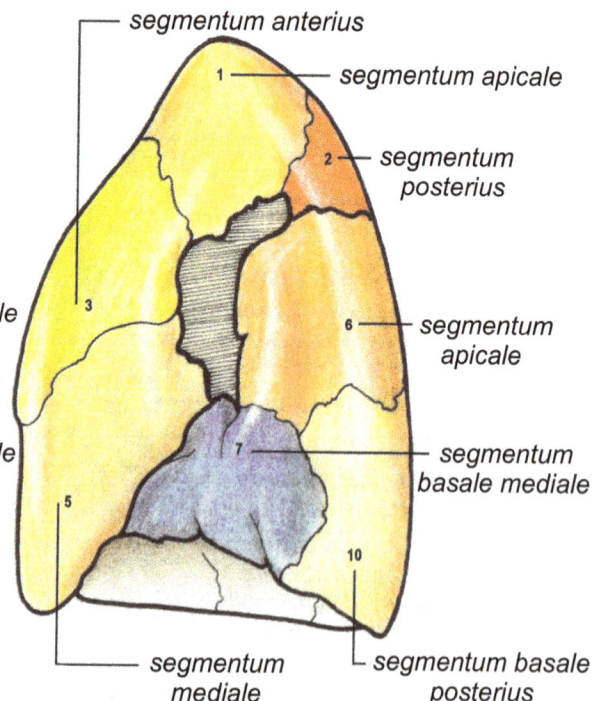

Fig. 77. Pulmonary segments (medial surface, right lung).

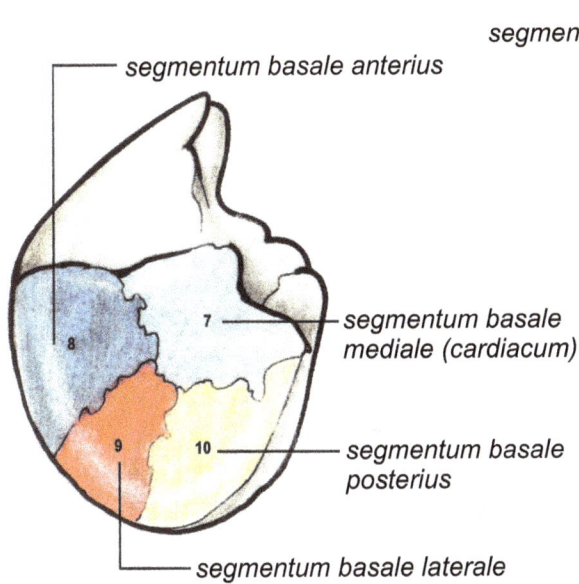

segmentum basale anterius

segmentum basale mediale (cardiacum)

segmentum basale posterius

segmentum basale laterale

Fig. 78. Pulmonary segments (diaphragmatic surface of the right lung).

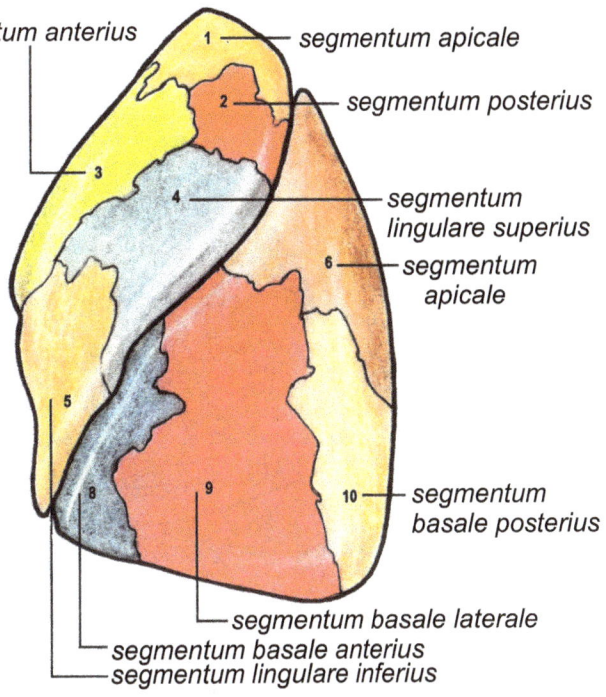

segmentum anterius

segmentum apicale

segmentum posterius

segmentum lingulare superius

segmentum apicale

segmentum basale posterius

segmentum basale laterale

segmentum basale anterius

segmentum lingulare inferius

Fig. 79. Pulmonary segments (lateral view, left lung).

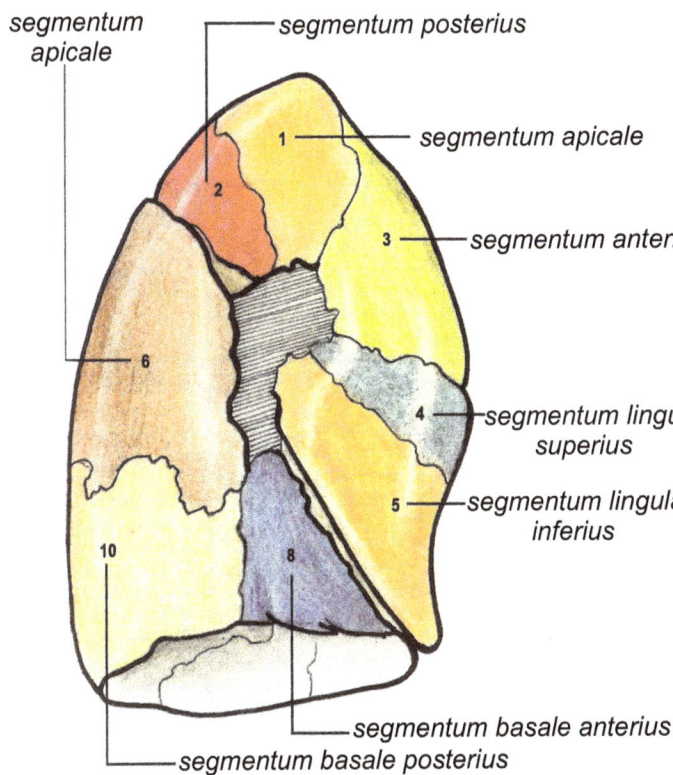

segmentum apicale

segmentum posterius

segmentum apicale

segmentum anterius

segmentum lingulare superius

segmentum lingulare inferius

segmentum basale anterius

segmentum basale posterius

Fig. 80. Pulmonary segments (medial view, left lung).

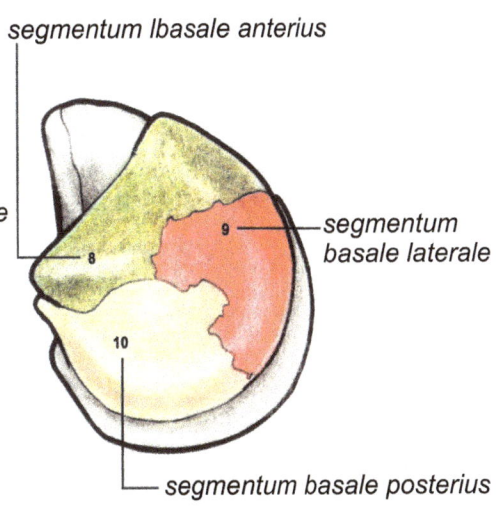

segmentum lbasale anterius

segmentum basale laterale

segmentum basale posterius

Fig. 81. Pulmonary segments (diaphragmatic surface of the left lung)

89

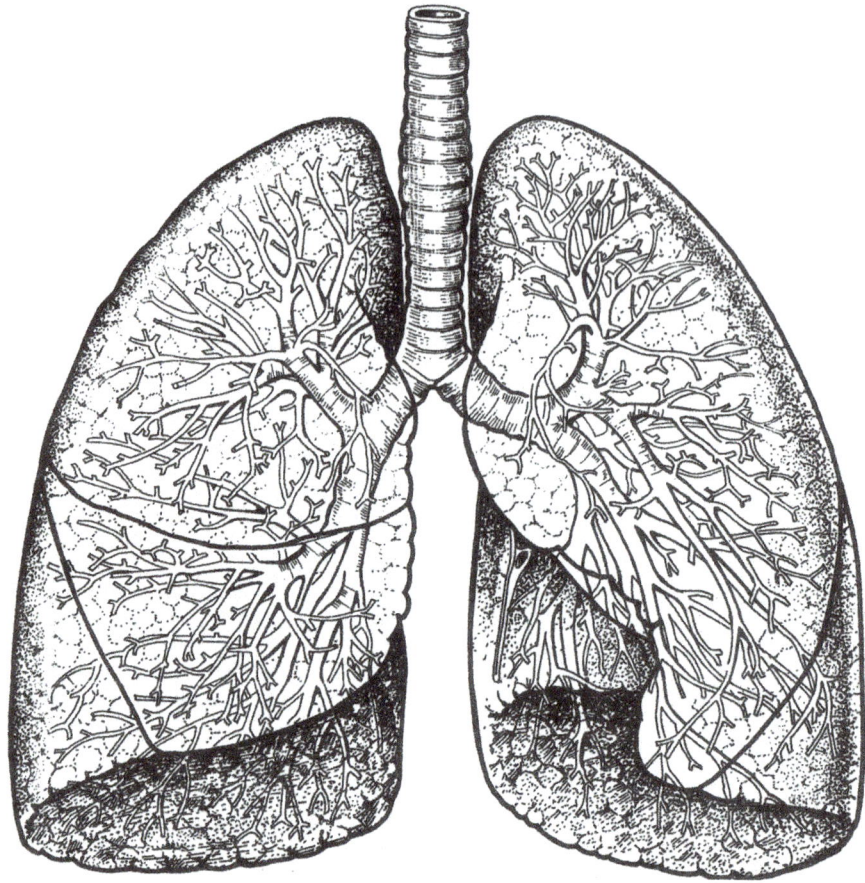

Fig. 82. The lungs and the tracheobronchial tree (overall view).

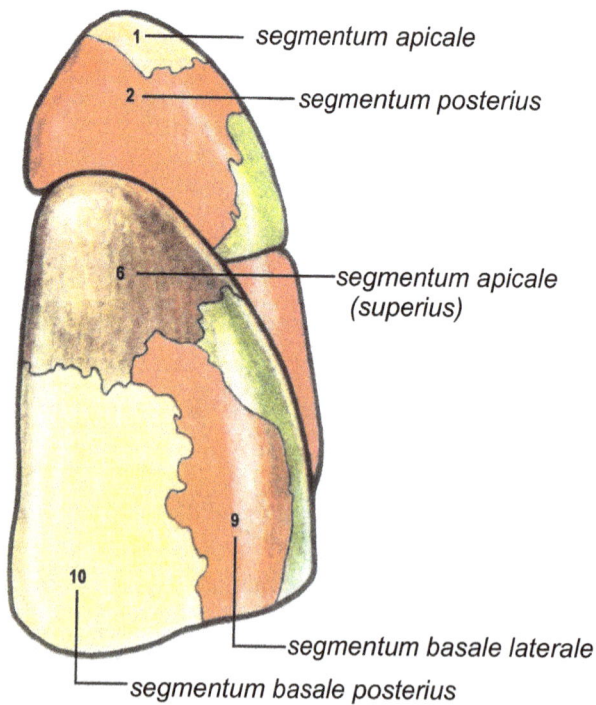

1 — segmentum apicale

2 — segmentum posterius

6 — segmentum apicale (superius)

9 — segmentum basale laterale

10 — segmentum basale posterius

Fig. 83. Pulmonary segments (posterior surface of the right lung).

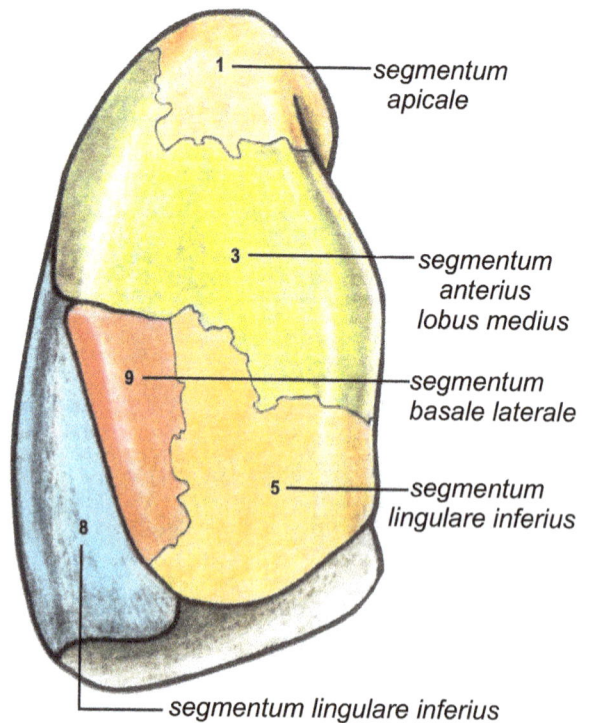

1 — segmentum apicale

3 — segmentum anterius lobus medius

9 — segmentum basale laterale

5 — segmentum lingulare inferius

8 — segmentum lingulare inferius

Fig. 84. Pulmonary segments (anterior view of the right lung).

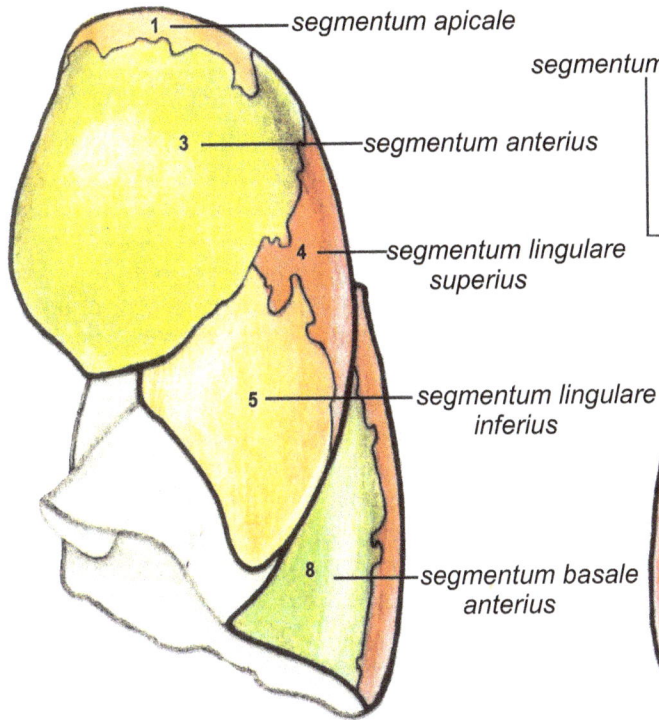

**Fig. 85. Pulmonary segments
(anterior view of the left lung).**

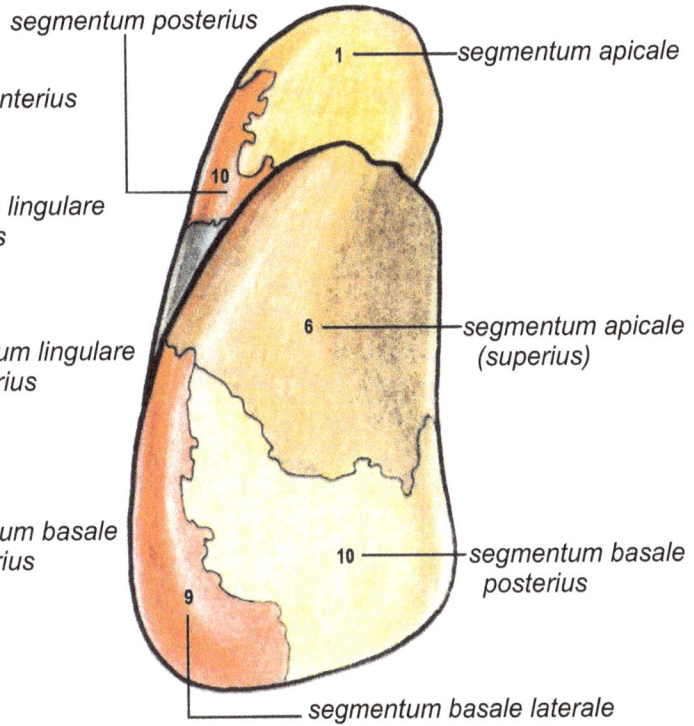

segmentum apicale

segmentum posterius

segmentum apicale

segmentum anterius

segmentum lingulare
superius

segmentum lingulare
inferius

segmentum apicale
(superius)

segmentum basale
anterius

segmentum basale
posterius

segmentum basale laterale

**Fig. 86. Pulmonary segments
(posterior surface of the left lung).**

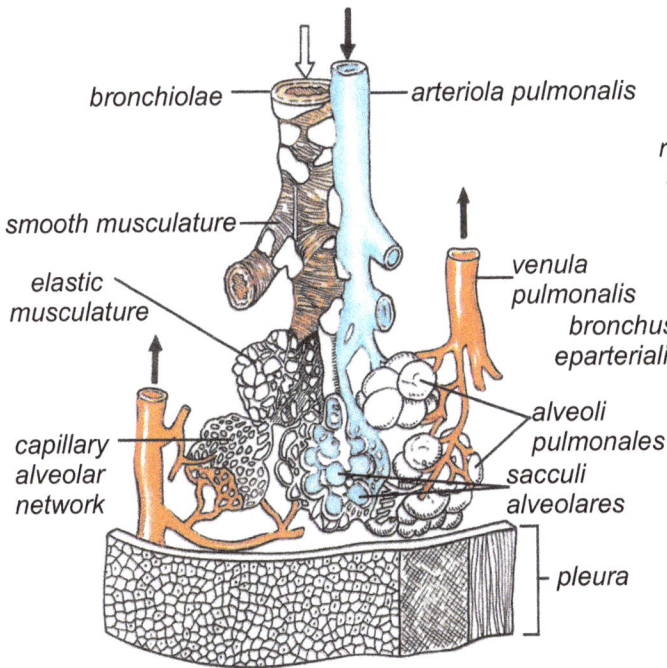

bronchiolae

arteriola pulmonalis

smooth musculature

elastic
musculature

venula
pulmonalis

capillary
alveolar
network

alveoli
pulmonales

sacculi
alveolares

pleura

**Fig. 87. Structure of the
alveolopulmonary complex.**

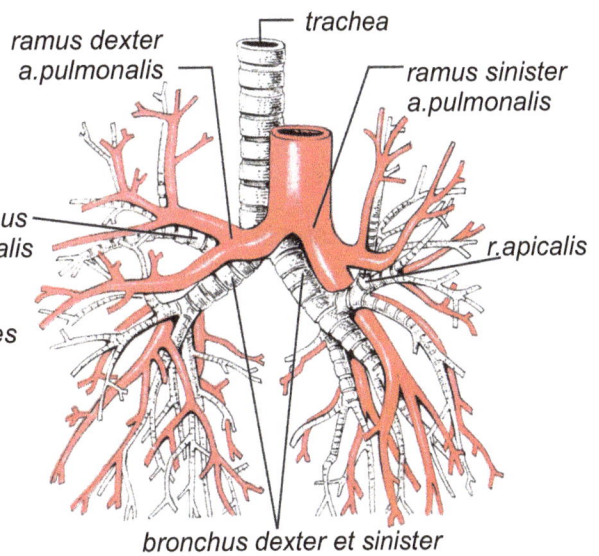

trachea

ramus dexter
a.pulmonalis

ramus sinister
a.pulmonalis

bronchus
eparterialis

r.apicalis

bronchus dexter et sinister

**Fig. 88. The tracheobronchial
free and the pulmonary artery
(scheme).**

91

The left pulmonary artery divides into three main branches: one to the superior lobe, which on turn divides into five segmental branches, one to the apical segment of the inferior lobe and one inferior, which divides into four branches, to the other segments of the inferior lobe.

The segmental branches further divide, following the bronchial divisions up to the level of alveoli, where they undergo capillarization.

The veins arise initially from the perialveolar capillary network. From this level begins the formation of trunks which increase progressively in size and run independently of the course of bronchi. Ultimately, two pulmonary veins, one superior and one inferior, are formed on the right and two on the left, which will empty into the left atrium of the heart. The nutrient circulation is assured by the bronchial arteries and veins. *The bronchial arteries,* two in number, right and left, have their origin at the level of the aorta, from which they reach the posterior surface of the respective bronchus and penetrate into the lung through the pulmonary hilum, giving off branches to the pulmonary vessels and to the bronchial tree, up to the level of the pulmonary lobules.

The bronchial veins are anterior and posterior. The posterior veins run behind the bronchi, are not satellites of the bronchial arteries and empty, on the right, into the azygos vein and, on the left, into the superior hemiazygos vein; the anterior veins empty either into the pulmonary veins or into the azygos vein, on the right, and into the superior hemiazygos vein, on the left.

The lymphatics of the lung start at the level of a perilobular and subpleural network, from where they drain into the peritracheobronchial lymph nodes, then into the anterior and posterior mediastinal lymph nodes. In each lung, three lymph territories may be distinguished. In the right lung there are a superior territory, which drains the lymph into the right laterotracheal lymph nodes, an inferior territory, which drains into the intertracheobronchial lymph nodes, and a middle territory, which drains into the right laterotracheal and the intertracheobronchial lymph nodes. In the left lung there are a superior territory, which drains into the laterotracheal lymph nodes on the left, an inferior territory, which drains into the intertracheobronchial lymph nodes, and a middle territory, which drains into the anterior mediastinal chains and into the lymph nodes at the level of the tracheal bifurcation *(inter-tracheobronchial lymph nodes). The nerve supply* is assured by the pulmonary autonomic plexuses. The lungs and the mediastinum. The lungs have relations with the mediastinum through their medial surfaces, covered with the pleura, which Sines the mediastinal fibrous lamina; they are connected to the mediastinum through the mesopneumonium and the pulmonary root *(radix pulmonis).* Anteriorly, the mediastinum is separated from the retrosternal extrapleural space, situated between the two symmetrical parietal pleurae, and has the shape of a clepsydra, consisting of two triangles joined at the top: above the thymic triangle and below the cardiopericardial triangle. Posteriorly, the thoracic mediastinum has the dimensions of the vertebral column, at the level of which lie the mediastino-vertebral sinuses and, respectively, on the right the interazygo-oesophageal recess and, on the left, the interaortico-oesophageal recess.

The Mediastinum

The space between the two pleuropulmonary regions makes up the mediastinum *(spatium mediastini).* It is situated between the sternum anteriorly, the vertebral column behind, the upper thoracic aperture above and the diaphragm below, bounded by the lamina propria mediastini, a connective formation, component of the intrafhoracic fascia. It contains viscera, vessels, the heart and the pericardium. The mediastinum is divided, by a conventional plane passing anteriorly to the bifurcation of the thorax and corresponding to the insertion of the pulmonary ligament, into the anterior and posterior mediastinum.

The Anterior Mediastinum

The anterior mediastinum is subdivided into two regions: upper (thymovascular) and lower (cardiorespiratory).

The Thimovascular Region

The upper region of the anterior mediastinum contains the thymus and the large vessels situated at the base of the heart and of the neck.

The Thymus
(Thymus)

The thymus is a lympho-epithelial organ playing a role in immunological processes. It develops up to puberty, after which begins its involution, the organ being progressively replaced by fibrous and fatty tissue.

Situated in the upper region of the mediastinum, it has a variable shape, being formed of two unequal lobes, right *(lobus dexter)* and left *(lobus sinister)*, connected with each other in their middle portion, conferring the aspect of the letter "H"(fig. 90).

It has relations, in the mediastinum, anteriorly with the sternum and the first *4-5* pairs of costal cartilages; posteriorly from above downwards, with the brachiocephalic veins, the superior vena cava, the aortic arch, the ascending aorta, the pulmonary trunk, the heart and the pericardium, and laterally with the mediastinal pleurae and the two phrenic nerves. Under maximum devebpement conditions, from the neonatal period to the age of 12-15 years, it reaches also the cervical region, entering into relations anteriorly with the subhyoid muscles, posteriorly with the trachea and laterally with the vasculonervous bundle (common carotid artery, vagus nerve and internal jugular vein).

The thymus is enveloped by a fibrous capsule, from which arise inside septa, delineating lobuli *(lobuli thymi)*, and then follicles that, at their level, present a medullar and a cortical zone, both consisting of reticular tissue, delineating the thymocytes (small lymphocytes). In the medulla lie also Hassal's corpuscles, consisting of concentric reticulo-epithelial cells, which have the aspect of onion coats and vary in number according to the age and to the state of activity.

The involution of the thymus starts concomitantly with the beginning of the gonadal function, but the gland may sometimes persist in the thymico-lymphatic state and can give rise to accidents of the heart arrest type (sometimes under minimum stress conditions or as a consequence of general anaesthesia); in this case the excision of the thymus (thymectomy) is compulsory.

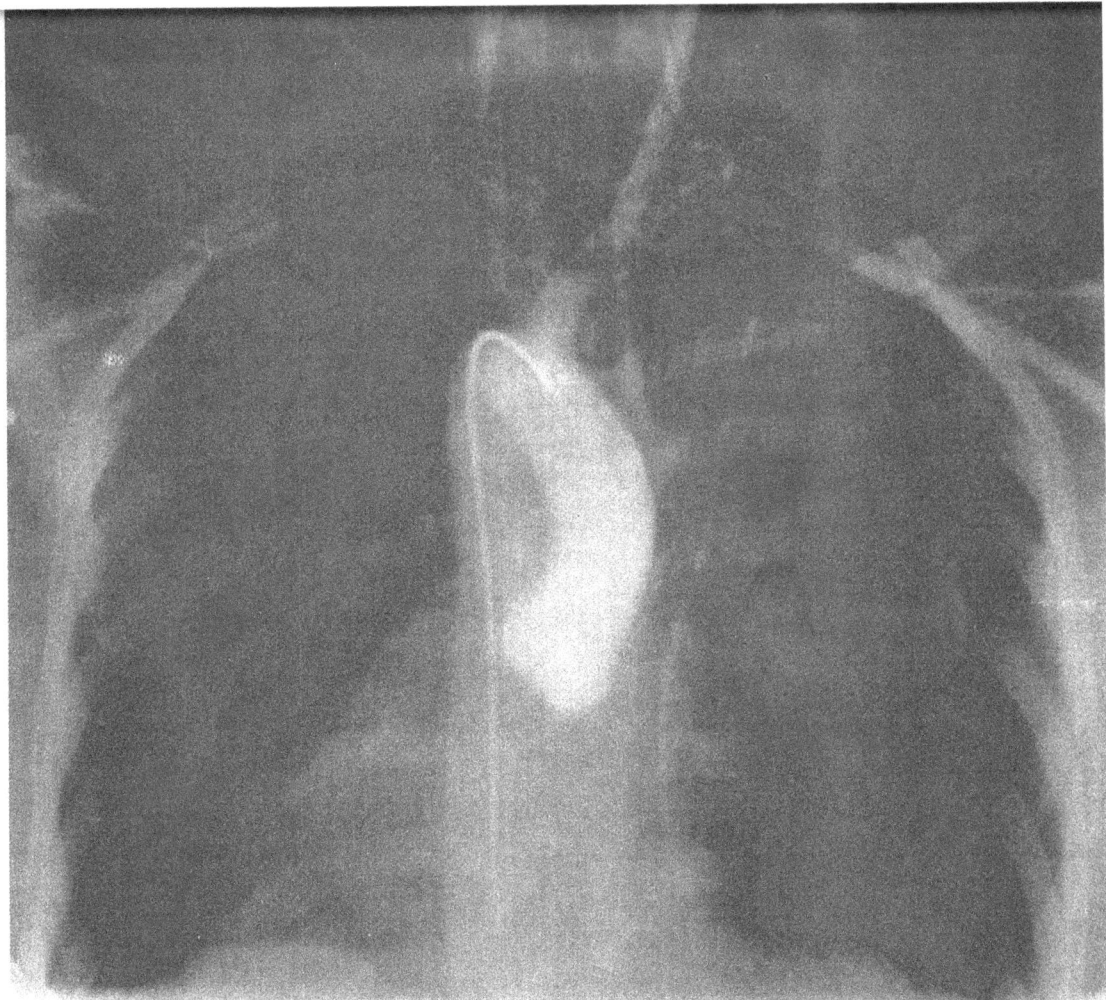

Fig. 89. Aortography (collection of the Central Military Hospital).

We mention that the tumours of the thymus generate disturbances by compression of the anatomical formations with which it is in relations, especially of the adjacent large vessels and nerves.

Vessel *and ne*rve supply

The arterial supply is provided by branches of the inferior thyroid artery, the brachiocephalic trunk, the left subclavian and the internal thoracic arteries.

The veins are satellites of the arteries and tributaries of the left brachiocephalic vein.

The lymph is drained into the inferior jugular parasternal and bronchomediastinal lymph nodes, and then, mostly, into the thoracic duct.

The nerve supply is assured by branches of the vagus nerve, the sympathetic thoracocervical chain and the phrenic nerve.

Behind the thymus are the large vessels from the base of the heart, the superior vena cava, the aortic arch and the pulmonary artery, which in their inferior part are contained in the pericardial sac, whereas the superior part is extrapericardial. These vessels occupy the inferior part of the superior region of t*he* anterior mediastinum. In its superior part lie the large vessels from the base of the neck: the right and left brachiocephalic venous trunks, which collect the internal jugular and subclavian veins; posteriorly, the arteries emerging from the aortic arch, the arterial brachiocephalic trunk (formed of the common carotid artery and the subclavian artery); on the left, the common carotid artery and subclavian artery, which emerge separately from the aortic arch. Hence, we emphasize that the vessels of the superior part of the anterior mediastinum are arranged on two planes: inferiorly, *the arterial plane* (aortic arch and pulmonary artery), situated anteriorly, and *the venous plane* (superior vena cava), situated posteriorly and to the right; superiorly, *the venous plane* (the brachiocephalic trunks with their veins of origin), situated anteriorly, and *the arterial plane* of the carotid and subclavian arteries, situated posteriorly. At this level lie also the anterior mediastinal lymph nodes (Botallo's arterial ligament) between the aortic arch and the left branch of the pulmonary artery, from where the respective recurrent emerges, the phrenic nerves descending on the right, at the level of the superior vena cava and on the left, laterally to the vagus nerve and the cardio-aortic nerve plexus with the cardiac ganglia. The aortic arch, which joins the ascending to the descending aorta, is situated in a parasagittal plane, crossing on the left the pulmonary root and coursing on the right in a direction similar to that of the azygos vein arch (fig. 91-96).

The Thoracic Aorta

The aorta is the trunk of origin of all the arteries of the systemic circulation. The thoracic aorta is made up of two segments:

- the arch of the aorta: the initial portion of the aorta, from the aortic orifice *(ostium aortae)* to the left flank of the fourth thoracic vertebra;

- the thoracic aorta, from the fourth thoracic vertebra to the aortic hiatus of the diaphragm, respectively T12 (fig. 89).

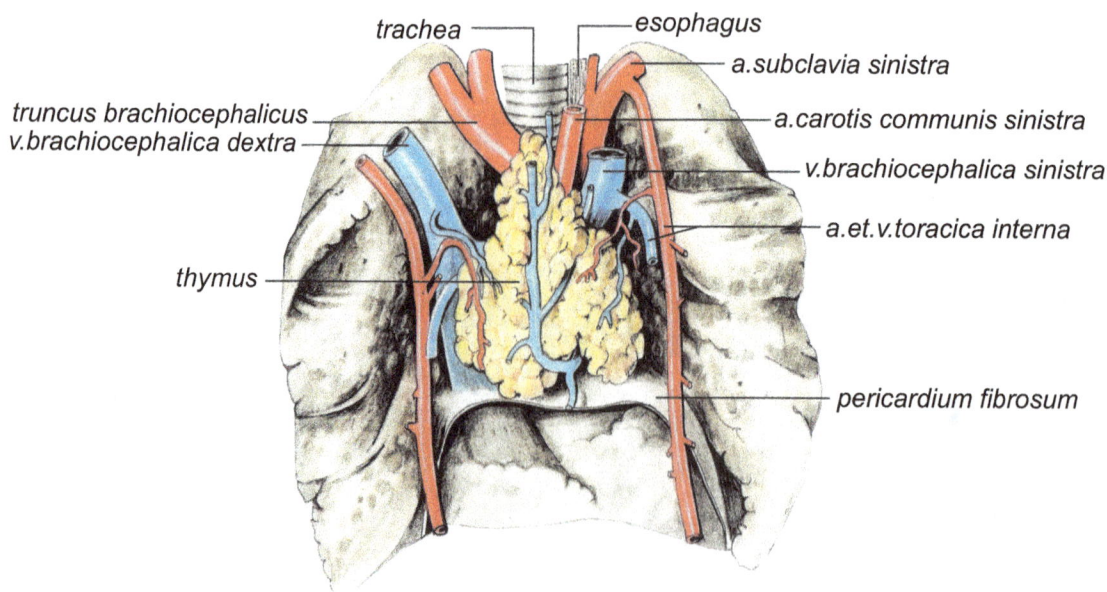

Fig. 90. The anterior mediastinum, superior port (thymus, relations).

The Aortic Arch

The aortic arch describes an arc of circle situated in a nearly sagittal, vertical plane, slightly inclined obliquely from before backwards and from the right to the left, and consisting of two portions:
an ascending portion, which initially runs obliquely upwards, forwards and to the right, of 3-4 cm in length, then becomes suddenly vertical, also of 3 cm in length, up to the level of the first left chondrosternal articulation;

- a horizontal portion, running obliquely backwards and to the left, so that its left surface is directed forwards and its right surface, backwards; it bestrides, with its inferior concavity, the left pulmonary root and then descends, becoming the descending thoracic aorta and ending at the level of the left lateral surface of the fourth thoracic vertebra (fig. 91, 92, 96, 97).

The ascending portion of *the* aortic arch joins the pulmonary arterial trunk, with wich it forms the arterial pedicle of the heart, and is nearly in totality intrapericardial.

Relations with the pericardium. The arterial pedicle is enveloped in a serous pericardial sheath: the serous pericardium, lining the fibrous pericardial sac which ascends the pedicle of the heart, fusing with the adventitia of the vessels.

The pulmonary arterial trunk arises from the right ventricle anteriorly to the orifice of origin of the aorta, which lies on its left side, and gives off, immediately after coming out, the emergence of the coronary arteries, being in relation also with the nerve filaments of the cardiac plexuses.

In its extrapericardial portion the ascending aorta has the following relations:

- on the right: with the extrapericardial superior vena cava, the right phrenic nerve and the mediastinal pleura, which covers also the fibrous pericardium;

- on the left: with the bifurcation of the pulmonary artery, the arterial ligament and the left mediastinal pleura;

- behind: with the right pulmonary artery and the intertracheobronchial lymph nodes;

- anteriorly: from the surface to the depth it has relations with the sternum, the internal mammary vessels, the anteriorpleurocostomediastinal culs-de-sac and the thymus.

The horizontal portion of the aortic arch has the following relations:

The left anterolateral surface corresponds to the mediastinal pleura, the internal surface of the left lung and the following structures: the left phrenic nerve, the left vagus nerve and the sympathetic cardiac nerves.

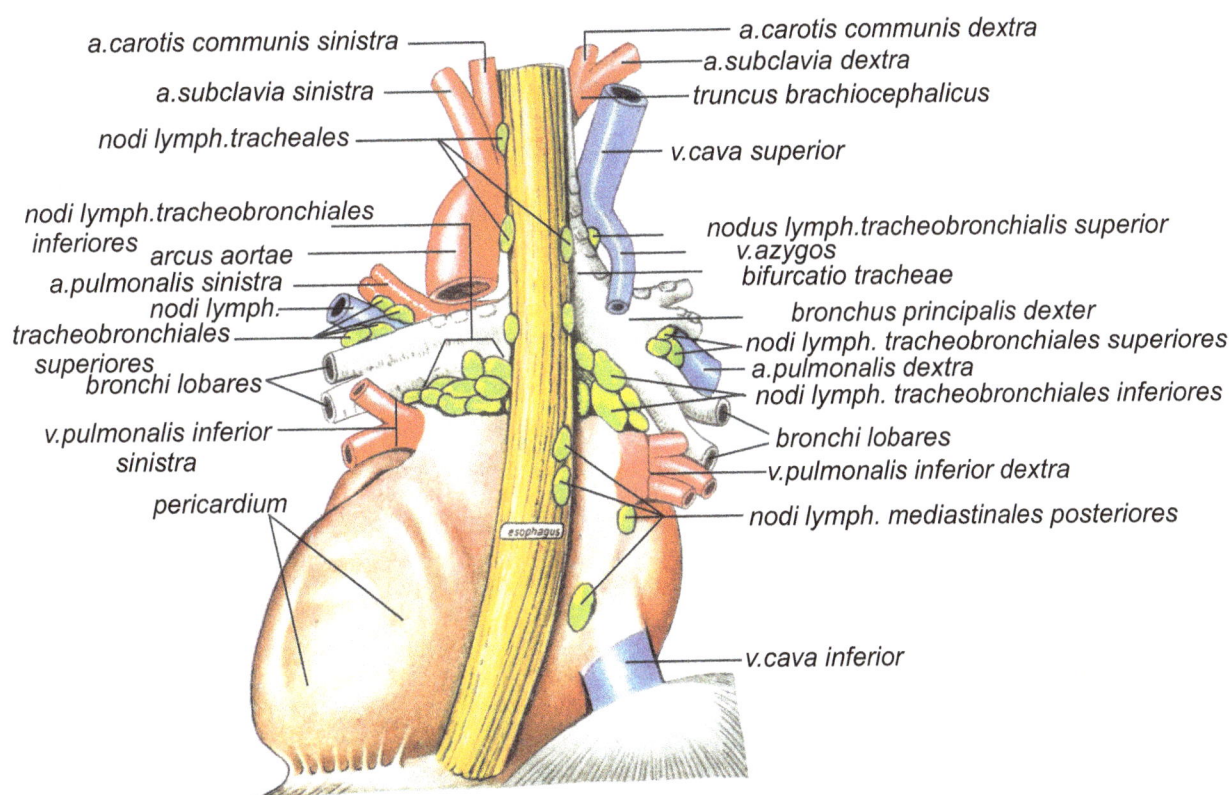

Fig. 91. The heart and the great vessels at its base (posterior view - relations).

The right posterolateral surface is in relation, from before backwards, with: the superior vena cava, into which the azygos vein arch empties; the trachea (the aortic impression); the left recurrent nerve; the cardiac sympathetic nerves and the vagus nerve; the thoracic oesophagus, which passes beyond the left border of the trachea (aorta narrowing); the retroesophageal thoracic duct, which becomes retro-subclavicular; the left flank of the fourth thoracic vertebra; the small superior azygos vein and the left thoracic sympathetic chain.

The inferior surface crosses above, successively, from before backwards and from the right to the left, the following elements: the right branch of the pulmonary artery, the bifurcation of the pulmonary artery on the left of the median line, the left pulmonary artery, the left bronchus.

The aorta is connected to the pulmonary bifurcation through the arterial ligament which is a fibrous cord, a remnant of the arterial canal

From the superior surface arise the three large arterial trunks: from before backwards, anteriorly to the trachea, the pretracheal brachiocephalic arterial trunk; laterally to the trachea, the left common carotid artery; laterally to the oesophagus, the left subclavian artery.

Between the brachiocephalic arterial trunk and the left primitive carotid artery lies the pretracheal vascular triangle: Neubauer's lowest thyroid artery (thyreoidea ima arteria) (when it is present), a tracheobronchial lymph node and, sometimes, an aberrant thyroid artery.

Between the left common carotid artery and the subclavian artery lies Bourgery's quadrilater, bounded: in front, by the left common carotid artery; behind, by the left subclavian artery; below, by the aorta; above, by the left superior intercostal vein (in the area of this quadrilater, the left vagus nerve crosses the left phrenic nerve .

The Descending Thoracic Aorta

Its origin is continuous with the aortic arch, at the left border of the fourth thoracic vertebra (fig. 94, 96). Its course is vertical, descending in the posterior mediastinum along the thoracic vertebral column. It traverses the diaphragm at the level of the twelfth thoracic vertebra, becoming the abdominal aorta.

In the posterior mediastinum it is the most posteriorly situated organ together with the thoracic duct and has the following relations: anteriorly, from above downwards, with the left root of the lung, the left vagus nerve, the thoracic oesophagus and the interaortico-oesophageal pleural sinus; posteriorly, the thoracic aorta is in relation with the intercostal arteries, which arise on its posterior surface, with the left thoracic sympathetic chain, the splanchnic nerve, the superior and inferior hemiazygos veins, which cross it behind, at the level of the sixth thoracic vertebra for the superior vein and at the level of the seventh or eighth vertebra for the inferior vein and, through these elements, with the vertebral column; laterally, on the left, it is covered by the mediastinal pleura and on the right, above, is the left border of the oesophagus; along its whole course it is in relation with the thoracic duct, which from retroaortic becomes downwards retroesophageal, and with the azygos vein, separated from the aorta by the thoracic duct.

The thoracic aorta crosses the diaphragm through the fibrous, unextensible aortic orifice, lying between the main medial pillars of the diaphragm, situated at the level of the twelfth thoracic vertebra.

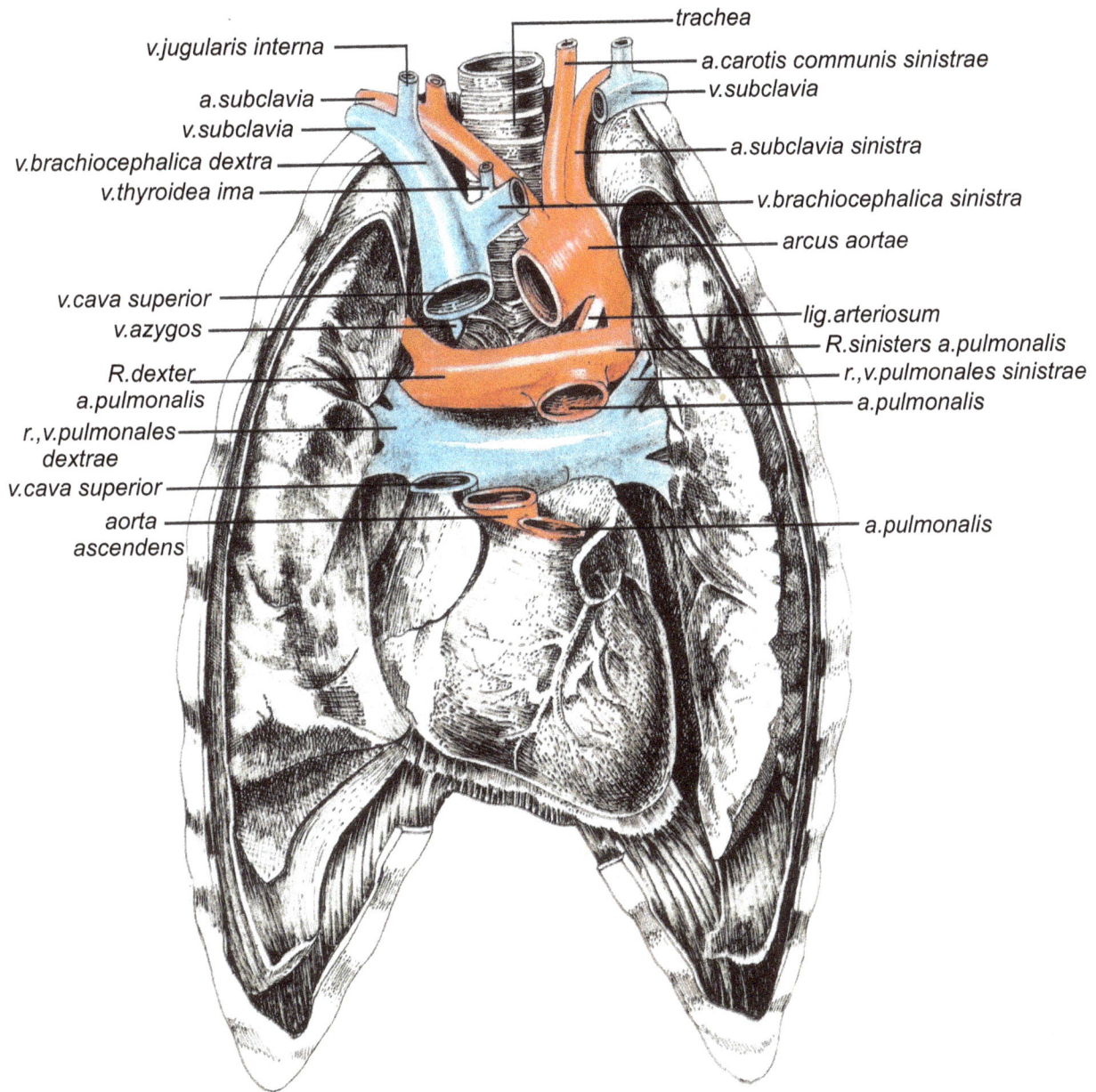

v.jugularis interna

trachea

a.carotis communis sinistrae

a.subclavia

v.subclavia

v.subclavia

v.brachiocephalica dextra

a.subclavia sinistra

v.thyroidea ima

v.brachiocephalica sinistra

arcus aortae

v.cava superior

lig.arteriosum

v.azygos

R.sinisters a.pulmonalis

R.dexter a.pulmonalis

r.,v.pulmonales sinistrae

a.pulmonalis

r.,v.pulmonales dextrae

v.cava superior

aorta ascendens

a.pulmonalis

Fig. 92. The anterior mediastinum (vascular part and heart - relations).

97

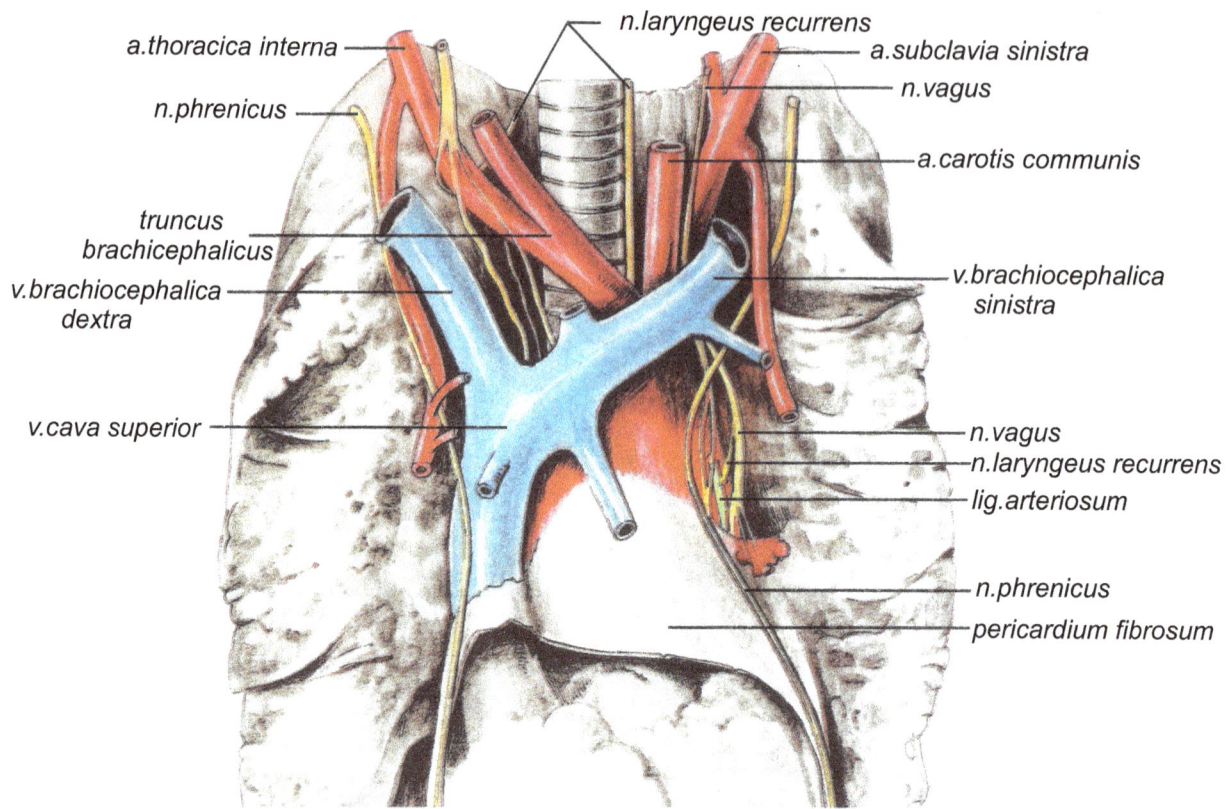

Fig. 93. The anterior mediastinum, superior part, the vascular plane, relations. Trachea

a.thoracica interna
n.phrenicus
truncus brachicephalicus
v.brachiocephalica dextra
v.cava superior
n.laryngeus recurrens
a.subclavia sinistra
n.vagus
a.carotis communis
v.brachiocephalica sinistra
n.vagus
n.laryngeus recurrens
lig.arteriosum
n.phrenicus
pericardium fibrosum

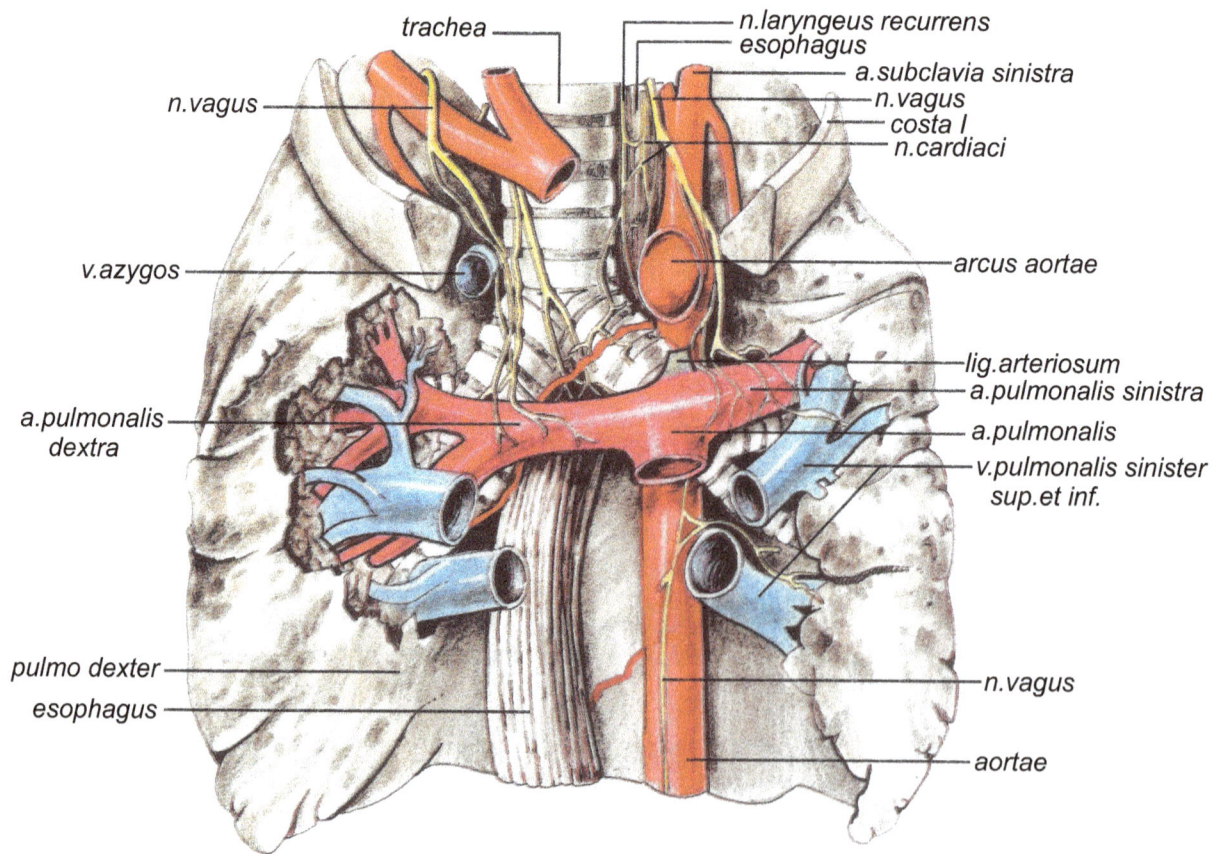

Fig. 94. The anterior mediastinum after the section of the vessels and the removal of the heart (relations).

trachea
n.vagus
v.azygos
a.pulmonalis dextra
pulmo dexter
esophagus
n.laryngeus recurrens
esophagus
a.subclavia sinistra
n.vagus
costa I
n.cardiaci
arcus aortae
lig.arteriosum
a.pulmonalis sinistra
a.pulmonalis
v.pulmonalis sinister sup.et inf.
n.vagus
aortae

98

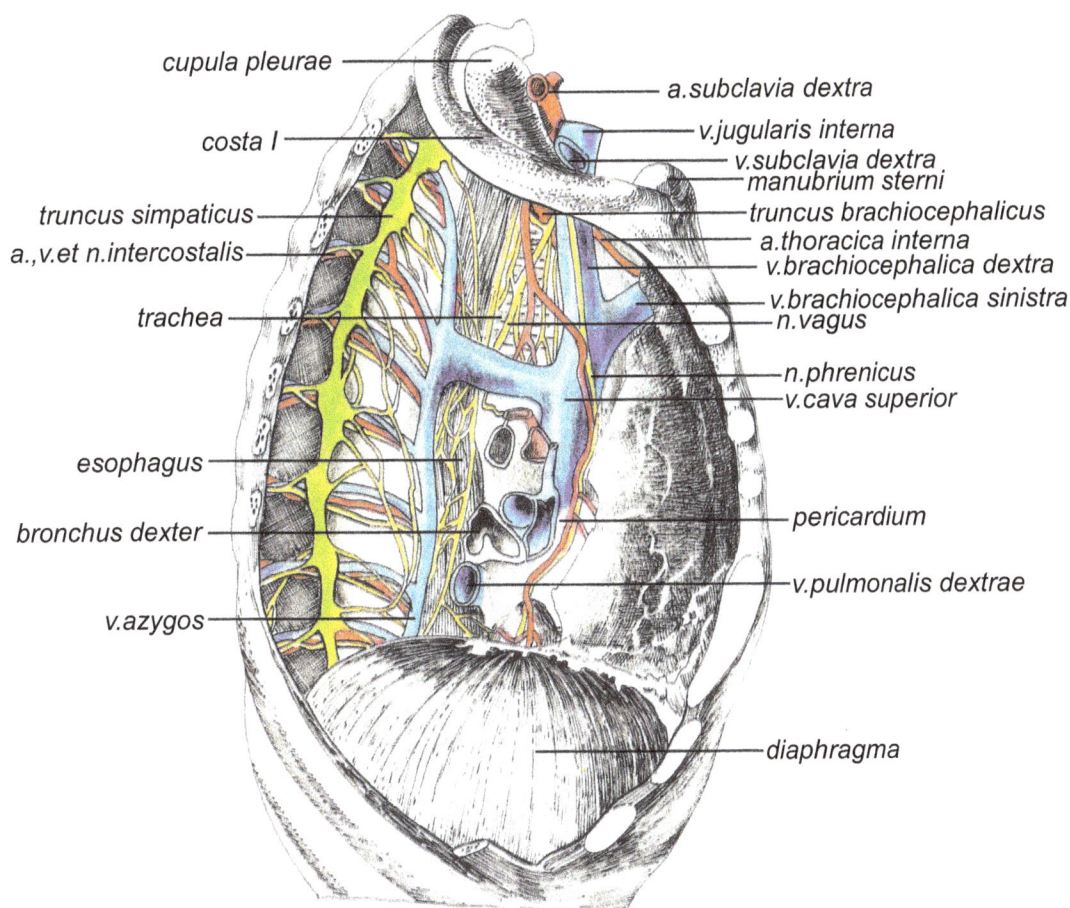

Fig. 95. The mediastinum (right lateral view, after removal of the right lung).

Anteriorly, it is in relation with the oesophagus, which crosses the diaphragm through the oesophageal hiatus projected at the tenth thoracic vertebra, and posteriorly with the twelfth thoracic vertebra (see also fig. 110 and 113).

The branches of *the* thoracic aorta are divided into two groups: visceral and parietal.

The visceral branches are the bronchial, oesophageal and mediastinal arteries.

The bronchial arteries, usually two in number, right and left, sometimes three, arise from the upper part of the descending aorta (or from the internal surface of the aortic arch), reach the anterior surface of the bronchus, sometimes retrobronchially, and end in the lungs, where they ramify.

The oesophageal arteries, two up to four, arise from the anterior surface of the aorta. They anastomose upwards with the oesophageal branches of the inferior thyroid arteries and with the bronchial arteries and downwards with the branches of the inferior diaphragmatic arteries and of the left gastric artery.

The mediastinal arteries are slender anterior arterioles, destined to the pericardium, the pleura and the adjacent lymph nodes.

The parietal branches are the aortic intercostal arteries.

These arteries, usually eight or nine in number, for the last eight or nine intercostal spaces (the arteries of the first two or three spaces arise from the cervico-intercostal trunk, a branch of the subclavian artery), have their origin on the posterior surface of the aorta, on either side of the median line, are accompanied by the intercostal veins, which empty into the system of the azygos veins, and end by anastomosis with the anterior intercostal arteries, branches of the internal thoracic artery (internal mammary artery).

The Pulmonary Artery

It is a functional artery, which conveys the venous blood of the right ventricle to the lungs, where it is oxygenated. It consists of a trunk and two branches resulted from division: the right and the left pulmonary arteries (fig.71, 88, 92, 94, 107, 113).

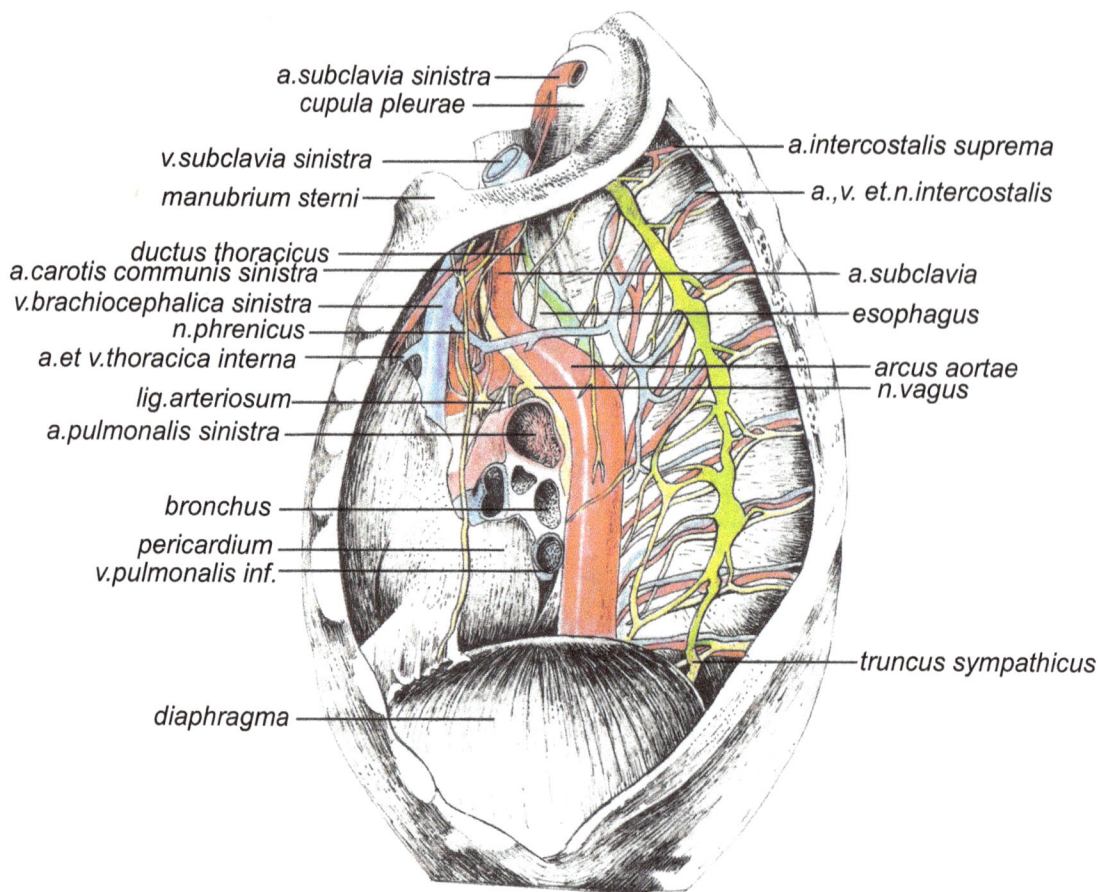

Fig. 96. The mediastinum (left lateral view, after removal of the left lung).

Labels on figure:
- a.subclavia sinistra
- cupula pleurae
- v.subclavia sinistra
- manubrium sterni
- ductus thoracicus
- a.carotis communis sinistra
- v.brachiocephalica sinistra
- n.phrenicus
- a.et v.thoracica interna
- lig.arteriosum
- a.pulmonalis sinistra
- bronchus
- pericardium
- v.pulmonalis inf.
- diaphragma
- a.intercostalis suprema
- a.,v. et.n.intercostalis
- a.subclavia
- esophagus
- arcus aortae
- n.vagus
- truncus sympathicus

The Pulmonary Arterial Trunk

It has its origin at the level of the pulmonary orifice, situated on the vertex of the arterial cone of the right ventricle. Its course is oblique backwards, to the left and upwards, describing a curve with the right posterolateral concavity, which bestrides the anterior and left surfaces of the ascending aorta.

It ends by bifurcation, the resulting branches forming a very obtuse angle; this bifurcation is situated on the left of the median line, at 2 cm below the tracheal bifurcation.

It joins the ascending aorta, with which it forms the arterial pedicle.

Within the pericardial sac it has relations with: the trunk of the ascending aorta (initially situated posteriorly and then anterolaterally on the right); the coronary arteries, which bound its two sides, and the nerve branches of the cardiac plexus.

Through the pericardium it has the following extrapericardial relations: anteriorly, from the surface into the depth, with the thoracic wall (sternum, internal port of the second costal cartilage and of the second left intercostal space, in which it projects), the left internal thoracic vessels, the anterior costomediastinal pleural culs-de-sac, the anterior border of the lungs and the lower portion of the thymus; behind and below, with the distal pulmonary veins; on the right, with the extrapericardial superior vena cava, the right phrenic nerve and the right mediastinal pleura; on the left, with the left mediastinal pleura, the left lung and the left phrenic nerve; above, the relations are those of the pulmonary bifurcation with the tracheal bifurcation, on the one hand, and with the inferior surface of the aortic arch, to which it is connected by the arterial ligament, on the other hand.

The Right Pulmonary Artery

It is the right branch resulted from the division of the pulmonary artery, which conveys the venous blood to the right lung (functional artery). Its origin is nearly continuous with that of the left pulmonary artery, on the left of the median line.

It consists of two segments: a preradicular segment, directed transversally and to the right, which crosses behind the aorta and the superior vena cava, and a radicular segment, which enters in the structure of the right pulmonary root.

100

The Left Pulmonary Artery

It is the left branch resulted from the division of the pulmonary artery, which conveys the venous blood into the left lung, to which it is a functional artery.

Its origin is nearly continuous with that of the right pulmonary artery and it is connected to the aorta through the arterial ligament.

It consists of two segments: an extrahilar segment, which reaches the upper border of the left bronchus without passing beyond it, and a hilar segment, which describes a concave course anteriorly and medially and passes behind the upper lobar bronchus.

The Superior Vena Cava

If is the collector trunk of the venous blood from the supradiaphragmatic portion of the human body; if is devoid of valves (fig. 91,93,95,98,107,116,117).

Its origin, situated behind the lower border of the first right costal cartilage, along the sternum, at the level of the second thoracic vertebra, results from the union of the right and left brachiocephalic venous trunks: the right almost vertical, the left very oblique downwards and to the right

Its course is nearly vertical and ends in the upper wall of the right atrium, through an orifice devoid of valves, which projects at the anterior extremity of the second right intercostal space; posteriorly if corresponds to the sixth thoracic vertebra or to the upper border of the seventh thoracic vertebra.

The superior vena cava is situated in the upper portion of the anterior mediastinum, paramedially on the right. It is extrapericardial in its upper two thirds and intrapericardial in its lower third.

The Extrapericardial Segment

If has the following relations:

- in front, from the surface to the depth, with the sternocostal plastron, the right infernal thoracic vessels, the anterior costomediasfinal cul-de-sac, the thymic region (in the upper mediastinal triangular space).

- behind, if is divided info two portions, superior and inferior, by the azygos vein arch, which opens into if on its posterior surface. The upper portion is in relation with the right laferotracheal lymph node chairs (the nodule of the azygos vein arch). The lower portion corresponds to the superior part of the right pulmonary root (the right extraradicular bronchus, the right pulmonary artery, the right superior pulmonary vein), which is crossed posteriorly by the right vagus nerve;

- on the right, if has relations with the right phrenic nerve, the right thoracic sympathetic chain and the right mediastinal pleura;

- on the left, with the ascending aorta and with the supraarterial part of the cavo-aorfic recess, between: the superior vena cava on the right, the aorta on the left, the left brachiocephalic venous trunk above and the serous pericardium below.

The Intrapericardial Segment

If has successively relations within the pericardium and relations through the pericardium.

Through the visceral leaflet of the pericardium if has relations anteriorly with the right atrium and posteriorly with the intrapericardial segment of the right superior pulmonary vein and with the inferior surface of the right pulmonary artery.

The extrapericardial relations are: in front, with the second right intercostal space; on the left, with the cavo-aortic recess, between the two vessels; on the right, with the right phrenic nerve, the right pleura and the right lung; behind, with the right pulmonary root (right pulmonary artery, right bronchus, right superior pulmonary vein).

The Cardio-Pericardial Region

If is bounded: anteriorly, by the inferior thoracic wall; below, by the diaphragm; laterally, by the right and the left mediastinal pleurae; posteriorly, by a conventional plane which passes anteriorly to the tracheal bifurcation; superiorly, by a conventional plane which passes above the heart.

This mediastinal region contains the pericardium and the heart.

The Pericardium

The pericardium is a fibroserous sac, which covers the heart and its vessels and consists of two components:

✎ an external, fibrous component, the fibrous pericardium (pericardium fibrosum) and

✎ a deep component, made up of two leaflets, the serous pericardium (pericardium serosum) (fig. 95, 96, 97).

The Fibrous Pericardium It has the shape of a frustum of a cone with the inferior base attached to the tendinous centre of the abdominal diaphragm *(pars diaphragmatica)* and with the summit directed upwards, where it adheres to the adventitia of the great vessels from the base of the heart The lateral surfaces are in relation with the mediastinal parietal pleurae which, covering the fibrous pericardium, cover also the phrenic nerves in their course to the diaphragm, together with the pericardial-phrenic vessels. The posterior surface of the fibrous pericardium is in contact with the oesophagus and the aorta *(pars vertebralis)*. Its anterior surface *(pars sterno - costalis)* is in relation with the pleura and the lungs *(pars retropleuro -pulmonaris)* and, in part, directly with the costal wall *(pars extrapleuralis)*, through the inferior triangle on the sternocostal plastron. The fibrous pericardium is connected to the skeleton and the adjacent organs through ligaments, the ligaments of the pericardium, which can be systematized info three main groups: sternopericardial, phrenicopericardial and vertebropericardial ligaments. There are also accessory ligaments (fig. 90, 93, 98).

The sternopericardial ligaments, They are two in number: superior and inferior.

The superior sternopericardial ligament is inserted on the manubrium of the sternum, below the sternothyroid muscle, and ends on the pericardium at the level of the great arterial trunks, being actually a cervicopericardial ligament

The inferior sternopericardial ligament extends from the lower extremity of the sternum and from the xiphoid process to the lower part of the pericardium.

The phrenicoperieardial ligaments are components of the endothoracic fascia, which consists of a subpleural fatly connective layer and a fibrous layer that merges with the pericardium, forming three resistant fibrous bands, the phrenicopericardial ligaments, the description of which varies according to the authors:

- *the anterior phrenicopericardial ligament* corresponds to the anterior region of the diaphragmatic base of the pericardium;

- *the right phrenicopericardial ligament* covers the posterolateral surface of the inferior vena cava;

- *the left phrenicopericardial ligament,* inconstant;.-;is situated on the left posterolateral side of the base of the pericardial cone.

The vertebropericardial ligaments. They have their origin on the prevertebral aponeurosis and extend from the sixth cervical vertebra to the fourth thoracic vertebra, ending below on the upper part of the pericardium: on the right, above the pulmonary roof and on the left, by two laminae surrounding the aortic arch.

The accessory ligaments. They connect the pericardium to the adjacent organs: the tracheopericardial, bronchopericardial and oesophagopericardial ligaments. The thyropericardial aponeurosis, a component of the middle cervical fascia *(fascia colli media)*, is also described.

The relations of the fibrous pericardium are those of the heart, except the superior ones, where the pericardium passes beyond the heart, covering the great vessels:

- anteriorly, with the thoracic wall, the pleura and the lung;

- above with the thymus;

- behind, with the structures of the posterior mediastinum and especially with the oesophagus;

- laterally, with the mediastinal pleurae and with the loose cell tissue, in which descend the phrenic nerves and the superior diaphragmatic vessels;

- below, with the phrenic centre of the diaphragm and, through the diaphragm, with the abdominal viscera, especially with the stomach, a relation which, under certain conditions, can cause the gastrocardiac syndrome, characterized by heart pain that may simulate an infarction in the case of gastric disturbances;

- in the superior area, the posterior, sternocostal and lateral portions meet in the form of a capsule *(capsula pericardii)*, where the aorta and the pulmonary artery leave the pericardial sac and the superior vena cava penetrates into it. At this level is formed, by prolongation of the pericardium on the aorta, the aortic recess *(recessus aorticus)*.

The Serous Pericardium

Like all the serosae, it consists of two leaflets: a visceral leaflet or epicardium, which covers the heart, and a parietal leaflet (pericardium serosum), which lines the fibrous pericardium.

Between the two leaflets lies a capillary cavity, the pericardial cavity (cavum pericardii), which contains a small amount of albuminous serous fluid, that facilitates the movements of the cardiac muscle (fig. 97, 98, 99).

The two components of the pericardial serosa are directly continuous with each other at the level of the pericardial culs-de-sac or of the pericardial sinuses.

Thus, after covering the heart and the coronary vessels with their branches, situated in the grooves between the ventricles and the atria, the visceral serous pericardium, the epicardium, courses from the apex of the heart to its base, i. e. to the vascular pedicle, arterial in front - formed by the aorta and the pulmonary artery -and venous behind - formed by the superior vena cava, situated above in the pericardial sac, and the inferior vena cava, situated below at the level of its passage orifice through the diaphragm -, but also to the pulmonary veins (two right and two left veins), situated on the posterior surface of the base of the heart.

The epicardium, i.e. the visceral pericardium, ascends the anterior surface of the heart and reaches the anterior surface of the aorta, situated at the base of the heart on the right, and the anterior surface of the pulmonary artery, situated on the left. It encloses both in a common serous sheath and ascends both up to the insertion of the fibrous pericardium on their adventitia, where it reflects from the anterior surface of the aorta on the deep surface of the fibrous pericardium, forming a preaortic cul-de-sac (preaortic sinus); if behaves in the same way also at the level of the pulmonary artery, giving rise to the prepulmonary cul-de-sac (prepulmonary sinus) and becoming at this level the parietal serous pericardium. By the common covering of the two arteries by the visceral serosa, they are enclosed in a serous muff. Behind the arteries arises Theile's transverse sinus (the index, introduced behind the pulmonary artery, appears on the right of the aorta). However, the anterior visceral serosa of the heart reaches also the anterior surface of the superior vena cava, at the insertion site of the fibrous pericardium into the adventitia of the vein, where if reflects on the fibrous pericardium and gives rise to the cul-de-sac in front of the vena cava. Covering the right atrium, it descends along the course of the right pulmonary veins, to the place where they penetrate into the left atrium, invests them and reflects on the fibrous pericardium, on the vertical line descending up to the inferior vena cava. If covers also the latter at its penetration site through the diaphragm info the right atrium and reflects also at the level of this vein on the fibrous pericardium, becoming thus a parietal fibrous pericardium also at this level.

However, the epicardium - the visceral pericardium - covers also the posterior or diaphragmatic surface of the heart and reaches the venous opening of the heart, which is posterior. After covering the posterior surface of the left atrium, if insinuates between the right and the left pulmonary veins and, at their penetration site into the atrium, it reflects on the fibrous pericardium; here it gives rise to a deep, oblique, upwards and rightwards directed cul-de-sac, termed oblique cul-de-sac or Hollers oblique sinus; then if passes on the anterior surface of the left superior pulmonary vein-on the anterior surface of the left atrium - and reaches the superior vena cava, forming in this way the serosa of the posterior wall of Theile's transverse sinus.

The exploratory finger, introduced into Theile's transverse sinus, has in front the arterial wall formed by the pulmonary artery and the aorta; behind, the anterior wall of the left atrium; above, the right pulmonary artery and below, the angle formed by the joining of the arterial pedicle with the anterior wall of the left atrium.

The right orifice of Theile's transverse sinus is a vertical slit between the aorta, medially, and the right auricle, the right atrium and the superior vena cava, laterally; the right coronary artery corresponds to its base. The finger introduced through this orifice goes round the arterial pedicle.

The left orifice is an elongated slit between the left auricle and the left atrium laterally, the left pulmonary artery above and the left coronary artery below.

The reflection line of the visceral serous pericardium on the parietal pericardium is T-shaped. Between the reflection on the inferior vena cava and the right pulmonary veins, on the one side, and the left pulmonary veins, on the other side, lies the recessus inferior pericardii *(sinus obliquus Halleri).*

The exploratory finger, introduced info Halleri's oblique sinus, should first lift the heart from its apex, then penetrate behind the heart, which remains in the palm, and enter into the sinus, bounded on either side by the two right and left pulmonary veins, anteriorly by the posterior wall of the left atrium and posteriorly by the wall of the fibrous pericardium lined by the pericardial parietal serosa.

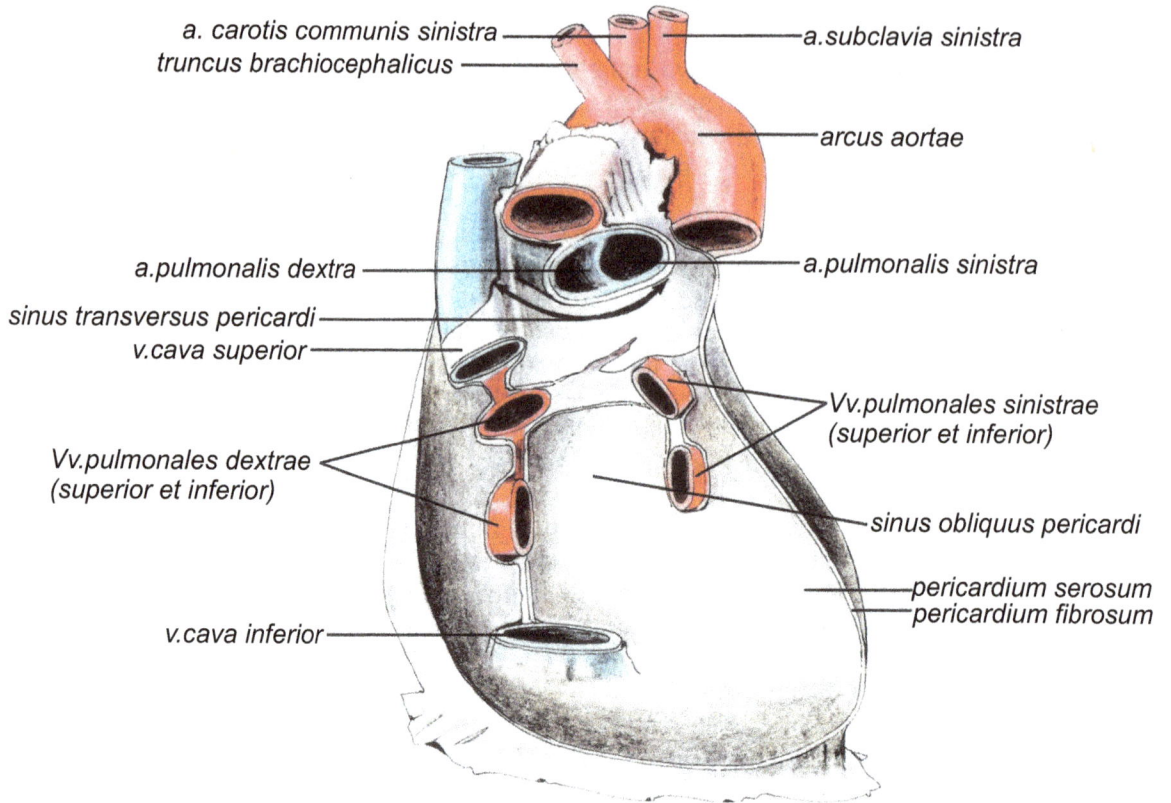

Fig. 97. The pericardium (interior view, after removal of the heart).

a. carotis communis sinistra — a.subclavia sinistra
truncus brachiocephalicus
arcus aortae
a.pulmonalis dextra — a.pulmonalis sinistra
sinus transversus pericardi
v.cava superior
Vv.pulmonales sinistrae (superior et inferior)
Vv.pulmonales dextrae (superior et inferior)
sinus obliquus pericardi
pericardium serosum
pericardium fibrosum
v.cava inferior

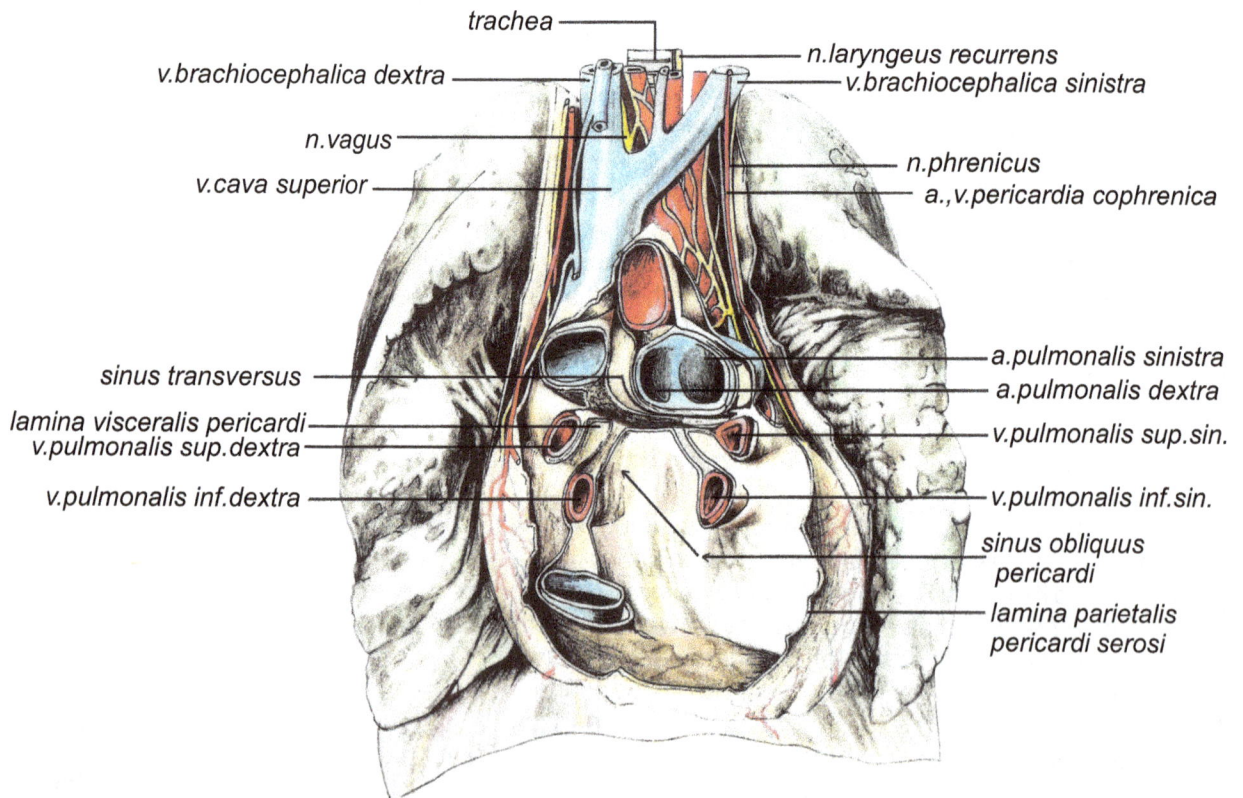

Fig. 98. The fibrous pericardium (extrinsic relations).

trachea
v.brachiocephalica dextra — n.laryngeus recurrens
v.brachiocephalica sinistra
n.vagus
v.cava superior — n.phrenicus
a.,v.pericardia cophrenica
sinus transversus — a.pulmonalis sinistra
a.pulmonalis dextra
lamina visceralis pericardi
v.pulmonalis sup.dextra — v.pulmonalis sup.sin.
v.pulmonalis inf.dextra — v.pulmonalis inf.sin.
sinus obliquus pericardi
lamina parietalis pericardi serosi

Beyond the fibrous pericardium, hence posteriorly to it, is situated the thoracic oesophagus.

Holler's oblique sinus has 10 cm in depth, is situated on the posterior surface of the left atrium, between the pulmonary veins (between the two venous pedicles), communicates behind with the great pericardial cavity and has the mentioned relation with the thoracic oesophagus.

At the level of the venous pedicle are also a cul-de-sac which corresponds to the left pulmonary veins and a cul-de-sac between the left superior pulmonary vein and the left branch of the pulmonary artery (left pulmonary recess), bounded medially by Marshall's vestigial fold.

Vessel and nerve supply of the pericardium

The vessel supply to the serous pericardium is provided by branches of *the coronary arteries* and *veins. The lymphatics* run to the subpericardial network of the heart. *The nerves* derive from the subpericardial cardiac plexus.

To the fibrous pericardium:

- *the arteries* are represented by branches of the superior diaphragmatic, of the bronchial and of the oesophageal arteries;
- *the satellite veins* empty into the azygos veins and the superior diaphragmatic veins;
- *the lymphatics* drain into the anterior and posterior mediastinal, diaphragmatic and intertracheobronchial lymph nodes;
- *the nerves* are represented by branches of the phrenic, vagus, recurrent and sympathetic cervical nerves.

The classical authors and especially Luschka have described numerous pericardial nerve branches of vagal, sympathetic, recurrent or phrenic origin. Recently these branches have been contested. However, the reflex incidents which occur at the opening or at the simple excitation of the pericardium prove the existence of a rich innervation, probably of microscopic order.

The envelopes of the pericardium are probably innervated from the same source as the heart. Nerve branches of the cardiac plexus give off rami which perforate the vascular sheaths or even the pericardium at the level of the reflection lines (Hovelaque), while others, deeper, ramify below the epicardium, forming an intrinsic plexus before entering into the myocardium.

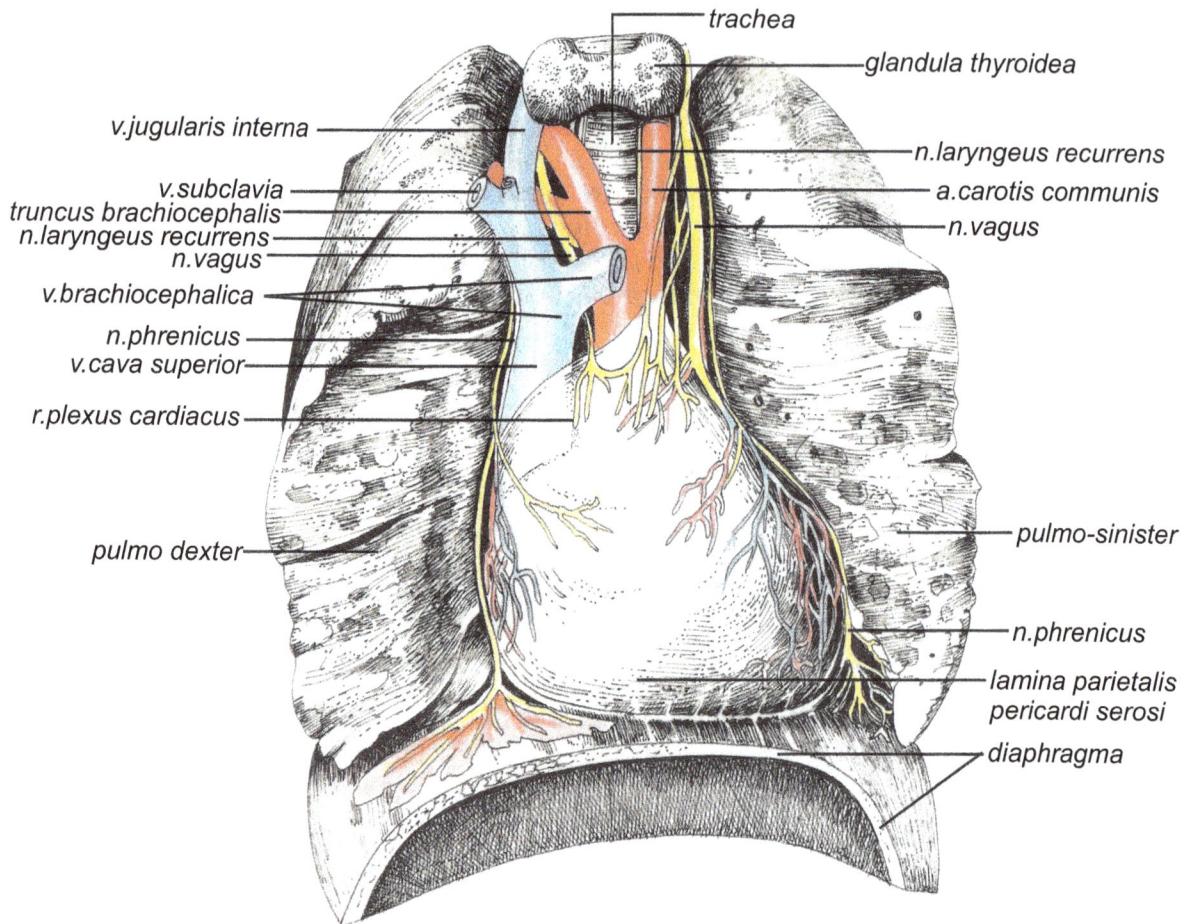

Fig. 99. The lungs and the anterior mediastinum after removal of the thymus (anterior overall view; the left vagus nerve is drawn anteriorly).

The Heart
[Cor]

The heart is a hollow muscle consisting of four chambers: two atria and two ventricles (fig. 100-112).

It has the shape of a frustum of a cone, with the base posterior and the apex anterior. The.great axis is obliquely directed forwards, downwards and to the left, practically horizontal in brevilineal subjects and nearly vertical in longilineal subjects *(cor pendulum).*

It weighs about 270 g in the male and less in the female; its colour is reddish.

Owing to its form, one distinguishes three surfaces, three borders, a base and an apex.

The grooves of the heart mark, on its external surface, the boundaries between the four cardiac chambers. Through the grooves run the coronary vessels, hidden by fat. The following grooves are distinguished: an interatrial *(sulcus interatrial),* an interventricular *(sulcus inferventricularis)* and an atrioventricular *(sulcus coronarisj* groove.

The surfaces are three in number: anterior *(sternocostal),* inferior *(diaphragmatic),* separated antero-inferiorly by a sharp border *(margo acutus)* and on the left and slightly posteriorly by a thicker border *(margo obtusus),* which may be actually considered as a left lateral mediastinal surface.

As the heart is situated asymmetrically, two fifths of its mass lie on the right of the median line, enclosing the greatest part of the right atrium and only a small part of the right ventricle, whereas the remaining three fifths are on the left of this line. The right, venous heart is anterior and the left, arterial heart is situated posteriorly. If is attached to the mediastinum only by the venous and arterial pedicles, which easily permits its pulsations, the more so as if is situated in a serous cavity which facilitates its contraction and relaxation movements.

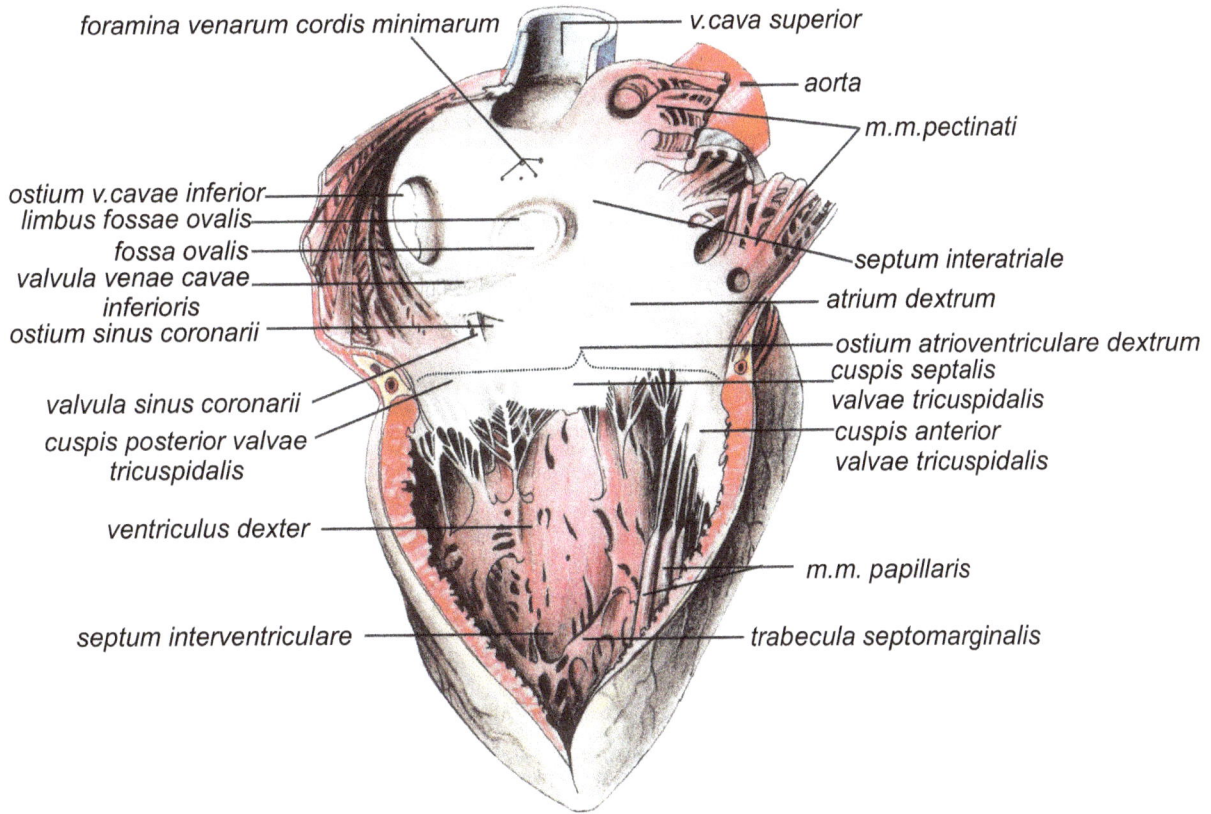

foramina venarum cordis minimarum — v.cava superior

aorta

m.m.pectinati

ostium v.cavae inferior
limbus fossae ovalis
fossa ovalis
valvula venae cavae
inferioris
ostium sinus coronarii

septum interatriale

atrium dextrum

ostium atrioventriculare dextrum
cuspis septalis
valvae tricuspidalis

cuspis anterior
valvae tricuspidalis

valvula sinus coronarii
cuspis posterior valvae
tricuspidalis

ventriculus dexter

m.m. papillaris

septum interventriculare

trabecula septomarginalis

Fig. 100. Open right heart (inside).

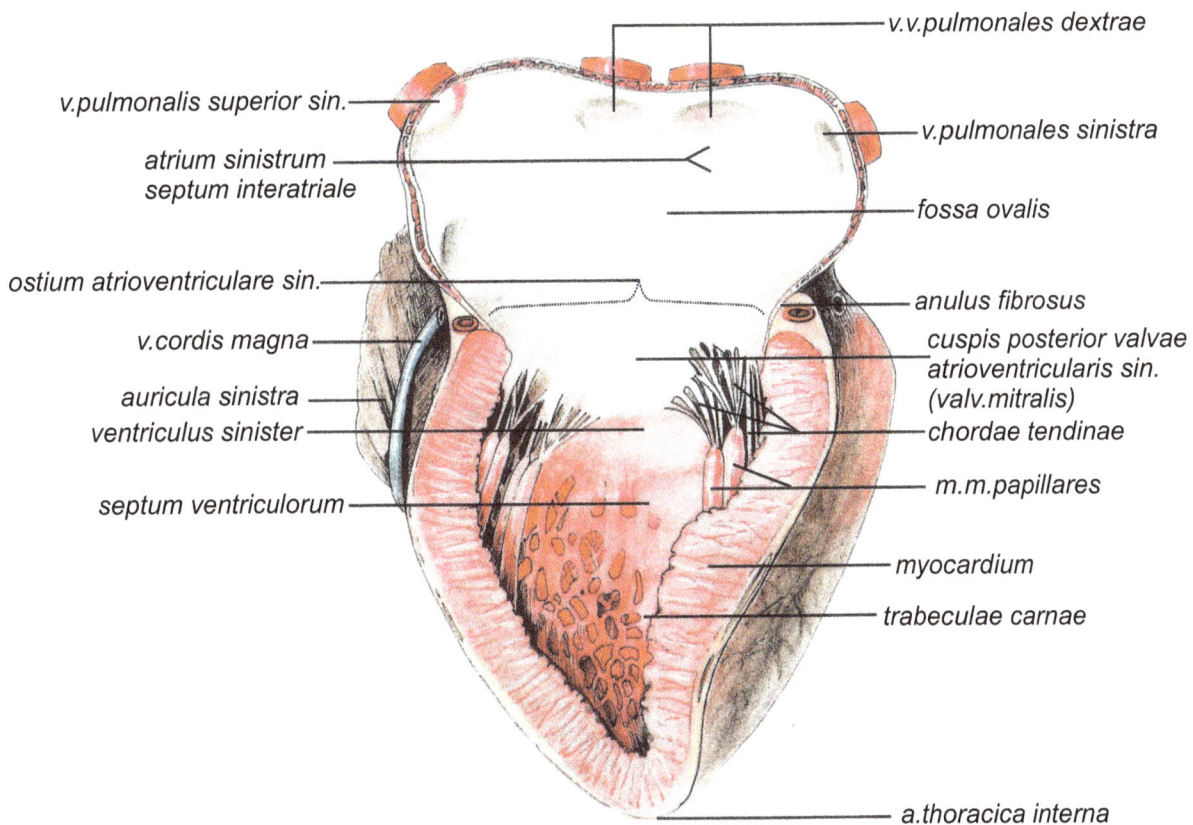

v.v.pulmonales dextrae

v.pulmonalis superior sin.

v.pulmonales sinistra

atrium sinistrum
septum interatriale

fossa ovalis

ostium atrioventriculare sin.

anulus fibrosus

v.cordis magna

cuspis posterior valvae
atrioventricularis sin.
(valv.mitralis)

auricula sinistra
ventriculus sinister

chordae tendinae

septum ventriculorum

m.m.papillares

myocardium

trabeculae carnae

a.thoracica interna

Fig. 101. Open left heart (inside).

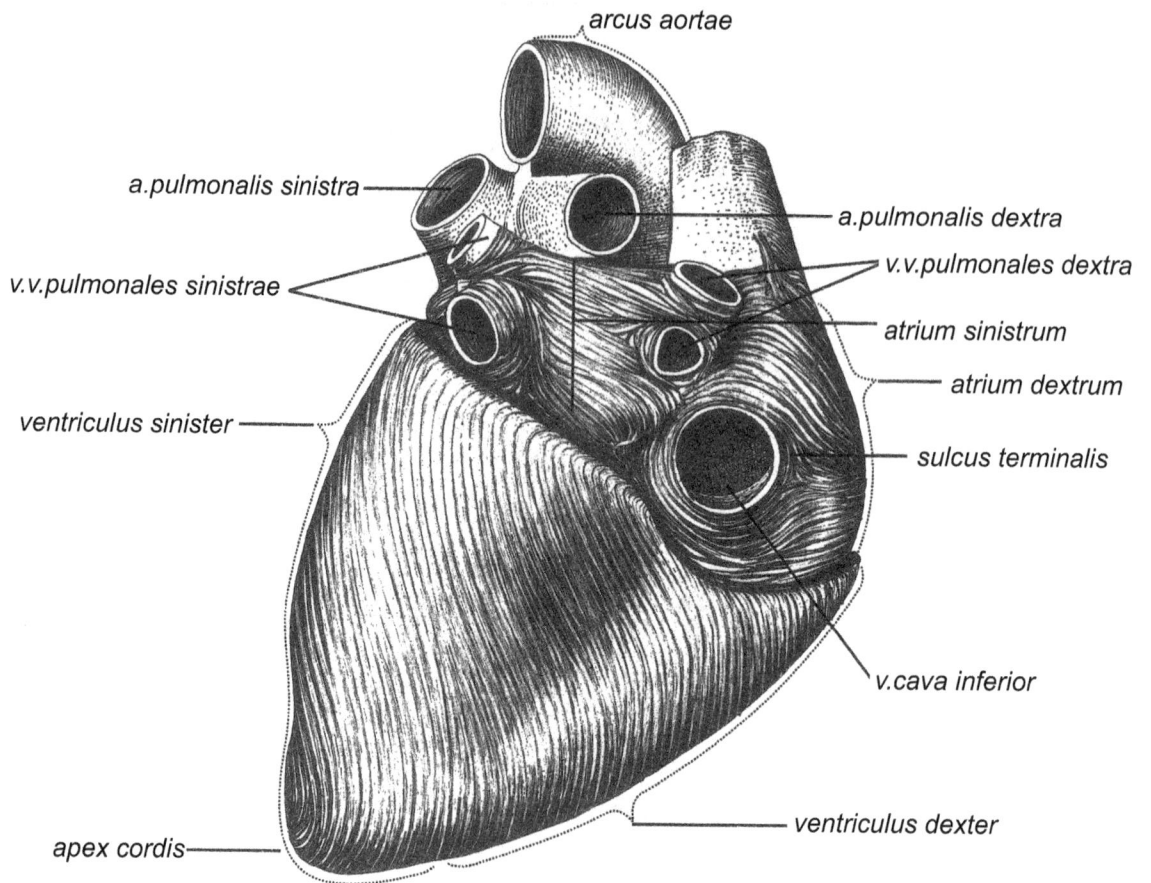

arcus aortae

a.pulmonalis sinistra

a.pulmonalis dextra

v.v.pulmonales sinistrae

v.v.pulmonales dextra

atrium sinistrum

atrium dextrum

ventriculus sinister

sulcus terminalis

v.cava inferior

ventriculus dexter

apex cordis

Fig. 102. The heart (structure of the myocardium).

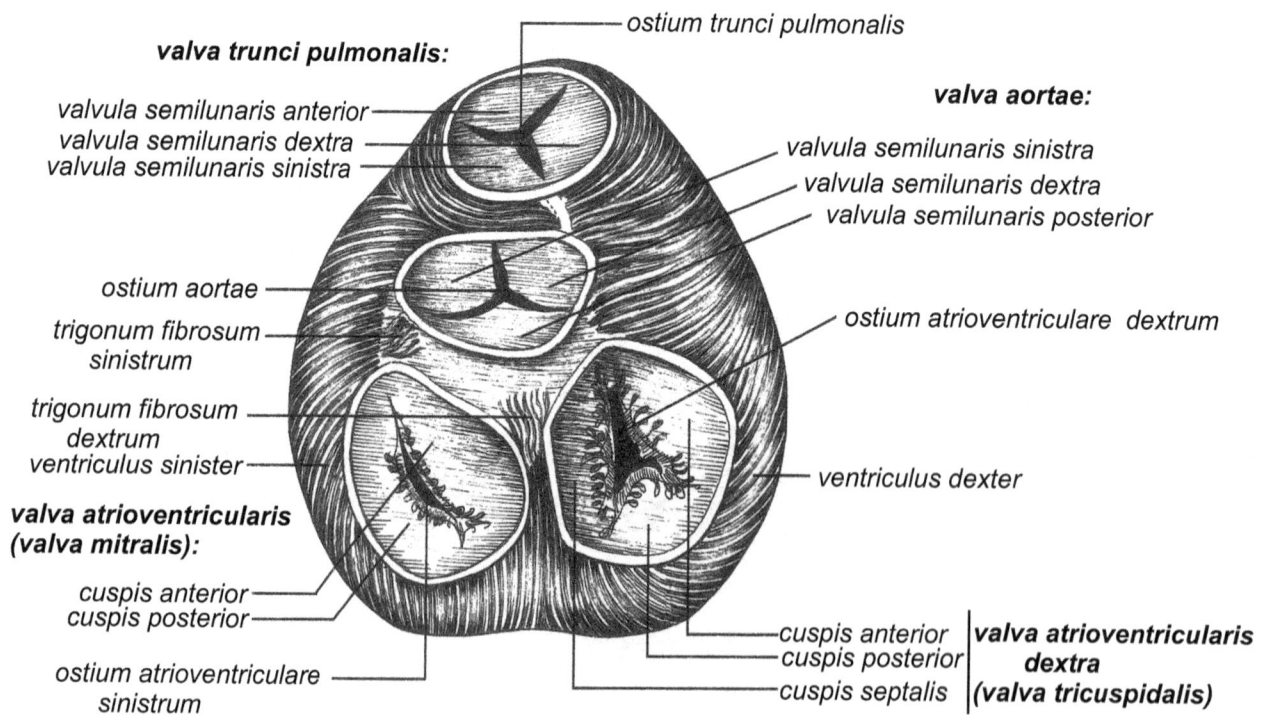

ostium trunci pulmonalis

valva trunci pulmonalis:

valva aortae:

valvula semilunaris anterior
valvula semilunaris dextra
valvula semilunaris sinistra

valvula semilunaris sinistra
valvula semilunaris dextra
valvula semilunaris posterior

ostium aortae

trigonum fibrosum sinistrum

ostium atrioventriculare dextrum

trigonum fibrosum dextrum
ventriculus sinister

ventriculus dexter

valva atrioventricularis (valva mitralis):

cuspis anterior
cuspis posterior

cuspis anterior
cuspis posterior
cuspis septalis

valva atrioventricularis dextra (valva tricuspidalis)

ostium atrioventriculare sinistrum

Fig. 103. The orifices of the heart.

108

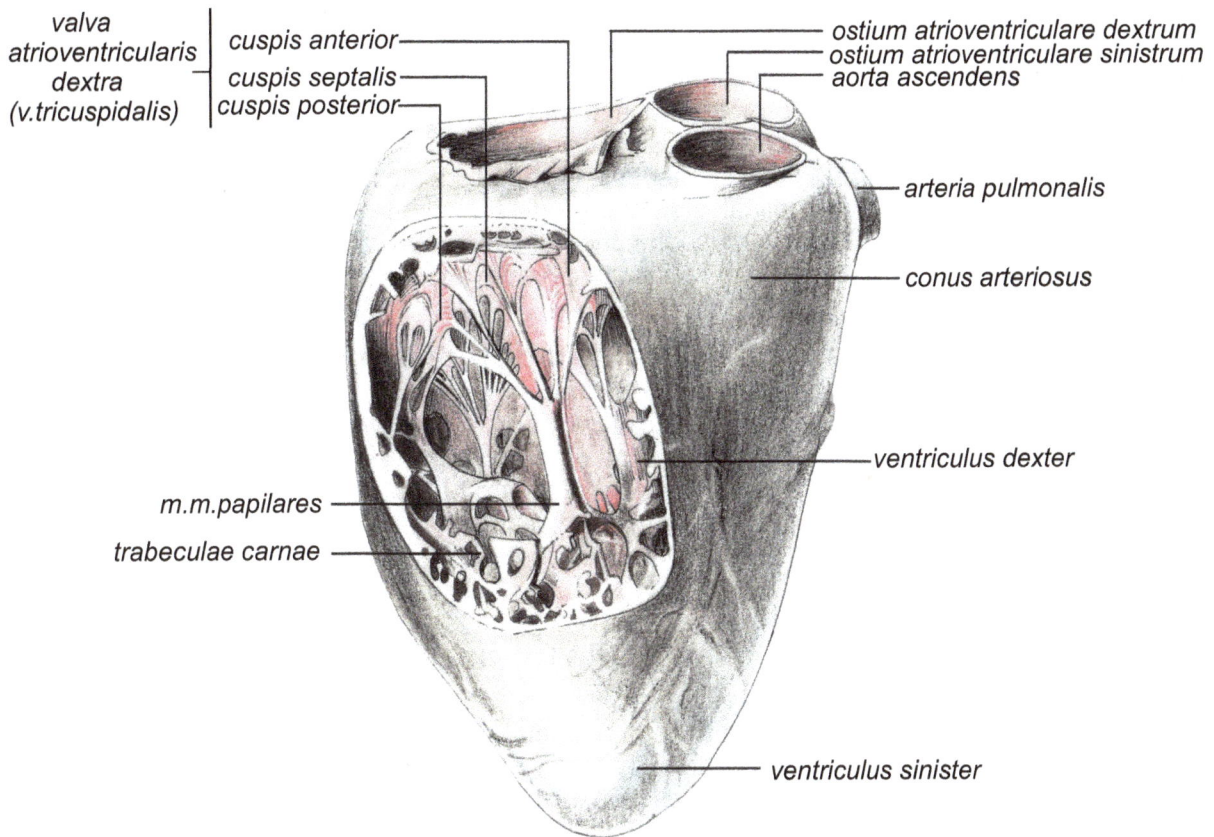

valva
atrioventricularis
dextra
(v.tricuspidalis)

cuspis anterior
cuspis septalis
cuspis posterior

ostium atrioventriculare dextrum
ostium atrioventriculare sinistrum
aorta ascendens

arteria pulmonalis

conus arteriosus

ventriculus dexter

m.m.papilares

trabeculae carnae

ventriculus sinister

Fig. 104. The right ventricle (inside) and the atrioventricular orifices after removal of the atria.

The base of the heart is formed by the posterior part of the atria and the afferent veins, respectively by the two venae cavae (superior and inferior), on the one hand, and the pulmonary veins, on the other.

The relations of the heart are the following (see also fig.92): *Anteriorly,* from the surface into the depth: with the sterno-costal plastron, to which it is attached by the sternopericardial ligaments and on which it can be projected on a zone called cardiac area; with the internal thoracic blood vessels and lymphatics; with the anterior costomediastinal pleural culs-de-sac which are nearly adjacent to each other mediastinally, from the second to the fourth intercostal space, forming the two triangles joined by their summits (superior-corresponds to the thymic recess; inferior-corresponds to the pericardium); with the anterior border of the lungs

Interiorly, it enters into a very wide relation with the diaphragm, to which it is fastened indirectly by means of the phrenicopericardial ligaments.

Through the diaphragm it is in relation with the vault (fornix) of the stomach and the left lobe of the liver.

The antero-inferior border *(margo acutus)* is in relation with the angle formed by the thoraco-abdominal wall with the anterior surface of the pericardial sac.

The right surface of the right atrium corresponds to the prehilar mediastinal surface of the right lung, anterior to the mesopneumonium, and to the right phrenic nerve, which descends on the fibrous pedicle, beneath the mediastinal pleura, along the venae cavae.

The left surface of the right atrium corresponds to the left T.mediastinal pleura, to the cardiac impression on the left lung, to the left phrenic nerve, which descends along the fibrous pericardium, penetrating into the diaphragm at the level of the heart apex, and to the left superior diaphragmatic vessels, more anterior than on the right

Posteriorly, the heart is in relation with the oesophagus, the vagus nerves, the pleural (interoesophago-pericardial, interaortico- oesophageal, interazygo-oesophageal) culs-de-sac, the descending aorta, the azygos and hemiazygos veins and the thoracic duct.

109

The relation of the left atrium with the oesophagus through Holler's transverse sinus explains the pain in the back, at the level of the cardiac vertebrae (on which project the constituent formations of the heart), as well as the discomfort in deglutition in the diseases of the left heart, especially in the case of a pericarditis with effusion. The dagger pushed by Brutus into Caesars chest, which caused his death, has penetrated through this vertebral zone.

In essence, the relations of the chambers of the heart are the following: the right atrium, with the sternocostal wall and the mediastinal pleura through the pericardium; the left atrium, with the bifurcation of the trachea and the oesophagus; the right ventricle, with the sternocostal wall and the costomediastinal sinus; the left ventricle, with the diaphragm and the left lung.

The internal Configuration of the Heart

The cone-shaped heart is divided into compartments, respectively into two atria and two ventricles; morphophysiologically it is made up of two compartments: the right heart and the left heart, resulted from the partition of the heart by septa (interventricular and interatrial septa).

The atria and ventricles communicate with each other through the atrioventricular orifices provided with a valvular apparatus, consisting of atrioventricular valves.

Each ventricle communicates with the corresponding artery through an arterial orifice furnished with three valves, the sigmoid valves.

The Septa of the Heart

The interventricular septum is triangular; anteriorly, its summit descends towards the apex of the heart, reaching its right side, so that the summit belongs to the left ventricle; its posterior base is continuous with the interatrial septum; its anterior and inferior borders join with the corresponding walls of the heart, along the interventricular grooves (fig 100,101,105,109). Its right, convex surface belongs to the right ventricle and its left, concave surface, to the left ventricle. The posterior part of the interventricular septum is sinuous, corresponding on the right to the tricuspid and pulmonary orifices and on the left, to the mitral and aortic orifices. The interatrial septum, thin and membranous, separates the two atria, forming their medial wall. Its right surface corresponds to the right atrium and presents on its postero-inferior part a depression; the oval fossa (fossa ovalis) (Bofallo's foramen), bordered upwards and forwards by a prominent arched margin, the limb of the oval fossa (limbus fossae ovalis) or Vieussen's ring.

The Ventricles

They are two pyramidal cavities, anterior and inferior to the atria and separated by the interventricular septum. On the walls of the ventricles, thicker than those of the atria, are present muscular ridges of the first, second and third order, termed trabeculae carneae.

The right ventricle.It has three walls, which present in their structure fleshy trabeculae under the form of columns in relief of the three above-mentioned orders, as well as the three papillary muscles (musculi papillares) (fig. 100,104,107,108).

Its base is traversed by two orifices, the atrioventricular and the arterial orifices, the latter being the origin of the pulmonary artery. The right atrioventricular (tricuspid) orifice is situated on the inferior part of the base of the right ventricle; through it the right ventricle communicates with the right atrium. This orifice is furnished with a tricuspid valve, made up of three triangular cusps, homologous with the three walls: anterior, inferior and medial. The pulmonary orifice is situated on the antero-superior part of the base and follows after the pulmonary infundibulum; the arterial cone, which forms the smooth portion of the ventricular walls [pars glabra), is situated above, in front and on the left of the tricuspid orifice and is furnished, like the aortic orifice, with three valves (in the shape of a swallow's nest), termed sigmoid valves(anterior, right posterolateral and left posterolateral). The left Ventricle. It has two walls, also presenting muscular columns of three orders, but they are much thicker than those of the right ventricle (fig. 101,102,107,109).

The atrioventricular orifice is situated on the lower part of the base and is furnished with two valves which have the aspect of an episcopal mitre, called therefore mitral valves.

The aortic orifice is provided with three sigmoid valves resembling those of the pulmonary artery, but more resistant and with a reverse arrangement: one posterior, one left anterolateral and the third right anterolateral.

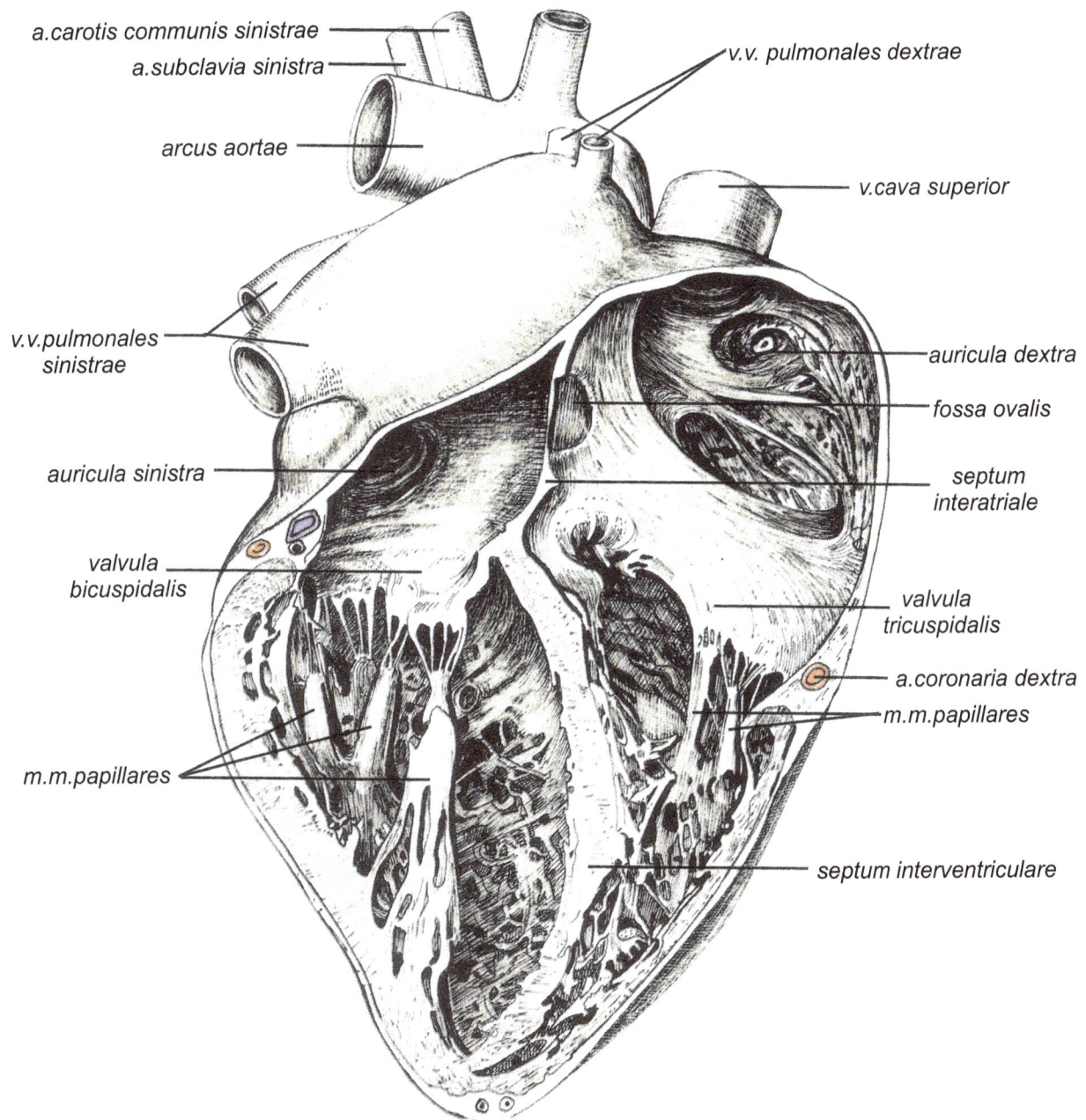

Fig. 105. The heart (inside).

a.carotis communis sinistrae

a.subclavia sinistra

arcus aortae

v.v.pulmonales sinistrae

auricula sinistra

valvula bicuspidalis

m.m.papillares

v.v. pulmonales dextrae

v.cava superior

auricula dextra

fossa ovalis

septum interatriale

valvula tricuspidalis

a.coronaria dextra

m.m.papillares

septum interventriculare

111

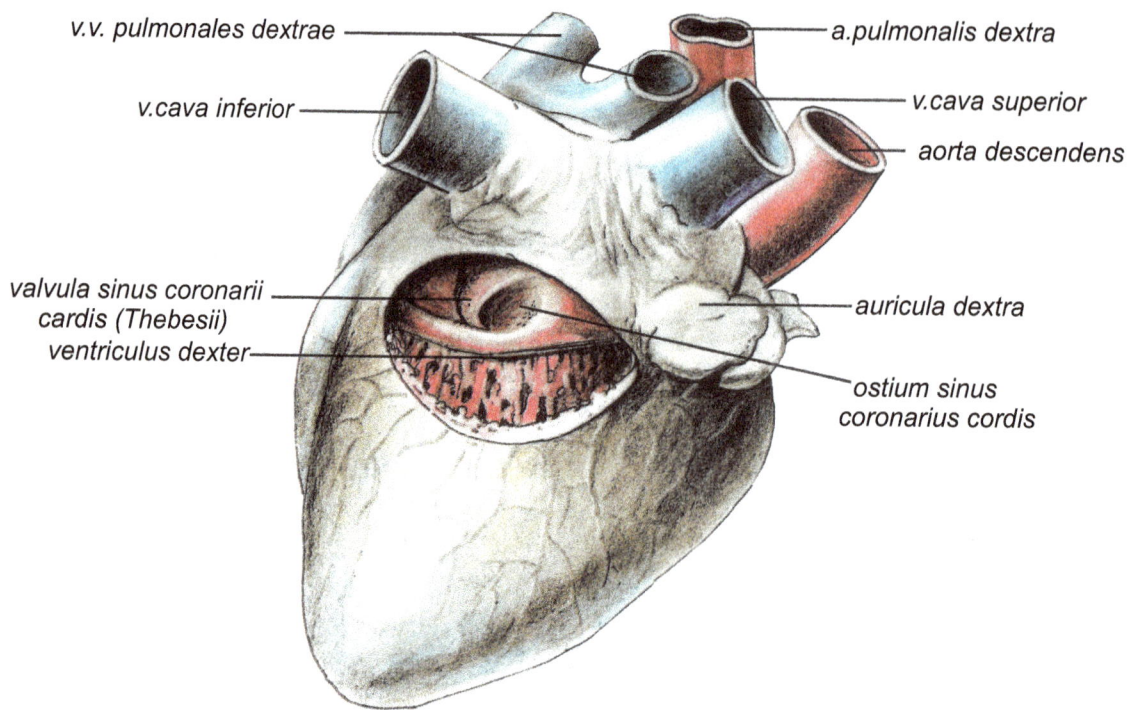

Fig. 106. The great vessels at the base of the heart; open right atrium and ventricle

v.v. pulmonales dextrae

a.pulmonalis dextra

v.cava inferior

v.cava superior

aorta descendens

valvula sinus coronarii
cardis (Thebesii)

ventriculus dexter

auricula dextra

ostium sinus
coronarius cordis

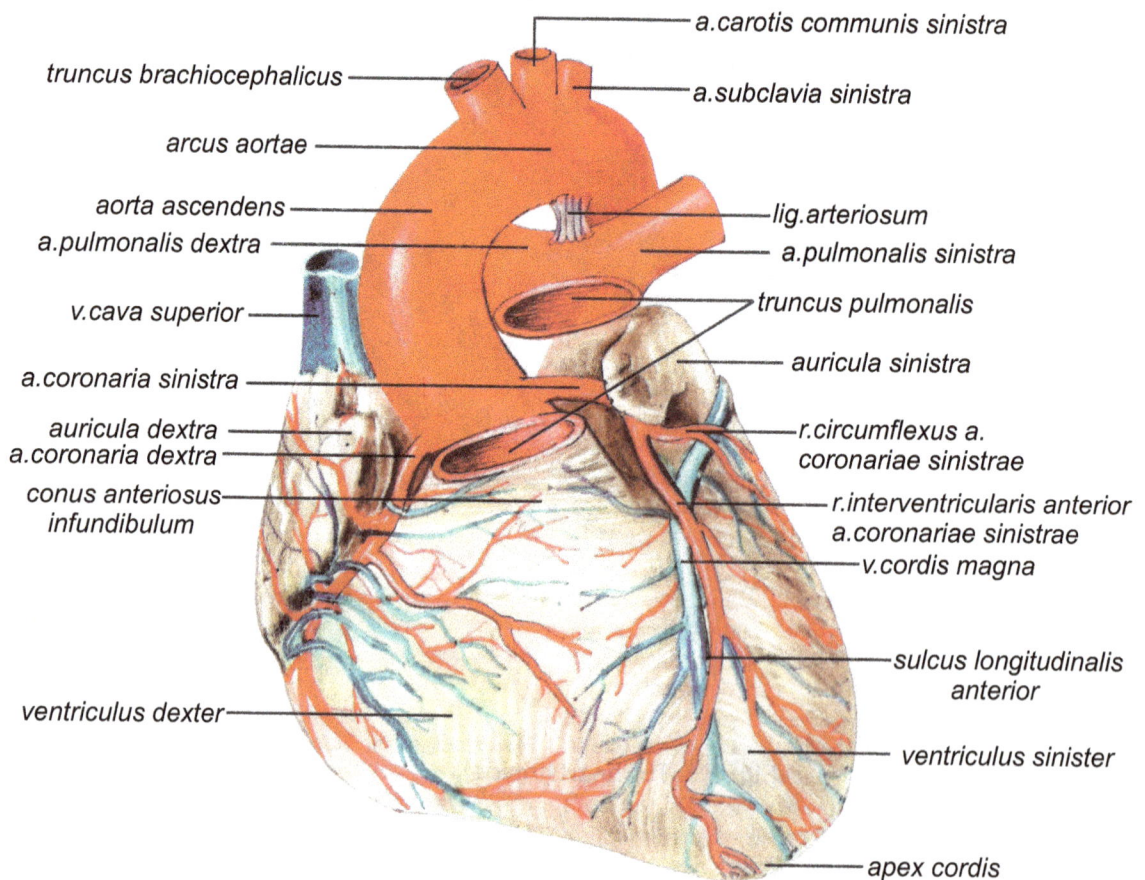

a.carotis communis sinistra

truncus brachiocephalicus

a.subclavia sinistra

arcus aortae

aorta ascendens

lig.arteriosum

a.pulmonalis dextra

a.pulmonalis sinistra

v.cava superior

truncus pulmonalis

a.coronaria sinistra

auricula sinistra

auricula dextra

a.coronaria dextra

r.circumflexus a.
coronariae sinistrae

conus anteriosus
infundibulum

r.interventricularis anterior
a.coronariae sinistrae

v.cordis magna

sulcus longitudinalis
anterior

ventriculus dexter

ventriculus sinister

apex cordis

Fig. 107. Blood vessel supply of the heart; sectioned pulmonary artery (anterior view).

The Atria

They are separated by the interatrial septum; their walls, which confer them a somewhat cuboid form, are thinner than those of the ventricles (fig. 100-103,108,109).

The right atrium . It has the following characters:
- on the anterior wall are present muscular columns termed pectinate muscles;
- the medial wall forms the right surface of the interatrial septum;
- on the upper wall is situated the opening orifice of the superior vena cava;
- on the lower wall lie: the orifice of the inferior vena cava furnished with a valve (the Eustachian valve) and the orifice of the coronary sinus, situated more anteriorly and more internally, also provided with a valve (the Thebesian valve) on the lower anterior wall is situated the right atrioventricular or tricuspid orifice;
- at the right margin of the orifice of the two venae cavae is situated the terminal crest (crista terminalis), which corresponds to His's terminal groove (sulcus terminalis).

The left atrum. It has also a cuboid shape with rounded edges; on its posterior wall lies the opening of the four orifices of the pulmonary veins and on the anterior wall, the mitral orifice.

Structure of the Heart

The heart, enveloped in a fibroserous coat, the pericardium, is formed by three coats:
an external (serous) coat: the epicardium;
- a middle coat: the myocardium and
- an internal coat: the endocardium.
The external serous coat corresponds to the visceral serous pericardium.

The Myocardium

If is a thick muscular coat, consisting of striated cardiac muscle fibres, proper to each cavity, and of fibres shared by the two ventricles (fig. 101,102). its fibres are inserted into a fibrous apparatus, located at the base of the heart, between the atria and the ventricles, and forming the cardiac skeleton (plane of the valves). The cardiac skeleton consists of four fibrous rings, forming the outline of the four orifices of the heart and having the same shape and orientation. At the place where the fibrous rings of the ventricular orifices meet the fibrous ring of the aortic valvular apparatus arise two fibrous triangles (trigone fibrosa), the right triangle corresponding to the nodal point of the heart. The upper membranous part of the interventricular septum is an integral part of the fibrous skeleton of the heart (fig. 103).

The Endocardium

It lines the cavities of the heart and is continuous with the endothelium of arteries (fig. 105). The valves with their fibrous skeleton are also covered with the endocardial leaflet. The inside of the heart is differently structured on the right and on the left; normally, the two halves of the heart are completely separated by the interatrial and, respectively, the interventricular septum. In the right atrium, the two venae cavae open into the venous sinus; their streams laminate, separating them from each other in the following way: the stream of the inferior vena cava is directed by its valve (the Eustachian valve) medially, towards the interatrial septum, whereas the stream of the superior vena cava is directed by the lateral intervenous torus (Lowers tubercle) (tuberculum intervenosum Loweri), giving rise to a vortex towards the right ventricle. On the septal wall are also located the valve of the coronary sinus (Thebesian valve), the oval fossa and the limb of the oval fossa, the smooth septal wall being continuous with the areolar wall of the right auricle. The blood stream, entered into the ventricle through the tricuspid valve, runs in an angular direction towards the arterial cone and penetrates during the ventricular systole into the pulmonary artery, opening its semilunar valves, while the tricuspid valve closes. The blood stream in the left branch of the pulmonary artery is less deviated than that in the arterial trunk; it flows at first through this branch and then, slower, through the right branch. The cardiac orifices (ostia) may be distinctly seen on the anatomical preparations on which the atria have been separated from the ventricles.

The atrioventricular orifices lie one beside the other, whereas the arterial orifices are situated one anteriorly to the other. On the same anatomical preparation may also be seen the fibrous skeleton, which is located between the atria and the ventricles and is disposed among their muscular fibres, no other connection existing between them than the neuromuscular one through the bundle of His, as well as the two - right and left - fibrous trigones of the skeleton, which separate the ventricular fibrous rings from the aortic ring, at its origin. The two orifices of the right ventricle, atrioventricular *(tricuspid)* and pulmonary, are separated inside the heart by the supraventricular crest *(crista supraventricularis)*, conferring to the two ventricular compartments (descending, *pars ventricularis,* and ascending, *pars infundibularis)* on this side a U-shaped aspect, in comparison with the Jeff ventricle, where the separation is achieved by the ventral cusp *(cuspis aorticus)* of the bicuspid valve and therefore the blood stream in the left ventricle is V-shaped.

The cusps of the tricuspid valve are: anterior, posterior and medial (septal) and those of the bicuspid valve: dorsal (parietal) and ventral (aortic). Owing to the oblique position of the heart, the location of the orifices corresponds to an oblique plane, perpendicular on its longitudinal axis.

The cardionector (conducting)system of the heart represents a neuromuscular structure, which assures the generation of the intermittent contraction impulse and its transmission from the atria into the ventricles, it begins in the sinuatrial node (Keith and Flack node), located in the terminal groove *(sulcus terminalis),* in its upper part, between the venous sinus, at the emptying site of the superior vena cava, and the right atrium, below the epicardium. The excitation originating in this node is conducted into the atrial musculature, passing to the atrioventricular (Aschoff and Tawara) node, situated in the interatrial wall, below the limb of the oval fossa and anteriorly to the orifice of the coronary sinus. From this node starts the atrioventricular (His) bundle, which passes through the fibrous skeleton, located between the atrial and the ventricular musculature, under the form of a common trunk *(crus commune),* on the posterior border of the membranous septum, and then divides into a thin right branch *(crus dexter),* to the ventricle on the same side, and a broader, ramified left branch *(crus sinister, ramus anterior* and *ramus posterior),* to the left ventricle. Hence, the left arm of the bundle is more developed than the right arm, development conditioned by the stronger musculature of the left ventricle owing to the wider distribution territory of the arterial system of the aorta (fig. 109).

The Keith and Flack node is supplied by an anterior atrial artery, a branch of one of the two coronary arteries (most frequently the right branch).

The Aschoff-Tawara node and the trunk of the bundle of His are supplied by the first posterior septal artery, a branch of the right coronary artery (less frequently, of the left branch).

The right branch of the bundle of His is supplied by the second anterior septal artery (left coronary); the left branch is supplied by the anterior and posterior septal arteries.

This blood supply explains the severity of septal infarctions (rhythm disturbances: atrioventricular block, bundle-branch block).

Vessel and nerve supply of the heart

Arteries of the heart. The arteries supplying the heart are the coronary arteries, two in number, right and left (fig. 107,108).

The left or anterior coronary artery (a. coronaris sinistra) arises from the aorta, immediately above the middle portion of the left sigmoid valve; the ostium of origin of the coronary arteries is more or less covered by the sigmoid valve during the systole; the penetration of blood info them is maximum during the diastole. Covered by the serous pericardium and surrounded by fat, it passes through the depression which separates the pulmonary artery, situated medially, from the left auricle, situated laterally, reaches the anterior interventricular groove and descends in if up to the apex of the heart, which it surrounds. It ends in the posterior interventricular groove at a few centimeters away from the apex of the heart.

*The right coronary artery (a. coronaris dextra)*arises at the level of the right sigmoid valve, runs between the pulmonary artery, situated medially, and the right auricle, situated laterally, reaches the left atrioventricular groove, the inferior surface of the heart, and penetrates info the posterior interventricular groove. If ends in this groove, where if anastomoses, in the neighbourhood of the apex, with the left coronary artery.

Anastomoses are present in 97% of cases, but they are most often insufficient, the circulation being of the terminal type, which explains the severity of infarctions.

114

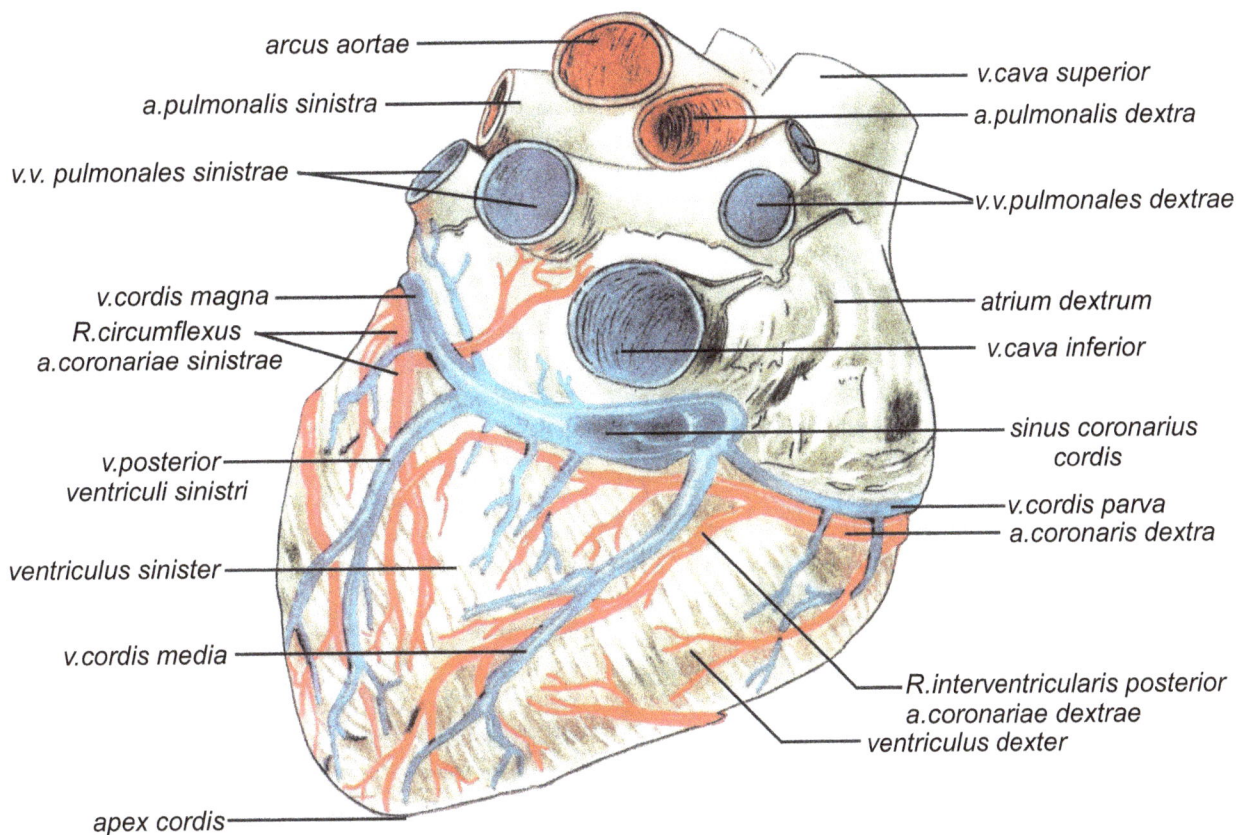

arcus aortae

a.pulmonalis sinistra

v.v. pulmonales sinistrae

v.cordis magna

R.circumflexus
a.coronariae sinistrae

v.posterior
ventriculi sinistri

ventriculus sinister

v.cordis media

apex cordis

v.cava superior

a.pulmonalis dextra

v.v.pulmonales dextrae

atrium dextrum

v.cava inferior

sinus coronarius
cordis

v.cordis parva
a.coronaris dextra

R.interventricularis posterior
a.coronariae dextrae
ventriculus dexter

Fig. 108. Blood vessel supply of the heart (posterior view).

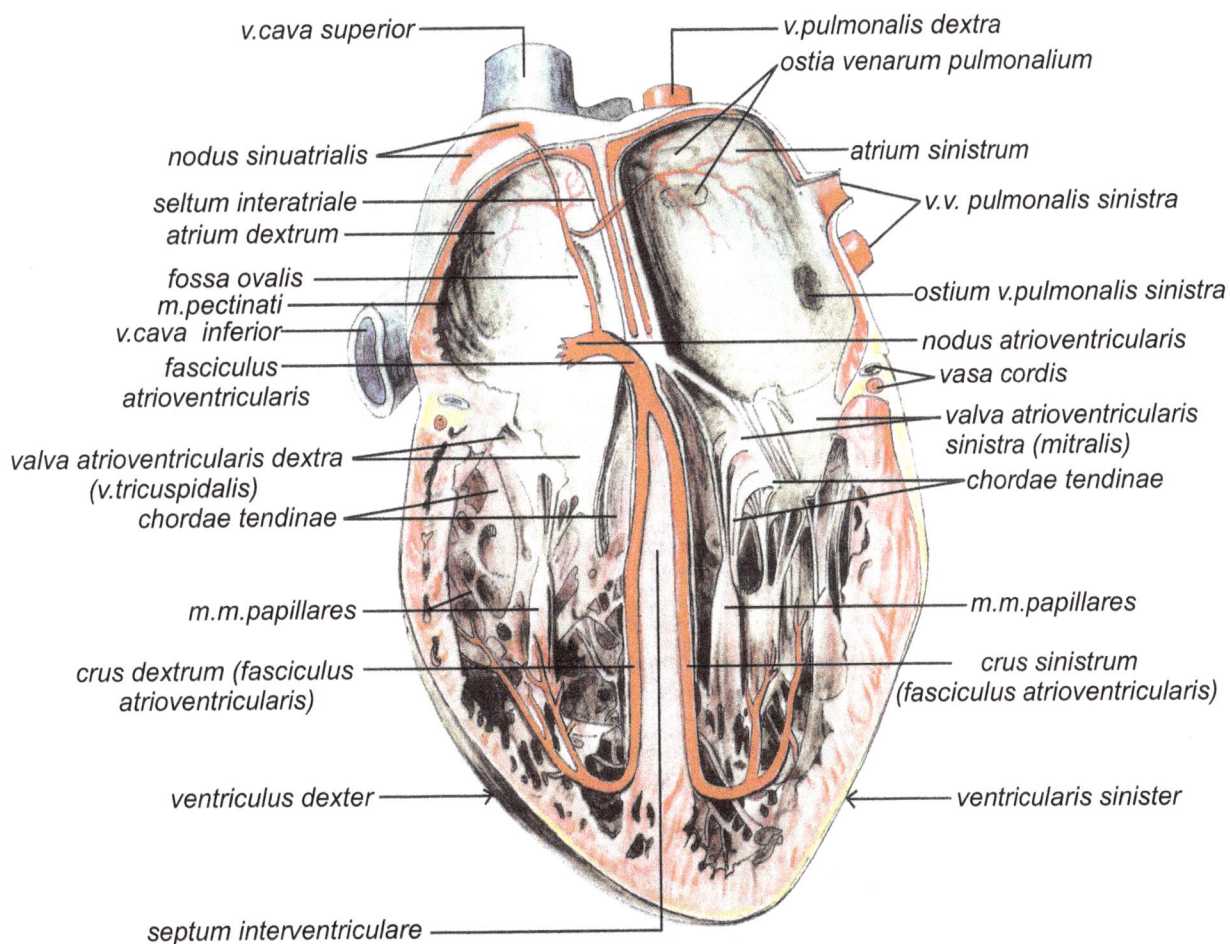

v.cava superior

nodus sinuatrialis

seltum interatriale

atrium dextrum

fossa ovalis

m.pectinati

v.cava inferior

fasciculus
atrioventricularis

valva atrioventricularis dextra
(v.tricuspidalis)

chordae tendinae

m.m.papillares

crus dextrum (fasciculus
atrioventricularis)

ventriculus dexter

septum interventriculare

v.pulmonalis dextra

ostia venarum pulmonalium

atrium sinistrum

v.v. pulmonalis sinistra

ostium v.pulmonalis sinistra

nodus atrioventricularis

vasa cordis

valva atrioventricularis
sinistra (mitralis)

chordae tendinae

m.m.papillares

crus sinistrum
(fasciculus atrioventricularis)

ventricularis sinister

Fig. 109. Autonomic nerve supply of the heart.

115

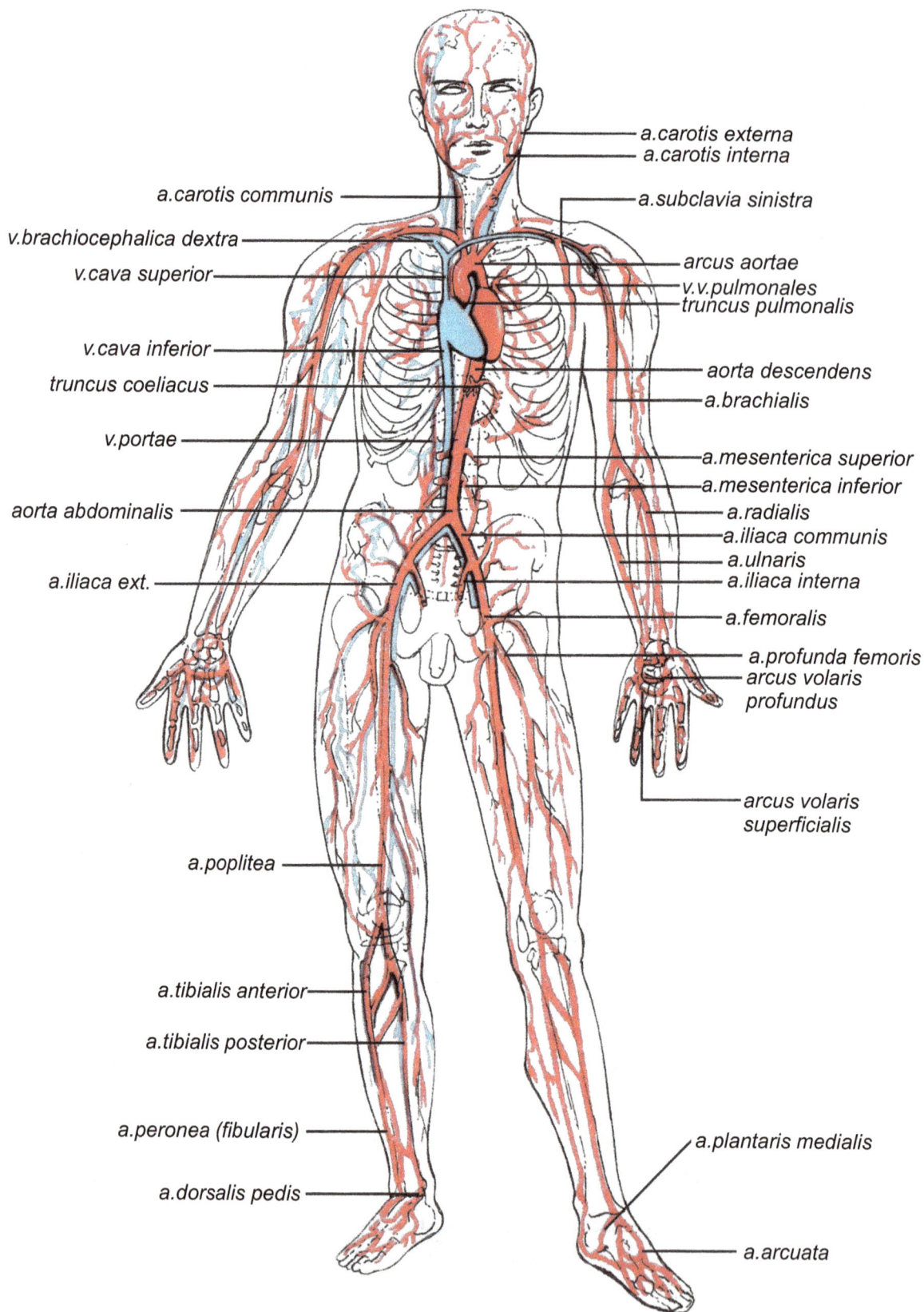

a.carotis externa
a.carotis interna
a.carotis communis
a.subclavia sinistra
v.brachiocephalica dextra
v.cava superior
arcus aortae
v.v.pulmonales
truncus pulmonalis
v.cava inferior
truncus coeliacus
aorta descendens
a.brachialis
v.portae
a.mesenterica superior
a.mesenterica inferior
aorta abdominalis
a.radialis
a.iliaca communis
a.ulnaris
a.iliaca interna
a.iliaca ext.
a.femoralis
a.profunda femoris
arcus volaris
profundus
arcus volaris
superficialis
a.poplitea
a.tibialis anterior
a.tibialis posterior
a.peronea (fibularis)
a.plantaris medialis
a.dorsalis pedis
a.arcuata

Fig .110. Blood circulation scheme.

116

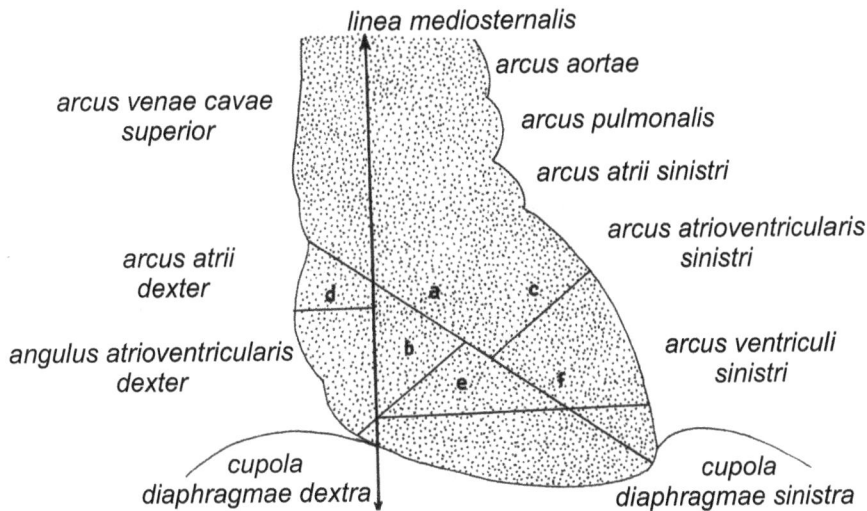

Fig. 111. Orthodiagram of the heart.

The veins of the heart are characterized by the presence of a large collecting system which drains most cardiac veins, the coronary sinus, to which, however, the small cardiac (thebesian) veins are not tributary.

The system of the coronary sinus. It is formed by the great coronary vein, which opens directly into the coronary sinus, and by the oblique vein of the left atrium (Marshall). The sinus is endowed with two valves: one at the opening of the great coronary vein, the Vieussens' valve, and the other at its emptying into the right atrium, the thebesian valve.

The small cardiac veins arise from the anterior and right sides of the right ventricle; they open info the upper wall of the right atrium through orifices *(foraminae);* the most important is the right marginal or Galien's vein.

The thebesian veins are small parietal veins, which open info the cardiac cavities through Vieussens' pores or foraminula.

The lymphatics of the heart consist of a subepicardial network, which receives the lymphatics of the myocardial and endocardial networks through periarterial valvulated collectors.

The subpericardial network drains through two main collecting trunks, left and right.

The left collector drains the left side of the network, ascends along the left surface, then posteriorly to the pulmonary artery and ends in the intertracheobronchial lymph nodes, which drain into the right laterotracheal chain.

The right collector drains the right side of the network, the atrio-ventricular groove, then reaches the anterior surface of the aorta and ends in a lymph node of the left anterior mediastinal (precarotid) chain.

Thus, the drainage territories are "cross".

The nerves of the heart derive from the cardiac plexus, made up of branches of the sympathetic trunk and of the vagus nerve.

The branches of the sympathetic trunk, three in number on each side, superior, middle and inferior, arise from the corresponding cervical ganglia.

The branches of the vagus, also in number of three, superior, middle and inferior, arise as follows: the superior branch directly from the vagus nerve; the middle branch from the recurrent laryngeal nerve and the inferior branch also from the vagus, from the lower zone of the recurrent nerve. The cardiac plexus is formed by the anastomosis of these sympathetic and vagal branches around the great vessels at the base of the heart. It consists of an anterior arterial plexus and of a posterior venous plexus.

The Posterior Mediastinum

The boundary between the anterior and the posterior mediastinum is represented by a conventional pretracheal plane, so that the trachea belongs to the posterior mediastinum. It contains also the oesophagus with the vagus nerves, the right vagus on its right and posterior side and the left vagus on its left and anterior side.

Behind the oesophagus lies the thoracic duct (ductus thoracicus), on its left side and then, behind, the descending aorta, on the right the azygos veins and, on either side, the thoracic orthosympathetic ganglionic trunk with the respective branches to the thoracic or abdominal organs (splanchnic nerves) (fig. 113).

The Thoracic Trachea

It consists of 16-20 tracheal rings (cartilagines tracheoles), incomplete on their dorsal (horseshoe-shaped) part which maintain its its lumen open and which are connected in the length by elastic ring-shaped ligaments (ligamenta annularia). The posterior surface of the tracheal tube is completed by the tracheal membrane (membrana trachealis), consisting of transversally arranged smooth muscle fibres. The internal surface is lined with a respiratory mucosa, rich in glands, and the external surface is covered by an adventitia, which fastens it to the oesophagus by fibromuscular tracts. It is situated slightly towards the right, whereas the oesophagus passes slightly towards the left, behind the trachea.

In its inferior part, the bifurcation of the trachea has relations anteriorly with the bifurcation of the pulmonary artery and the right branch of the pulmonary artery, since the bifurcation of the pulmonary artery is situated at 0.5-1 cm on the left of the bifurcation of the trachea.

Anteriorly to the bifurcation of the pulmonary artery lies the arch of the aorta, which crosses the left anterolateral part of the trachea, on which it forms an area termed aortic depression. Between the trachea and the aorta lies the posterior cardiac plexus(fig. 113,114).

Above the aorta, the anterior surface of the trachea is in relation, from the right to the left, with the arterial brachiocephalic trunk, the left common carotid artery and the mediastinal ganglionic chain. Between the brachiocephalic trunk and the left common carotid artery runs the artera thyreoidea ima (Neubauer), when if exists.

These arterial trunks are crossed anteriorly by the left venous brachiocephalic trunk. The superior vena cava descends anteriorly to the right border of the trachea, from which it is separated by lymph nodes. In the adult, anteriorly to the left venous brachiocephalic trunk are the thymic remains.

The posterior surface of the thoracic trachea lies anteriorly to the oesophagus. Laterally, the relations are different on the right side in comparison with the left side.

On the left, the thoracic trachea is in relation with: the arch of the aorta; the left common carotid artery, the left vagus nerve, the left subclavian artery, the thoracic duct, the left recurrent laryngeal nerve, the left sympathetic ganglionic chain and the left mediastinal pleura.

On the right if is in relation with: the arch of the azygos vein, the arterial brachiocephalic trunk, the right vagus nerve, lymph nodes and the right mediastinal pleura.

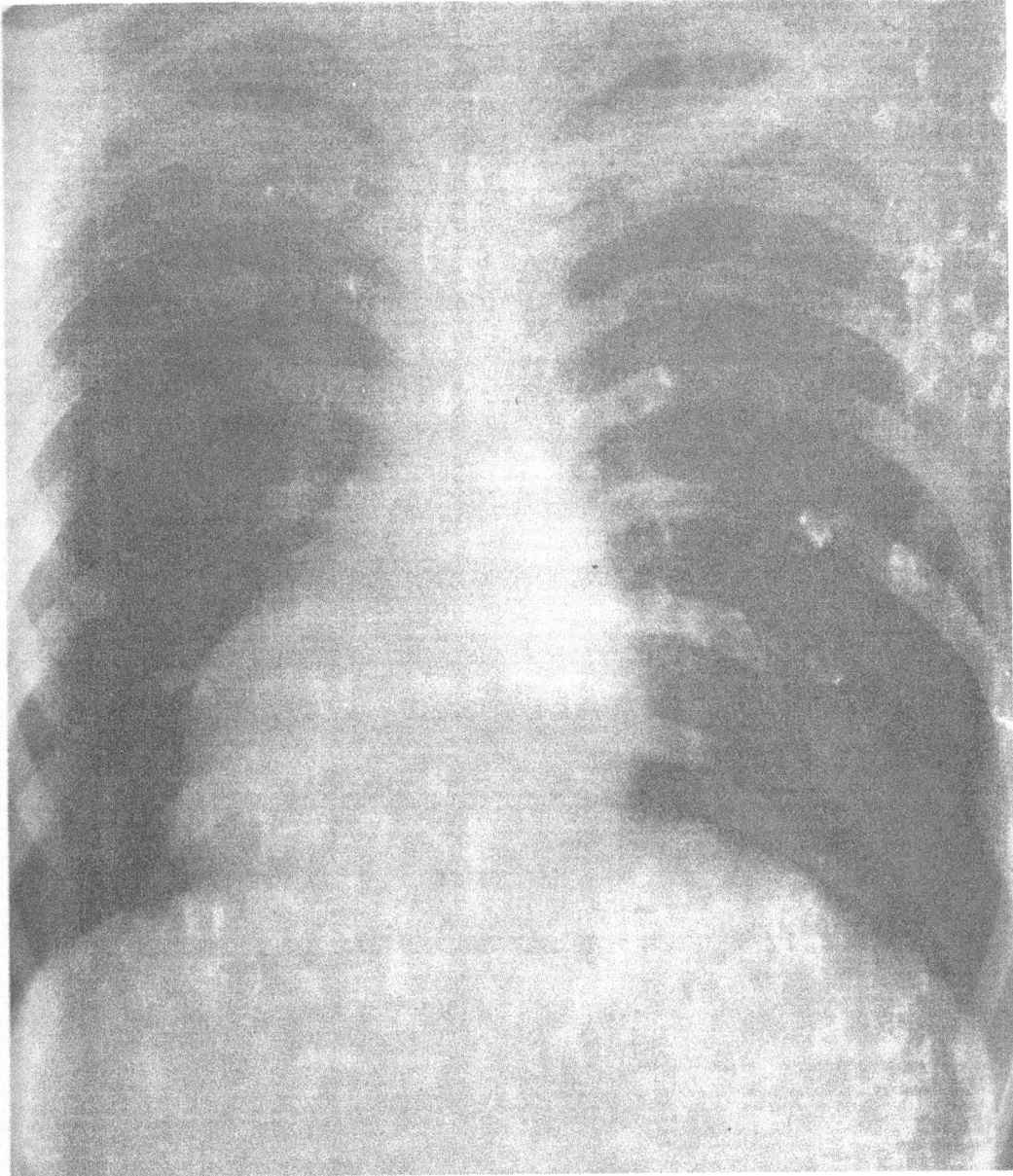

Fig. 112. Radiography of the heart.

The Thoracic Oesophagus

It crosses, in its lower part, the descending aorta, passing to the left of the median line and then towards the cardia, which projects to the left of the eleventh thoracic vertebra. The oesophagus has a total length of 22-25 cm and, if we add the distance of 15 cm from the rima oris to the mouth of the oesophagus, situated at the level of Chassaigriac's carotid tubercle (C6 vertebra), we find the minimum distance of 40-45 cm which an oesophageal catheter (Faucher) has to cover to enter the stomach (fig. 113-114).

The caliber of the oesophagus is almost uniform, presenting on the whole the following narrowings: at the mouth of the oesophagus (cricoid narrowing), at the arch of the aorta (aortic narrowing), at the left bronchus (bronchial narrowing) and at the cardia (cardiac narrowing), where sometimes a spasm may be present due to the hypertrophy of the musculature, the section of this sphincter (Heller's cardiotomy) being recommended in this case. Sometimes, a narrowing at the level of the diaphragm (diaphragmatic narrowing) can also exist, the orifice of the diaphragm being projected at the level of the tenth thoracic vertebra. The oesophagus presents in its upper part, continuous with the pharynx, striated muscle fibres and in its lower part, smooth muscles.

The oesophageal mucosa has longitudinal folds and is mobile on the muscular coat through the submucosa.

119

Vessel and nerve supply

The arteries of the thoracic oesophagus derive from the neighbouring branches (inferior thyroid, bronchial and superior phrenic arteries).

The veins form submucous plexuses which, in the inferior part, unite with the gastric veins, entering in the territory of the hepatic portal vein, and in the superior part, with the azygos vein, thus entering in the territory of the superior vena cava. The veins situated in the distal portion of the oesophagus become varicose in cases of compression of the trunk of the portal vein (cirrhoses, liver tumours), being the source of upper digestive haemorrhages.

The lymph vessels of the thoracic oesophagus drain into the lymph nodes scattered from the cardia up to the supraclavicular group, explaining the lymph node metastases which may involve even the Troisier-Virchow's supraclavicular lymph node.

The vagus nerves approach the oesophagus at the level of the bifurcation of the trachea. Up to here the oesophagus receives branches of the recurrent nerves and of the orthosympathetic cervicothoracic chain.

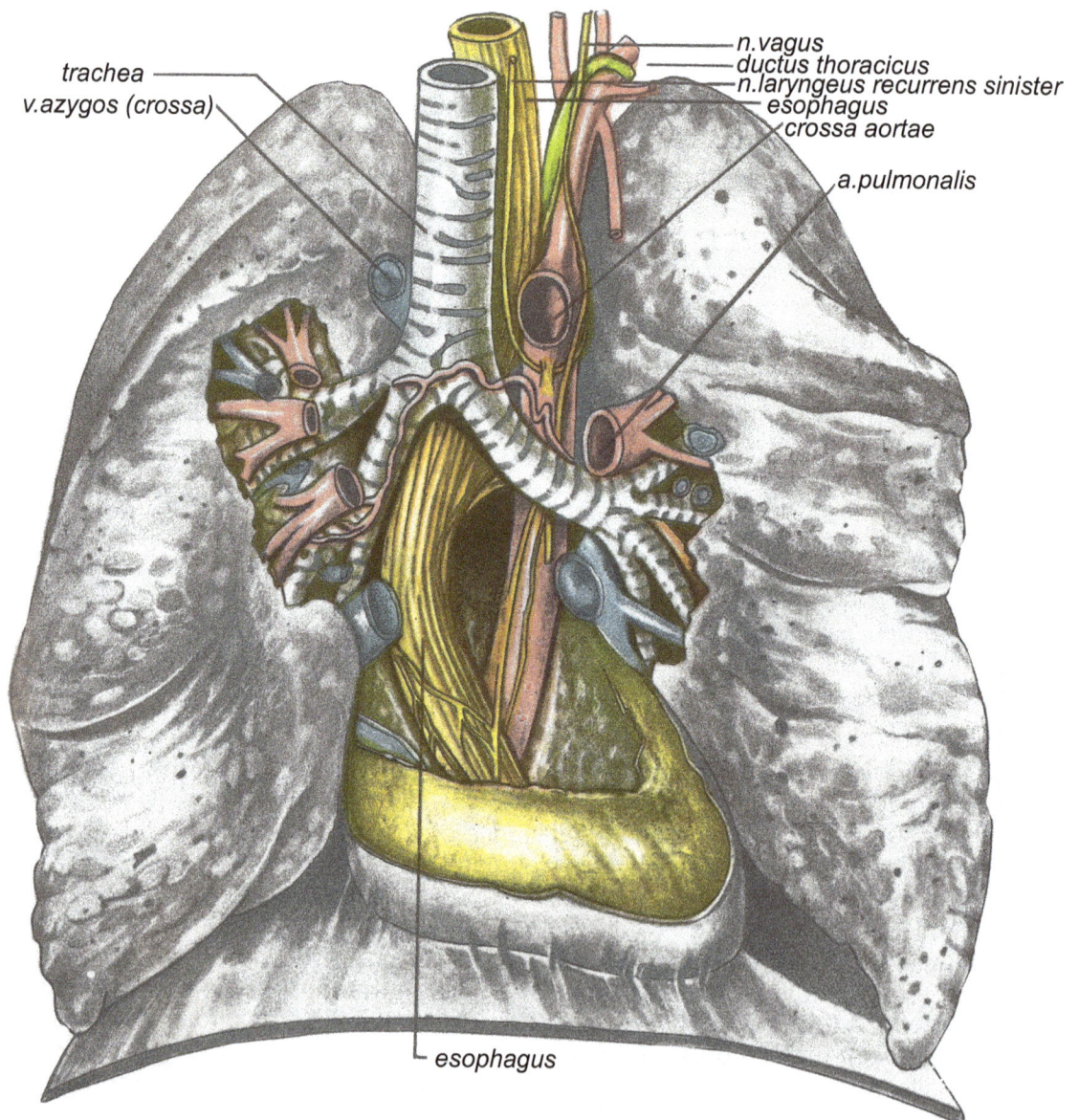

Fig. 113. Region of the tracheal bifurcation after removal of the heart (relations).

The Thoracis Duct
(Ductus thoracicus)

The thoracic duct is the largest lymph trunk of the organism. It collects the lymph from the subdiaphragmatic zone, except a part of the liver and the supraumbilical lymph vessels of the abdominal wall. It receives also the lymph from the level of the posterolateral wall of the thorax and from a part of the base of the neck (fig. 115).

It results from the junction of two lumbar lymph trunks, at the level of the first two lumbar vertebrae or of the last two thoracic-vertebrae, hence it originates either in the abdomen or in the thorax. Classically it is considered that the thoracic duct begins with a dilated segment, called Pecquet's cistern *(cisterna chyli).* However, this segment exists only if the thoracic duct has an abdominal origin and if the intestinal and lumbar lymph trunks open directly at the level of its lower extremity.

From its origin the thoracic duct ascends the right side of the aorta, reaching the base of the neck, where it describes a concave curve downwards and forwards (the arch of the thoracic duct), then it courses forwards and to the left up to the juguloclavian confluence, where it ends by opening either in this place or into the internal jugular vein or into the left subclavian vein.

The left jugular and subclavian lymph trunks empty into the duct anteriorly to its opening or directly into the venous angle.

The relations of the thoracic duct are the following: In the abdomen, if it has an abdominal origin, if is situated behind the right flank of the aorta and medially to the right crus of the diaphragm *(crus mediale).*

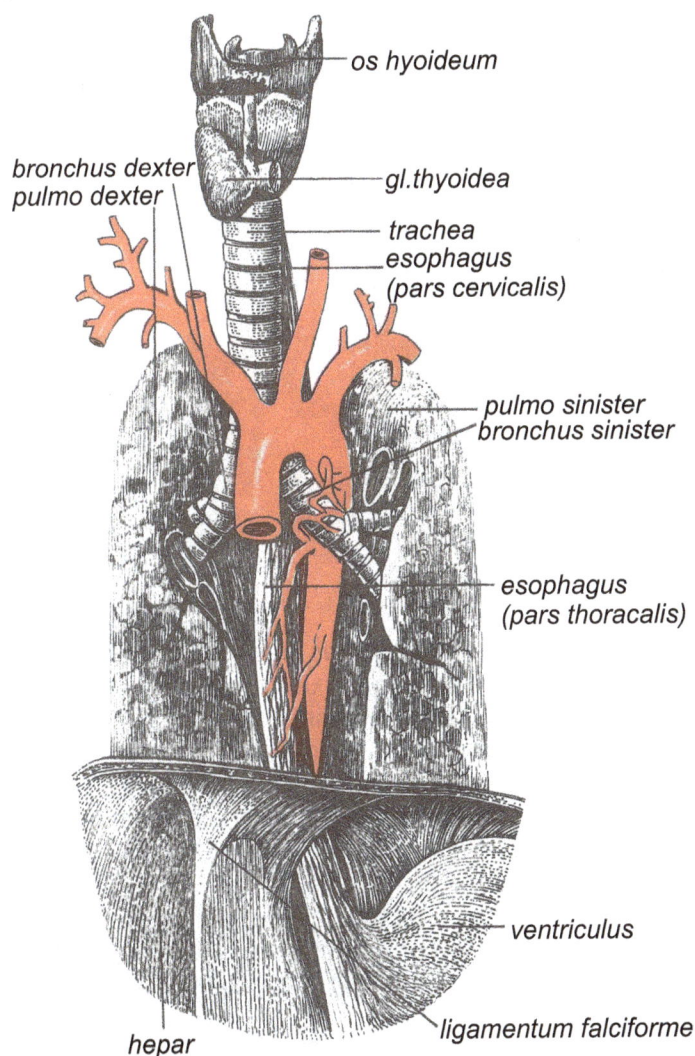

os hyoideum

bronchus dexter
pulmo dexter

gl.thyoidea

trachea
esophagus
(pars cervicalis)

pulmo sinister
bronchus sinister

esophagus
(pars thoracalis)

ventriculus

hepar

ligamentum falciforme

**Fig. 114. Aorta, trachea and oesophagus
(relations).**

In the thorax we distinguish an interazygo-aortic and a supraaortic segment. In the interazygo - aortic segment the thoracic duct is situated on the right side of the arch of the aorta; above, if crosses the internal surface of the arch of the aorta.

On the right of the thoracic duct courses the azygos vein and, behind, the duct is in relation with the right intercostal arteries.

In the supraaortic segment, the thoracic duct ascends the posteromedial surface of the subclavian artery and is in relation anteriorly with the left common carotid artery and the vagus nerve, posteriorly with the long muscle of the neck *(longus colli muscle),* medially with the oesophagus and the recurrent laryngeal nerve and laterally with the subclavian artery and the pleura.

At the level of the neck, the arch of the thoracic duct passes through a triangular space termed trigone of the vertebral artery, comprised between the longus colli muscle and the oesophagus medially, the anterior scalenus muscle laterally and the first rib inferiorly. It courses posterolaterally to the vasclonervous bundle of the neck, anteromedially to the vertebral artery and vein, as well as medially to the phrenic nerve.

The thoracic duct has collateral afferences from the infercostal, juxta-vertebral and posterior mediastinal nodes.

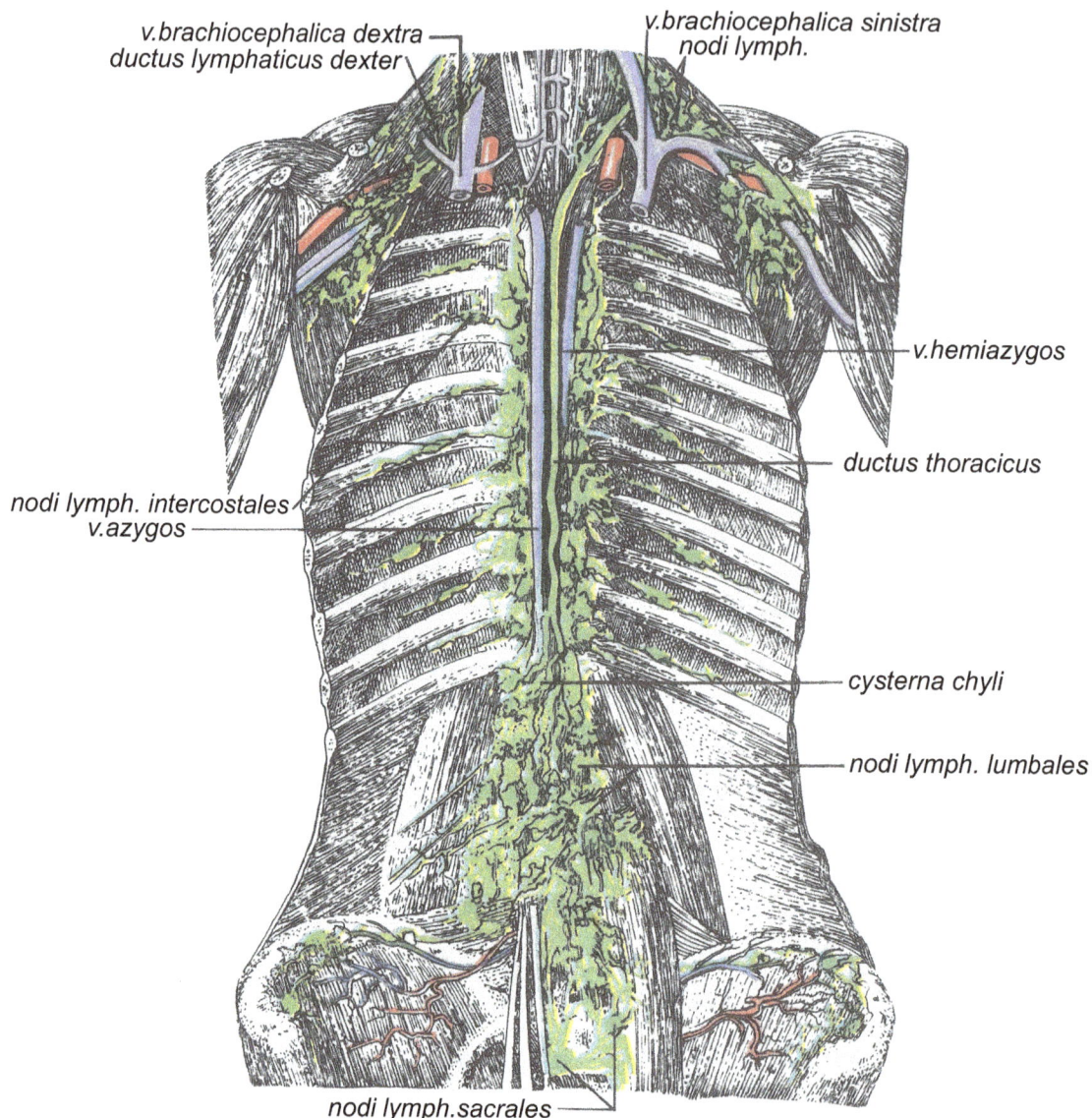

v.brachiocephalica dextra
ductus lymphaticus dexter

v.brachiocephalica sinistra
nodi lymph.

v.hemiazygos

ductus thoracicus

nodi lymph. intercostales
v.azygos

cysterna chyli

nodi lymph. lumbales

nodi lymph.sacrales

Fig. 115. Lymph circulation.

The Azygos Vein System The azygos vein *(vena azigos major)* has its origin in the thoracic cavity, at the level of the right eleventh intercostal space; it results from the union of two branches of origin, respectively an external branch, formed of the right ascending lumbar vein and the twelfth intercostal vein, and an inconstant internal branch, either formed on the surface of the inferior vena cava or (less frequently) derived from the right renal vein (fig. 116, 117).

In the thorax, the azygos vein ascends the anterior surface of the vertebral column, on the right of the median line up to the T_4 level, where it arches forward above the right roof of the lung and opens through its arch at the level of the posterior wall of the superior vena cava.

In its ascending, thoracic portion, if has the following relations: medially with the thoracic duct and farther off with the aorta, laterally with the right mediastinal pleura, posteriorly with the vertebral column and the right intercostal arteries, anteriorly with the root of the lung and lower down with the oesophagus through the interazygo - oesophageal cul-de-sac.

The arch of the azygos vein passes above the right bronchus and pulmonary artery and below the right laterotracheal lymph nodes and is enclosed by the right pleura laterally and by the oesophagus, the trachea and the right vagus nerve medially.

122

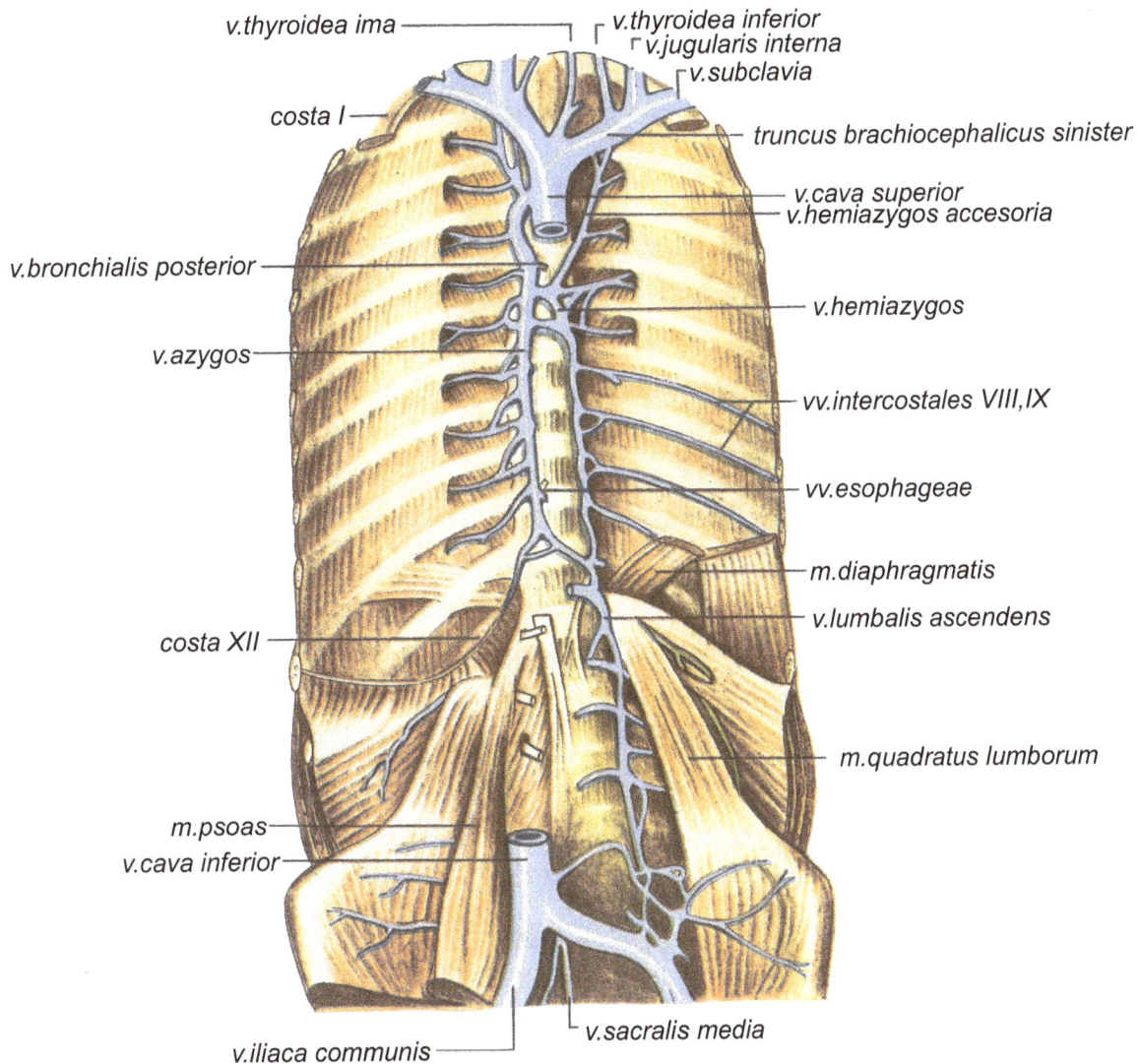

v.thyroidea ima
v.thyroidea inferior
v.jugularis interna
v.subclavia
costa I
truncus brachiocephalicus sinister
v.cava superior
v.hemiazygos accesoria
v.bronchialis posterior
v.hemiazygos
v.azygos
vv.intercostales VIII,IX
vv.esophageae
m.diaphragmatis
v.lumbalis ascendens
costa XII
m.quadratus lumborum
m.psoas
v.cava inferior
v.iliaca communis
v.sacralis media

Fig. 116. The azygos vein system.

The azygos vein receives the posterior right bronchial vein, the oesophageal and pericardial veins, the intercostal veins and the small azygos veins (hemiazygos and accessory hemiazygos veins) (which descend the left border of the vertebral column, outside the aorta, up to the seventh rib, where they arch to the right, pass behind the aorta, the thoracic duct and end in the azygos vein; they collect blood from the level of the first 6-7 intercostal spaces and sometimes from the left posterior bronchial vein).

The Orthosympatheic Ganglionic *Chain (Truncus sympathicus)* It is the last anatomic structure also situated symmetrically, paravertebrally at the level of the heads of ribs and anteriorly to the intercostal vessels which they cross. The ganglia of the thoracic sympathetic chain, unlike the cervical ones, are arranged segmentally. Only the first (and sometimes also the second) thoracic ganglion is fused with the last cervical ganglion in the ganglion stellatum, with which it forms the subclavian loop (Vieussen's ansa). The thoracic orthosympathetic chain gives off cardiac *(nervus cardiacus imus),* bronchopulmonary, oesophageal and aortic branches. From the last 6 (sometimes 7-8) ganglia emerge the greater splanchnic nerve *(nervus splanchnicus major)* and the lesser splanchnic nerve *(nervus splanchnicus minor),* which pass through the diaphragm to the abdominal viscera, in the nerve supply of which they participate. The thoracic sympathetic chain contains generally 11 ganglia (actually, the first ganglion is generally united with the inferior cervical ganglion, forming the ganglion stellatum), which are situated at the level of the costovertebral joints.

123

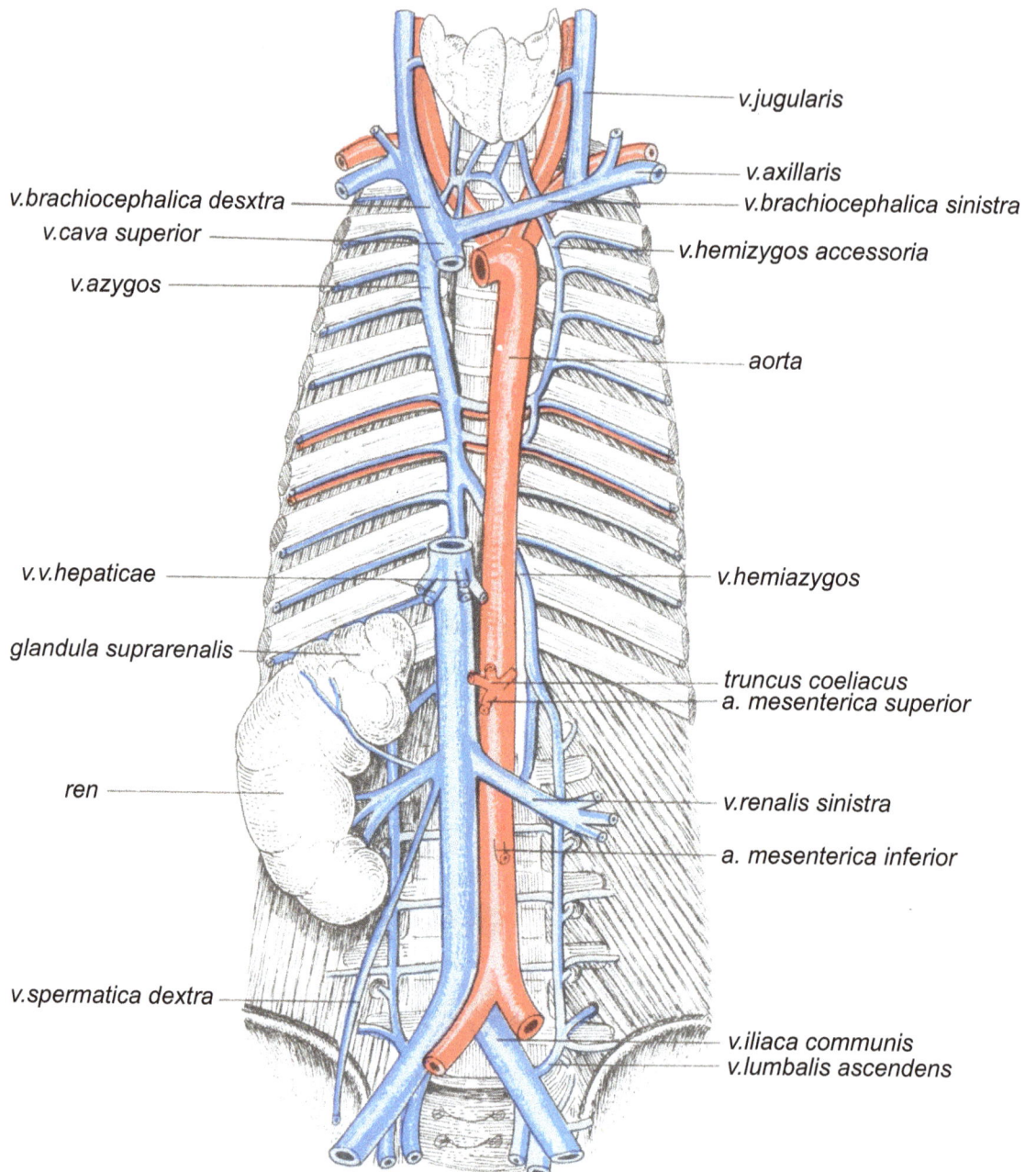

Fig. 117. The cava and azygos vein system.

The ganglionic chain has the following relations: posteriorly with the intercostal vessels; laterally with the pleura; medially with the vertebral column; anteriorly with the aorta, which covers only the upper part of the thoracic chain and the azygos veins.

The splanchnic nerves are generally two in number: the greater and the lesser splanchnic nerves.

The greater splanchnic nerve results from the union of the branches arising constantly from the thoracic ganglia 6, 7, 8 and 9 and inconstantly from the thoracic ganglia 5 and 10.

The lesser splanchnic nerve is formed by the branches given off by the thoracic ganglia 10, 11 and 12. It descends laterally to the greater splanchnic nerve, passes through the diaphragm, usually with the sympathetic chain, less frequently with the greater splanchnic or the aorta. Arrived in the abdomen, it divides into three groups of branches: superior, which goes to the convexity of the semilunar ganglion; middle, which course to the superior mesenteric ganglion; inferior, which courses to the renal plexus.

The Abdomen

Chapter 3

Preliminaries

The abdomen is that portion of the trunk which lies between the thorax, from which it is separated by the diaphragmatic muscle, and the pelvis, with which it communicates, forming together a common cavity, the peritoneal cavity (cavum peritoneale). The boundary between the abdominal cavity and the pelvic cavity is the superior aperture of the pelvis or pelvic inlet (aditus pelvis), whereas the perineal area corresponds to the inferior aperture or pelvic outlet.

Boundaries and Walls

The upper boundary of the abdominal cavity is formed by the diaphragmatic muscle, which separates it from the thoracic cavity.

The diaphragmatic muscle (the diaphragm). It has the shape of a transverse, elongated dome, with the convexity upwards, which by its base attaches itself on the anatomic elements that delineate the inferior thoracic aperture. It is made up of two parts: a central, tendinous part, called phrenic centre or tendinous centre *(centrum tendineum),* and a peripheral, muscular part, formed of fascicles inserted into various areas of the inferior thoracic aperture. The diaphragm may be considered as made up of a series of digastric muscles, the bellies of which insert themselves on two opposite points of the thorax, whereas their intermediate tendons interweave, forming the phrenic centre. The fascicles may be systematized into three parts: the vertebral lumbar part *(pars lumbalis),* the costal part *(pars costalis)* and the sternal part *(pars sternalis).*

The lumbar vertebral portion *(pars lumbalis)* has an internal and an external part. The internal part is attached to the lumbar column, the right side on the first four lumbar vertebrae and the left side on the first three, through two symmetrical pillars: the internal pillar *(crus mediale)* and the external pillar *(crus laterale).* A middle or accessory pillar *(crus intermedius)* is also described. The external part is a muscle plate, the fibers of which are attached to the arch of the psoas muscle *(arcus lumbocostalis medialis)* and to the arch of the quadratus lumborum muscle *(arcus lumbocostalis lateralis).*

The costal portion *(pars costalis)* is attached through 6 digitations to the inner surface of the ribs VII-XII and interweaves with the corresponding digitations of the transverse muscle of abdomen.

The sternal portion *(pars sternalis)* is attached to the posterior aspect of the xiphoid process.

On both sides of the median line there is a space called trigonum fibrosum sternocostale (Larrey's space), between the two muscle fascicles attached to the xiphoid process; posterolaterally, another space, called trigonum fibrosum lumbocostale, connects the renal area to the subpleural tissue (Bochdalek's hiatus), through which sometimes occur diaphragmatic hernias. The diaphragm has three large openings through which pass: the inferior vena cava (through the foramen venae cavae), the oesophagus, the thoracic duct and the two vagus nerves (through the hiatus esophageus), as well as the aorta (through the hiatus aorticus). Likewise, it is traversed by the sympathetic trunk, in the interstice between the lateral pillar *(crus laterale)* and the muscle plate attached to the arch of the psoas muscle, by the greater and lesser splanchnic nerves, the origin trunk of the azygos vein and the internal mammary artery, through Larrey's space.

Vessel and nerve supply of the diaphragm

Arteries. The arterial supply is provided in the upper part by the musculophrenic artery (branch from the internal thoracic artery) and the superior phrenic artery (branch from the descending aorta) and in the lower part, by the inferior phrenic artery (from the abdominal aorta).

The venous system is to a great extent satellite to the arteries.

The diaphragm is abundantly supplied with lymphatic vessels. A subperitoneal and a subpleural network may be distinguished. Through intrathoracic aspiration, the lymph circulation is directed more towards the thoracic cage. Thus, a "pleuritis" associated to diseases of the subdiaphragmatic space occurs much more frequently than a secondary involvement of the subdiaphragmatic peritoneum in pleural, pulmonary or mediastinal suppurations. The regional lymph nodes, into which the diaphragm lymph drains more frequently, are situated in the anterior and posterior mediastinum.

The inferior boundary of the abdominal cavity is formed by the internal iliac fossae, which contain the iliopsoas muscles. They are situated laterally, enclosing, with the promontory of sacrum (angulus lumbosacralis) and the pubic bone, the pelvic inlet (superior aperture of the pelvis).

The anterolateral wall of the abdomen is formed by broad muscles (the external oblique muscle, the internal oblique muscle and the transverse muscle), by their aponeuroses and by the fasciae which invest them.

Posteriorly to these muscles and to the fascia investing the transverse muscle (fascia transversalis) lie the subperitoneal tissue and the peritoneum.

The subperitoneal tissue is abundant and loose in the lower part of the wall, where it can be easily detached; the remainder is well represented and achieves a more adherent connection of the peritoneum to the abdominal wall.

On the inner part of the abdominal wall, peritoneal folds start from the umbilicus, representing mostly rudiments of the vessels of embryonic circulation. Upwards, towards the diaphragm and the superior aspect of the liver, starts the falciform ligament, enclosing in its border the oblitera-ted umbilical vein, of which the round ligament of the liver (ligamentum teres hepatis) has resulted. Downwards, on either side descends an umbilical fold (plica umbilicalis medialis), containing the obliterated umbilical vessels (ligg. umbilicale mediale), and medially - the median umbilical fold (plica umbilicalis mediana) containing the obliterated urachus (ligamentum umbilicale medianum). The epigastric folds (plicae epigastricae), which contain the inferior epigastric vessels, are placed laterally to the median umbilical folds and are not connected with the umbilicus, but penetrate paramedially-posteriorly into the rectus abdominis sheath.

These folds delineate the supravesical inguinal fossa (fossa inguinalis supravesicalis), the medial inguinal fossa (fossa inguinalis medialis) and the lateral inguinal fossa (fossa inguinalis lateralis), placed laterally to the epigastric fold. At its level lies the deep ring of the inguinal canal.

The layers of the abdominal wall are:
- lateral: skin; subcutaneous fatty tissue; the external oblique muscle of abdomen; the internal oblique muscle of abdomen; the transverse muscle of abdomen; the transverse fascia; the peritoneum;
- paramedian: skin; subcutaneous cell tissue; the anterior leaflet of the rectus abdominis sheath; the rectus abdominis; the posterior leaflet of the rectus abdominis sheath; the transverse fascia and the peritoneum.

The inguinal region. The inguinal region *(regio inguinalis)* corresponds in depth to the inguinal canal *(canalis inguinalis).*

It is triangular, the lower boundary being represented by the femoral arch, the upper boundary by a horizontal line tangent to the anterior superior iliac spine and the inner one, by the external border of the rectus abdominis.

The superficial plane of the region is formed by the integument and the subcutaneous fatty connective tissue, in the depth of which lie branches of the superficial iliac circumflex artery (the permanent pulsations of this artery bring about the condensation of the subcutaneous cell tissue under the form of Thomson's fascia),

The satellite veins and the fibrils of the abdominogenital nerve. Beneath this plane lies the aponeurosis of the external oblique muscle.

The plane underlying the aponeurosis encloses an interstice called inguinal canal, crossed by the spermatic cord in the male and the round ligament of uterus in the female. The inguinal canal has four walls (anterior, posterior, superior and inferior) and two orifices (the superficial and the deep ring).

The inguinal canal *(canalis inguinalis).* It is 4-5 cm long and begins at the deep inguinal ring *(annulus inguinalis profundus),* in the lateral inguinal fossa. It is situated laterally to the epigastric fold, which contains the inferior epigastric vessels.

It runs obliquely from above downwards and medially through the layers of the abdominal wall and ends at the superficial inguinal ring *(annulus inguinalis superficialis),* which lies at the level of the medial inguinal fossa (fig. 11 8,119).

The inguinal canal is bounded as follows:
- below: by the inguinal ligament *(ligamentum inguinale)* and the reflex (Colles') ligament *(ligamentum reflexum);*
- above: by the transverse muscle of the abdomen and the interna! oblique muscle of the abdomen in the lateral half and the conjoined tendon *(tendo conjunctivus aut falx inguinalis)* in the medial half;
- behind: by the transverse fascia, the conjoined tendon and the reflex (Colles') ligament;
- in front: by the aponeurosis of the external oblique muscle of the abdomen with the medial and lateral parts, the intercrural fibres of the aponeurosis of the external oblique muscle and the superficial abdominal fascia.

127

The superficial inguinal ring *(annulus inguinalis superficialis)*, situated subcutaneously, is enclosed by three pillars, the crura: the medial crus *(crus mediate sive superius)*, the lateral crus *(crus laferale sive inferius)*, both arising from the aponeurosis of the ipsilateral external oblique muscle of the abdomen, and the posterior crus *(crus. posterius)*, which arises form the aponeurosis of the contralateral muscle and is identical with Colles' reflex ligament. This ring acquires a falciform border through the intercrural *arciform)*, inguino-abdominal (from the extra-abdominal fascia) and inguinofemoral (from the femoral fascia) fibres, which penetrate the linea alba, preventing the removal of the pillars.

The deep inguinal ring *(annulus inguinalis profundus)* lies in the transverse fascia, above and medially to the middle of the femoral arch. The free border is formed by reflection of the transverse fascia, which invaginates within the canal; below it is bounded by Hesselbach's ligament (lig. interfoveolare) and the iliopectineal fasciae.

The inguinal canal contains:

- *in the male*: the cremaster muscle, fibrils of the ilio-inguinal and iliohypogastric nerves, the ductus deferens, the deferential artery, a branch of the inferior vesical artery, a branch of the hypogastric artery, the spermatic artery, a branch of the aorta, the funicular artery, a branch of the epigastric artery; the bulky and flexuose anterior venous plexus accompanies the spermatic artery; it empties on the right directly into the inferior vena cava and on the left, into the left renal vein, which flows, in its turn, into the inferior vena cava. The dilatation of the anterior venous plexus, on the left, occurs as a consequence of emptying of the left spermatic vein into the left renal vein, which in some cases forms a barrier to the venous flow to the heart; it is known under the name of varicocele (it is operated by means of the "AMARRAL" or "IVANISIEVITCH" cure, by removal of the veins in which the blood is stopped in its way to the heart); the posterior venous plexus, considerably smaller,is situated around the funicular artery; the respective veins flow into the hypogastric veins; the lymphatics drain into the lumbar lymph nodes;

-*in the female*, the inguinal canal contains the round ligament of uterus (lig. teres uteri). It begins at the angle of the uterus, passes across the pelvis on the anterior aspect of the broad ligament and penetrates the deep ring of the inguinal canal Less thick than the spermatic cord, it crosses the canal, comes out through the superficial ring and attaches itself to the spine of pubis, the symphysis pubis and the adipose tissue of labia majora pudendi.It contains branches of the greater and lesser abdominogenital and genitofemoral nerves, like in the male. It is accompanied by an arteriole, derived from the epigastric artery, and by veins. The inguinal region is an area at the level of which hernias may occur. Clinically, two main forms of hernias are distinguished: direct and indirect. The direct hernias pass across the abdominal wall, medially to the epigastric vessels of the epigastric (lateral) fold, and therefore they are also called medial inguinal hernias. The indirect hernias penetrate the inguinal canal and begin laterally to the epigastric vessels (lateral inguinal hernias). Lateral hernias may be congenital or acquired, medial hernias are always acquired.

Of the total number of inguinal hernias, 80% are lateral and 20% medial. In the congenital inguinal hernias, the vaginal process (peritoneovaginal canal) remains patent.

Direct hernias perforate vertically the abdominal wall, indirect hernias perforate it obliquely. Both emerge at the level of the superficial inguinal ring, but direct hernias remain usually at the root of the scrotum and are of a lesser size.

Interparietal hernias are those indirect hernias which do not emerge, but remain in the inguinal canal and at pressure, lodge themselves between the layers of the abdominal wall

As *cryptorchism* or *cryptorchidism* is designated an arrest in development and descent of the testis. If the testis still remains after birth in the abdominal cavity, the result is an abdominal cryptorchism (abdominal testis, retentio ahdominalis). If it remains in the inguinal canal, the cryptorchism is inguinal (inguinal testis, retentio inguinalis). The complete descent of the testis into the scrotum is a sign of maturation of the newborn.

The testis are situated in the scrotum in 96% of the newborn children, but only in 68% of the immature infants. In the testicular ectopia, the testis is lodged in atypical sites. Dystopic testes degenerate, as a rule, rapidly, with destruction only of the germinal, but not of the secretory epithelium. In 10% of cases, a malignant degeneration occurs. Fifty percent of the tumours are seminomas. Therefore if is indicated to perform the early surgical repair of testicular anomalies, inclusively the extirpation of the atrophic testicle, when if cannot be descended in the bursa.

Hernias of the umbilical cord are rare and develop following an incomplete reposition of organs during foetal life.

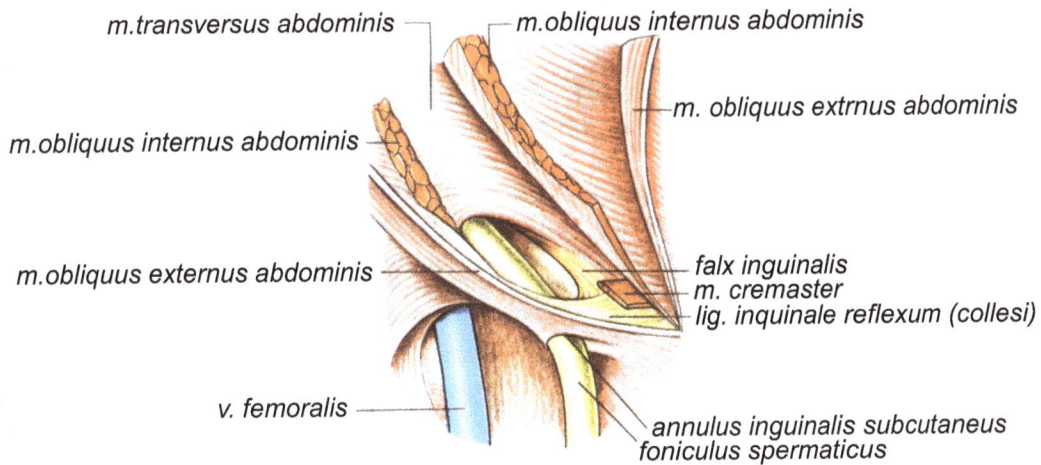

Fig. 118. Superior wall of the inguinal canal

Labels in figure:
m.transversus abdominis — m.obliquus internus abdominis — m. obliquus extrnus abdominis — m.obliquus internus abdominis — m.obliquus externus abdominis — falx inguinalis — m. cremaster — lig. inquinale reflexum (collesi) — v. femoralis — annulus inguinalis subcutaneus — foniculus spermaticus

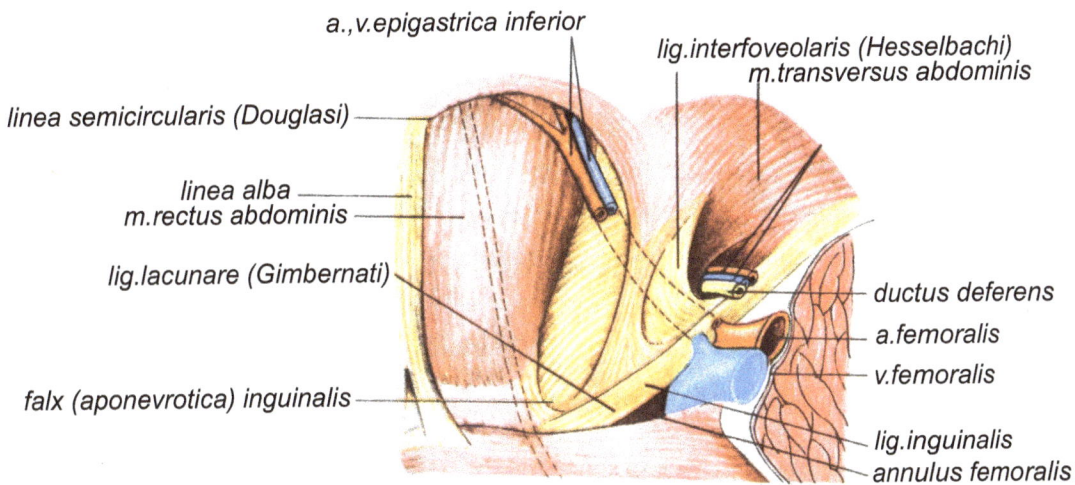

Fig. 119. Posterior wall of the inguinal canal.

Labels in figure:
a.,v.epigastrica inferior — lig.interfoveolaris (Hesselbachi) — m.transversus abdominis — linea semicircularis (Douglasi) — linea alba — m.rectus abdominis — lig.lacunare (Gimbernati) — ductus deferens — a.femoralis — v.femoralis — falx (aponevrotica) inguinalis — lig.inguinalis — annulus femoralis

Thus, they occur owing to the persistence of the physiological umbilical hernia which, normally, subsides very early, before birth. Unlike the umbilical hernia, the hernia of the umbilical cord is not covered with epidermis, but with the thin, transparent amnion and the gelatinous tissue of the umbilical cord.

The posterior wall of the abdomen is formed by the rachidian region and the lumbo-iliac regions.

The rachidian region contains the vertebral column and the soft parts which cover if posteriorly (skin, subcutaneous adipose tissue and muscular planes of the posterior region of the trunk).

The lumbo-iliac region is bounded superiorly by the twelfth rib (more exactly, by the arches of the psoas muscle and of the *quadratus lumborum muscle),* laterally by the lateral border of the quadratus lumborum, inferiorly by the iliac crest and medially by the vertebral bodies.

Muscles and aponeuroses present themselves sfrafigraphically, from before backwards, as follows: the quadratus lumborum muscle, the iliopsoas muscle, the posterior insertion aponeurosis of the transverse muscle of abdomen, the posterior muscles of the trunk at this level *(spinal muscles, serratus posterior inferior muscle, internal oblique muscle, latissimus dorsi muscle).*

Between the posterior border of the infernal oblique muscle (laterally), the lateral border of the spinal muscles (medially), the serratus postero-inferior muscle and the twelfth rib (superiorly) lies a space called lumbo-costo-abdominal space ("Grynfeltt's tetragon"), where the aponeurosis of the transverse muscle is directly situated on the latissimus dorsi muscle, representing a hernial area of the abdominal wall. Another similar area is the "Pefif's trigone" *(trigonum tumbale)* witch lies between the latissimus dorsi muscle and the external oblique muscle and has the base formed by the iliac crest; this trigone, too, represents an area of weak resistance of the posterior abdominal wall, through which hernias may occur.

129

The Abdominal Cavity

Since the abdomen and the pelvis are lined with the peritoneal serous membrane, the term abdominopelvic cavity is synonymous with peritoneal cavity (fig. 120). The abdomino-pelvic cavity is located between the diaphragm and the perineum. The boundary between the abdominal and the pelvic cavities is at the level of the pelvic inlet.This boundary has only a didactic value, because actually suppurations and haematomas may spread in the whole cavity, regardless of the site of their genesis. In addition, some organs, for example the uterus, become intraabdominal during pregnancy, whereas others, in case of ptosis, may be found in the pelvis, in spite of their abdominal site (fig. 120-199).

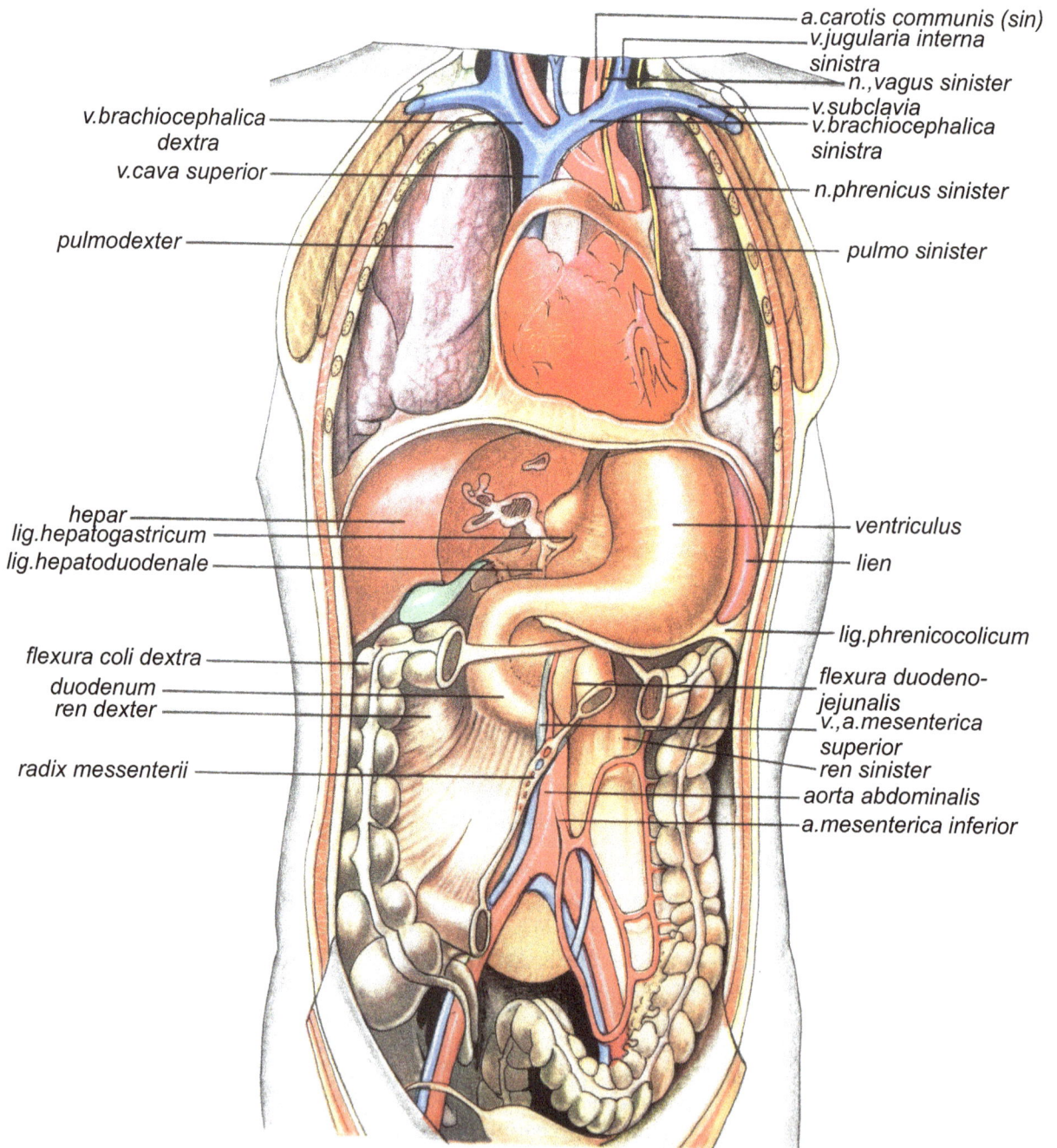

Fig. 120. Overall view of the thoracic and abdominopelvic cavities.

The Peritoneum

The walls and viscera of the abdominopelvic cavity are lined or invested by the peritoneal serous membrane. If has an area of about 1,500 cm^2.

Like the other serous membranes, the peritoneum is formed of a parietal and of a visceral leaflet. The parietal peritoneum *(peritoneum parietale)* is the inner investing layer of the walls of the abdominopelvic cavity, to which it is attached by a connective tissue, the so-called *tela subseroso (fascia endoabdominalis),* that is continuous with the endothoracic fascia, which lines the pleura (fig. 126). The visceral leaflet *(peritoneum viscerale)* covers the abdominopelvic organs. The parietal leaflet is thicker and more resistant, but less adherent then the visceral leaflet.

The two leaflets are continuous with each other and enclose the peritoneal cavity *(cavum peritonei),* a potential cavity that becomes real only in pathological cases. In this cavity with smooth and glistening walls lies a thin lamina of albuminous fluid, which facilitates the gliding of organs. In the male, the peritoneal cavity is completely closed, whereas in the female if communicates with the outside through the orifices of the uterine tubes, the tubes, the uterus and the vagina (fig. 126,127).

Some organs are almost completely invested by peritoneum and are called intraperitoneal organs, others are only partially covered and are termed extraperitoneal organs. The visceral leaflet of the peritoneum extends to the level of complicated reflexion lines, that generate peritoneal folds, divided into mesenteries, ligaments and omenta (epiploons). They represent both attachement means of the viscera and the place where the neurovascular pedicles penetrate or leave organs.

A mesentery is a peritoneal fold which attaches a hollow organ of the alimentary canal to the abdominal wall (the mesentery proper, the mesocolon). The mesenteries contain the neurovascular pedicles which run from the abdominal wall to the organs invested by peritoneum. Each mesentery is formed of two leaflets separated by a thin lamina of connective-adipose tissue, crossed by vessels and nerves. The two leaflets begin at the parietal peritoneum, which they prolong towards the organs of the abdominal cavity, with the visceral peritoneum of which they are continuous, at the level where the vessels and nerves penetrate the organs for which the respective neurovascular pedicles are destined.

The omenta are peritoneal formations with an own individuality, which connect to each other the organs of the supramesocolic portion of the abdominal cavity.

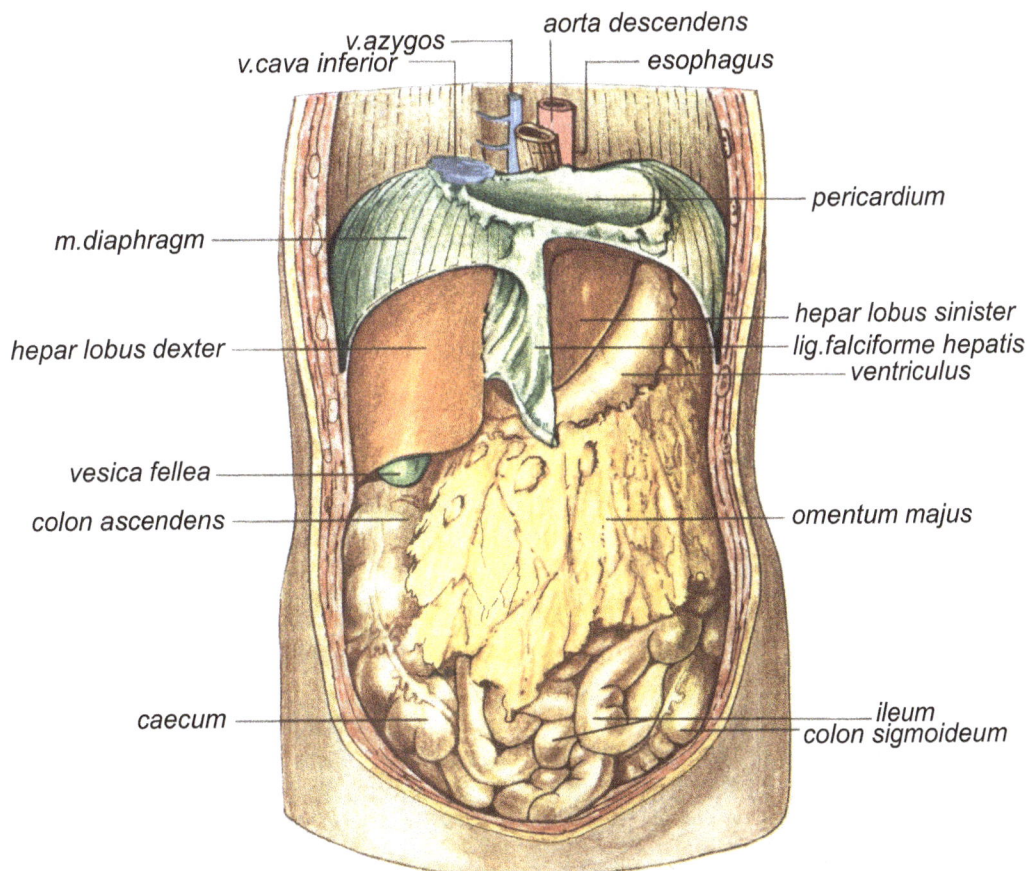

Fig. 121. Anterior view of the abdominal viscera after raising the anterior wall

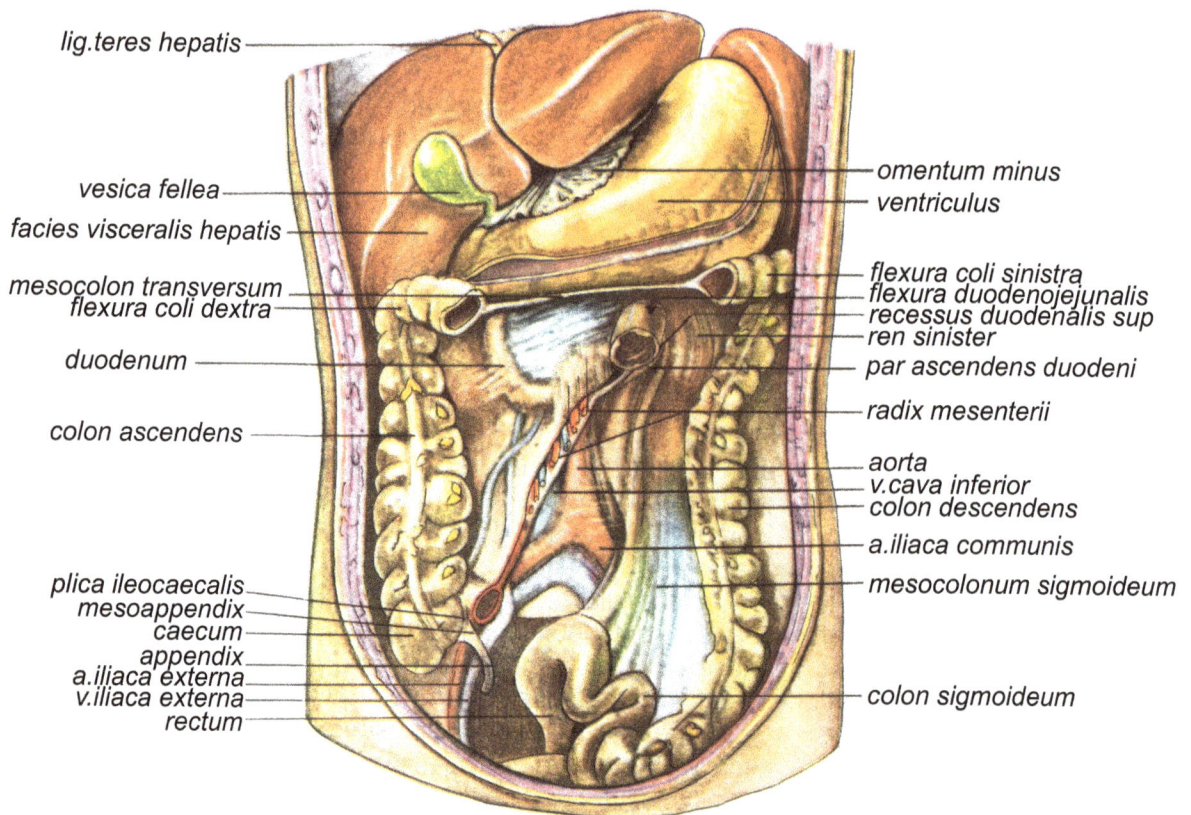

Fig. 122. The supramesocolic part - anterior view; the colic areas of the inframesocolic part.

The lesser or oesogastroduodenohepatic omentum (omentum minus), derived from a portion of the anterior mesogastrium, is represented by a quadrilateral peritoneal lamina,which attaches the liver to the abdominal oesophagus, the lesser curvature of the stomach and the first part of the duodenum. It has two aspects (anterior and posterior) and three borders (oesogastroduo-denal, hepatic and free) (fig. 131, 132).

It is made up of three areas: left or superior, median and right. The left or superior area, which contains vascular branches and nervous rami arising from the left vagus nerve, is termed pars densa, owing to the fact that it is thicker. The median area is called pars flaccida, because at this level the lesser omentum is very thin and transparent. The right area, which is very thick, corresponds to the free border of the lesser omenfum,containing the hepatic pedicle, and is called pars vasculosa.

The greater or gastrocolic omentum *(omentum majus),* derived from a portion of the posterior mesogastrium, is represented by an approximately quadrilateral peritoneal fold, which begins at the greater curvature of the stomach, passes in front of the transverse colon, to which it adheres, and descends to the pelvis, anteriorly to the jejuno-ileal loops and to the colic frame which surrounds these loops.

It presents several elements (fig. 121,123).

An anterior lamina inserts above on the greater curvature of the stomach, from the pylorus to the gastrosplenic omentum, with which it is continuous, and reflects below so as to be continuous with its posterior leaflet. The anterior lamina of the greater omentum is constituted of two leaflets: an anterior leaflet, from the visceral peritoneum of the anterior aspect of the stomach, and a posterior leaflet, arising from the visceral peritoneum of the posterior aspect of the stomach which, when reaching the greater curvature, adheres to the above mentioned anterior leaflet. After having reflected upwards, both leaflets form the posterior lamina, which passes superiorly over the transverse colon and its mesocolon. At this level the two leaflets separate from each other. That which descended from the posterior aspect of the stomach ascends above the pancreas, becoming parietal and forming at the same time the posterior wall of the omental bursa (hence, of this small peritoneal cavity which lies behind the stomach and the lesser omentum). The other leaflet, after having attached to the posterior abdominal wall, is directed ventrally as a leaflet of the transverse mesocolon, invests the transverse colon, becoming thus visceral peritoneum, and leaves it, forming the inferior leaflet of the transverse mesocolon. After reaching the posterior wall of the abdomen, it reflects downwards on the wall, becoming parietal peritoneum.

132

As a consequence of the coalescence processes that occur during the development of the peritoneum, the posterior lamina blends with the transverse mesocolon and the anterior lamina, so that the greater omentum is formed of four leaflets of peritoneal serous membrane to which, in the gastrocolic ligament, the two leaflets of the mesocolon are added, this ligament being thus formed of six serous leaflets.

From the above mentioned details it would appear that the greater omentum Is not one and the same with the gastrocolic omentum and thai the greater omentum is made up of two parts: a superior part, represented by the gastrocolic omentum, and a larger inferior part, the greater omentum proper. The researches performed, among which also those of. I. Filip, have demonstrated the absence of coalescence of the two laminae and their role in the occurrence of the gastric volvulus. On the basis of these data, Gh. Niculescu has introduced a personal operative technique, which consists in the restoration of the coalescence, with good clinical results.

The gastrosplenic omentum *(ligamentum gastrolienal)* is a quadrilateral peritoneal lamina, which connects the fornix of the stomach to the hilum of the spleen and forms the anterior wall of the left projection of the omental bursa (the splenic or lienal recess of the omental bursa). The gastrosplenic omentum is consti¬tuted of two leaflets: an anterior one, derived from the greature curvature of the stomach, and a posterior one, arising from the peritoneum of the omental bursa, which comes to the spleen from the posterior surface of the stomach. Between the two leaflets run short vessels and the left gastroepiploic artery (fig. 130).

The pancreaticosplenic omentum *(ligamentum pancreaticolienale)* has a quadrilateral shape and extends from the hilum of spleen to the tail of pancreas. It is made up of two peritoneal leaflets: an anterior leaflet, arising from the peritoneum of the omental bursa, and a posterior one, which comes from the peritoneum of the great abdominal cavity. Between the two leaflets lie the tail of the pancreas and the splenic vessels.

A ligament is a peritoneal fold which attaches to the abdominal wall the intraperitoneal, usually parenchymal organs of the alimentary canal or organs which do not belong to the latter (the coronary ligament, the falciform ligament, the triangular ligaments, the round ligaments, the broad ligaments etc.) (fig. 128).

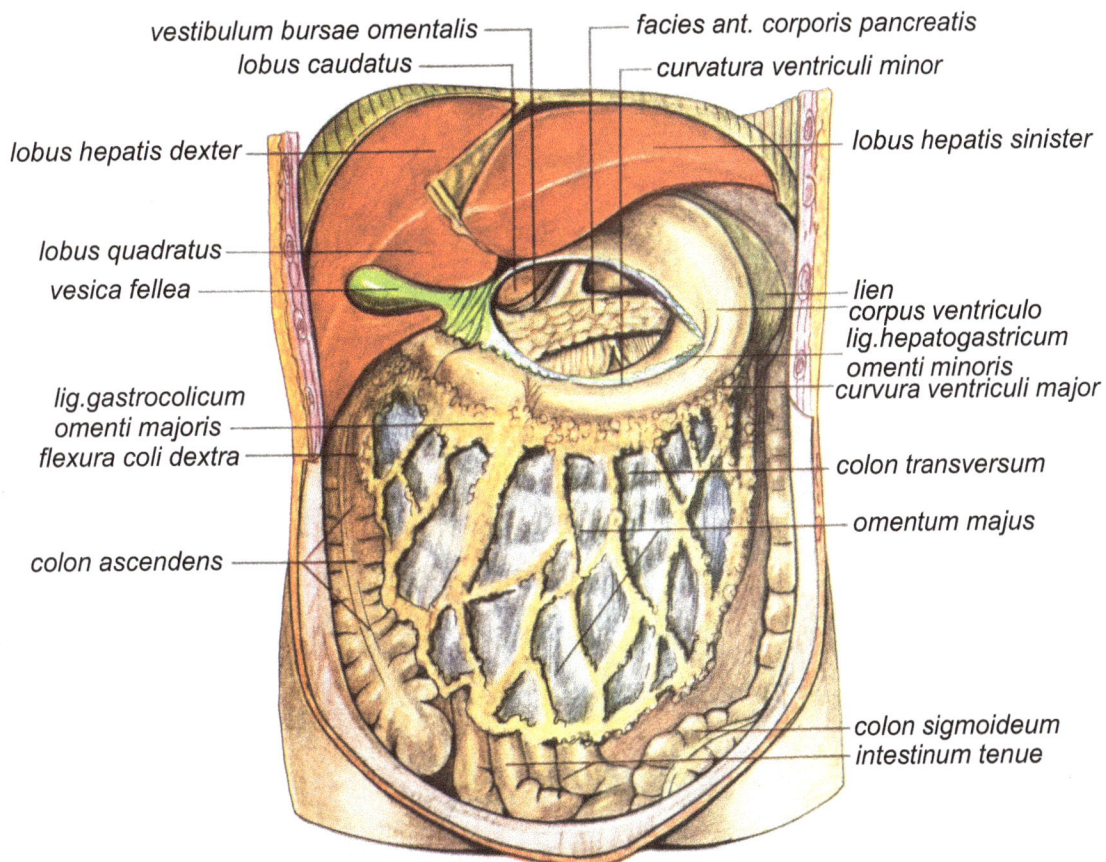

Fig. 123. The suprainesocolic part after section of the lesser omentum and marked fraction towards the left of the stomach; display of the pancreas; the greater omentum.

133

Between the abdominal wail and an organ which in the course of its development changes from intraperitoneal to retroperitoneal, lies the coalescence fascia (Treitz's fascia - refropancreatic; Toldt's fascia -retrocolic). This fascia results from the fusion of the primitive mesentery with the primary parietal peritoneum through the coalescence phenomenon (fig. 132).

The abdominal cavity is divided by the presence of the transverse colon into two large portions: the supramesocolic portion *(cavum supraomentale)* and the inframesocolic portion *(cavum infraomentale)*.

The supramesocolic (or glandular) portion is bounded: above by the diaphragm, below by the transverse colon and mesocolon with the two phrenocolic ligaments and the greater omentum *(omentum majus),* through its anterosuperior surface.

The supramesocolic portion is divided into three compartments: the liver bed, the stomach bed and the spleen bed, which enclose the omental bursa (fig. 122,123,125).

The Omental Bursa
(The Epiploic or Retrogastric Cavity)

The omental bursa (the lesser peritoneal cavity) is a diverticulum of the greater peritoneal cavity, situated behind the stomach. If communicates with the greater peritoneal cavity through the Winslow's hiatus *(foramen epiploicum)* (fig. 124,129).

The omental bursa presents for study two portions: the vestibule *(vestibulum bursae omentalis)* and the omental bursa proper, which communicate with each other through an orifice called *foramen bursae omentalis.*

The walls of the vestibule are formed: in front, by the lesser omentum; behind, by the parietal peritoneum which lines the posterior abdominal wall between the inferior vena cava and the peritoneal folds determined by the left gastric and hepatic arteries; above, by the inferior surface of the liver; below, by the lesser curvature of the stomach (fig. 131).

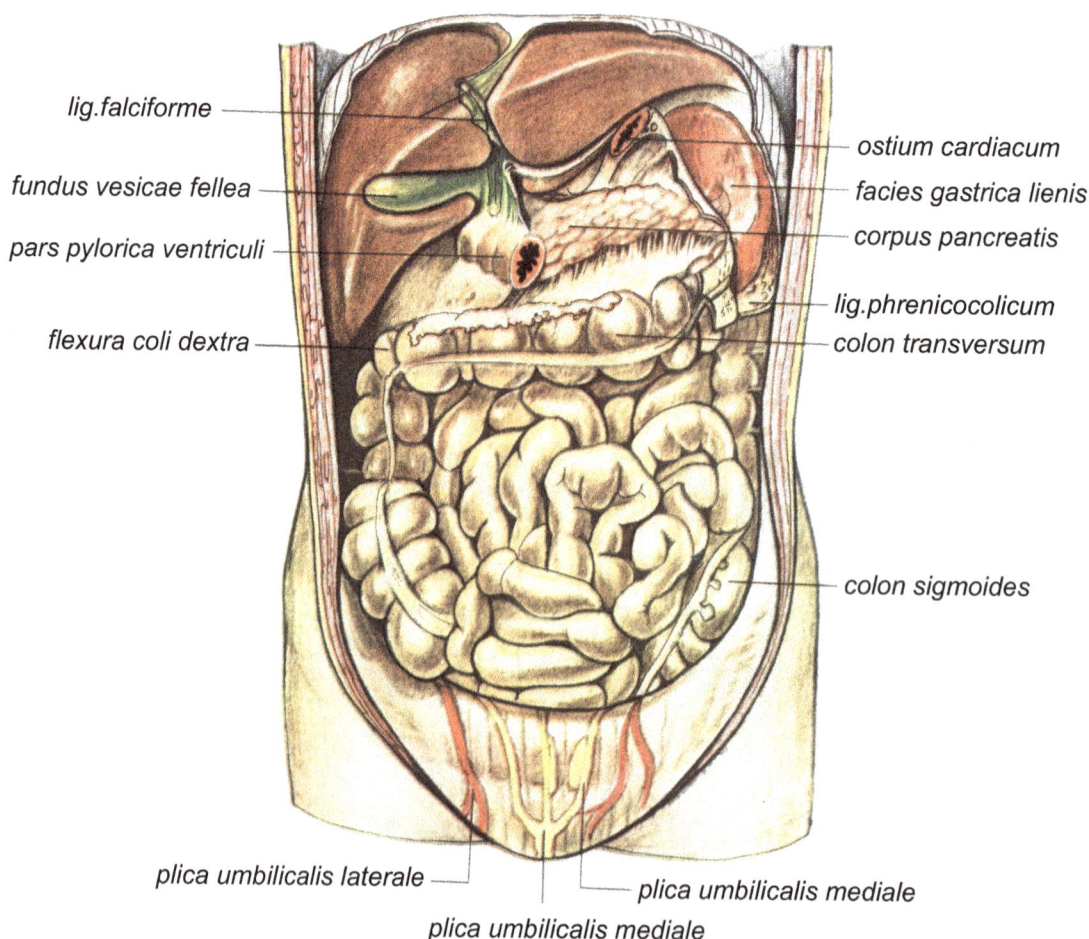

Fig. 124. The inframesocolic part and the omental bursa open anteriorly, through removal of the stomach

134

The vestibule widens gradually from below upwards. Its superior portion sends out a projection behind the liver, between the spigelian lobe and the diaphragm, up to the coronary ligament.

Winslow's hiatus is the right opening of the vestibule of the omental bursa, through which the communication with the greater peritoneal cavity takes place. This opening, elongated from above downwards, is bounded: in front, by the free right border of the lesser omentum *(pars vasculara),* where lies the liver pedicle (posteriorly, the vena porta situated in the ligament; in front and to the left, the hepatic artery: to the right, the common bile duct); behind, by the outline of the inferior vena cava; above, by the caudate tubercle of the spigelian lobe and below, by the first part of the duodenum.

The foramen bursae omentalis is the left opening of the vestibule of the omental bursa, through which the vestibule communicates with the omental bursa proper. This opening is bounded: anteriorly, by the lesser curvature of the stomach; posterosuperiorly, by the gastropancreatic ligament (the peritoneal fold at the level of the left gastric artery); postero-inferiorly, by the duodenopancreatic ligament.

The omental bursa extends from the right to the left, between the *foramen bursae omentalis* and the hilum of the spleen. It is bounded above by the reflexion of the parietal peritoneum on the visceral peritoneum, which invests the posterior surface of the stomach, and below by the transverse mesocolon and the fusion of the constituent leaflets of the greater omentum, in which it may extend under the form of the omental recess. The anterior wall is represented superiorly by the posterior surface of the stomach and inferiorly, by the anterior leaflet of the greater omentum; the posterior wall is in relation with the body and tail of pancreas, the left kidney and adrenal gland and the diaphragm.

The omental bursa sends out three projections: the superiorprojection, represented by the oesophageal recess, the left or lienal projection and the inferior or omental projection.

The left projection (recessus lienalis) extends up to the spleen and is bounded: posteriorly, by the pancreaticosplenic omentum, across which run the splenic vessels, and anteriorly, by the gastrosplenic omentum, in which the short vessels and the left gastroepiploic artery are situated. At the level of the hilum of the spleen, the anterior leaflet of the pancreatico-splenic omentum is continuous with the posterior leaflet of the gastrosplenic omentum.

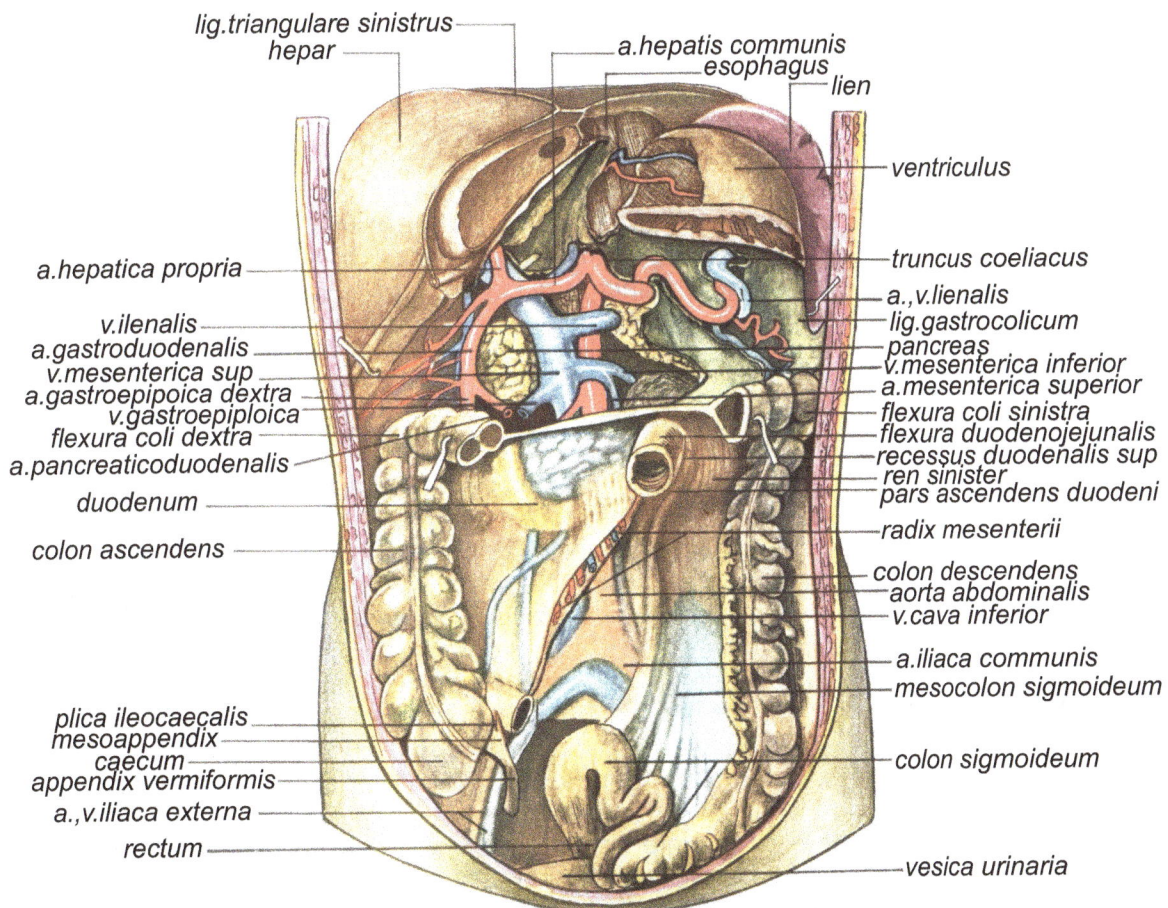

Fig, 125. Vassals of the suprasnesocolic part (stomach and pancreas partly removed; liver and spleen drawn upwards; mesocolon drawn downwards) and colic areas (after section of the mesentery and removal of the intestinal loops).

135

The superior projection (recessus superior) ascends to the diaphragm and forms a blind pouch on the left side of the oesophagus.

We mention that the peritoneal serous membrane attaches to the posterior wall the supramesocolic portion of the second part of the duodenum and may be traced above the right kidney, in *the hepatoduodeno-renocolic pouch* (Rutherford-Morison's hepatorenal pouch). This space, enclosed - as its name shows - by the duodenum, the upper pole of the right kidney, the liver and the hepatic angle of the colon, establishes a communication between the subhepatic space, the vestibule of the omental bursa and the right parietocolic space (the inframesocolic part), above the hepatic angle of the colon.

The *inframesocolic* or *intestinal* portion is bounded: superiorly, by the transverse colon with its mesocolon, the left colophrenic ligament and the greater omentum (by its postero-inferior surface): inferiorly, by the superior aperture of the pelvis; anteriorly, laterally and posteriorly, by the corresponding walls of the abdominal cavity. It is divided by the mesentery, the ascending and the descending colon into five spaces (fig. 124).

-the *right and left parietocolic spaces,* between the ascending, respectively descending colon and the right and left lateral walls of the abdomen;

- the *right mesentericocolic space (the right colic pouch),* between the right side of the mesentery and the ascending colon, in the shape of a triangle with the base upwards;

- the *left mesentericocolic space (the left colic pouch),* also triangular in shape, with the base downwards, having a large communication with the pelvis;

- the *pelvis,* situated in the lower part of the peritoneal cavity, separated from it by a conventional plane which passes from the promontory, along the arcuate lines, towards the pubis.

Course of the Peritoneum in the Various Regions of the Abdominal Cavity

The topography of the peritoneum in the different areas of the abdominal cavity contributes to the understanding of the uninterrupted continuity of the peritoneal leaflet and affords a general view of it (fig. 126,127,129-134).

From the deep surface of the umbilicus, the parietal peritoneum lines the posterior aspect of the anterior abdominal wall and then the inferior surface of the diaphragm, up to the coronary and triangular ligaments of the liver. Here it forms the superior leaflet of the coronary and triangular ligaments and is continuous with the visceral peritoneum of the liver, which covers the superior aspect of the liver, reaches the anterior border, then passes on the inferior aspect of the liver, up to its hilum, where it is continuous with the anterior leaflet of the lesser omentum *(omentum minus).*

At the lesser curvature of the stomach, the peritoneum covers the anterior aspect of the stomach up to the greater curvature, becoming the tunica serosa of the latter, then it forms the anterior leaflet of the gastrocolic omentum, which is continuous with the anterior sheet of the greater omentum *(omentum matus).* At the level of the free border of the greater omentum, which may occasionally extend as far as into the pelvis, the peritoneum is reflected upwards and is continuous with the posterior sheet of the greater omentum.

This sheet ascends up to the posterior abdominal wall, where it attaches on it, and then it is continuous with the superior leaflet of the transverse mesocolon and reaches the transverse colon, investing it and becoming its tunica serosa, after which it comes to the posterior abdominal wall. At this level, the visceral peritoneum is continuous with the parietal peritoneum, which covers the posterior and duodenal abdominal walls. However, as we have already mentioned, the stomach is invested also on its posterior aspect by the peritoneal serous membrane, which comes from the lesser omentum. This peritoneal leaflet makes up the anterior wall of the omental bursa and, at the same time, when it reaches the greater curvature, it forms the posterior leaflet of the greater omentum. It descends with the anterior leaflet of the latter in proximity of the pubis, occasionally it ascends towards the posterior abdominal wall, covers the pancreas and becomes the posterior wall of the omental bursa. It lines this wall and the diaphragm, then the inferior surface of the liver and forms laterally, on the left, the posterior abdominal wall - up to the level of the inferior vena cava, where if is continuous with the parietal peritoneum of the greater peritoneal cavity - and then the anterior aspect of the right Kidney, after which if passes on the right lateral abdominal wall and then on the anterior abdominal wall, up to the right leaflet of the falciform ligament of the liver.

136

From the left surface of the falciform ligament, the peritoneum covers the liver up to the hilum, where it is continuous with the anterior leaflet of the lesser omentum; it reaches the lesser curvature of the stomach, passes on the anterior aspect of the latter, becomes tunica serosa and gains the greater curvature, where it forms the gastrosplenic omentum; the latter extends up to the hilum of the spleen, covers the spleen and reaches again the hilum, where if is continuous with the posterior leaflet of the pancreaticosplenic omentum, up to the tail of the pancreas; then it passes on the posterior abdominal wall, where the visceral peritoneum becomes parietal peritoneum. It covers the left kidney and the posterior, left lateral and anterior abdominal walls, up to the left leaflet of the falciform ligament of the liver.

Following its course in the transverse plane, beneath the transverse mesocolon and beginning from the anterior abdominal wall, it may be seen that the peritoneal serous membrane covers the right posterior lateral wall in the right parietocolic space, reaches the ascending colon and covers it. Then it lines the right mesenfericocolic space up to the roof of the mesentery, where the parietal peritoneum becomes the right leaflet of the mesentery and then visceral peritoneum. It invests the jejunum and the ileum and then, becoming their tunica serosa, hence visceral peritoneum, if is continuous with the left leaflet of the mesentery up to the posterior abdominal wall, where if becomes again parietal peritoneum, lining the left mesenterico-colic space up to the descending colon. It covers also the latter (partially if is half-retroperitoneal, lining the left parietocolic space), passes on the left lateral abdominal wall and reaches the posterior aspect of the anterior abdominal wall.

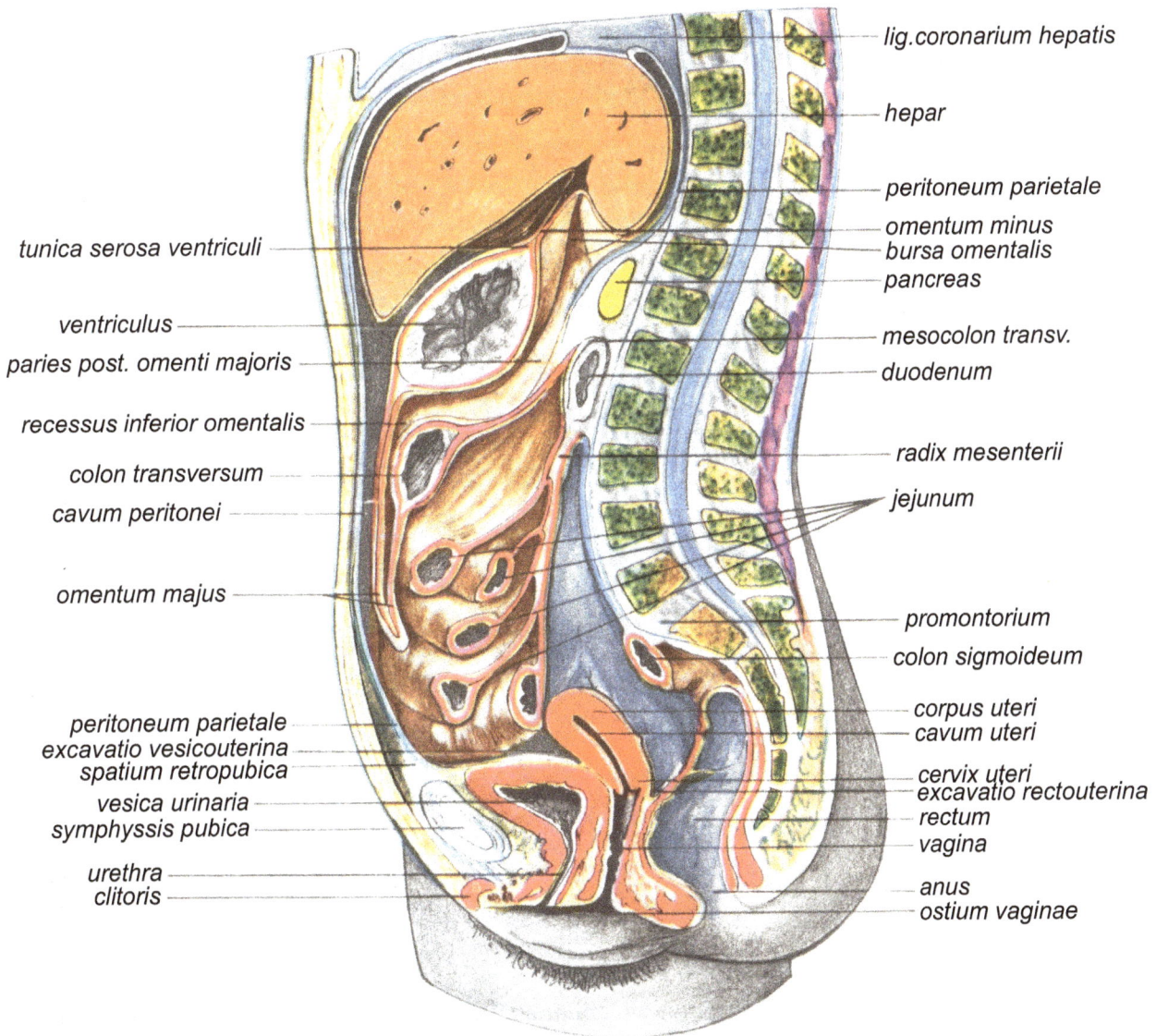

Fig. 126. Disposition of the peritoneum at the level of the obdominopelvic cavity (sagittal section).

137

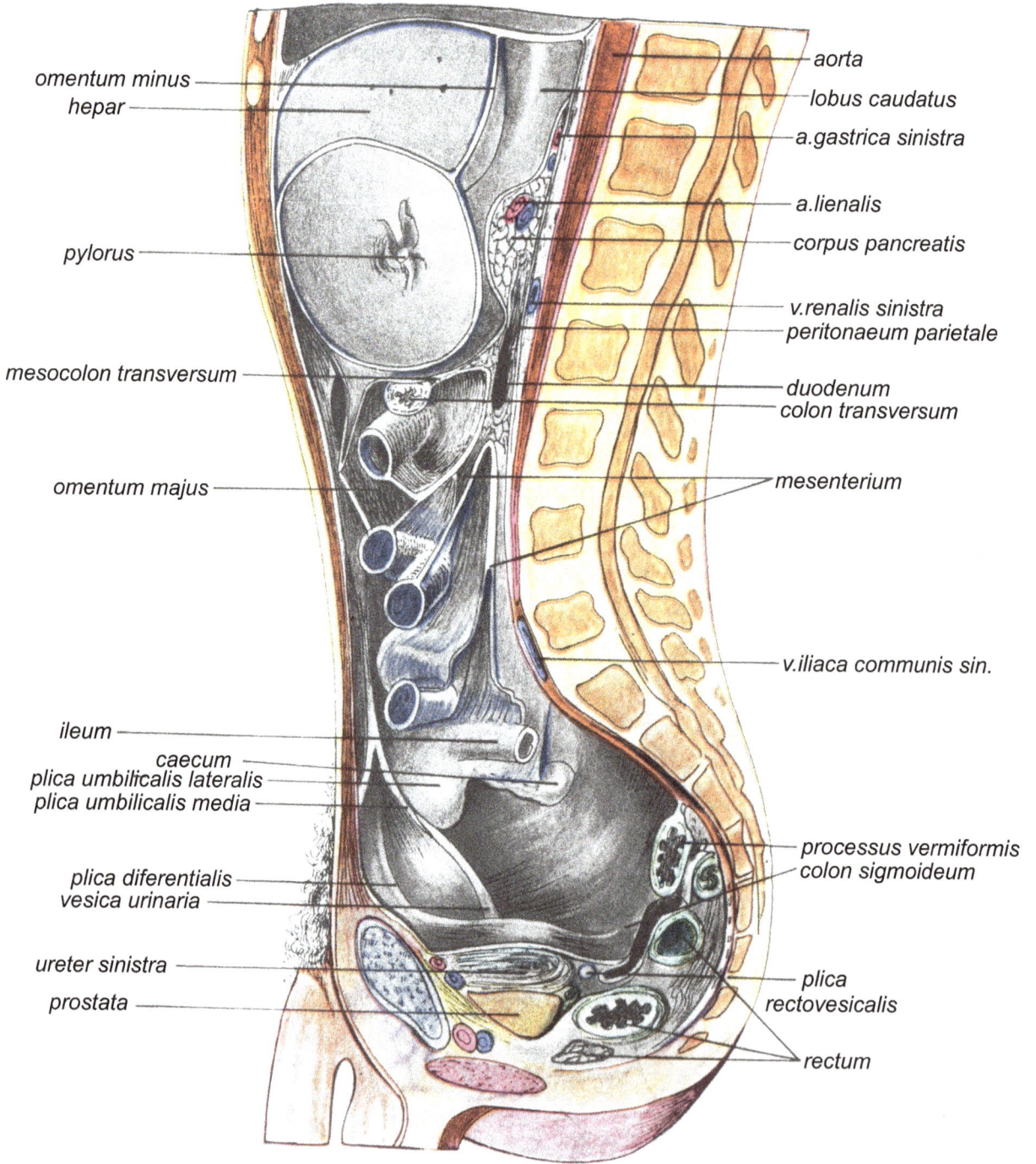

omentum minus

hepar

pylorus

mesocolon transversum

omentum majus

ileum

caecum
plica umbilicalis lateralis
plica umbilicalis media

plica diferentialis
vesica urinaria

ureter sinistra

prostata

aorta

lobus caudatus

a.gastrica sinistra

a.lienalis

corpus pancreatis

v.renalis sinistra
peritonaeum parietale

duodenum
colon transversum

mesenterium

v.iliaca communis sin.

processus vermiformis
colon sigmoideum

plica
rectovesicalis

rectum

Fig. 127. Sagittal section through the abdominal cavity - disposition of the peritoneum.

138

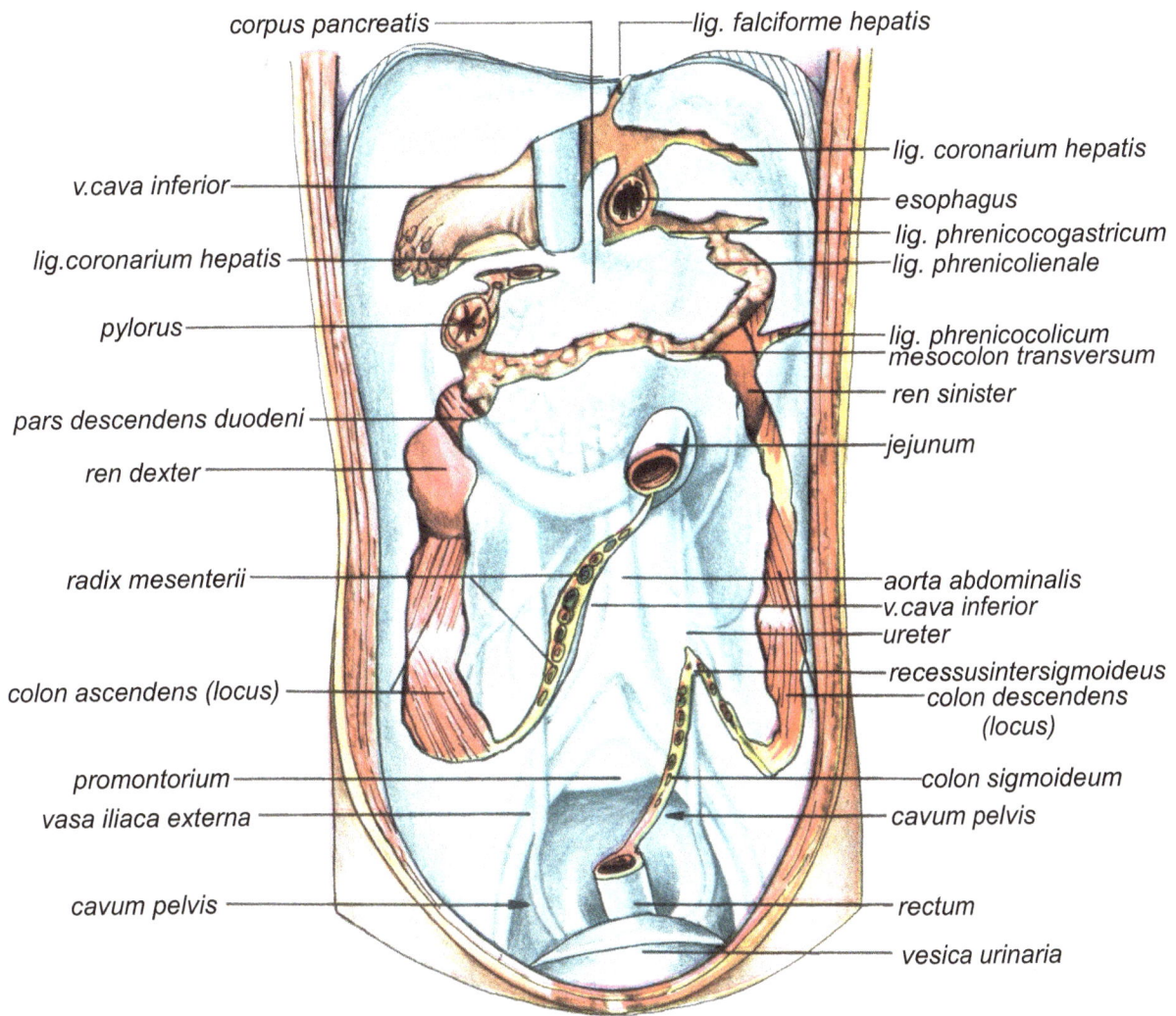

Fig. 128. Posterior wall of the abdominopelvic cavity

corpus pancreatis

lig. falciforme hepatis

lig. coronarium hepatis

v.cava inferior

esophagus

lig. phrenicocogastricum

lig. coronarium hepatis

lig. phrenicolienale

pylorus

lig. phrenicocolicum

mesocolon transversum

pars descendens duodeni

ren sinister

ren dexter

jejunum

radix mesenterii

aorta abdominalis

v.cava inferior

ureter

colon ascendens (locus)

recessusintersigmoideus

colon descendens
(locus)

promontorium

colon sigmoideum

vasa iliaca externa

cavum pelvis

cavum pelvis

rectum

vesica urinaria

139

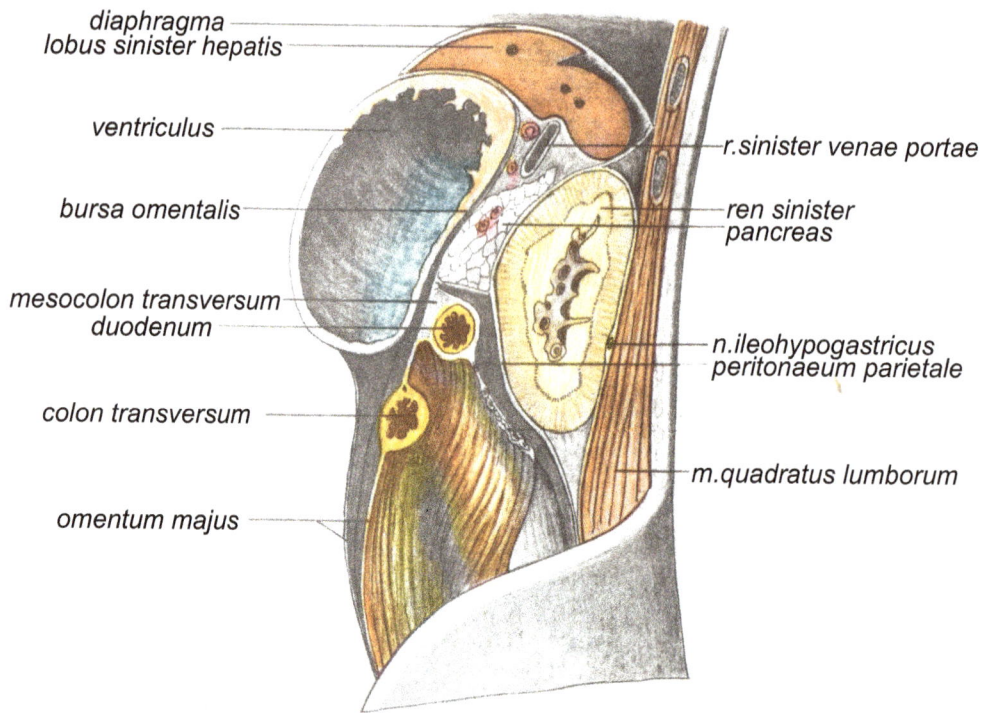

Fig. 129. Parasagittal section at the level of the omental bursa.

Labels, clockwise from top left:
- diaphragma
- lobus sinister hepatis
- ventriculus
- bursa omentalis
- mesocolon transversum
- duodenum
- colon transversum
- omentum majus
- r.sinister venae portae
- ren sinister
- pancreas
- n.ileohypogastricus
- peritonaeum parietale
- m.quadratus lumborum

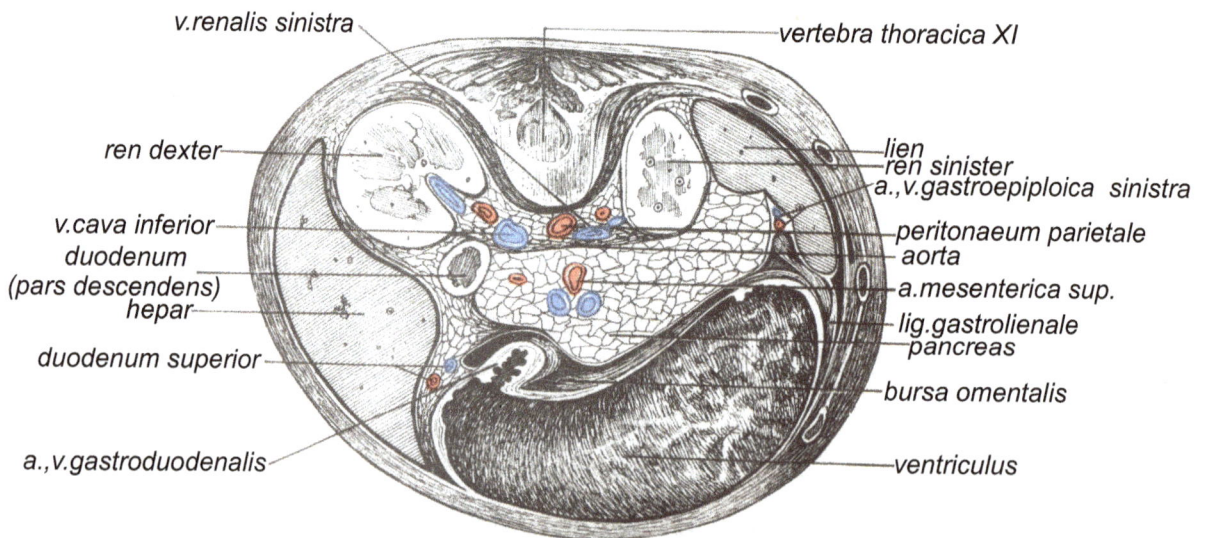

Fig. 130 Transversa section of the level of the eleventh thoracic vertebra

Labels:
- v.renalis sinistra
- vertebra thoracica XI
- ren dexter
- lien
- ren sinister
- a.,v.gastroepiploica sinistra
- v.cava inferior
- peritonaeum parietale
- duodenum
- aorta
- (pars descendens)
- a.mesenterica sup.
- hepar
- lig.gastrolienale
- pancreas
- duodenum superior
- bursa omentalis
- a.,v.gastroduodenalis
- ventriculus

140

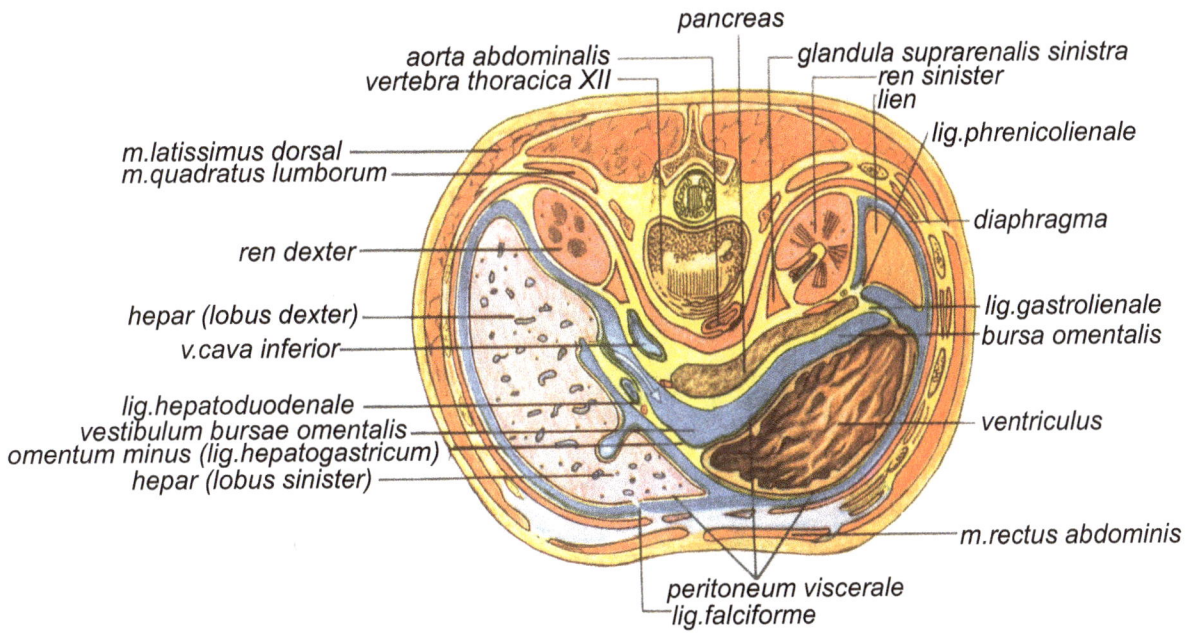

Fig. 131. Transverse section through the abdominal cavity at the level of the twelfth thoracic vertebra.

Labels (Fig. 131):
pancreas
aorta abdominalis
vertebra thoracica XII
glandula suprarenalis sinistra
ren sinister
lien
lig.phrenicolienale
m.latissimus dorsal
m.quadratus lumborum
diaphragma
ren dexter
hepar (lobus dexter)
lig.gastrolienale
bursa omentalis
v.cava inferior
lig.hepatoduodenale
vestibulum bursae omentalis
omentum minus (lig.hepatogastricum)
hepar (lobus sinister)
ventriculus
m.rectus abdominis
peritoneum viscerale
lig.falciforme

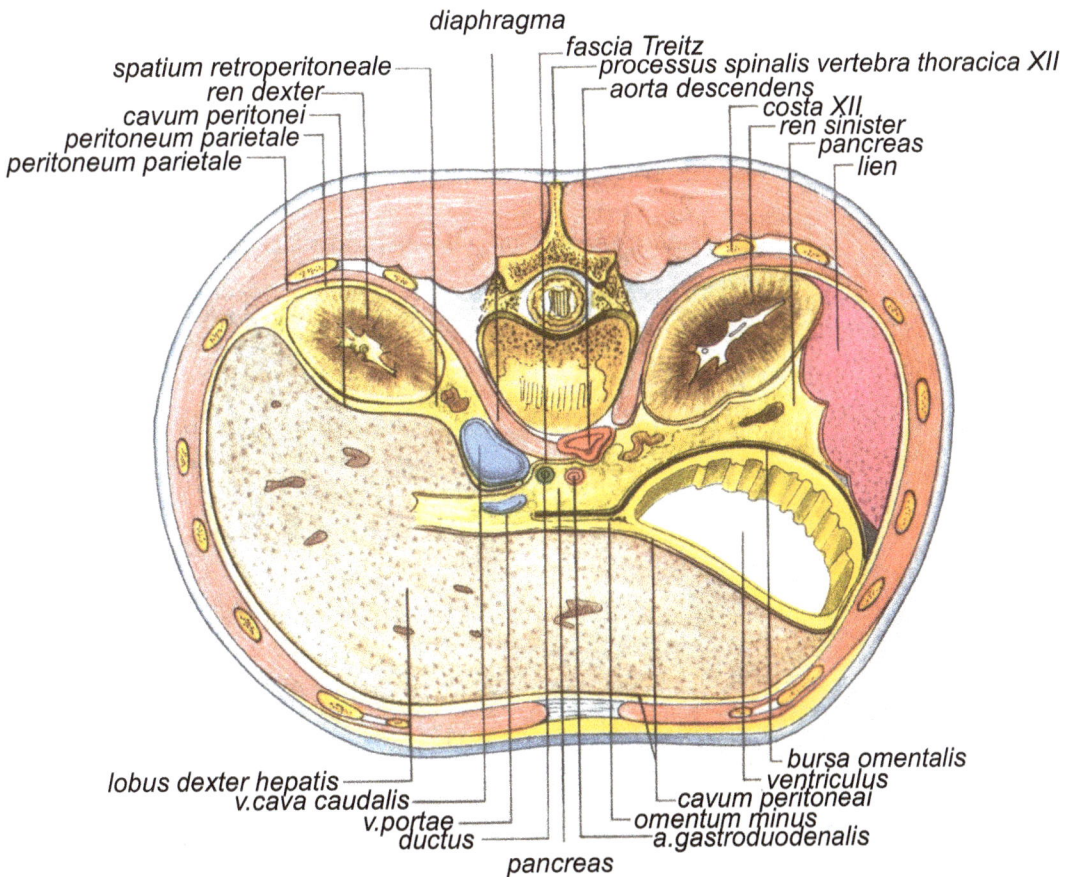

Fig. 132. Section at the level of the space between the twelfth thoracic and the first lumbar vertebrae.

Labels (Fig. 132):
diaphragma
fascia Treitz
processus spinalis vertebra thoracica XII
aorta descendens
spatium retroperitoneale
ren dexter
costa XII
ren sinister
pancreas
lien
cavum peritonei
peritoneum parietale
peritoneum parietale
lobus dexter hepatis
v.cava caudalis
v.portae
ductus
pancreas
bursa omentalis
ventriculus
cavum peritoneal
omentum minus
a.gastroduodenalis

141

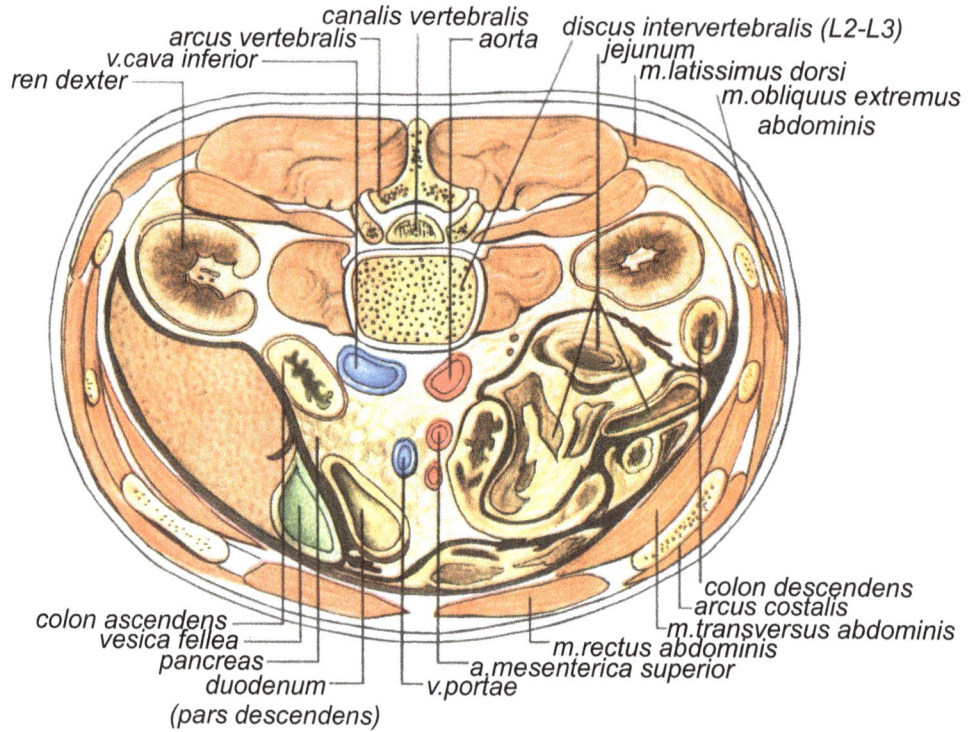

Fig. 133. Transverse section at the level of the third lumbar vertebra.

Labels (Fig. 133):
ren dexter
v.cava inferior
arcus vertebralis
canalis vertebralis
aorta
discus intervertebralis (L2-L3)
jejunum
m.latissimus dorsi
m.obliquus extremus abdominis
colon descendens
arcus costalis
m.transversus abdominis
m.rectus abdominis
a.mesenterica superior
v.portae
duodenum (pars descendens)
pancreas
vesica fellea
colon ascendens

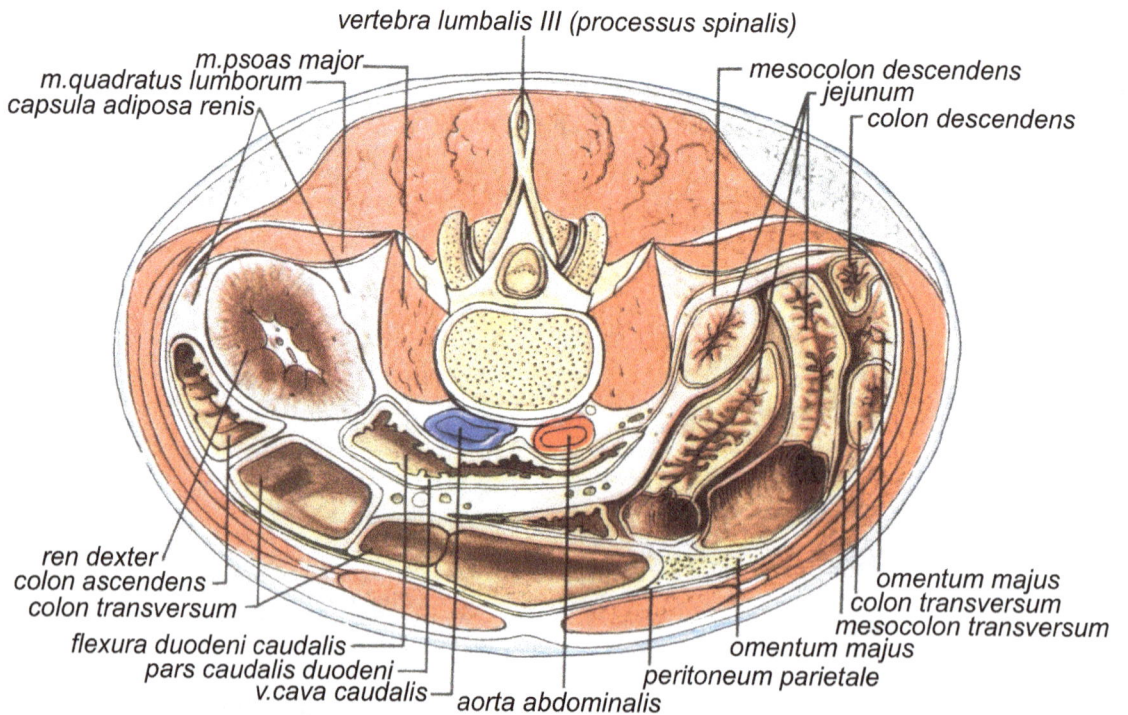

Fig. 134. Transverse section of the level of the fourth lumber vertebra.

Labels (Fig. 134):
m.quadratus lumborum
m.psoas major
capsula adiposa renis
vertebra lumbalis III (processus spinalis)
mesocolon descendens
jejunum
colon descendens
ren dexter
colon ascendens
colon transversum
flexura duodeni caudalis
pars caudalis duodeni
v.cava caudalis
aorta abdominalis
peritoneum parietale
omentum majus
omentum majus
colon transversum
mesocolon transversum

142

The Supramesocolic Part

The organs contained in this part are presented, in order to facilitate their comprehension, not only from the topographical point of view, but also by taking into account functional considerations.

The Gastric Region

If contains the abdominal part of the oesophagus and the stomach.

The Abdominal Oesophagus

It is a short segment, which is continuous with the thoracic oesophagus or, more exactly, with the diaphragmatic oesophagus and opens into the stomach through the orifice called cardia. This oesophago-cardial region plays an important physiological role, as if opposes the reflux of gastric juice towards a mucous membrane which is unable to neutralize the peptic effects.

The abdominal oesophagus is very short, of about 3 cm in length, after which if implants itselfs info the stomach; its right border is continuous with the lesser curvature and its left border forms with the fornix or vault of the stomach an acute angle, open upwards towards the diaphragm.

It is situated in the left subphrenic space, at the level of the tenth thoracic vertebra, the cardia being projected to the left of the eleventh thoracic vertebra.

The anterior surface is covered with the anterior leaflet of the peritonea! serous membrane, which is then continuous with the lesser omentum, beneath which descend the branchings of the left vagus nerve, and enters in relation also with the liver, on which it leaves an impression under the form of a vertical groove.

The posterior surface is in relation with the right vagus nerve, the pillars of the diaphragmatic muscle and the vertebral column.

The right border corresponds to the liver (the spigelian lobe) and to the lesser omentum, the anterior leaflet of which passes in front of the oesophagus, whereas the posterior leaflet is reflected to be continuous with the diaphragmatic peritoneum.

The left border forms, as mentioned above, an angle with the fornix of the stomach.

Vessel supply

The arterial supply is provided by branches of the left gastric artery, which give off the oesophago-cardiogastric arteries, and by branches of the inferior phrenic artery.

The veins make up a submucous network, which drains into the left gastric vein, level at which portacaval anastomoses are formed.

The lymphatics drain chiefly into the coronary gastric nodular chain.

The Stomach
(Gaster)

The stomach is a dilated organ of the digestive tube, situated between the oesophagus and the duodenum, in the subphrenic space, in the stomach bed, which communicates widely with the liver bed and the spleen bed. The stomach bed is bounded: below, by the transverse colon and the transverse mesocolon; above, by the diaphragmatic muscle; behind, by the parietal peritoneum which forms the posterior wall of the omental bursa and by the organs which if covers (pancreas, left kidney, left adrenal gland); in front, by the anterior wall of the trunk and by the liver; to the left, by the spleen bed; to the right, by the liver bed (see also fig. 120, 122).

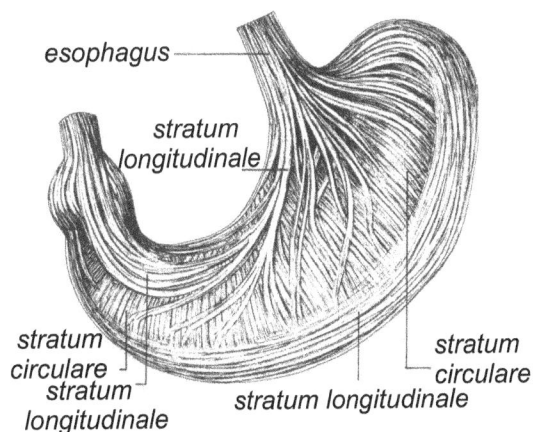

esophagus

stratum longitudinale

stratum circulare
stratum longitudinale

stratum circulare

stratum longitudinale

Fig. 135. The stomach - disposition of the muscle fibres (external view).

143

The shape and size of the stomach show extremely numerous individual variations. Examined in vivo, the stomach has generally the shape of a capital "J", with a longer vertical descending and a shorter horizontal portion. The vertical portion is oblique, directed anfero-inferiorly and formed of two segments: the vault of the stomach (fundus ventriculi sive fornix) and the body of the stomach (corpus ventriculi), separated by a horizontal conventional plane, tangent to the upper border of the cardia or carried through the cardiac notch (incisura cardiaca).

The horizontal portion (pars pylorica), in turn, is formed of two other segments: the pyloric antrum (antrum pyloricum), situated on the right of the body of the stomach, slightly dilated, and the pyloric canal (canalis pyloricus), a narrow and short cylindrical segment, which is continuous with the duodenum; the place of its continuity with the duodenum is marked by the duodenopyloric groove, in which lies the prepyloric vein (v. praepylorica).

The boundary between the body of the stomach or pars digestoria and the antral portion of the stomach is formed by a conventional plane carried perpendicularly on the tangent of the greater curvature, from the angular incisure, which is situated at the place where the lesser curvature changes its direction, from vertical becoming horizontal (fig. 141).

The stomach has two surfaces, anterior and posterior, which are separated by the two curvatures: the lesser curvature (curvatura minor), directed to the right and upwards, on which the lesser omentum is attached, and the greater curvature (curvatura major), directed to the left and downwards, attached to the transverse colon through the gastrocolic ligament and to the spleen through the gastrolienal ligament.

If communicates upwards, through the cardia, with the oesophagus and downwards, through the pyloric orifice, with the duodenum (fig. 136, 139).

In dependence on the degree of tonicity, the stomach may be: orthotopic, j-shaped, as it was described above, with a steerhorn appearance; hypertonic, in which the difference between the two portions is no more obvious; hypotonic, in which the two curvatures get close to each other and the horizontal part tends towards verticalization; atonic.

Anatomofunctionally, the stomach is considered to consist of two portions; a digestive portion *(pars digestoria),* which is vertical and formed of the fundus and corpus ventriculi, and a portion of food evacuation *(pars egesforia),* formed of the pyloric antrum and the pyloric canal.

Inner Configuration

The inner surface of the stomach presents numerous folds of the mucous membrane, some of which follow the long axis of the organ, along the greater curvature, whereas other, small folds are vertical, transversal or oblique and irregular. Along the lesser curvature lies a furrow, between two longitudinal folds, called gastric canal *(canalis ventriculi)* or Waldeyer's "Magenstrasse" (gastric road).

There are also some finer grooves, which mark off polygonal zones called gastric areas *(areae gastricae),* at the surface of which some elevations *(plicae villosae)* may be seen, separated by other depressions, into which the gastric glands open at the level of the gastric pits *(foveolae gastricae).* The cardiac orifice is provided with the cardio-oesophageal valve which is, as a matter of fact, the result of the acute angle formed by the oesophagus with the fornix of the stomach (some authors consider that the cardiac orifice has no valve, neither an anatomical sphincter).

The separation between the oesophageal and the gastric mucosa is very clear-cut. The pyloric orifice has a sphincter *(m. sphincter pylori)* and a valve (formed by a fold of the mucosa, it is an inconstant formation, of a functional nature, and therefore it is not recognized in the anatomic nomenclature) (fig. 136-138).

Relations of the Stomach

The anterior surface has relations with two regions of the wall of the trunk: the thoracic wall and the abdominal wall.

The thoracic region is delineated by the left chondrocostal arch. The relations of the stomach with the thoracic viscera are achieved through the diaphragm.

The stomach projects upwards as far as the fifth left intercostal space.

It corresponds to the left lung and pleura, the heart and the pericardium.

The left lobe of the liver interposes more or less between the diaphragm and the anterior surface of the stomach, but, the latter lies in immediate contact with the wall of the trunk at the level of Traube's semilunar space, enclosed between the dullness of the liver on the right, of the spleen on the left, of the heart and the costal arch downwards. It is a tympanic area of percussion costal arch downwards. It is a tympanic area of percussion

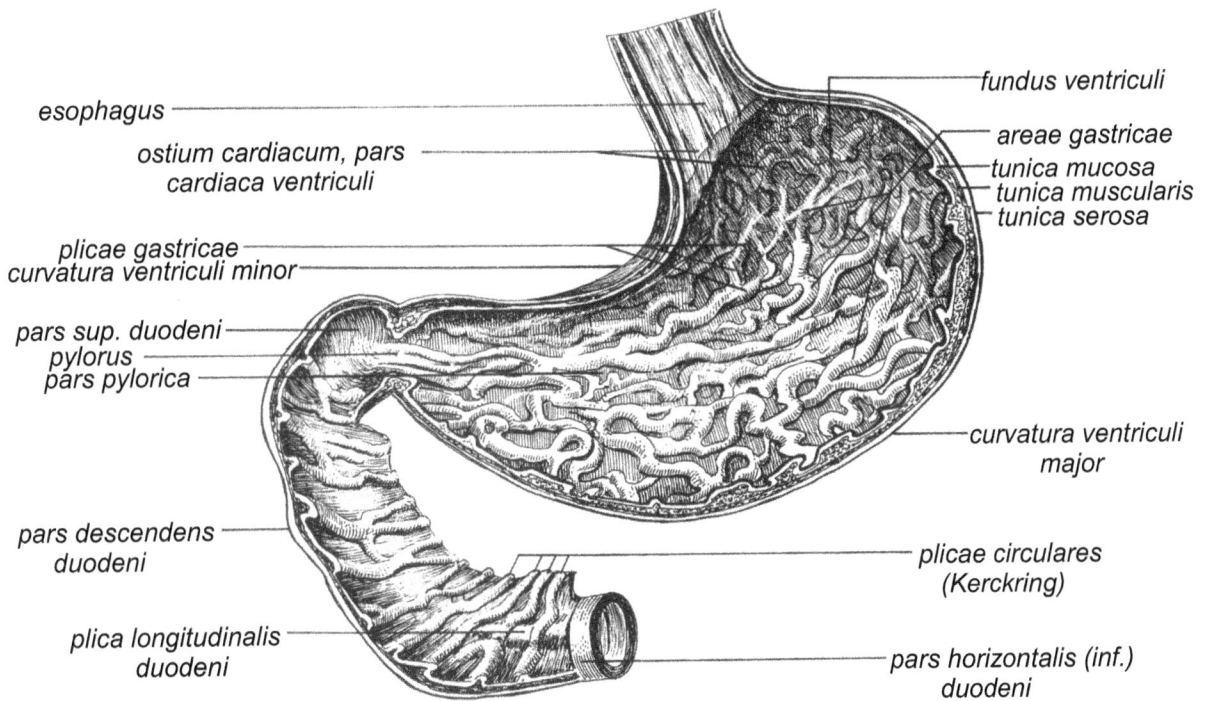

esophagus

ostium cardiacum, pars
cardiaca ventriculi

plicae gastricae
curvatura ventriculi minor

pars sup. duodeni
pylorus
pars pylorica

pars descendens
duodeni

plica longitudinalis
duodeni

fundus ventriculi

areae gastricae
tunica mucosa
tunica muscularis
tunica serosa

curvatura ventriculi
major

plicae circulares
(Kerckring)

pars horizontalis (inf.)
duodeni

Fig. 136. Internal structure of the stomach

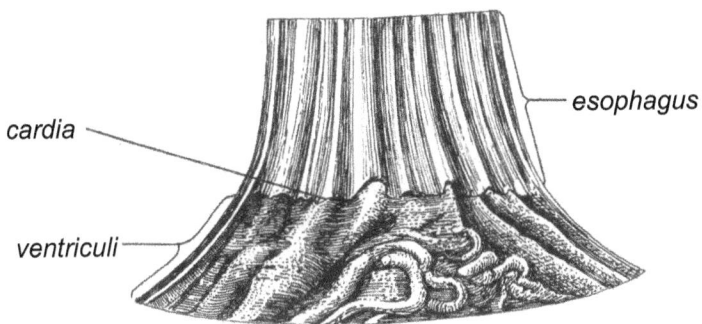

cardia

ventriculi

esophagus

Fig. 137. Structure of the cardia.

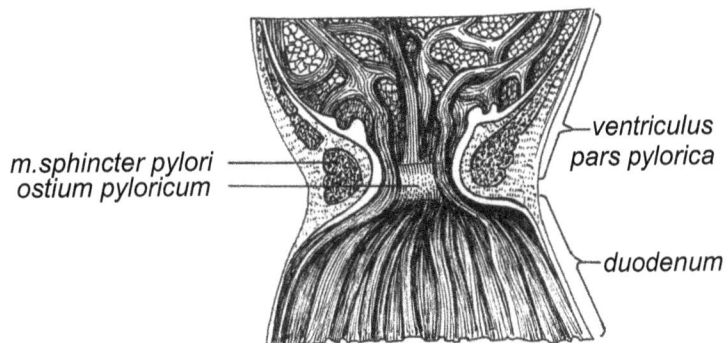

m.sphincter pylori
ostium pyloricum

ventriculus
pars pylorica

duodenum

Fig. 138. The pylorus – Internal structure

145

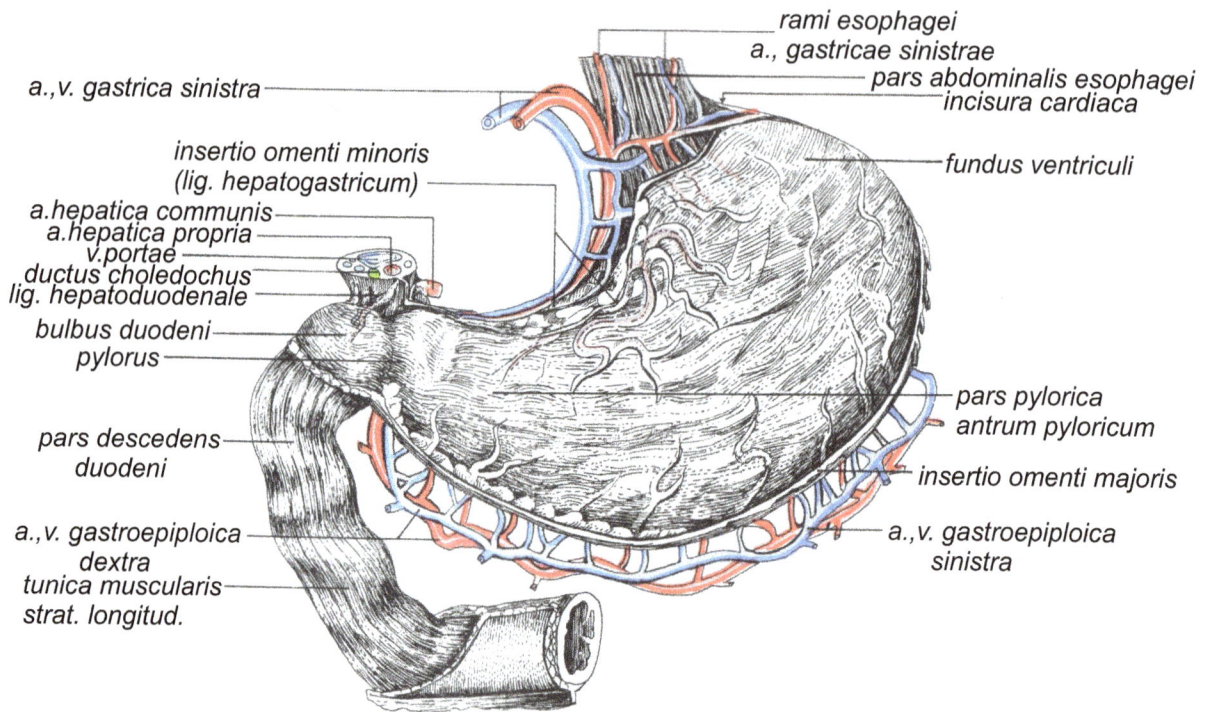

Fig. 139. Blood supply of the stomach.

Labels (clockwise):
- a.,v. gastrica sinistra
- rami esophagei
- a., gastricae sinistrae
- pars abdominalis esophagei
- incisura cardiaca
- fundus ventriculi
- insertio omenti minoris (lig. hepatogastricum)
- a.hepatica communis
- a.hepatica propria
- v.portae
- ductus choledochus
- lig. hepatoduodenale
- bulbus duodeni
- pylorus
- pars pylorica
- antrum pyloricum
- insertio omenti majoris
- a.,v. gastroepiploica sinistra
- pars descedens duodeni
- a.,v. gastroepiploica dextra
- tunica muscularis strat. longitud.

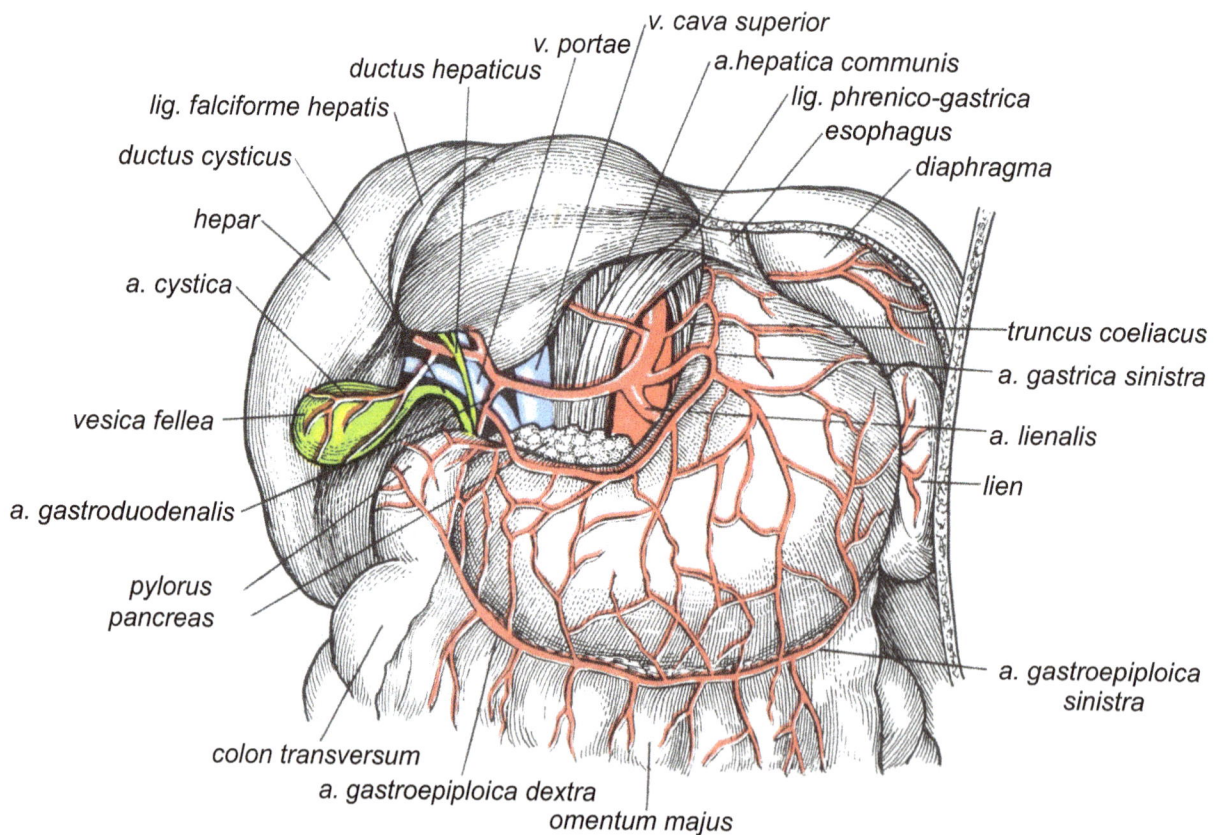

Fig. 140. The coeliac trunk.

Labels:
- ductus hepaticus
- v. portae
- v. cava superior
- a.hepatica communis
- lig. phrenico-gastrica
- esophagus
- diaphragma
- lig. falciforme hepatis
- ductus cysticus
- hepar
- a. cystica
- truncus coeliacus
- a. gastrica sinistra
- a. lienalis
- lien
- vesica fellea
- a. gastroduodenalis
- pylorus
- pancreas
- a. gastroepiploica sinistra
- colon transversum
- a. gastroepiploica dextra
- omentum majus

146

The abdominal region. The stomach projects also on the anterior abdominal wall, in the Labbe's triangle area, enclosed between the inferior border of the liver, in the epigastrium, the left costal arch and the transverse colon downwards (or the line which connects the extremities of the ribs of the tenth pair).

The posterior surface comprises three portions:

The superior portion, at the level of which is situated the phrenicogastric ligament; the stomach is in relation with the diaphragmatic muscle.

The median portion, which corresponds to the retrogastric projection of the omental bursa, through which the stomach is in relation: with the body and the tail of the pancreas, which is free in the pancreaticosplenic ligament, that contains the splenic vessels; with the anterior surface of the left kidney, enclosed by the adrenal gland, the spleen and the tail of the pancreas.

The inferior portion, situated beneath the lower border of the pancreas, where the stomach descends towards the transverse mesocolon and corresponds to Riolan's arch (the left superior colic pedicle). Through the mesocolon it has relations with the duodenojejunal angle and the first loops of the small intestine.

The lesser curvature (curvatura minor) gives attachement to the lesser omentum which, at this level, presents two segments: pars flaccida and pars condensa, higher situated, towards the oesophagus. The vessels and branches of the vagus nerves descend close to the lesser curvature. Beneath the lesser omentum lies the vestibule of the omental bursa.

The lesser curvature forms, to its right side, through the pyloroduodenal incisure, the boundary to the first part of the duodenum and to the coeliac region (on the deep plane of which, beneath the parietal peritoneum of the vestibule of the omental bursa, run the inferior vena cava, the abdominal aorta with the coeliac trunk and the solar plexus), from where emerge, at the level of the lower border of T12,on the anterior surface of the aorta, the coeliac trunk and its three branches enclosed by the solar plexus.

The greater curvature (curvatura major) is made up of two portions: a vertical and a horizontal portion.

The vertical portion corresponds, from above downwards, to the phrenicogastric ligament and the gastrosplenic omentum, which is very narrow and contains the short arteries, branches of the splenic artery.

The horizontal portion gives attachement to the gastrocolic ligament which contains, at a certain distance from the greater curvature, the arterial circle made up of the gastro-epiploic arteries.

The pylorus is a true channel, the musculature of which is structured as a sphincter. It achieves the communication between the gastric antrum and the first part of the duodenum and is directed obliquely upwards, backwards and to the right.

The pylorus is projected on the L1 vertebra. Deep sited, it has the following relations: in front and above, with the quadrate lobe of the liver and the neck of the gallbladder; below, with the pancreas, the portal vein, the hepatic artery, the gastroduodenal artery, the origin of the right gastro-epiploic artery, as well as with the transverse colon and the gastrocolic ligament; behind, the pylorus corresponds to the vestibule of the omental bursa and, through it, to the neck of the pancreas and the gastroduodenal artery; below, it bounds also the coeliac region and enters in relation with the transverse mesocolon and Riolan's arch; above, it is in relation with the lesser omentum and the liver pedicle.

The relations of the cardia are the same as those of the abdominal oesophagus.

Structure of the Stomach

The stomach presents the following tunics or coats (fig. 135-138):

- the mucous tunic, made up of a monostratified columnar epithelium, mucinous glands, which secrete mucus, in the antropyloric zone, and fundic glands, with an acidopeptic secretion;

- the submucous tunic, containing blood vessels, lymphatics and Meissner's submucous plexus;

- the muscular funic, which consists of three layers: a superficial longitudinal layer, beneath the serosa, a second,median layer, formed of circular fibres, and a third, deep layer, made up of oblique fibres.

- the serous tunic, represented by the visceral peritoneum; it lacks on a small area at the level of the fornix, where a fibrous tissue which connects the fornix to the diaphragmatic muscle (gastrophrenic ligament) is present.

147

The peritoneum is reflected in the region of the lesser and greater curvature, forming the lesser and the greater omentum; likewise, from the fornix (fornix ventriculi) and the greater curvature the peritoneum is reflected towards the hilum of the spleen, forming the gastrosplenic ligament.

Vessel and nerve supply (fig. 139,140)

The nutrient arteries arise from the coeliac trunk (the left gastric or coronary gastric artery, the hepatic artery and the splenic artery), forming two perigastric arterial circles, situated along the two curvatures.

The vascular circle of the lesser curvature is constituted of two anastomosed pedicles: the left gastric artery (a. gastrica sinistra) and the pyloric artery *(a. gastrica dextra)*.

The left gastric artery *(a. gastrica sinistra)* arises from the coeliac trunk, representing the superior branch of its trifurcation (it appears occasionally as a collateral branch of the splenic artery).

Its course comprises three portions: the posterior parietal portion, in which the artery is adherent to the posterior planes, fixed and surrounded by branches of the vagus nerve and by preaortic lymph nodes; the median portion, in which the artery, at first almost vertical, is directed towards the lesser curvature and raises a peritoneal fold, which separates the vestibule from the omental bursa proper; the gastric portion, in which the artery reaches the lesser curvature, at the junction of the upper third with the lower two thirds, and divides into two branches, anterior and posterior, which run very close to the lesser curvature and anastomose with the homologous branches of the pyloric artery (a. gastrica dextra).

The left gastric vein *(v. gastrica sinistra)* arises in proximity of the pylorus, ascends along the lesser curvature and becomes satellite of the artery which it accompanies in its falx. Arrived below the posterior parietal peritoneum, it is directed behind the pancreas and empties into the left flank of the portal vein.

The pyloric artery has, in most cases, its origin in the common hepatic artery or, occasionally, in the gastroduodenal artery. It is much thinner than the left gastric artery.

In the liver pedicle, the pyloric artery is situated in front and to the left of the elements of the pedicle (the portal vein behind, the hepatic duct on the left, the hepatic artery on the right). It is thus contained in the free border of the lesser omentum and ascends along the lesser curvature to anastomose with the left gastric artery.

The pyloric vein, satellite of the artery, empties into the portal trunk.

The vascular circle of the greater curvature is constituted of the anastomosis of the right and left gastro-epiploic arteries.

The right gastro-epiploic artery arises through the bifurcation of the gastroduodenal artery at the inferior border of the first part of the duodenum. It is thicker than the left. It ascends along the curvature in the greater omentum.

The right gastro-epiploic vein is satellite of the artery and comes beneath the pylorus, after which it bends and runs in front of the body of the pancreas, faking part in the formation of the anterior pancreatic arch. It ends by joining the right superior colic vein and empties info the superior mesenteric vein.

The left gastro-epiploic artery issues from the splenic artery or from one of its branches. It enters the stomach through the gastrosplenic omentum and has a retrogastric and a subgastric segment.

The left gastro-epiploic vein is satellite of the artery and opens into the splenic vein in the hilum of the spleen.

The short vessels arise from the splenic artery or its branches and supply the upper part of the greater curvature and the fundus of the stomach.

The lymph vessels of the stomach drain info three groups of lymph nodes, which should be extirpated in the course of gastrectomies performed for cancer (fig 142):

- the group of lymph nodes of the left gastric artery, from the lesser curvature to the origin of the artery;

-the group of lymph nodes satellite of the splenic artery, situated in the hilum of the spleen;

-the group of lymph nodes satellite of the right gastro¬epiploic artery, with retropyloric nodes of the gastroduodenal chain and subpyloric nodes.

The second lymphatic relay is represented by the hilar and preaortic nodes. In addition, there are connexions of the gastric lymph vessels with the supraclavicular lymph node (Troisier's node, in gastric cancer).

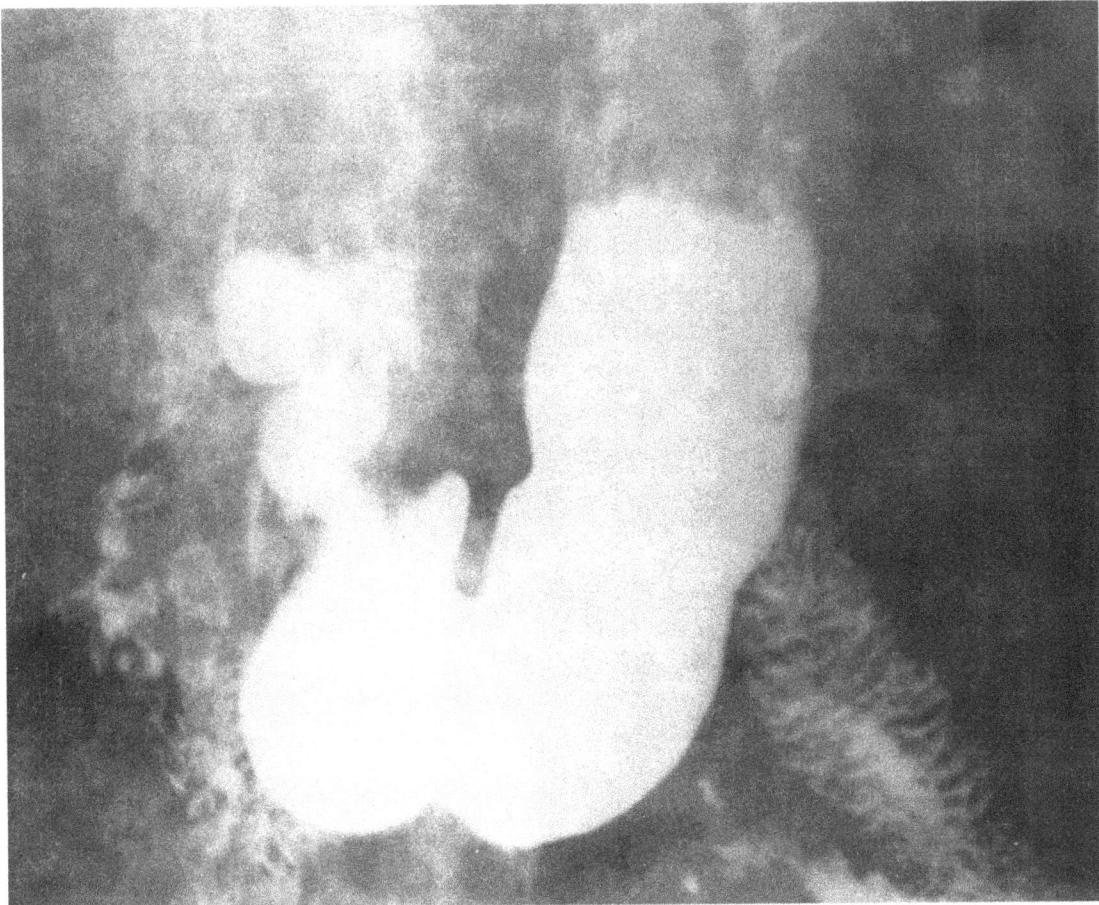

Fig. 141. Radiography of the stomach.

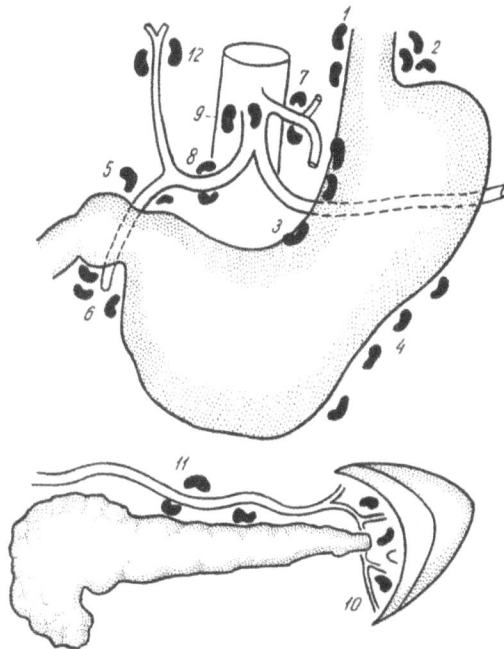

Fig. 142. Lymph nodes of the stomach.
First set: 1 - right paracardiac; 2 - left paracardiac; 3 - of the lesser curvature;
4 - of the greater curvature; 5 - suprapyloric; 6 - subpyloric.
Second: 7- group of the left gastric artery; 8 - group of the common hepatic artery;
9 - group of the coeliac trunk; 10 - group of the hilus of the spleen;
11 - group of the splenic artery.
Third set: 12 - group of the hepato-duodenal ligament;
13 - retropancreatic group (covered on the scheme).

149

The gastric nerve supply is double:
- sympathetic, derived from the solar plexus through the periarterial plexuses;
- parasympathetic, derived from the vagus nerves. Their physiological significance in gastric motility, secretion and sensibility should be emphasized, which explains the necessity of their surgical section (Dragstedf's vagotomy) in gastroduodenal ulcers.

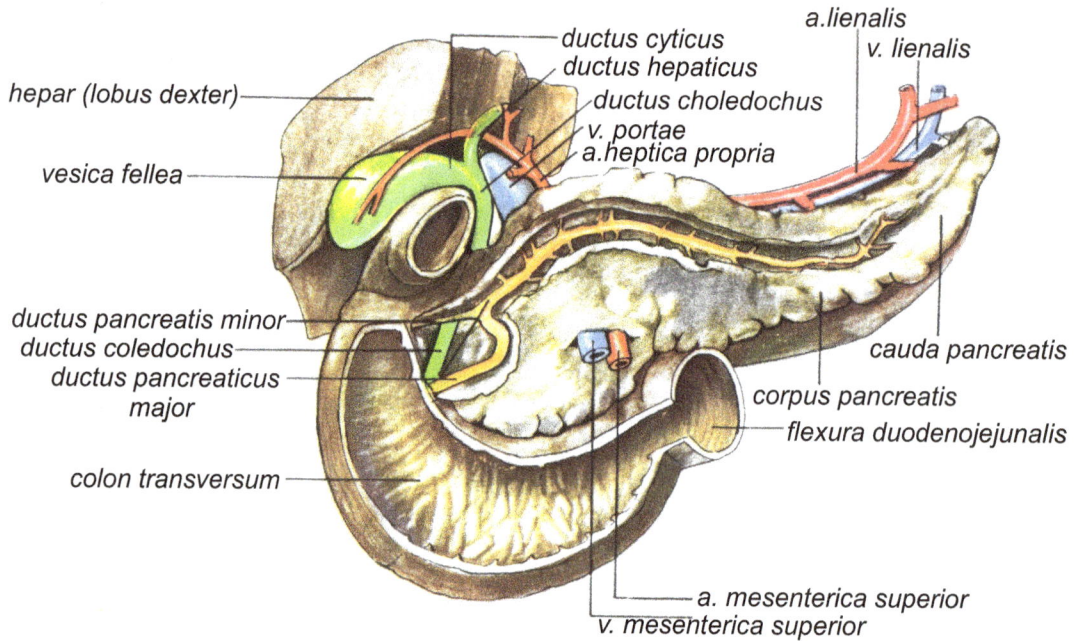

Fig. 143. The pancreas (excretory ducts - arrangement).

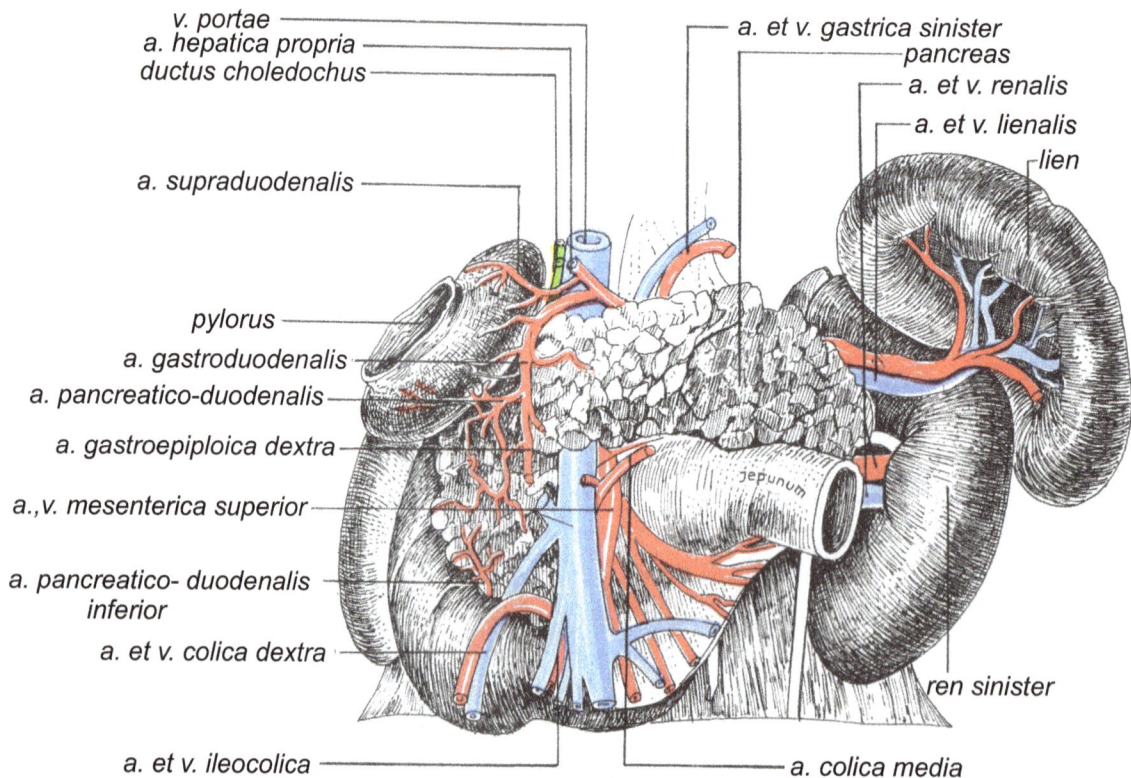

Fig. 144. Duodenum and pancreas - anterior view, vessel supply

150

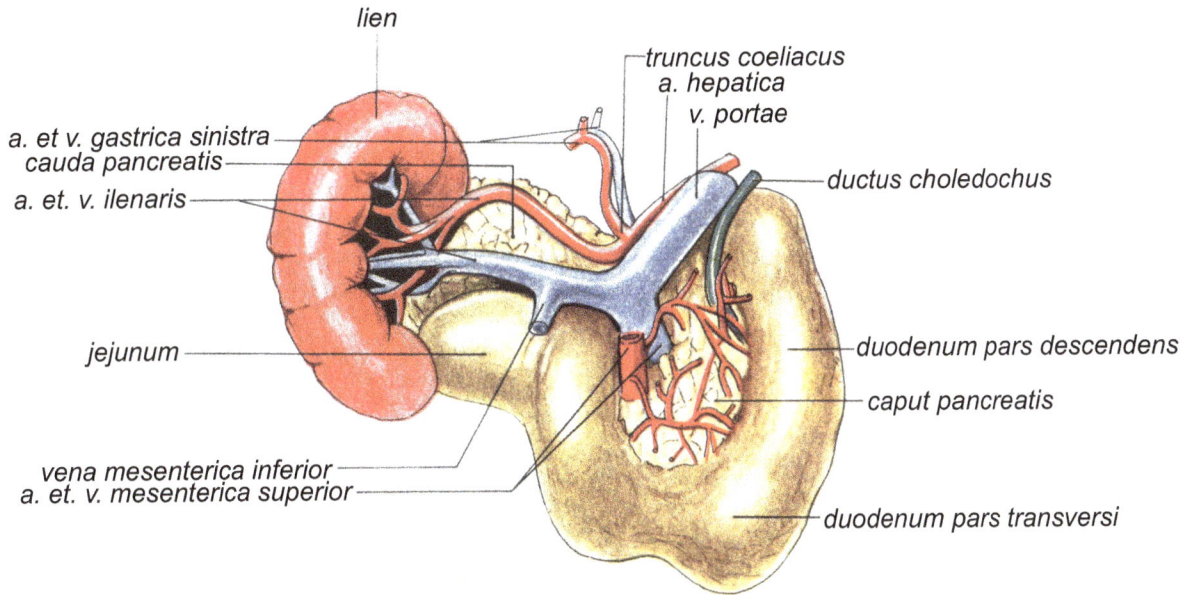

Fig. 145. Duodenum, pancreas and bile duct, viewed from behind.

lien

truncus coeliacus
a. hepatica
v. portae

a. et v. gastrica sinistra
cauda pancreatis

a. et. v. ilenaris

ductus choledochus

jejunum

duodenum pars descendens

caput pancreatis

vena mesenterica inferior
a. et. v. mesenterica superior

duodenum pars transversi

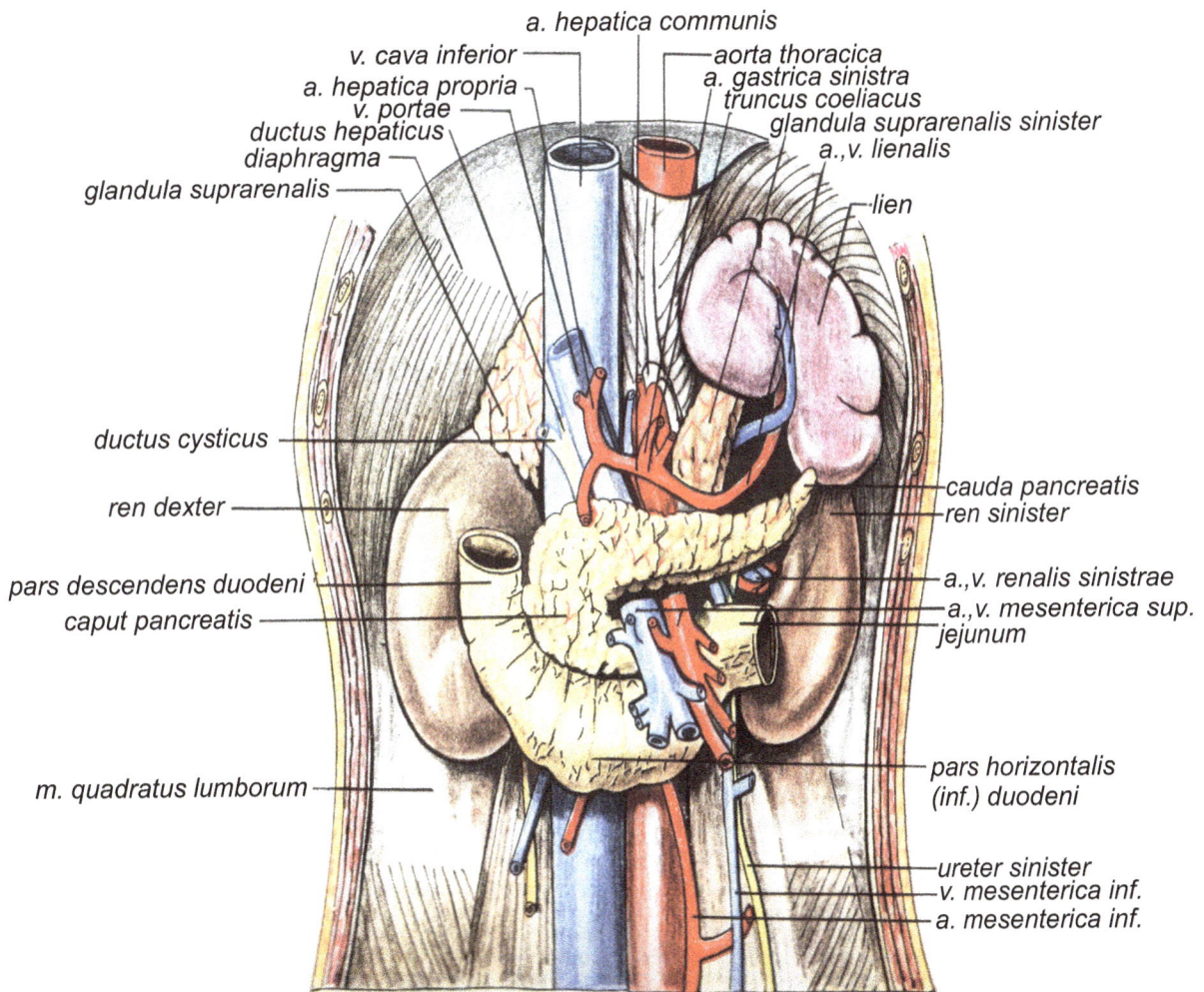

Fig. 146. Ths coeliac region.

a. hepatica communis

v. cava inferior
a. hepatica propria
v. portae
ductus hepaticus
diaphragma
glandula suprarenalis

aorta thoracica
a. gastrica sinistra
truncus coeliacus
glandula suprarenalis sinister
a.,v. lienalis

lien

ductus cysticus

ren dexter

pars descendens duodeni
caput pancreatis

cauda pancreatis
ren sinister

a.,v. renalis sinistrae
a.,v. mesenterica sup.
jejunum

m. quadratus lumborum

pars horizontalis
(inf.) duodeni

ureter sinister
v. mesenterica inf.
a. mesenterica inf.

151

The Duodenopancreatic Region

The duodenopancreatic region is made up of two organs, the duodenum and the pancreas, closely interdependent embriologically, anatomically (especially the head of the pancreas and the duodenum, to a lesser extent the tail of the pancreas, which is in relation with the hilum of the spleen) and pathologically. We shall describe them concomitantly with the supramesocolic part, although they are situated in the retroperitoneal space (they are secondarily retroperitoneal organs), since through the omental bursa they have close relations with the organs of this part and, in addition, the vascular pedicles, also derived from the coeliac trunk, increase this interdependence. Moreover, the duodenopancreatic region represents topographically a transitional area between the supra- and the inframesocolic part (fig. 143,151).

The Duodenum

It is the first part of the small intestine, as it is situated between the pyloric portion of the stomach and the jejunum and ends at the level of the duodenojejunal angle *(duodenojejunal flexure)* (the place where the intestine becomes mobile owing to the existence of the mesentery). It is horseshoe-shaped with the concavity directed upwards, surrounds the head of the pancreas and has very close relations with the pancreas, as well as with the pancreatic and biliary excretory ducts. It is made up of four parts: the first part *(pars superior),* which extends from the duodenopyloric groove to the neck of the gall bladder and is also called duodenal bulb; the second *(pars descendens),* from the neck of the gall bladder, where the duodenum changes its direction, becoming vertical, to the inferior pole of the right kidney; the third *(pars horizontalis),* which changes again its direction and extends from the inferior pole of the right kidney to the left side of the vertebral column, more exactly to the root of the mesentery; the fourth *(pars ascendens),* which ascends along the left surface of the body of the L_2 vertebra, beginning from the left aspect of the mesentery. In this course, it forms three flexures: the superior duodenal flexure *(flexura duodeni superior),* the inferior duodenal flexure *(flexura duodeni inferior)* and the duodenojejunal flexure *(flexura duodenojejunalis)* (fig. 143, 145). Inside the duodenum are circular folds (Kerckring's folds) (which are lacking in the superior part) and intestinal villi. At the level of the descending part lies a longitudinal fold of mucosa *(plica longifudinalis duodeni),* in the medial wall of the organ, formed as the result of the descent of the common bile duct in the thickness of the wall; inferiorly to the fold lies a prominence, the major duodenal papilla *(papilla duodeni major),* that contains the Voter's ampulla, into which open, together, the common bile duct and the main pancreatic duct (Wirsung' canal); 3 cm above is situated the minor duodenal papilla *(papilla duodeni minor),* on the summit of which opens the accessory pancreatic duct (Santorini's duct) (fig. 143,145,150). The duodenal wall has four tunics; serous, muscular, submucous and mucous. The submucous tunic contains Brunner's duodenal glands *(glandulae duodenales),* of tubulo-alveolar type, densely grouped, especially in the first part of the duodenum, up to 3 mm in size, macroscopically visible, typical of the duodenum. Duodenal relations: As regards the relations with the peritoneum, in the adult the duodenum becomes retroperitoneal, as the result of a process of coalescence which forms the Treitz's fascia, through which the posterior surface of the duodenum blends with the peritoneum of the posterior abdominal wall. The mesentery persists at the level of the proximal segment of the first part of the duodenum, while the distal segment and the second, third and fourth parts of the duodenum are covered with peritoneum only on their anterior surface. The second part is crossed by the root of the transverse mesocolon and the third part, by the root of the mesentery (fig. 147). The peritoneum of the proximal segment of the duodenal bulb, the prepyloric segment, is continuous upwards with the duodenal portion of the gastrohepatic omentum, which is also called duodenohepatic ligament and contains in its depth the elements of the liver pedicle. The peritoneum of the fourth part of the duodenum and of the duodenojejunal angle forms, in most cases, a number of crescentic folds *(plicae semilunares),* beneath which lie more or less discernible depressions, called duodenal fossae or recesses, resulted from a coalescence defect - either an excessive coalescence or the presence of vessels which raise the peritoneum under the form of these folds. At their level may occur infernal hernias, through penetration of the intestinal loops into the respective recesses. They are the following (fig. 148,149):
- *the superior duodenal recess,* at the level of the upper segment of the ascending part of the duodenum;

- *the inferior duodenal recess,* at the level of the lower segment of the ascending part of the duodenum;
    - *the superior or venous paraduodenal recess,* situated between the superior and inferior duodenal recesses;
    - *the inferior ar arterial paraduodenal recess,* very rarely present, formed by the superior left colic artery;
    - *the superior duodenojejunal recess* (Thoma Ionescu), formed by the fold of the inferior mesenteric vein, situated on the left of the vertebral column, between the upper border of the duodenojejunal angle and the root of the transverse mesocolon;
    - *the superior retroduodenal recess* (Gruber-Landzert), situated behind the fourth part of the duodenum.

In addition, we mention that the duodenum surrounds -above, laterally and below - the head of the pancreas.

The superior part (first part) has relations: in front and above with the liver and the gall bladder; behind, with the common bile duct, the portal vein and the gastroduodenal artery; below, with the head of the pancreas.

The descending part (second part) is crossed anteriorly by the transverse mesocolon. The supramesocolic portion has relations with the liver and the gall bladder, to which it is occasionally connected through the cysticoduodeno-colic ligament; the inframesocolic portion has relations with the loops of the small intestine. Posteriorly, this portion has relations with the anterior surface of the right kidney, with the renal pedicle (therefore it is also called "Cruveilhier's prerenal portion") and with the inferior vena cava. Between the duodenum and these organs lies the Treitzs coalescence fascia. Medially, it has relations with the pancreas and laterally, with the liver and the ascending colon.

The inferior, horizontal part (third part) is situated in the submesocolic or intestinal region and is anteriorly in relation with the root of the mesentery, with the superior mesenteric vessels (the artery on the left, the vein on the right) which cross if, emerging from below the pancreas, and with the superior jejunal loops. Behind, through Treitz's fascia, it has relations with the psoas muscle, the inferior vena cava, the right ureter, the aorta and the origin of the right internal spermatic artery or of the right ovarian artery. Above, it is in relation with the pancreas, from which it is separated through the superior mesenteric vessels and, below, with the loops of the small intestine.

The ascending part (fourth part) has the following relations: above and in front, with the stomach-through the transverse mesocolon and with the loops of the small intestine (jejunal coils); behind, with the psoas muscle, the left renal pedicle and the aorta; medially, with the aorta and the superior origin of the mesentery which crosses it, and laterally, with the medial border of the left kidney, from which it is separated through the ureter and Treitz's vascular arch (formed superiorly by the inferior mesenteric vein, medially by the inferior mesenteric vein accompanied by the left colic artery and inferiorly, by the left colic artery); the peritoneum may insinuate beneath Treitz's arch and form a recess, with the concavity directed medially, liable to give rise to an internal hernia.

Vessel and nerve supply (fig. 144,146)

The arterial supply is provided by the pancreatico-duodenal arteries which arise as follows: the two superior arteries, from the gastroduodenal artery, a branch of the hepatic artery, and the two inferior, from the superior mesenteric artery.

The veins are satellites of the arteries and tributaries of the portal vein.

The lymph vessels are united with those of the pancreas and drain into the hepatic and coeliac lymph nodes.

The nerve supply is sympathetic and parasympathetic.

The Pancreas

The pancreas is a mixed exocrine and endocrine gland, situated behind the dorsal parietal peritoneum, hence it is a retroperitoneal organ, like the duodenum. It is divided by the root of the mesocolon, like the duodenum, into a supramesocolic and an inframesocolic region. It is a deep organ, solidary with the duodenum (fig. 143-147,150).

Elongated transversally, it has somewhat the shape of a hammer with the handle to the left (Mackel) and is made up of four portions:

The head (caput pancreatis), somewhat circular in shape, sited within the duodenal horseshoe, presents a projection directed medially and inferiorly, situated, behind the superior mesenteric vessels and called uncinate process *(processus uncinatus).*

The neck, a narrower segment, is situated between the first part of the duodenum and the superior mesenteric vessels.

The body (corpus pancreatis), elongated, prismatic-triangular in shape on section, has three surfaces (anterior, posterior and inferior) and three borders (superior, anterior and inferior). On the anterior surface, near the head, a prominence is present, called omental tubercle (tuber omentale), and the posterior surface contains two grooves for the splenic vein and artery.

The tail (cauda pancreatis) lies in the pancreatico-splenic omentum and may have various shapes, as it can be long or short (fig. 146).

It is altached to the posterior abdominal wall, at the level of the head, through Treitz's fascia, corresponding to the coalescence of the mesoduodenum, at the level of the body of the pancreas, through the coalescence of the posterior mesogastrium.

At the level of the head, the root of the transverse mesocolon defines the position of the pancreas in the two parts of the peritoneal cavity. There are two portions of the peritoneal serosa which cover the pancreas on its anterior surface: a supra- and a submesocolic portion.

At the level of the body, the pancreas corresponds to the omental bursa, thus it is supramesocolic (fig. 147).

The relations of the pancreas are the following:

The head of the pancreas is crossed transversally, in front, by the root of the transverse mesocolon, which divides it into a supramesocolic portion, in relation with the pyloric portion of the stomach and situated in the vestibule of the omental bursa, and an inframesocolic portion, in relation with the transverse colon and the loops of the small intestine.

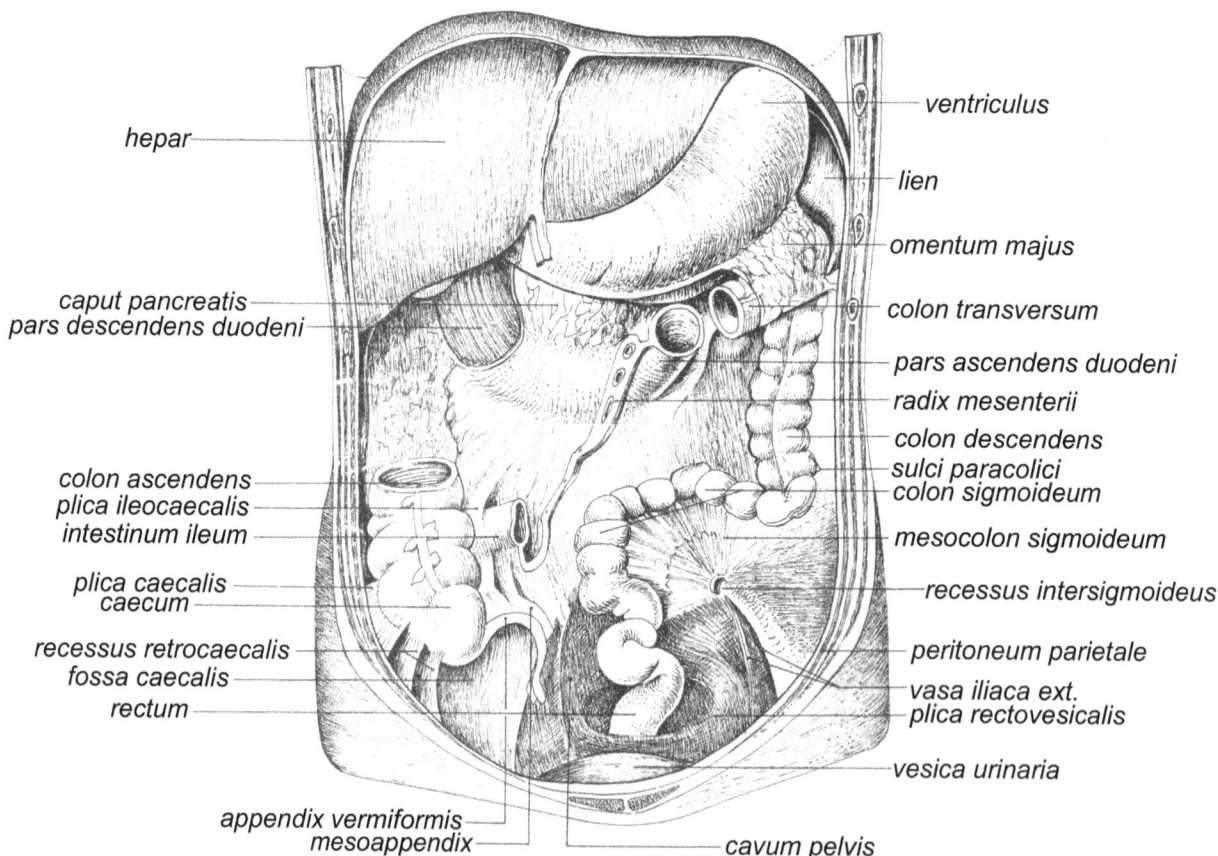

Fig. 147. The supramesocolic part - anterior view - and the inframesocolic part, after section of the ascending colon, the transverse colon and the mesentery.

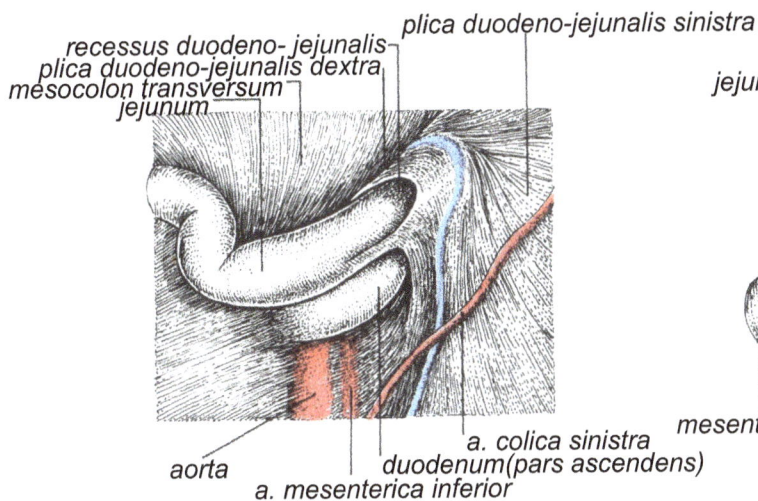

recessus duodeno-jejunalis
plica duodeno-jejunalis dextra
mesocolon transversum
jejunum

plica duodeno-jejunalis sinistra

aorta
a. mesenterica inferior
duodenum(pars ascendens)
a. colica sinistra

Fig. 148. The duodenojejunal recesses.

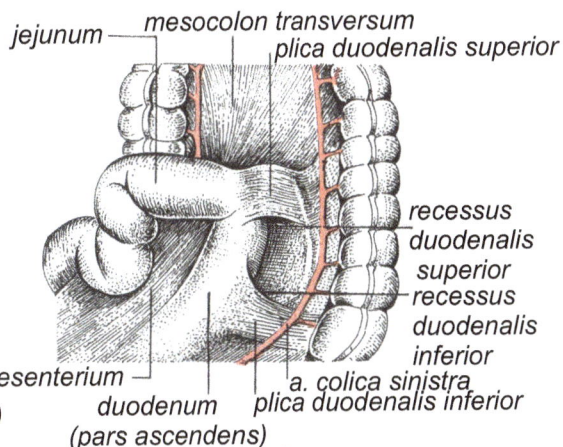

jejunum
mesocolon transversum
plica duodenalis superior

recessus
duodenalis
superior
recessus
duodenalis
inferior
a. colica sinistra
plica duodenalis inferior

mesenterium
duodenum
(pars ascendens)

Fig. 149.The duodenal recesses

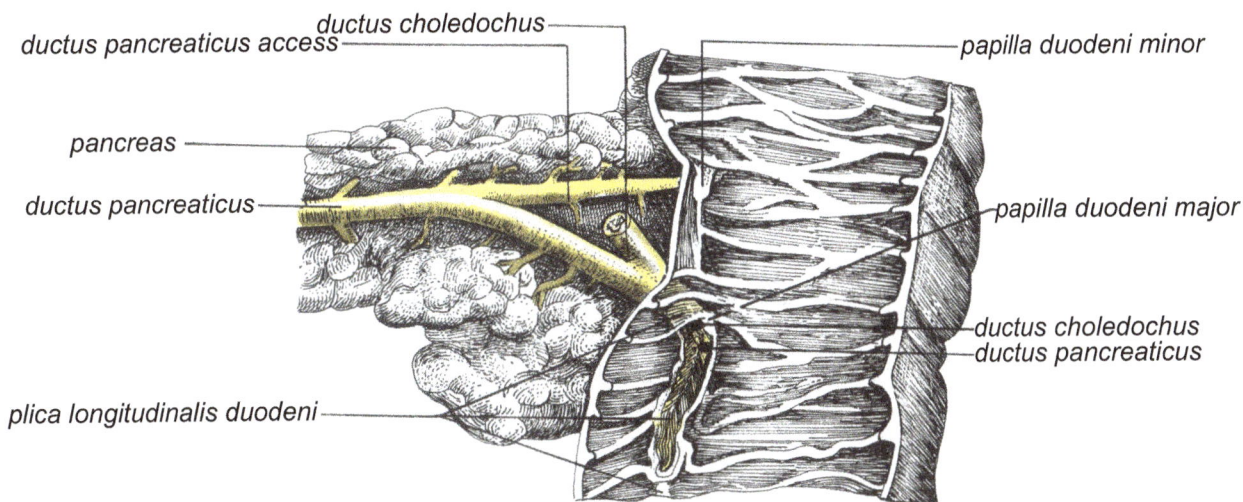

ductus pancreaticus access
ductus choledochus
papilla duodeni minor

pancreas

ductus pancreaticus

papilla duodeni major

ductus choledochus
ductus pancreaticus

plica longitudinalis duodeni

Fig. 150. Structure of the duodenum - opening of the pancreatic ducts.

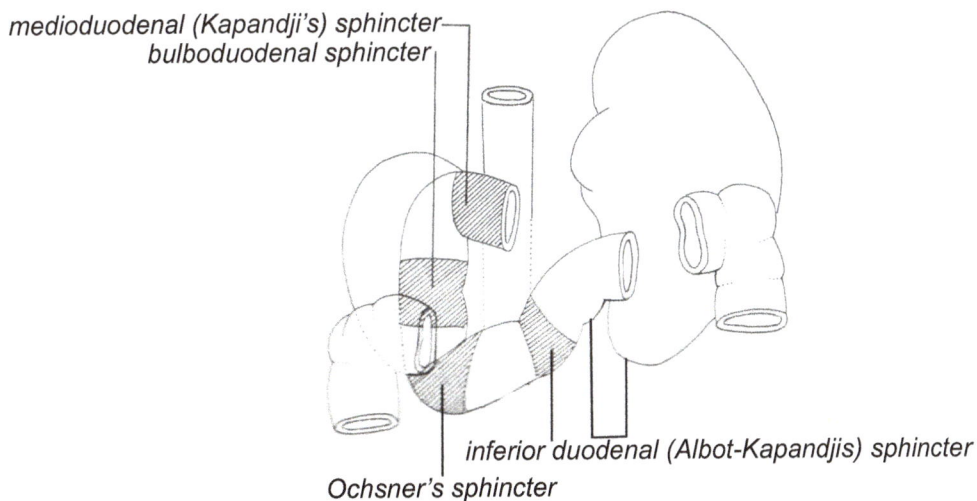

medioduodenal (Kapandji's) sphincter
bulboduodenal sphincter

inferior duodenal (Albot-Kapandjis) sphincter

Ochsner's sphincter

Fig. 151. Sphincters of the duodenum

155

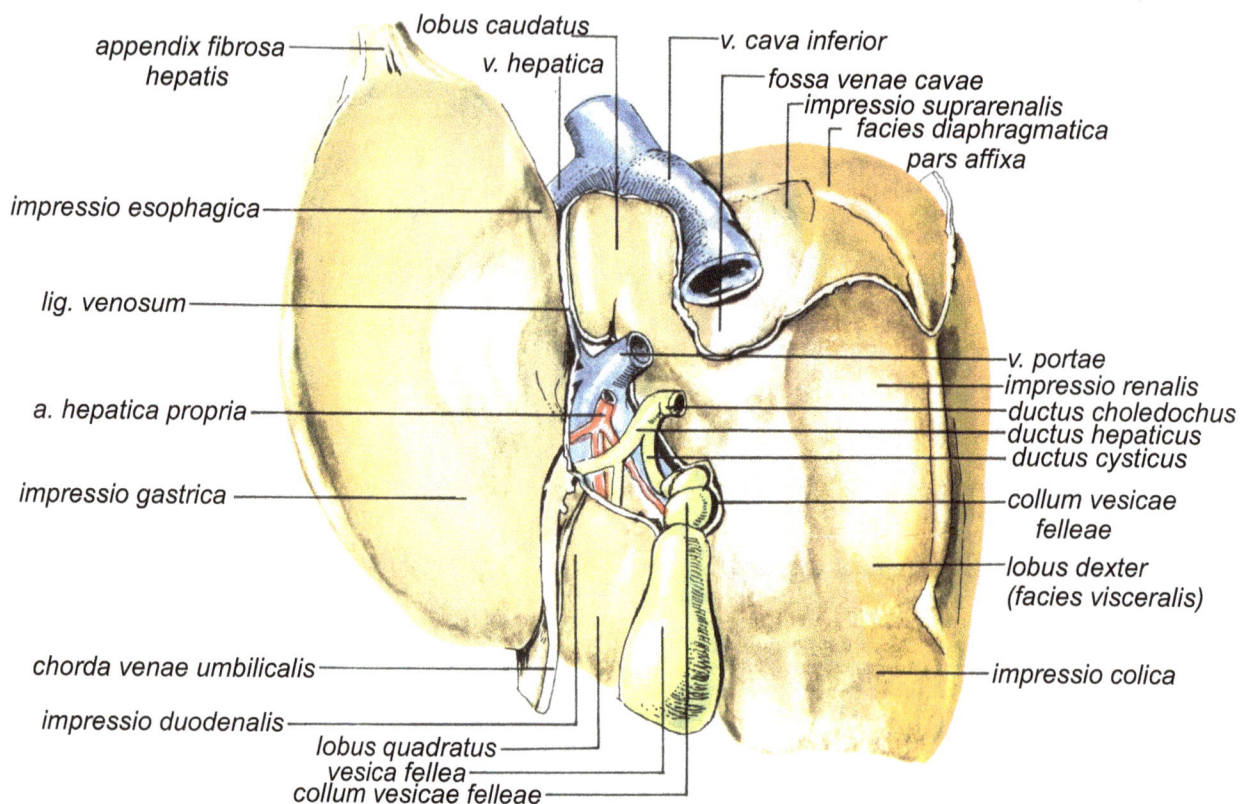

Fig. 152. Inferior surface of the liver.

- appendix fibrosa hepatis
- lobus caudatus
- v. hepatica
- v. cava inferior
- fossa venae cavae
- impressio suprarenalis
- facies diaphragmatica
- pars affixa
- impressio esophagica
- lig. venosum
- a. hepatica propria
- impressio gastrica
- v. portae
- impressio renalis
- ductus choledochus
- ductus hepaticus
- ductus cysticus
- collum vesicae felleae
- lobus dexter (facies visceralis)
- chorda venae umbilicalis
- impressio duodenalis
- lobus quadratus
- vesica fellea
- collum vesicae felleae
- impressio colica

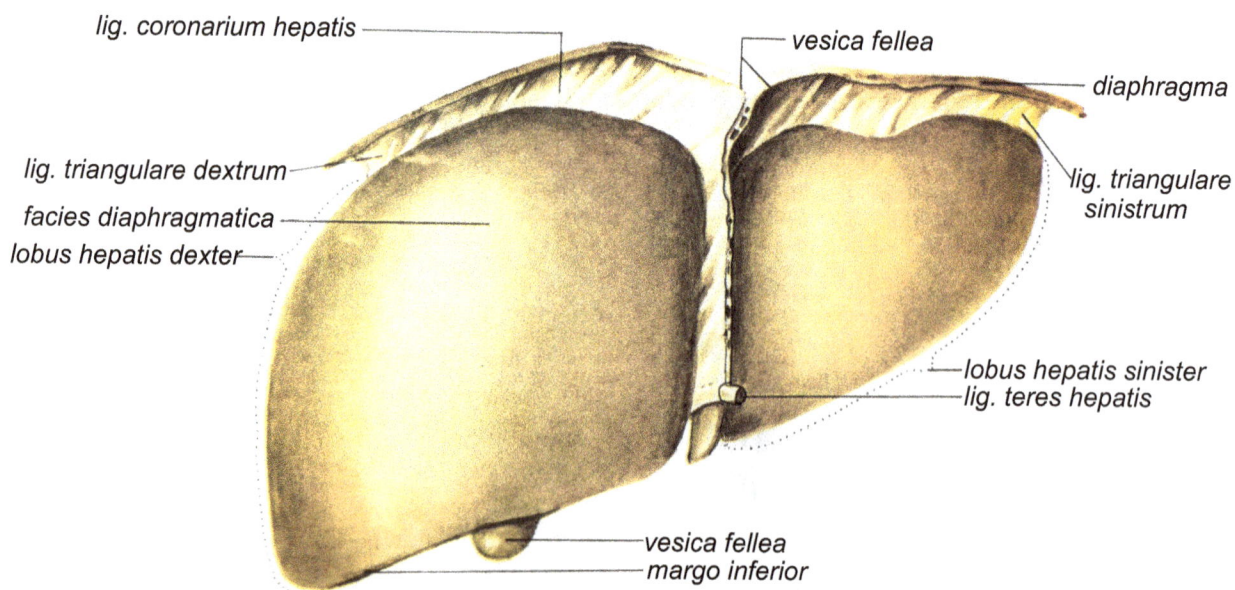

Fig. 153. Disposition of the peritoneum at the level of the liver (hepatic ligaments).

- lig. coronarium hepatis
- vesica fellea
- diaphragma
- lig. triangulare dextrum
- facies diaphragmatica
- lobus hepatis dexter
- lig. triangulare sinistrum
- lobus hepatis sinister
- lig. teres hepatis
- vesica fellea
- margo inferior

On the anterior surface of the head run the right gastroepiploic vessels and sometimes the pancreaticoduodenal vessels.

The posterior surface is in contact with the common bile duct, which turns away from the portal vein and gives rise to a groove or tunnel in the pancreatic tissue, applied on the pancreas through Treitz's fascia, then with the duodenopancreatic vascular arches; through Treitz's fascia it corresponds to- the inferior vena cava, the internal border of the kidney, the right ureter and the right renal pedicle.

The superior border is in relation with the first part of the duodenum.

The right border is enclosed by the second part of the duodenum, which surrounds it, and is in relation with the biliopancreatic confluent.

The inferior border corresponds to the third part of the duodenum, with which if lies in a close contact.

The left border is in relation with the superior mesenteric vessels.

The neck of the pancreas is a constricted portion of the pancreas, enclosed between the first part of the duodenum and the superior mesenteric vessels which, as a matter of fact, determine it.

Its anterior surface is in relation with the peritoneum of the posterior wall of the omental bursa at the level of the gasroduodenal artery, with the root of the transverse mesocolon which attaches on it, with the pylorus and the transverse colon.

The posterior surface is in relation, through Treitz's fascia, with the inferior vena cava and with the site of origin of the portal vein.

The superior border is in contact with the duodenum.

The inferior border corresponds to the superior mesenteric vessels which pass over it.

The body of the pancreas has three surfaces and three borders, it has an irregular shape, of a triangular prism.

The anterior surface *(fades anterior)* is covered with the parietal leaflet of the omental bursa and is in contact through it with the stomach.

The posterior surface *(facies posterior)* is in contact with Treitz's fascia and, through it, has relations, from the right to the left, with the following elements: with the aorta and the origin of the superior mesenteric artery; with the coeliac plexus; with the left renal vein; with the left renal pedicle and the left kidney; with the splenic artery and vein (which receives at this level the inferior mesenteric vein); with the pancreaticosplenic lymph nodes.

The inferior surface, of smaller extent, presents three impressions, from left to right: a duodenojejunal impression, made by the duodenojejunal flexure; a jejunal impression, left by the superior jejunal loops; a colic impression, made by the transverse colon.

The superior border corresponds, on the right, to the coeliac trunk and solar plexus and on the left, to the splenic pedicle.

The anterior border is the site of insertion of the transverse mesocolon.

The posterior border corresponds to the duodenojejunal angle and the left kidney.

The tail of the pancreas lies in the pancreaticosplenic omentum.

Its anterior surface forms part of the posterior wall of the omental bursa and is in relation with the splenic vessels.

The posterior surface corresponds to the left kidney.

The inferior surface gives attachement to the transverse colon.

The free border corresponds to the hilum of the spleen.

Structure of the Pancreas

It is a mixed, exocrine-endocrine gland, enveloped at the periphery in a connective capsule.

The exocrine pancreas consists of spherical acini, resembling those of the parotid gland.

The excretory canalicular system begins at the level of the acini through the intercalary ducts which unite into collector ducts, forming, ultimately, the main excretory duct, called also Wirsung's canal or pancreatic duct *(ductus pancreaticus),* and the accessory excretory duct, called also Santorini's duct *(ductus pancreaticus accesorius),* which open into the duodenum, between the two papillae, inferior *(major)* and superior *(minor).*

The endocrine pancreas is represented by the islets of Langerhans, diffusely scattered in the tissue of the exocrine gland.

They contain beta cells, which secrete insulin, and alpha cells, which secrete glucagon and the lipocaic factor.

Vessel and nerve supply (fig. 144,145)

The arterial supply is provided by the pancreaticoduodenal arteries, branches of the gastroduodenal arteries, which derive from the hepatic and superior mesenteric arteries, as well as by the pancreatic arteries, branches of the splenic artery.

The common hepatic artery sends sometimes directly to the pancreas a medial hepatic artery.

The veins, satellites of the arteries, drain the blood into the portal vein.

The lymph vessels end in the peripancreatic and especially in the retropancreatic lymph nodes, as well as in the pancreatscolienal superior and inferior pancreatic and coeliac lymph nodes.

The autonomic sympathetic and *parasympathetic innervation* is represented by the coeliac plexus, the superior mesenteric plexus and the splenic plexus.

The Hepatic Region

The hepatic region *(regio hepatica),* situated in the right part of the supramesocolic portion of the abdominal cavity, contains the liver (fig. 152-170).

It is situated in the right hypochondrium (subphrenic space) and in the superior area of the left epigastrium. It is in relation above, with the diaphragmatic muscle, with which it comes in contact through its convex surface *(fades diaphragmatica),* and below, with the organs of the superior abdominal part (oesophagus, stomach, supramesocolic duodenum) and, through its inferior surface *(fades visceralis),* with some delimiting organs of this part (the hepatic angle of the colon and the right half of the transverse colon). Below, the hepatic bed is related to the adjacent spaces of the two parts: supramesocolic and inframesocolic. Thus, through the Winsiow's hiatus *(foramen epiploicum)* it communicates with the omental bursa; through the hepatoreno-duodenocolic space *(spatium hepatocolicum),* with the right parietocolic space *(spatium parietocolicum dexter);* above the cardiac notch *(incisura cardiaca)* and the fundus of stomach *(fornix ventriculi),* with the splenic bed *(loja lienalis)* and in front of the greater omentum *(omentum ma/us),* with the inferior abdominal part.

It is almost completely invested with parietal peritoneum and the organs contained in it are covered with visceral peritoneum, with the exception of the coalescence area of the liver through *pars affixa* and, naturally, of the insertion lines of the following ligaments; *ligamentum falciforme, ligamentum coronare, ligamentum triangulare, ligamentum coronare ventriculi, plica gastropancreatica, ligamentum hepato-esophagicum-pars tensa et ligamentum hepatogastricum pars flaccida, ligamentum hepatoduodenale, ligamen-tum hepatorenale, ligamentum renocolicum,* respectively *ligamentum phrenocolicum dexter.*

The hepatic peritoneum. The visceral peritoneum invests the liver and is continuous with the parietal peritoneum through the ligaments and with the peritoneum of the other organs of the digestive tube through an omentum: the lesser omentum *(omentum minus).*

The hepatic ligaments are peritoneal formations which connect the liver to the diaphragm and to the anterior abdominal wall (fig. 153-155).

The coronary ligament (ligamentum coronare), extending transversal from right to left, situated in a transverse plane, has two leaflets: superior and inferior. Each represents the reflexion on the liver of the parietal peritoneum, which becomes visceral peritoneum, so that in the area of the ligament the liver comes in contact with the posterior elements, without peritoneal interposition *(pars affixa hepatica).*

The superior leaflet is continuous with the falciform ligament.

The inferior leaflet passes on the anterior surface of the inferior vena cava and ascends along its left border, enclosing the spigelian lobe. It runs along the superior border, then along the left border of this lobe and unites with the posterior leaflet of the lesser omentum. The left segment of the inferior leaflet of the coronary ligament is continuous, on the right side, with the anterior leaflet of the lesser omentum.

Its two extremities contribute to the formation of the triangular ligaments, the left one being better individualized.

The left triangular ligament (ligamentum triangulare sinistrum) represents the left extremity of the coronary ligament, the leaflets of which are coalescent. It is an almost horizontal lamina, which connects the left lobe to the diaphragm.

Thus, it has a superior surface, an inferior surface and three borders: posterior-diaphragmatic, anterior-hepatic and lateral-free.

The right triangular ligament (ligamentum triangulare dextrum) has an identical disposition, but is shorter and less well individualized than the left one.

The falciform ligament (ligamentum falciforme) passes transversally from the anterior abdominal wall and from the diaphragm to the liver and is situated in a plane which looks backwards and to the left. It is triangular in shape and has the following borders: an anterosuperior border, extending from the diaphragm to the umbilicus; a postero-inferior border, from the superior leaflet of the coronary ligament to the round ligament; a free border, from the umbilicus to the round ligament.

The falciform ligament contains the round ligament in its free border (a fibrous vestige of the obliterated umbilical vein) and the paraumbilical vein.

The lesser omentum (omentum minus) is a vertical and transverse fold extending from the liver to the abdominal oesophagus, the stomach and the first part of the duodenum. It is constituted of two leaflets and has, consequently, two surfaces and four borders.

The hepatic border is variously disposed, in dependence on the topographical segment. Thus, at the level of the oesophagus, the anterior leaflet covers the anterior surface of the oesophagus and the posterior leaflet reflects on its right border; at the level of the lesser curvature of the stomach, the two leaflets are continuous with the visceral, peritoneum of the stomach, becoming thus the serous tunic of the latter (the vascular circle of the lesser curvature). At the level of the mobile duodenum (duodenal bulb), on the left of the gastroduodenal artery, the disposition is identical At the level of the fixed duodenum, the anterior leaflet covers the duodenum, while the posterior leaflet reflects and is continuous with the parietal peritoneum, forming the duodenocaval ligament.

The diaphragmatic border is very short. The two surfaces are continuous with the diaphragmatic peritoneum.

The free border contains the hepatic pedicle and bounds anteriorly the hiatus of Winslow. There is sometimes a projection of the lesser omentum towards the right side of the region, forming the cysticoduodeno-colic ligament.

The anterior surface of the lesser omentum is covered by the liver (subhepatic space).

The posterior surface forms the anterior wall of the vestibule of the omental bursa and correponds to the coeliac region.

The lesser omentum is made up of three parts:
- the right part *(pars vascularis):* which contains the hepatic pedicle (the posterior portal vein, the hepatocholedochus and anteriorly and on the right, the hepatic artery) in front and on the left.
- the median part *(pars flaccida):* transparent and very thin;
- the superior left part *(pars tensa):* thicker, containing the hepatic branch of the left gastric artery and the hepatic branch of the vagus nerve.

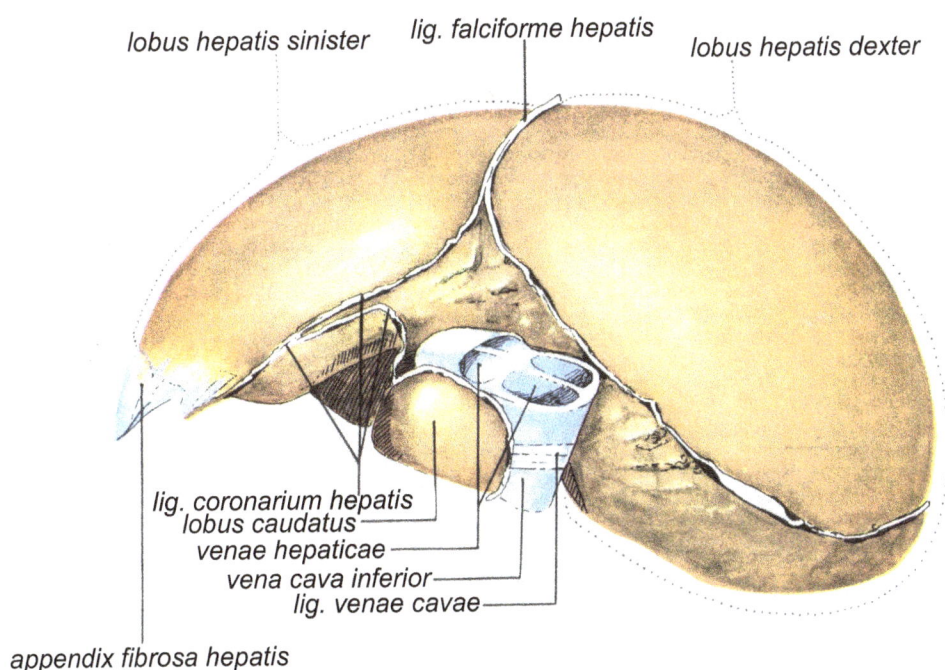

Fig.154. Diaphragmatic surface of the liver (posteriorview).

The Liver
(Hepar)

The liver *(hepar)* is a bulky gland, situated in the right subphrenic space in the supramesocolic part of the abdomen. It has two surfaces: superior *(facies diaphragmatica)* and inferior *(fades visceralis)* separated, in their anterior portion, by an inferior border *(margo acutus)*.

Posteriorly, the two surfaces are continuous with each other, through a rounded marge *(margo obtusus)*, which is not a separative boundary, but may be considered also as a posterior surface, corresponding to the hepatic area that is not covered with peritoneum, the so-called "bare area" *(pars affixa)*.

The superior or diaphragmatic surface *(fades diaphragmatica)*is convex and broad beneath the diaphragm; it is divided into two lobes by an anteroposterior peritoneal ligament, called falciform ligament (already described), which connects the liver to the diaphragmatic muscle. The delimitation into the two lobes is completed, on the inferior surface, by the left sagittal groove. The right lobe, on the diaphragmatic surface, bears often the impression of the costal arch and the left lobe shows the cardiac impression (fig. 153,154).

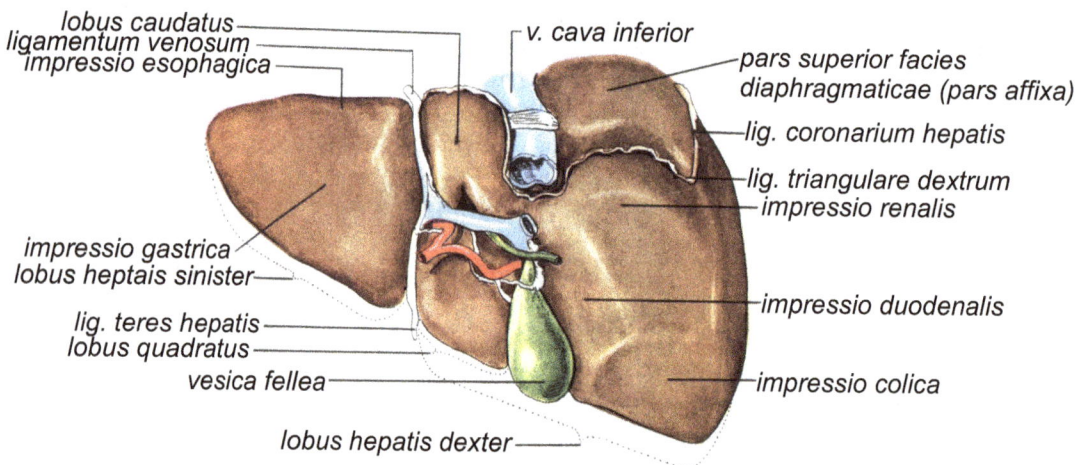

Fig. 155. Inferior surface of the liver - porta hepatis (transverse fissure).

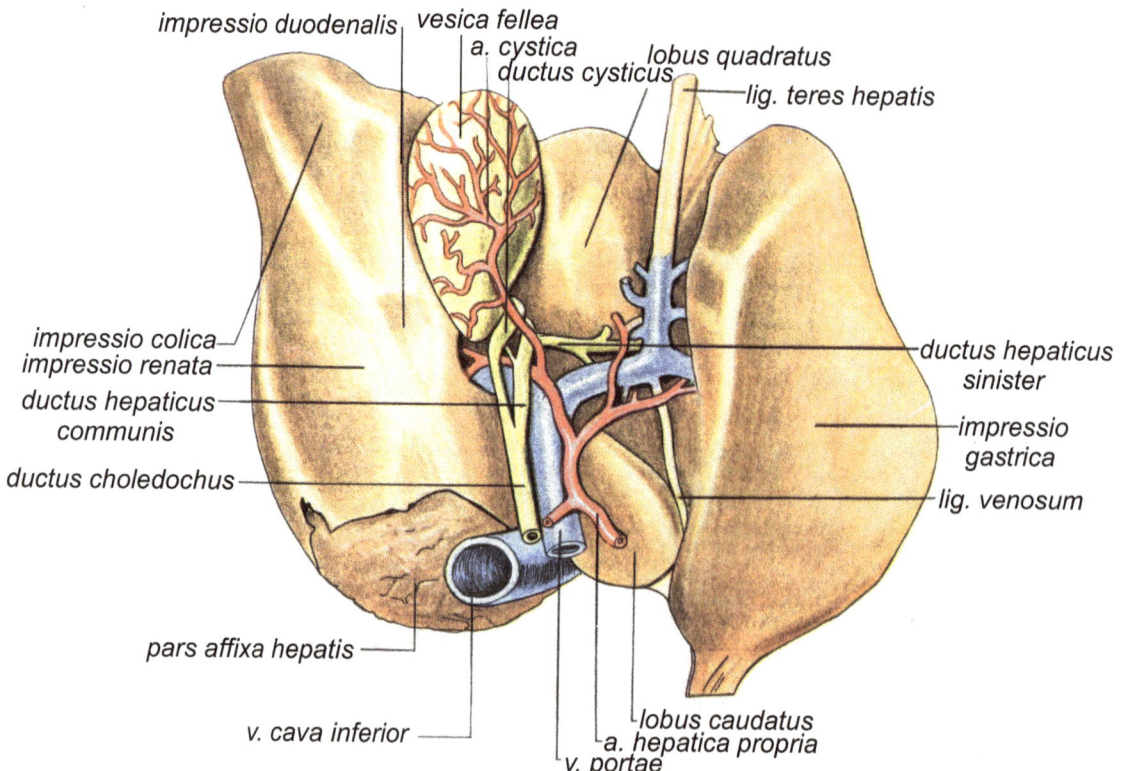

Fig. 156. Visceral surface of the liver (disposition of the portal vein, the hepatic artery and the choledochus; blood supply of the gall bladder).

160

The inferior surface *(fades visceralis)* looks downwards, backwards and to the left and has three grooves (fig. 152,155,157):

- the left anteroposterior (left sagittal) groove *(sulcus sagittalis sinister)* begins at the pit of the anterior border and reaches posteriorly the incisure of the venous ligament. It is made up of two components: an anterior component, which corresponds to the round ligament *(ligamentum teres hepatis)*, the obliterated portion of the umbilical vein, and a posterior component, due to the venous ligament *(ligamentum venosum)*, the obliterated vestige of the canal of Arantius *(ductus venosus)*;

- the right anteroposterior (right sagittal) groove *(sulcus sagittalis dexter)* presents in front the fossa for the gall bladder *(fossa vesicae felleae)* and behind, the fossa for the vena cava *(sulcus venae cavae)*. The anterior segment is separated from the posterior segment by an area of hepatic parenchyma, called caudate process *(processus caudatus)* of the spigelian caudate lobe;

- the transverse groove *(sulcus transversus)* lodges the hilum of the liver *(porta hepatis)*, made up of the elements of the hepatic pedicle (behind, the portal vein; in front on the left, the hepatic artery; in front on the right, the hepatic duct).

The three sulci divide the visceral surface of the liver into three areas, of which two lateral- one on the right and the other on the left of the sagittal sulci-and the third medial, situated between the two grooves.

The right lateral area corresponds to the right hepatic lobe and has the following impressions, produced by the viscera with which it comes in contact: the colic impression *(impressio colica)* in front; the renal impression *(impressio renalis)* behind, with the suprarenal impression *(jmpressio suprarenalis)*; the duodenal impression *(impressio duodenalis)* medially, between both.

The left lateral area corresponds to the left hepatic lobe and is in contact with the fornix and with a portion of the anterior surface of the stomach *(impressio gastrica)* anteriorly and with the oesophagus *(impressio esophagea)* posteriorly, where lies the groove due to the presence of this organ, medially to which the omental tuber *(tuber omentale)* is situated.

The medial area is divided by the hilum of the liver into two lobes:

- the quadrate lobe *(lobus quadratus)*, situated anteriorly to the hilum (in front of the transverse sulcus), which is in contact with the pylorus and the duodenal bulb;

- posteriorly, the second lobe, the spigelian caudate lobe *(lobus caudatus)*, situated in the omental bursa: it has relations posteriorly with the diaphragm, on the right with the vena cava, on the left with the oesophagus and below, with the coeliac trunk, the superior border of the pancreas and the lesser curvature of the stomach.

Its diaphragmatic surface, owing to its convex shape, is almost entirely in contact with the diaphragmatic muscle. Only its anterior and middle parts are in contact with the abdominal wall, in the triangular space of the infracostal angle, in the epigastrium, where they form the hepatic triangle, in which the liver is accessible to palpation.

The diaphragmatic surface has an area which is not covered with peritoneum *(pars affixa)*, bounded by the peritoneal leaflets of the coronary ligaments, from which start, on either side, the right and the left triangular ligaments *(ligamentum triangularis dexter et sinister)* and, towards the superior surface, the suspensory ligament of the liver *(ligamentum falciforme)*.

The liver presents on this diaphragmatic surface a depression *(area cardiaca)*, the remainder of its surface corresponding to the relation with the pleuropulmonary regions *(areae pleuralae et pulmonalae)*. In the area which is not covered with peritoneum *(pars affixa)*, the liver is in contact with the diaphragm, on either side of the inferior vena cava.

The peritoneal cavity forms above and in front of the pars affixa, bounded by the coronary ligament of the liver, two peritoneal spaces: the anterior phrenicohepatic peritoneal space, divided by the falciform ligament into two areas, right and left, and the posterior phrenicohepatic peritoneal space, in relation with the diaphragmatic surface, which corresponds to the visceral surface of the liver. In these spaces occur the subphrenic abscesses.

Structure of the liver.

The structure of the liver includes branches of the portal vein, of the hepatic artery, the intrahepatic bile ducts, lymph vessels and nerves.

Segmentation of the liver.

The liver is described as being divided into two main lobes, right and left which, however, are not of a special surgical interest, as they do not correspond to the vascular reality of the liver. As a matter of fact, the liver is divided into a number of territories, dependent either on the portal system or on the suprahepatic system.

This fact is very important, because in case of hepatic tumours, cysts etc., extirpations of these segments may be performed after a previous ligation of the respective vessels. These territories are separated by planes with a lesser vessel supply, namely by scissures, which are virtual (fig. 160,161, 163,164).

There are three scissures:

The main portal scissure, which extends from the gall bladder to the left border of the inferior vena cava, forming a sloping plane of 70 degrees backwards and to the right, and crosses the hilum at the level of the vasculobiliary bifurcations;

The right portal scissure, extending from the right extremity of the hilum to the anterior border of the liver, on the inferior surface, and ending, on the superior surface, at the right border of the vena cava;

The left portal scissure, which forms a dihedral angle of 90 degrees and encloses two segments: a sagittal segment, which follows the canal of Arantius, and a frontal one, which prolongs the hilum to the left.

The scissures lodge the suprahepatic veins. They enclose five areas:
- *the right paramedian area,* located between the main and the right scissures;
- *the right lateral area,* situated to the right of the right scissure;
- *the left paramedian area,* situated between the main and the left scissures;
- *the left lateral area,* situated to the left of the left scissure;
- *the dorsal area,* the spigelian lobe respectively, a separate entity supplied by the vessels of the two pedicles.

The areas are each divided into segments, supplied by a branch of the portal vein. The segments are numbered from 1 to 8, clockwise:
- *the dorsal area* corresponds to segment 1;
- *the left lateral area* corresponds to segment 2;
- *the left paramedian area,* to segments 3 and 4;
- *the right paramedian area,* to segments 5 and 8; the latter is the only visible on the superior surface of the liver;
- *the right lateral area* corresponds to segments 6 and 7.

As a matter of fact, there is no plane of separation between these various segments. The distribution of the pedicle is of a terminal type, the ligation bringing about the necrosis of the corresponding territory. The suprahepatic systematization of the liver overlaps, in the main, the portal systematization.

The suprahepatic areas are three in number: the right suprahepatic area, drained by the right suprahepatic vein; the medial suprahepatic area, drained by the sagittal suprahepatic vein, which follows the main portal scissure; the left suprahepatic area, drained by the left suprahepatic vein.

The portal vein is the venous trunk which collects blood from the subdiaphragmatic digestive tract and carries it back to the liver.

It emerges through three main venous roots:

-*the inferior mesenteric vein,* which drains the left colon; it empties into the splenic vein, forming a venous confluent, which unites with the *superior mesenteric vein,* that drains the small intestine, the pancreas and the right colon; it is continuous, towards the liver, with the portal vein.

These three veins join at the level of the second lumbar vertebra, behind the neck of the pancreas, to form the trunk of the portal vein (fig. 165,166).

The portal vein ascends towards the hilum of the liver, where it ends by bifurcation. It is 8-10 cm long and its calibre is of 15 mm.

It has the following relations:

At the level of the posterior surface of the pancreas: in front, with the neck of the pancreas; behind, through Treitz's fascia, with the inferior vena cava; on the right, with the retropancreatic choledochus; on the left, with the origin of the superior mesenteric artery.

In the hepatic pedicles, three segments may be distinguished in the portal vein: at the base of the pedicle, at the free border of the lesser omentum, in the hilum of the liver.

At the base of the pedicle, the portal vein is related:

-in front, with the lesser omentum (the anterior leaflet) and the epiploic tubercle of the pancreas, which separates it from the first part of the duodenum; on the left, with the hepatic artery, which describes its arch and gives oft the gastroduodenal artery; on the right, with the choledochus, which delimits the interporto-choledochal triangle; behind, with the duodenocaval ligament:

162

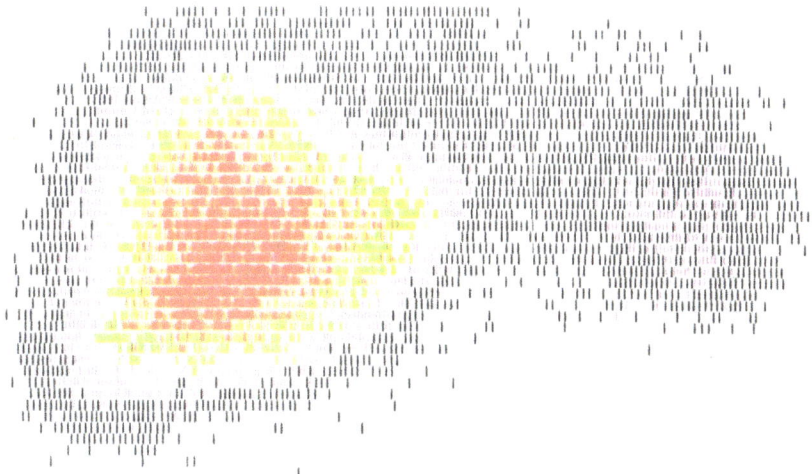

Fig. 157. Normal hepatosplenic scintigram

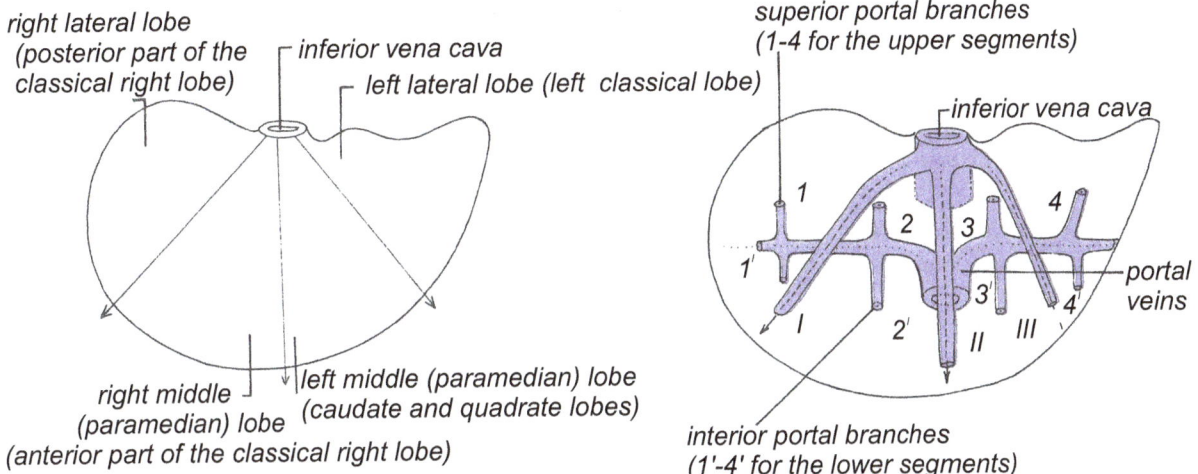

right lateral lobe
(posterior part of the classical right lobe)

inferior vena cava

left lateral lobe (left classical lobe)

superior portal branches
(1-4 for the upper segments)

inferior vena cava

portal veins

right middle (paramedian) lobe
(anterior part of the classical right lobe)

left middle (paramedian) lobe
(caudate and quadrate lobes)

interior portal branches
(1'-4' for the lower segments)

Fig. 158. Intrahepatic topography of the portal vein and suprahepatic veins (schematic representation)

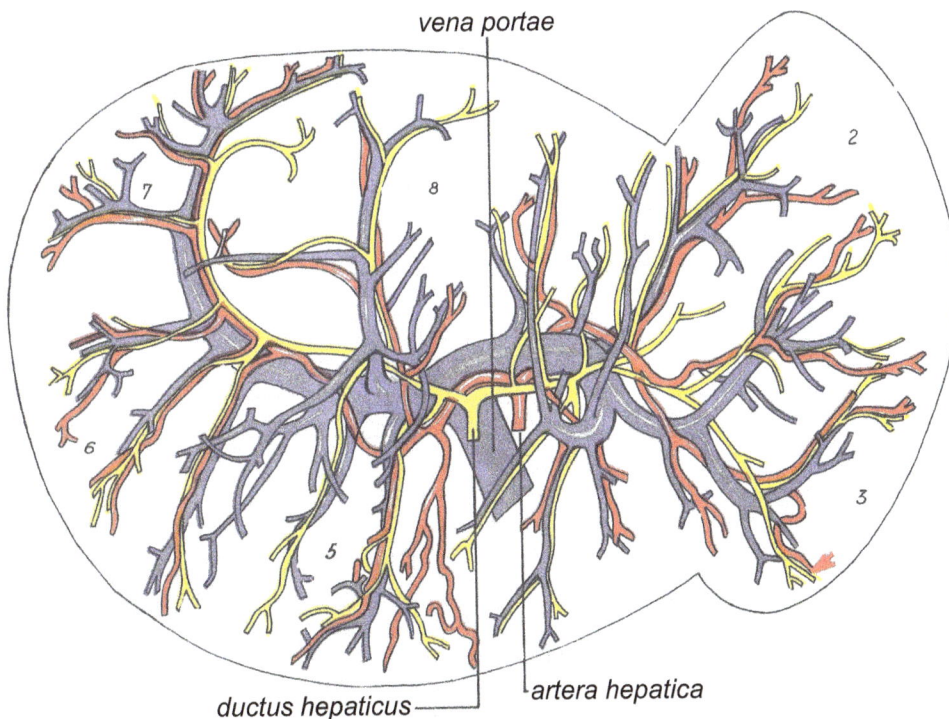

vena portae

ductus hepaticus

artera hepatica

Fig. 159. Disposition of the portal vain, the hepatic artery and the bile ducts in the liver - schematic representation

163

In the free border of the lesser omentum, the relations of the elements of the hepatic pedicle are the following:

- the portal vein lies posteriorly in the pedicle; the hepatic artery ascends along its left flank; the hepatic duct is situated on its right flank and receives, at a variable level, the cystic duct; the lymph vessels conceal the portal vein; the nerve plexuses are situated in front and behind;

- the relations with the neighbouring organs are achieved through the free border of the lesser omentum and are as follows: in front, with the liver; behind, with the hiatus of Winslow, which separates it form the inferior vena cava; on the right, is the communication with the greater peritoneal cavity.

In the hilum of the liver, the portal vein is the most posterior element and bifurcates on the right side of the hilum: its right, short branch seems to be continuous with the trunk and its left, long branch is arciform. The hepatic artery bifurcates on the anterior left flank of the vein:

- its right branch passes between the hepatic duct, in front, and the right branch of the portal vein, behind, and ends beneath the right hepatic duct;

- the left branch lies beneath the left hepatic duct, in front of the left portal vein.

The right and left hepatic ducts unite in front of the right portal vein, forming the common hepatic duct *(ductus hepaticus communis)*.

The portal trunk receives also the left gastric vein, the pyloric vein and the right superior pancreaticoduodenal vein.

Its right branch receives, in addition, the two cystic veins and the left branch is in relation with two obliterated veins: the canal of Arantius and the umbilical vein of the round ligament.

The terminal branches of the portal vein are (fig. 158, 159):

a. *The very short right branch* is continuous with the direction of the trunk and divides into two sectorial branches:

- the latero-inferior vein (right lateral area), which describes a curve with the concavity directed posteriorly and to the left; from the bend arise the branches of segment 6; from its ending emerge the branches of segment 7;

- the centrosuperior vein (right paramedian area), which describes a curve with the concavity posteriorly and runs towards the superior surface of the liver; from its convexity arise the branches of segment 5 and from its ending, the branches of segment 8;

b. *The left branch,* first directed transversally in the hilum, bends abruptly at the level of the fossa for the umbilical vein and ends at 2 cm from the anterior border of the liver; it will give off, in the hilum, the branches of segment 1; at the level of its angle, it gives rise to the angular vein, which courses towards segment 2; from its ending emerge two pedicles:

- the right pedicle, which is directed towards segment 4 and
- the left pedicle, towards segment 3.

Anomalies are possible: the right branch may lack and be replaced by a bifurcation.

There are portacaval anastomoses, which develop in case of portal obstruction. They are of five types:

1. *Esocardio-tuberosity anastomoses.* The veins of the upper pole of the stomach anastomose with the oesophageal veins (azygos system) and with the left inferior diaphragmatic vein, but in the case of a portal obstacle, the derivation formed by these veins is inadequate and they dilate and rupture, leading to severe digestive haemorrhages.

2. *Umbilical anastomoses.* They occur through the paraumbilical veins; the umbilical vein itself remains sometimes permeable.

3. *Rectal anastomoses.* They result from the union of the superior haemorrhoidal veins, branches of the inferior mesenteric vein, with the median and inferior haemorrhoidal veins, which are dependent on the inferior caval system.

4. *Perineoparietal anastomoses.* They are represented by Retzius' veins, the portorenal anastomosis and the ligamentous veins of the liver.

5. *Portosuprahepatic anastomoses.* They are very rare.

Besides the hepatic portal vein, there are accessory portal veins, situated in the thickness of the peritoneal ligaments, which connect the circulation of the intra-abdominal viscera - especially of the stomach, gall bladder and liver - to that of the abdominal wall and the diaphragmatic vault. We mention among these; the veins of the falciform ligament; the veins of the left triangular ligament; the veins of the round ligament; the cystic veins; the parabiliary veins, which form a true arch along the bile duct.

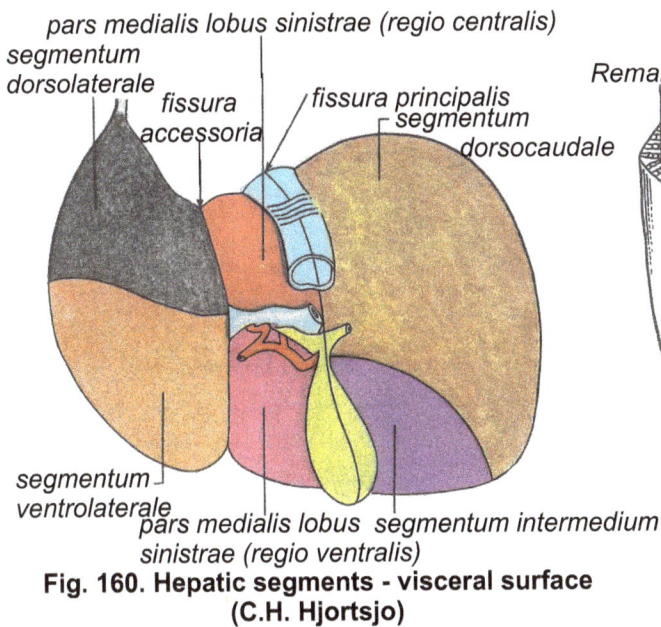

Fig. 160. Hepatic segments - visceral surface (C.H. Hjortsjo)

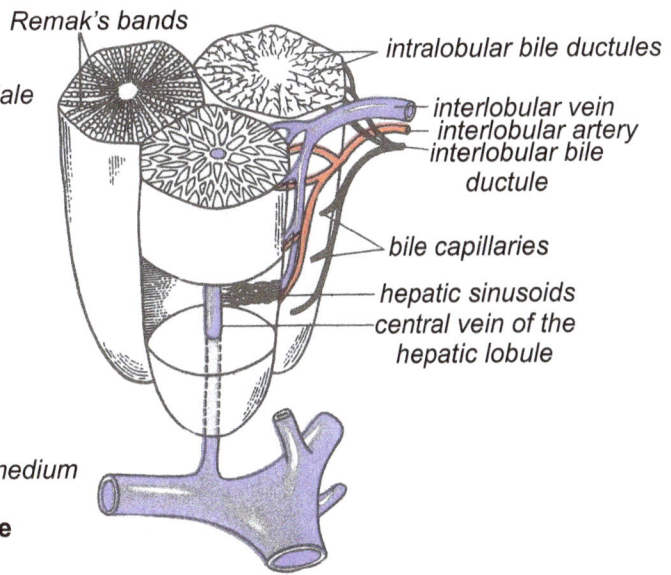

Fig. 162. Structure of the liver - hepatic lobule – scheme

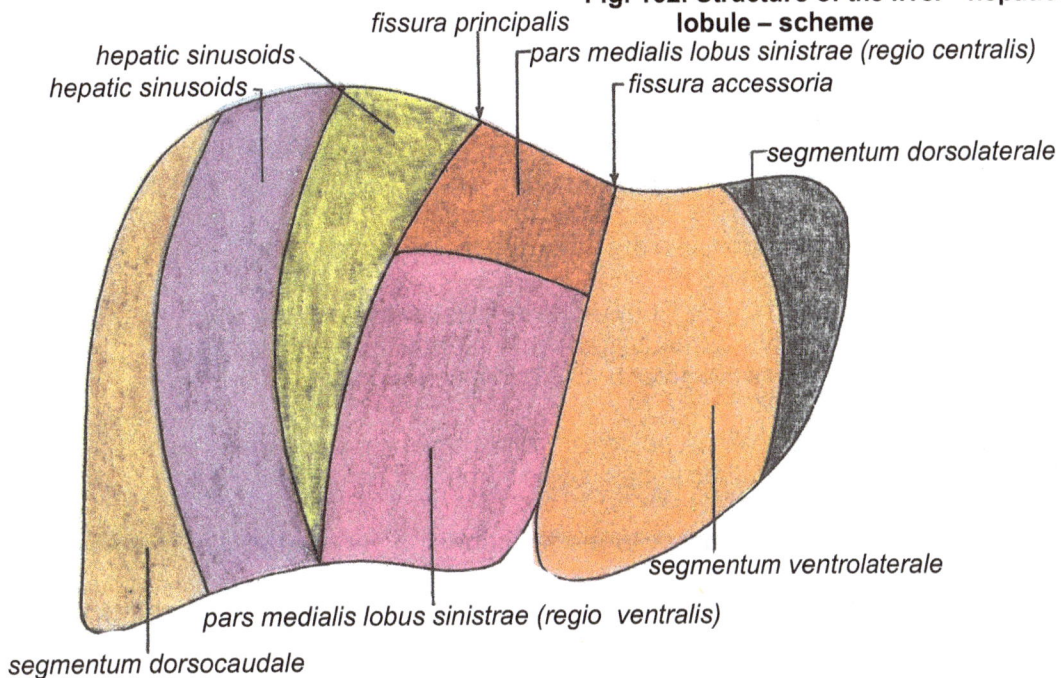

Fig. 161. Hepatic segments - diaphragmatic surface (C.H.Hjortsjo)

The hepatitc artery. It arises from the coeliac trunk, on the upper border of the epiploic tubercle of the pancreas, after the coeliac trunk has given off the left gastric artery (fig. 166).

This bulky artery is first continuous with the direction of the coeliac trunk, then runs obliquely forwards, to the right and slightly downwards. When it joins the portal vein, it divides into two terminal branches: the proper hepatic artery and the gastroduodenal artery.

The gastroduodenal artery is directed obliquely downwards, forwards and to the right, crosses the posterior surface of the first part of the duodenum, forms the boundary between the fixed and the mobile portions and bifurcates on its lower border into the right gastro-epiploic artery and the anterior and superior pancreaticoduodenal arteries.

The proper hepatic artery ascends along the left anterior flank of the portal vein, in the free branch of the lesser omentum and divides at the level of the hilum into its two branches, left and right. Its relations are those of the biliary pedicle, which were described concomitantly with the portal vein.

It gives off a number of collateral branches, such as the pyloric and the cystic arteries, as well as terminal branches. The branches of the hepatic artery follow, on the whole, those of the portal vein (fig. 159).

165

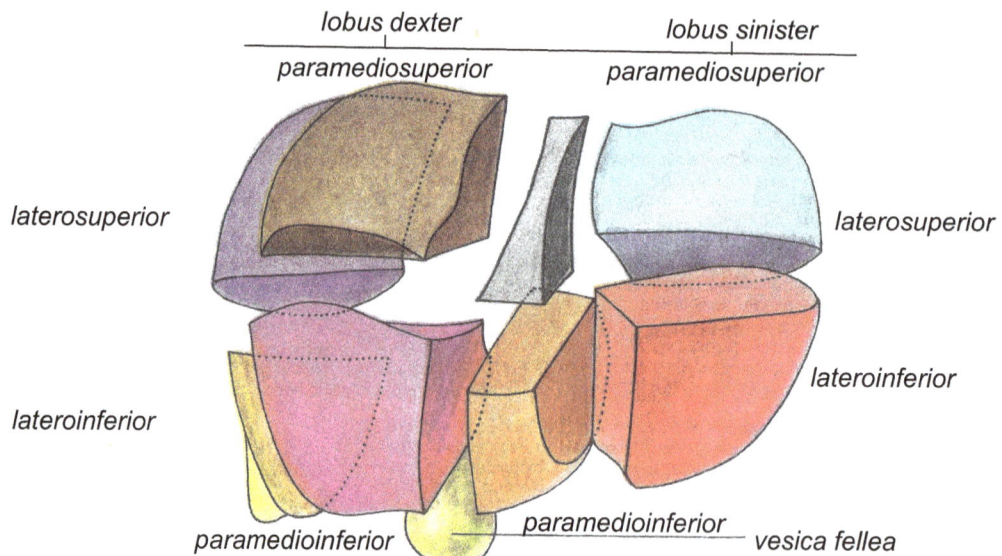

a. facies diaphragmatica

Fig. 163. Hepatic segments - diaphragmatic surface (Reiferscheid)

The left branch is directed forwards and beneath the left branch of the portal vein and divides into two terminal branches:

- the anterior branch, which is distributed to segment 3 and
- the posterior, transverse, branch, directed to segment 2.

Before bifurcating, the left branch gives off two collaterals:an anterior branch, which supplies segment 4, and a posterior branch, which crosses the portal vein and supplies segment 1.

The right, transverse branch runs laterally, passing between the right hepatic duct, in front, and the right branch of theportal vein, behind. Then it reaches the upper border of the rightportal vein, whose satellite it is, and divides, like the latter, into twobranches: paramedian, for segments 5 and 8, and right lateral, for segments 6 and 7.

Lymph vessels of the liver. They are superficial and deep.

The superficial lymph vessels lie beneath the hepatic peritoneum:

- the lymph vessels of the anterior half of the superior surface and those of the inferior surface run to the nodes of the hilum;
- the lymph vessels of the posterior half of the superior surface and those of the posterior surface course to the latero-aortic nodes; some of them accompany the inferior vena cava and end in the thorax;
- the lymph vessels adjacent to the falciform ligament run to the intrathoracic, retroxiphoid lymph nodes;

The deep lymph vessels follow the venous branches and are grouped into two flows: a portal flow, which is directed towards the nodes of the hilum, and a suprahepatic flow, which ends around the vena cava, within the thorax.

The suprahepatic veins. Four groups are distinguished: the suprahepatic main (right, median and left) and accessory veins, which reach the posterior surface of the liver on the borders of the vena cava (fig. 152,154).

The left suprahepatic vein is formed by the union of a bulky, sagittal branch, which drains segment 3, with a transverse, thin branch for segment 2 and with an intermediate, inconstant branch.

Thus constituted, the trunk of the left suprahepatic vein, long of 1-3 cm, unites with the median suprahepatic vein, to empty into the left anterior flank of the inferior vena cava.

The median suprahepatic vein is formed of branches coming from segments 3 and 4. It receives, in addition, the vein of segment 8. The trunk, 12 cm long, courses along the main portal scissure and ends through a trunk common with the left vein.

The right suprahepatic vein is formed by the branches draining the venous blood of the right lateral area. It is lodged in the right portal scissure, is 12 cm long and its calibre is very large. It describes a curve concave downwards following the hepatic dome.

It has a short extrahepatic course, running in the retrohepatic connective space, and empties into the right anterolateral flank of the inferior vena cava, at 1.2 cm above the confluent of the two preceding veins.

166

The posterior suprahepatic veins are the accessory suprahepatic veins or the veins of the right border and the veins of the spigelian lobe.

The intrahepatic bile ducts. The intrahepatic bile ducts are segmental channels, which join to form the right and left hepatic ducts.

The duct of segment 3, directed first obliquely forwards and to the right, bends to become sagittal. It unites with the duct of segment 2, to form a transverse trunk, which is directed to the right and receives, on its anterior surface, one or several ducts of segment 4 and on its posterior surface, one or several ducts of segment 1. In this way is usually constituted the left hepatic duct.

The ducts of segments 6 and 7 unite to form the right lateral duct.

The ducts of segments 5 and 8 unite to form the right paramedian duct.

The right paramedian and *the right lateral duct* join to form the right hepatic duct. This way of formation undergoes numerous variations.

The liver is invested with a connective capsule, called Glisson's capsule, lined with peritoneal serosa. At the level of the hilum, the Glisson's capsule surrounds the vessels that penetrate the liver and has, with them, a fan-like distribution, dividing the liver parenchyma into lobular units.

The classical hepatic lobule is represented by a morphological unit separated from the similar units by a periacinous stroma. In man, this hepatic lobule has been defined by analogy with the hepatic lobule of the pig, in which the perilobular connective tissue is very precisely and well developed.

These lobules have the shape of a pyramid or of a cone with a hexagonal base, the base being usually directed towards Glisson's capsule.

At the periphery of these formations is outlined, under the shape of thin bands, the perilobular stroma which, at the level of the vertices of the polygones and especially in the contact area of several lobules, has the shape of a triangular connective space, called Kieman's space, which contains the four tubular elements: the branch of the portal vein, the branch of the hepatic artery, the bile duct and the lymph vessel.

This, actually, the pig's hepatic lobule is separated from the similar structures by a perilobular stroma, the angles of the polygone being marked by portal spaces (the spaces of the perilobular stroma which contain the four elements of the hilum) and the centre, by the centrolobular vein, towards which converge both the sinusoidal capillaries and the hepatic cords.

In man, the classical hepatic lobule is structurally less systematized, because of the absence of a complete perilobular stroma. This lobule is defined as being formed of the whole cordal parenchyma, which is tributary to a centrolobular vein and delimited at the periphery by the arbitrary line which unites the portal spaces. In man, the portal spaces are not connected to each other by connective trabeculae, but only by perilobular circular vessels.

The portal hepatic lobule. In man we meet actually the portal hepatic lobule or Sabourin's inverted hepatic lobule, which is centred by a true portal space and, more exactly, by a bile duct. This portal lobule is made up of the totality of secretory units, regardless of their site, which are tributary to the same bile duct. The periphery of the portal lobule is represented by the arbitrary line which unites with each other three centrolobular vessels situated in closer proximity and is triangular in shape.

From the vessels lodged in these portal spaces arise the perilobular arteries and veins, which surround the periphery of the lobule and supply the lobular parenchyma through the sinusoidal capillaries. Parallel to these perilobular vessels run the perilobular bile canaliculi, into which the bile from the intralobular canaliculi is drained and which empty in the portal spaces into the interlobular canaliculi.

The hepatic acinus. In man is present the hepatic acinus described by Rappaport, made up of the totality of morphofunctional units which are tributary to the same vascular supply and to the same bile ductule; in other words, the hepatic acinus is constituted of the totality of cells which are supplied by the same vessel and which discharge the bile into the same bile ductule.

The hepatic acini are smaller units than the lobules, since they enter as constituent elements both into the structure of the classical lobule and into that of the portal lobule of the liver.

The classical hepatic lobule should be regarded as made up of several morphofunctional subunits, the hepatic acini.

Each hepatic acinus is located in two neighbouring classical lobules. Within the same classical hepatic lobule, the acinar units are autonomous structures, both from the viewpoint of the vascular supply and of the bile drainage. The existence of the hepatic acini demonstrates also the presence of a "segmental" circulation in the lobule, which explains the unequal injury to the various lobular areas.

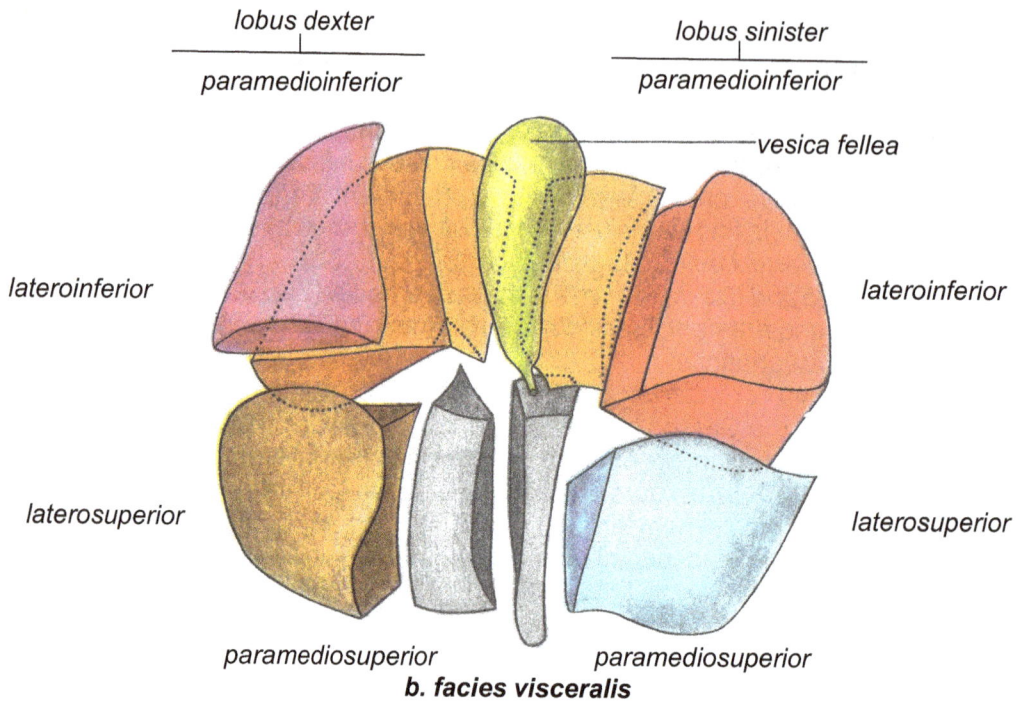

lobus dexter   lobus sinister

paramedioinferior   paramedioinferior

vesica fellea

lateroinferior   lateroinferior

laterosuperior   laterosuperior

paramediosuperior   paramediosuperior

b. facies visceralis

Fig. 164. Hepatic segments - visceral surface (Reiferscheid)

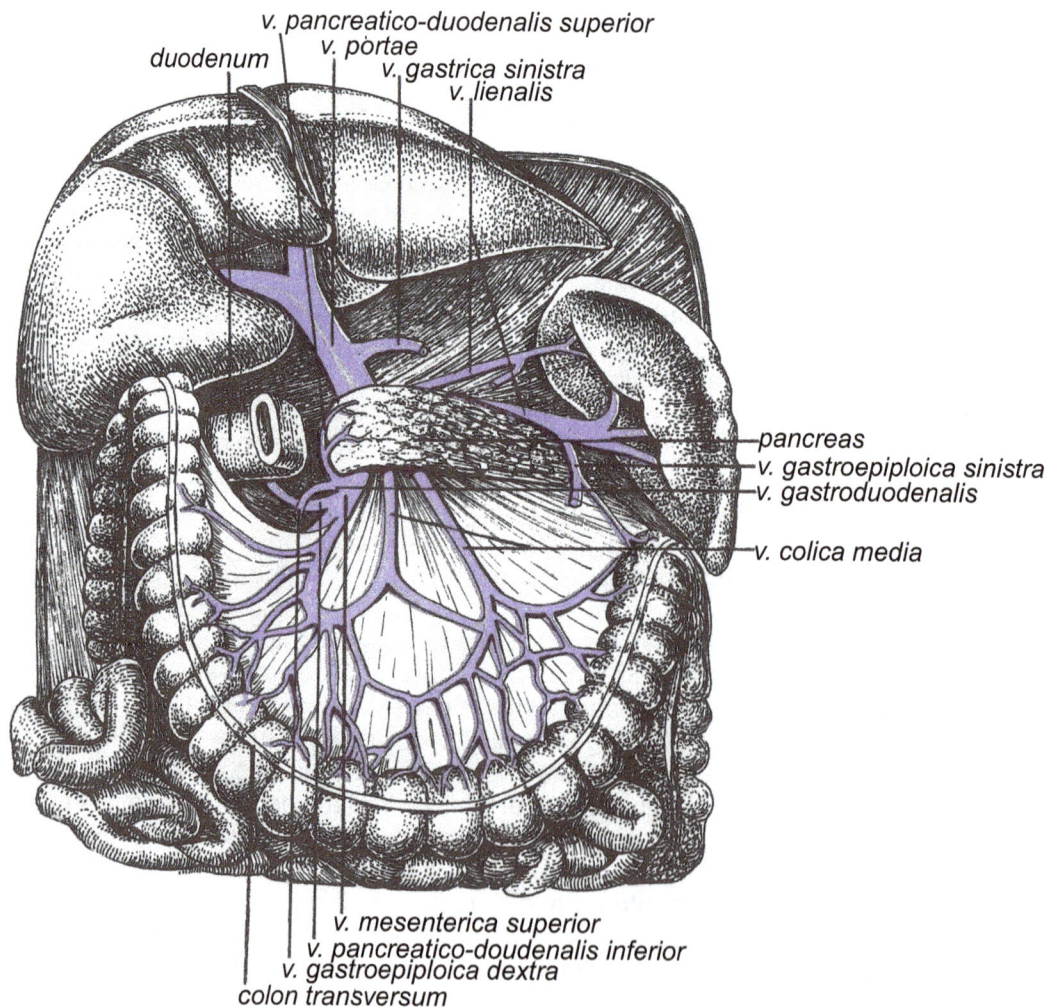

duodenum   v. pancreatico-duodenalis superior
v. pòrtae
v. gastrica sinistra
v. lienalis

pancreas
v. gastroepiploica sinistra
v. gastroduodenalis

v. colica media

v. mesenterica superior
v. pancreatico-doudenalis inferior
v. gastroepiploica dextra
colon transversum

Fig. 165. Portal, splenic and middle colic veins

These acinar structures explain many of the aspects met in the hepatic pathology, since in such cases the unequal lesion of the various areas situated in the same hepatic classical lobule may be observed. The lobule contains also cell cords, the so-called Remak's hepatic cords, which have a radiate disposition relative to the centrolobular vein, towards which they converge. The hepatic cords are surrounded by sinusoidal capillaries, with a discontinuous wall. Between the sinusoidal capillaries and the hepatic cords lie Disse's spaces, in the wall of which are lodged tridimensional networks of reticulin fibres. The hepatocytes may vary considerably in shape. They are constituent elements of the hepatic cords. They are cubical cells with two polyhedral sides, one of which represents the vascular pole and the other the biliary pole.

The biliary pole is formed of the surface of the cell presenting a linear groove which delimits, together with that of the neighbouring cell, the biliary capillary. The vascular pole (or surface) comes in an immediate contact with the blood from the sinusoidal capillaries. Between the vascular pole and the capillary wall of discontinuous type lies the Disse's space.

There is no contiguity relation between the biliary and the sinusoidal blood capillaries, as the body of the hepatocyte is interposed between them; the change of this relation in certain pathological conditions results in the passage of bile into blood and the appearance of icterus.

The Extrahepatic Biliary Tract

The extrahepathic biliary tract represents a topographically independent structure, which includes also the vasculonervous elements of the hepatic pedicle (the hepatic artery, the portal vein, the lymph vessels and the autonomic plexus). It is situated in the hepatoreno-duodenocolic region, bounds anteriorly the hiatus of Winslow *(foramen bursae omentalis)* and is related above with the visceral leaflet of the liver, below with the right kidney and the duodenal region, laterally and on the left with the hepatogastric area of the lesser omentum *(pars flaccida)*. It is made up of a main biliary passage and an accessory biliary passage (fig. 150, 156, 167-170).

The main biliary passage is represented by the hepatocholedochus.

The common hepatic duct (ductus hepaticus communis) is formed by the junction of the two hepatic ducts, right and left *(ductus hepaticus proprius, dexter et sinister)*. It descends into the free border of the lesser omentum and fuses, at a variable level, generally at the upper border or behind the first part of the duodenum, with the cystic duct *(ductus cysticus)*, which comes from "he gall bladder *(vesica fellea)*. It forms the choledoch duct *(ductus choledochus)*, which describes a curve with the concavity directed towards the right and anteriorly and ends at the median third of the second part of the duodenum. It is on the average 5 cm long and its calibre is of 5 mm (fig. 156).

The choledochus is composed of four segments: supraduodenal, very short, often lacking; retroduodenal; retropancreatic; intraparietal (within the wall of the duodenum). It ends with an orifice which if shares with the pancreatic duct *(ductus pancreaticus major)* and the two ducts unite, forming the hepatopancreafic ampulla of Voter *(ampulla hepatopancreatica)* (fig. 150).

As regards the relations of the main biliary passage (the hepatocholedochus), we mention below the more important.

In the retroduodenal segment, it passes behind the first part of the duodenum. It is crossed by the superior right pancreaticoduodenal artery.

In the retropancreatic segment, it descends on the posterior surface of the head of pancreas, in which it digs a groove or even a canal. Behind are Treitz's fascia, the inferior vena cava, the right renal vein, the right spermatic vein and the right suprarenal vein.

In the intraparietal segment, it unites with the Wirsung's canal and both penetrate the dilation called ampulla of Voter *(ampulla hepatopancreatica)*. In other cases, more rarely, they open separately. The two canals (separated), as well as the ampulla of Voter are surrounded by oblique and circular muscle fibres, which make up the Oddi's sphincter.

Vessel and nerve supply (fig. 167)

The arteries of the main biliary passage are represented by the recurrent branch of the cystic artery, branches of the hepatic artery and branches of the pancreatico-duodenal arteries, especially of the superior one.

The veins are tributaries of the portal vein and are mainly satellites of the hepatic artery, as they open into the lymph nodes of the hepatic pedicle.

The nerves are represented by the anterior hepatic plexus, satellite of the hepatic artery, and the posterior hepatic plexus, satellite of the portal vein.

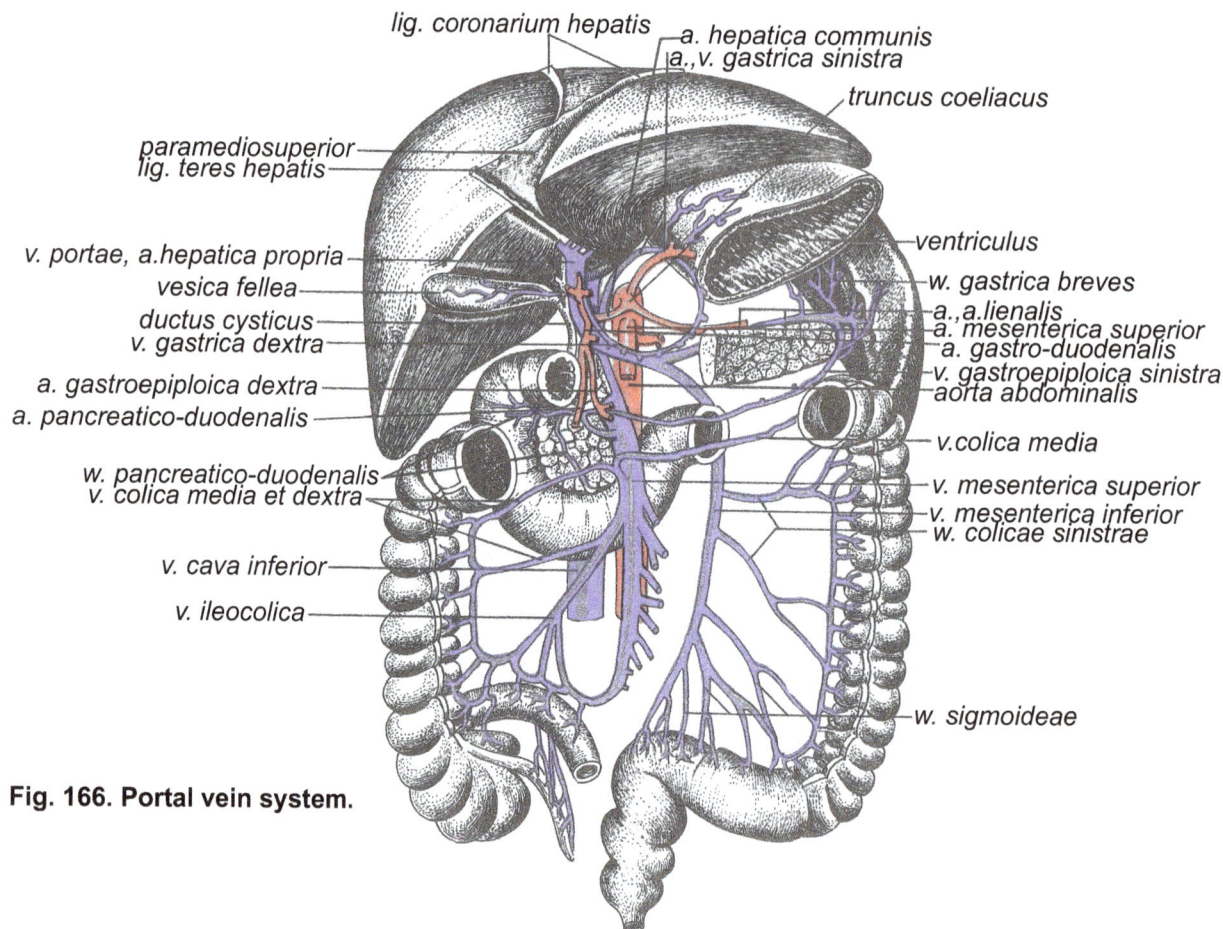

Fig. 166. Portal vein system.

Labels for Fig. 166:
- lig. coronarium hepatis
- a. hepatica communis
- a.,v. gastrica sinistra
- truncus coeliacus
- paramediosuperior
- lig. teres hepatis
- ventriculus
- v. portae, a.hepatica propria
- vesica fellea
- ductus cysticus
- v. gastrica dextra
- w. gastrica breves
- a.,a.lienalis
- a. mesenterica superior
- a. gastro-duodenalis
- a. gastroepiploica dextra
- a. pancreatico-duodenalis
- v. gastroepiploica sinistra
- aorta abdominalis
- w. pancreatico-duodenalis
- v. colica media et dextra
- v.colica media
- v. mesenterica superior
- v. mesenterica inferior
- w. colicae sinistrae
- v. cava inferior
- v. ileocolica
- w. sigmoideae

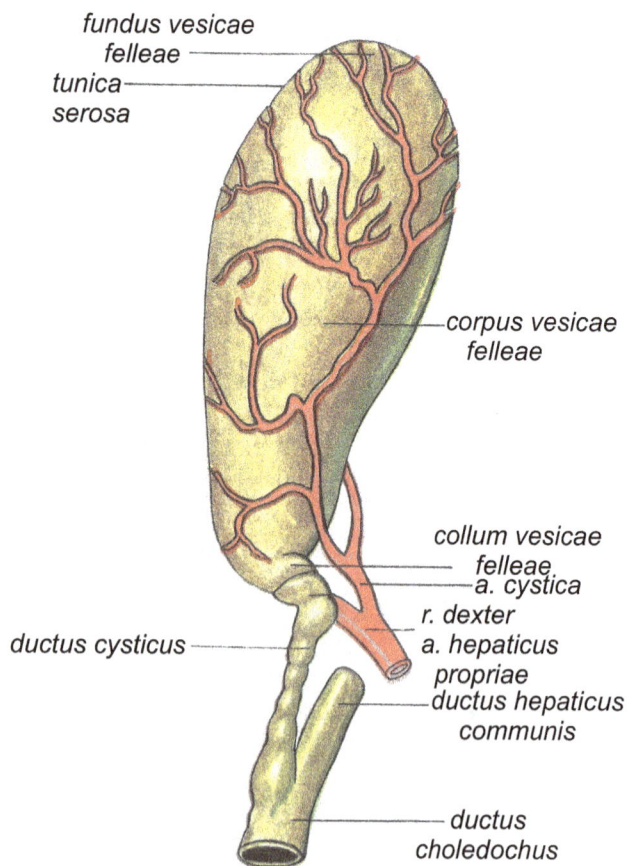

Fig. 167. Gall bladder - blood supply.

Labels for Fig. 167:
- fundus vesicae felleae
- tunica serosa
- corpus vesicae felleae
- collum vesicae felleae
- a. cystica
- r. dexter
- a. hepaticus propriae
- ductus cysticus
- ductus hepaticus communis
- ductus choledochus

The **accesory biliary passage** is formed of the gall bladder *(vesica fellea)* and the cystic duct *(ductus cysticus)*.

The *gall bladder (vesica fellea)* is pear-shaped and situated in the cystic fossa of the inferior surface of the liver. It is formed of the following portions: a fundus, which usually projects beyond the anterior border of the liver; a body *(corpus),* which lies under the liver, in the gall bladder fossa, in relation with the right angle of the colon and the duodenal bulb (between the superior surface of the body of the gall bladder and the inferior surface of the liver there is a fibrous tissue, in which run accessory portal veins); a neck *(collum)* which is continuous with the cystic duct *(ductus cysticus).* As regards its structure, three tunics are described; the peritoneal serosa, which covers its inferior surface, but which may sometimes completely invest it and form a mesentery, that connects if to the inferior surface of the liver; the muscular tunic, which s very thin; the mucous tunic, which forms permanent folds, bounding the polygonal depressions, and other folds which disappear on its distension; when it is full, its capacity is of about 50-60 cm^3.

170

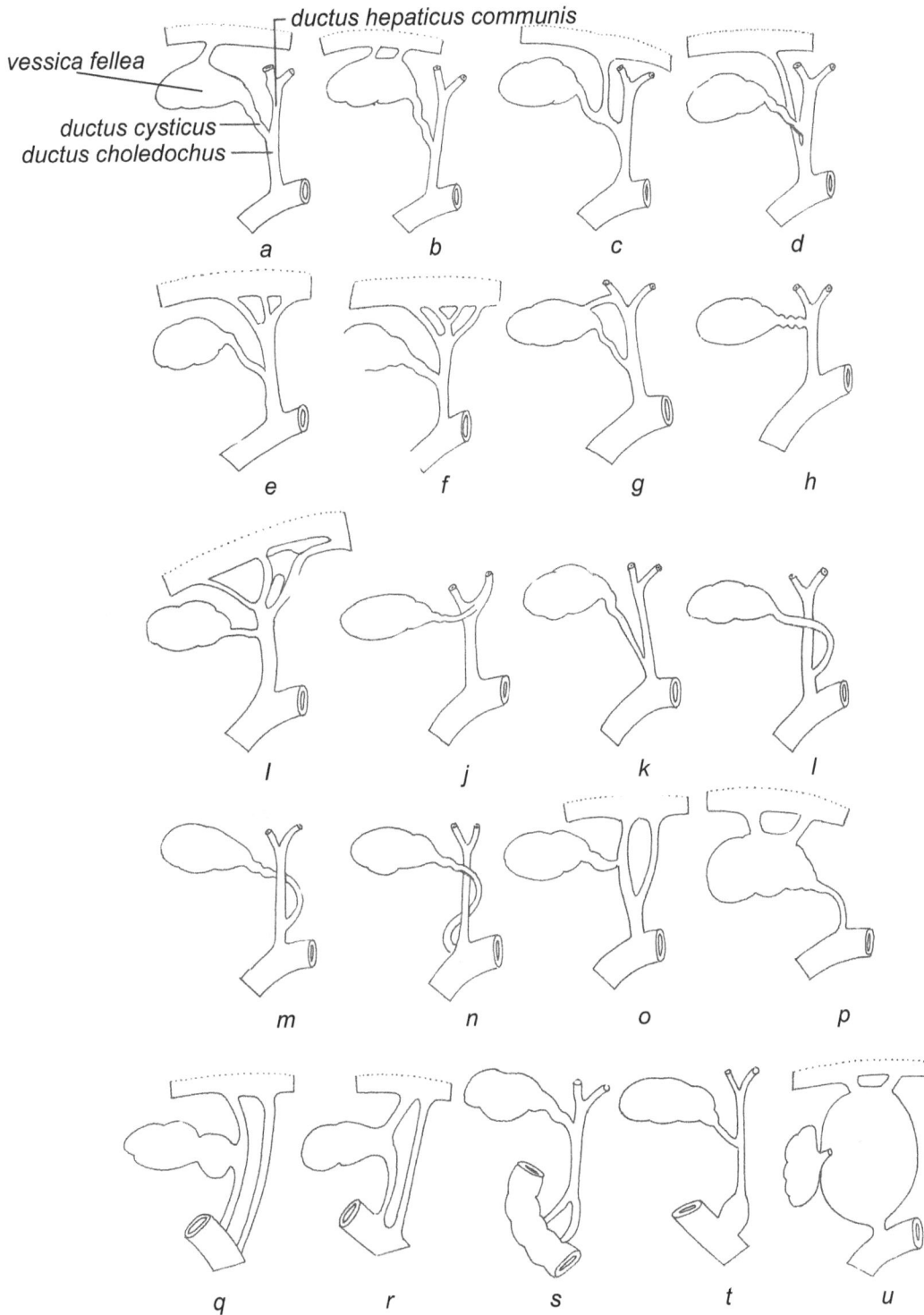

Fig. 168. Variants of the extrahepatic biliary passages: a. single cholecystohepatic duct; b. double cholecystohepatic duct; c. cysticohepatic duct; d.choledochohepatic duct; e. hepatic duct with three branches; f. hepatic duct with four branches; g. right cholecystohepatic duct; h. high opening of the cystic duct; i. variants of the origin of the main biliary passage; j. left cysticohepatic duct; k. low emptying of the cystic duct; l. anterior spiral cystic duct; m. posterior spiral cystic duct; n. double spiral cystic duct; o. absence of the common hepatic duct; p. hepatic ducts opened into the cystic duct and cystic duct opened into the duodenum; q. double emptying of the hepatic ducts into the duodenum; r. similar to q; s. bifurcation of the terminal choledochus; t. distal diverticulum of the choledochus; u. congenital cyst of the hepatocholedochus (according to B. Kourias)

The cystic duct (ductus cysticusj is bent on the neck of the gall bladder; it is directed downwards, to the left and backwards and ends in an acute angle in the right flank of the hepatic duct, in which it opens or with which it joins, forming the choledochus. The confluent projects generally at the inferior border of the L$_1$ vertebra. It is on the average 3 cm long and its calibre diminishes from the choledochus towards the gall bladder. Seldom it appears externally nodulated, aspect corresponding inside to numerous transverse and oblique folds, which form Heister's spiral valve.

As the level of the free border of the lesser omentum, the cystic duct has more important relations with: the portal vein, situated behind; the hepatic duct which, with the cystic duct and the liver, forms Budd's triangle, crossed by the right branch of the hepatic artery; the hepatic artery, on the left of the hepatic duct.

At the base of the hepatic pedicle, where the cystic duct usually adheres to the hepatic duct, it is related, in front, to the first part of the duodenum, from which it is separated through the lesser omentum (the posterior leaflet), and on the left, to the portal vein and the bifurcation of the common hepatic artery. At this level,through the lesser omentum, it is in relation, in front, with the liver and the first part of the duodenum and behind, with the hiatus of Winslow and the inferior vena cava (see also fig. 169 and 170).

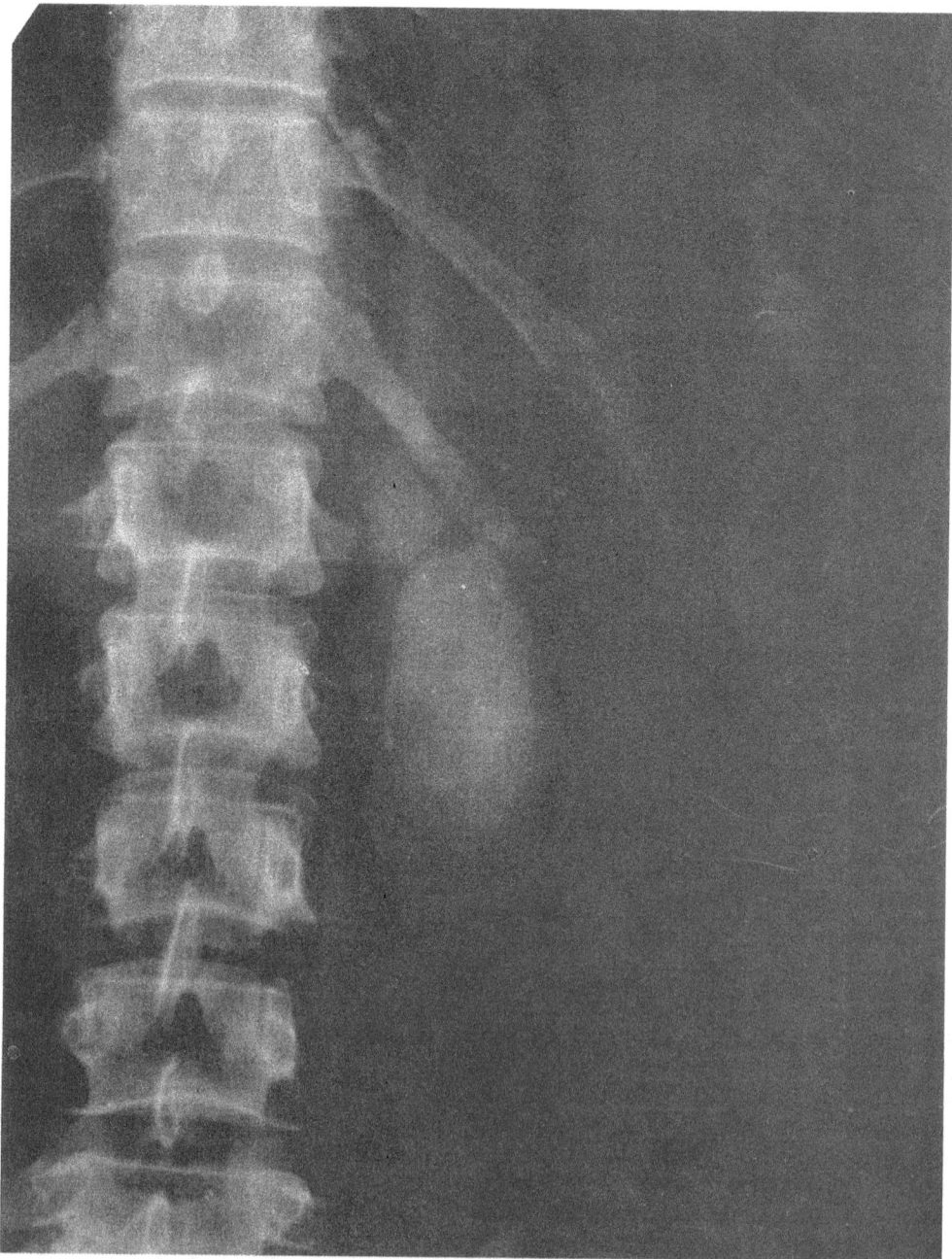

Fig. 169. Cholecystography - image of the cholecyst before its contraction.

Vessel and nerve supply (fig. 167)

The arterial supply of the accessory biliary passage is provided by the cystic artery, which either arises as a branch of the proper hepatic artery, or issues from the right branch of the hepatic artery, but in both cases it penetrates the gall bladder at the level of the neck and divides into two branches: right and left.

The cystic veins, superficial, two in number, are satellites of the cystic artery and empty into the right branch of the portal vein. The deep veins, which enter the liver directly, are accessory portal veins.

The lymph vessels drain into the lymph nodes of the neck of the gall bladder and into the lymph nodes of the anterior border of the hiatus of Winslow, from where they reach the retroduodeno-pancreatic lymph nodes.

The nerves are derived from the anterior hepatic plexus.

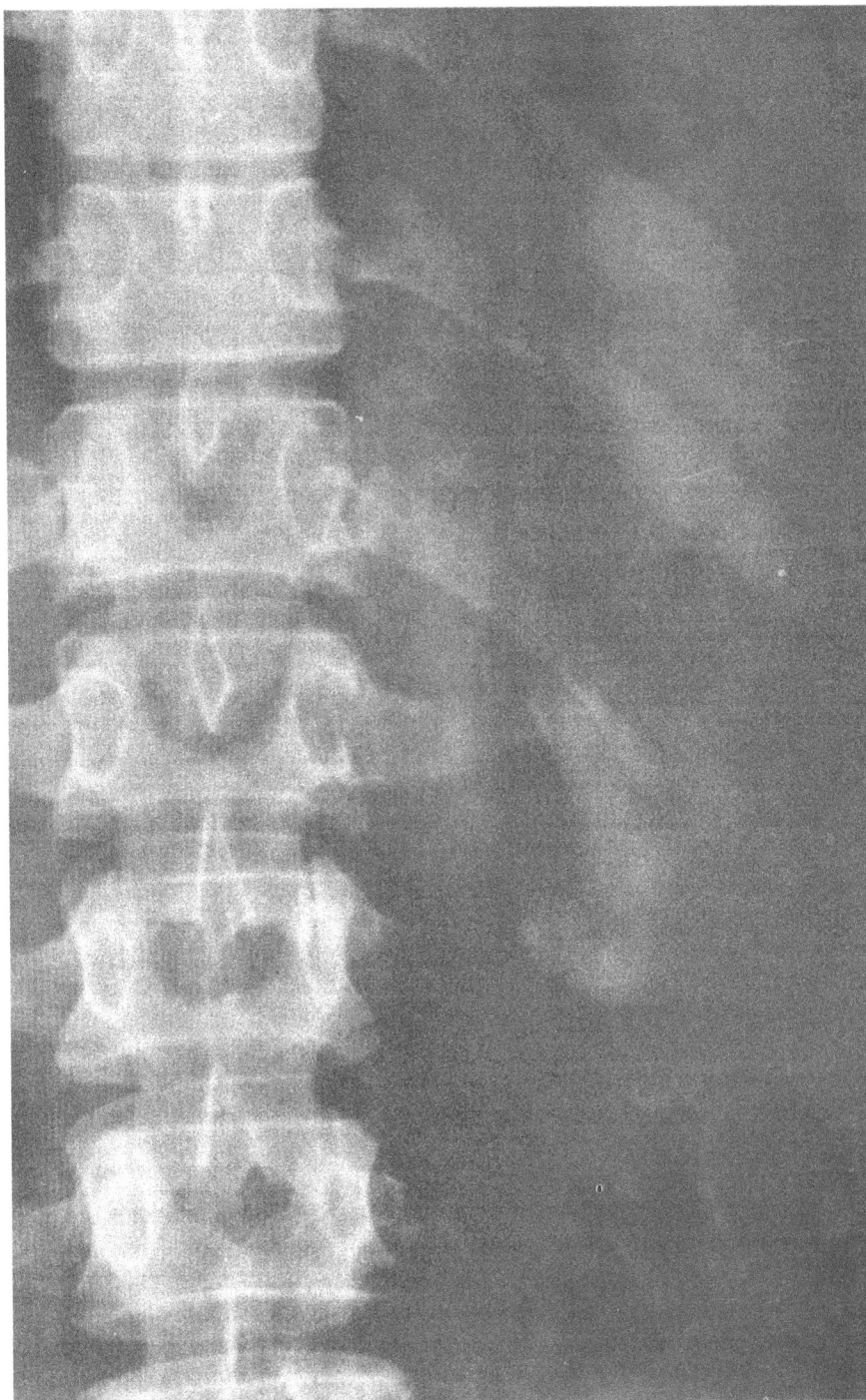

Fig. 170. Cholecystography - image of the contracted cholecyst.

173

The Splenic Region

The splenic region contains the spleen (lien), a lymphoid organ.

The walls of the splenic bed are the following:

- superior, posterior and lateral - the left diaphragmatic dome, above which lie the pleural recess (sinus), the base of the left lung, the ninth, tenth and eleventh ribs and the corresponding infercostal spaces;

- inferior - the upper half of the anterolateral surface of the left kidney, the left adrenal gland and the left phrenicocolic ligament (sustentaculum lienis);

- medial - the body of the eleventh thoracic vertebra, the upper half of the body of the fwelfth thoracic vertebra and the greater curvature of the stomach, through the gastrosplenic ligament;

- anterior - the gastrocolic ligament and, on an anterior plane, the left thoracoabdominal wall

The spleen is a lymphoid organ of a tetrahedral form, generally variable, with four sides and two poles. It is lodged in the left subphrenic space and projects on the anterior and lateral walls of the abdomen, into the left hypochondrium, at the level of the T11 vertebra, and laterally, at the level of the ninth, tenth and eleventh ribs, so that it is situated obliquely below the left costal arch and, under normal conditions, it cannot be palpated. Its long axis lies along the tenth rib (fig. 171-173).

In vivo, as it is highly vascular, the spleen has a lateral surface in relation with the diaphragm, a medial surface subdivided, through the presence of the hilum, into three contact areas with the adjacent organs (medially and posteriorly, the area with the impression of the left kidney; medially and anteriorly, the gastric area; medially and inferiorly, the area of the left aortic angle), two borders and two extremities.

In the corpse, it is somewhat bloodless and the neighbouring organs leave their impression on it, as shown above.

The diaphragmatic or lateral surface (fades diaphragmatica) is smooth and convex owing to the diaphragm. Through the latter it is in relation with the left pleural sinus and the left lung. It is covered with peritoneum, as the spleen is an intraperitoneal organ. At this level may also be sometimes found the end of the left hepatic lobe, which insinuates between the spleen and the diaphragm, when the left lobe of this organ is more developed.

The gastric or anteromedial surface (facies gastrica) is concave and behind it lies the hilum of the spleen, which divides it into a prehilar portion, in relation with the stomach, and a retrohilar portion, in contact with the left kidney.

At the level of the hilum of the spleen, it may have relations with the tail of the pancreas, in the pancreaticosplenic ligament, when the tail of the pancreas is longer.

The renal surface (fades renalis) is directed poster - medially and has a somewhat lesser extent; it has a smooth surface in relation with the anterolateral surface of the left kidney and with the adrenal gland.

The colic surface (facies colica) or the base of the spleen is the contact area of the spleen with the left colic angle and the left phrenicocolic ligament (sustentaculum lienis).

The notched or anterosuperior border (margo crenatus sive acutus) separates the diaphragmatic from the gastric surface.

The thick postero - inferior border (margo inferior sive margo obtusus) is the boundary between the diaphragmatic and the renal surface.

The infernal border (margo medialis) separates the gastric from the renal surface.

The ventral pole (or antero-inferior pole or ventral end) projects into the left tenth intercostal space and rests on the phrenicocolic ligament.

The posterior or posterosuperior pole is situated in proximity to the vertebral column at the level of T10, therefore it is also called vertebral end.

The spleen is invested by a proper fibrous capsule, quite resistant This fact is of the utmost clinical importance, since in case of severe injuries sited in the left hypochondrium or at the base of the left hemithorax, ruptures of the parenchyma may occur, leaving nevertheless the capsule undamaged (subcapsular rupture). In other cases, a subcapsular haematoma develops, which ruptures the capsule, producing an extracapsular intraperitoneal haemorrhage (two-stage haemorrhage). Therefore, these injured patients should be carefully followed-up and, in case of breakdown or lowering of the arterial pressure, acceleration of the pulse, appearance of thirst and cold sweat, hence in the presence of haemorrhage signs, emergency surgery should be performed to save them.

The spleen is entirely invested by visceral peritoneum, which reflects at the level of the hilum on the splenic vessels and gives rise to the pancreaticosplenic and gastrosplenic ligaments.

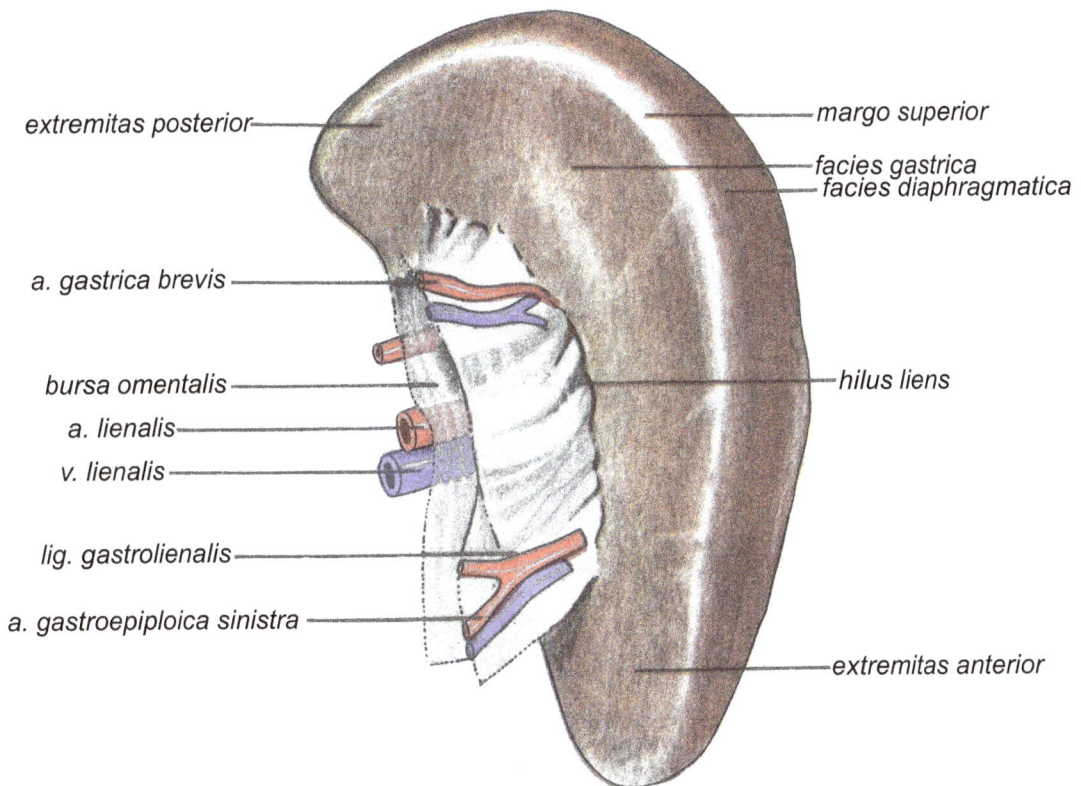

Fig. 171. The spleen - splenic vessels.

extremitas posterior

margo superior

facies gastrica
facies diaphragmatica

a. gastrica brevis

bursa omentalis

a. lienalis

v. lienalis

hilus liens

lig. gastrolienalis

a. gastroepiploica sinistra

extremitas anterior

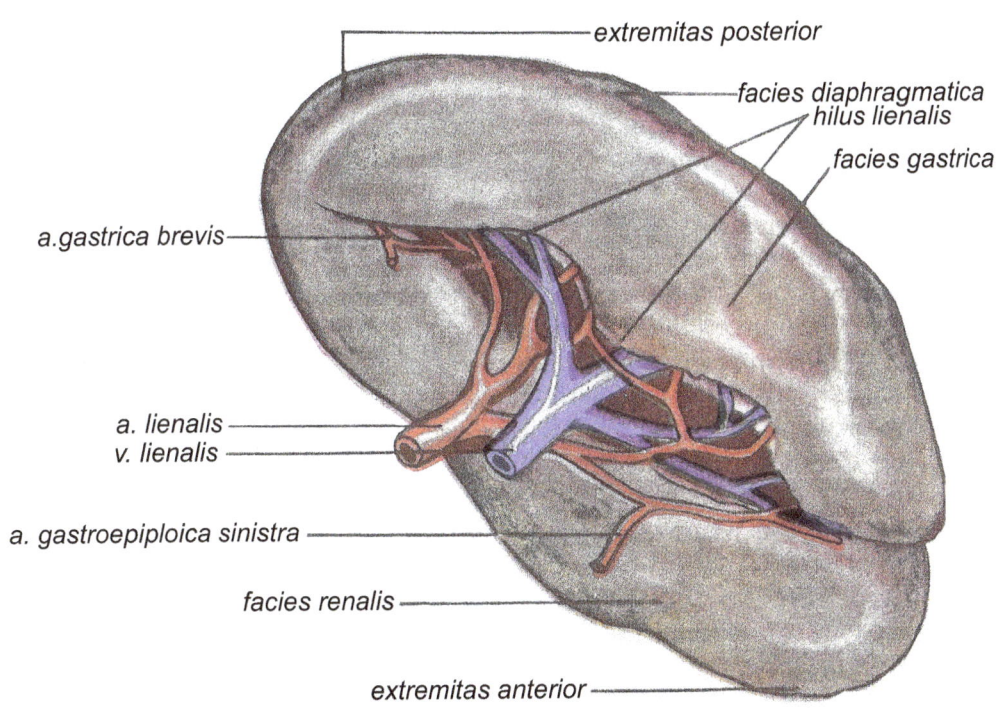

Fig. 172. The spleen - hilus of the spleen.

extremitas posterior

facies diaphragmatica
hilus lienalis

facies gastrica

a.gastrica brevis

a. lienalis
v. lienalis

a. gastroepiploica sinistra

facies renalis

extremitas anterior

175

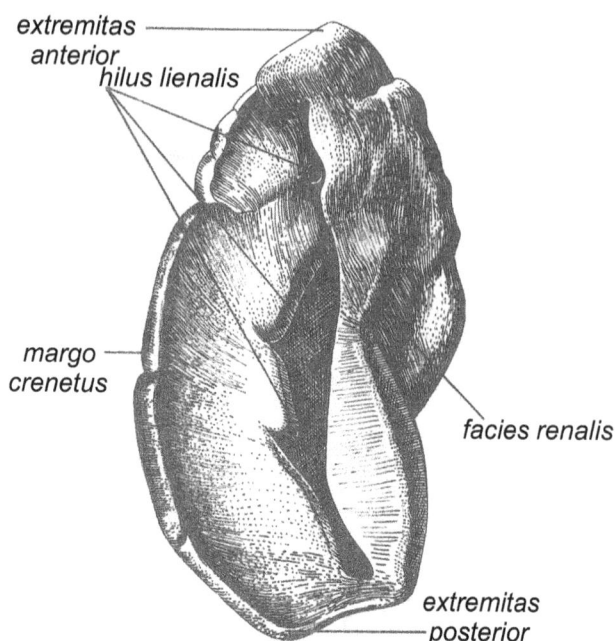

extremitas anterior

hilus lienalis

margo crenetus

facies renalis

extremitas posterior

Fig. 173. The spleen - overall view

Vessel and nerve supply

The arterial supply of the spleen is derived from the splenic artery, a branch of the coeliac trunk, which reaches the hilum of the spleen along the upper border of the pancreas, where it divides into two branches, which ramify into 6-8 arterioles that penetrate the splenic parenchyma. The splenic artery gives rise to pancreatic branches, to the left gastro - epiploic artery and to the short branches for the fornix of the stomach.

The splenic vein, of larger size than the homonymous artery, is situated behind it, receives the inferior mesenteric vein, then unites with the superior mesenteric vein and empties into the portal vein.

The lymph vessels drain into the retropancreatic and coeliac lymph nodes.

The autonomic nerve supply derives from the solar plexus.

The Inframesocolic Part

In this abdominal space, situated below the transverse mesocolon, lie the jejunum and the ileum, surrounded by the framework of the large intestine, respectively the caecum with the vermiform appendix, the ascending colon, the transverse colon, the descending colon and the sigmoid colon continuous with the rectum - the latter, however, situated in the pelvis. The viscera of the inframesocolic part are covered anteriorly with the greater omentum *(omentum majus)*. Attached above to the greater curvature of the stomach, the greater omentum descends over the transverse colon, covering it at the level of the taenia omentalis, and passes also over the loops of the small intestine, covering them, too. It is quadrilateral in shape, with an irregular surface, crossed by vessels and masses of fat; it has two leaflets, one anterior and the other posterior, presenting great individual variations.

Vessel and nerve supply of the greater omentum

The arterial supply is very rich and is achieved through the anastomose of the right and left gastro - epiploic arteries, for the anterior leaflet, and through branches of the splenic and inferior pancreaticoduodenal arteries, for the posterior leaflet

The veins are satellites of the arteries; those of the anterior leaflet empty into the left gastro-epiploic vein or into the splenic vein and those of the posterior leaflet, usually into the superior mesenteric vein.

The lymph vessels are very numerous. The anterior leaflet drains the lymph into the right gastroepiploic lymph nodes and then into the subpyloric and hepatic lymph nodes, which shows the necessity of removing the greater omentum in gastric cancer. The posterior leaflet drains the lymph into the lymph nodes of the pancreaticosplenic chain.

The nerve supply is very rich, too, the nerves being disposed in plexuses. They arise especially from the hepatic and splenic plexuses.

The Jejunum and Ileum
(Jejunum et ileum)

The jejunum and ileum represent the mobile segment of the small intestine, unlike the duodenum, already studied, which represents its fixed portion.

The jejuno-ileum extends from the duodenojejunal angle to the ileocaecal valve (or ileocaecal angle). The duodenojejunal angle lies on the left of the vertebral column, at the level of the intervertebral L1L2 disk, and the ileocaecal angle is in the right iliac fossa, in front of the interline of the right sacroiliac joint, midway between the two borders of the iliopsoas muscle. It is on the average 6.5 m long and its diameter diminishes from the jejunum towards the ileum from 2.5-3 cm to 2 cm.

176

It is cylindrical in shape and the mesentery attaches it to the posterior wall of the trunk. From its origin to its end, it forms a number of coils called intestinal loops. The superior, jejunal loops are disposed horizontally over each other, in the superior, median and left portions of the submesocolic part. The inferior, ileal loops are arranged vertically, beside each other, with the exception of those of the last segment, of a few centimeters in length, disposed perpendicularly on the caecum, into which it opens. The boundary between the jejunum and the ileum is not clear-cut, but it is estimated that the proximal two fifths of the small intestine constitute the jejunum and the distal three fifths, the ileum. There are, however, some elements of differentiation, among which the following: the jejunum has a greater number of circular folds and is more consistent on palpation, but Payer's patches are absent, while in the ileum they are very numerous; the blood vessel supply is more abundant then that of the ileum; the musculature of the jejunum is stronger than that of the ileum.

The relations of the jejuno- ileum are the following: in front, with the greater omentum, which separates it from the anterior abdominal wall; behind, with the posterior abdominal wall and the existent retroperitoneal organs (duodenum, kidney, ureters, the large arterial, venous and lymph trunks); above, with the transverse colon and mesocolon; in the right mesentericocolic space, with the horizontal portion of the duodenum, the head of the pancreas, the inferior pole of the left kidney, the colic vessels and the right ureter; in the left mesentericocolic space, with the ascending part of the duodenum, the duodeno-jejunal angle, the left colic flexure, the inferior mesenteric vessels and the left ureter; below, the intestinal loops descend up to the level of the iliac fossae and the pelvis, where they come in contact with the urinary bladder, the rectum and in the female the uterus; laterally, with the ascending and descending portions of the colon.

The Meckel's diverticulum represents the remnant of the omphalo-enteric duct, existent in 2 per cent of the subjects. It is situated at about 80 cm from the ileocaecal valve and is glove-finger-shaped; it is 5-6 cm long and lies free in the peritoneal cavity. Less frequently, instead of it may be present a fibrous band, extending up to the deep cicatrix of the umbilicus.

The Mesentery
(Mesenterium sive mesostenium)

It is the serous connection which attaches the jejuno-ileal loops to the posterior abdominal wall It contains the vasculonervous pedicle of the intestine.

Whereas the parietal border (the root) is 15 cm long, the intestinal border has the length of the small intestine. The height is variable, from 12 to 15 cm, but much lesser at the two ends: at the duodenojejunal angle and the caecum. It is directed obliquely from left to right and from above downwards, from the level of the left side of the L_1 vertebra to the right sacro-iliac joint and divides the inframesocolic part into two colic spaces: right and left: it has two surfaces, anterior right and posterior left, and two borders: intestinal and parietal.

The root of the mesentery *(radix mesenterii)* has the shape of an elongated S and presents: a superior segment, directed obliquely downwards and to the right, from the duodenojejunal angle to the lower border of the third part of the duodenum, which crosses the aorta and the inferior vena cava; a median, vertical segment, which descends in front of the superior mesenteric artery up to the disk between the L_4 and L5vertebrae; an inferior segment, directed obliquely downwards ana to the right, in front of the psoas, which reaches the caecum, after having passed over the ureter and the right primitive iliac vessels.

In the root of the mesentery run, one after another, the superior mesenteric artery and its ileocaeco-colic branch.

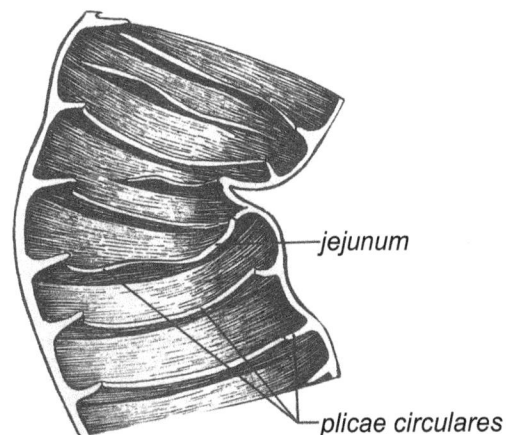

jejunum

plicae circulares

Fig. 174. Structure of the jejunum (circular folds).

Internal configuration. The inner surface of the small intestine has the following elements: the circular folds *(plicae circulares),* also called connivent valves of Kerkring, formed of transverse circular folds of the mucous membrane, numerous in the jejunum and fewer in the ileum, which contains, however, lymphoid aggregations: Peyer's patches. The role of the circular folds of Kerkring is to increase the absorption surface of the intestine (fig. 174,177,178).

In the structure of the intestine, are present the following tunics:
- the serous tunic, formed of the visceral peritoneum, which reflects on the posterior wail of the abdominal cavity and constitutes the mesentery;
- the muscular tunic, formed of an outer longitudinal and an inner circular layer, between which lies Auerbach's nerve plexus;
- the submucous tunic, which contains Meissner's nerve plexus;
- the mucous tunic, consisting of an epithelial component and corium. The epithelial component contains, on the one hand, monostratified columnar epithelium, made up of enterocytes and goblet cells, and on the other hand, a glandular apparatus *(glandulae intestinales),* formed of Lieberkuhn's glands (fig. 175,176,180).

The corium contains numerous lymphocytes, either diffusely scattered (solitary lymphatic follicles), or grouped into lymphoid follicles (aggregated lymphatic follicles), known especially under the name of Peyer's patches (or plaques).

The corium is separated from the submucosa by a layer of smooth muscle *(muscularis mucosae).*

The intestinal villi *(villi intestinales)* are cylindrical formations of the intestinal mucous membrane, about 0.5-1.5 mm high, glove-finger-shaped, adapted to the absorption function which, besides that of digestion of food, is the main function of the small intestine.

They characterize, specifically, the small intestine; they are lacking in the superior duodenum and disappear in the terminal ileum.

They are structures of the mucous membrane, formed of a vasculoconnective central axis, expansion of the corium, lined by asurface epithelium. The epithelium of the villi is columnar, monostratified, with a plateau, constituted especially of enterocytes; goblet cells are infrequent virtually absent; endocrine cells are rare too (fig.179).

The stroma of villi, an "emanation" of the corium, is formed of a connective reticular tissue, containing a variable number of tine collagen and elastic fibres. The cell component of this stroma is rich and polymorphous, represented by autochthonous cells, fibroblasts and histiocytes (the latter are often of macrophage type), as well as by allogenic elements, of haematic origin.

The allogenic population is made up of neutrophils and eosinophils (tissular or haematic), lymphocytes and plasmocytes etc. At the level of the stroma of the villi is situated, centrally, a lymph vessel, the central lacteal (chyle) vessel, which begins with a " blind" (closed) end and opens info the lymphatic network of the corium.

The blood supply of the villi is provided by an arteriovenous capillary. The arteria. segment, derived from the arteries of the corium, has an ascending course, from the base to the tip of the villi; at this level it bends (forming an arch), capillarizes abundantly beneath the basemen membrane of the absorptive epithelium and is continuous with venules which empty into the plexus of the underlying corium.

The stroma of the villi contains, in addition, nerve fibres derived from Meissner's nerve plexus, fibres which end either at the level of blood and lymph vessels, or at the level of the contractile elements. The contractile elements are represented either by smooth muscle fibres, derived from the muscularis mucosae, or by spindle cells (called also fusiform cells)
. The contraction of these muscle fibres produces the shortening of villi and furthers the passage of the absorbed substances into the blood capillaries and lymph vessels; their relaxation is accompanied by lengthening of the villi,which augments their surface and promotes absorption. In this way, the villi do not represent only an absorption apparatus, but also a "propulsive" device, that ensures the passage of absorbed substances into the blood and the lymph.

Vessel and nerve supply

The arterial supply of the jejuno- ileum and the mesentery is ensured by the superior mesenteric artery, which nourishes the whole jejuno-ileum, the ascending caeco- colic segment and the right two thirds of the transverse colon, up to the Bohm - Cannon's point.

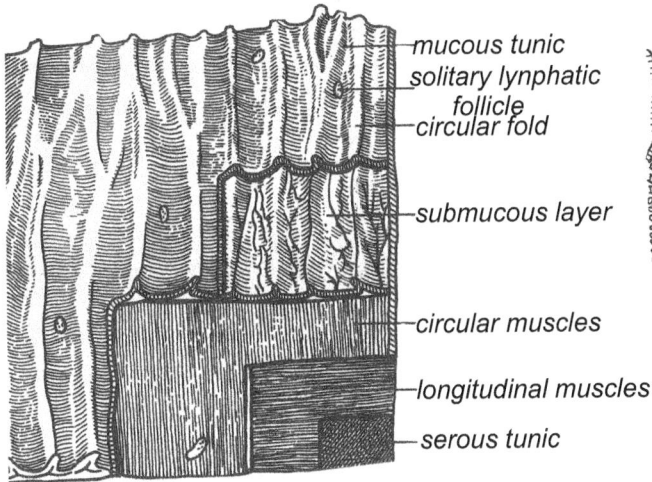

Fig. 175. Structure of the small intestine
(stratigraphic view).

Labels: mucous tunic, solitary lynphatic follicle, circular fold, submucous layer, circular muscles, longitudinal muscles, serous tunic

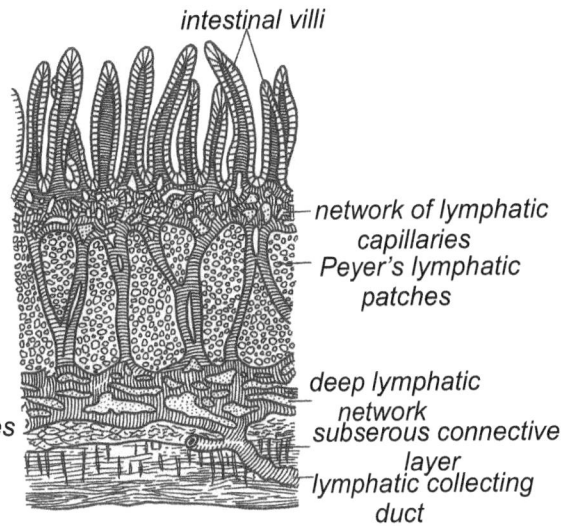

Fig. 176. Structure of the small intestine -
vertical section.

Labels: intestinal villi, network of lymphatic capillaries, Peyer's lymphatic patches, deep lymphatic network, subserous connective layer, lymphatic collecting duct

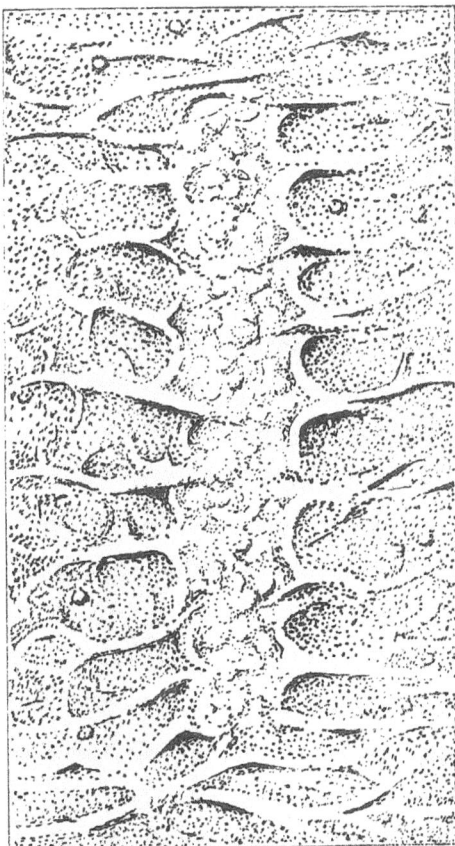

plicae ileum

Fig. 177. Structure of the ileum.

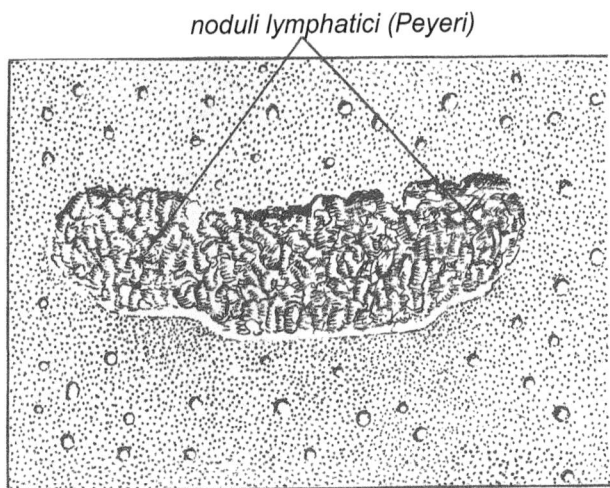

noduli lymphatici (Peyeri)

Fig. 178. Peyer's patches.

179

The superior mesenteric artery has its origin in the aorta, about 2 cm beneath the coeliac trunk and, after descending behind the pancreas and in front of the duodenum, reaches the root of the mesentery, where it has the following relations: on the right, with the superior mesenteric vein; behind, with the vein of the first jejunal loop; in front, in the transverse mesocolon, with the middle colic artery and on the left, with the first jejunal loop.

It gives off collateral branches to the retropancreatic region and to the right colon, after which, in the jejuno-ileal segment, it sends 14-15 branches to the jejunum and the ileum (jejunal and ileal arteries), that form arches of various degrees, through which, finally, emerge the vasa recta.

Between the superior mesenteric artery on the left, the ileobicaeco-appendiculocolic trunk on the right and the terminoileal anastomotic arch below, is enclosed a triangular, less vascularized area, called "Treves' vascular area".

The veins form a submucous network, from which arise the jejunal and ileal veins, tributaries of the superior mesenteric vein.

The lymph vessels of the small intestine and the mesentery emerge from the central lacteals of the villi, form a submucous network and are satellites of the veins.

They have a first relay in the mesenteric lymph nodes, about 200 in number, from where the afferent vessels pass either directly, or through the coeliac lymph nodes and the intestinal collector trunk, to the Pecquet's cistern *(cisterna chyli)*.

The autonomic nerve supply of the small intestine is derived from the coeliac plexus and the superior mesenteric plexus (fig.182).

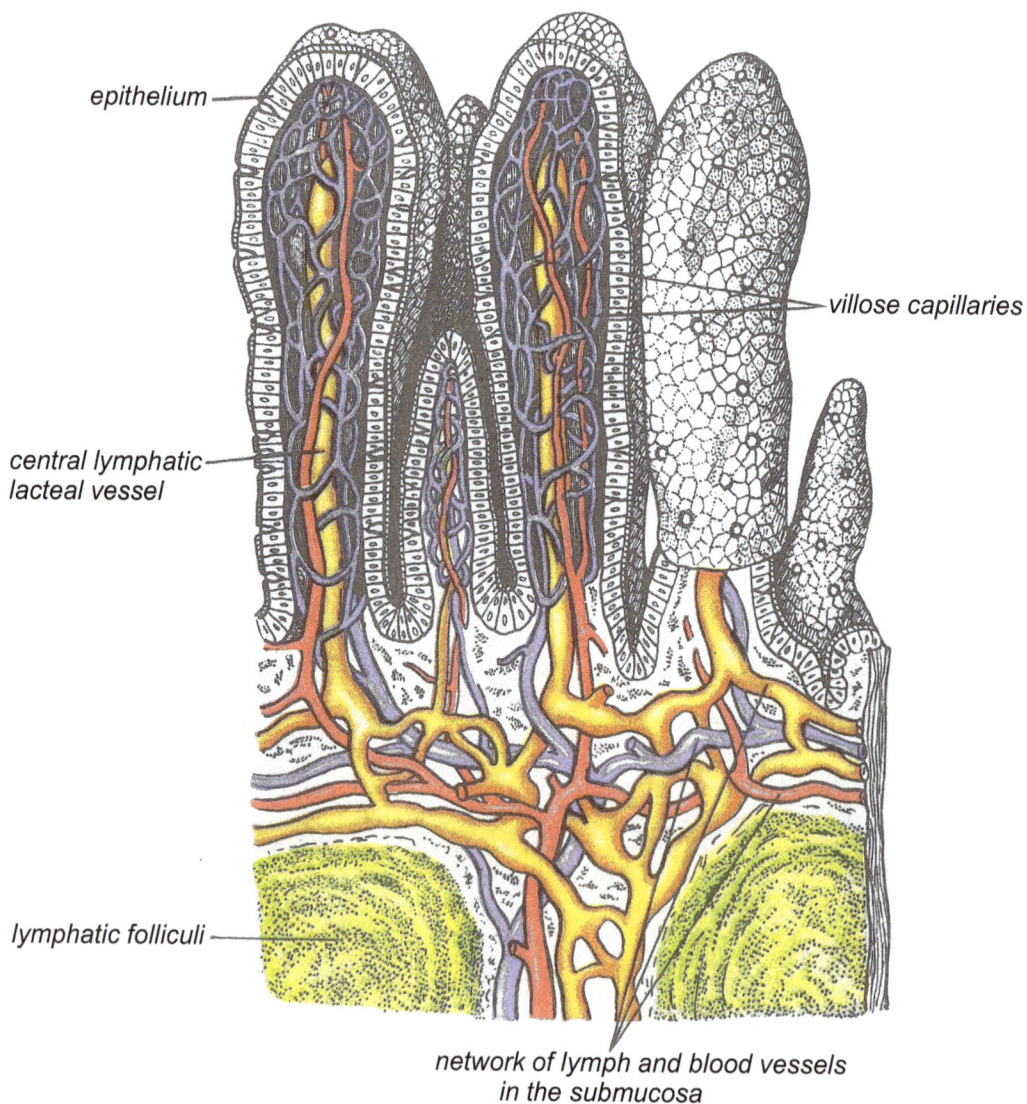

epithelium

villose capillaries

central lymphatic lacteal vessel

lymphatic folliculi

network of lymph and blood vessels in the submucosa

Fig. 179. Intestinal villi – structure

180

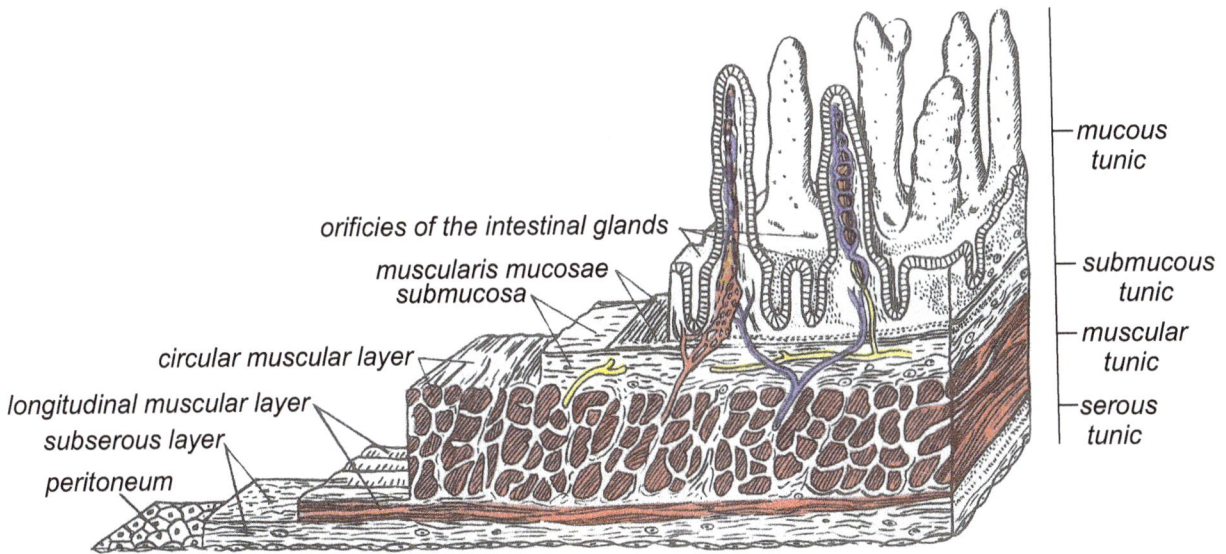

Fig. 180. Schematic section through the wall of the small intestine

- mucous tunic
- submucous tunic
- muscular tunic
- serous tunic

orifices of the intestinal glands
muscularis mucosae
submucosa
circular muscular layer
longitudinal muscular layer
subserous layer
peritoneum

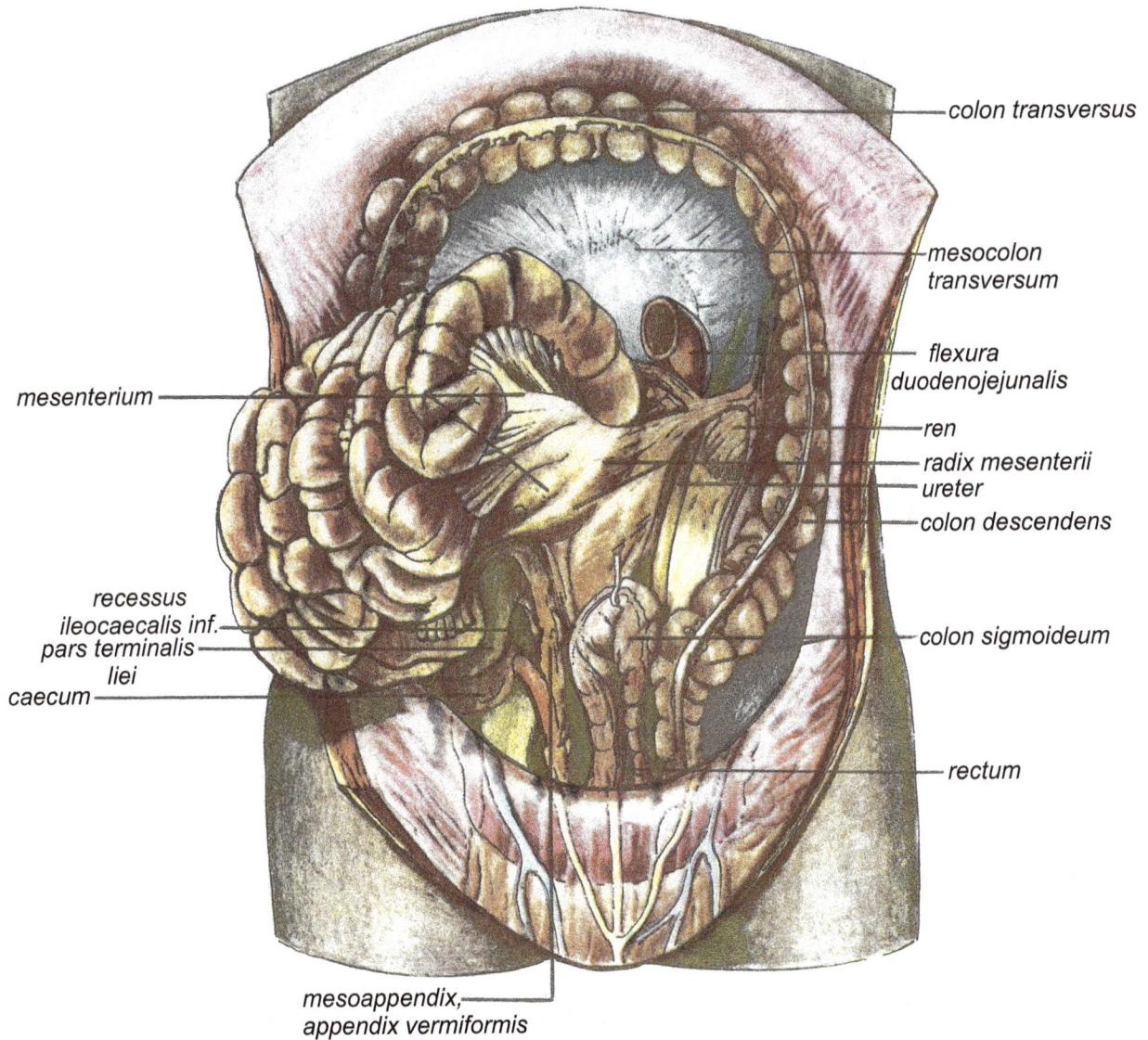

- colon transversus
- mesocolon transversum
- flexura duodenojejunalis
- ren
- radix mesenterii
- ureter
- colon descendens
- colon sigmoideum
- rectum

mesenterium
recessus ileocaecalis inf.
pars terminalis liei
caecum
mesoappendix, appendix vermiformis

Fig. 181. Inframasocolic part (the transverse colon and the sigmoid are pulled upwards).

181

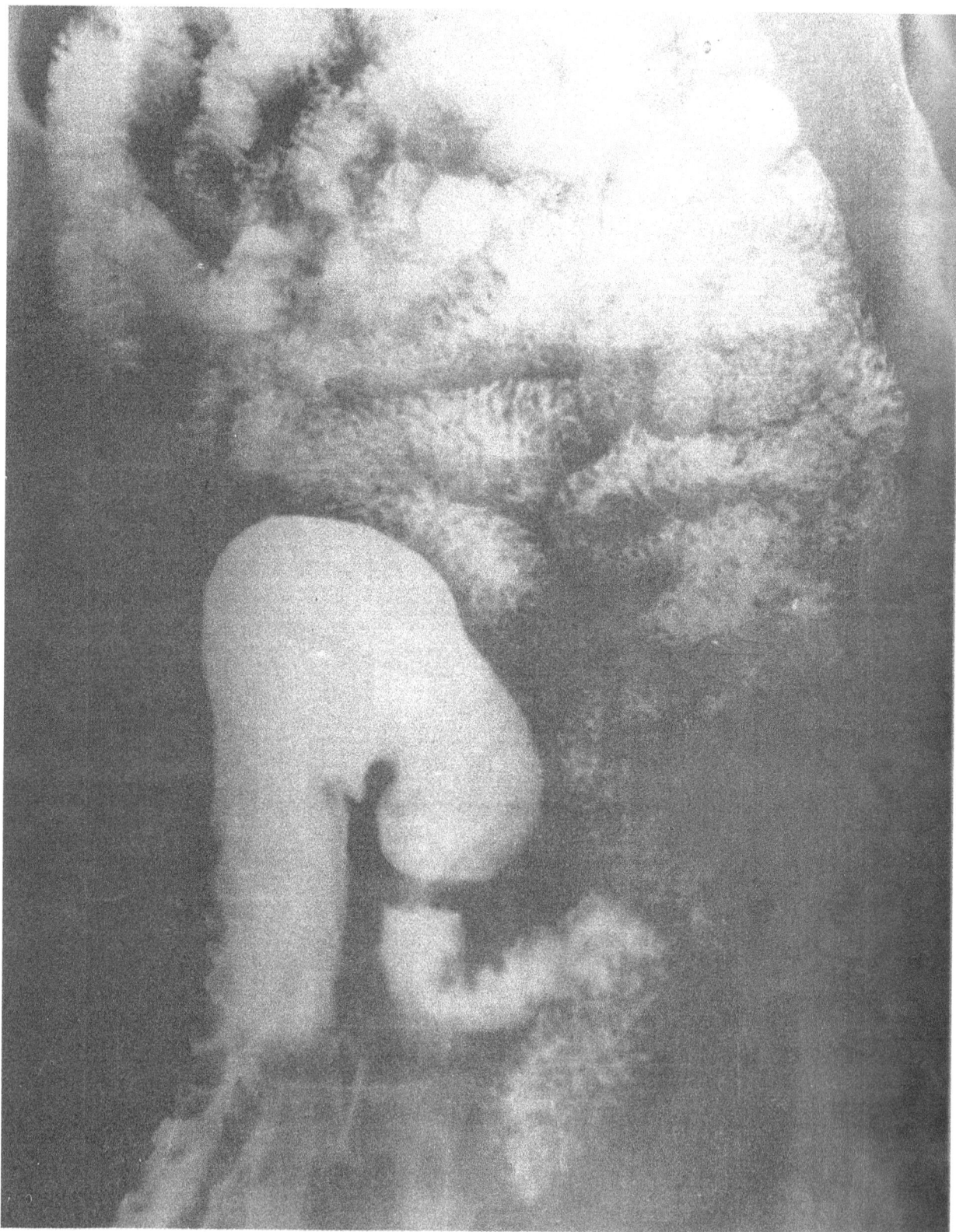

Fig. 182. Roentgenography of the stomach, the duodenum and the small intestine
(collection of the Central Military Hospital)

The Large Intestine
(Intestinum crasium)

The large intestine, the colon, is continuous with the small intestine and is situated between the ileocaecal valve and the origin of the rectum, at the level of the S_3 vertebra.

It is 1.60-1.85 m long and has a diameter of about 7-8 cm at its origin, which diminishes towards the terminal portion up to 3-3.5 cm.

The anatomofopographical elements and the disposition of the peritoneum, which ensures a certain degree of mobility to some segments and of fixity, through coalescence fasciae, to others, lead to the division of the large intestine into the following portions: caeco- appendicular, ascending colon, transverse colon, descending colon, sigmoid colon and rectum, which opens outside through the anal orifice (fig.184,185).

The large intestine has the following morphological peculiarities, which distinguish it from the small intestine (fig. 183).

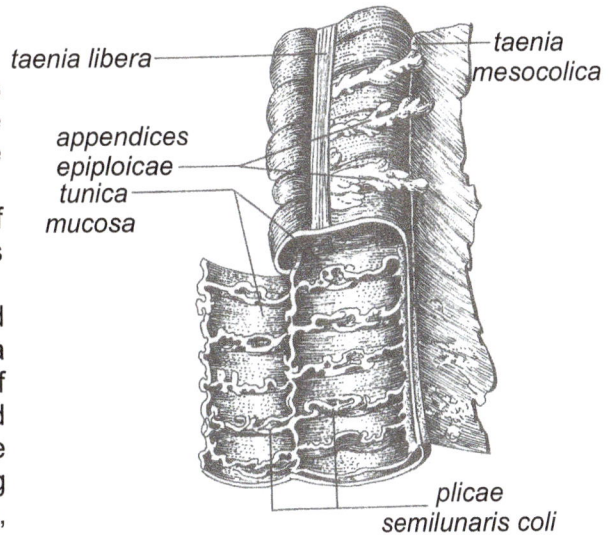

Fig. 183. Structure of the large intestine.

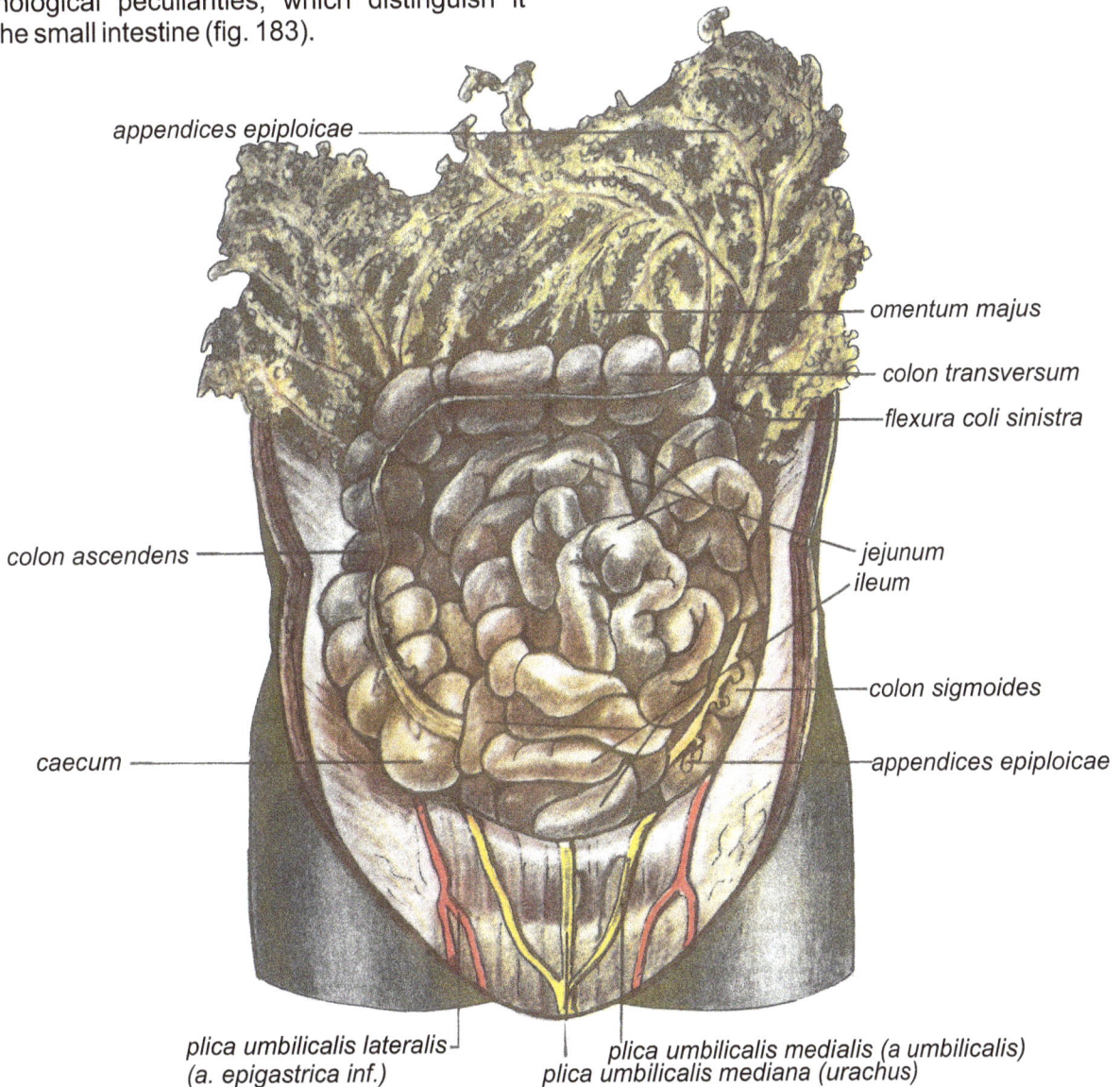

Fig. 184. Inframesocolic port (the greater omentum is pulled upwards).

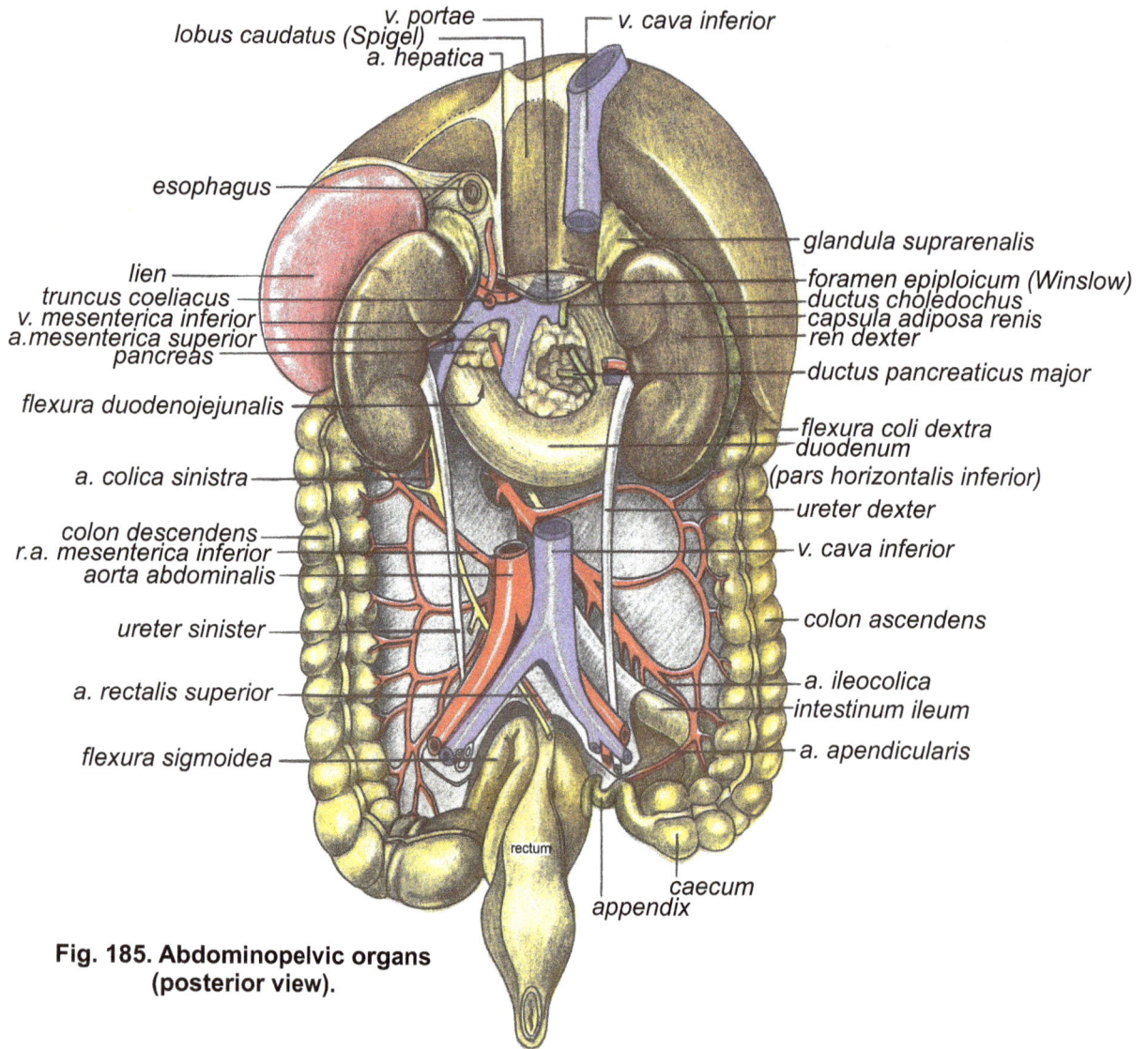

Fig. 185. Abdominopelvic organs (posterior view).

v. portae
lobus caudatus (Spigel)
a. hepatica
v. cava inferior
esophagus
glandula suprarenalis
foramen epiploicum (Winslow)
ductus choledochus
capsula adiposa renis
ren dexter
lien
truncus coeliacus
v. mesenterica inferior
a.mesenterica superior
pancreas
ductus pancreaticus major
flexura duodenojejunalis
flexura coli dextra
duodenum
(pars horizontalis inferior)
a. colica sinistra
ureter dexter
colon descendens
r.a. mesenterica inferior
aorta abdominalis
v. cava inferior
ureter sinister
colon ascendens
a. rectalis superior
a. ileocolica
intestinum ileum
flexura sigmoidea
a. apendicularis
rectum
caecum
appendix

The teniae of the colon (taeniae coli) are three muscular bands which arise from the site of attachment of the vermiform appendix on the caecum. One of them is visible, that which gives no attachement to any peritoneal formation and which is called free tenia (taenia libera); the second corresponds to the insertion of the embryonic mesocolon and is termed mesocolic tenia (taenia mesocolica); the third, which gives attachment to the greater omentum on thetransverse colon, is called omental tenia (taenia omentalis).

The hausfrations of the colon (hausfra coli) are prominent areas of the intestinal wall, separated through deep transverse furrows bulging into the lumen of the intestine under the form of folds, called semilunar folds (plicae semilunares coli). Unlike the small intestine, where the folds are formed only by mucous membrane, in the large intestine all the layers of the wall contribute to their structure.

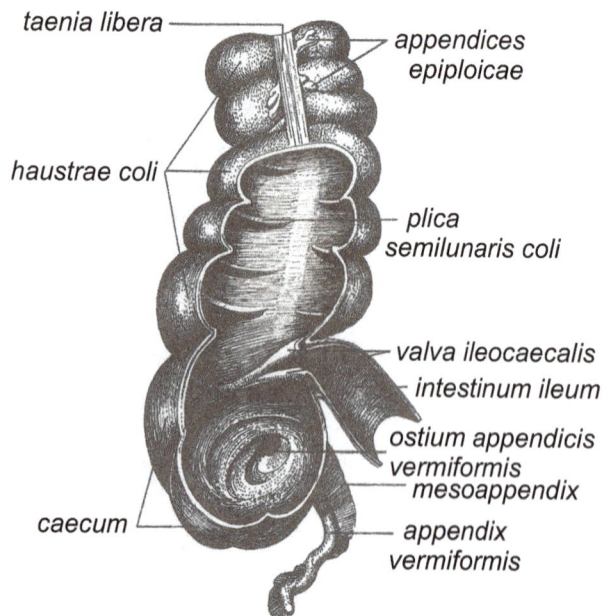

taenia libera
appendices epiploicae
haustrae coli
plica semilunaris coli
valva ileocaecalis
intestinum ileum
ostium appendicis vermiformis
mesoappendix
caecum
appendix vermiformis

Fig. 186. The caecum.

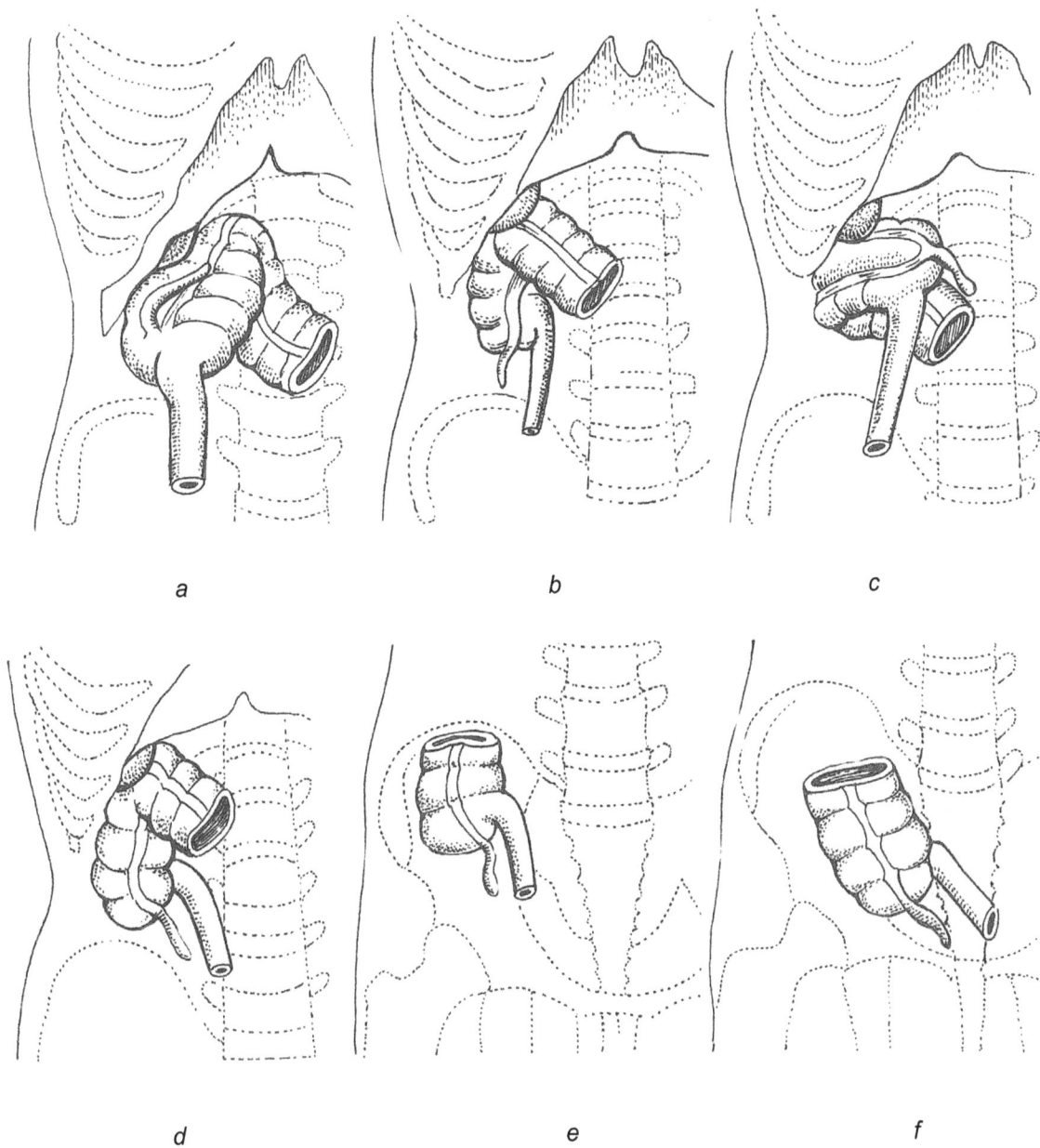

a *b* *c*

d *e* *f*

Fig. 187. Variants of the position of caecum, a. subhepatic horizontal caecum, the vermiform appendix in immediate relation with the gall bladder; b. subhepatic vertical caecum; c. reversed subhepatic horizontal caecum; d. caecum in lumbar position; a. caecum in iliac position; f. caecum in pelvic position.

The epiploic appendices (appendices epiploicae) represent the third characteristic of the large intestine, which differentiates if from the small intestine. They are yellow, tassel-like formations, constituted of a peritoneal coat which contains a mass of fat: the attachement of the epiploic appendices at the level of the teniae.

The following coats may be distinguished in the structure of the large intestine:

- the serous tunic *(tunica serosa sive tela serosa)* formed of peritoneum, at the level of all the segments of the large intestine - although not complete in some segments of the colon -, with the exception of the last portion of the rectum, which presents a fibrous, thick adventitia;

- the mucous funic *(tunica mucosa),* formed of two layers: an outer layer, consisting of longitudinal fibres grouped in the three teniae, and an inner layer, made up of circular fibres;

- the submucous tunic *(tela submucosal,* which contains neurovascular and lymphatic elements;

- the muscular tunic *(tunica muscularis),* constituted of a columnar epithelium with many Lieberkuhn's glands and a corium, in which numerous lymphoid elements are present.

The Caecum and the Appendix
(Caecum et appendix vermiformis)

The caecum is the portion of the large intestine situated below the opening of the ileum, from the level of the ileocaecal valve to that of the opening of the appendix into it. It is saccular in shape and is continuous upwards with the ascending colon; from the site of implantation of the appendix into the caecum start the three muscular teniae (anterior, free posterolateral and adherent posteromedial); in addition, a fundus and a body of the caecum are described.

It is lodged in the right iliac fossa, but in dependence on the embryologic development abnormalities it may be also high situated (horizontal caecum in subhepatic position, the appendix entering in relation with the gall bladder, or vertical subhepatic caecum, with a short segment of the ascending colon) or low situated (caecum in lumbar, iliac or pelvic position) (fig. 187).

When it lies in the right iliac fossa, it may display the following four morphological types:

- the true infundibular type, with the appendix developed inferiorly;

- the type inflected in obtuse angle, with the appendix directed downwards and medially;

- the type inflected in right angle, with the appendix situated behind the caecum;

- the type inflected in acute angle, with the appendix directed upwards and medially (pre- or retroileal).

Relations of the caecum and appendix. Anteriorly, the caecum is directly related to the anterior abdominal wall, in front of if passing sometimes the greater omentum. Posteriorly it is in relation with the elements lodged in the right iliac fossa: peritoneum, subperitoneal connective-adipose tissue- which at this level is continuous with that situated deeply in relation to the femoral arch and is called Bogros' space-, iliac fascia, iliopsoas muscle and branches of the lumbar plexus (the femoral nerve and the femorocutaneous nerve).

Laterally is situated the iliac fascia, below which are the iliac muscle and the femorocutaneous nerve. Superiorly, if is in relation with the iliac crest and the broad muscles of the abdomen. Between the caecum and the infernal iliac fossa is enclosed the right parietocolic space, prolonged downwards. Medially, the caecum is related to the terminal ileal loops and the right inferior end of the greater omentum; on a deeper plane, to the external iliac vessels. Below, if is in a more remote contact with the femoral arch and the deep ring of the inguinal canal.

As regards the inner configuration, the caecum is devoid of hausfrations and is constituted of a rounded cavity; the teniae are longitudinal, smooth and start from the site of implantation of the appendix. In rare cases there may be present a few haustrations, separated through circular folds. Inside the caecum lies the ileocaecalvalve and, below this, on its median surface, the appendicular orifice (fig. 186).

The ileocaecal valve (valva ileocaecalis) (Bauhin's valve) has the shape of a horizontal slit, with a superior' lip *(labium superius)* and on inferior lip *(labium inferius).* The two lips close the ileocaecal orifice *(ostium ileocaecalis)* and coalesce at the ends info the anterior commissure and the posterior commissure. From each commissure starts a frenulum of the commissure *(frenulum valvae ileocaecalis).* Around the orifice is formed a muscular apparatus with circular smooth fibres, called Keith's sphincter (fig. 188).

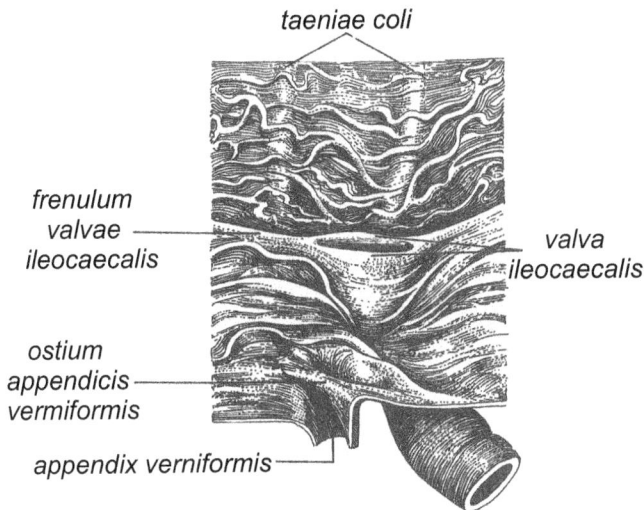

Fig. 188. Ileocaecal valve.

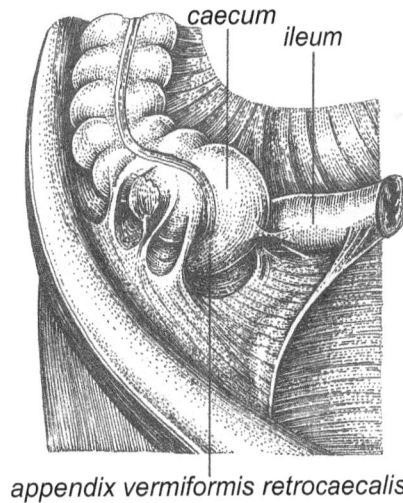

Fig. 189. Retrocaecal and paracolic recess.

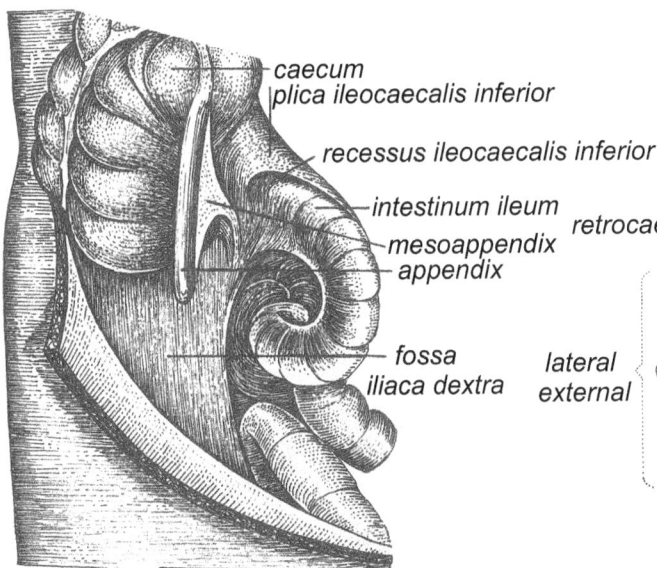

Fig. 190. Lower ileocaecal recess

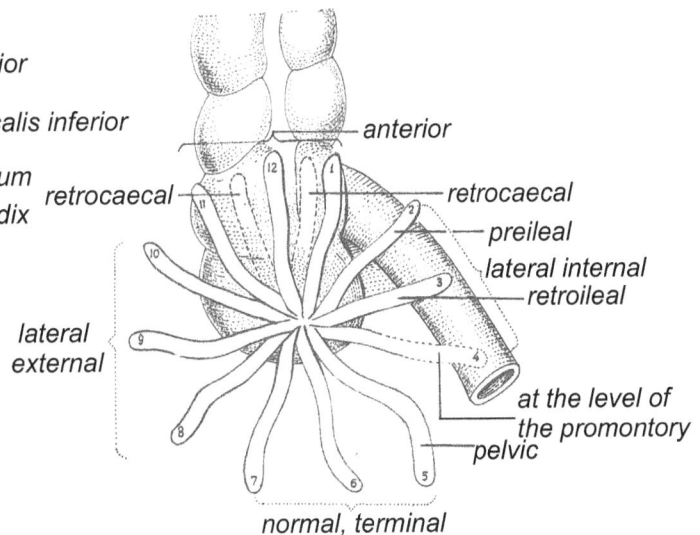

Fig. 191. Position of the vermiform appendix
(according to Gerota).

The structure of a valve includes a muscular lamina which belongs to the small intestine and a lamina which depends on the caecum. They are constituted of a mucous, a submucous and a muscular tunic, the latter formed of circular fibres.

The ileocaecal valve resembles a funnel, with the small opening directed towards the caecum and the large towards the ileum, so that the ileum seems to invaginate within the caecum. If allows the passage of the intestinal contents into the large intestine and prevents its reflux from the colon to the ileum.

The appendicular orifice (ostium appendicis vermiformis) is situated at the junction of the inferior wall with the postero-infernal wall of the caecum, at 1.5-2 cm below the ileocaecal valve. It is infundibular or circular in shape. The mucous membrane forms, at the site of implantation, a bulging fold, called Gerlach's valve.

In dependence on the behaviour of the peritoneum, different morphological varieties of the caecum mayoccur:

- a mobile caecum, when the peritoneum envelops if entirely and thus the caecum is not fixed to the posterior wall of the right iliac fossa;

187

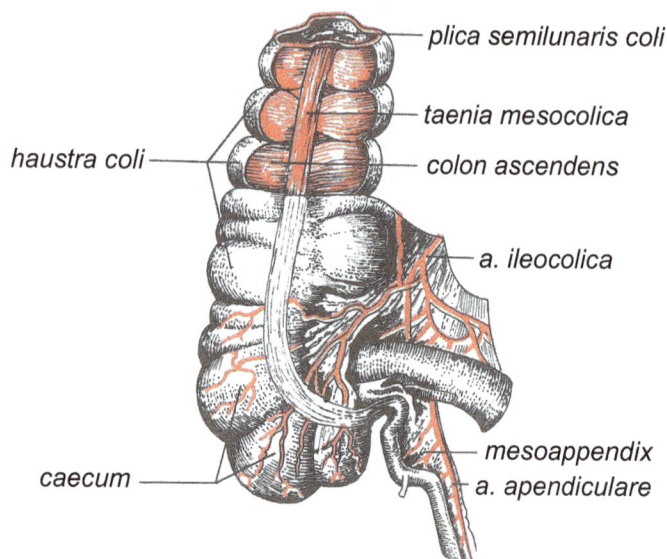

Fig. 192. Caecum and appendix - external aspect and blood supply.

Labels on figure:
- plica semilunaris coli
- taenia mesocolica
- colon ascendens
- a. ileocolica
- haustra coli
- mesoappendix
- a. apendiculare
- caecum

- a fixed caecum with a free fundus, as a result of the process of coalescence of its posterior surface with the peritoneum of the right iliac fossa;

- if the coalescence is incomplete, if results in a fixed caecum corresponding to a retrocaecal recess;

- if the coalescence is complete, the caecum is entirely fixed. Occasionally, in this situation, the appendix is free in the abdominal cavity or if is retrocaecal and the process of coalescence occurs sometimes only on the posteromedian line of the caecum, resulting in a mesocaecum.

As a consequence of the disposition of the peritoneum towards the caecum and the appendix, in the caeco-appendicular region appear folds and fossae.

Thus, between the lateral wall of the caecum and the parietal peritoneum lies the paracaecal fold, which bounds the paracaecal fossa; medially, lies the ileocaecal fold, which arises owing to a branch of the ileocolic artery, situated on the anterior surface of the caecum; if forms the superior ileocaecal fossa.

Between the ileum, the caecum and the appendix is the ileo-appendicular fold, which forms the inferior ileocaecal fossa, where sometimes the appendix may be situated. Behind the caecum lies frequently the retrocaecal fossa, in which, too, may be situated the appendix (fig.189,190).

The positions of the appendix vary concomitantly with the caecum and may be subhepatic, iliac and pelvic.

The appendix (processus vermiformis). It is a cylindrical tube, up to 9 cm long, which arises from the fundus or the medial surface of the caecum. Its base lies at the site of confluence of the surface of the caecum. Its base lies at the site of confluence of the three muscular teniae of the caecum and at 2 cm below the ileocaecal valve: it is sinuous, mostly with a short, radicular portion attached to the caecum, and a long, floating portion, which is movable. On its medial side is the appendicular mesentery. Usually, when the mesentery is short, the appendix is sited medially relative to the caecum; when it is long, the appendix may assume various clockwise positions.

In dependence on the direction of the appendix we mention, according to Testut, the following positions: *the descending appendix* (42% of cases), which is situated in the medial portion of the iliac fossa and the apex of which may reach the pelvis (it has relations: behind, with the psoas muscle; in front, with the abdominal wall and the ileal loops; medially, with the ileal loops; laterally, with the fundus of the caecum, and its apex comes in contact with the internal and external iliac vessels and the testicular vessels; its inflammation spreads to the pelvic organs); *the lateral appendix* (26%) which has lateral relations with the inguinal ligament; *the medial appendix* (17%), in relation with the intestinal loops; *the retrocaecal ascending appendix* (13%), situated behind the caecum and even behind the ascending colon. A very complete description of these positions was given by Gerota (fig. 191).

The structure of the caecum and appendix is characterized by the existence of four tunics (coats):

- the serous funic, formed of the caeco-appendicular peritoneum;

- the muscular funic, consisting of longitudinal and circular fibres, with the difference that, unlike in the caecum, in the appendix, in the longitudinal layer, instead of forming further teniae, these fibres unite into a continuous layer;

- the submucous tunic;

- the mucous funic in the caecum is identical to that of the whole large intestine; in the appendix, it presents a great number of lymphatic follicles and this is the reason why the appendix is also called abdominal tonsil.

188

Vessel *and nerve* supply (fig.192,193)

The arterial supply of the caecum and appendix is derived from the superior mesenteric artery through the ileocolic branch, from which arise, in the neighbourhood of the colic angle, the following branches: the anterior caecal artery, the posterior caecal artery and the appendicular artery.

The anterior caecal artery (sometimes double) gives off a branch to the inferior area of the colon and is distributed to the anterior wail of the caecum and the base of the appendix.

The posterior caecal artery gives rise to a branch for the inferior area of the colon and then gives off caecal branches, which supply the bottom of the caecum.

The appendicular artery lies in the mesoappendix and ramifies within the appendix.

The veins, satellites of the arteries, ore tributaries of the ileocolic veins that empty info the superior mesenteric vein and then info the portal vein, which explains the hepatic abscesses consecutive to an acute appendicitis.

The lymph vessels drain info the following lymph node groups:

- the ileocaecal lymph nodes, situated at the base and in the thickness of the mesoappendix, in the caeco-appendicular angle, on the anterior and posterior surface of the caecum;

- the mesenteric lymph nodes, situated on the course of the mesenteric vessels;

- the duodenopancreatic lymph nodes.

The nerve supply of the caecum and appendix derives from the superior mesenteric plexus.

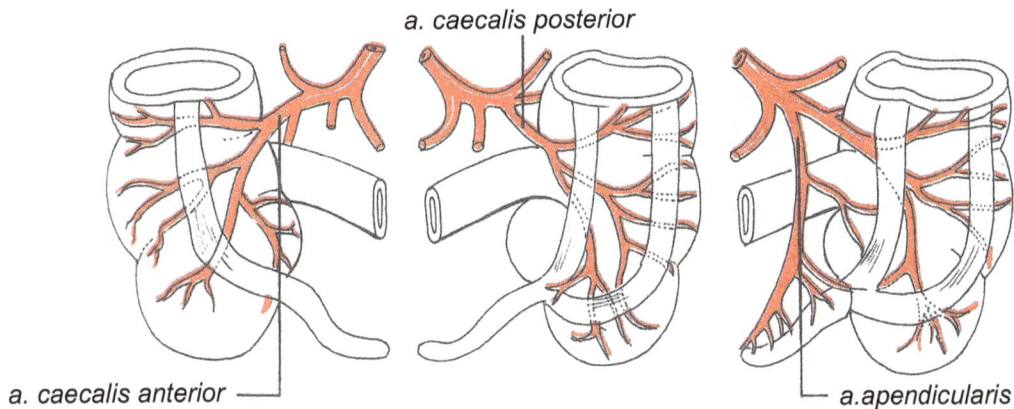

a. caecalis posterior

a. caecalis anterior

a.apendicularis

Fig. 193. Blood supply of the caecum

The Colon

The colon is that portion of the large intestine which extends from the caecum (the ileocaecal valve, respectively) to the rectum fat the level of the S_3 vertebra, respectively). If is divided into the following segments: the ascending colon *(colon ascendens),* from the origin up to below the visceral surface of the liver, at the right flexure of the colon *(flexura coli dextra)* or the hepatic angle; the transverse colon *(colon transversum),* from the right flexure to the left flexure, situated at the level of the spleen *(flexura coli sinistra)* or of the splenic angle, in relation with the spleen; the descending colon *(colon descendens),* from the left colic flexure to the level of the iliac crest; the sigmoid colon *(colon sigmoideum),* from the level of the iliac crest to that of the S_3 vertebra, where if is continuous with the rectum (fig. 185,194).

The ascending colon *(colon ascendens)* is lodged in the upper part of the right iliac fossa and in the right lumbar region; it is covered anteriorly with peritoneum and has the following relations: behind, with the iliopsoas muscle and the anterior surface of the lower pole of the right kidney, through Toldt's coalescence fascia; laterally, with the lateral wall of the abdomen, from which if is separated through the right parietocolic space; anteriorly and medially, if is covered by the loops of the small intestine.

Sometimes a more or less apparent mesocolon may be present.

The right flexure of the colon *(flexura coli dextra)* has relations, behind, with the descending portion of the duodenum and the right kidney and in front, with the visceral surface of the liver *(impressio colica)* and the fundus of the gall bladder. The right flexure is maintained in its position by the right phrenocolic ligament It may form an acute angle, when the transverse colon is longer, and a right or obtuse angle, when the transverse colon ascends towards the splenic angle.

189

The transverse colon *(colon transversum)* extends between the right and the left flexures, in an oblique ascending direction, from the right to the left. It has a long mesentery, the transverse mesocolon, which endows it with a great motility.

The roof of the mesocolon is attached transversally on the posterior abdominal wall and crosses, from the right to the left, the second portion of the duodenum (the descending part) and the head of the pancreas. Its ends are, on the right, above the lower pole of the right kidney and, on the left, above the upper pole of the left kidney. The transverse mesocolon contains Riolan's arch and the middle colic artery *(artera colica media);* it has two surfaces, anterosuperior and postero-inferior, the latter in relation with the loops of the small intestine. The transverse colon is described as consisting of two portions: a right portion, with a short, rather fixed mesentery, extending from the right colic flexure to the crossing with the superior mesenteric vessels, and a longer, more movable, left portion, represented by the remainig transverse mesocolon.

The relations of the transverse colon are: in front, with the anterior abdominal wall, on its anterior surface being attached the greater omentum; above, with the inferior surface of the liver and the greater curvature of the stomach, to which if is connected through the gastrocolic ligament; below, with the duodenojejunal flexure and the loops of the small intestine; behind, with the right kidney, the descending part of the duodenum, the head and the body of the pancreas.

The left flexure of the colon *(flexura coli sinistra)* is related, behind, with the left kidney and the left adrenal gland; superolaterally, with the spleen *(impressio colica);* in front, with the stomach, which on the left passes beyond it. It is deep, high, situated in the left hypochondrium and much acuter than the right flexure. If is connected to the diaphragmatic muscle through the left phrenicocolic ligament (which forms the sustentaculum lienis).

The descending colon *(colon descendens),* extending up to the iliac crest, is somewhat longer than the ascending portion. It is parietalized like the ascending colon through Toldt's fascia of coalescence.

If is related, behind, with the quadratus lumborum muscle and the lateral border of the left kidney, with the iliohypogastric and ilio-inguinal nerves, in front and medially with the small intestine and laterally, with the left parietocolic space.

The anterior, medial and lateral surfaces are invested by peritoneum and posteriorly lies the left Toldt's fascia of coalescence.

The sigmoid colon *(colon sigmoideum)* extends from the level of the iliac crest to the third sacral vertebra and is continuous with the descending colon up to the level of the rectum. It is also called iliopelvic or terminal colon. It has the following characteristics: the haustrations are less clear-cut, there are only two teniae instead of three, the epiploic appendices are very numerous and disposed in two rows and it has a great mobility, as the result of the existence of a rather long mesentery. The sigmoid colon shows great variations as regards its position and degree of mobility, individual variations which are related to its length and to the length of its mesentery.

In dependence on the degree of mobility if has three portions, delimited by two angles; a first, *fixed portion,* extending from the iliac crest to the inner border of the left psoas ("iliac colon"); a second, *movable portion,* formed of a more or less long loop, in dependence on the height of the individual or on the existence of congenital malformations-dolicho-sigma- with the concavity upwards or downwards, known also as "sigmoid loop"; the third portion, with a short mesentery and an oblique direction from the right to the left; called also "rectosigmoid segment". The last two portions make up the pelvic colon, with a small mesentery, and vary from 25 to 30 cm.

The line of insertion of the mesosigmoid info the posterior wall of the pelvis begins at the iliac crest, runs medially up to the lateral border of the left psoas muscle, crosses the left ureter, where it forms the intrasigmoid recess *(recessus intrasigmoideusj,* in the fundus of which it may be palpated, and descends towards the median line, up to the level of the S_3 vertebra (fig.195).

The relations of the sigmoid colon are the following: in the iliac segment, anteriorly, with the anterolateral wall of the abdomen, the loops of the small intestine and the greater omentum and posteriorly, with the posterior wall of the abdomen in the left parietocolic space, the Toldt's fascia of coalescence, the iliac fascia with the iliac muscle, the testicular or ovarian vessels, the genitofemoral nerve, the external iliac vessels; in the pelvic segment, anteriorly, with the urinary bladder (in the female, with the uterus and the adnexa) and the loops of the small intestine, and posteriorly, with the ureter, the rectal ampulla, the left internal and external iliac vessels, inferiorly, with the urinary bladder, the ureters (in the female, the broad ligaments, the uterine tubes, the ovaries), the rectal ampulla and the pouch of Douglas (or *Douglas' cul de sac).*

190

Vessel and nerve supply (fig.196-199).

The blood supply of the large intestine is provided by two arteries: the superior mesenteric artery and the inferior mesenteric artery. Practically, from the anatomicosurgical point of view, taking info account the blood supply criterion, the colon is divided into the right and the left colon.

A. *The arterial supply of the right colon* is provided by the superior mesenteric artery; three of its main branches are here interesting:

1. *The ileocolic artery* (the inferior right colic artery), which arises from the superior mesenteric artery, slightly below the right superior colic artery, runs within the roof of the mesentery, obliquely downwards and to the right, crosses the superior mesenteric vein, the right ureter and the spermatic vascular trunk and ends in front of the ileocaecal angle, giving off the following branches: the ascending colic artery, which tabs part in the formation of the colic vascular arch; the anterior caeca! artery, which supplies the caecum and the initial portion of the ascending colon; the posterior caeca! artery, which supplies the posterior surface of the caecum; the appendicular artery and the ileal artery.

2. *The right superior colic artery* arises from the superior mesenteric artery in front of the third portion of the duodenum, is directed upwards and to the right, crosses the submesocolic segment of the duodenopancreas and gives off: an inferior branch, which courses towards the ascending colon, anastomoses with a ramus of the middle colic artery and forms the right paracolic arch; a superior branch, which lies in the depth of the transverse mesocolon, anastomoses with the left colic artery and forms the Riolan-Haller's arch. It supplies usually the right two thirds of the transverse colon, up to the Bohm-Cannon's point

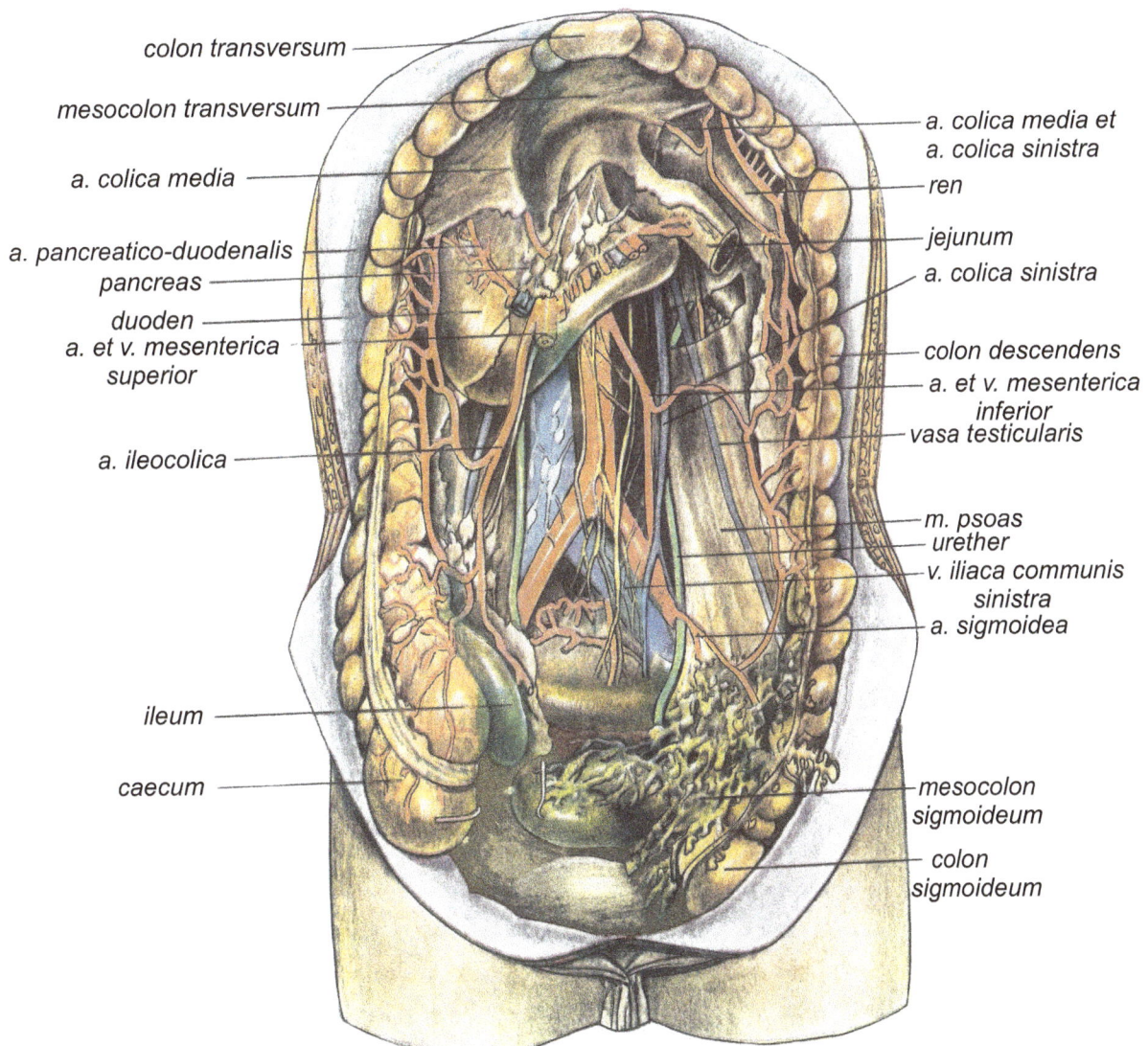

Fig. 194. Large intestine with the caecum and the transverse colon pullrd upwards; the retroperitoneal space is displayed

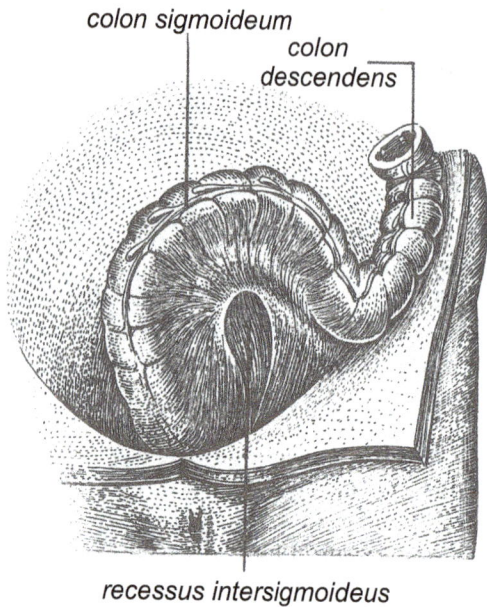

Fig. 195. The intersigmoidal recess.

colon sigmoideum
colon descendens
recessus intersigmoideus

2. *The middle colic artery* arises beneath the superior right colic artery and courses towards the inner border of the ascending colon, where it divides into an ascending and a descending branch. These branches anastomose with the corresponding branches of the superior and inferior right colic arteries. This is the proper artery of the ascending colon.

The veins are generally satellites of the arteries. The right superior colic vein is an exception to this rule, since if receives, in front of the head of the pancreas, the right gastro-epiploic vein and forms the gastrocolic trunk.

B. *The arterial supply of the left colon* is derived from the inferior mesenteric artery, which arises on the anterior surface of the abdominal aorta, at 5 cm above its bifurcation, at the level of the intervertebral L_3-L_4 space.

Its course includes four segments: retroduodenal, lumbar, iliac and pelvic.

In the retroduodenal segment, the artery is covered by Treitzs fascia, the pancreas and the duodenum; in the lumbar segment, if is situated between the left psoas muscle, on the right of the inferior mesenteric vein and of the ureter; in the iliac segment, if is sited in the mesosigmoid and crosses the common iliac vessels; in the pelvic segment if lies also in the mesosigmoid and ends in front of the third sacral vertebra, behind the superior end on the rectum, bifurcating info the superior hemorrhoidal arteries.

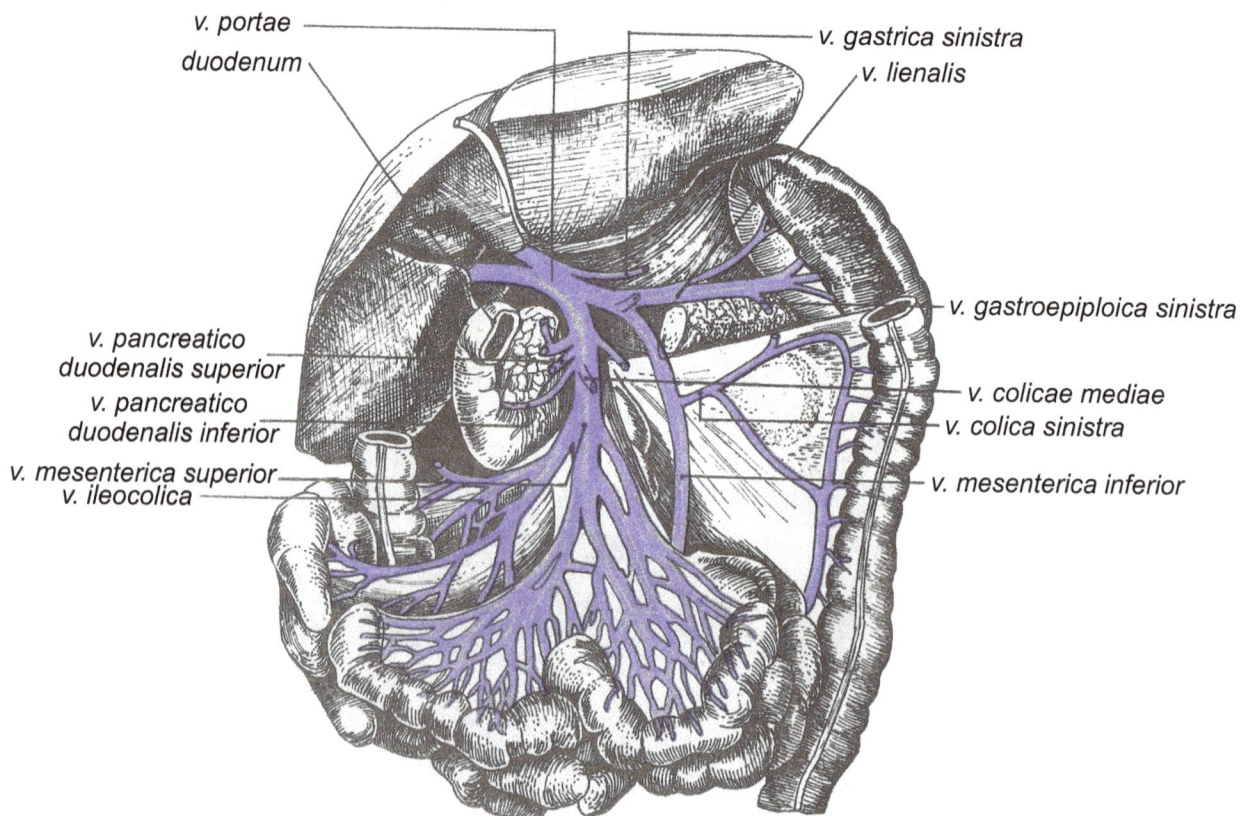

v. portae
duodenum
v. gastrica sinistra
v. lienalis
v. pancreatico duodenalis superior
v. pancreatico duodenalis inferior
v. mesenterica superior
v. ileocolica
v. gastroepiploica sinistra
v. colicae mediae
v. colica sinistra
v. mesenterica inferior

Fig. 196. The superior mesenteric vein.

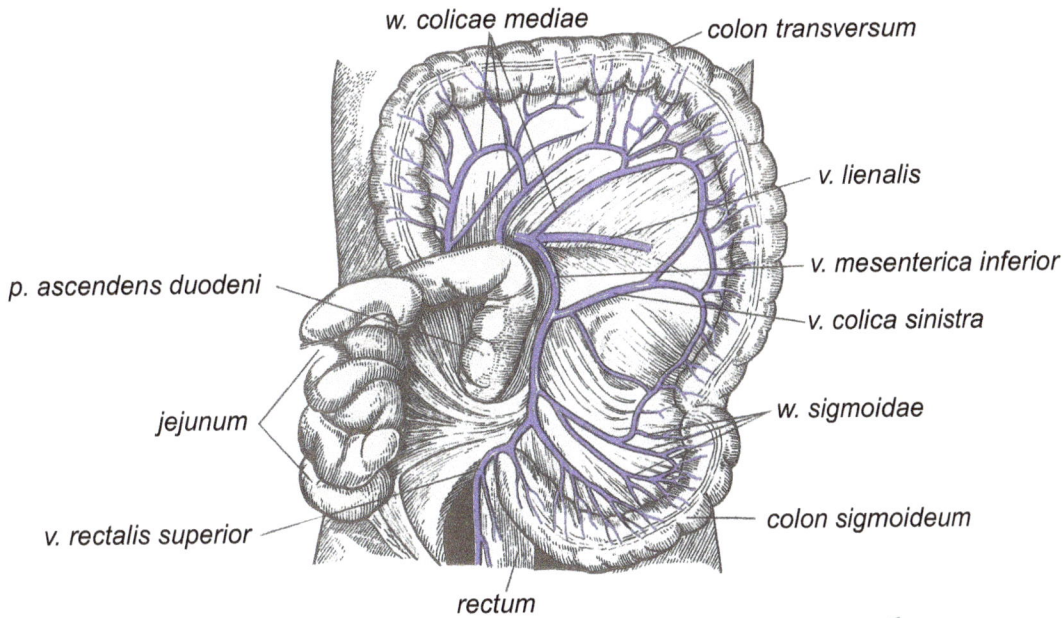

Fig. 197. The inferior mesenteric vein.

w. colicae mediae
colon transversum
v. lienalis
v. mesenterica inferior
v. colica sinistra
p. ascendens duodeni
w. sigmoidae
jejunum
colon sigmoideum
v. rectalis superior
rectum

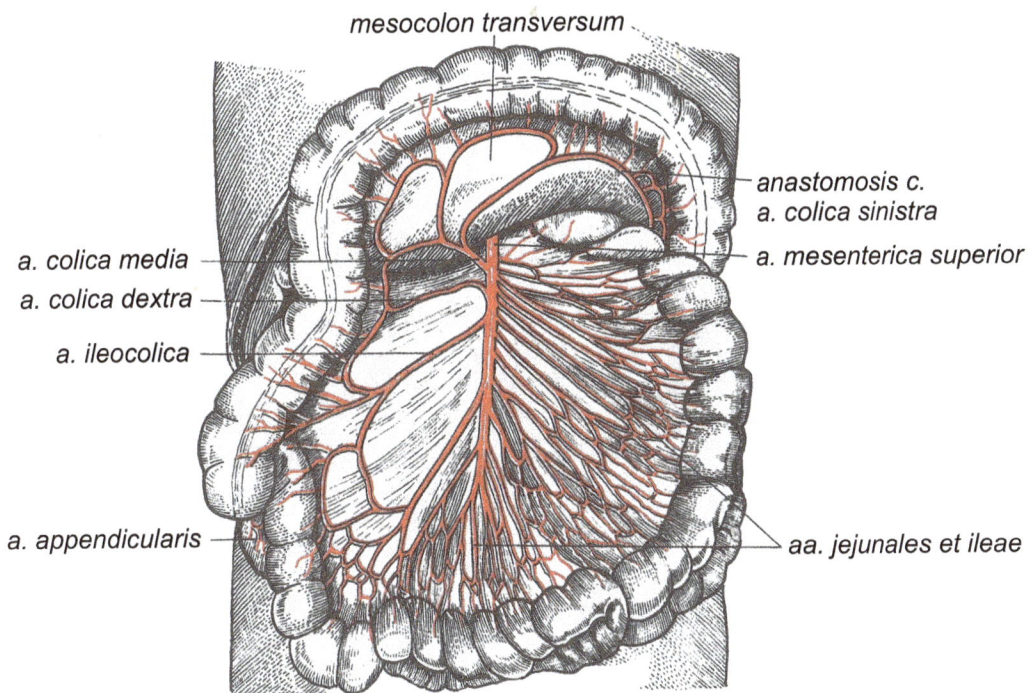

Fig. 198. The superior mesenteric artery.

mesocolon transversum
anastomosis c. a. colica sinistra
a. mesenterica superior
a. colica media
a. colica dextra
a. ileocolica
a. appendicularis
aa. jejunales et ileae

It gives off some collateral branches, which supply the colon and are described below, and two terminal branches, which will be studied in the chapter regarding the rectum.

1. *The left superior colic artery,* which arises at 2-3 cm below the duodenum, has an ascending course, passes through the left colic space between the duodenum, on the right, and the kidney, on the left; it crosses the inferior mesenteric vein, completing Treitz's vascular arch (formed of the superior left colic artery and the inferior mesenteric vein), then the left ureter, the left spermatic or ovarian vessels and the left kidney and, at the splenic angle of the colon, it divides into two branches: an ascending branch, destined to the transverse colon, which will form Riolan's arcade through its anastomosis with the superior right colic artery, and a descending branch, to the descending colon, which will anastomose with the middle left colic artery, if it exists, or with the superior sigmoid artery, forming the left paracolic arcade.

193

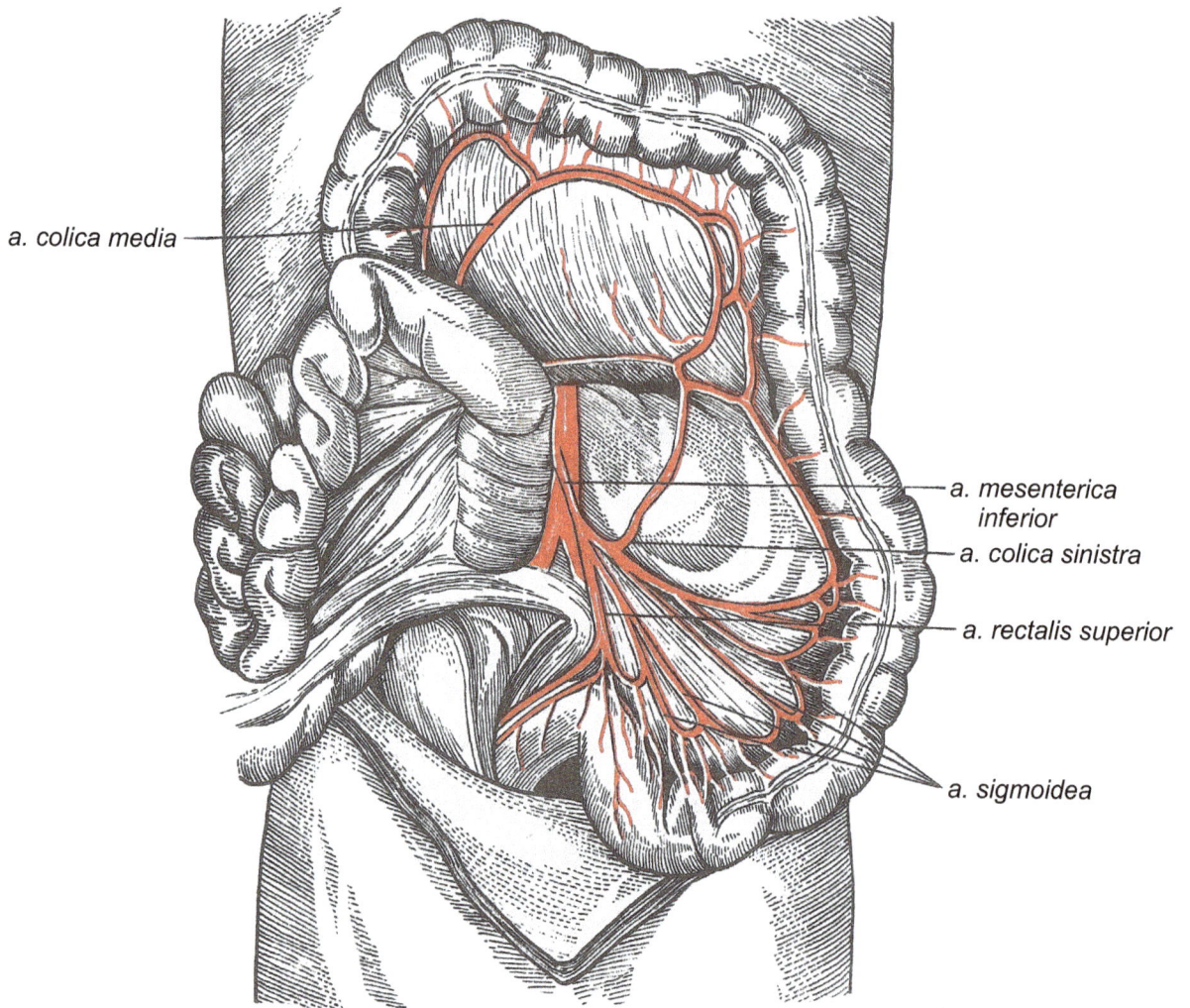

a. colica media

a. mesenterica inferior

a. colica sinistra

a. rectalis superior

a. sigmoidea

Fig. 199. The inferior mesenteric artery

194

2. *The trunk of the sigmoids,* which divides into three branches: the superior sigmoid artery, the middle sigmoid artery and the inferior sigmoid artery. These three arteries are anastomosed through a paracolic arcade, which is united with that of the descending colon.

The veins are only partly satellites of the arteries; their network of origin empties into the sigmoid veins and into the left superior colic vein, which drains the blood into the portal system through the inferior mesenteric vein, that forms a common trunk with the splenic vein.

The sigmoid veins are satellites of the arteries. They run into the sigmoid mesocolon and empty into the inferior mesenteric vein at the level of the pelvic inlet.

The left superior colic vein is not satellite of its homonymous artery and empties into the upper part of the inferior mesenteric vein.

The lymph vessels of the colon form networks in the mucous, muscular and subserous coats, after which they enter the paracolic (ileocolic, right, middle and left colic) lymph nodes and men the superior and inferior mesenteric lymph nodes, from where they drain into the portal and retropancreatic groups of lymph modes.

The nerve supply is autonomic; sympathetic and parasympathetic The caecum, the appendix, the ascending colon and the right two thirds of the transverse colon receive sympathetic fibres from the coeliac and superior mesenteric ganglia and the parasympathetic nerve supply from the vagus nerve. As regards the remainder of the colon, its sympathetic supply is derived from the inferior mesenteric plexus and its parasympathetic supply, from the sacral parasympathetic plexus through the pelvic splanchnic nerves.

The Rectum
[Rectum]

The rectum is the last portion of the large intestine. Its upper boundary lies at the end of the sigmoid, at the level of the S_3 vertebra, and its lower boundary, at the level of the anocutaneous line, situated in the junction area of the anal mucous membrane with the perineal skin.

It consists of two portions: a pelvic portion, the rectal ampulla *(ampulla recti),* situated at the level of the concavity of the sacrum, and a perineal portion or the anal canal *(canalis analis),* which surrounds the coccyx, describing a curve with the concavity directed anteriorly.

The rectum lies in a well delimited space, which allows to use the cleavage planes in surgery or in its fixation, in case of prolapse, and even in extirpations performed for rectal cancer.

External configuration. In the pelvic portion, it resembles a more or less cylindrical tube, devoid of the characteristics of the colon. The longitudinal musculature is not disposed under the form of teniae, but it appears scattered in a uniform layer (fig. 200).

On its outer surface there are two or three transverse grooves that correspond to the Houston's valves situated inside.

In the perineal portion, the rectum is surrounded by the levator ani muscles and by sphincters.

The rectum deserves its name only in animals, in humans it presents two sagittal curvatures. In its course through the pelvic excavation it is situated posteriorly' and has a route somewhat parallel to the sacral and coccygeal concavities, presenting thus an upper curve concave forwards and a lower curve convex forwards. The first is the sacral flexure *(flexura sacralis)* and the second, resulted from turning round the coccyx, is the perineal flexure *(flexura perinealis),* after which the rectum opens outside the organism, through the anal orifice.

It is 15-20 cm long, of which 12-16 cm belong to the sacral portion and about 3-4 cm to the perineal portion.

The cavity of the empty rectum is nearly capillary, its walls being in contact; only at the level of the ampulla, the calibre is 2-5 cm in diameter. In fullness it has a considerable volume, compressing the neighbouring organs.

Internal configuration. In the ampullar region, the mucous membrane presents three transverse folds *(plicae tronsversales recti),* called also Houston's valves. The first valve is situated on the left side, at 6-7 cm from the anus, the second, the so-called Kohlrausch's valve, lies on the right side, 2 cm higher, and the third valve is situated on the left anterolateral wall, 11 cm from the anus.

In the perineal portion there are small longitudinal folds, Morgagni's anal columns *(columnae anales),* 7-10 in number, which join above the anal orifice, where they give rise to formations of swallow's nest aspect, the so-called Morgagni's anal sinuses *(sinus anales sive rectales)* (fig. 201).

195

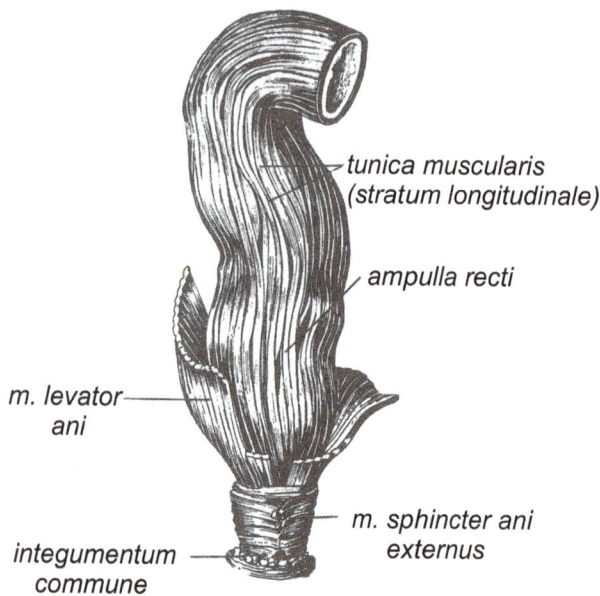

Fig. 200 The ractum - overall view.

At the external anal orifice lie a number of radial folds. The haemorrhoidal area is a circular zone of the anal canal, which corresponds posteriorly to some bulgings of the anal columns and where, in the submucous membrane, dilations of the rectal venous plexus may be seen.

Structure off the rectum. The rectum is made up of the following tunics; an outer serous, respectively fibrous connective tunic, a middle, muscular tunic and an inner, mucous tunic (fig. 202).

The outer tunic (tunica serosa et fibrosa) is formed, in the anterior and superior portions of the rectal ampulla, by the peritoneal serosa, which is reflected from the rectum on the urinary bladder in the male and on the uterus in the female. In this way is formed the Douglas[5] cul-de-sac, the most declivous point of the peritoneal cavity, in which purulent collections may eventually accumulate in the case of peritonitis. The remainder of the rectum is invested by an adventitia of fibrous connective tissue.

The muscular tunic (tunica muscularis) is made up of external longitudinal fibres and internal circular fibres.

The longitudinal musculature is no more arranged under the form of bands, but constitutes a continuous coat around the rectum and may be divided into three layers:
- the external layer, which ends on the superior fascia of the levator ani muscle;
- the middle layer, interspersed with the fibres of the levator ani;
- the internal layer, situated between the external striated sphincter of the anus and the internal smooth sphincter, made up by circular musculature.

The circular musculature forms a layer in the depth, internal to the longitudinal musculature and to the external striated sphincter of the anus *(sphincter ani externus)*. Its lower portion forms the internal smooth sphincter *(sphincter ani internus)*.

The submucous tunic (tela submucosa)

The mucous tunic (tunica mucosa) presents a columnar monostratified epithelium, at the level of the pelvic rectum, and a pluristratified pavement epithelium at the level of the anal canal.

Relations. The rectum is lodged in the rectal space.

The rectal space is formed, in front, by the prostatoperitoneal aponeurosis in the male and the parametrium in the female; behind, by the sacrum and the coccyx, the pyramidal muscles and the ischiococcygeal muscles; laterally, by the levator ani muscle; below, by the angle between the rectum and the levator ani muscle, the space being closed; above, by the peritoneum.

Around the rectum are: the prerectal space, between the rectum and the prostato-peritoneal aponeurosis in the male or the parametrium in the female; the retrorectal space, between the rectum and the sacro-coccyx; the laterorectal spaces between the sagittal sacrorecto-genitopubic aponeuroses and the walls of the pelvis and, in the frontal plane, some plates of connective tissue, containing the middle haemorrhoidal arteries.

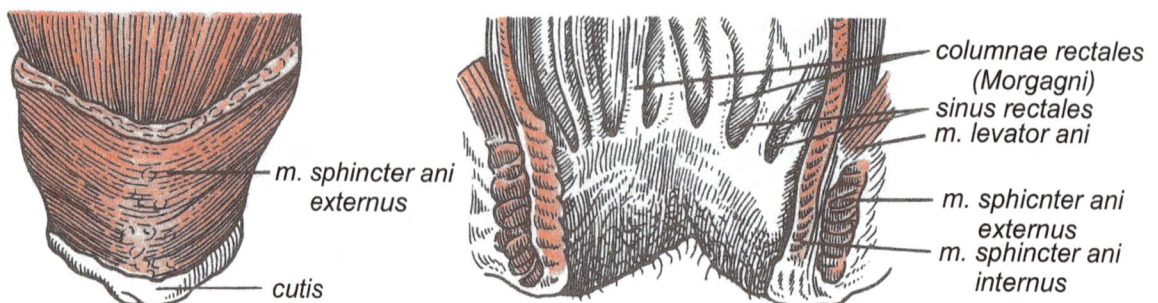

Fig. 201. Structure of the rectal sphincters.

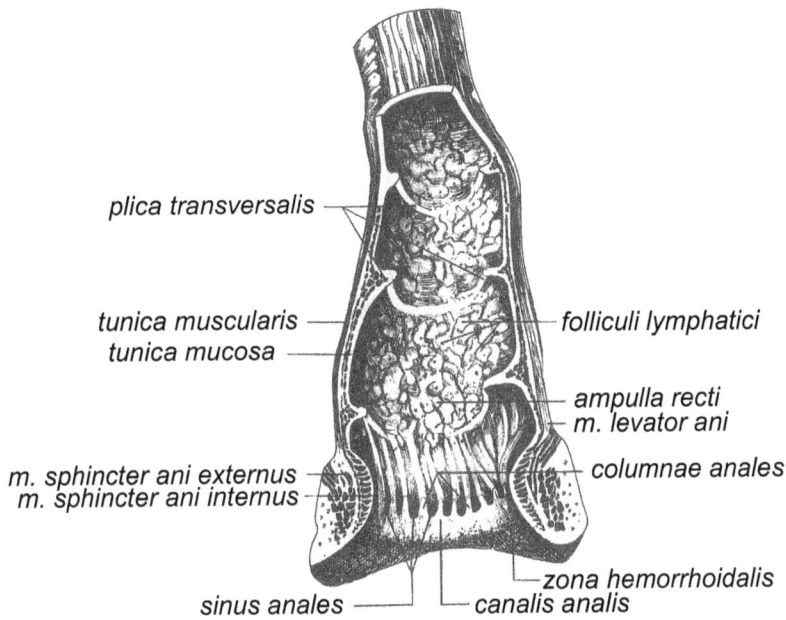

plica transversalis

tunica muscularis
tunica mucosa

m. sphincter ani externus
m. sphincter ani internus

folliculi lymphatici

ampulla recti
m. levator ani

columnae anales

zona hemorrhoidalis

sinus anales
canalis analis

Fig. 202. Structure of the rectum.

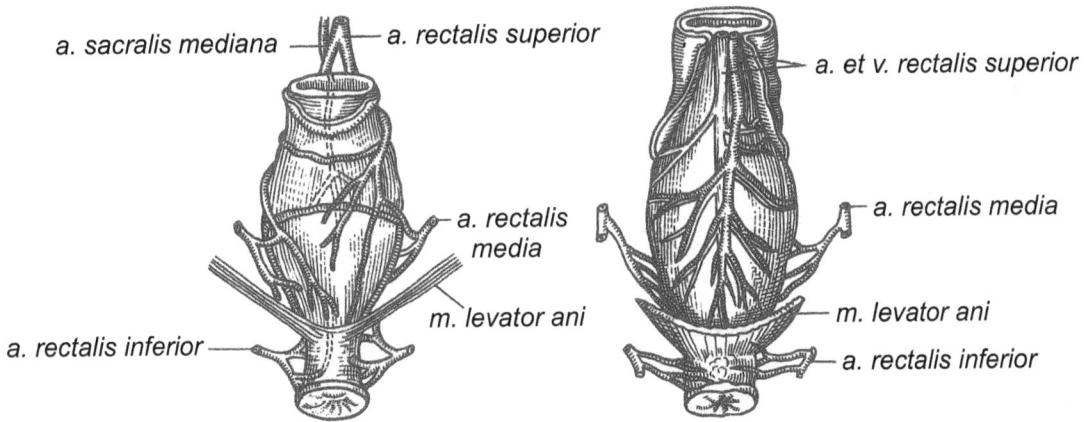

a. sacralis mediana
a. rectalis superior

a. et v. rectalis superior

a. rectalis media

a. rectalis media

m. levator ani

m. levator ani

a. rectalis inferior

a. rectalis inferior

Fig. 203. Vassal supply of the rectum.

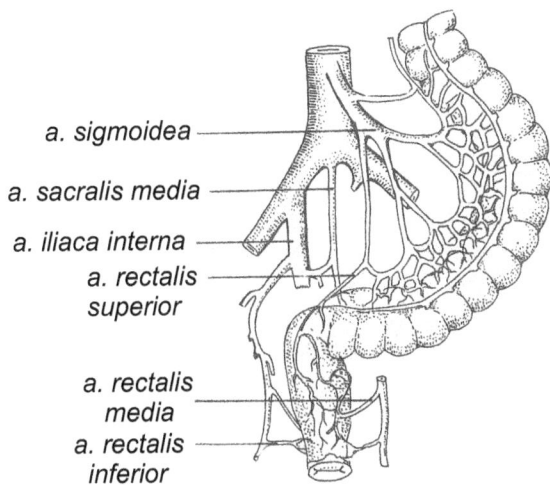

a. sigmoidea

a. sacralis media

a. iliaca interna

a. rectalis superior

a. rectalis media

a. rectalis inferior

Fig. 204. Arterial supply of the rectum (scheme).

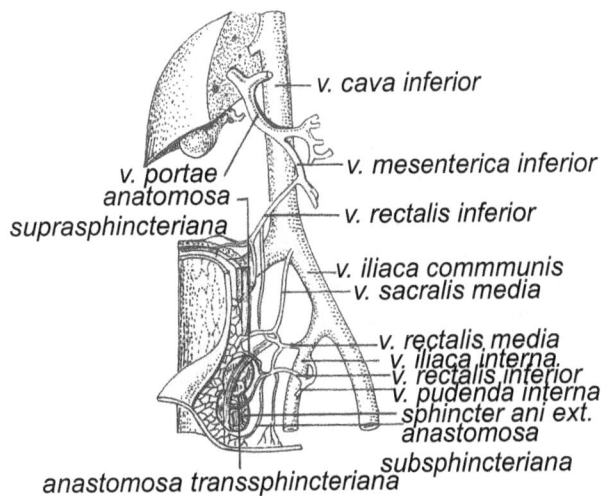

v. cava inferior

v. portae
anatomosa suprasphincteriana

v. mesenterica inferior

v. rectalis inferior

v. iliaca commmunis
v. sacralis media

v. rectalis media
v. iliaca interna
v. rectalis inferior
v. pudenda interna
sphincter ani ext.
anastomosa subsphincteriana

anastomosa transsphincteriana

Fig. 205. Venous drainage of the rectum (scheme)

197

Means of fastening. The rectum is fastened through the peritoneum which covers it, the vascular pedicles (the superior and the middle haemorrhoidal vessels), the levator ani muscles, theadhesions and the connective connections between the rectum and the surrounding urogenital formations.

The relations of the pelvic rectum are the following: in front, with the cul-de-sac of Douglas (the rectovesical pouch in the male, the recto-uterine pouch in the female, which is deeper), the fundus of the urinary bladder, the interdeferential triangle, the ampullae of the deferent ducts, the seminal vesicles and the prostate in the male and the vagina in the female; behind, in the retrorectal space, with the median sacral artery, the lymph nodes, the sacral sympathetic chains, the sacrum and the coccyx; laterally, if is covered with peritoneum; through the reflexion of the peritoneum on the lateral walls of the pelvis (pararectal recesses),

Tthe lateral surfaces are divided into an upper segment, invested by peritoneum, which is in relation with the sigmoid colon and the loops of the small intestine, and a lower segment, related to the sacrorectogenitopubic line and, through it, to the ureter and the internal iliac vessels. When the rectum is distended, the upper segment comes in contact with the internal iliac vessels and, in the female, also with the ovaries and the uterine tubes.

The relations of the perineal rectum are the following: anteriorly, they differ between sexes, namely in the male, with the bulb of the urethra, the bulbo-urethral glands (Cowper's glands).

The superficial transverse muscle of the perineum, the membranous urethra, the apex of the prostate through the prostato-peritoneal aponeurosis, and in the female, with the posterior wall of thevagina through the rectovaginal space; posteriorly, with the strap of the levatores ani, the anococcygeal raphe and the posterior projections of the ischiorectal fossae; laterally, with the ischiorectal fossa and the pudendal vasculonervous bundle.

The anus (anus) is the orifice through which the rectum opens to the outside; it is situated in the posterior perineum. It has the shape of a sagittal slit and an integument with radiating folds, rich in sebaceous and sudoriferous (sudoriparous) glands. The anal musculature is of sphincterian type and will be described in the chapter regarding the perineum.

Vessel and nerve supply of the rectum

The arterial supply of the rectum is derived from the following three vascular trunks (fig. 203, 204):

- *the superior haemorrhoidal artery (a. rectalis superior),* a ramus of the inferior mesenteric artery, supplies the pelvic rectum and anastomoses with the middle haemorrhoidal artery and the inferior haemorrhoidal artery;

-*the middle haemorrhoidal artery (a. rectalis media),* a ramus of the hypogastric (internal iliac) artery, has a small volume;

- *the inferior haemorrhoidal artery (a. rectalis inferior),* a branch of the internal pudendal artery *(a. pudenda interna),* which is a branch of the hypogastric artery, supplies the perineal rectum and anastomoses with the other haemorrhoidal arteries.

The venous network is formed of (fig. 205):

-*the superior haemorrhoidal vein (vena rectalis superior),*which unites with the sigmoid veins, forming the mesenteric vein, that empties into the portal vein;

-*the middle haemorrhoidal veins (venae rectales mediae),*which collect the venous blood from the lower part of the rectal ampulla and empty into the hypogastric vein;

-*the inferior haemorrhoidal veins (venae rectales inferiores),*which open into the internal pudendal vein, that drains the blood into the internal iliac vein and the inferior caval system.

The anastomoses between the haemorrhoidal veins achieve a connection between the portal system (through the superior haemorrhoidal veins) and the inferior caval system (through the middle haemorrhoidal veins and the inferior haemorrhoidal veins).

The lymph vessels of the rectum are grouped in:

- *superior lymph vessels,* that accompany the superior haemorrhoidal vein and reach the lymph nodes of the colon;

- *middle lymph vessels,* which open into the hypogastric and sacral lymph nodes;

- *inferior lymph vessels,* which convey the lymph into the inguinofemoral lymph nodes (superior and medial groups).

- *The nerve supply* of the rectum is derived from the pudendal plexus and the sacrococcygeal plexus.

The Retroperitoneal Space
(Spatium retroperitoneals)

The retroperitoneal space is an area which comprises connective tissue and is situated between the posterior genital peritoneum and the posterior abdominal wall, the intra-abominal fascia *(fascia abdomini interna)* respectively, at the level of the iumbo-iliac region
. It is actually the posterior portion of a much larger space outside the peritoneum *(spatium extraperitoneale)*.

It contains the kidneys, the ureters, the adrenal glands, partially the ascending and descending colon, the duodenum, the pancreas, the abdominal aorta, the inferior vena cava, abdominal autonomic nervous trunks and lymphatics (trunks and nodes), all lodged in an abundant connective tissue (fig. 206,207).

Some of these anatomical elements are primitively retroperitoneal (the adrenal glands, the kidneys, the large abdominal vessels, the renal pelves with the ureters, the nervous autonomic plexuses and the lumbo- aortic lymph nodes), a fascia being differentiated around them, called internal urogenital tunic, whereas others are secondarily retroperitoneal (the last portions of the duodenum, the pancreas, the ascending and descending colon) consecutively to the formation of the fasciae of coalescence, resulted from the union of the mesenteries with the respective parietal peritoneum (the retroduodenopancreatic or Treitz's fascia, the right retrocolomesocolic or Toldt I fascia and the left retrocolomesocolic or Toldt II fascia).

Posteriorly, the retroperitoneal space is related to the anatomical elements that make up the posterior abdominal wall and form the lumbo-iliac region, constituted stratigraphically of integument the latissimus dorsi muscle, the lumbodorsal aponeurosis *(aponevrosis lombodorsalis)*,

The quadratus lumborum muscle and the psoas muscle, and laterally, to the broad muscles of the abdomen the external and internal oblique muscles of abdomen *(mm. obliqui abdominis externus et internus)*, the transverse muscle of abdomen *(m. transversus abdominis)* and the internal abdominal fascia *(fascia abdominis internus)*.

Anteriorly, through the peritoneum, the retroperitoneal space comes in contact with the anatomical elements of the peritoneal cavity. Inferiorly, it is continuous with the subperitoneal space.

The retroperitoneal space is constituted of an adipose tissue *(corpus adiposum pararenale)*, in which lies the perirenal fascia *(fascia perirenalis)*, that surrounds an adipose capsule *(capsula adiposa perirenalis)*, containig the adrenal glands, the kidneys, the abdominal artery and the inferior vena cava. The perirenal fascia has an anterior plate *(lamina pararenalis)* and a posterior plate (lamina retrorenalis), which join laterally.

The perirenal fascia corresponds to the internal urogenital tunic. It is continuous in the pelvis with the "visceral retinacula", that form a fibrous intrapelvic supporting apparatus *(corpus fibrosum infrapelvinum)*, and anteriorly, with the umbilicovesical fascia *(fascia umbilico-vesicalis)*.

These elements form the inner urogenital fibrous tunic *(tunica fibrosa urogenitalis interna)*, which starts from the umbilicus, descends into the pelvis and ascends posteriorly at the level of the retroperitoneal organs (fig. 219).

The inner urogenital tunic merges into the retroperitoneal connective tissue. Inferiorly, it is continuous with the connective tissue of the internal iliac fossa up to the inguinal arch, passes into the pelvis along the neurovascular formations, encloses posteriorly the ureter, laterally the ovarian vessels, in the inguinal canal the spermatic cord and, continuous with it, the vessels that run along the external border of the muscle.

Then it accompanies the external iliac vessels, which have become femoral through their passage into the lacuna vasorum. In the lesser pelvis, it is continuous, along the internal iliac vessels and their visceral and parietal branches, withthe "retinacula" (neurovascular septa or sheaths), which support the pelvic organs.

Anteriorly, the inner urogenital tunic extends from the urinary bladder upwards, occupying the space between the two fibrous folds of the umbilical arteries, up to the umbilical cicatrix, on the peritoneal surface.

Above, towards the abdominal diaphragm, the inner urogenital tunic merges gradually into the connective tissue of the subphrenic space and constitutes actually a means of fixation of some retroperitoneal organs.

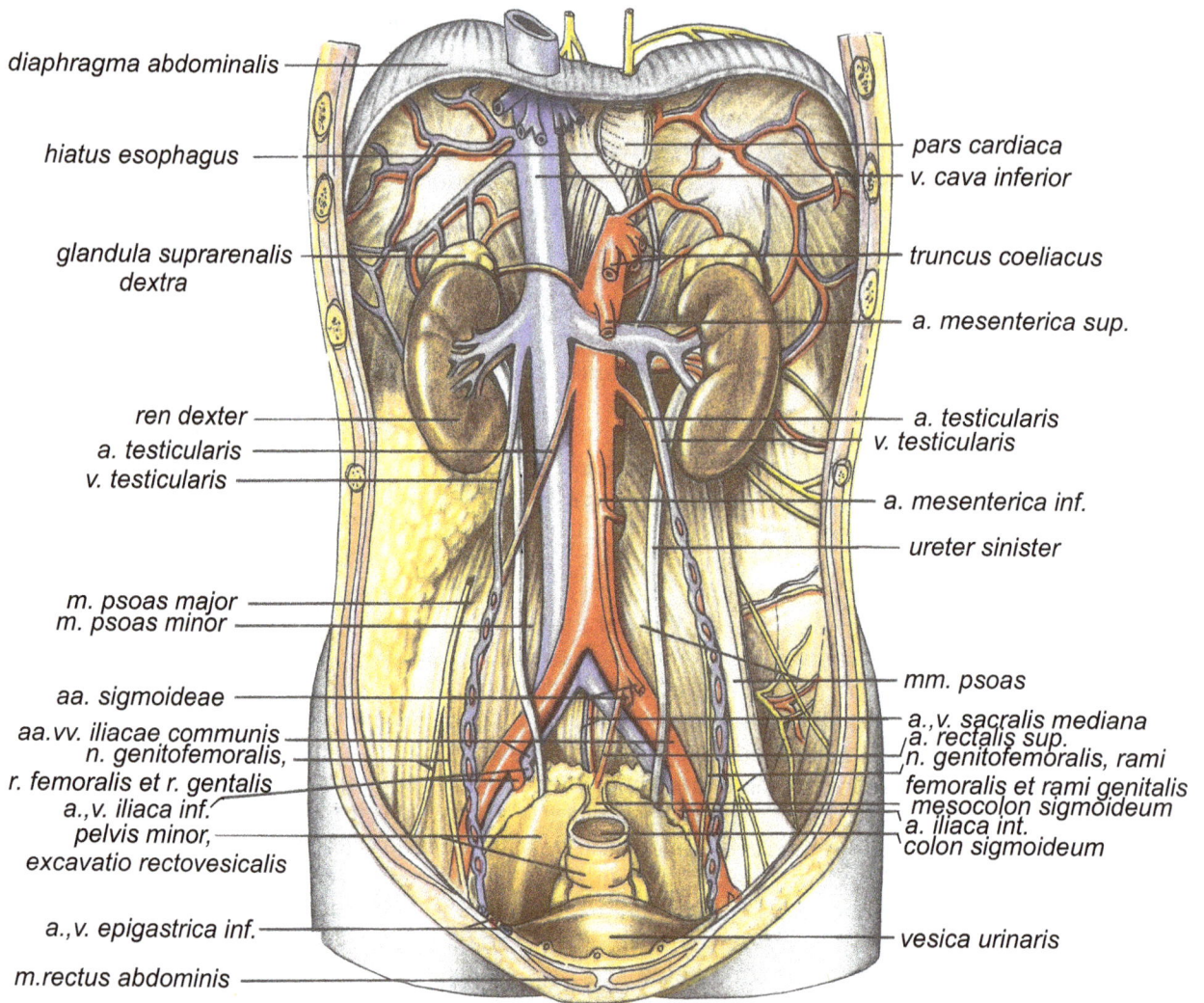

Fig. 206. The retroperitoneal space

diaphragma abdominalis

hiatus esophagus

glandula suprarenalis dextra

ren dexter
a. testicularis
v. testicularis

m. psoas major
m. psoas minor

aa. sigmoideae
aa.vv. iliacae communis
n. genitofemoralis,
r. femoralis et r. gentalis
a.,v. iliaca inf.
pelvis minor,
excavatio rectovesicalis

a.,v. epigastrica inf.

m.rectus abdominis

pars cardiaca
v. cava inferior

truncus coeliacus

a. mesenterica sup.

a. testicularis
v. testicularis

a. mesenterica inf.

ureter sinister

mm. psoas

a.,v. sacralis mediana
a. rectalis sup.
n. genitofemoralis, rami
femoralis et rami genitalis
mesocolon sigmoideum
a. iliaca int.
colon sigmoideum

vesica urinaris

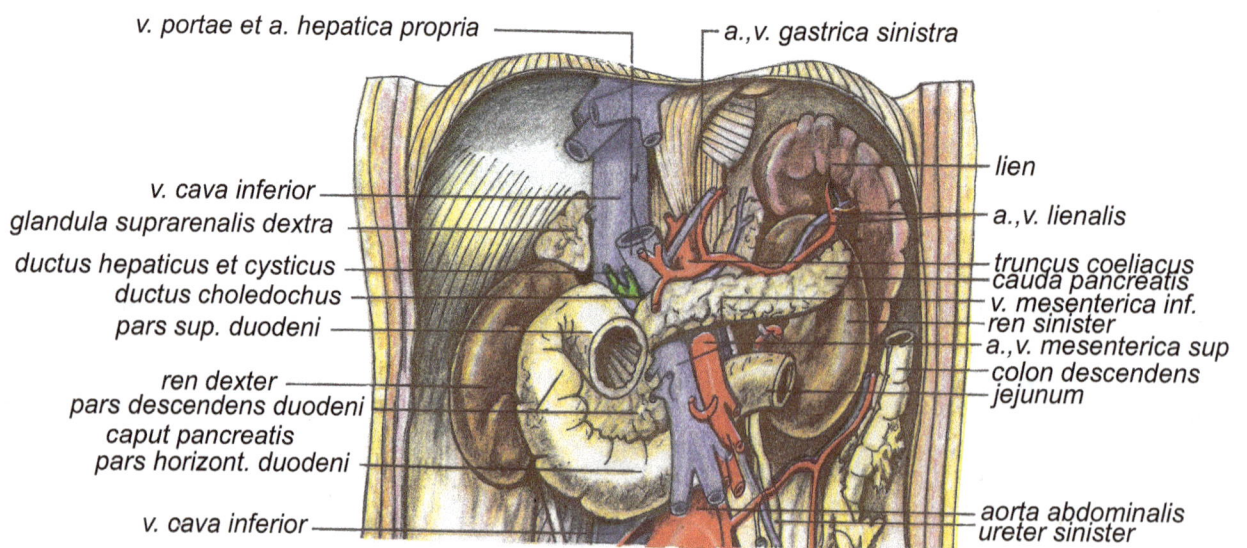

Fig. 207. The retroperitoneal space (duodenopancreatic region).

v. portae et a. hepatica propria

a.,v. gastrica sinistra

v. cava inferior
glandula suprarenalis dextra

ductus hepaticus et cysticus
ductus choledochus
pars sup. duodeni

ren dexter
pars descendens duodeni
caput pancreatis
pars horizont. duodeni

v. cava inferior

lien

a.,v. lienalis

truncus coeliacus
cauda pancreatis
v. mesenterica inf.
ren sinister
a.,v. mesenterica sup
colon descendens
jejunum

aorta abdominalis
ureter sinister

200

The Suprarenal (or Adrenal) Glands (Glandulae suprarenales)

The suprarenal glands are situated above the superior pole of each kidney; the right gland is *triangular* and the left, *semilunar* They have three surfaces: an anterior *(fades anterior)*, a posterior *(fades posterior)* and a basal surface *(fades renalis)*, which is in relation with the kidney. As the glands are devoid of hilum, the numerous neurovascular elements penetrate through their anterior surface and through the borders. The right suprarenal gland is lower situated, at the level of the T_{12} vertebra, whereas the left lies at the level of the T11 vertebra *(regio suprarenalis)* (fig. 208, 209).

Both are lodged in the suprarenal bed, bounded by the perirenal fascia. They are firmly fastened by the vasculonervous pedicles and by the connective links with the diaphragmatic muscle and the liver. The relations of the suprarenal glands are the following: behind, with the abdominal diaphragm, which separates them from the costodiaphragmatic sinus and, through this sinus, with the two last ribs. Between the gland and the diaphragmatic muscle are the thoraco-abdominal sympathetic chain and the splanchnic nerves. Behind, the right suprarenal gland is also related to the inferior vena cava, to the superior flexure of the duodenum and to the liver, to its right lobe respectively. In addition, the left suprarenal gland has relations with the spleen, the tail of the pancreas, the posterior surface of the fundus of stomach, from which it is separated by the omental bursa.

Their medial borders are in contact with the coeliac plexus and have relations with the semilunar ganglia, the aorta on the left and the inferior vena cava on the right. The superior pole is related to the diaphragm and the inferior pole is situated on the upper pole of the kidney and has more remote relations even with the renal pedicle.

The suprarenal gland is constituted of the suprarenal medulla and the suprarenal cortex. The cortex consists of cells arranged in cords, forming three zones: the glomerular zone *(zona glomerulosa)*, the fascicular zone *(zona fasciculata)*, with cells aranged in parallel columns, and the reticular zone *(zona reticularis)* with cells having the appearance of irregular structures. The medulla is made up of chromaffin cells, which are brown or yellow stained by chromium salts; between them are sinusoid capillaries, sympathetic nerve fibres and even sympathetic nerve cells.

Vessel and nerve supply

They have a very rich vessel and nerve supply, given their functional importance and the correlation of the adrenal medulla with the orthosympathetic system.

The arterial supply is derived from the three suprarenal arteries: the superior suprarenal artery *(artera suprarenalis superior)*, which is a branch of the inferior diaphragmatic artery, the middle suprarenal artery *(artera suprarenalis media)*, which arises directly from the aorta, and the inferior suprarenal arteries *(artera suprarenalis inferior)*, which arise from the renal artery.

The veins drain into the central vein *(vena centralis)*, which empties, on the right, into the inferior vena cava and, on the left, into the left renal vein.

The lymph vessels of the suprarenal glands drain into the lumbo-aortic lymph nodes and anastomose with the renal and pleural lymph vessels.

The nerve supply is provided by a rich suprarenal plexus, derived from the coeliac plexus, which forms a medial (solar) nerve pedicle, that has a sympathetic and parasympathetic component. The suprarenal glands are also supplied by branches of the splanchnic nerves, which form a posterior nerve pedicle, made up of the last six ganglia of the thoracic trunk.

The Kidneys
(Ren dexter et ren sinister)

The kidneys are the main organs of the urinary apparatus, which is formed of the two kidneys and their excretory passages, represented by the calix, the renal pelvis, the ureters, the urinary bladder and the urethra. (fig 210-222).

The kidneys are retroperitoneal organs: the left kidney lies higher than the right.

They are bean-shaped and present: two surfaces- an anterior surface (facies anterior) and a posterior surface (facies posterior); two borders – a lateral convex border (margo lateralis) and a medial border (margo medialis), buried in the hilum (hilum renalis); two poles – a superior pole (extremitas superior) and an inferior pole (extremitas inferior), their vertical axis being directed obliquely downwards and outwards and the transverse axis , obliquely backwards and outwards.

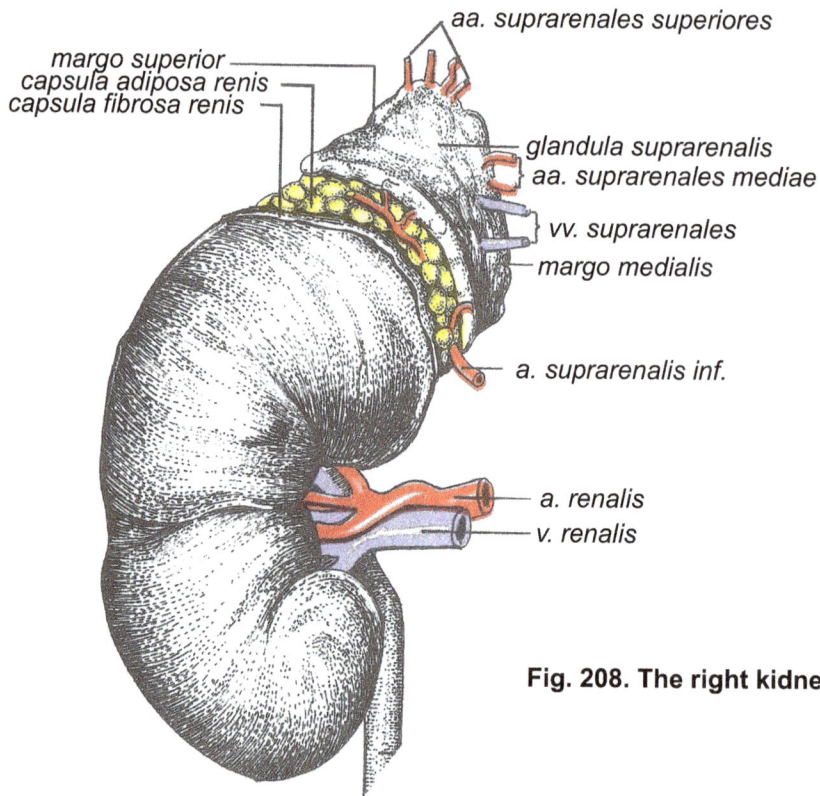

aa. suprarenales superiores

margo superior
capsula adiposa renis
capsula fibrosa renis

glandula suprarenalis
aa. suprarenales mediae

vv. suprarenales
margo medialis

a. suprarenalis inf.

a. renalis
v. renalis

Fig. 208. The right kidney - anterior view.

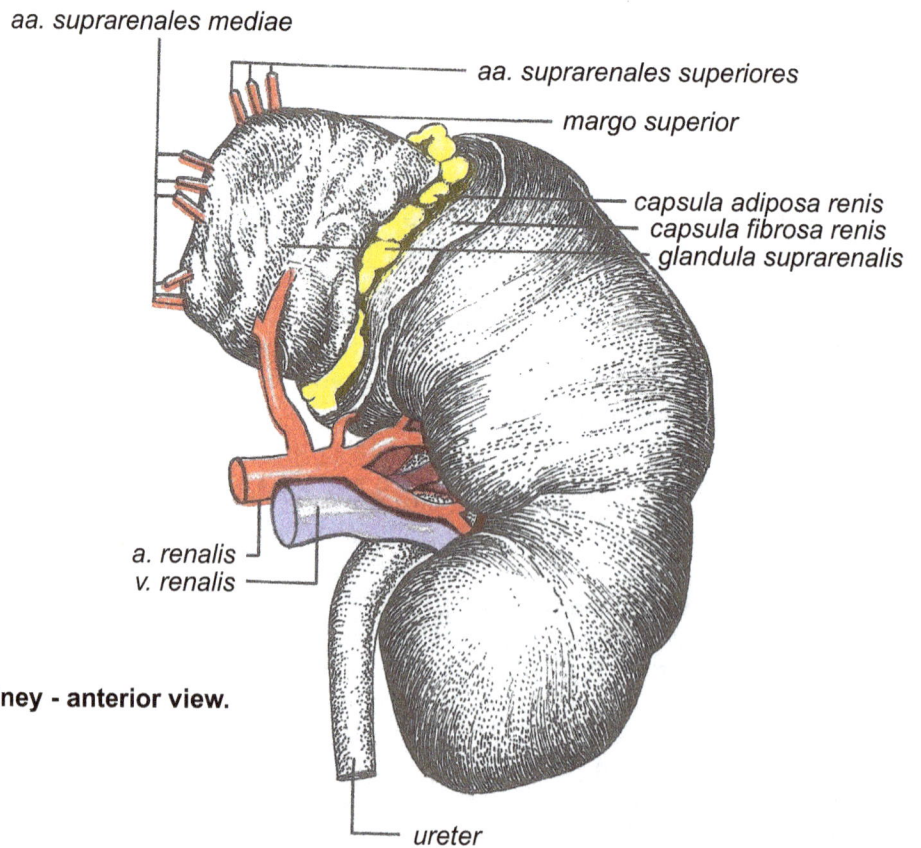

aa. suprarenales mediae

aa. suprarenales superiores

margo superior

capsula adiposa renis
capsula fibrosa renis
glandula suprarenalis

a. renalis
v. renalis

Fig. 209. The left kidney - anterior view.

ureter

202

They measure about 12 x 6 x 3 cm, weigh from 120 to 140 g and are red-brown colored and of a firm consistence.

The kidneys are constituted of a parenchyma, in the centre of which is buried the renal sinus.

The renal sinus (sinus renalis), about 3 cm in depth, is a cavity which lodges the renal pedicle formed of the excretory passages (calix and renal pelvis), vessels (renal artery and vein), nerves and a mass of fat. After removing its contents, the renal sinus, parallelepipedal in shape if it is sectioned frontally, presents on the bottom of its surface the renal papillae, around which lie the minor calices; among these appear the interpapillary eminences, formed of cortical substance.

The renal parenchyma is surrounded by the dense renal proper capsule (capsula fibrosa), which may be decorticated, and consists of two zones: the medullar (medulla renis) and the cortical (cortex renis) zone.

The medulla, situated deeply, consists of the renal pyramids of Malpighi or malpighian pyramids (pyramides renales), in number from 7 to 14, which have a base (basis pyramidis) , parallel to the lateral border of the kidney, and an apex, the renal papilla (papilla renalis), that bulges within the renal sinus. On each papilla is attached a minor calix (calix minor) and there may be seen 15-20 papillary openings (foraminae papillares), situated on the intercaliceal surface of the pyramid, surface called area cribrosa.

The pyramid may be considered as formed of a papillary zone and a limiting or external striated zone, the striations of which are formed of the collecting tubules and the blood vessels.

The renal cortex (cortex renis), situated externally at the periphery, towards the capsule, is determined by the existence of the renal glomeruli and extends, like a band, between the base of the pyramids and the renal capsule. The renal cortex, in turn, penetrates among the pyramids, forming the Bertin's columns (columnae renales) which, at he level of the sinus, form the interpapillary eminences. Likewise, the medulla at the level of the bases of the pyramids of Malpighi, which form the renal lobes, penetrates the cortex under the form of pale striations, that constitute the medullary rays, the "pyramids of Ferrein", each striation being made up of 50-100 uriniferous tubules. Between the pyramids of Ferrein which, according to the PNA, are the renal lobules [although as lobules have been designated both the convoluted part, centered by an (inter) lobular artery, and the adjacent portion of the medullary rays (pyramids of Ferrein)] lies the convoluted part (pars convoluta), darker-colored, constituted of the renal corpuscles, with their afferent vessels and the canaliculi of the nephrons to which they are related. The renal corpuscles appear like red dots, the cortical substance is yellow-reddish and the medullary substance is pale blue-reddish in color.

Thus, the kidneys are constituted of lobes (lobi renalis), which comprise in their structure a renal pyramid of Malpighi with all the afferent pyramids of Ferrein and the whole cortical substance that surrounds them. The lobes are made up of cortical lobules (lobuli corticales), represented by a medullary ray, and the surrounding convoluted cortical part. In the middle of the ray lies the collecting canaliculus and the convoluted part contains all the tributary nephrons of the respective collecting tubule. (fig. 214)

The morphological and functional unit of the kidney is the nephron.

Structure and ultrastructure of the nephron

The nephrons, morphofunctional units of the kidney, represent their secretory portion. They are constituted of a renal corpuscle of Malpighi (or malpighian corpuscle) and a uriniferous tubule (the "tubular nephron"). They are in number of 2-2.5 millions in both kidneys (fig.215).

The renal corpuscle of Malpighi

The renal corpuscle, a complex capillaro-epithelial structure, is spherical in shape, with an average diameter of 200 microns (ranging between 150 and 300 microns), visible to the naked eye as a red dot in the convoluted part.It is formed of a capillary coil, the renal glomerulus, situated between two arterioles (hence, an arterial rete mirabile),contained within a double-walled epithelial capsule (which belongs to the uriniferous tubule), Bowman's capsule; between two walls lies the Bowman's urinary space.

Each renal corpuscle has two poles: a vascular pole, which belongs to the renal glomerulus, and a urinary pole, that belongs to Bowman's capsule. The vascular pole is represented by the glomerular hilum, through which penetrates the afferent arteriole, which capillarizes into 30-50 capillary loops non-anastomosed with each other, that unite into afferent arterioles, which leave the glomerulus via the same vascular pole. This pole comes in contact with the proximal convoluted segment of the uriniferous tubule. The urinary pole represents the continuation of Bowman's capsule with the proximal convoluted segment of the uriniferous tubule.

Bowman's capsule. It arises through penetration of the capillary glomerulus at the "closed" ("blind" or "glove-finger-shaped") end of the uriniferous tubule; thus results a cavity, bounded by two monostratified epithelial leaflets: a parietal leaflet, disposed on the basement membrane of the capsule, and a visceral leaflet, which "moulds" on all capillary loops of the glomerulus.

The two leaflets continue at the level of the vascular pole, between them lying the urinary (or glomerular) space, which communicates, at the level of the urinary pole, with the proximal convoluted tube.

The renal glomerulus. The renal glomerulus presents a rete mirabile, formed of a tuft of capillaries situated between two arterioles: an afferent arteriole and an efferent one, the lumen of which is lesser. At the level of the vascular pole, these two arterioles are (nearly) adjacent, bounding an angle with the opening outwards (towards the parenchyma).

The calibre difference between the two glomerular arterioles (the size of the diameter of the efferent arteriole is lesser than of the afferent arteriole) brings about an increased hydrostatic pressure (similar to that of the venous stasis) in the glomerular capillaries, which promotes the glomerular filtration.

Through both kidneys pass daily 1,500 l of blood, of which an amount of 180 l of water is filtered, together with glucosis, amino acids and some salts, which represents the primary urine (fig. 216).

The uriniferous tubule. The uriniferous tubule is made up of a proximal tubule, the intermediate segment, the distal tubule and, continuous with the latter, of the collecting tubules.

The proximal tubule (first segment). The proximal tubule or first segment represents the initial portion of the uriniferous tubule. It is 14 mm long and 60 microns in diameter. It is made up of two segments: a tortuous segment and a straight one.

At the level of the proximal convoluted tubule, about 98.5-99% of the water present in the primary urine, as well as glucose and a part of the amino acids and of the salts which are necessary to the organism, *ore* resorbed; in this way is formed the quantity of water contained in the final urine (1.5 l).

The intermediate segment is represented by the thin descending limb of Henle's loop.

Functionally, this segment takes part in the urine concentration and dilution processes.

The distal tubule (second segment) is also formed of two segments: the straight segment or tubule (the former thick ascending limb of Henle's loop), situated in the renal medulla (about 9 mm long), and the distal convoluted tubule (4.5-5 mm long), situated in the juxtaglomerular cortex, in relation with the vascular pole, consequently with the glomerular arterioles.

The cells of the distal tubule are metabolically active; they have a rich enzymatic endowment; among the most active enzymes rank the alkaline phosphatase and the succinic dehydrogenase. The epithelium of the distal tubule takes part both in the processes of water resorption and in those of secretion and synthesis (fig. 217).

The juxtaglomerular complex deserves a special description.

The juxtaglomerular complex is represented by three structures of a particular type: the juxtaglomerular apparatus, located at the level of the afferent glomerular arteriole; the pole cushion ("Polkissen"), situated on the afferent arteriole, in the angle formed by the two glomerular arteries; the macula densa (Zimmermann), an epithelial structure situated at the boundary between the straight (ascending) and the convoluted part of the distal tubule, in close relation with the afferent arteriole.

It is considered that, in addition to its secretory function, of renin elaboration, the juxtaglomerular complex has also the significance of a barosensitive and/or chemosensitive (I. Diculescu) receptor, taking part in the regulation of the glomerular circulation.

The collecting tubules, situated mostly in the renal medulla and in the intermediate striated part (pyramids of Ferrein) of the cortex, derive from the branching of the ureteral bud and their main function is that of collection and excretion of the final urine.

According to the diameter, the cytologic structure and the topographic site, there are three main types of collecting tubules: the intermediate segments, which begin in the renal cortex, receive some distal tubules and pass into the medulla, where they are continuous with the straight collecting tubules *(tubuli renales recti)* which make up the major part of the medulla, and then with Bellini's papillary ducts, that are situated in the papillary portion of the renal pyramids (of Malpighi) and open through the papillary orifices at the level of the area cribrosa. In this way, from thousands of collecting channels (4,000-6,000) at the base of the pyramid of Malpighi, on the tip of the papilla will result 15-20 collecting channels.

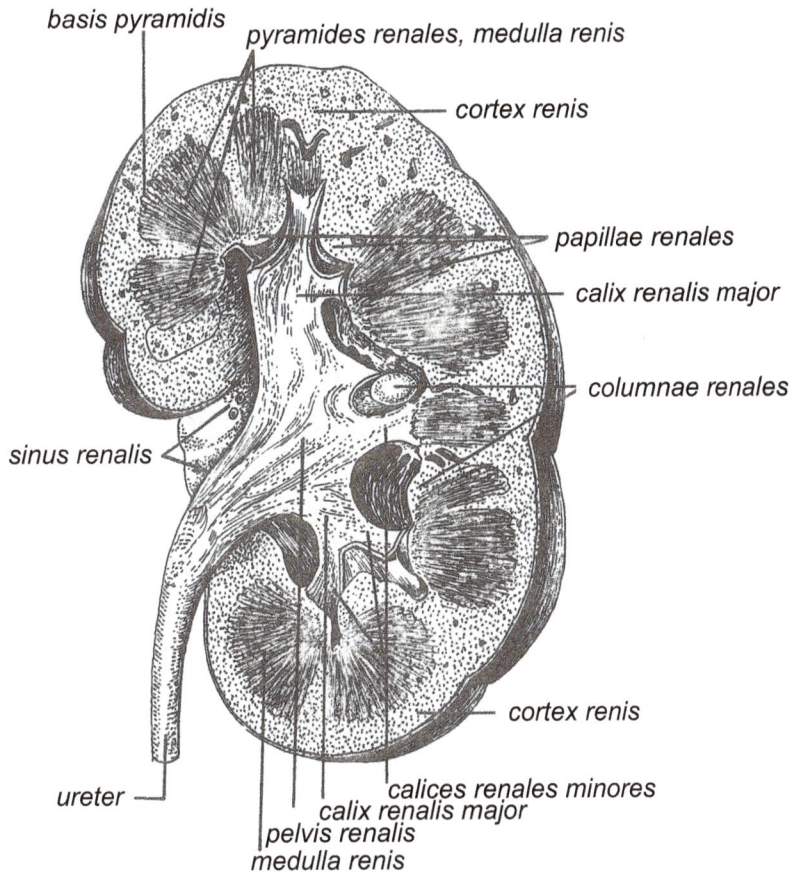

Fig. 210. Structure of the kidney

basis pyramidis

pyramides renales, medulla renis

cortex renis

papillae renales

calix renalis major

columnae renales

sinus renalis

cortex renis

calices renales minores

calix renalis major

pelvis renalis

medulla renis

ureter

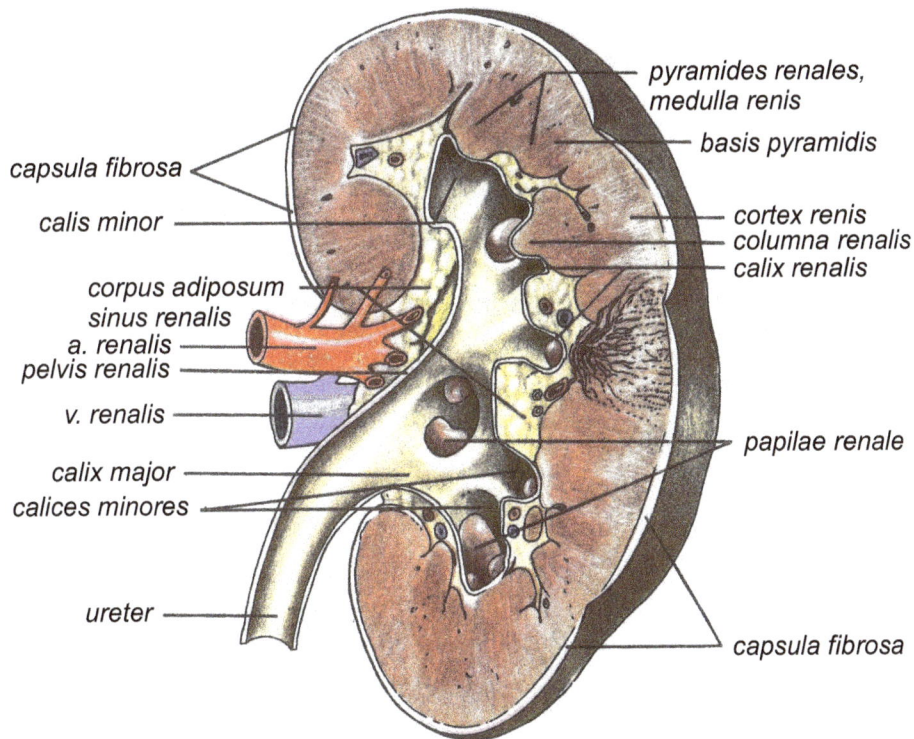

Fig. 211. The kidney, the renal pelvis and the ureter (longitudinal section).

pyramides renales, medulla renis

basis pyramidis

capsula fibrosa

cortex renis

columna renalis

calix renalis

calis minor

corpus adiposum

sinus renalis

a. renalis

pelvis renalis

v. renalis

papilae renale

calix major

calices minores

ureter

capsula fibrosa

205

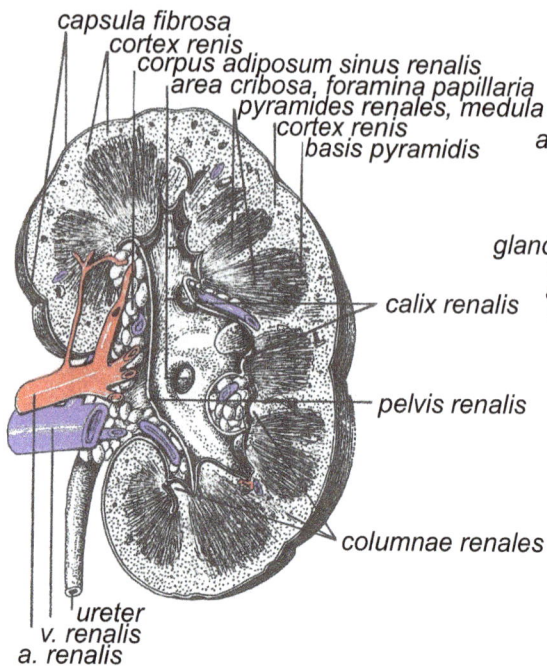

Fig. 212. The kidney - macroscopic structure.

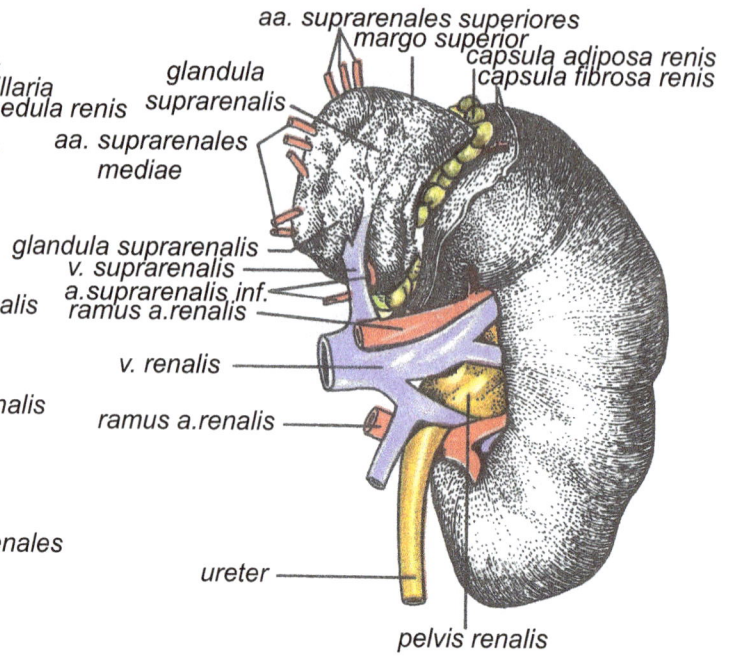

capsula fibrosa
cortex renis
corpus adiposum sinus renalis
area cribosa, foramina papillaria
pyramides renales, medula renis
cortex renis
basis pyramidis

calix renalis

pelvis renalis

columnae renales

ureter
v. renalis
a. renalis

aa. suprarenales superiores
margo superior
capsula adiposa renis
capsula fibrosa renis
glandula
suprarenalis
aa. suprarenales
mediae

glandula suprarenalis
v. suprarenalis
a.suprarenalis inf.
ramus a.renalis

v. renalis

ramus a.renalis

ureter

pelvis renalis

Fig. 213 The left kidney - anterior view (renal pedicle).

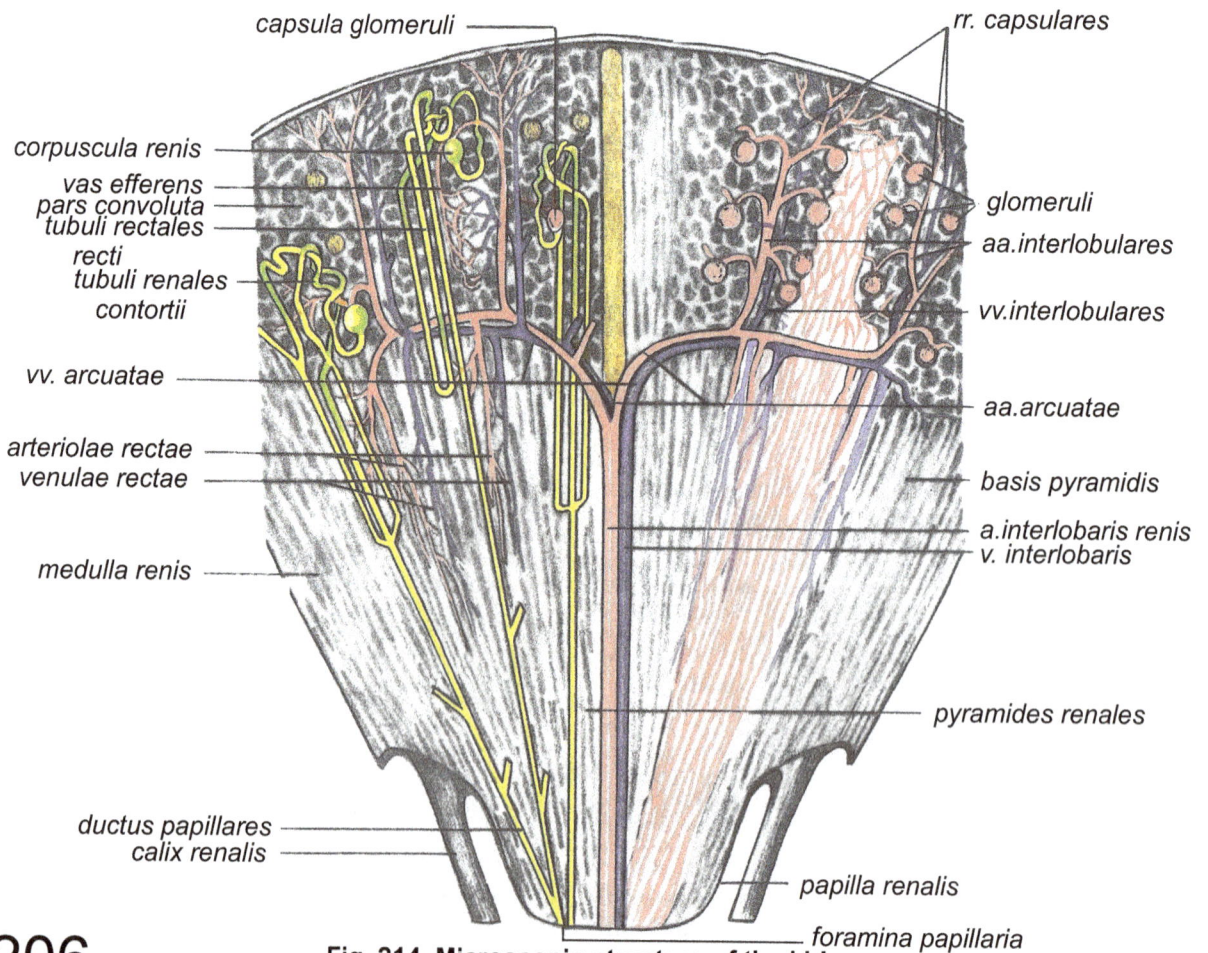

capsula glomeruli
rr. capsulares

corpuscula renis
vas efferens
pars convoluta
tubuli rectales
recti
tubuli renales
contortii

vv. arcuatae

arteriolae rectae
venulae rectae

medulla renis

ductus papillares
calix renalis

glomeruli
aa.interlobulares

vv.interlobulares

aa.arcuatae

basis pyramidis

a.interlobaris renis
v. interlobaris

pyramides renales

papilla renalis
foramina papillaria

Fig. 214. Microscopic structure of the kidney.

206

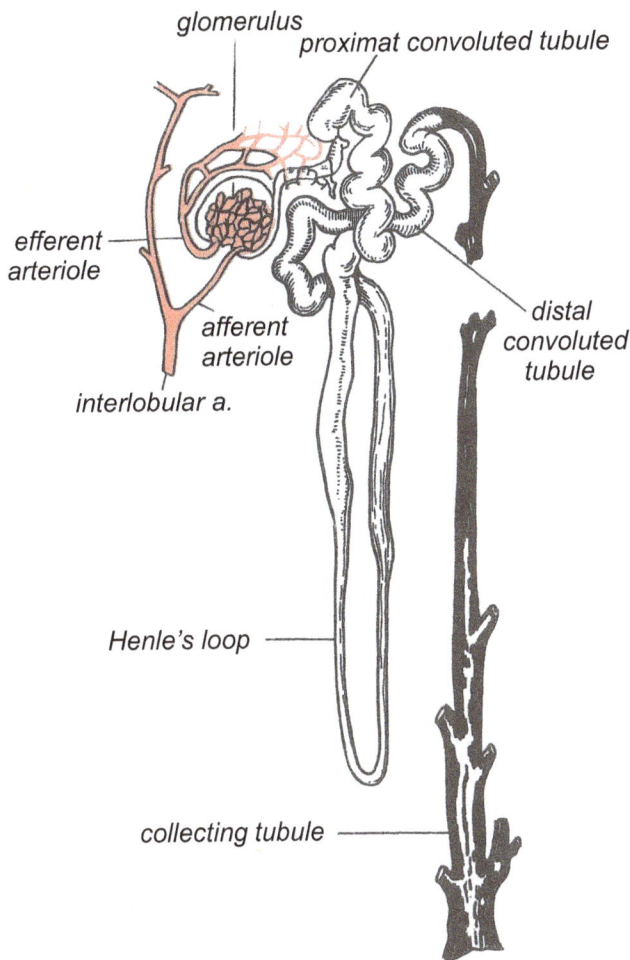

Fig. 215. The nephron.

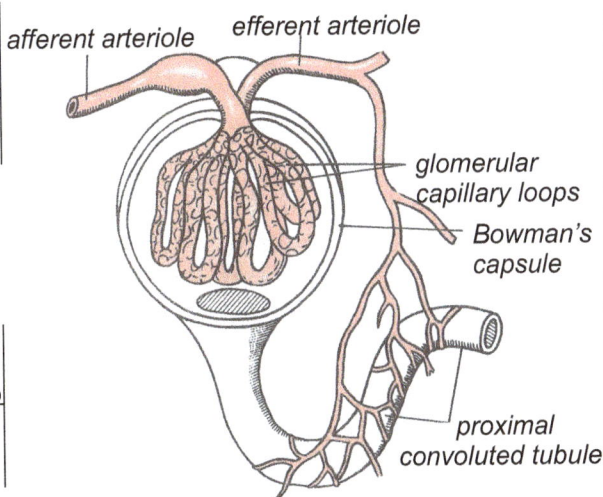

Fig. 216. The renal glomerulus.

Situation and relations. The kidneys are situated retroperitonealy, in the lumbodiaphragmatic region, within the renal bed, which has the following boundaries: above, a transverse plane which passes through the T11 vertebra; below, a similar plane, that passes at the level of the L3 vertebra; laterally, a parasagittal plane through the tip of the twelfth rib; medially, the paravertebral line. The superior renal pole lies in the subphrenic space and the inferior pole, in the sublumbar space.

The kidney is invested by a segment of the infernal urogenital tunic which, at this level, is termed renal fascia and presents an anterior (prerenal) lamina, that passes in front of the kidney, of the renal pedicle, of the abdominal aorta and of the inferior vena cava and is continuous with the anterior contralateral lamina and a posterior lamina (Zuckerkandl's retrorenal fascia), which covers the quadratus lumborum muscle and the iliopsoas muscle and than inserts into the lumbar column. Above, the two laminae invest the suprarenal glands and continue up to the diaphragm to which they attach; below, they are continuous with the pelvic segments of the internal urogenital fibrous tunic (fig. 219).

The renal beds communicate with each other through their anterior part and are open below, along the two ureters.

Between the kidney and the internal urogenital tunic there is the perirenal fat. Likewise, between the posterior surface of the internal urogenital funic (the retrorenal lamina, respectively) and the muscles quadratus lumborum and psoas lies the pararenal fat or adipose capsule (of Gerota).

Outside the renal bed, the kidney has relations with the skeleton, the peritoneum, the adjacent viscera, the neighbouring vessels and nerves and the abdominal walls, especially with the posterior wall.

The relations with the skeleton are the following: the medial border of the kidney is situated at a distance varying between 3 cm (at the level of the superior pole) and 6 cm (at the level of the inferior pole) from the midline of the vertebral column; the hilum of the kidney lies at the level of the transverse process of the L_2 vertebra and is situated at the bottom of the renal sinus, place where the elements of the pedicle penetrate within the kidney; the dorsal surface of the upper half of the kidneys comes in relation with the eleventh and twelfth ribs, the diaphragm and the triangular hiatus of Bochdaleck; between the ribs and the posterior surface of the kidney lies the costodiaphragmatic pleural sinus (which should not be opened in the course of lumbotomies) that.

207

Towards the midline, descends by about 2 cm below the twelfth rib and crosses, at about 12 cm laterally from this line, the twelfth and eleventh ribs. When the twelfth rib is long, the pleural sinus remains hidden and protected beneath if.

When this rib is short or absent, the pleura, which descends considerably below the eleventh rib, is in relation only with the lumbocostal ligament.

The peritoneum has important relations with the two kidneys.

From the anterior surface of the left kidney the peritoneum is continuous above with the superior leaflet of the transverse mesocolon, that crosses the kidney at the junction of the upper with the middle third; below, the anterior surface is in relation with the inferior leaflet of the transverse mesocolon; at the lateral border of the kidney, the peritoneum is continuous with the peritoneum which covers the descending colon, with the posterior parietal peritoneum which invests the medial part of the iliac fascia and, medially, with the peritoneum that covers the left surface of the duodenojejunal loop and that forms the two fossae (upper duodenal and lower duodenal).

The peritoneum which covers the anterior surface of theright kidney is continuous: above, with the peritoneum of the inferior surface of the liver, forming the hepatorenal ligament; medially, with the peritoneum of Winslow's hiatus and of the anterior surface of the first portion of the duodenum, forming the duodenorenal ligament; laterally, with the parietal peritoneum and below with the superior leaflet of the transverse mesocolon.

The anterior surface of the right kidney is in relation; in the upper third and the lateral half of the middle portion, with the inferior surface of the right lobe of the liver, through the peritoneum; in the medial part of the middle third, with the descending portion of the duodenum, directly, without interposition of the peritoneum; in the inferior area, with the right colic flexure, the Toldt I fascia and the inferior intestinal loops.

The anterior surface of the left kidney, in the upper third, is in immediate relation with the pancreas, above which run the splenic vessels, and, through the peritoneum, with the base of the spleen, with the posterior surface of the stomach, with the transverse mesocolon, that crosses it, with the left flexure of the colon and with the loops of the small intestine.

The mesocolon divides the kidney into a supra- and an inframesocolic portion. The supramesocolic portion of the right kidney is situated below the liver, in the greater peritoneal cavity, whereas that of the left kidney corresponds to the omental bursa.

The posterior surface of the kidneys in the thoracic portion is related, through Gerota's pararenal fat: medially, to the crura of "he diaphragm and laterally, to the right fascicles of the diaphragm, which attach on the arches of the psoas and quadratus lumborum muscles.

At this site, there is a space between the muscle fibres of the diaphragm, through which the posterior surface of the kidneys comes in relation with the costodiaphragmatic pleural sinus, which explains the possibility of spreading of the perirenal inflammatory processes to the pleura.

The posterior surface of the kidneys in the abdominal portion is related: medially, to the psoas muscle and laterally, to the quadratus lumborum muscle. It is also in relation with the ilioinguinal and iliohypogastric nerves, which at this level pierce the transverse muscle, to situate themselves between the latter and the internal oblique muscle of the abdomen, which explains, in the various renal pathological processes, the referred pain in the inguinal region, in the genital area and towards the root of the thigh.

Behind the transverse muscle, the kidney is related to "Grynfeltt's tetragon" and "Petit's triangle", that represent areas through which suppurations arising from the kidney may drain outwards.

The lateral border has relations, on the right, with the inferior surface of the liver and the ascending colon and on the left, with the renal surface of the spleen and with the descending colon.

The medial border, on the right, is covered by the descending portion of the duodenum and is related to the inferior vena cava; on the left, it has relations with the duodenojejunal flexure and the abdominal aorta.

In the middle of the medial border of the kidney are located the renal pedicles with their elements, the disposition of which, looked at from above downwards and from before backwards, is the following: vein, artery, pelvis. The lower half of the medial border of the kidneys is in relation with the upper portion of the ureter.

The superior extremity of the kidney is in relation with the suprarenal glands, from which it is, however, separated by a loose tissue, owing to which the suprarenal glands are held in position in the course of the renal ptosis.

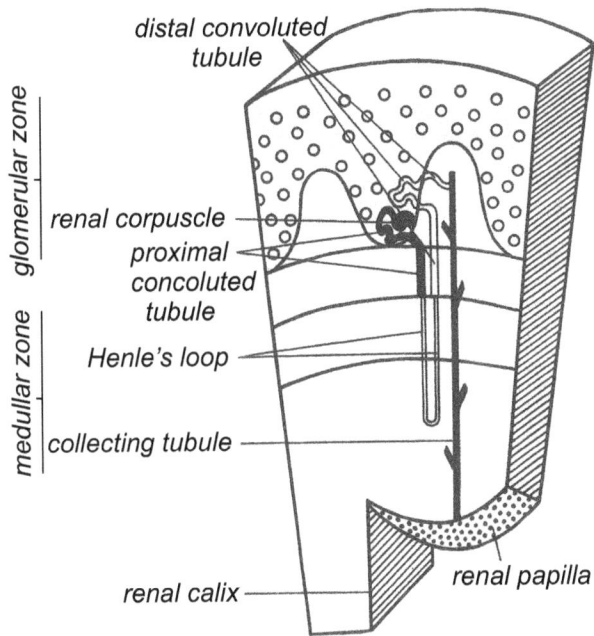

distal convoluted tubule

glomerular zone

renal corpuscle

proximal concoluted tubule

Henle's loop

medullar zone

collecting tubule

renal calix

renal papilla

Fig. 217. Scheme of the structure of the glomerular and medullar zones of the kidney.

The inferior extremity rests on the quadratus lumborum muscle.

The kidney is fastened in its bed by the walls of the bed themselves, which form below its inferior pole a hammock, through the perirenal fat, which connects it to the walls of the bed, and through the renal vessels, which constitute an important suspension factor. However, the most important means of fastening the kidney is the abdominal pressure, which applies it to the lumbar wall.

The left kidney is belter fixed than the right, since at this level may be observed: the presence of a tighter rectocolic fascia; the body of the pancreas situated on its anterior surface; owing to the disposition of the suprarenal vein, which empties into the left renal vein, the vascular pedicle of the left kidney is better fastened (fig. 220,221).

Vessel and nerve supply of the kidney

The arterial supply of the kidney is derived from the renal arteries *(aa. renales).*

Usually two in number, they arise at the level of the L1 vertebra from the abdominal aorta, below the origin of the superior mesenteric artery.

They are situated retroperitoneally and are obliquely directed inferiorly, laterally and posteriorly, the left renal artery being shorter than the right; they end at the level of the renal hilum, where they usually ramify into five terminal branches, of which four situated in front of the renal pelvis (prepyelic) and one behind the renal pelvis (retropyelic).

The renal arteries have a series of relations at their origin, along the prepedicular course, in the renal pedicle, in the sinus and at the level of the renal hilum.

Thus, at their origin they are related: above, to the coeliac trunk, situated in the centre of the coeliac region; below, to the origin of the genital arteries (spermatic or ovarian artery); medially, to the superior mesenteric artery; behind and on the right, to the origin of the thoracic duct; in front, to the body of the pancreas and the left renal vein.

The prepediculate segment exists only for the right renal artery, since the left renal artery is from the very beginning pediculate. The right renal artery, in the prepediculate segment, is retrocaval and has the following relations: in front, with the corticocaval Interstice and the inferior vena cava, which receives at this level the left renal vein is surrounded by pre- and retrocaval ganglia and by the duodenopancreatic visceral complex; behind, with the right border of the body of the L1 vertebra, with the right crus of the diaphragm, the splanchnic nerves and the internal root of origin of the azygos vein.

The pediculate segment of each kidney is constituted of; the renal vein, situated in front of the artery (the left, longer, less oblique, which has a prepediculate segment); the renal artery, the renal pelvis (in the posterior plane), the lymph vessels disposed in three planes: pre- inter and retrovascular; the nerves, arranged in two planes around the artery. Each pedicle is contained in the renal bed and, through it, is in relation: behind, with the posterior lumbar wall (at the level of the L$_1$ vertebra); medially, the left with the aorta and the right, with the inferior vena cava; laterally, with the hilum of the kidney; above, with the adrenal glands; below, with the ureter and the genital vessels (spermatic artery and ovarian artery). In front, on the right, the pedicle is supramesocolic and has relations, through the Treitz's fascia, with the duodenopancreatic region (the head of the pancreas, the second portion of the duodenum and the common bile duct); on the left, the pedicle is crossed by the root of the transverse mesocolon, which contains Riolan's arch. Above the root of the mesocolon, the pedicle has relations with the pancreas and the splenic vessels and beneath the root of the mesocolon, with the duodeno-jejunal angle and the left colic artery. In the hilar end segment, the renal arteries divide into two trunks: pre- and retropyelic, which leads to the formation of two vascular planes; an anterior, prepyelic, bulky plane and a posterior plane, practically reduced to the retropyelic artery, which is hidden by the posterior lip of the hilum, so that the posterior surface of the renal pelvis is free and allows the pyelotomy by lumbar route.

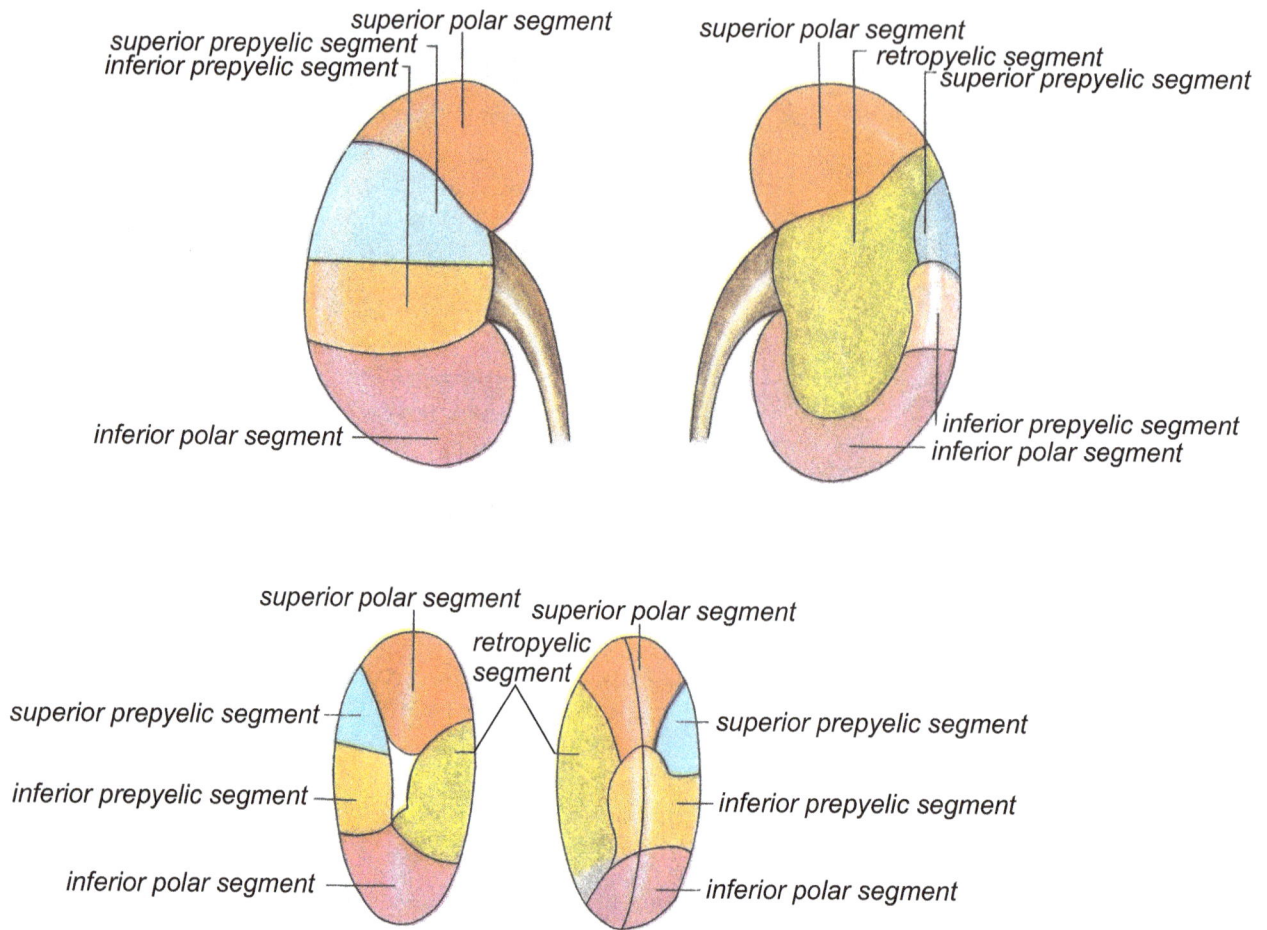

superior prepyelic segment
inferior prepyelic segment
superior polar segment

superior polar segment
retropyelic segment
superior prepyelic segment

inferior polar segment

inferior prepyelic segment
inferior polar segment

superior polar segment
retropyelic segment
superior polar segment

superior prepyelic segment

superior prepyelic segment

inferior prepyelic segment

inferior prepyelic segment

inferior polar segment

inferior polar segment

Fig. 218. Renal segments.

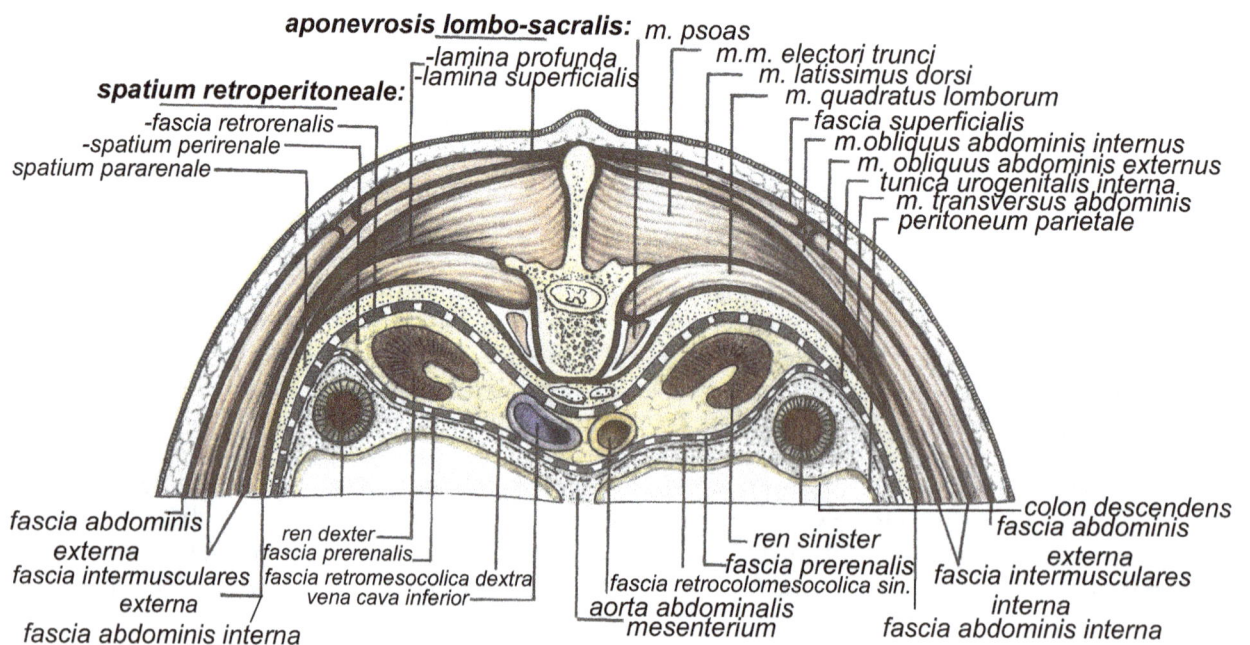

aponevrosis lombo-sacralis: m. psoas
-lamina profunda
-lamina superficialis
m.m. electori trunci
m. latissimus dorsi
m. quadratus lomborum
fascia superficialis
spatium retroperitoneale:
-fascia retrorenalis
-spatium perirenale
spatium pararenale
m.obliquus abdominis internus
m. obliquus abdominis externus
tunica urogenitalis interna
m. transversus abdominis
peritoneum parietale

fascia abdominis
externa
fascia intermusculares
externa
fascia abdominis interna
ren dexter
fascia prerenalis
fascia retromesocolica dextra
vena cava inferior
fascia retrocolomesocolica sin.
aorta abdominalis
mesenterium
ren sinister
fascia prerenalis
colon descendens
fascia abdominis
externa
fascia intermusculares
interna
fascia abdominis interna

Fig. 219. Internal urogenital fascia - disposition at the level of the kidneys.

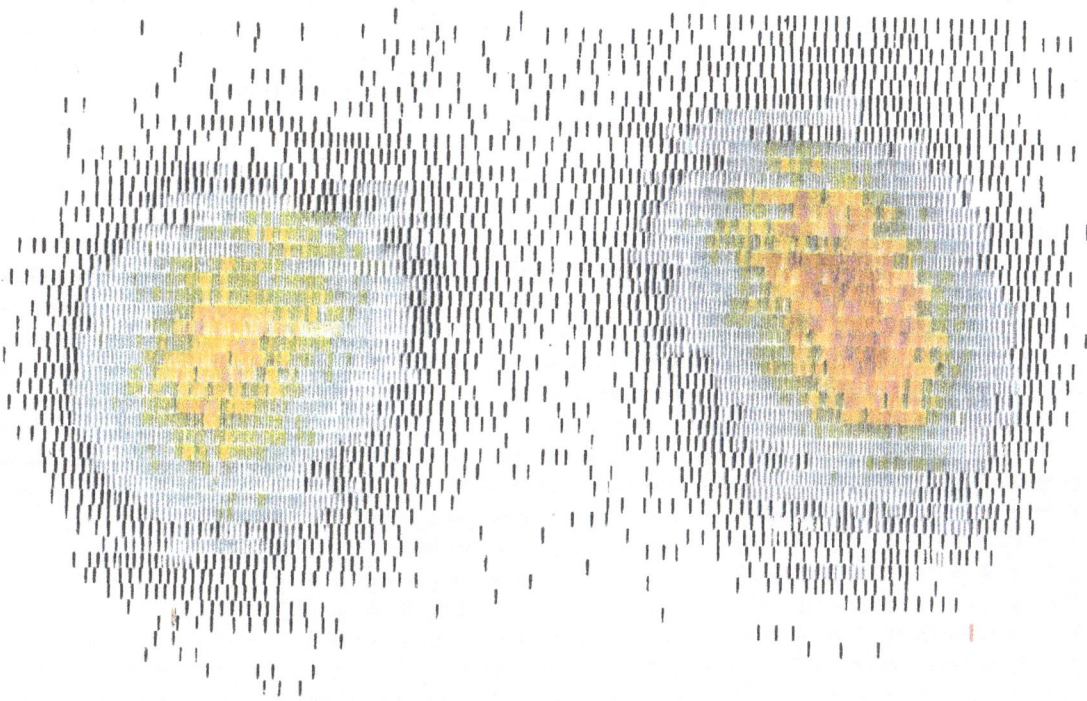

Fig. 220. Normal renal scintigram (collection of the Central Military Hospital).

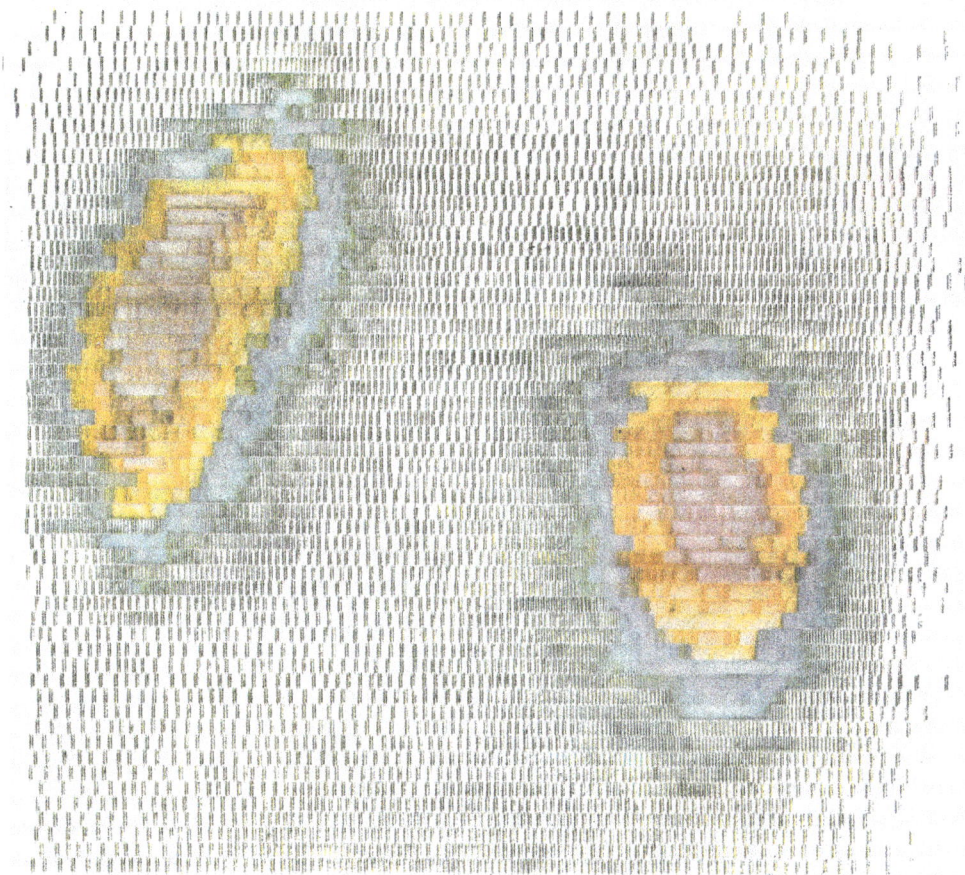

Fig. 221. Renal scintigram showing a slight ptosis of the right kidney
(collection of the Central Military Hospital)

Before penetrating the hilum, the renal arteries give off ganglionic branches, anterior and posterior ureteral branches, capsuloadipose branches, which empty into the exorenal arterial arch. The end branches are represented, in most cases, by four prepyelic branches and one retropyelic branch. They are segmental branches which mark the boundaries, at the level of the kidney, of five segments, surgically important for performing partial nephrectomies; these segments, characterized by a more individualized vessel and nerve supply, are the following (fig. 218):

- the upper segment *(segmentum superius)* (superior polar), supplied by the superior segmental (polar) artery *(a. segmenti superioris)*;

- the lower segment *(segmentum inferius)* (inferior polar), supplied by the inferior segmental (polar) artery *(a. segmenti inferioris)*;

- the antero-superior segment *(segmentum anterius superius)* (superior prepyelic), supplied by the anterior - superior segmental (polar) artery *(a. segmenti anterioris superioris)*;

- the antero-inferior segment *(segmentum anterius inferius)* (inferior prepyelic), supplied by the antero-inferior segmental (prepyelic) artery *(a. segmenti anterioris inferioris)*;

- the posterior segment *(segmentum posterius)* (retropyelic), supplied by the posterior segmental (retropyelic) artery *(a. segmenti posterioris)*.

Between the territories of the anterior segmental (prepyelic) arteries and those of the posterior segmental (retropyelic) artery lies a less vascularized zone, called "Hyrtl's avascular zone", at the level of which nephrotomies for extraction of renal calculi are performed.

The renal segmentation is of an utmost practical value, since it permits to perform segmentectomies (partial nephrectomies) in case of limited diseases of the kidney (tuberculosis, renal tumours etc.). At the level of the renal sinus, the segmental arteries divide into interlobar arteries *(arteriae interlobares renis)*, which penetrate at Bertin's columns and have a peripyramidal course, from the papilla to the base of the pyramid of Malpighi. At the level of the base of pyramids they divide into the arcuate arteries *(arteriae arcuafae)*, which ramify into interlobular arteries *(arteriae interlobulares)*. The latter are also, like the before-mentioned branches, of the terminal type; they are disposed radially and divide into capsular branches *(rami capsulares)*, which anastomose with the arteries of the perirenal adipose capsule. From the interlobular arteries break up the afferent arterioles, which penetrate the renal capsule to form the renal glomerulus, constituted of 25-50 capillaries non-anastomosed with each other, united into the efferent arteriole (forming thus an arterial capillary *rete* mirabile). After leaving the corpuscle at the level of the arterial pole, the efferent arteriole capillarizes around the renal tubules, forming the whole peritubular capillary -this time terminal - network. These efferent arterioles of the glomeruli, situated in proximity to the medulla, supply also the medulla through the straight arterioles *(arteriolae rectae)*, formerly called false *(spuriae)* straight arterioles. However, the renal medulla is supplied, too, by proper vessels, also termed straight arterioles, formerly considered as true *(verae)* straight arterioles, arising directly from the arcuate artery and the interlobular artery. It follows that, in the arterial circuit the cortical zone is supplied before the medullary zone. As regards the perirenal adipose capsule, we mention that its arterial supply is provided by the branch derived directly from the renal artery and by branches of the inferior phrenic arteries (branches of the aorta), of the superior mesenteric artery, of the lumbar and suprarenal arteries. All these branches form at this level an arterial arch.

In conclusion, it may be emphasized that the arterial distribution is terminal in type and that, as a matter of fact, there may be delimited an anterior, more extensive area, a posterior area, bounded by Hyrtl's avascular line, passing 1 cm behind the convex border of the kidney, and two polar areas, of which the inferior is less individualized. Consequently, partial nephrectomies are possible and the obliteration of a branch of the renal arteries brings about an infarction of the corresponding area.

The venous supply originates from the parenchyma under the form of the arcuate veins, satellites of the arteries, which anastomose and form a suprapyramidal venous arch, called there fore arciform veins, situated at the level of the base of the medullary pyramids, to which comes, through ascending and descending veins, the venous blood from the kidney.

Descending veins are the interlobular veins *(venae interlobulares)*, satellites of the homonymous arteries. Ascending veins are the straight veins *(venulae rectae)*, satellites of the homonymous arteries, which derive from the medullary substance.

The suprapyramidal venous arch is reached by the interlobar veins *(venae interlobares)*, that descend at the level of the renal columns, from where also other venous branches converge within the renal sinus, uniting at the level of the calices and forming the extremely varied venous branches of the renal sinus, that constitute the renal vein.

The renal veins open into the inferior vena cava and are retroperitoneal; the vein on the right side is shorter than the left.

The veins constitute the anterior element of the pedicle.

The prepediculate segment exists only for the left renal vein, which has in this segment relations: anteriorly, with the duodenopancreatic region, through Treitz's fascia; posteriorly, with the aorta, superiorly and anteriorly, with the origin of the superior mesenteric artery. Into the left renal vein empty the left spermatic vein or ovarian vein, the left suprarenal vein and, occasionally, the left inferior phrenic vein. At the level of the adipose capsule there is a very abundant venous network, that converges towards an exorenal venous arch, parallel to the outer border of the kidney, arch which receives venous blood from the kidneys through the perforating veins and drains into the veins of the colon, of the posterior abdominal wall, of the suprarenal glands, the ureter, the testicles or the ovaries. Actually, this system of veins represents an important route of derivation in case of thrombosis of the renal vein.

Arteriovenous anastomoses are present on the renal capsule, as well as at the level of the calices, around the papillae. Likewise, it should be mentioned that on the convex surface of the kidneys run the stellate veins of Verheyen, resulted from the confluence of several capsular veins towards the sites where begin, below the capsule, the interlobular veins, into which they empty after having passed across the capsule.

Lymph vessels. Three groups of lymph vessels are described in the kidney: deep lymph vessels, that accompany the blood vessels and on the course of which may be seen small nodes, a lymphatic plexus beneath the fibrous capsule and, in **the** perirenal adipose capsule, a lymphatic plexus, anastomosed with the subcapsular plexus. These three plexuses drain, to the right, into the right latero - aortic lymph nodes and into the retrocaval nodes and, to the left, into the left latero-aortic lymph nodes from the lumbar region.

The nerve supply is autonomic, provided by the sympathetic fibres of the coeliac, superior mesenteric and corticorenal ganglia. In addition, the kidneys receive fibres from the lesser splanchnic nerve and the vagus nerve. The nerve fibres come along the renal artery, form the renal plexus, penetrate the kidney and supply the various renal structures, especially the neuro-uro-arterial juxtaglomerular apparatus.

The excretory passages of the kidney. The urine formed is discharged outwards through the excretory passages, represented by the minor renal calices *(calices renales minores)*, the major renal calices *(calices renales ma/ores)*, the renal pelvis *(pelvis renalis)*, the ureter *(ureter)*, the urinary bladder *(vesica urinaria)* and the urethra. The calices, the renal pelvis and the ureter constitute also a topographic region, called pyelo- ureteral region, bounded: above, by a transverse plane that passes beneath the twelfth thoracic vertebra; below, by a conventional transverse plane drawn through the anterio-superior iliac spines; medially, by a parasagittal plane which passes at 2-2.5 cm from the median line of the body; laterally, by a parasagittal plane, too, drawn through the transverse processes of the lumbar vertebrae (fig. 222).

The minor renal calices *(calices renales minores)* are musculomembranous formations, lodged in the renal sinus, that attach with their proximal extremity around the renal papillae and, through their distal extremity, unite with each other, forming the major calices (however, not always). There are from 6 to 12 minor calices, the same number as that of the renal papillae, but fewer than the pyramids, of which two may open at the level of the same minor calix, through the union of their papillae (compound pyramids). The inner surface of the calices, which comes in contact with urine, is smooth; the outer surface is in relation with the fat and the vessels of the renal sinus.

The major calices *(calices renales majores)* are two or three in number and result from the union of the 6-12 minor calices. In this way arises a dendritic pelvis. When the minor calices open directly into the pelvis, which in this case is more voluminous, arises an ampullar pelvis.

The renal pelvis *(pelvis renalis)* has the shape of an anteroposteriorly flattened funnel and a musculomembranous constitution; at its base open the major calices and through its apex it is continuous with the ureter. Its morphological picture is very diverse, but generally there are two types of renal pelvis: the ampullar renal pelvis, when the minor calices are short and the pelvis large and bulky, and the ramified or dendritic pelvis, when the minor calices unite, forming the major calices, which are long, whereas the pelvis is smaller.

The renal pelvis has two surfaces, anterior and posterior, three borders, inferior, superior and superolateral or the base, which receives the major calices, and an apex, which is continuous with the ureter. The renal pelvis is made up of two portions: an intrarenal (intrasinusal) and an extrarenal (extrasinusal) portion.

213

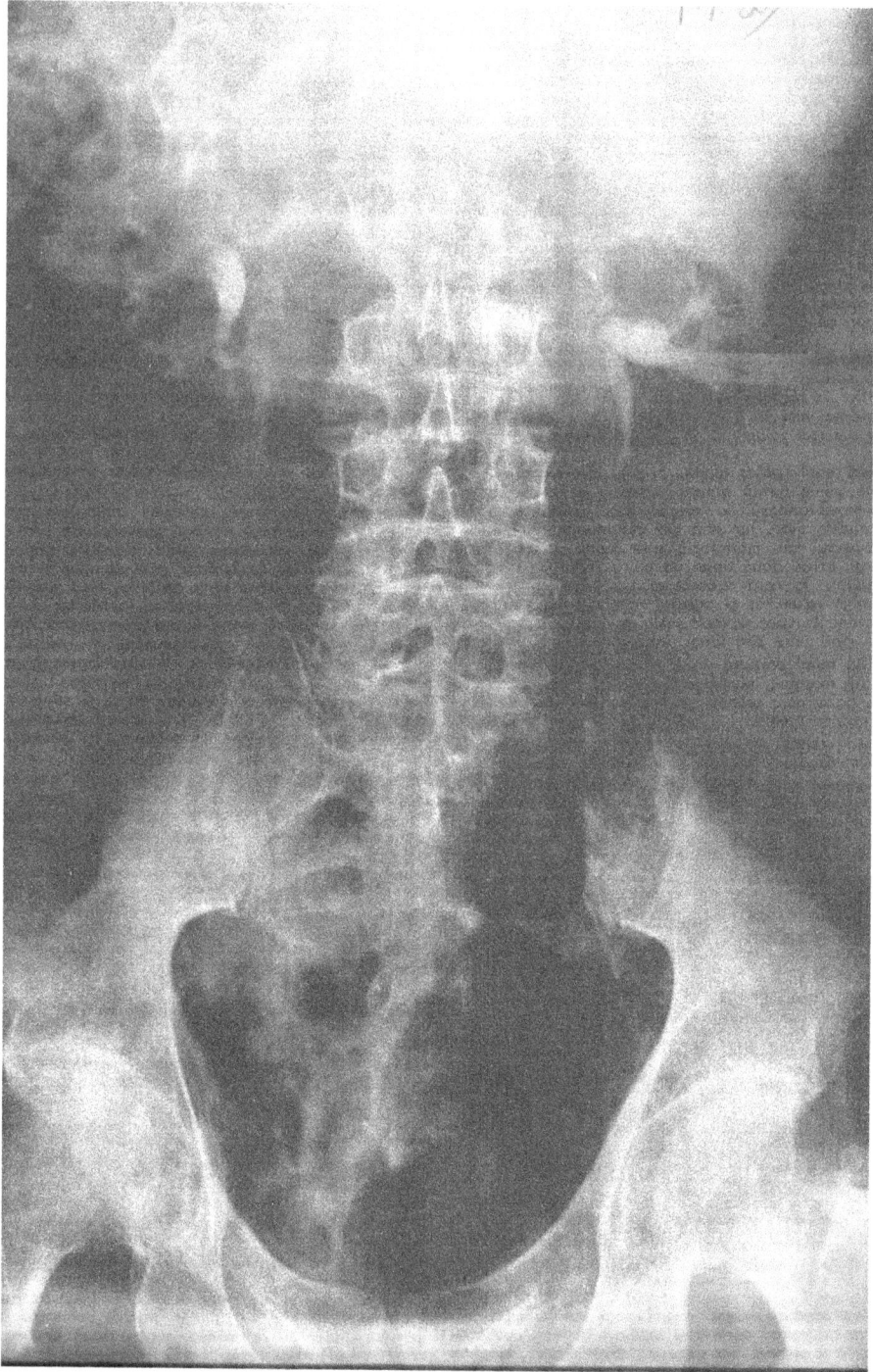

Fig. 222. Normal urography.

The intrarenal portion is usually limited and the renal pelvis, which occupies the posterior plane of the renal pedicle, is related to: the prepyelic vessels (the veins being situated in front of the arteries). The posterior surface of the renal pelvis is in relation only with the retropyelic artery, so that the route of acces in pyelotomy is posterior.

The extrarenal (extrasinusal) part of the renal pelvis is related, in the renal pedicle, anteriorly to the prepyelic segmental arterial branches and to the prepyelic venous branches.

In addition to the relations with the vascular elements in the pedicle, we mention the following general relations: above, with the suprarenal glands; below, with the ureter, with which it is continuous; laterally, with the hilum of the kidney; behind, with the various planes of the posterior wall, the psoas muscle and the perirenal fat respectively; in front, the relations are different on the right and on the left: on the right, through Treitz's fascia, the pelvis is related to the second portion of the duodenum and to the vena cava and on the left, it is crossed by the root of the transverse mesocolon and has relations with the body of the pancreas.

214

The Ureter
[Ureter]

The ureter is a musculomembranous tube which unites the renal pelvis with the urinary bladder and crosses longitudinally the abdominal and pelvic cavities. It is extraperitoneal and measures about 30-35 cm in length.

It arises from the renal pelvis at the level of the second lumbar vertebra and descends obliquely downwards and medially, since it is not rectilinear:

It has an abdominal portion *(pars abdominalis),* almost vertical, which extends up to the superior pelvic aperture (the terminal line of the bony pelvis), and a pelvic portion *(pars pelvina),* which descends within the pelvis up to the level of the urinary bladder. The ureter shows three bendings: one at the level of the kidney *(renal flexure),* the second at the terminal line (marginal flexure) and the third in the pelvis *(pelvic curvature).*

It presents also a succession of constricted and dilated areas: immediately after the renal pelvis, a dilated area, called infundibulum, then a constricted area, termed ureteral neck, again a dilatation, the lumbo-iliac spindle, followed at the level of the terminal line by the marginal constriction, after which it dilates again, forming the pelvic spindle, to constrict finally at the level of the orifice of entrance into the urinary bladder.

The fixity of the ureter is relative: sometimes, a ureterolumbar ligament is present, but most frequently it adheres to the peritoneum with which it may be mobilized, fact which should be taken into account in operations. It is surrounded by a vascular nerve plexus that should be spared in surgery (extraction of calculi, ureterotomies, ureterostomies).

The relations of the ureter vary in dependence on its portions.

The abdominal portion (pars abdominalis) is divided into a lumbar segment, form the renal pelvis to the sacrum, and an iliac segment, from the sacrum to the terminal line.

The lumbar segment of the right ureter has relations: anteriorly, with the posterior parietal peritoneum to which it adheres, being related in the upper part, through the peritoneum, to the second portion of the duodenum and to the root of the mesentery, which passes over it, after which it is crossed by the genital vessels (spermatic vessels in the male and ovarian vessels in the female) and by the caliceal vessels; over the lumbar segment of the right ureter passes the mesosigmoid, in the recess of which it may be palpated; posteriorly, with the psoas muscle and, through the iliac fascia, with the genitofemoral and lateral femoral cutaneous nerves; medially, with the aorta, to the left with the inferior vena cava, to the right with the lumbo-aortic lymph nodes and the sympathetic trunks; laterally, with the ascending colon to the right and the descending colon, to the left.

The iliac segment, 3-4 cm in length, descends in front of the iliac vessels and crosses on the right, the external iliac artery, 1.5 cm below the bifurcation of the common iliac artery, and on the left, the common iliac artery, 1.5 cm above the bifurcation; behind, it is related to the psoas muscle; in front, the right ureter is crossed by the mesentery and has relations with the terminal ileum, the ileocolic vessels and the vermiform appendix (when it has a descending position); the left ureter is crossed, in front, also in this segment, by the sigmoid mesocolon, that forms, at this level, the intersigmoidal recess, landmark of the ureter; laterally, the ureters are related to the genital vessels (spermatic in the male and ovarian in the female).

The pelvic portion (pars pelvina) begins at the terminal line and has a fixed parietal segment and a visceral segment, which have different relations in the male versus the female.

In the male, *the parietal segment* descends vertically along the lateral wall of the pelvis and is intimately related to the internal iliac artery and vein; *the visceral segment,* directed transversally, is situated in the pelvic-subperitoneal space, runs between the urinary-bladder and the rectum, crosses the deferent duct over which it passes, courses in front of the base of the seminal vesicle and penetrates the urinary bladder through a very narrow orifice.

In the female, it should be mentioned that, in *the parietal segment,* the lateral surface of the ureter is crossed by the uterine artery, which is a branch of the infernal iliac artery; the medial side is covered with peritoneum and bounds the ovarian fossa, having thus relations with the ovary.

The visceral segment, also directed transversally has a first part in the base itself of the broad ligament of the uterus, where if has, too, an important relation with the uterine artery, that crosses if in front, after which it passes in front of the vagina and opens into the urinary bladder.

This relation is very important and should be taken info account in case of hysterectomy (removal of the uterus), because of the true danger of infra - operative lesion of the ureter.

215

The *intravesical portion* is short, has an oblique course and the opening is provided with a fold of the vesical mucous membrane, which prevents the reflux of urine from the bladder into the ureter. The ureteral openings and the urethral orifice bound, at the level of the urinary bladder, the so-called Lieutaud's trigone *(trigonum vesicae).*

Structurally, the excretory urinary tract has the following coats: *the adventitia (tunica adventitial,* made up of fibrous and elastic connective tissue; *the muscular tunic (tunica muscularis),* which in the renal pelvis is constituted of a single circular layer, but at the level of the ureter, in the upper third, is formed of three layers(outer longitudinal, median circular and inner longitudinal), like in the lower third, whereas in the middle third only two layers (outer longitudinal and inner circular) are present; *the mucous tunic (tunica mucosa),* made up of a stratified epithelium, called transitional epithelium.

Vessel and nerve supply

The arteries supplying the calices and the renal pelvis are derived from the renal artery. The ureter is supplied by branches deriving from multiple sources, which from above downwards are the following: branches of the renal artery and lumbar arteries (for the upper portion of the ureter), of the testicular or ovarian artery and the common iliac artery (for the middle third), of the hypogastric artery (internal iliac artery), the inferior vesical artery, the superior vesical artery and the deferential artery (in the male) (for the lower third).

The veins are satellites of the arteries.

The lymph vessels of the abdominal ureter drain into the lumbar lymph nodes and those of the pelvic ureter, info the internal iliac lymph nodes.

The nerve supply is autonomic and derives from the renal, aortic and hypogastric nerve plexuses, the nerve branches following the course of the arteries.

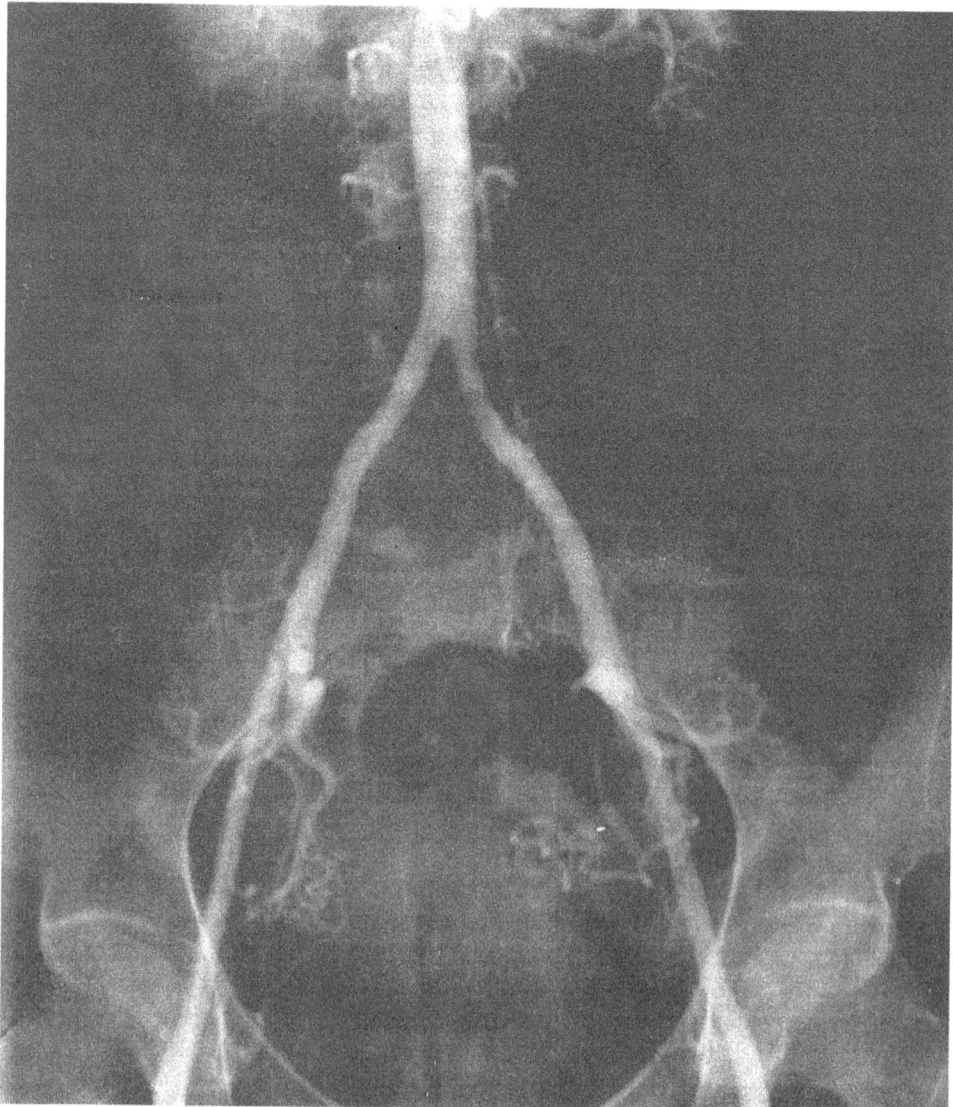

Fig. 223. Aorto-iliac arteriography
(showing the abdominal aorta, the extenal iliac arteries and the internal iliac arteries)

The Retroperitoneal Vessels

The Abdominal Aorta

The abdominal aorta extends from the diaphragmatic orifice, situated at the level of the intervertebral disk T11 -T12 up to the level of the fourth lumbar vertebra, where it trifurcates into its end branches: the common iliac arteries and the median sacral artery (fig. 223, 225).

The aorta descends nearly vertically, slightly to the left of the midline, in front of the lumbar column, in the retroperitoneal space.

It measures on the average from 15 to 18 cm, Its calibre diminishes from above downwards.

In its course, the abdominal aorta is surrounded by a loose connective tissue, by the aortic sympathetic nerve plexus and by the lymph nodes of the lumbo-aortic chain: the preaortic, the right latero - aortic, the left latero - aortic and the retroaortic lymph nodes.

If is related behind, to the lumbar vertebral column, the cistern of Pacquet *(cisterna chyli)* and the origin of the thoracic canal; laterally and behind, are the lumbar sympathetic trunk and the ascending lumbar veins.

In front, the relations are differentiated:

In the coeliac area, it gives rise to the inferior phrenic arteries and to the coeliac trunk, from which arise the hepatic artery, the splenic artery and the left gastric artery *(a. gastrica sinistra);* in front of them lie the omental bursa, the pars flaccida of the lesser omentum, the inferior surface of the liver and the lesser curvature of the stomach.

In the duodenopancreatic area it has relations with the pancreas and the third portion of the duodenum, applied on the aorta through the parietal peritoneum. Here is the origin of the superior and inferior mesenteric arteries, of the genital arteries and of the renal arteries. The third portion of the duodenum and the pancreas are caught in the aortomesenteric fold. The aorta is situated on the midline of Regie's tetragon, bounded by the portal vein, the left suprarenal vein, the splenomesenteric trunk and the renal vein.

In the subduodenal area lies the terminal aortic region. At this level it has relations, through the mesosigmoid, with the loops of the small intestine.

On the left, the aorta is related, from above downwards, to: the diaphragm, the left suprarenal gland, the inner border of the left kidney, the left ureter, the anterior surface of the psoas muscle and the left genital vessels.

On the right, laterally to the aorta runs the inferior vena cava. In the space between the aorta and the vena cava lies the chain of lumbo-aortic lymph nodes.

The bifurcation of the aorta has the following relations: in front, with the lumbar sympathetic chains, the presacral nerve, the sigmoid mesocolon and the origin of the superior haemorrhoidal artery; behind, with the lumbar nerve plexus: below, with the promonto-iliac triangle, the lateral borders being formed by the common iliac arteries and the base by the promontory - at this level lie the left common iliac vein and the iliac lymph nodes; above, with the origin of the inferior mesenteric artery, situated *4-5* cm higher; on the right, with the origin of the inferior vena cava and the precaval or right latero - aortic lymph nodes; on the left, with the intersigmoid fossa, the origin of the trunk of the sigmoid arteries and the inferior mesenteric vein.

The abdominal aorta gives off parietal and visceral branches: the parietal branches are the inferior diaphragmatic and the lumbar arteries; the visceral branches are paired [the middle suprarenal arteries, renal arteries, genital arteries (spermatic in the male, ovarian in the female), middle ureteral arteries] and unpaired (the coeliac trunk with the hepatic artery, splenic artery, left gastric artery, superior mesenteric artery and inferior mesenteric artery).

To sum up:

1. The aorta divides into the two common iliac arteries.

2. The common iliac artery *(arteria iliaca communis)* bifurcates into a right branch and a left, bulky branch. They result from bifurcation of the aorta at the fourth lumbar vertebra. They descend obliquely downwards and outwards up to the sacro-iliac joint, where each divides into an external and an internal iliac (hypogastric) artery.

1. The external iliac artery *(arteria iliaca externa)* courses from the sacro-iliac joint to the inguinal ligament, where it becomes the femoral artery. It has two collateral branches; the inferior epigastric artery *(arteria epigastrica inferior),* which gives off the anastomotic artery *(ramus pubicus)* with the obturator artery (arch of death) and the deep circumflex iliac artery *(arteria circumflexa ilium profunda),* with the ascending (abdominal) branch and the transverse (iliac) branch.

217

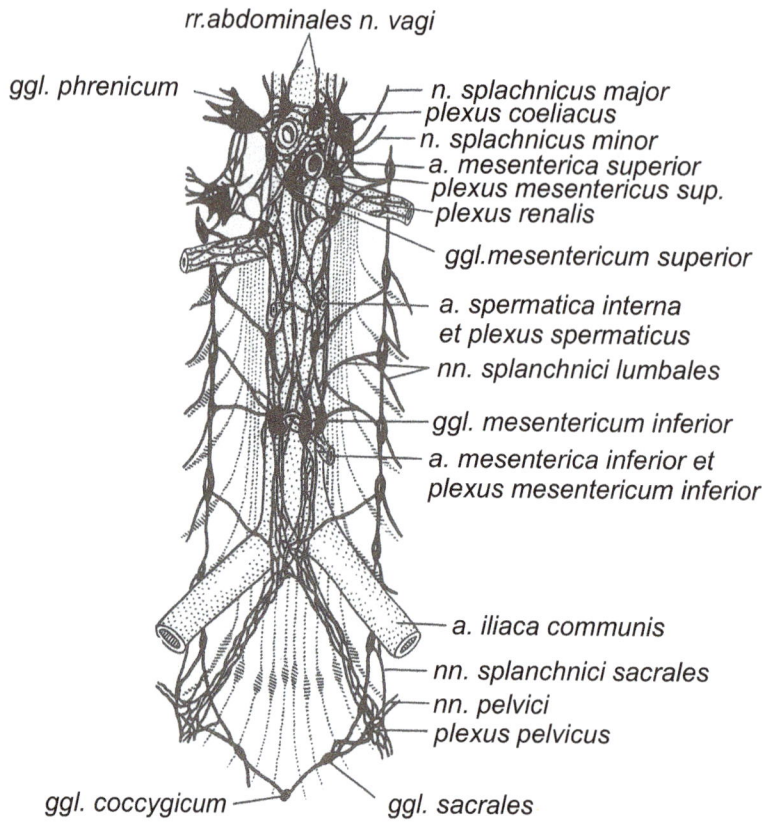

Fig. 224. Nerve plexuses of the abdominopelvic region

rr.abdominales n. vagi

ggl. phrenicum

n. splachnicus major
plexus coeliacus
n. splachnicus minor
a. mesenterica superior
plexus mesentericus sup.
plexus renalis

ggl.mesentericum superior

a. spermatica interna
et plexus spermaticus

nn. splanchnici lumbales

ggl. mesentericum inferior

a. mesenterica inferior et
plexus mesentericum inferior

a. iliaca communis

nn. splanchnici sacrales
nn. pelvici
plexus pelvicus

ggl. coccygicum

ggl. sacrales

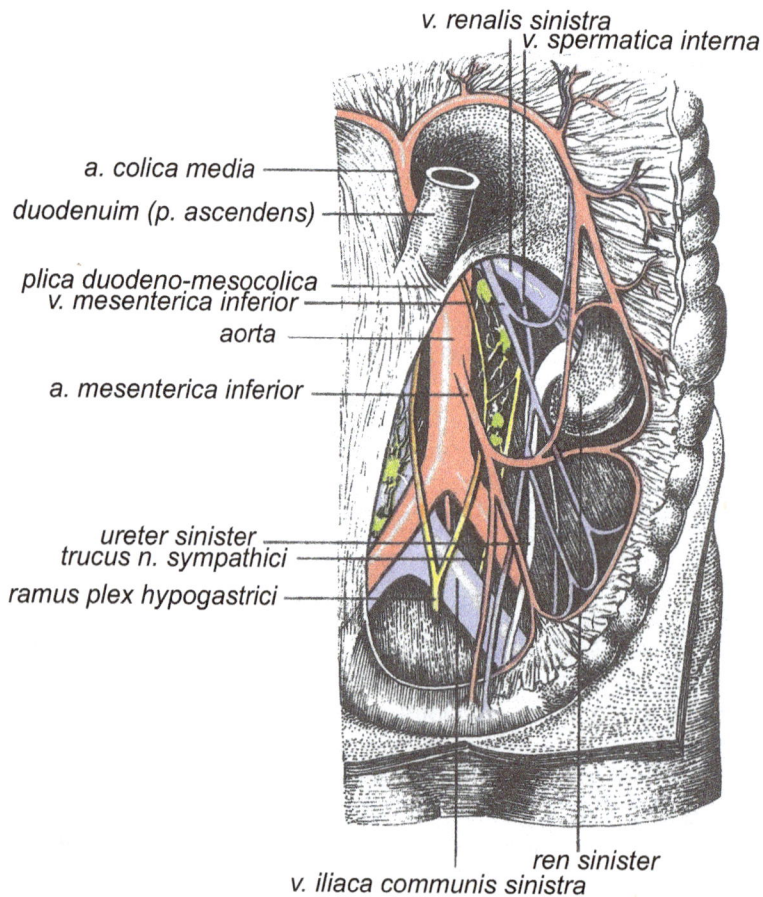

Fig. 225.The retroperitoneal space (left half).

v. renalis sinistra
v. spermatica interna

a. colica media

duodenuim (p. ascendens)

plica duodeno-mesocolica
v. mesenterica inferior

aorta

a. mesenterica inferior

ureter sinister
trucus n. sympathici

ramus plex hypogastrici

v. iliaca communis sinistra

ren sinister

218

1. The internal iliac or hypogastric artery *(arteria iliaca interna)* is the main artery of the pelvis, with the origin at the sacroiliac joint. It courses downwards and forwards, up to the level of the greater ischiadic incisure *(incisura ischiadica major)*. The extrapelvic branches are: the obturator artery, the superior gluteal artery *(arteria glutaea superior)* and inferior gluteal artery *(arteria gluteaea inferior)* and the internal pudendal artery *(arteria pudenda interna)*.

The parietal intrapelvic branches are the iliolumbar and the lateral sacral arteries.

The visceral intrapelvic branches are the umbilical, inferior vesical and inferior haemorrhoidal arteries (in the female are present, in addition, the uterine and vaginal arteries).

The Inferior Vena Cava

If conveys to the heart the venous blood from the lower limbs, the lesser pelvis and the abdomen and is satellite of the aorta (fig. 226).

If is formed by union of the two common iliac veins, right and left, at the level of the intervertebral disk between the fourth and the fifth lumbar vertebra, from where it ascends vertically on the right of the vertebral column and, reaching the first lumbar vertebra, it slightly bends to the right and digs a groove on the posterior surface of the liver it crosses the diaphragm, enters the thoracic cavity, passes through the pericardium and opens into the right atrium.

In the abdominal cavity, where if is situated on the lateral surface of the lumbar vertebral column, the inferior vena cava has the following relations; anteriorly, with the posterior parietal peritoneum and, through it, with the roof of the mesentery, with the right spermatic or ovarian vessels, with the duodenum (third portion), the head of the pancreas and the portal vein, from which if is separated through the hiatus of Winslow, and with the posterior surface of the liver, being here situated in the groove of the inferior vena cava between the right lobe and the spigelian caudate lobe; posteriorly, with the vertebral column, the right lumbar arteries, the right renal artery, the right inferior phrenic artery, the greater and lesser splanchnic nerves, as well as with the right sympathetic chain; medially, with the abdominal aorta; laterally, with the psoas muscle, the right ureter, the medial border of the right kidney, the right suprarenal gland and the left portion of the right lobe of the liver.

Into the inferior vena cava open the inferior phrenic veins *(venae phrenicae inferiores)*, the lumbar veins *(venae lumbales)*, the suprarenal veins *(venae suprarenales)*, the genital veins *(vena spermica interna et vena ovarica)*, the renal veins *(venae renales)* and the upper group of hepatic veins *(venae hepaticae)*.

The Abdominal
Retroperitoneal Lymphatics
[Systema lymphatica]

The abdominal retroperitoneal lymphatics are formed of lymph nodes and lymph channels (fig. 227).

The lymph nodes *(nodi lymphafici)* are contained within the retroperitoneal cell tissue and constitute two groups: the iliac group and the lumbo-aortic group.

The iliac group *(nodi lymphafici iliaci)* is made up of the external iliac, infernal iliac (hypogastric) and common iliac subgroups.

The lumbo-aorfic group *(nodi lymphafici lumbales)* has the lymph nodes disposed in lumbo-aortic and viscera! subgroups.

According to their topography, the lumbo- aortic lymph nodes are: left latero - aortic, right latero-aortic, preaortic and retroaortic.

The visceral lymph nodes are the following: gastric lymph nodes *(nodi lymphafici gastrici)*; splenic lymph nodes *(nodi lymphatici lienales)*; hepatic lymph nodes *(nodi lymphatici hepatici)*; mesenteric lymph nodes *(nodi lymphatici mesenterici)* and colic lymph nodes *(nodi lymphafici colici)*.

The lymphatics afferent to the abdominal nodes arise from the external and interna! iliac nodes, from the lumbar, spermatic or ovarian, renal and suprarenal, gastric, hepatic, splenic, intestinal and colic nodes.

The efferent lymphatics run to: the cistern of Pecquet *(cisterna chyli)*, which lies in front of the vertebral column, behind the aorta, at the L1-L2 level, collects most of the subdiaphragmatic lymphatics and is continuous with the thoracic duct; the thoracic duct *(ductus thoracicus)*, which has its origin in the cistern of Pecquet, has a short abdominal course, after which it penetrates the posteriormediastinum and ends in the left venous confluent between the internal jugular vein and the left subclavian vein (Pirogoff's trigone).

The Lumbar Sympathetic Chain

The lumbar sympathetic chain is situated retroperitoneally, laterally to the vertebral bodies; if is continuous with the thoracic sympathetic chain from the diaphragm to the level of the promontory.

It is constituted of a nerve chain or cord, on the course of which are interposed sympathetic nerve ganglia.

The lumbar sympathetic trunk (*truncus sympathicus*) is thiner than the thoracic trunk and the ganglia, in number of 4-5, lie at the level of the vertebral bodies. The lumbar sympathetic chain has the following relations: to the right and in front, with the inferior vena cava and the right common iliac vein; to the left and in front, it is only partially related to the aorta; medially and behind, with the lumbar vertebral column; laterally and behind, with the insertions of the psoas muscle.

The main collateral branches of the lumbar sympathetic trunk are: the periaortic vascular branches; the connecting branches between the two lumbar sympathetic chains; the pelvic splanchnic nerve made up of the viscera! branches arising from the 4-5 lumbar ganglia, which appears completely formed at the level of the fifth lumbar vertebra or of the promontory (anteriorly to the promontory, the two splanchnic nerves - right and left - form the "presacral nerve", from which arise, in the pelvis, the right and left hypogastric nerves); the communicating branches (*rami communicantes*) of the lumbar sympathetic chain; the anastomotic branches which unite to form the lumbo-aortic plexus, that is continuous above with the solar plexus and below, with the hypogastric plexus (fig. 224).

The Solar Plexus
(Plexus coeliacus seu solaris)

The solar plexus is complex and extends on the anterior surface of the abdominal aorta, from the coeliac trunk to the renal arteries (T12-L1); it is made up of three pairs of nerve ganglia and nerve fibres that solidarize it (fig. 224).

The nerve ganglia (semilunar, superior aorticomesenteric and aorticorenal) are disposed from above downwards and lie. near the origin of the coeliac trunk, of the superior mesenteric artery and of the renal arteries.

The semilunar ganglia (ganglia coeliaca) are situated paramedially, to the right and the left of the coeliac trunk, in the posterior plane of the coeliac region (Luschka). They are crescent-shaped, with the concavity upwards, the right being about 2 cm in length and the left 1.5 cm. They are placed in the retroperitoneal connective tissue and have relations: in front, with the posterior peritoneum of the omental bursa, the right being partially covered by the inferior vena cava and by the head of the pancreas; behind, with the crura of the diaphragm and the aorta; below, with the pancreas on the right and the arcuate artery on the left; medially, with the coeliac trunk and the aorta; laterally, with the suprarenal glands.

The superior aorticomesenteric ganglia (ganglia mesenteric) are placed near the origin of the superior mesenteric artery and are connected through anastomotic filaments with the semilunar and the aorticorenal ganglia. *The aorticorenal ganglia (ganglia renales)* are situated near the origin of the renal arteries; they are covered by the inferior vena cava on the right and the renal vein on the left.

The afferent nerve branches of the solar plexus are: the abdominal pneumogastric nerve *(nervus vagus),* which passes across the diaphragmatic muscle under the form of a perioesophageal plexus, from which arises a posterior abdominal trunk, that gives off the "solar branch"; the greater splanchnic nerve *(nervus splanhnicus major),* which is formed of the anterior branches of the thoracic sympathetic ganglia (5-9), penetrates the abdomen through the diaphragmatic muscle and divides into a main branch, which runs to the semilunar ganglion and forms with the branch of the vagus nerve the Wrisberg's loop, and into suprarenal filaments; the lesser splanchnic nerve *(nervus splanchnicus minor);* the right phrenic nerve (abdominal branch).

The efferent branches of the solar plexus are periarterial and form the inferior diaphragmatic plexus, the pericoeliac plexus, the hepatic plexus, the coronary gastric plexus, the splenic, renal, superior mesenteric, inferior mesenteric, spermatic or ovarian, suprarenal plexuses etc.

It should be mentioned that *the coeliac region* is situated retroperitoneally and bounded: below, by the horizontal portion of the lesser curvature of the stomach, the pyloric canal and the superior border of the duodenum; above, by the visceral surface of the liver (the spigelian lobe), on the left, by the vertical portion of the lesser curvature of the stomach; on the right, by the portal vein, the proper hepatic artery and the hepatocholedochus, which form the hepatic pedicle.

The lesser omentum *(omentum minus)* separates the coeliac region from the greater peritoneal cavity.

In the depth, the region presents an osteofibromuscular plane, which is formed of the vertebral levels T11-L1 and is bounded by the suprarenal glands.

In the coeliac region are situated the coeliac trunk with the left gastric artery *(arteria gastrica sinistra)*, the aorta, the coeliac plexus, the abdominal branch of the vagus nerve, lymph nodes, the cistern of Pecquet *(cisterna chyli)*, behind and to the right of the abdominal aorta; into the cistern open the intestinal and lumbar lymph trunks.

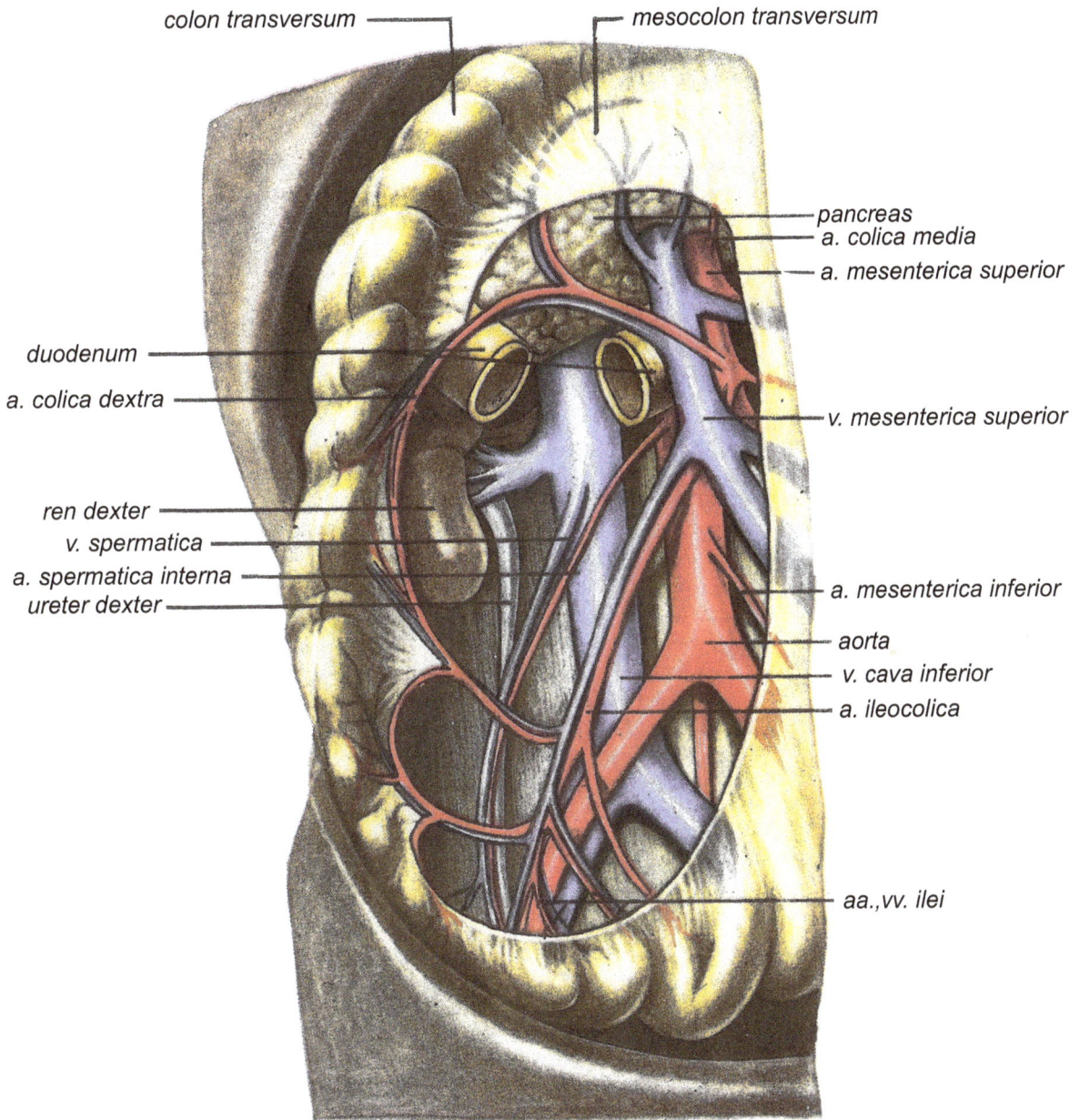

Fig. 226. The retroperitoneal space (right half).

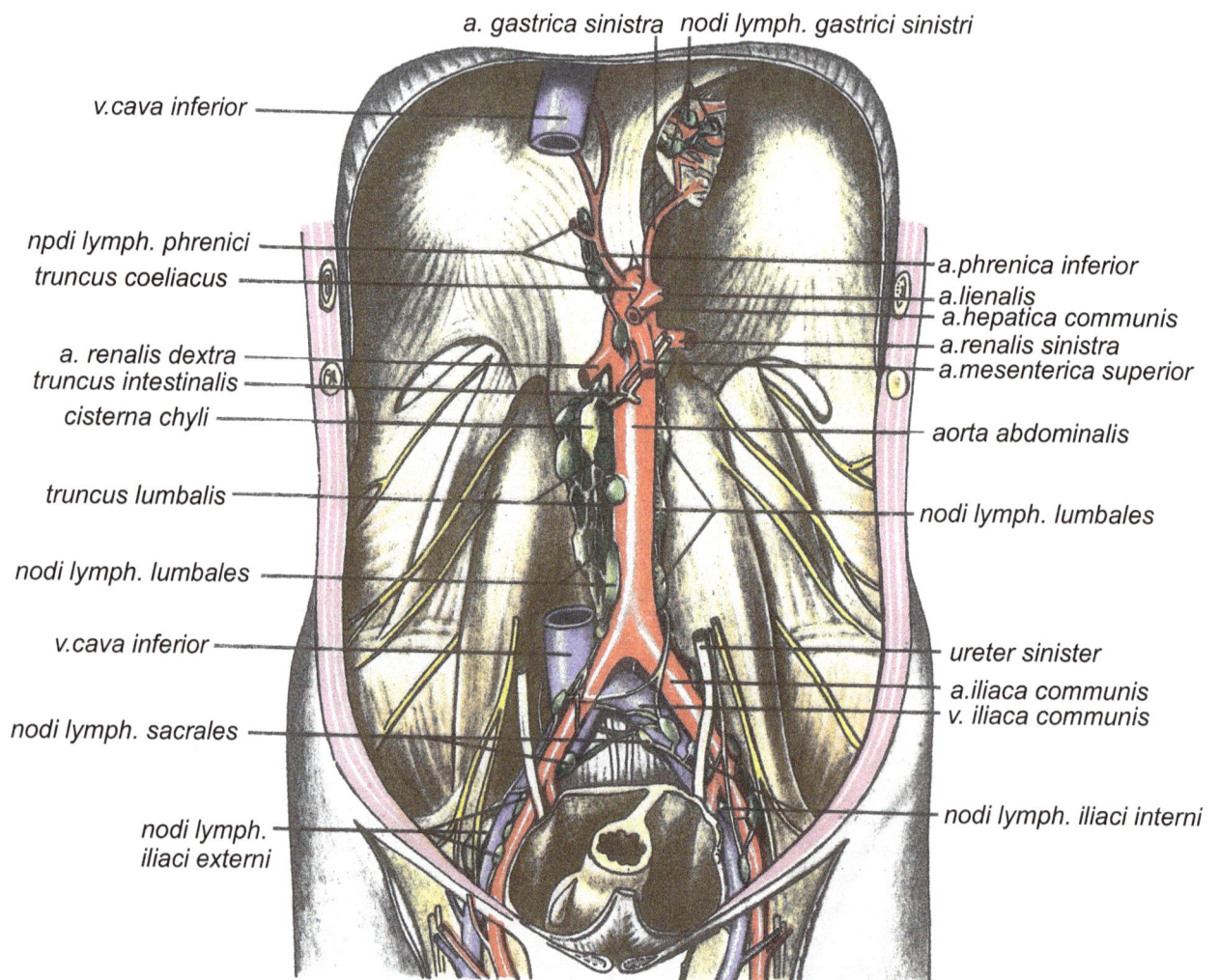

Labels on the figure:

- a. gastrica sinistra
- nodi lymph. gastrici sinistri
- v.cava inferior
- npdi lymph. phrenici
- truncus coeliacus
- a. renalis dextra
- truncus intestinalis
- cisterna chyli
- truncus lumbalis
- nodi lymph. lumbales
- v.cava inferior
- nodi lymph. sacrales
- nodi lymph. iliaci externi
- a.phrenica inferior
- a.lienalis
- a.hepatica communis
- a.renalis sinistra
- a.mesenterica superior
- aorta abdominalis
- nodi lymph. lumbales
- ureter sinister
- a.iliaca communis
- v. iliaca communis
- nodi lymph. iliaci interni

Fig. 227. Lymph vessels of the retroperitoneal space.

222

The Pelvis

Chapter 4

PRELIMINARIES

The pelvis is situated inferiorly, continuous with the abdominal cavity. We distinguish the greater pelvis and the lesser pelvis. The greater pelvis is enclosed by the walls of the abdominal cavity corresponding to the infernal iliac fossae. The lesser pelvis is situated between the terminal line constituted by the promontory and the arcuate line of the ilium and pubis, which bound the pelvic inlet or superior pelvic aperture *(aditus pelvis)* and the pelvic outlet or inferior pelvic aperture *(exitus pelvis)*, formed of smooth walls, of peritoneum respectively, and bounded by four bony points: anteriorly the lower border of the pubic symphysis, laterally the ischial tuberosities and posteriorly the apex of the coccyx.

In the lesser pelvis or pelvic proper cavity there is described also a median aperture, which corresponds to a plane passing through the third sacral vertebra, the ischial spines and the most prominent point of the posterior surface of the pubic symphysis.

At the level of the inferior aperture, the pelvic floor is formed of the pelvic diaphragm and the urogenital diaphragm.

The pelvic diaphragm, funnel-shaped with the base up-wards, is formed of the levatores ani muscles *(musculi levatores ani)* with their pubococcygeal (medially) and iliococcygeal (laterally) fascicles and posteriorly, by the ischiococcygeal muscles *(musculi coccygei)*. This diaphragm, in addition to the lateral insertions, attaches also on the fascia of the internal obturator muscle through the tendinous arches *(arcus tendineus fasciae obturatorae)* and medially, through the anococcygeal raphe *(raphe anococcygea)*, on the coccyx.

Medially to the levatores ani muscles lies the perineal hiatus with an anterior zone for the urethra and the vagina *(hiatus urogenitalis)* and a posterior zone, crossed by the anorectal canal *(hiatus analis)*. The two hiatuses are separated through the fibrous centre of the perineum, the resistant part of the pelvic floor.

Anteriorly and inferiorly to the pelvic diaphragm lies the urogenital diaphragm, triangular in shape, also formed of two symmetrical zones, which bounds the urogenital hiatus, being thus crossed by the urethra and vagina in the female and by the urethra in the male. The urogenital diaphragm is a triangular musculoaponeurotic formation, constituted of an aponeurosis (Carcassone's urogenital aponeurosis), invested on the inferior surface by the deep transverse muscle of the perineum *(musculus transversus perinei profundus)*. This is medially entangled with the muscular tunic of the vagina, on which is placed, posteriorly and in a more superficial plane, the superficial transverse muscle of the perineum *(musculus transversus perinei superficialis)*, that forms the anobulbar raphe in the male and the anovaginal raphe in the female. Medially and anteriorly, the urogenital diaphragm together with the inferior arcuate ligament of the pubis *(ligamentum arcuatum pubis)*, form the preurethral ligament *(ligamentum transversum pelvis)*, bounding the openings through which the vessels and nerves pass to the erectile apparatus. Around the urogenital canals, the urogenital diaphragm forms sphincters: the striated sphincter of the urethra in the male *(sphincter urethrae externus)* and the urethrovaginal sphincter in the female, constituted of the sphincter of the urethra *(sphincter urethrae externus)* and the vaginal sphincter *(sphincter vaginae)*.

We mention that the two diaphragms overlap only in their anterior part and are separated through an area represented by the anterior projections of the ischiorectal fossae *(spatium interdiaphragmata)*. In this way, the urogenital diaphragm corresponds to the urogenital trigone of the perineum, which has thus two diaphragms; the pelvic diaphragm above and the urogenital diaphragm below. Behind, in the anal trigone of the perineum there is only the pelvic diaphragm, under which lies the pelvic-subcutaneous space *(fossa ischiorectalis)*, separated into two lateral parts, in the superficial area, by the presence of the external anal sphincter muscle, attached behind on the coccyx through the anococcygeal raphe.

The pelvic floor formed in this way represents the inferior boundary of the pelvis, which supports the pelvic viscera, and through its complex structure, especially through its sphincteric formations, it takes part in the functionality of the terminal segments of the urogenital and digestive systems.

The walls of the pelvis are represented by the bony pelvis, constituted of the two iliac bones, and by musculo-aponeurotic formations which are attached on the bony elements (see vol. I). The bony frame which bounds the pelvic cavity is formed; laterally, by the inferior part of the medial surface of the hip bone which presents at this level the obturator foramen, closed by the obturator membrane; anteriorly, by the posterior surface of the pubic symphysis; posteriorly, by the anterior surface of the sacrum and of the coccyx.

In the space between the two hip bones and the lateral borders of the sacrum are situated the sacro-sciatic ligaments, which divide this space into two orifices: a superior orifice, the greater ischiadic incisure, called also the greater sciatic notch, and an inferior orifice, the lesser ischiadic incisure, called also the lesser sciatic notch.

Two muscles are attached to the lateral walls of the pelvis: the internal obturator muscle *(musculus obturator internus)* and the piriform muscle of the pelvis *(musculus piriformis)*. The obturator internus is attached on the bony frame of the obturator foramen and reflects into the lesser ischiadic incisure. The piriform muscle inserts into the anterior surface of the sacrum and passes across the greater ischiadic incisure. In addition, there are the muscles which attach at the level of the iliac fossae and on the external surface of the hip bones.

Among the aponeuroses we mention the aponeurosis of the obturator internus, which covers the deep surface of the muscle.

Laterally, the pelvis presents the following more important orifices through which the communication with the external regions takes place: the obturator foramen *(foramen obturatum)*, crossed by the obturator pedicle; the greater sciatic notch *(incisura ischiadica major)*, crossed by the neurovascular elements destined to the gluteal region, which form two neurovascular pedicles (superior and inferior gluteal pedicles), separated through the piriform muscle of the pelvis; the lesser sciatic notch *(incisura ischiadica minor)*, which contains the internal pudendal pedicle.

The Pelvic Peritoneum

The pelvic peritoneum lines the walls of the pelvis and reflects on the median viscera of the pelvic cavity.

in the male, the peritoneum covers anteroposteriorly the urinary bladder and the rectum (fig.228, 229, 237, 243).

At the level of the urinary bladder, it covers only the upper part of the posterior surface of the bladder. It forms laterally, when the bladder is distended, two pouches (the paravesical fossae).

At the level of the rectum, if covers the anterior surface and the lateral surfaces of the pelvic rectum.

Between the urinary bladder and the rectum lies the cul-de-sac of Douglas (the rectovesical pouch); its bottom lies 7 cm away from the border of the anus; if is elevated by the bottom of the seminal vesicles, which form two secondary pouches: an anterior, vesicoseminal, and a posterior, seminorectal pouch.

Between the rectum and the lateral wall of the pelvis, the peritoneum forms the pararectal fossae.

In the female, the peritoneum covers anteriorly the superior surface of the urinary bladder and posteriorly, the anterior and lateral surfaces of the rectum (fig.230, 231, 232, 250).

Between the urinary bladder and the rectum, it covers the uterus and forms laterally to the uterus, on either side, the broad ligaments.

The uterus and the broad ligaments divide the peritoneal segment of the pelvic cavity info two parts: *an anterior part,* which corresponds - on the midline - to the vesico-uterine pouch and -laterally - to the paravesical fossae and the preovarian fossa, and *a posterior part,* which corresponds - laterally - to the ovarian fossae and the pararectal fossae and - on the midline - to the cul-de-sac of Douglas or vagino-uterorectal pouch (its bottom lies 7 cm away from the anal orifice, like in the male, and its anterior leaflet covers the first 2 cm of the posterior surface of the vagina and is continuous with the peritoneum of the posterior surface of the uterus and with the posterior leaflet of the broad ligaments).

The cul-de-sac off Douglas is the most declivous region of the peritoneal cavity and is consequently the site of the eventual serosity or pus collections that may form in the peritoneal cavity.

Constituted of the peritoneal serous membrane, it is different in the two sexes.

In the male, the cul-de-sac of Douglas separates the urinary bladder from the rectum and is made up of the peritoneum which reflects from the anterior aspect of the rectum on the bottom of the seminal vesicles (the end of the deferent ducts), at 1.5 cm above the base of the prostate, on the posterior surface of the urinary bladder.

The prominence of the seminal vesicles divides if info two secondary pouches: a small, preseminal, and a large, retroseminal pouch, which is the true cul-de-sac of Douglas.

In this way, the cul-de-sac of Douglas, wide-open in its upper part, may lodge the loops of the small intestine or the sigmoid, but if is narrower in its Sower part.

225

Its most important relations are: in front, with the trigone of the bladder, the terminal ureters, the seminal vesicles or vesicular glands and the deferent ducts and posteriorly, with the rectum, which   in this   area is accessible to the rectal touch.

In the female, the cul-de-sac of Douglas separates the uterus from the rectum.

The peritoneum   lines the posterior surface of the uterus, then the upper part of the posterior surface of the vaginal fornix, descending up to 7 cm above the anus; posteriorly, if reflects on the anterior surface of the rectum and laterally, on the pelvic wall

Actually, it is bounded laterally by the prominence of the uterosacral ligaments, forming a space resembling a posteriorly open U.

Its inferior, retrovaginal segment is free; the superior segment contains loops of the small intestine, if may lodge an adnex or the retroverted uterus. The cul-de-sac of Douglas is accessible to vaginal and rectal touch.

The peritoneum divides the pelvic cavity info two parts: the peritoneal and the subperitoneal part.

The peritoneal part *(cavum pelvis peritoneale)* contains the portion of pelvic viscera covered with peritoneum: the loops of the small intestine, the caeco-appendix in low position and the sigmoid colon; the uterine tubae *(salpinx)* open info the peritoneal cavity; the cul-de-sac of Douglas forms the most declivous part of the peritoneal cavity.

The subperitoneal part *(cavum pelvis subperitoneale)* is encircled by the pelvic walls, the pelvic floor and the peritoneum. This segment contains the extraperitoneal part of the pelvic viscera and the intrapelvic fibrous body, represented by their retinacula, which constitute the connecting elements between the pelvic walls (the parietal leaflet of the intrapelvic fascia) and the viscera invested by the visceral portion of the same intrapelvic fascia.

These retinacula are called, for the rectum, pararectum *(paraproctium),* including also the retrorectal lamina; for the uterus, parauterum; for the vagina, paravaginum *(paracolpium);* for the prostate and the urinary bladder, paracystium, the latter including also the umbilico-prevesical fascia.

The Romanian school cosiders, on the basis of the studies performed, that the retrorectal lamina and the umbilico-prevesical fascia have appeared as the result of coalescence processes, through the disappearance of the mesorectum and of the vesical ligaments, the remains of which are the pubovesical ligament in the female and the puboprostatic ligament in the male.

The subperitoneal part if is divided into three areas by the connective-fibrous sacrorecto-genifopubic laminae: a median area and two lateral areas.

The sacrorecto-genifopubic laminae are appendages of the sheath of the internal iliac (hypogastric) artery, which form two paramedian septa extending from the sacrum to the pubes and which cover the lateral surfaces of the pelvic viscera. They contain the arteries of the viscera and the hypogastric plexus.

The median area is enclosed between the two sacrorecto-genitopubic laminae and forms the visceral space.

The two lateral areas are comprised between the pelvic walls and the fibrous sacrorecto-genifopubic laminae and form the lateropelvi-visceral space.

The visceral space. Both in the male and in the female, it is divided info two main spaces by a transverse wall,, which in the male is the prosfafoperineal aponeurosis (retroprostatic septum) and in the female is made up by the urethrovaginal fascia (urethrovaginal septum) and the rectovaginal fascia (rectovaginal septum).

The anterior space contains the genito-urinary apparatus. If is, on its turn, subdivided by the umbilico-prevesical aponeurosis into an anterior portion, which is the prevesical space of Retzius, and a posterior portion (the retrovesical space), which in the male is the vesicoprostatic region (that contains the urinary bladder, the prostate, the genital space with the deferent ducts and the seminal vesicles) and in the female, the vaginal region (which contains the urinary bladder, the uterus and the vagina).

The posterior, rectal space contains the rectum and is separated from the pelvic wall by the presacral laminae which enclose thus a retrorectal space.

The lateropelvi-visceral space which, owing to its relation with the rectum, is called pelvirectal space, has the following boundaries; above, the reflection of the peritoneum from the rectum on the organs lying in front of if; below, the intrapelvic fascia (the deep perineal fascia), which covers the pelvic diaphragm obliquely and medially downwards; medially, the pelvic viscera from which if is separated through the sacrorecto-genitopubic laminae; laterally, the pelvic walls.

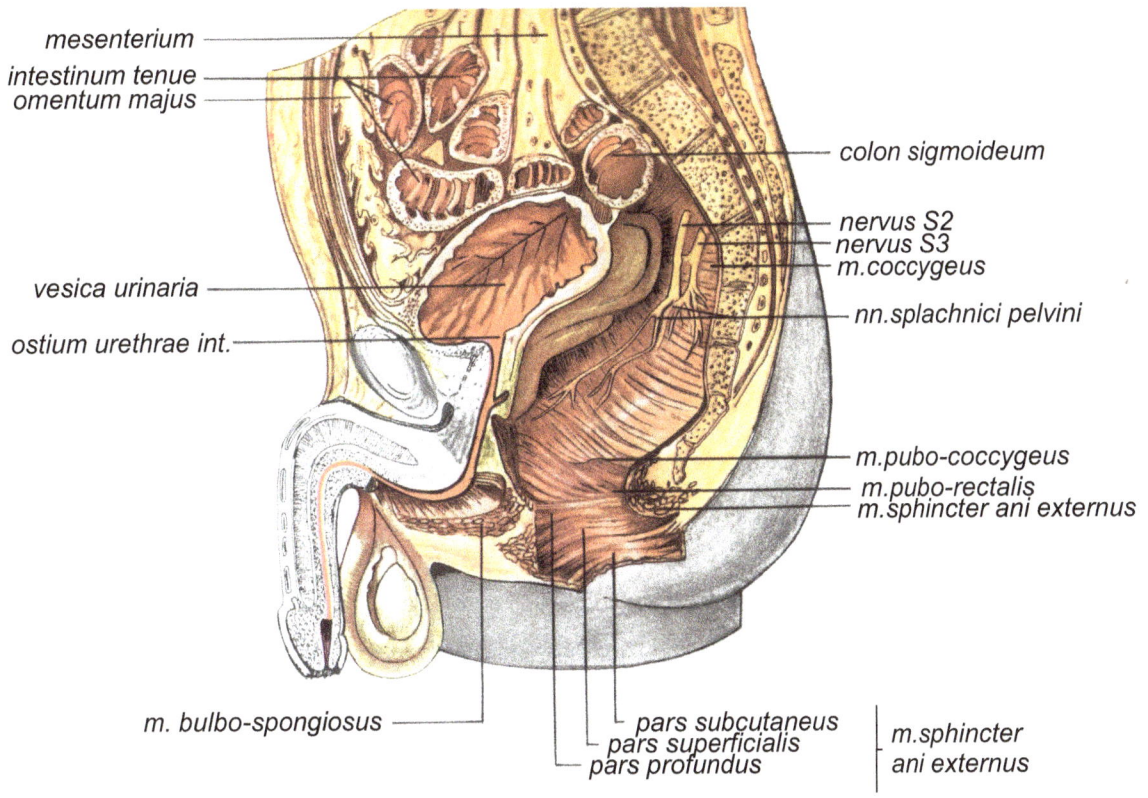

mesenterium

intestinum tenue
omentum majus

colon sigmoideum

nervus S2
nervus S3
m.coccygeus

vesica urinaria

nn.splachnici pelvini

ostium urethrae int.

m.pubo-coccygeus
m.pubo-rectalis
m.sphincter ani externus

m. bulbo-spongiosus

pars subcutaneus
pars superficialis
pars profundus

m.sphincter
ani externus

Fig, 228. The external sphincter muscle of onus in the male.

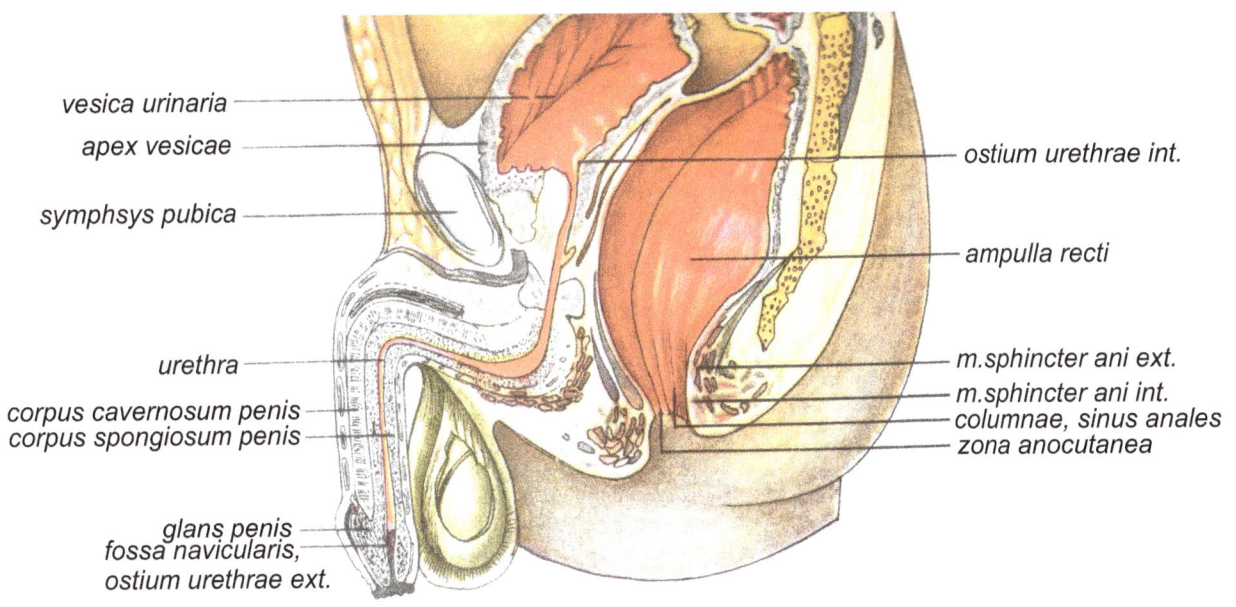

vesica urinaria

apex vesicae

symphsys pubica

ostium urethrae int.

ampulla recti

urethra

corpus cavernosum penis
corpus spongiosum penis

m.sphincter ani ext.
m.sphincter ani int.
columnae, sinus anales
zona anocutanea

glans penis
fossa navicularis,
ostium urethrae ext.

Fig. 229. Mediosagittal section through the pelvis in the male.

227

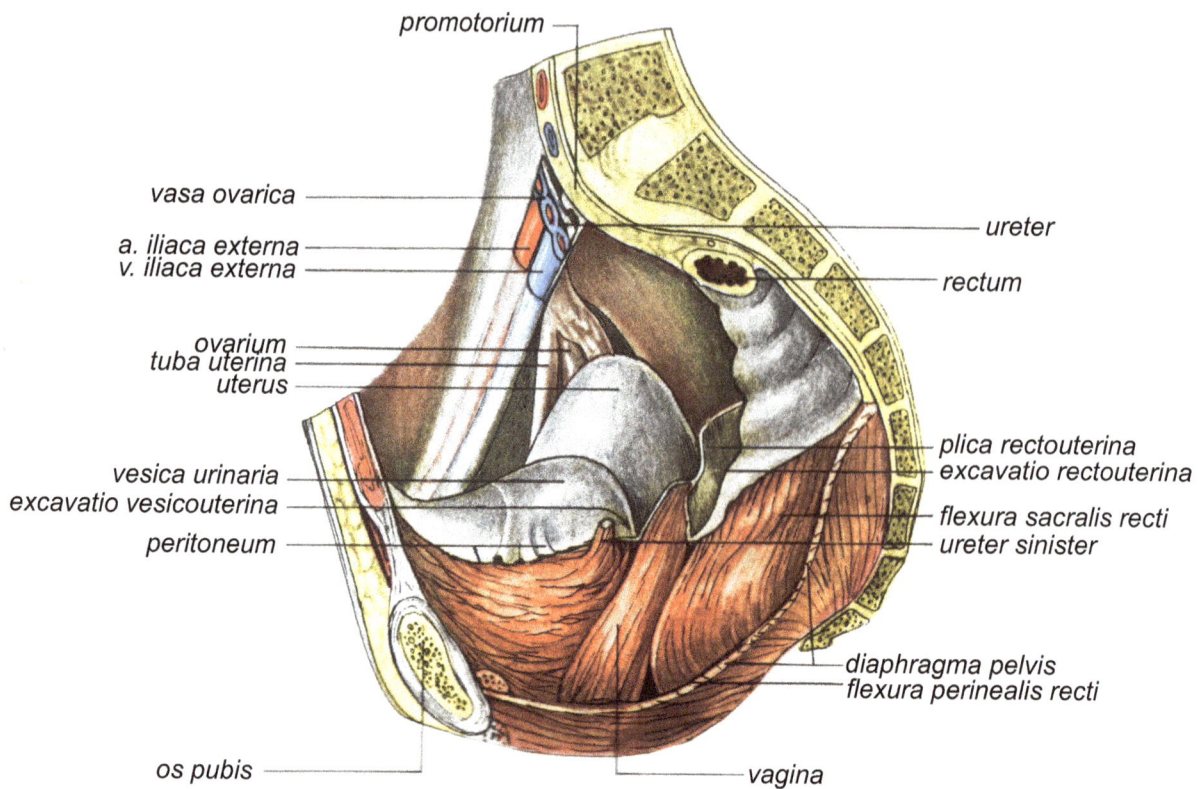

Fig. 230. The pelvis in the female; disposition of the peritoneum.

promotorium

vasa ovarica

a. iliaca externa
v. iliaca externa

ovarium
tuba uterina
uterus

vesica urinaria
excavatio vesicouterina

peritoneum

os pubis

ureter

rectum

plica rectouterina
excavatio rectouterina

flexura sacralis recti
ureter sinister

diaphragma pelvis
flexura perinealis recti

vagina

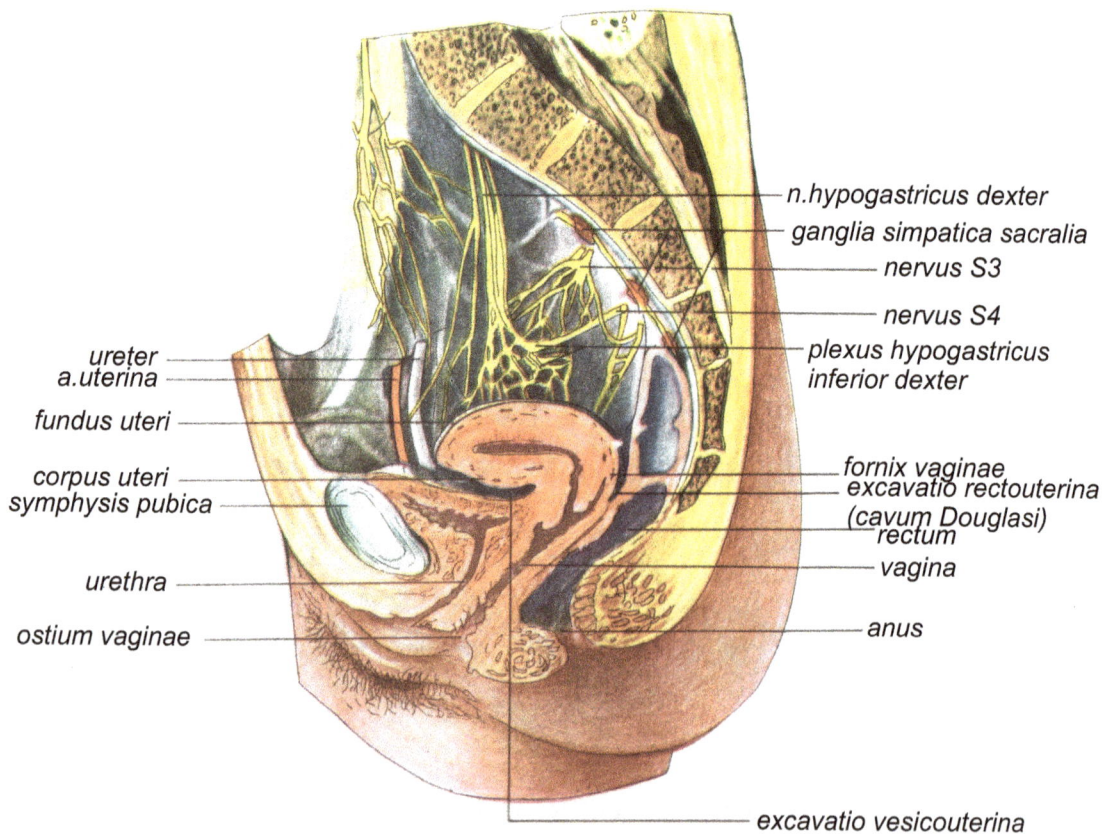

Fig. 231. Nerve supply to the pelvis in the female (mediosagital section).

n.hypogastricus dexter
ganglia simpatica sacralia
nervus S3

nervus S4
plexus hypogastricus
inferior dexter

ureter
a.uterina

fundus uteri

corpus uteri
symphysis pubica

urethra

ostium vaginae

fornix vaginae
excavatio rectouterina
(cavum Douglasi)
rectum

vagina

anus

excavatio vesicouterina

228

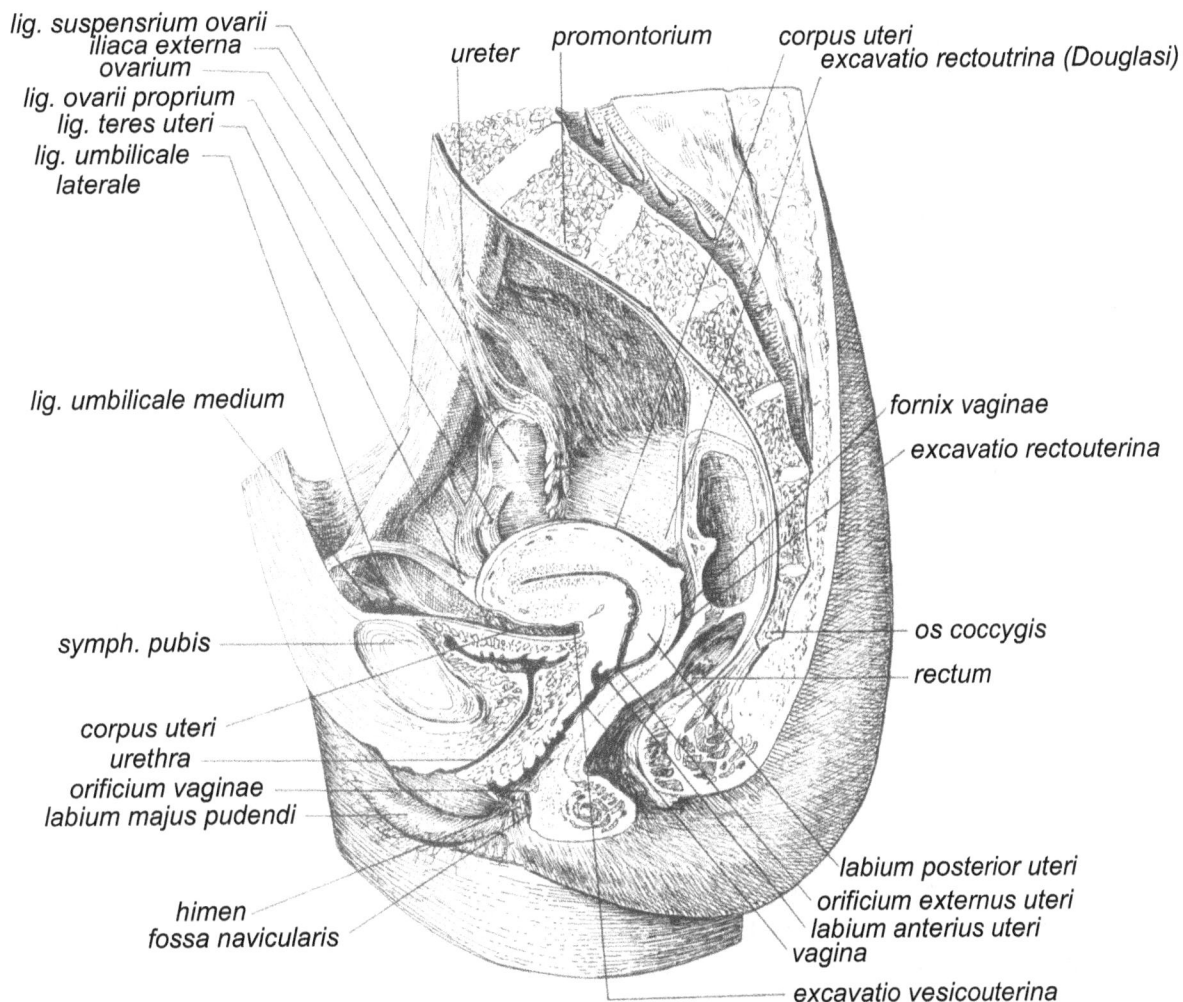

Fig. 232. Mediosagittal section through the pelvis in the female.

The contents of the lateral space are represented by a connective-adipose tissue, the obturator pedicle, the infernal iliac vessels with their branches, especially visceral, lymphatics, the hypogastric nerve plexus, the parietovisceral segment of the ureters and the transverse septa, which are made up of the connective laminae containing vessels and of the peritoneal joining fasciae, among which we mention; the connective coat of the infernal iliac artery, the umbilico-prevesical aponeurosis, the prostatoperitoneal aponeurosis and the retrorectal lamina.

The Romanian school of anatomy has demonstrated that this fibrous apparatus is, in this morphological manner, proper to man, representing an adaptation to orthostatism. This fibrous apparatus of the pelvis has the role of supporting the pelvic organs and represents the pelvic component of the internal urogenital tunic, to which - as we have already seen - belong the perineosuprarenal fascia and the umbilico-prevesical fascia. This tunic may be compared to a hammock and represents a unitary connective formation that assures the statics of the various organs with which if comes in contact, under the conditions of adaptation of their function to orthostatism.

229

The Urnary Bladder
(Vesica urinaria)

The disposition of the fasciae and of the pelvic peritoneum leads to the formation of a visceral sac of a particular clinical and surgical significance - the urinary bladder sac.

The urinary bladder *(vesica urinaria)* is a musculomembranous reservoir info which open the ureters, through which it receives the urine excreted by the kidneys, that is discharged through the urethra outside the organism, the urine being retained in the bladder between mictions.

The bladder has a summit *(apex vesicae)* and a bottom *(fundus vesicae),* opposite to the apex as orientation, from which starts the urethra. The portion between the apex and the fundus is the body of the bladder *(corpus vesicae);*

The shape of the bladder varies according to its filling degree, so that when it is empty, it is tetrahedral in shape, with a triangular base directed postero-inferiorly and with an apex directed anterosuperiorly, from which starts, towards the deep cicatrix of the umbilicus, the urachus, contained in the median umbilical fold *(ligamentum umbilicale medianum).*

Laterally lie medial umbilical ligaments *(ligamentum umbilicale medianum),* resulted from the obliteration of the umbilical arteries. When the bladder is filled, it assumes an ovoid form, with the axis directed obliquely posterosuperiorly (fig. 233).

The mean capacity of the bladder is from 250 to 350 ml, marked variations being recorded according to the age, the sex, the pathological conditions etc. (fig. 238-240).

The urinary bladder is located in the pelvis, within a sac the walls of which are formed: in front, by the umbilico-prevesical fascia and, through it, by the two pubic bones joined by the pubic symphysis; behind, by the prostatoperitoneal fascia, the rectovesical septum in the male and the vesicovaginal septum in the female; laterally, by the internal obturator muscles and the anal levatores; below, by the prostate in the male and the vagina, the urogenital diaphragm respectively in the female (we mention that in the female the bladder is lower situated, coming in contact with the urogenital diaphragm itself); above, by the peritoneum.

Around the bladder lie the pelvi-subperitoneal connective tissue and the circumscribed (prevesical, retrovesical) spaces described above.

The bladder is fixed in its sac: below, by the perineum; above, by the peritoneum which connects it to the neighbouring organs and to the median and medial umbilical ligaments; in front, by the pubovesical ligaments *(ligamentum pubovesicale),* which contain smooth muscle fibres, and by the puboprostatic ligaments *(ligamentum puboprostaticum);* behind, by a group of muscle fibres that run towards the rectum *(m. rectovesicalis)* in the male or to the uterus *(m. uterovesicalis)* in the female.

Through the walls of the sac in which it lies, the bladder has relations which are different in the male and in the female, variable to a certain extent according to whether the organ is empty or full

As regards the empty bladder, which presents three surfaces (anterior, posterior and inferior), two lateral borders and an apex, it has the following relations:

the anterior surface is related to the umbilico-prevesical fascia, the space of Retzius, the pubovesical ligaments, Santorini's venous plexus, lymphatics, the neurovascular obturator fascicle and the pubic symphysis; the posterior surface is covered with peritoneum and hasrelations with the rectum in the male and the uterus in the female, with loops of the small intestine, with the sigmoid colon and sometimes with the vermiform appendix, if the latter is normally sited;

- the inferior surface (fundus of the bladder), extending from the neck of the bladder to the vesicorectal pouch in the male and to the vesico - uterine pouch in the female, has relations: in the male, with the prostate, the seminal vesicles, the ampullae of the deferent ducts through the latter, with the rectum; in the female, with the cervix uteri and the anterior wall of the vagina;

- the lateral borders are in relation with the levatores ani and the internal obturator muscles; the peritoneum, which invests their upper portion, forms the laterovesical pouches, which contain coils of the small intestine; below the reflection of the peritoneum, the bladder is in relation with the pelvi-subperitoneal connective tissue and in the male, in this region, the ureter crosses the deferent duct;

- at the apex of the bladder lies the urachus and at the level of the two postero-inferior angles are implanted the ureters.

The full bladder changes somewhat its relations. At the upper part of the anterior surface lies the prevesical peritoneal pouch, which should be pushed upwards in the course of cystotomy, in order to avoid the opening of the peritoneal cavity (fig. 243, 244).

Inside, the urinary bladder presents the following structural elements: at the level of the fundus of the bladder, there is a smooth area, devoid of folds, triangular in shape, containing in its posterior angles the ureteral orifices and in its anterior angle, the urethral orifice; it is called Lieutaud's vesical trigone *(trigonum vesicae)*.

The ureteral orifices are each bounded by a mucous fold *(plica ureterica)*; between the two orifices is sited a transverse prominence, called interureteral fold *(plica interureterica)*, behind which lies a depression, termed interureteral fossa.

The internal orifice of the urethra *(ostium urethrae internum)*, which is round in the child and the female, is a transverse opening in the male owing to the prostate; it is the most declivous point of the bladder, surrounded by a submucous venous plexus, which plays a part in the closing mechanism of the orifice. The body of the bladder is smooth in the child, but with advancing age it folds, forming more or less marked columns and recesses.

Structure of the bladder. The bladder is made up of a serous tunic *(tunica serosa)*, a fibrous tunic *(tunica fibrosa)*, a muscular tunic *(tunica muscularis)* and a mucous tunic *(tunica mucosa)*.

The serous tunic is made up of peritoneum. At the level of the anterior abdominal wall, the vesical peritoneum forms a prevesical pouch, when the bladder is full, and at the same time it moves away from the upper border of the symphysis.

Behind, the peritoneum, attached to the vesical adventitia through connective tissue, covers the posterior surface of the bladder and descends to the level of the seminal vesicles, where it reflects on the rectum, forming the vesicorectal pouch (cul-de-sac of Douglas) in the man.

In the female, it descends less on the posterior surface of the bladder and reflects on the uterus up to the boundary between the corpus and the cervix uteri, without reaching the anterior fornix of vagina, from which it is about 2-3 cm away, so that the lower portion of the bladder and the cervix uteri are extraperitoneal.

At the level of the posterior surface of the bladder, the peritoneum covers also the terminal portion of the deferent duct, which descends medially towards its ampullar portion, so that between the deferential ampullae, laterally, and the vesicorectal pouch is formed on the posterior surface of the bladder, in the male, the interdeferential trigone.

Laterally to the bladder, the peritoneum forms the paravesical fossa *(fovea paravesicalis)*, which extends up to the lateral umbilical fold *(ligamentum umbilicale lateralis)* formed by the epigastric artery and is continuous above with the supravesical fossa *(fovea supravesicalis)*, situated above the symphysis, on either side of the median umbilical fold.

On the posterior aspect of the bladder, the peritoneum forms one fold more, the transverse vesical fold - plica vesicalis transversa - when the bladder is empty.

The fibrous tunic is made up of connective tissue, a component of the visceral leaflet of the intrapelvic fascia.

The musculature is formed of three layers of smooth fibres.

The external layer is made up of longitudinal fascicles, which start from the urachus and scatter on the ventral and dorsal walls, constituting the pubovesical muscles. The middle layer consists ofcircular, irregularly orientated fascicles. The internal layer consiste of plexiform fibres, directed longitudinally in the upper part and transversally in the lower part.

The muscle fibres of the plexiform layer are continuous with the muscular stratum of the initial portion of the urethra, which it covers circularly, forming the sphincter of the bladder *(m. sphincter vesicae)*. The three layers form an architectonic unit, called detrusor muscle of the bladder *(musculum detrusor vesicae)* (the vesical muscle).

The submucosa is constituted of loose connective tissue, but it is absent at the level of the trigone of the bladder, so that at this level the mucous membrane loses its mobility. This fact explains the location of choice of tuberculous lesions in this area.

The mucous membrane is made up of a connective-elastic corium and of a stratified pavement epithelium.

Vessel and nerve supply

The bladder has four *arterial pedicles,* formed of: the superior vesical arteries, branches of the umbilicovesical artery; the inferior vesical arteries, branches of the internal iliac arteries; an anterior pedicle formed of a branch of the internal pudendal artery and of the inferior epigastric artery; a posterior pedicle, formed of branches of the middle rectal artery in the male and of the uterine and vaginal arteries in the female.

The arteries of the bladder anastomose to a great extent with each other and are distributed so as to form three networks: submucous, subserous and subepithelial.

The veins of the bladder are numerous and very bulky; they arise from three vesical plexuses; submucous, muscular and subserous; they accompany the arteries and run towards the internal iliac veins and the internal pudendal veins. The submucous plexus forms, around the orifices at the level of the trigone of the bladder, a kind of venous cushion, which drains the blood into the vesicopudendal plexus (fig. 243-245).

The bladder is supplied by a well developed network of *lymph vessels* in the muscular stratum. At this level start perforating branches, which empty into the perivesical lymphatic network, also well developed, The efferent vessels follow three pathways: an anterior pathway, towards the lymph nodes of the external iliac chain; a posterior pathway, towards the lymph nodes of the common iliac chain and the internal iliac lymph nodes; an inferior pathway, towards the lymph nodes of the bifurcation of the aorta.

The nerves of the bladder arise from the pelvic (sacral) sympathetic and parasympathetic system. The nerves of sympathetic origin arise from the hypogastric plexus, which on its turn is related to the lumbo-aortic plexus through the presacral nerve. The parasympathetic branches arise from the sacral parasympathetic system through the pelvic nerves. The nerves form in the wall of the organ two plexuses: one in the muscular layer and the other in the submucosa. From these plexuses start filaments which end in the epithelial layer.

The parasympathetic nerves are distributed mainly in the body of the bladder *(m. detrusor);* those of sympathetic origin are destined mainly to the neck of the bladder. The striated sphincter receives filaments from the pudendal nerve.

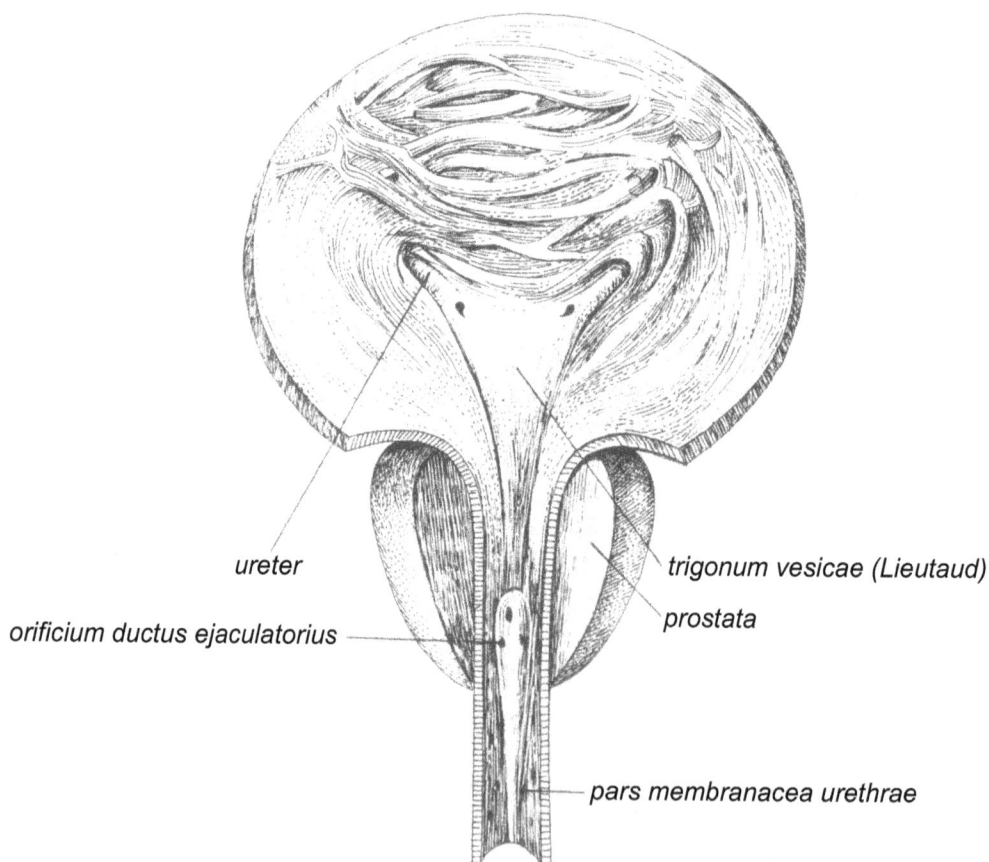

Fig, 233. The urinary bladder and the urethra structure

The Internal Male Genital Organs

In the pelvis is situated the region of the prostate and of the seminal vesicles, region which includes also segments of the urethra and of the pelvic portion of the deferent ducts - with which the prostate has intimate intrinsic relations - and the ejaculatory ducts, to which the seminal vesicles are closely related.

For didactic purposes, in order to offer a complete view of the male genital apparatus, in this chapter will be described also the testis, the epididymis and the other spermatic passages, organs which belong topographically to other regions.

For the same reasons, the order of dealing with the internal components of the male genital apparatus is the following: testis, epididymis, deferent ducts, ejaculatory ducts, prostate, seminal vesicles and urethra (fig. 234).

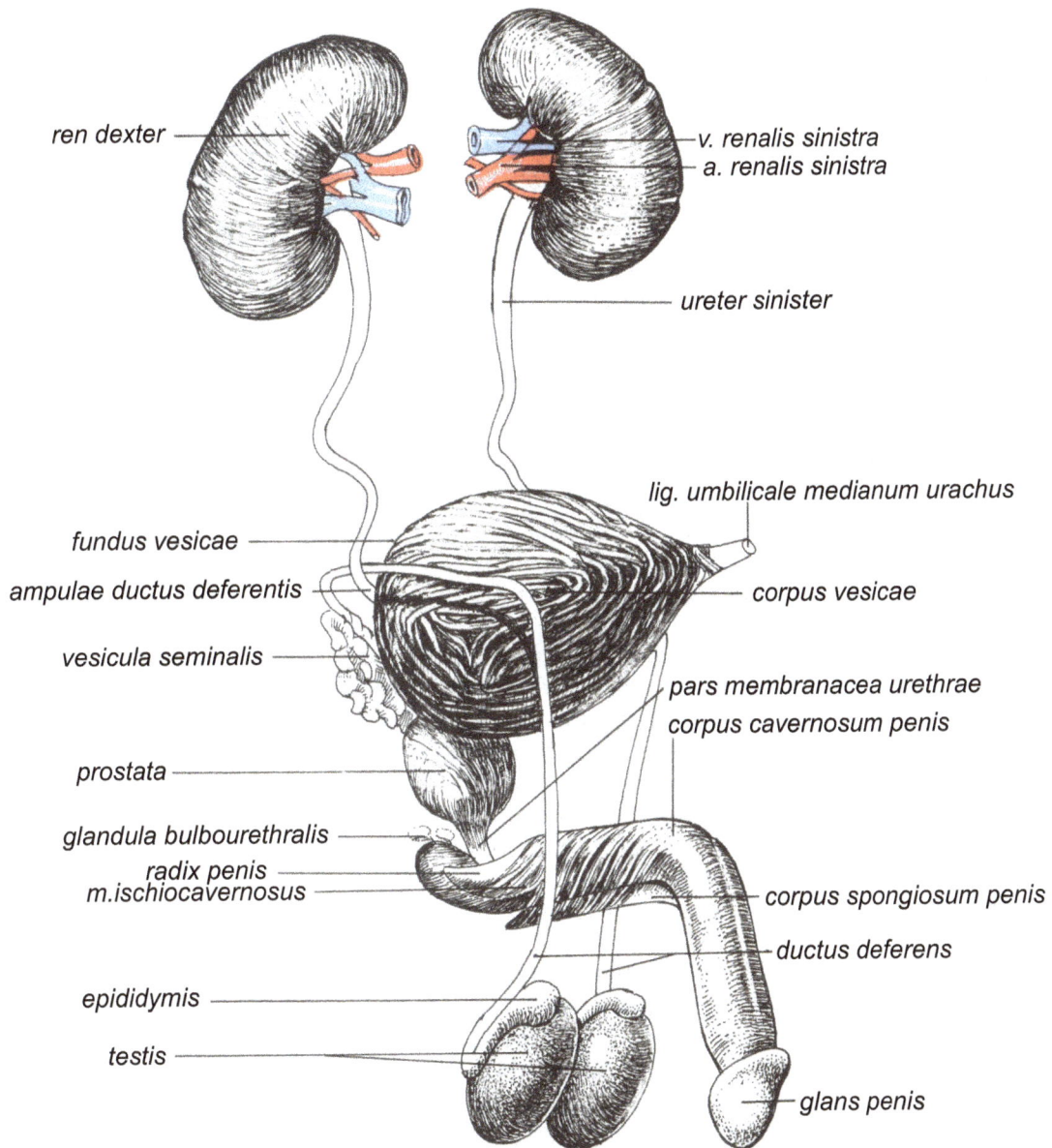

Fig. 234. The urogenital apparatus in the male (overall view)

The Testis
[Testis, orchis]

The testes, two in number, are organs producing spermatozoa and, at the same time, endocrine glands elaborating androgenic hormones.

They develop in the lumbar region, from where they descend (descensus testis) along the dorsal wall of the abdominal cavity, pass through the inguinal canal and reach the scrotum, where we find them normally at birth, separated through the scrotal septum (septum scroti).

However, this process of descent of the testes may be arrested at various levels, namely in the abdominal cavity, in the inguinal canal or at the external orifice of this canal and in this case we speak about ectopia testis, which may be unilateral or bilateral. It requires surgery or medical treatment, otherwise the atrophy of testes may occur.

The testes are ovoid in shape, transversal!!)/ flattened, with the axis directed obliquely from above downwards and anteroposteriorly. Their average dimensions are 4-5 cm in length and 2.5 cm in breadth; they are white-bluish in colour, of a firm, elastic consistency, similar to that of the eyeball, and are completely covered with the vaginal serosa.

They have two surfaces - medial and lateral, two borders - a supero-anterior, free border, and an inferoposterior border, and two extremities - superior (extremitas superior seu capitalis) and inferior (extremitas inferior).

The medial surface is slightly convex and covered with the vaginal serosa.

The lateral surface is free and invested, like the medial, by the serosa derived from the peritoneum (called epiorchium); if is more vaulted than the medial surface.

In proximity of the posterior border of the testis, the serosa insinuates between the body of the epididymis and the lateral surface of the testis, forming the vaginal interepididymo-testicular cul-de-sac called testicular bursa or sinus of the epididymis (sinus epididymidis); the upper border is in relation with the epididymis and with the vessels of the spermatic cord.

The superior extremity of the testis is covered by the head of the epididymis. Between the head of the epididymis and the testis lies a small ovoid body, called sessile hydatid (appendix testis), containing a gelatinous substance (it is a remnant of the upper extremity of the mullerian duct); the inferior extremity is situated outside the vaginal tunic and gives attachment to the inferior ligament of the epididymis. Slightly below it lies a bundle of connective elastic and smooth muscle fibres, which extends from the inferior extremity of the testis and epididymis to the deep surface of the scrotum, attaching the testis to the scrotum (Hunter's gubernaculum testis).

Beneath the vaginal serosa, i.e. beneath the epiorchium, the testis is invested by a fibrous, dense membrane, called tunica albuginea.

The breadth of the albuginea, which is on the average of 1 mm, increases at the posterior border of the testis, especially in the upper half, where it is much wider and forms the mediastinum testis or Highmore's body, which has the shape of a triangular pyramid.

Through the apex of the latter, blood vessels penetrate the testis and 10-15 canaliculi (ductuli efferentes) leave it; these efferent ductules, belonging to the spermatic passages, cross the tunica albuginea and enter the head of the epididymis.

From the apex and the lateral sides of Highmore's body start numerous connective septa, which divide the testis into five zone, called lobuli.

The lobuli, pyramidal or conic in shape, with the base against the albuginea, opposite to the mediastinum, are 250-300 in number and contain the convoluted seminiferous tubules (tubuli seminiferi contorti). The length of these very flexuose tubules varies from 70 to 80 cm and their number from 1 to 4 and in each lobule they form anastomoses with one another (fig. 235).

They constitute the parenchyma of the testis (parenchyma testis) and represent the glandular portion of the testis, made up of a basement membrane, on which lie two cell types: germinative cells of various ages (the more aged, the spermatogonia, generate towards the lumen of the tubes increasingly young cells, whereas within the lumen of the convoluted tubules are the spermatozoa or spermia) and Sertoli's supporting or sustentacular cells.

Among the convoluted tubules, in the connective tissue, are Leydig's interstitial cells, that accomplish the function of internal secretion.

The convoluted tubules run towards the mediastinum testis, where they generate Holler's rete testis, from which arise the efferent ductules that open into the head of the epididymis.

234

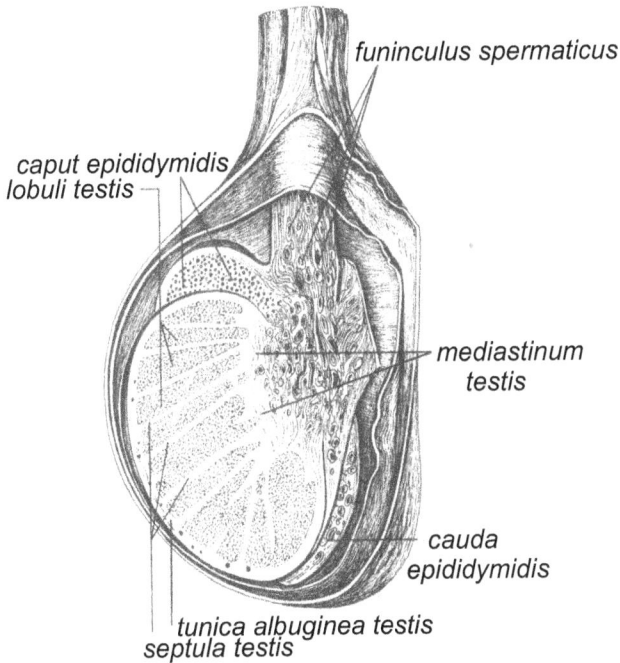

caput epididymidis
lobuli testis
funinculus spermaticus
mediastinum testis
cauda epididymidis
tunica albuginea testis
septula testis

Fig. 235. The testicle - structure.

The Epididymis

The epididymis is a tubular, elongated organ, situated on the posterior border of the testis, with the concavity directed anteroinferiorly, being thus adapted to the convex orientation of the posterior border of the testis (fig. 235).

It consists of three parts: a head *(caput)*, a body *(corpus)* and a tail *(cauda epididymis)*. It has a length of about 5 cm and a breadth of 12 mm at the level of the head, its thickness diminishing from the head towards the tail, which gives it the shape of a large comma.

The body is bulky and connected with the testis through the efferent cones, fibrous connective tissue and vaginal serosa.

On the head of the epididymis is implanted a vesicle, which is inconstant and represents a remnant of the superior extremity of the wolffian duct. The body is prismatic - triangular and above it may lie some small vesicles which constitute the organ of Giraldes, a remnant of the wolffian duct corresponding to the paraoophoron; the tail is flattened from above downwards.

The vaginal serosa covers the epididymis and reflects on the testis, forming the inter-epididymotesticular pouch *(bursa of testis* or sinus of epididymis). The epididymis is in relation, medially, with the elements of the spermatic cord and its inferior surface is adherent to the testis through a connective tissue. The inferior extremity of the epididymis is continuous with the deferent duct and forms with it an acute angle of the shape of a hairpin open upwards. Like the testis, the epididymis is connected to the scrotum through the scrotal ligament *(gubernaculum testis)* and two epididymal ligaments: one superior and the other inferior; between them is formed the sinus of the epididymis through insinuation of the serosa between the testis and the epididymis. The epididymis is actually a very sinuous duct, measuring unravelled about 6 cm in lenght, that are packed in the 5 cm which it measures in situ. These windings are held togheter by the connective tissue, which forms a so-called epididymal albuginea, but this is far from being as dense as that of the testis, so that it may tumefy in the case of an infection.

The Deferent Duct

The deferent duct begins at the tail of the epididymis and ends at the junction of the seminal vesicles with the ejaculatory duct. It is a cylindrical tube, save the portion in the neighbourhood of its end, where its calibre increases and its surface becomes irregular, covered with bosses, to which correspond inside the diverticles that form the ampulla of the deferent duct *(ampulla ductus deferentis)*. It has a firm consistency, a length of about 60 cm and a diameter of 34 mm (fig. 236, 239). The deferent duct consists of the following portions: epididymotesticular, funicular and iliopelvic.

The epididymotesticular portion begins at the tail of the epididymis and runs upwards and anteriorly on the posterior border of the testis, along the medial surface of the epididymis; it is separated from the epididymis by the spermatic veins of the posterior plexus.

The funicular portion begins at the superior part of the body of the epididymis, from where the deferent duct runs vertically up to the external ring of the inguinal canal. In this route, the deferent duct belongs to the spermatic cord.

The spermatic cord (funiculus spermaticus) is the pedicle that suspends the testis and the epididymis. It contains the deferent duct, the internal spermatic (testicular), deferential and cremasteric arteries, nerve filaments accompanying these arteries, the anterior and the posterior spermatic venous plexuses, richly anastomosed, which form the spermatic vein, lymph vessels and Cloquet's ligament, resulted from the obliteration of the perineovaginal canal. All these elements are united by fibrous connective tissue.

235

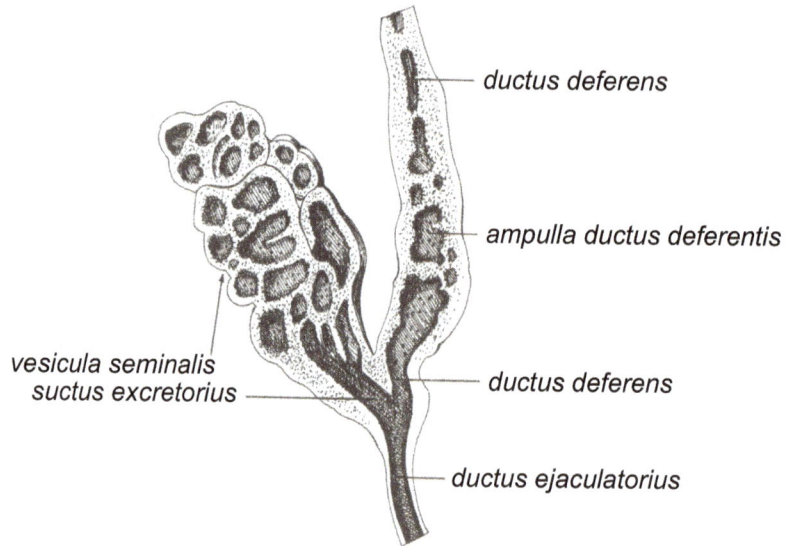

ductus deferens

ampulla ductus deferentis

ductus deferens

vesicula seminalis
suctus excretorius

ductus ejaculatorius

Fig. 236. Seminal vesicle, deferent ducts, excretory ducts, ejaculatory ducts.

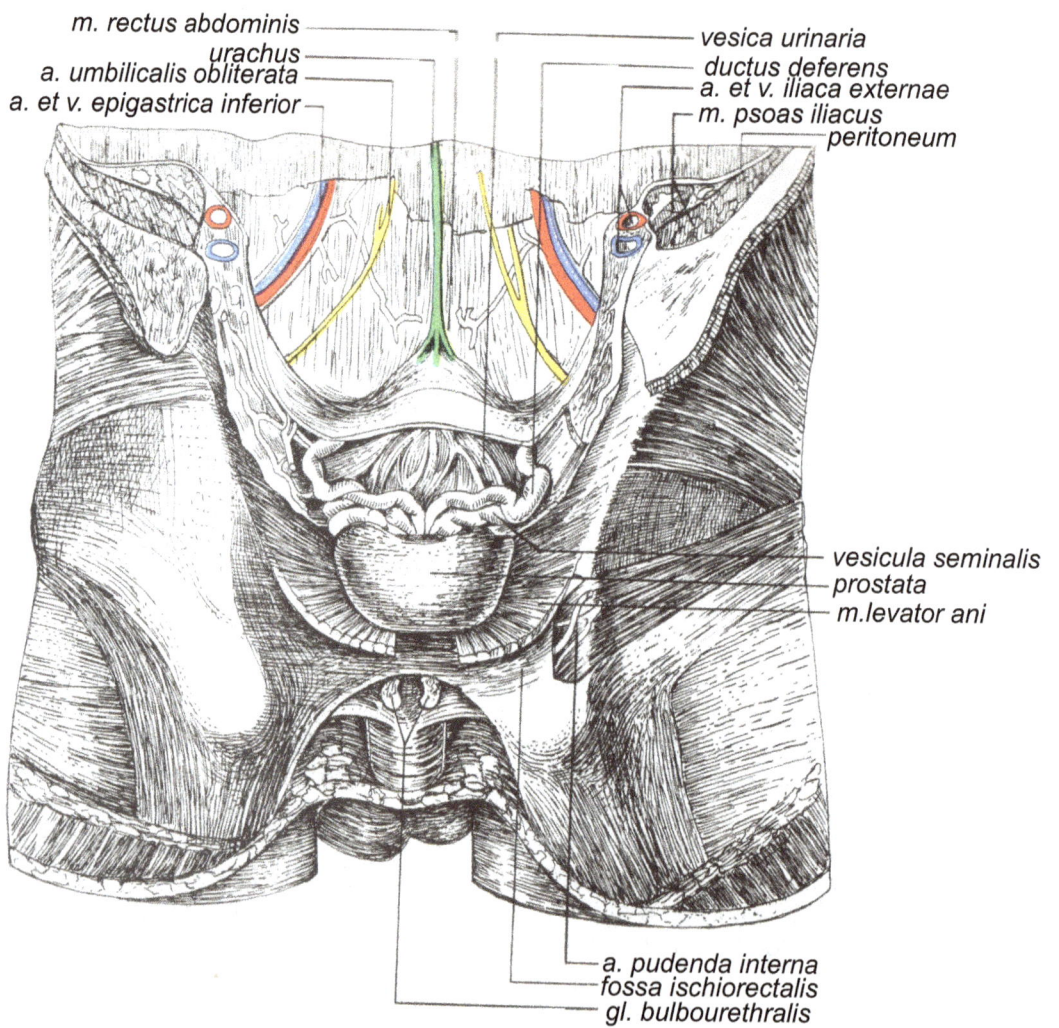

m. rectus abdominis

urachus

a. umbilicalis obliterata

a. et v. epigastrica inferior

vesica urinaria

ductus deferens

a. et v. iliaca externae

m. psoas iliacus

peritoneum

vesicula seminalis

prostata

m.levator ani

a. pudenda interna

fossa ischiorectalis

gl. bulbourethralis

**Fig. 237. Frontal section through the pelvis so the male, in front of the
rectum (view of the visceral space from behind)**

The *inguinal portion,* situated in the inguinal canal, extends from the superficial inguinal ring to the deep inguinal ring, at the level of which it enters the abdominopelvic cavity. It has also relations with the elements of the spermatic cord, as well as with the branches of the abdominogenital and genitofemoral nerves and with the funicular artery. At the level of the deep inguinal ring, the elements of the spermatic cord separate from each other. The vessels of the anterior spermatic venous plexus form the spermatic vein, which ascends on the posterior abdominal wall, emptying differently on the right in comparison with the left side; the veins of the posterior spermatic plexus open into the epigastric vein; the deferent duct, accompanied by the deferential artery, penetrates the pelvic cavity, crosses the epigastric vessels, passes above them describing a curve, the inferior concavity of which is opposite the curve with the superior concavity, determined by the epigastric artery, then crosses the external iliac vessels and descends within the pelvic cavity. Here it runs on the lateral side of the urinary bladder, passes above the ureter and the umbilical artery, turns round the base of the seminal vesicles, becomes retrovesical and presents at this level an ampullar dilation *(ampulla ductus deferentis),* which is related in front to the fundus of the urinary bladder and behind to the rectal ampulla, circumscribes medially, with the contralateral duct, the interdeferential triangle and is related laterally to the homolateral seminal vesicle *(glandula vesiculosa).* In its structure we distinguish an external tunic, the adventitia, made up of connective tissue, a muscular coat, constituted of fibres with a spiral course, and a mucous membrane of epithelial type.

Vessel *and nerve* supply of the testis, epididymis ans deferent duct

The arterial supply of the testis is provided by the spermatic artery, a branch of the aorta. The epididymis is supplied by the spermatic artery, too, and by the deferential artery, a branch of the vesiculodeferential artery which, on its turn, is a branch of the internal iliac (hypogastric) artery. The deferent duct, in its epididymotesticular and funicular portions, is supplied by the deferential artery and by branches of the funicular artery, a branch of the inferior epigastric artery, which anastomoses with the spermatic and deferential arteries.

The veins of the testis and of the anterior part of the epididymis constitute the anterior venous group or the anterior spermatic plexus and the veins of the posterior part of the epididymis form the posterior spermatic plexus. The veins of the deferent duct drain, on the one hand, into the two mentioned venous plexuses and, on the other hand, in the pelvis, into the vesicoprostatic and seminal plexuses.

The lymph vessels of the testis and epididymis ascend along the spermatic vessels and drain into the abdomino-aortic lymph nodes, Sometimes, a part of the lymphatics of the testis drain into an external iliac node situated at the bifurcation of the common iliac artery, The lypmhatics of the deferent duct drain into the external and internal iliac lymph nodes.

The nerve supply of the testis and epididymis derives from the solar plexus through the spermatic plexus and from the hypogastric plexus through the vesiculodeferential plexus.

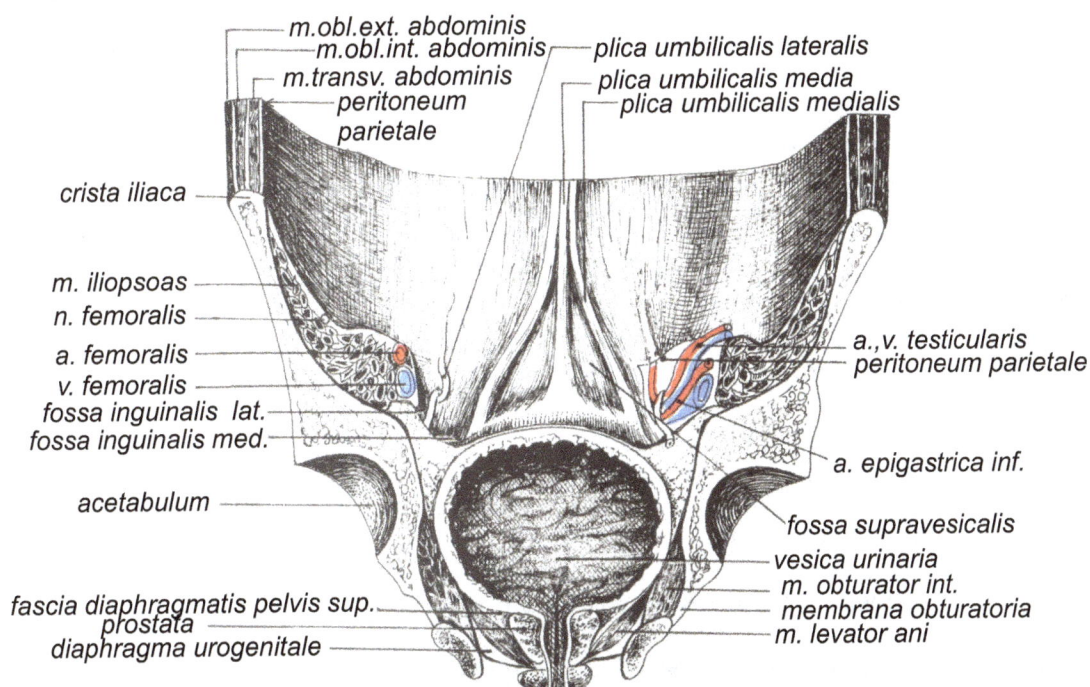

Fig. 238. Internal surface of the abdominal wall; the urinary bladder.

237

The Ejaculatory Ducts
(Ductus ejaculatorius)

The ejaculatory duct, continuous with the deferent duct, is formed above the base of the prostate by the union of the deferent duct with the excretory duct of the seminal vesicle, then enters the prostatic parenchyma and opens into the prostatic urethra through an orifice situated on the colliculus seminalis.

Its structure is similar to that of the deferent duct, with the difference that its lumen is narrower than that of the deferent duct, allowing the expulsion of sperm with an increased force and speed into the urethra, during the contraction of the musculature of the deferent duct

Three funics are described: the adventitia, situated externally, absent in the intraprostatic portion; the muscular funic, made up of a longitudinal and a circular layer, and the mucous funic, formed of a corium and columnar epithelium (fig. 236, 240 - 242).

The Prostate

The prostate is situated in a well cricumscribed splanchnic bed, which allows surgical procedures to be performed in good conditions.

The prostate is a glandular organ of the size of a chestnut, functionally attached to the male genital apparatus and situated in the pelvi-subperitoneal space, above the urogenital diaphragm and below the urinary bladder, to which if is closely connected. If develops around the initial portion of the urethra which crosses if and has with it very close anatomotopographical and clinico-operative relations.

In the adult male, the prostate measures 20-30 mm in length, 40 mm in breadth and 25-30 mm in thickness. Its normal weight is 20-25 g.

It has the shape of an anferoposteriorly flattened cone, with the base directed upwards, towards the fundus of the bladder and the apex directed downwards. If is crossed by the prostatic portion of the urethra *(pars prostatica urethrae);* the base is in relation with the urinary bladder, and the apex, with the urogenital diaphragm. Its anterior surface is in relation with the pubic symphysis and the posterior surface, with the rectum *(fades rectalis).*

Through rectal digital touch if may be palpated and massaged. The lateral surfaces are in relation with the levatores ani.

If is made up of two lobes connected posteriorly by the isthmus of the prostate and anteriorly, by the preurethral portion. In some aged sujects, the isthmus undergoes hyperthrophy and gives rise to a median lobe, which develops in the urethra and the bladder (in the uvula of the latter) and should be removed.

The prostatic urethra traverses the prostate on a distance of 3 cm. It presents on its dorsal wall the so-called *colliculus seminalis,* prolonged upwards and downwards through the urethral crest *(crista urethralis).* On the colliculus seminalis is sited the prostatic utricle, a glove-finger-shaped tube, 1-6 mm long and 9-10 mm deep. (If is a remnant of the mullerian duct) (fig. 237-240, 242).

On each side open the ejaculatory ducts and the orifices of the 30-50 tubulo-alveolar proper glands, which are enveloped by connective tissue and fascicles of smooth muscles, that form the so-called prostatic muscle.

The secretion of the prostate is viscous, of a characteristic odour, and accomplishes the function of contributing to the motility of spermatozoa, which travel at a rate of 25 microns per second.

The prostate is situated in a bed bounded by six walls:

The posterior wall is formed by the Denonvilliers' prostatoperitoneal aponeurosis (rectovesical septum).

The anterior wall is constituted of the pubis.

The lateral walls, right and left, are made up of the laminae of connective tissue which belong to the sacrorecta-genitopubic aponeurosis and of the levatores ani; in their thickness are situated the lateral venous plexuses of the prostate.

The superior wall is made up by the puboprostatic ligaments, the fundus of the urinary bladder, the deferent ducts and with the proper capsule of the gland. The prostate is attached the seminal vesicles (vesicular glands).

The inferior wall is formed of the urogenital diaphragm.

238

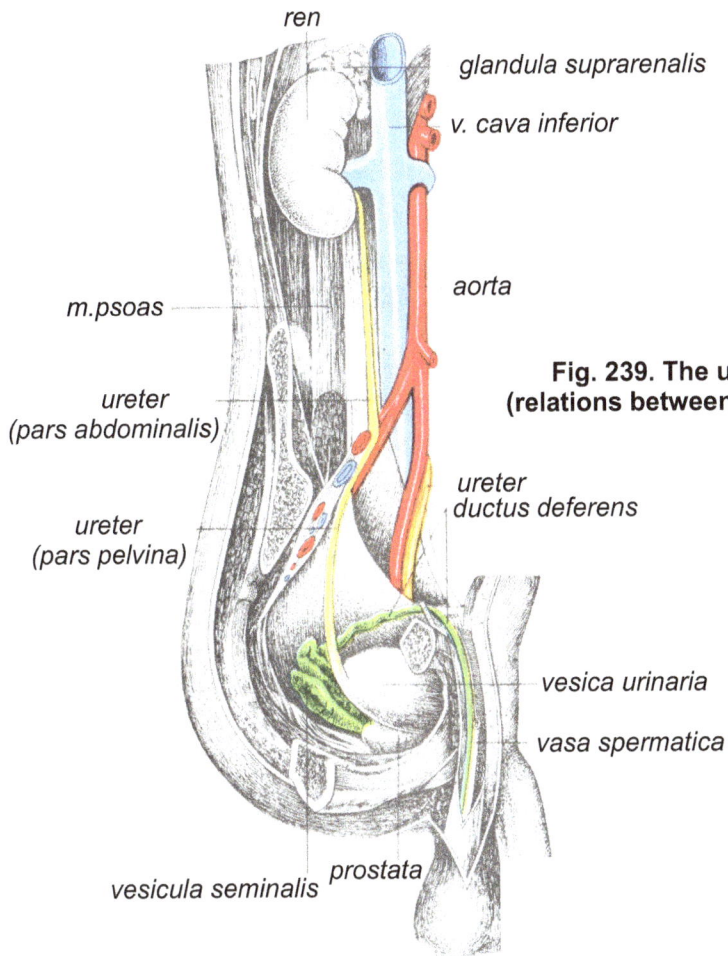

ren
glandula suprarenalis
v. cava inferior
aorta
m.psoas

**Fig. 239. The urogenital apparatus in the male
(relations between the ureter and the deferent ducts).**

ureter
(pars abdominalis)

ureter
(pars pelvina)

ureter
ductus deferens

vesica urinaria

vasa spermatica

vesicula seminalis   prostata

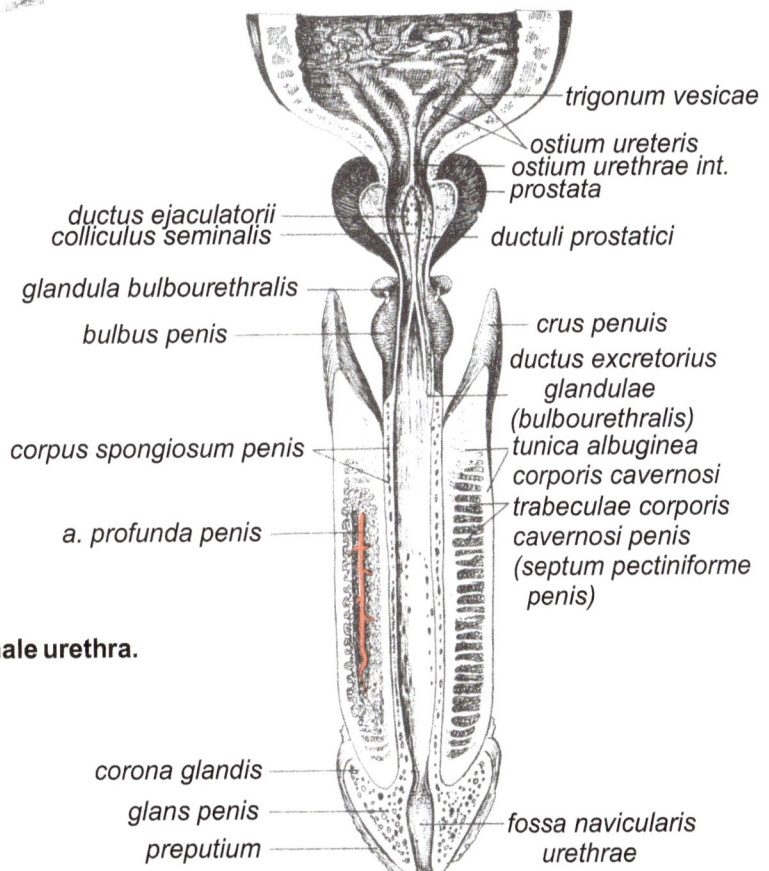

trigonum vesicae
ostium ureteris
ostium urethrae int.
prostata

ductus ejaculatorii
colliculus seminalis

ductuli prostatici

glandula bulbourethralis

crus penuis

bulbus penis

ductus excretorius
glandulae
(bulbourethralis)
tunica albuginea
corporis cavernosi
trabeculae corporis
cavernosi penis
(septum pectiniforme
penis)

corpus spongiosum penis

a. profunda penis

Fig. 240. The urinary bladder and the male urethra.

corona glandis

glans penis

preputium

fossa navicularis
urethrae

239

The prostatic bed communicates, through the spaces between the puboprostatic ligaments, with the prevesical space and posteriorly with the retrovesical space. Between the bed and the gland lies the periprostatic space, which contains loose connective tissue. Around the gland, this condensed tissue forms the periprostatic fascia *(fascia prostatae)*. Laterally to this tissue lie the vesicoprostatic venous plexuses. The periprostatic fascia should not be confounded with the proper capsule of the gland. The prostate is attached

through the peritoneum, the puboprostatic ligaments and the adhesions to the urethra and the urinary bladder. The relations of the prostate may be grouped into two categories: intrinsic relations, with the organs which cross it, and extrinsic relations, achieved through the walls of the prostatic bed.

The prostate has intrinsic relations with the prostatic urethra and the ejaculatory ducts.

The extrinsic relations of the prostate are the following: anteriorly, with the pubic symphysis, from which it is separated by a space containing connective tissue and the prostatic venous plexus, bounded above by the puboprostatic ligaments; posteriorly, with the rectal ampulla, from which it is separated by Denonvilliers' prostatoperitoneal aponeurosis; inferolaterally, with the levator ani muscle, the sacrorecto-genitopubic laminae and the anterior projections of the ischiorectal fossae; superiorly, the base of the prostate *(basis prostatae)* is related to the base of the urinary bladder, the seminal vesicles and the deferent ducts; inferiorly, the apex of the prostate *(apex prostatae)* is in a direct relation with the genital diaphragm on which it rests.

Structurally, the prostate is made up of a glandular substance *(substantia glandularis prostatae)* and a musculo-connective stroma, in which prevail the smooth muscle fibres, specific to this gland, which have the role of evacuating the prostatic secretion during ejaculation.

The glandular substance forms four lobes; two lateral lobes *(lobus dexter, lobus sinister)*, which are connected by the isthmus of the prostate *(isthmus prostatae)*, situated in front of the urethra, but sometimes the isthmus may lack; the median lobe *(lobus medius)*, situated in the posterosuperior part of the gland; the posterior lobe *(lobus posterius)*, situated in the postero-inferior portion in proximity to the rectum. The glands are of two kinds: periurethral glands of a mucous type, situated around the urethra, which open through small orifices into the urethra, and actual prostatic glands, of a tubuloalveolar type, 30 to 50 in number, the excretory ducts of which join to form the prostatic ducts *(ductuli prostatici)*, that open into the prostatic sinuses; they are related to the sphincter of the extrasphinteric urethra, contrary to the periurethral ducts, which are intrasphincteric.

The interglandular stroma is composed of smooth muscular, collagen and elastic fibres.

The prostate is enveloped by a proper capsule, made up of a dense connective tissue, with elastic and smooth muscle fibres, from the level of which start connective musculo-elastic septa, which divide the glandular parenchyma into lobules and converge on a central area, crossed by the ejaculatory ducts and the prostatic utricle, the urethra being situated anteriorly.

Vessel and nerve supply of the prostate

The arterial supply is derived from the inferior vesical arteries *(a. vesicalis inferior)* and the middle rectal arteries *(a. rectalis media)*.

The veins open info the prostatic venous plexus, which surrounds the gland and then drain into the internal pudendal vein *(v. pudenda interna)* (fig. 243-245).

The lymph enters the external and internal iliac and the sacral lymph nodes. *The nerve supply* is derived from the prostatic plexus, a branching of the inferior hypogastric plexus, which contains sympathetic and parasympathetic fibres.

The Seminal Vesicles
[Vesiculae seminales siveglandulae vesiculosae]

The seminal vesicles are two in number and have the role, on the one hand, of producing a secretion which is added to the seminal fluid and, on the other hand, of a reservoir in which the secretion of the spermatic ducts accumulates. They are conic in shape, directed obliquely, with an irregular surface owing to their twisting; they are situated above the prostate, between the rectum and the urinary bladder (fig. 236, 241, 242).

They have relations, in front, with the fundus of the urinary bladder, behind with the ampullar portion of the deferent ducts and laterally with the prostatic venous plexus. Their base is related to the rectovesical pouch (cul-de-sac of Douglas) and crossed by the ureter {before entering the urinary bladder); their apex is continuous with the excretory duct *(ductus excretorius)*, which will unite with the deferent duct, to form the ejaculatory ducts.

The inner aspect is areolar, owing to the presence of numerous folds and diverticula.

We distinguish in their constitution an adventitia situated externally, a muscular tunic made up of superficial longitudinal and deep circular fibres and a mucous coat *(tunica mucosa),* constituted inside of secretory cells.

Vessel and nerve supply

The arterial supply is provided by branches of the inferior vesical artery and of the middle rectal artery; *the veins* run to the vesical and prostatic plexus, *the lymph* drains into the internal iliac lymph nodes and *the nerves* derive from the inferior hypogastric, the vesical and the prostatic plexuses.

The Urethra
[Urethra]

The urethra begins at the level of the urethra! orifice of the urinary bladder *(orificium urethrae internum)* and ends at the urinary meatus *(orificium urethrae externum),* situated at the free extremity of the glans penis. The urethra is made up of four portions; the intramural portion, very short, at the base of the urinary bladder; the prostatic portion *(pars prostatica),* 3 cm long, which passes-through the prostate; the membranous portion *(pars membranacea),* which crosses the urogenital diaphragm; the spongiose portion *(pars spongiosa),* embedded in the corpus spongiosum penis (cavernous body of urethra) (fig.240-242).

The prostatic portion has variable relations with the prostate, forming a tunnel in the gland, which is very close to the anterior surface of the prostate and may even be uncovered anteriorly by glandular tissue.

The membranous portion, called so because its constituent funics are surrounded by the external striated sphincter and by the perineal fascia, has relations, in front, with the transverse ligament of the perineum and with the vesicoprostatic venous plexuses and behind, with the bulbo-urethral glands and the tendinous centre of the perineum.

The spongiose portion is situated in the corpus spongiosum penis, into which it penetrates on its superior surface, 0.5-1 cm anteriorly to its posterior end.

The shape of the urethra varies according to its functional state and to the aspect obtained by instrumental or radiological exploration. The empty urethra, in the intermictional rest stage, is a virtual canal, the walls of which are in contact and which on exploration shows areas of narrowing. In the normal state, it is S-shaped, presenting a posterior curve, concave anteriorly, by going round the pubic symphysis from behind forwards and from above downwards, and an anterior curve, convex anteriorly, of the portion situated in the pendulous penis, which hangs below the pubic symphysis. However, certain morphological features are also dependent on its calibre. Thus, the empty urethra has the walls in contact, whereas the full urethra has four areas of narrowing: at the level of the external orifice, in the corpus spongiosum penis up to the prepubic angle, in the membranous area owing to the striated sphincter, and at the internal orifice. Belween these narrowings may be seen dilated portions, among which also the intraprostatic region.

The interior aspect is consequently variable.

Thus, at the level of the prostatic urethra is present the urethral crest *(crista urethralisj,* which is sited on the posterior wall of the urethra and which, in its median portion, shows an elevation called verumontanum *(colliculus seminalis),* on the tip of which opens the prostatic utricle *(utriculus prostaticus),* a channel ended in cul-de-sac, that penetrates the prostate and represents a remainder of the mullerian ducts. On either side open the two orifices of the ejaculatory ducts, level at which the male urethra becomes a common uroseminal pathway. We mention that the verumontanum bounds laterally some vertical depressions, called prostatic sinuses *(sinus prostaticus),* into which open the orifices of the two ejaculatory ducts and about 10 narrower orifices of the 20-30 prostatic glands of the urethra. In the membranous part are situated the orifices of the urethral glands.

At the level of the spongiose urethra open the orifices of the bulbo-urethral glands (Cowper) and the Morgagni's lacunae *(lacunae urethrales),* which lie behind the bulb of the *corpus spongiosum* (spongiose body). They are larger sinuses *(foramina)* and smaller sinuses *(foraminula)* of the mucosa and are bounded by folds, into the bottom of which open the urethral glands (Littre). About 1-2 cm before the external urinary meatus lies a mucous transverse fold *(valvula fossae navicularis)* (Guerin), crescentic in shape, which descends from the posterior wall of the navicular fossa. The catheterization of the urethra should take into account these morphological details of the urethral walls, consequently it should be performed with a great care.

241

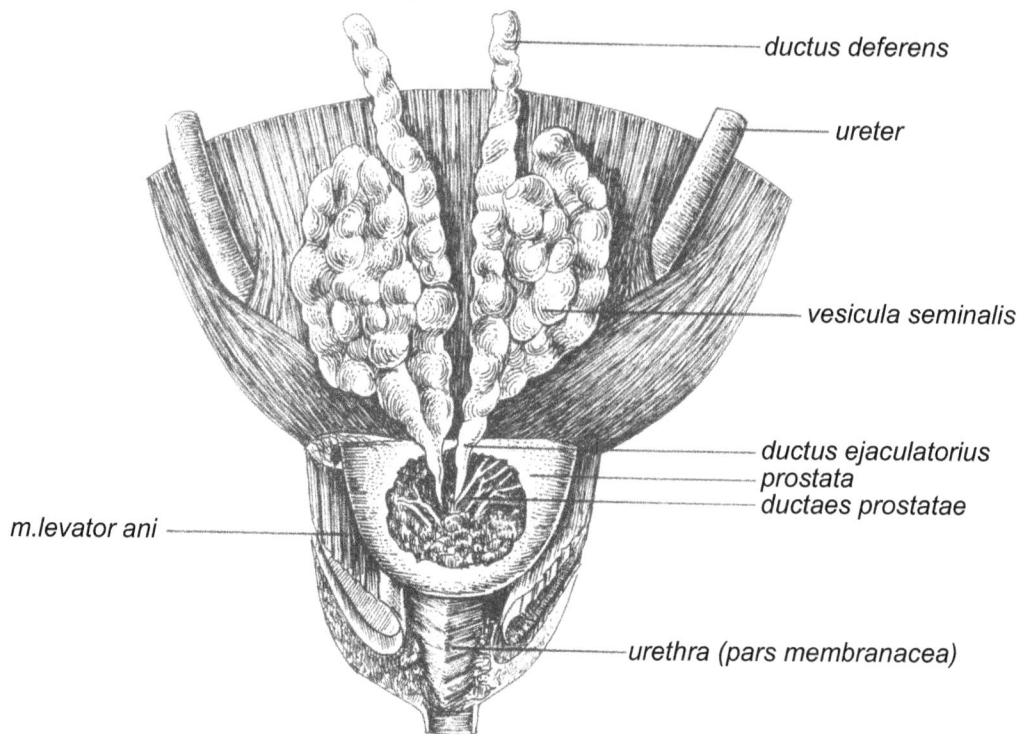

ductus deferens

ureter

vesicula seminalis

ductus ejaculatorius
prostata
ductaes prostatae

m.levator ani

urethra (pars membranacea)

Fig. 241. Dissected prostate and seminal vesicle, viewed from behind

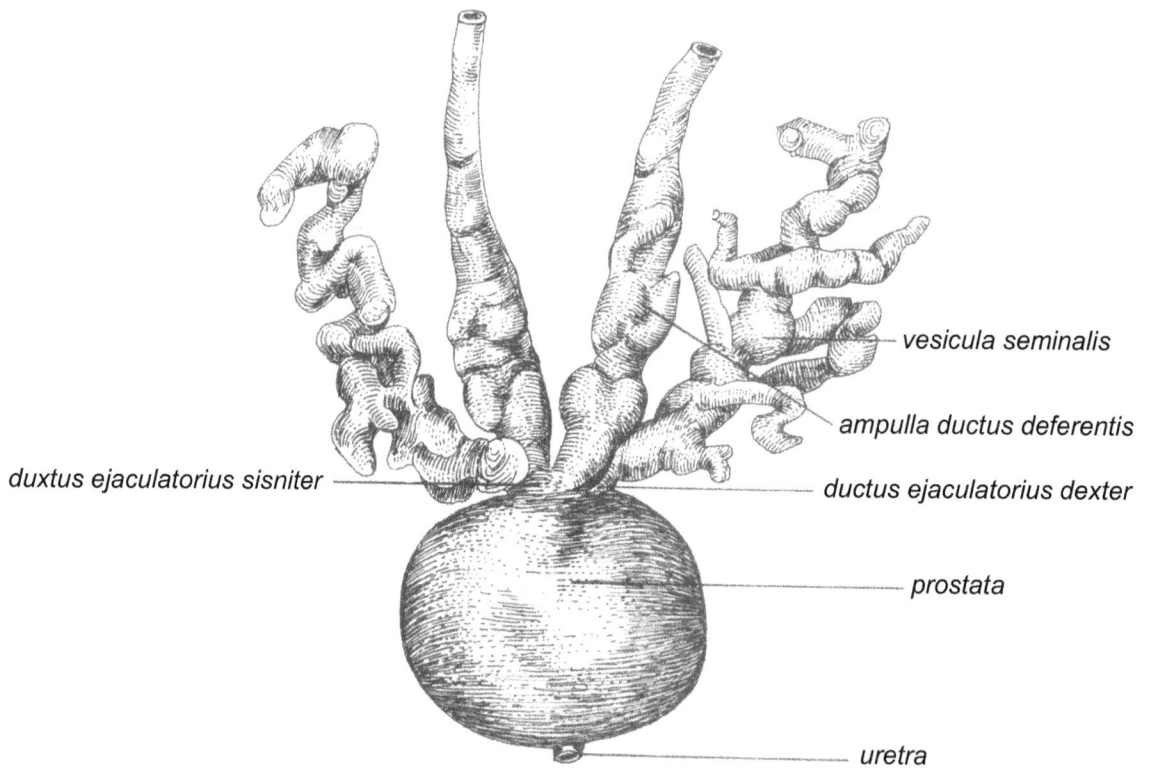

vesicula seminalis

ampulla ductus deferentis

duxtus ejaculatorius sisniter

ductus ejaculatorius dexter

prostata

uretra

Fig. 242. The prostate and seminal vesciles

242

As regards the structure, the urethra consists of a mucous tunic and a muscular funic. In the prostatic portion, the urethral wall is united with the glandular tissue and in the spongiose portion, whit the corpus spongiosum penis.

The mucous tunic is constituted of an epithelium of urinary type up to the colliculus seminalis, then of a columnar stratified epithelium up to the navicular fossa, of a non-keratinized pavement epithelium at the level of the meatus and of a fibro-elastic richly vascularized corium, which contains a venous plexus made up of blood lacunae. The mucosa contains numerous intraepithelial, intramucous and tubulo-acinous urethral glands *(glandulae urethrales)*, which secrete intermittently mucus that protects the urethral mucosa against the corrosive action of urine, often acid owing to the nature of food.

The muscular tunic contains smooth and striated fibres. The smooth fibres are longitudinal, continuous with those of the plexiform layer of the urinary bladder, and circular external which, at the level of the initial zone of the urethra, form the sphincter of the urinary bladder *(sphincter vesicae superior) (lissosphincter),* situated partly inside the prostate. The striated fibres form the striated sphincter of the urethra *(rhabdosphincter),* situated outside the prostate, at the level of the membranous urethra, called also musculus sphincter vesicae inferior and supplied by the pudendal nerve.

Vessels *and- nerve* supply of the urethra

The arterial supply is provided to the prostatic urethra by the middle rectal artery and the inferior vesical artery, to the membranous urethra by the artery of the bulb of penis and to the spongiose urethra, by the urethral artery and the dorsal artery of the penis.

The veins drain into the prostatic plexus for the prostatic urethra, into the internal pudendal vein for the membranous portion and into the deep dorsal vein of the penis for the spongiose portion, reaching finally the internal iliac vein (fig. 243-245).

The lymph flows into the external and internal lymph nodes for the prostatic and membranous portions and into the inguinal and external iliac lymph nodes for the spongiose urethra.

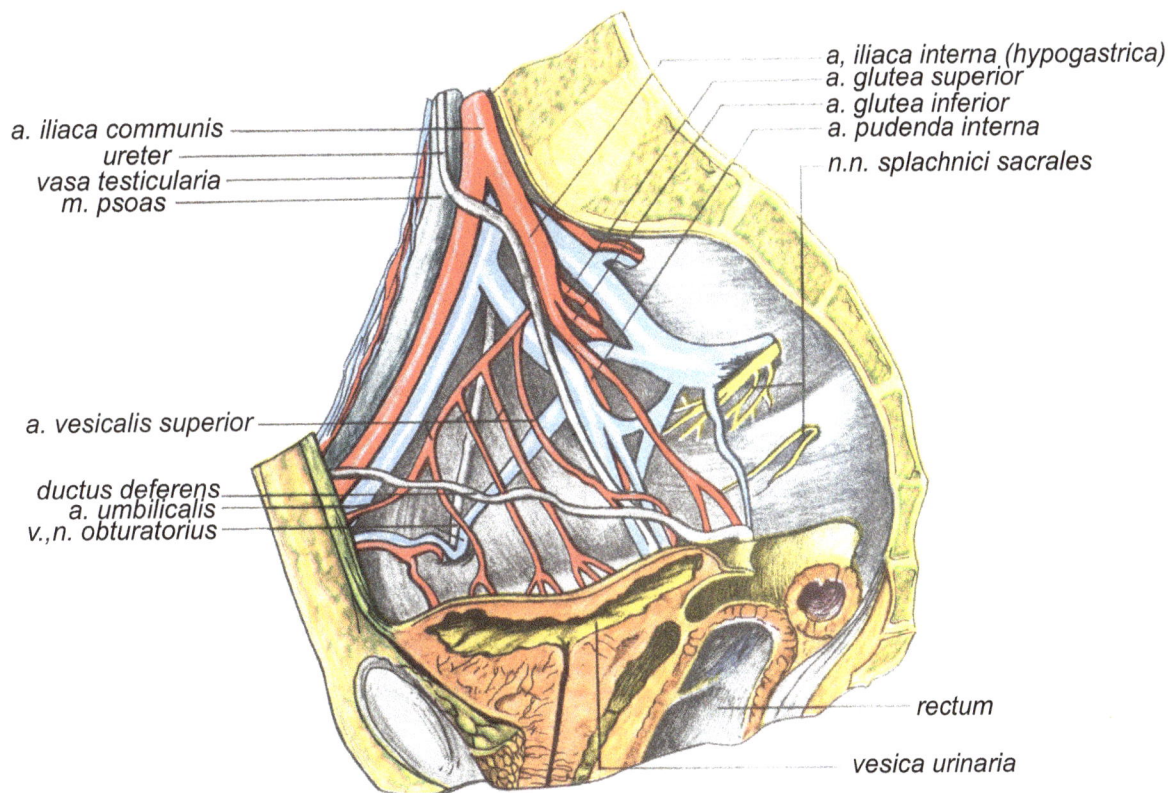

Fig. 243. Lateral wall of the pelvis in the male.

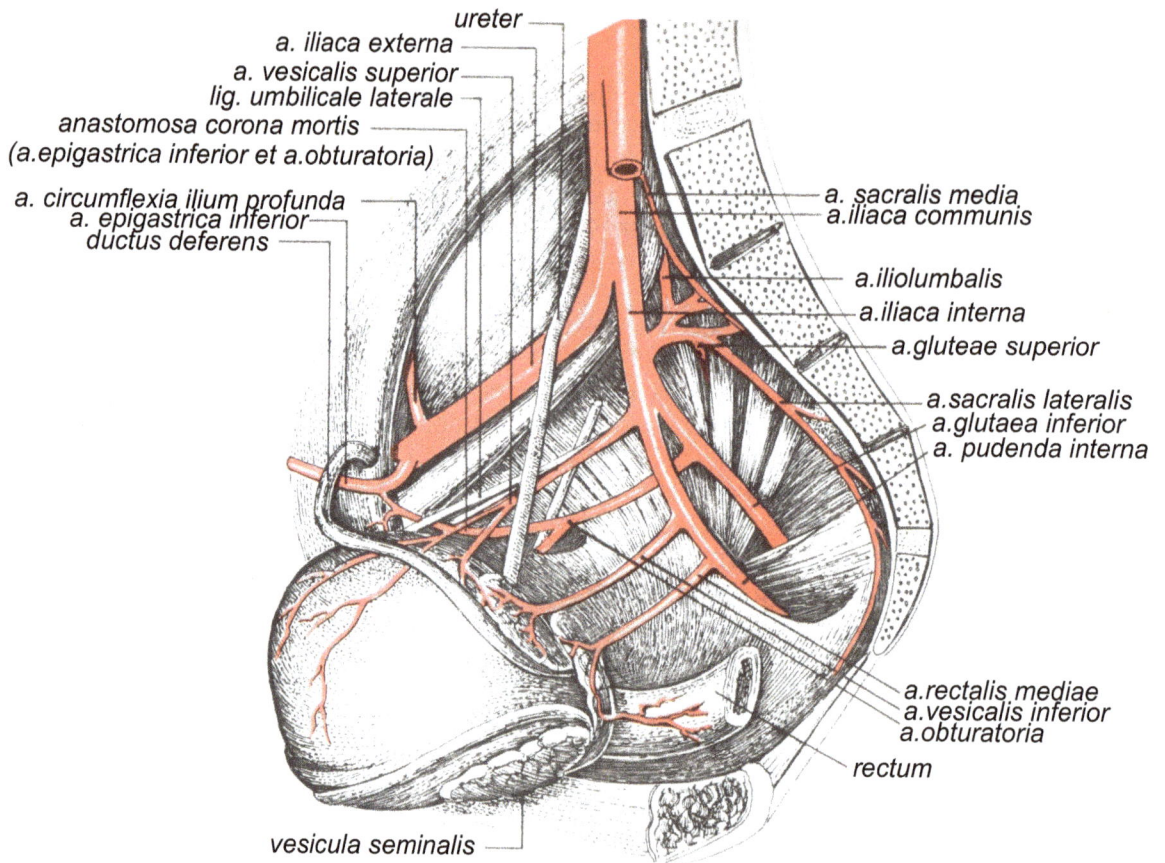

ureter
a. iliaca externa
a. vesicalis superior
lig. umbilicale laterale
anastomosa corona mortis
(a.epigastrica inferior et a.obturatoria)
a. circumflexia ilium profunda
a. epigastrica inferior
ductus deferens
a. sacralis media
a.iliaca communis
a.iliolumbalis
a.iliaca interna
a.gluteae superior
a.sacralis lateralis
a.glutaea inferior
a. pudenda interna
a.rectalis mediae
a.vesicalis inferior
a.obturatoria
rectum
vesicula seminalis

Fig. 244. The internal iliac artery in the male.

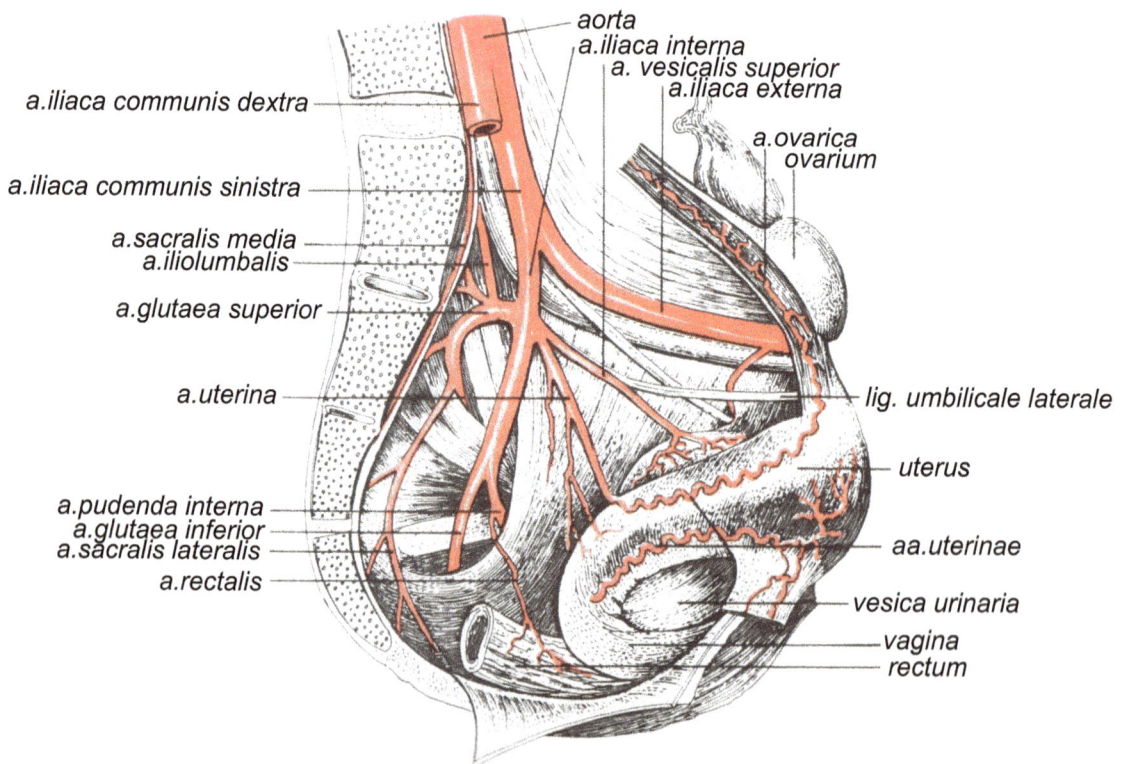

aorta
a.iliaca interna
a. vesicalis superior
a.iliaca externa
a.iliaca communis dextra
a.ovarica
ovarium
a.iliaca communis sinistra
a.sacralis media
a.iliolumbalis
a.glutaea superior
a.uterina
lig. umbilicale laterale
uterus
a.pudenda interna
a.glutaea inferior
a.sacralis lateralis
aa.uterinae
a.rectalis
vesica urinaria
vagina
rectum

Fig. 245. The internal iliac artery in the female.

The Female Internal Genital Organs

The female internal genital organs situated in the pelvis pelvis, but it is may be considered as making up, from the topographical stand- point, the region of the female internal genital organs.

This region is bounded anteriorly by the vesicourethral, the uterine region, region, posteriorly by the rectal region, laterally by the pelvic walls and extends interiorly up to the vulva. The upper boundary corresponds approximately to the plane of the aditus pelvis but is the subject to great variations, both under physiological (pregnancy) and pathological conditions. This region may be divided into the vaginal region, which corresponds to the vagina the utherine region which corresponds to the uterus, and the tubo-ovarian region, which corresponds to the uterine tubes and the ovary (fig. 246, 247).

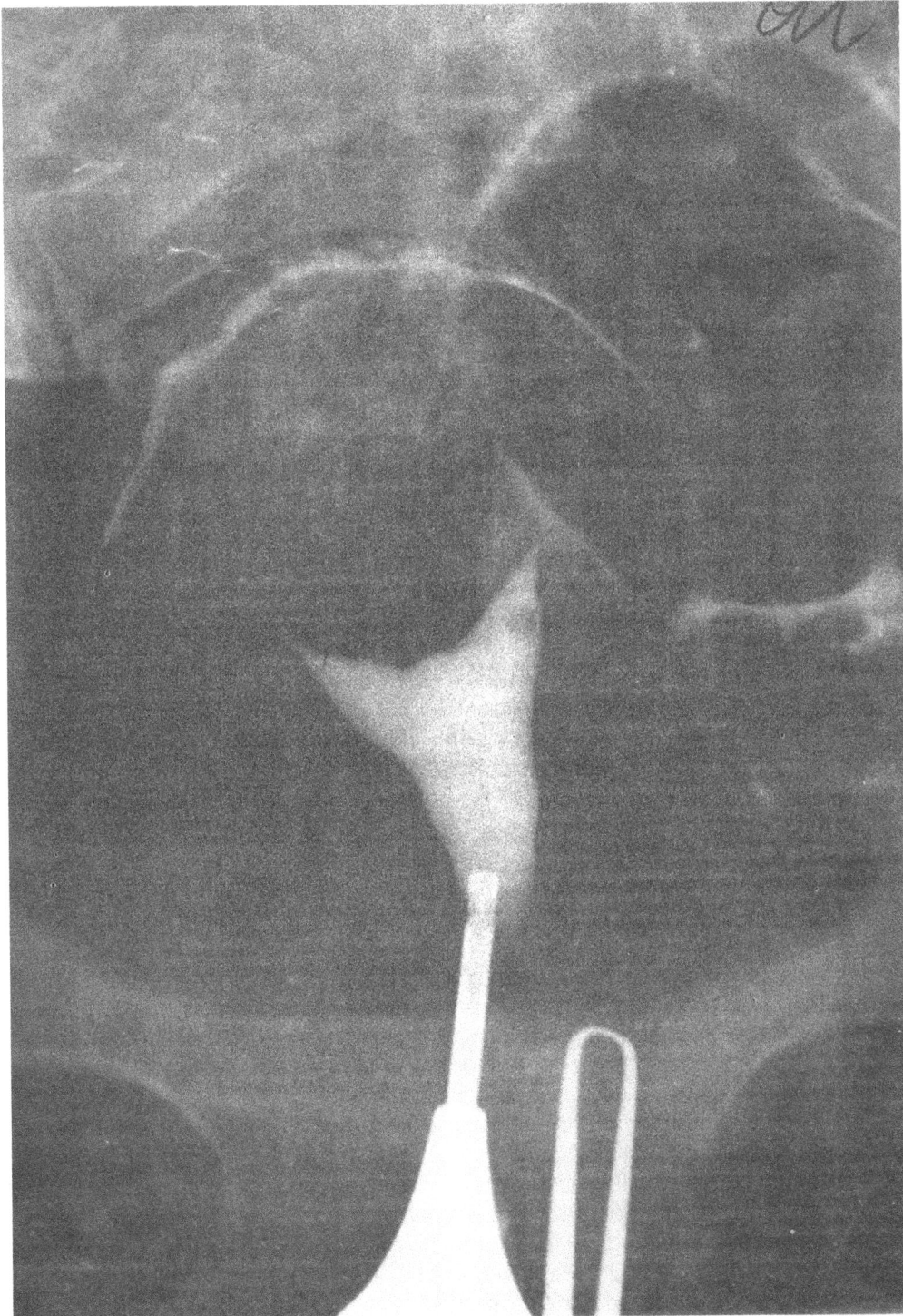

Fig. 246. Histerosalpingography.

The Vaginal Region

The Vagina
(Kolpos)

For didactic considerations, the whole pelviperineal vagina will be described here.

The vagina is a cylindrical musculomembranous extensible and elastic canal and has the role of copulation, as well as a role in the storage of spermatozoa near the uterine cervix and in the passage of the foetus and the afterbirth in labour.

It is cylindrical in the upper third, flattened anteroposteriorly in the middle third and flattened transversally in the lower third. The anterior wall of the vaginal cavity is in contact with the posterior wall

Its size varies, the average length being 8-9 cm. The posterior wall is longer than the anterior, because its attachment is higher situated than the uterine cervix.

The vagina, like the urethra, has an oblique supero-inferior and postero-anterior direction, forming with the uterus the uterovaginal angle, open towards the pubic symphysis. With the horizontal line, the vagina forms an angle of about 60°. In its course, it crosses the urogenital hiatus of the pelvic diaphragm, which it divides into a pelvic portion, situated above the diaphragm, in the vaginal bed, and a perineal portion, adherent to the deep transverse muscle of the perineum and to the middle perineal fascia. It is lodged in the pelvic portion within a space called vaginal bed and is bounded anteriorly by the urinary bladder and the urethra, posteriorly by the rectum, superiorly by the uterus and inferiorly if is closed by the adherence of the vagina to the perineum. Between the vagina and the walls of the vaginal bed lies the paracolpium, made up of connective tissue. It has a rather great mobility, owing to the laxity of the surrounding tissues, and follows the uterine cervix in its movements, except the lower extremity, which is fixed in the perineum. Its suspension is secured by the cervix uteri, the pelvisubperitoneal cell tissue condensed around the vessels (sacrorecto-genitopubic laminae), the connections with the urethra, the rectum and the urinary bladder. The support is assured by the tendinous centre of the perineum (the levatores ani muscles, although not. attached on the vagina, grip like in a strap the posterior wall of the latter).

Cylindrical in shape, the vagina has an upper end, a lower end, an outer surface and an inner surface (fig. 247-249, 251).

The upper end inserts into the cervix uteri somewhat at the junction of the middle third with the lower third. In this way, a kind of circular groove is formed, which surrounds the cervix and is called *vaginal fornix*. The vaginal fornix is subdivided info four culs-de-sac that correspond to the vaginal walls: the anterior fornix *(fornix anterior)*, the posterior fornix *(fornix posterior)* and two lateral fornices *[fornix lateralis dexter et sinister)*. As the vagina attaches much higher on the posterior surface of the cervix than on its anterior surface, the posterior fornix is deeper (20-25 mm) than the anterior, which is often reduced to a groove. The posterior fornix is covered with recto-uterine peritoneum (cul-de-sac of Douglas), which extends also on the posterior wall of the vagina, permitting the digital exploration or the examination through puncture of the collections formed in the recto-uterine excavation. The lateral fornices correspond to the parametria and have relations with the uterine vessels, the vaginal vessels and the anterior hypogastric plexus. Likewise, the lateral fornices and the anterior fornix are related to the ureter which passes into the area between the vagina and the urinary bladder, to open into the latter.

The lower end (the perineal portion) of the vagina is sited in the depth of the fibromuscular, only slightly extensible planes of the perineum and represents the least dilatable part of the vaginal tube. It is related anteriorly to the urethra and posteriorly to the anal canal, from which it is separated by the rectovaginal triangle, the vagina being connected, at the summit of the triangle, to the rectum by the rectovaginal muscle. The relations of the vagina are the following: the anterior surface *(pars anterior)* has various relations: the superior portion is related to the urinary bladder and the vesical trigone (Lieutaud's trigone) and the inferior portion to the urethra; the vesical trigone is related to Pawlick's vaginal trigone (bounded by the bifurcation of the anterior vaginal column and by a transverse fold situated above, in the vagina); the posterior surface corresponds above to the culs-de-sac of Douglas (the space between the posterior vaginal fornix and the rectum); in the median part it is in relation with the anterior aspect of the rectum and in the lower third it bounds, with the rectum, an anatomic region triangular in shape, with the base downwards and the apex upwards, known under the name of rectovaginal triangle *(trigonum rectovaginale)*.

246

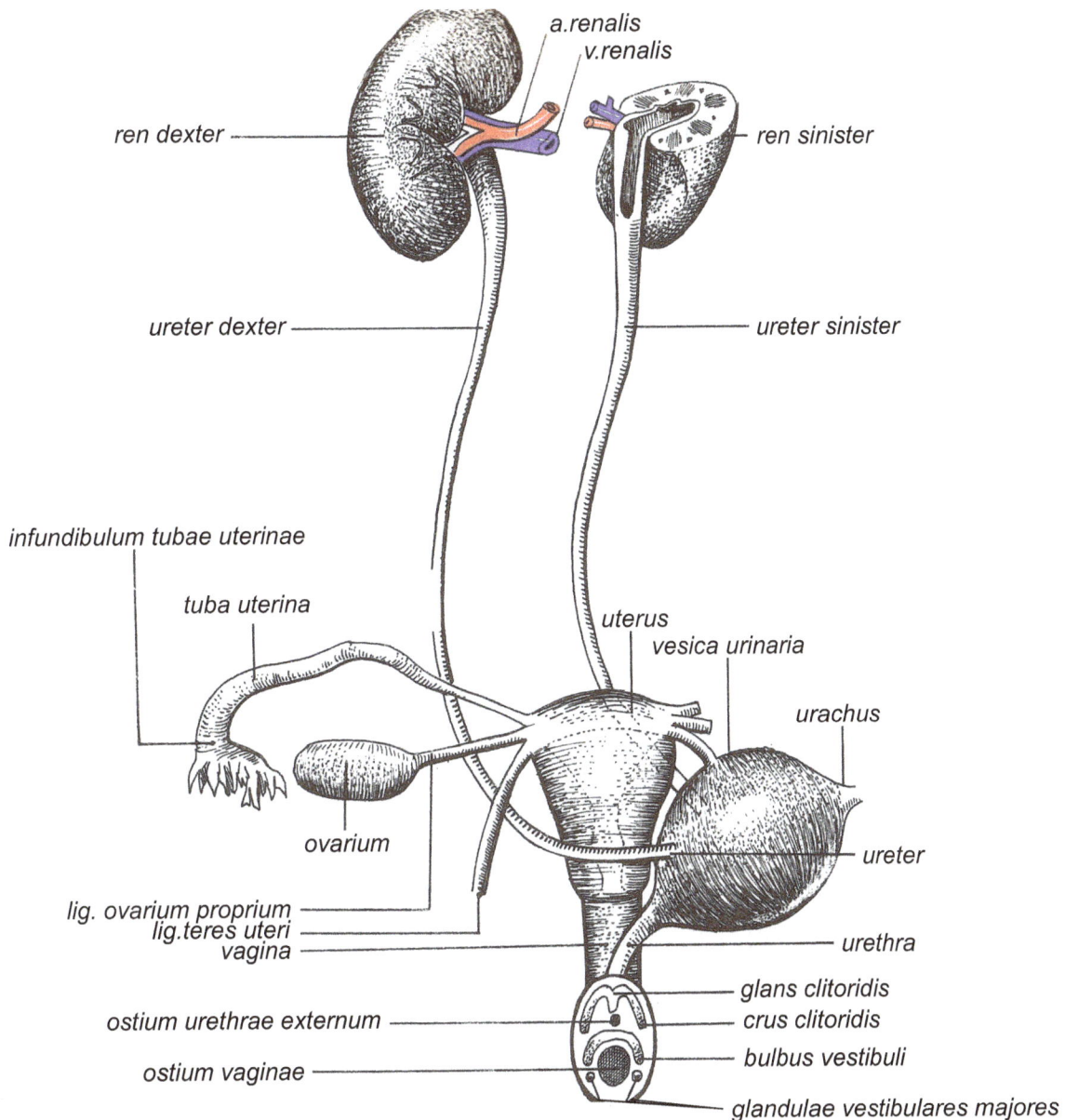

Fig. 247. The urogenital apparatus in the female (overall view).

The lateral walls are related, in the upper third, to the base of the broad ligaments and to the vessels and nerves which are situated in this space (the vesical and uterovaginal venous plexuses, the cervicovaginal arteries, the arch of the uterine artery, branches of the hypogastric plexus); in the median portion, to the perineal aponeurosis and the levatores ani muscles, important supporting elements of the organ, and in the lower portion, to the urogenital diaphragm.

The vaginal epithelium presents, all along it, a number of transverse folds and on the midline, there are two longitudinal, smooth, round ridges, one in front and other behind, called the anterior and posterior vaginal columns (columna rugarum anterior et posterior). The anterior column ends at the anterior border of the vulvar orifice of the vagina with a small thickening - the urethral tubercle of the vagina (carina urethralis vaginae), situated immediately below the urinary meatus. In the superior portion, the anterior column bifurcates, bounding, together with a transverse fold sited at the external orifice of the cervix uteri, a triangular region, known under the name of Pawlick's triangle, which corresponds to Lieutaud's triangle of the bladder. The folds are more marked in girls and in virgins (fig. 251).

As regards the structure, the vagina consists of three tunics: the external tunic (the adventitia), made up of connective elastic tissue, the muscular tunic (tunica muscularis), constituted of muscle fibres and connective elastic tissue, which in the area of the lower extremity of the vagina forms a smooth sphincter, and the mucous tunic (tunica mucosa), made up of a stratified pavement epithelium, in which glands are lacking, and of corium.

247

Vessel and nerve supply

The arteries derive mostly from the vaginal artery which, in turn, is a branch of the uterine artery. In addition, the vagina is supplied by branches of the inferior vesical, middle rectal and internal pudendal arteries.

The veins arise at the level of the mucous and muscular tunics, forming plexuses *(plexus venosus vaginalis),* developed especially on the lateral parts of the vagina.

They form with the uterine plexus, which is directed towards the uterine vein, the vesical plexus, that is continuous with the vesicovaginal veins, and the rectal plexus, that reaches the rectal veins. All these plexuses empty into the internal iliac vein.

The lymph vessels of the superior part end in the infernal iliac nodes and those of the inferior part, in the sacral and superficial inguinal nodes. *The lymph vessels of the vagina* anastomose with the lymph vessels of the cervix uteri and of the vulva; those which run on the posterior part; have connections with the rectal lymphatics. *The autonomic nerve supply* is derived from the inferior hypogastric plexus and the uterovaginal plexus, and the somatic nerve supply, from the sacral plexus, through the internal pudendal nerve.

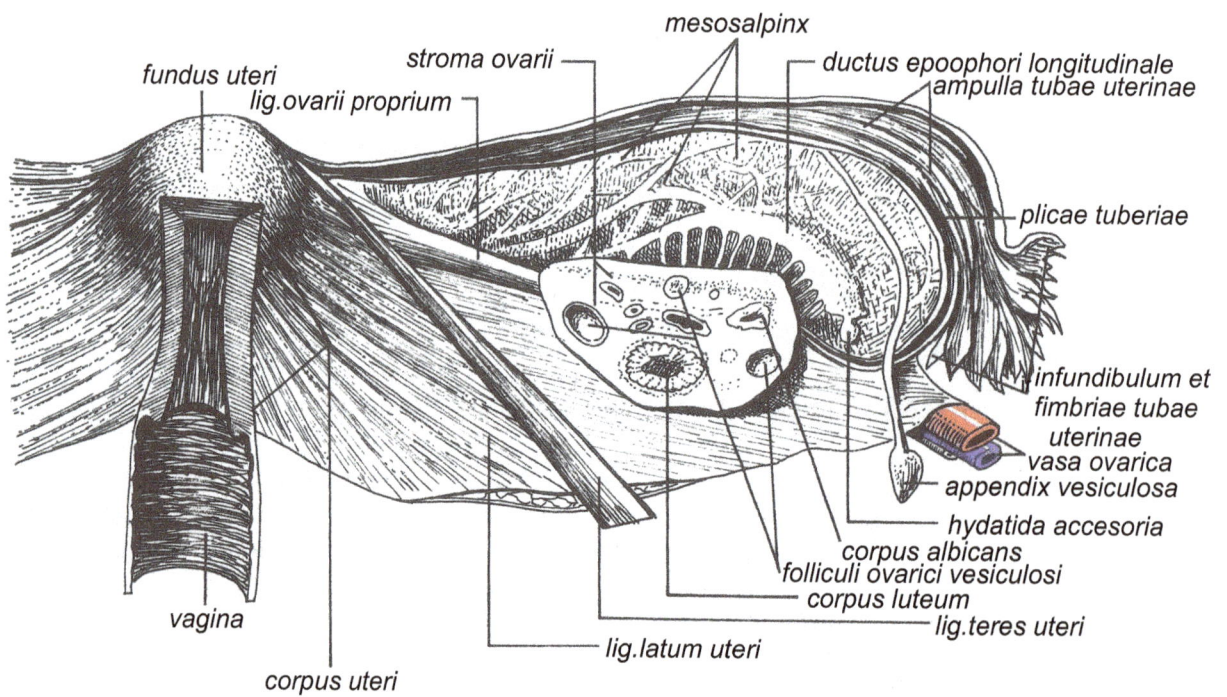

Fig. 248. The female genital apparatus.

248

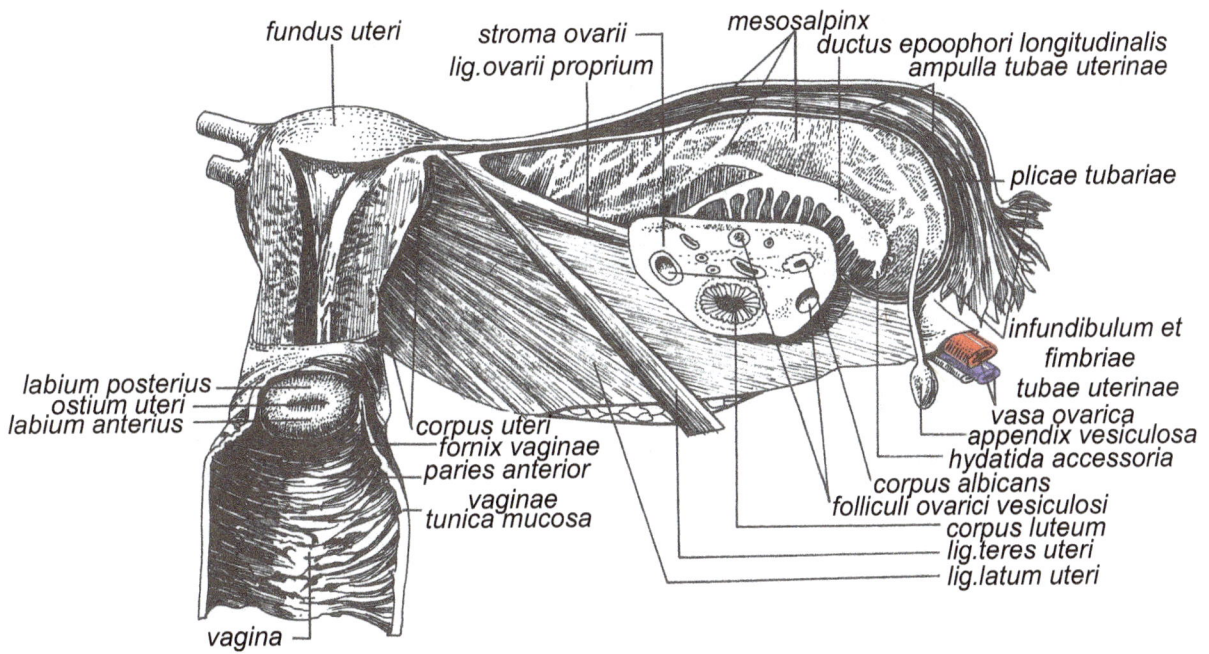

Fig. 249. The female genital apparatus (cervix and vagina).

fundus uteri

stroma ovarii
lig.ovarii proprium

mesosalpinx

ductus epoophori longitudinalis
ampulla tubae uterinae

plicae tubariae

infundibulum et
fimbriae
tubae uterinae

vasa ovarica
appendix vesiculosa
hydatida accessoria
corpus albicans
folliculi ovarici vesiculosi
corpus luteum
lig.teres uteri
lig.latum uteri

labium posterius
ostium uteri
labium anterius

corpus uteri
fornix vaginae
paries anterior
vaginae
tunica mucosa

vagina

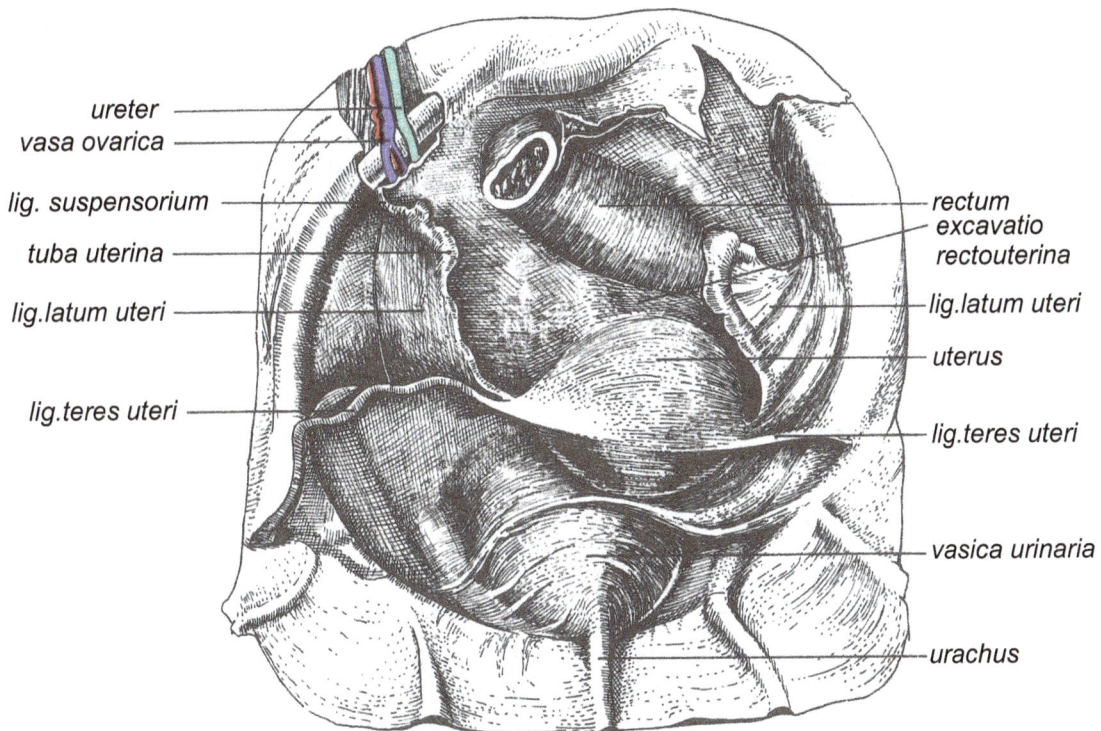

Fig. 250. The female pelvis viewed from above (uterine ligaments)

ureter
vasa ovarica
lig. suspensorium
tuba uterina
lig.latum uteri
lig.teres uteri

rectum
excavatio
rectouterina
lig.latum uteri
uterus
lig.teres uteri
vasica urinaria
urachus

249

The Uterine Region

The uterine region contains as the main organ the uterus *(uterus, hysteron)*.

The uterine region is related, in front, to the urinary bladder, behind to the rectum, laterally to the uterine tube and the ovary, below to the vagina. The uterus is situated between the bony walls of the pelvis, so that normally it is not palpable abdominally. During pregnancy it increases in size and, at the beginning of the third month, it passes beyond the boundaries of the pelvis, becoming an abdominal organ, so that it may be explored also by palpation of the anterior abdominal wall.

The Uterus
(Uterus, hysteron)

In the adult woman, the uterus is piriform, with the bulky part upwards and the apex downwards; it is embedded in the superior portion of the vagina (fig. 247).

The uterus is made up of the following portions, from above downwards: the body of the uterus *(corpus uteri)*, which is slightly flattened from before backwards and has a conoidal aspect with an anterior flat surface *(fades vesicalis)* and a posterior slightly bulging surface *(fades intestinalis)*. The right and left borders *(margo uteri dexter et sinister)* are concave in nulliparae and convex in multiparae. The superior part of the uterine body represents the fundus of the uterus *(fundus uteri)*, which presents laterally the uterine horns, that are continuous with the uterine tubes (fig. 249, 251).

The neck of the uterus *(cervix uteri)* is represented by the lower extremity of the uterus; it is cylindrical in shape, continuous with the corpus uteri, and is invaginated at the level of the upper extremity of the vagina. About 3 cm in length, it is subdivided into a supravaginal portion *(portio supravaginalis cervicis)* and an inferior vaginal portion, invaginated within the vagina *(portion vaginalis cervicis)*. The attachment line of the vagina on the cervix is circular, consequently higher situated posteriorly than anteriorly, so that the posterior vagina! cul-de-sac *(fornix vaginale posterior)* is deeper than the anterior. During pregnancy, in the second half, appears the isthmus of the uterus *(isthmus uteri)*, a semicircular groove, visible on the anterior surface, situated between the corpus and the cervix. The ostium uteri, which is the external orifice of the cervix, is different in shape in virgins, nulliparae and multiparae. It has generally the shape of a transverse depression, bounded by two lips, an anterior *(labium anterius)* and a posterior lip *(labium posterior)*, connected by two lateral commissures. In multiparae, the ostium is half-open. In nulliparae, the uterus measures 6-7 cm in length, 4-5 cm in breadth and 2-2.5 cm in thickness, whereas in multiparae these sizes increase by about 1 -2 cm and attain during pregnancy very high values, the uterus becoming an abdominal organ (the age of pregnancy may be established on the basis of the size of uterus). The position of the uterus is also of an utmost importance. Thus, the longitudinal axes of the corpus and cervix form an angle, called angle of flexion, open towards the pubic symphysis, varying between 140° and 170°, the uterus being in this way in anteflexion. The longitudinal axes of the cervix uteri and of the vagina form an angle, also open anteriorly, called angle of version, of 90-100°, the uterus being thus in anteversion, too. The uterus and the vagina present an axis that corresponds to the axis of the pelvic excavation and represents the axis along which the foetus descends during birth. The uterus being a mobile organ, the uterine body shifts against the cervix, at the level of the isthmus, either backwards, as the result of filling of the urinary bladder, or forwards, when the rectum is full, or in both directions, owing to the presence of the intestinal loops, after which it returns to its normal position. Under pathological conditions, the return to normal does no more occur and instead of anteversion and anteflexion appears the retroversion, associated to retroflexion or lateroversion.

The uterus is held in its normal position by suspension means, which fasten it to the walls of the pelvic excavation, and by supporting means, which hold it from below upwards.

Suspension means. The suspension means are represented by the peritoneum, the broad ligaments *(lig. latum uteri)*, the round ligaments *(lig. teres uteri)* and the uterosacral ligaments (fig. 250, 251).

The peritoneum, passing from the posterior surface of the urinary bladder towards the uterus, covers this organ beginning with the isthmus, then the anterior surface (the vesico-uterine excavation), the fundus and the posterior surface, up to the level of the supravaginal portion of the cervix, after which it passes on the anterior aspect of the rectum (the recto-uterine excavation) and the cul-de-sac of Douglas.

The broad ligaments (lig. latum uteri) represent the main means of uterine suspension. They are two quadrilateral peritoneal laminae which, after covering the anterior and posterior surfaces of the uterus, reach the borders, come close to each other and form at the level of the uterus a kind of small wings, directed towards the pelvic excavation.

The broad ligaments have two surfaces and four borders: the anterior surface passes over the round ligament and the lateral surface of the urinary bladder; the posterior surface is in relation with the ovary, to which it forms a mesentery *(mesovarium),* attached to the ovary at the level of Farre-Waldeyer's line, the ovary being the single intraabdominal organ devoid of peritoneal investment (it is attached to the pelvic wall by the suspensory ligament of the ovary); the superior, free border, contains the uterine tube, which is invested on its anterior, superior and posterior surfaces by the peritoneal serosa; the inferior, widened border constitutes the base of the broad ligament; this portion contains a great amount of loose and fatty connective tissue, which is continuous with the pelvisubperitoneal tissue *(parametrium);* at this level lie also the hypogastric plexus and the crossing between the uterine artery and the ureter; the medial border is situated along the uterine border, which supports the insertions of the broad ligament on the uterus; the ligament contains at this level the ascending portion of the uterine artery, venous plexuses, lymph vessels and nerves; the lateral border is thin, movable and made up of two portions:

a)   an upper portion, which corresponds to the free border of the mesosalpinx (in the upper portion, the anterior and posterior layers of the broad ligament invest the uterine tube, then join below the latter and constitute the mesosalpinx - the superior small wing of the broad ligament -, which contains the tubal vessels; the posterior surface of the mesosalpinx is continuous with the mesovarium - the posterior small wing of the broad ligament - into which run the ovarian vessels);

b)   a lower portion, which is attached on the lateral wall of the pelvis, at the level of the internal obturator muscle. The two layers of the broad ligament reflect on the anterior and posterior pelvic walls and are continuous with the parietal peritoneum.

Beneath the layers of the broad ligament may be observed three prominences, representing the so-called wings of the broad ligament; the superior wing, identical with the upper border, which contains the uterine tube *(mesosalpinx);* the anterior wing, along the round ligament; the posterior wing, which corresponds to the mesovarium.

Between the two layers of the broad ligament there is an amount of connective fatty tissue, which is continuous with that of the neighbouring regions, Inside this tissue are located the uterus and the ureter, the horizontal portion of the uterine artery, the ovarian artery, the ovarian plexuses, the lymph vessels and the tubal and utero-ovarian nerves. Some authors have mentioned the presence of smooth muscle fibres and connective fibres situated at the base of the parametria, which constitute Mackenrodt's transverse ligament.

The round ligaments (lig. teres uteri) are connective muscular bands, which have their origin in the tubal angle of the uterus, below the tube.

They are directed obliquely forwards and laterally [the anterior wing), cross the lateral surface of the urinary bladder, then the external iliac vessels, describe a curve with the concavity inferomediaily and penetrate through the deep inguinal ring into the inguinal canal, which they cross in its whole length; then they leave it through the subcutaneous orifice and end on the spine of the pubis and in the adipose connective tissue of the mons pubis *(mons Veneris).*

They are constituted of a connective elastic skeleton, muscle fibres, an arteriole, which is a branch of the inferior epigastric artery, the vein that drains into the inferior epigastric vein and into the femoral vein, lymphatics, nerve fibrils of the uterine plexus, the genitofemoral, iliohypogastric and iliac-inguinal nerves.

The uterosacral and *uterolumbar ligaments* are made up of smooth muscle fibres and connective fibres. The uterosacral ligaments begin at the level of the uterine isthmus and run partially towards the rectum (muscle fibres) and partially towards the anterior surface of the sacrum (connective fibres). They bring about the formation of falciform folds, which bulge under the peritoneum. Their inferolateral border is continuous with the sacrorecto-genitopubic aponeurosis.

The sopporting means are the sacrorecto-genitopubic ligaments (described before), the connexions with the adjacent organs and the perineum. With respect to the connexions with the adjacent organs, we mention that the supravaginal portion of the cervix uteri adheres, on the one hand, to the urinary bladder through the pelvi-subperitoneal connective tissue, situated in the vesico-uterine excavation, and on the other hand, to the rectum, through the sacrorecto-genitopubic laminae.

The peritoneum, although not directly, is the most important supporting means through the vagina.

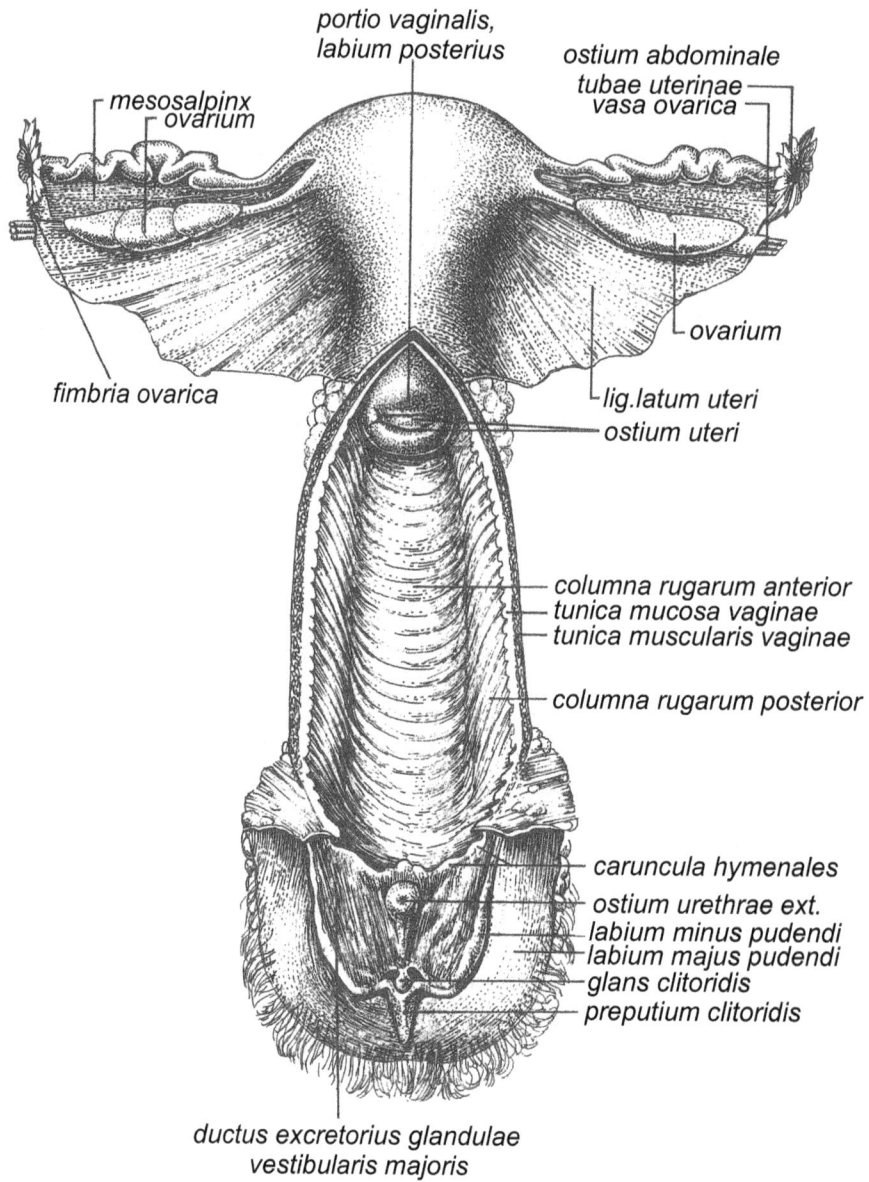

Fig. 251. Uterus, vagina and external genital organs (anterior view)

portio vaginalis,
labium posterius

ostium abdominale
tubae uterinae
vasa ovarica

mesosalpinx
ovarium

ovarium

fimbria ovarica

lig.latum uteri

ostium uteri

columna rugarum anterior
tunica mucosa vaginae
tunica muscularis vaginae

columna rugarum posterior

caruncula hymenales
ostium urethrae ext.
labium minus pudendi
labium majus pudendi
glans clitoridis
preputium clitoridis

ductus excretorius glandulae
vestibularis majoris

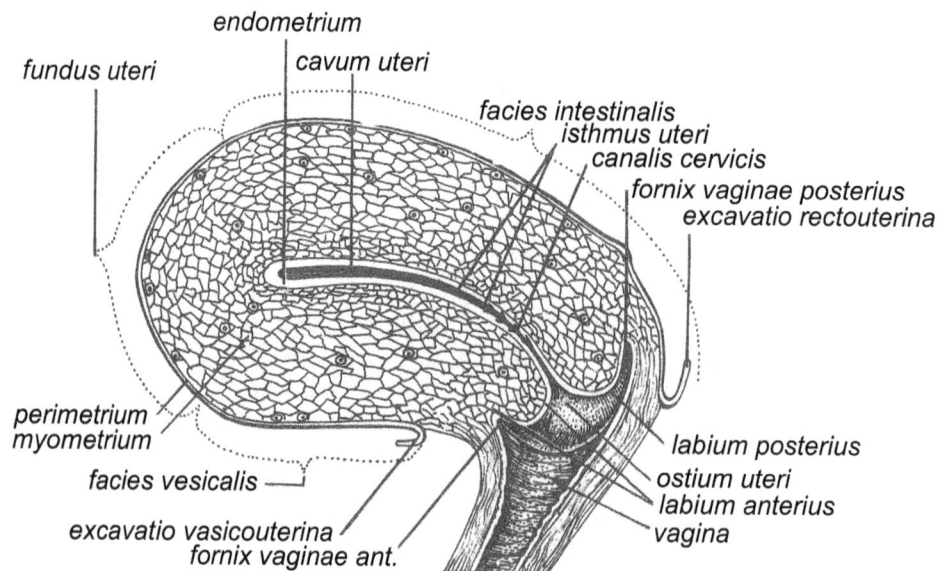

Fig. 252. The uterus - sagittal section

endometrium
cavum uteri

fundus uteri

facies intestinalis
isthmus uteri
canalis cervicis

fornix vaginae posterius
excavatio rectouterina

perimetrium
myometrium

facies vesicalis

excavatio vasicouterina
fornix vaginae ant.

labium posterius
ostium uteri
labium anterius
vagina

252

In conclusion, the uterus is maintained in position by the contribution of all these anatomical structures, of which: the vagina and the perineum are in the foreground in the fastening of the uterus and the round ligaments counteract the backward tilting; the uterosacral ligaments limit the forward inclination and the broad ligaments, together with the sacrorectopubic laminae, prevent the laterodeviation. The occurrence of pathological lesions of these structural complexes brings about position disturbances of the uterus and a prolapse of the uterus, requiring surgical treatment.

The relations of the uterus are complex.

The corpus uteri (body of the uterus), which is pear-shaped and flattened anteroposteriorly, has: two surfaces invested by peritoneum, of which an anterior surface, related to the urinary bladder and separated from it through the vesico - uterine pouch, and a posterior surface, related to the rectum and separated from it through the recto-uterine pouch, in which descend the loops of the small intestine and, sometimes, the sigmoid colon; two lateral borders, devoid of peritoneum, each corresponding to the homolateral broad ligament and related to the uterine vessels; above, the fundus of the uterus, covered with peritoneum, is in contact with the intestinal loops and with the sigmoid colon.

The cervix uteri (neck of the uterus) is cylindrical in shape and made up of two portions: supra-vaginal and intravaginal.

The supravaginal portion measures 15-20 mm in height, and is related, anteriorly to the urinary bladder, posteriorly to the cul-de-sac of Douglas and the rectum, laterally to the broad ligament, which contains, at this level, the uterine artery, venous plexuses and the terminal portion of the ureter. The uterine artery isdirected upwards and medially, describes an undulated curve and the ureter passes obliquely, directed downwards, medially and forwards, below the uterine artery and reaches the urinary bladder, into which it opens.

The relation between the ureter and the uterine artery is particularly important, since in hysterectomies, if the surgeon is not sufficiently attentive, he may injure the ureter. Therefore, the uterine artery should be carefully laid bare.

The intravaginal portion varies from 8 to 12 mm in length and from 20 to 25 mm in diameter; it has the shape of a frustum of cone, at the apex of which lies the external os of the cervix.

The inner conformation of the uterus has the following characteristics:

The uterine cavity (cavum uteri) has, on a mediofrontal section, the shape of an iscosceles triangle with a curved base. The anterior and posterior plane and smooth surfaces are applied on each other. At the level of the angles of this triangle are situated three openings: two upper openings - right and left -through which the uterine cavity communicates with the uterine tubes, and a lower opening, through which it communicates with the cervix uteri.

The cervical canal (canalis cervicis uteri) is fusiform in nulliparae and communicates, in the upper part, with the cavity of uterus through the internal os and in the lower part, with the vagina through the external os. On each of its two surfaces - anterior and posterior - a longitudinal ridge may be seen, from which start, to the right and the left, a number of longitudinal folds, that constitute, on the whole, the so-called arbor vitae uteri *(plicae palmatae)*. In multiparae, the cavity of the cervix has the shape of a frustum of cone. Structure of the uterus . The uterus consists of three tunics, which constitute its walls (fig. 252):

- the serous tunic *(tunica serosa)* is made up of the peritoneum which invests the uterus, lined in the depth by a layer of connective tissue that forms the subserous layer *(tela subserosa);*

the muscular tunic *(tunica muscularis)* is characterized by the functional orientation of the smooth muscle fibres, which have an arciform arrangement around the isthmus, the fundus and the uterine horns and are continuous with the muscle fibres of the broad ligaments. This muscular tunic or myometrium is constituted of smooth muscle fibres, disposed in three layers: an external subserous layer, consisting of longitudinal and transverse fibres; a middle layer, made up of richly anastomosed fibres that contain numerous blood vessels, chiefly veins; an internal layer, consisting of longitudinal and circular fibres. They are arranged in spiral bundles; between them lies an abundant connective tissue, which is imbibed with fluid during pregnancy, permitting the sliding of the muscle fibres as the pregnant uterus increases in size. The circular fibres represent 30% of the structure of the uterine body, whereas the isthmus and the cervix contain approximately 10-15% muscular elements;

- the mucous tunic *(tunica mucosa)* or endometrium is constituted of a ciliated columnar epithelium, the cilia being directed towards the vagina; it lines the whole internal surface of the uterus, inclusively that of the cervical canal, save the external surface of the intravaginal portion of the cervix, which is covered by a nonkeratinized stratified pavement epithelium. The endocervical mucous membrane, cylindrical glandular in type, is separated from the pavement mucous membrane of the exocervix by a demarcation area called cervicovaginal junction.

253

The mucous membrane of the uterine cavity, constituted of a simple cuboprismatic epithelium with ciliated cells, is provided with numerous glands (glandulae uterinae), of a simple or ramified tubular type, the fundus of which enters deeply up to the myometrium; it is strongly hormonodependent.

The mucous membrane of the cervical canal, unlike that of the uterine cavity, is glandular in type, with numerous muciparous cells and is less hormonodependent. The cervical glands are very ramified, acinous in type, with muciparous cells. The obliteration of the canals of the cervical glands leads to the formation of cysts called Naboth's ovules (ovula Nabothi).

Vessel and nerve supply (fig. 253)

The arterial supply of the uterus is assured by the uterine artery (a. uterina), a branch of the internal iliac artery, and, accessorily, by the ovarian artery (a. ovarica) and the artery of the round ligament.

The uterine artery arises from the anterior trunk of the internal iliac artery, either isolated, or through a common trunk with the umbilical artery, at the level of the ovarian fossa. In its course it presents three segments: retroligamentous, subligamentous and intraligamentous. It describes a curve with the superior concavity of 15 cm in length and ends below the ovary, by anastomosis with the ovarian artery.

In its parietal, retroligamentous segment, the uterine artery runs obliquely downwards and forwards, on the pelvic wall, and has relations: anteriorly, with the umbilical and obturator arteries; laterally, with the pelvic wall, from which it detaches at the level of the ischial spine; medially, with the satellite ureter, adherent to the peritoneum.

It its transverse subligamentous segment, the artery is directed medially, involving a connective condensation, emanation of the coat of the internal iliac artery, from which it penetrates the broad ligament. It passes transversally through the base of the broad ligament into the intraligamentous portion, runs towards the cervix uteri and then ascends the lateral border of the uterine body, divides into two terminal branches and anastomoses with the ovarian artery.

The uterine artery and the ureter have important relations: they cross under the form of an elongated X at the level of the supravaginal portion of the cervix. After crossing the ureter, the uterine artery describes its arch with the superolateral concavity and enters the broad ligament.

If gives off numerous collateral branches, of which we mention: peritoneal, ureteral and vesical branches and the vaginal artery (a. vaginalis), which is distributed to the cervix and the superior portion of the vagina.

In addition, if gives off branches with a tortuous course to the cervix and corpus. We mention that the uterine branches divide into anterior and posterior rami, which run on the respective surfaces of the uterus and diminish gradually their calibre as they approach the midline, so that this median area of the uterine body is paucivascular, permitting surgical incisions under more favourable conditions.

The terminal branches are the ovarian branch, which anastomoses with the corresponding ramus of the ovarian artery, forming the paraovarian arch, and the tubal branch, which enters the mesosalpinx and anastomoses with the tubal branch of the ovarian artery, constituting the subtubal arch.

To the arterial supply of the uterus contribute also a branch of the ovarian artery and the artery of the round ligament.

The veins collect first in the uterine sinuses (cavities with an endothelial wall, existent in the muscular funic), then they drain into the uterine venous plexuses (plexus venosus uterinus), situated between the layers of the broad ligament, where they accompany the uterine arteries; from here, the venous blood runs either towards the uterine veins (venae uterinae), which empty into the internal iliac vein, or towards the veins of the uterine tube and the ovary, with the ovarian vein which will empty, on the right, into the inferior vena cava and, on the left, info the renal vein.

The veins of the round ligament, of slight importance, open into the inferior epigastric vein.

The lymph of the uterus drains info various groups of lymph nodes.

Thus, the lymph of the uterine fundus joins the lymph of the ovary and flows through the suspensory ligament of the ovary (Jumbo-ovarian ligament), following the course of the ovarian vessels, to the lumbar lymph nodes. From the tubal angle, the lymph vessels follow the route of the round ligament through the inguinal canal, to the superficial inguinal lymph nodes.

The lymph of the uterine body and of the upper portion of the cervix flows through the vessels which accompany the uterine artery into the parauterine nodes and then info the internal iliac nodes, arranged around the internal iliac vessels.

254

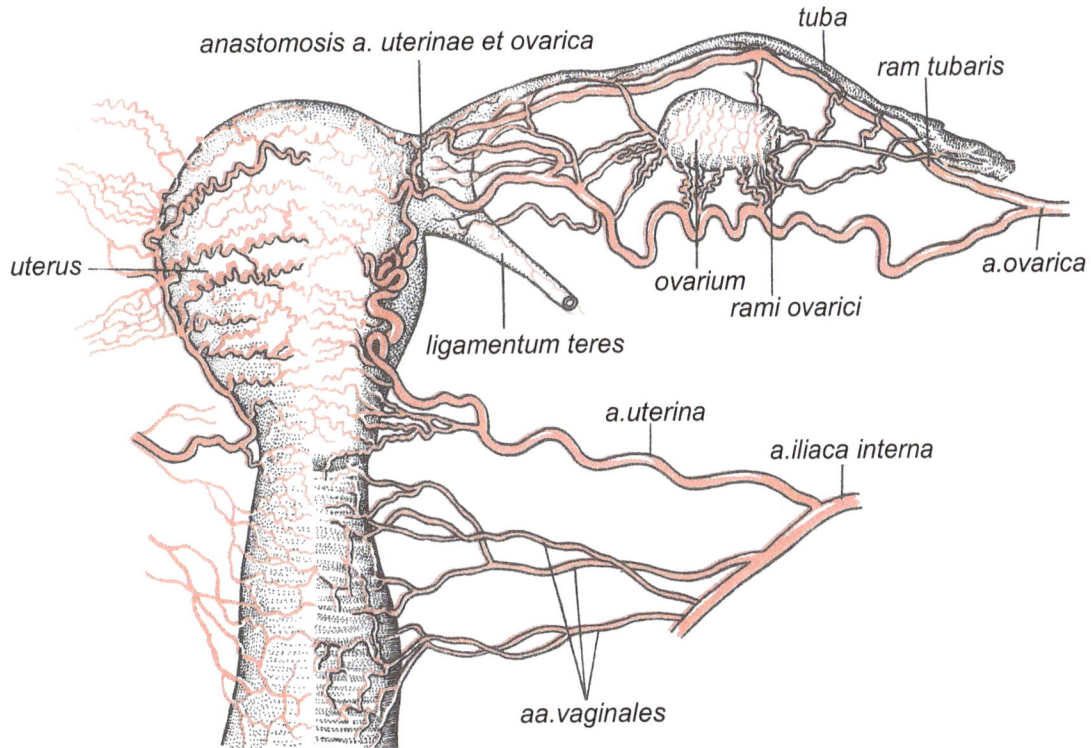

Fig. 253. Arterial supply of the ovary and uterus

The lymph of the lower cervical segment and of the upper part of the vagina drains into the infernal, external and common iliac and sacral lymph nodes (fig. 258).

The autonomic sympathetic and parysympathetic nerve supoly is derived from the caudal aortic plexus and from the third and fourth pair of sacral nerves. The caudal aortic plexus forms the uterovaginal plexus, which is reached also by parasympathetic fibres of the pelvic nerve. On the sides of the cervix are intercalated numerous ganglionic cells: Frankenhauser's cervical ganglion. From this ganglion, myelinic and amyelinic fibres run to the vagina, the uterus and the urinary bladder. The adnexa receive the sympathetic fibres from the ovarian plexus. The fibres of the pelvic nerve have "heir course in the recto-uterine fold.

The Tubo-Ovarian Region

The Uterine Tubes
(Tubae uterinae)

The uterine tubes or oviducts are two musculomembranous channels, 10-12 cm long, situated in the upper part of the broad ligament. They begin at the level of the uterine horns and are directed laterally, up to the median part of the ovary, where they bend and describe a loop with the concavity orientated medially, which encircles the ovary.

Each tube consists of the following portions:

The infundibulum of the uterine tube (infundibulum tubae uterinae), funnel-shaped with an enlarged base, is made up of a bunch of 10-15 fringes or fimbria *(fimbria tubae).* One fringe is longer, called ovarian fimbria *(fimbria ovarica),* and has a course parallel to the tubo-ovarian ligament, to which if adheres. In the centre of the infundibulum lies the abdominal opening of the tube *(ostium abdominale)* through which the peritoneal cavity communicates with the outside.

The ampullar portion (ampulla tubae uterinae), which constitutes two-thirds of the length of the tube and narrows close to the isthmus, has a tortuous course around the ovary (fig. 248, 249, 251, 255).

The isthmus (isthmus tubae uterinae), which is narrower, penetrates the uterine body between the round ligament and the proper ovarian ligament.

255

The uterine portion (pars uterina seu interstitialis) passes through the wall of the uterus and opens info it through the uterine orifice *(ostium uterinum),* in the superior angle of the uterine cavity.

The tube is firmer attached in the medial than in the lateral portion. To the attachment of the tube contribute also the suspensory ligament of the ovary and the tubo-ovarian ligament.

The tube is continuous with a narrow canal, of 1-2 mm, which enlarges at the level of the ampulla and presents longitudinal folds *(plicae tubariae),* from the uterine ostium to the abdominal ostium, favouring the progression of spermatozoa.

The longitudinal folds are very numerous and ramified at the level of the ampulla (labyrinth of the tube).

According to its course, there is described a transverse portion of the tube, from the uterus to the lower extremity of the ovary, made up of the isthmus and located in the superior border of the broad ligament, and a portion constituted of the ampulla of the tube, which surrounds the ovary and forms, with the infundibular portion, the ovarian bursa, a pouch within which the ovary is lodged.

In the thickness of the mesosalpinx, which is the peritoneum that connects the tube to the uterus and to the fundus of the pelvis, there are sometimes embryonic remnants of the primordial kidney *(epoophoron).*

The uterine tube, situated in the free border of the broad ligaments, has a rather variable position and relations, according to the mobility of the parametrium and uterus. The isthmic portion is usually transverse. In the neighbourhood of the ovary, where the ampulla begins, the tube ascends parallel to the axis of the ovaryup to the extremity of the latter, where it bends and is directed medially and downwards, on the medial aspect of the ovary, which if covers with its infundibular portion; the fimbria and the mesosalpinx form a coat which separates the ovary from the peritoneal cavity *(ansa tubae).*

This is actually a reception device of the ovary which, when if reaches the surface of the ovary, is taken over by the fimbrio-snfundibular anatomical complex and directed within the tube, after which if is sent towards the uterine cavity.

Structure of *the uterine* tube. The tube is made up of an external serous layer *(tunica serosa)* which belongs to the mesosalpinx, dependent on the peritoneum, beneath which lies a loose connective layer, through which run the vessels and nerves *(tela subserosa),* of a muscular layer *(tunica muscularis)* and, finally, of the tubal mucosa *(tunica mucosa).*

The muscular layer consists of smooth muscle fibres, arranged in three strata: a deep, thicker, longitudinal stratum, a median, circular one and a superficial one, made up of longitudinal fibres.

The three strata intermingle, forming a unitary system, in which interpenetrate also the smooth fibres that come from the broad ligament The musculature is more scantily represented at the level of the ampulla in comparison with the isthmus.

The mucosa of the tube is constituted of a monostratal columnar epithelium, with vibratile cilia and mucous secretory cells disposed on a corium.

The movements of the vibratile cilia, which are more numerous in the lateral half of the tube, promote the migration of the fecundated ovum towards the uterine cavity. The secretion of the mucous cells serves also as a mucous coat of the ovum and may have a nutritional role for the spermatozoa and the zygote.

The Ovary
[Ovarium]

The ovary, the female sex gland, is a paired organ - a right and a left ovary -, with a double secretory function: external and infernal. Thus, it is the ovule-producing organ *(gonad)* and, at the same time, an endocrine gland which, through the hormones produced, induces the secondary sex characters and plays an important part in the making up of the female constitutional type (fig. 248,249,251). The ovary is ovoid in shape, with the great axis vertical, and has two surfaces (medial and lateral), two borders (free and mesovarian) and two extremities (tubal and uterine). If is usually 3 cm in length, 2 cm in breadth and 1 cm thick and weighs 6-8 g; during the climacterium, its size diminishes. The aspect and consistency are variable, the consistency being usually elastic and slightly fluctuant; after menstruation the ovary becomes hard, fibrous and of a white-grey colour.

The external surface is smooth in the child and becomes more and more irregular with advancing age of the woman and increasing number of births.

After its development in the lumbar region, the ovary migrates towards the inferior region of the abdomen, so that at birth it is located in the lesser pelvis, in the retrouterine eavity, behind the brood ligament, below the tube and in front of the rectum. The ovary is the single organ of the abdominal cavity which is not covered by peritoneum, the latter ending at the hilum of ovary (margo mesovaricus), at the level of the Waldeyer-Farres peritoneo-ovarian line.

The ovary is thus devoid of a proper visceral serous membrane and is covered only by the germinal epithelium. Therefore, when describing the ovary, both the term of intra-abdominal organ (situated in the peritoneal cavity) and that of extraperitoneal organ (destitute of the visceral serous coat) are used.

In the subperitoneal space of this area run the gluteal vessels and nerves, which may be irritated in the course of the ovarian inflammatory process, inducing pain with gluteal irradiations.

Among the above mentioned relations, important from the surgical point of view are especially those of the ovary with the ureter, which may be injured during operations on the ovary (fig. 254).

The lateral (parietal) surface of the ovary corresponds, in nulliparous women, to Krause's groove or fossa, bounded behind by the internal iliac vessels and the ureter, in front by the pelvic attachment of the broad ligament, above by the external' iliac vessels and below, by the common origin of the umbilical and of the uterine artery. On the bottom of the fossa lies the obturator vasculonervous bundle. This relation explains the pain at the medial surface of the thigh, which women suffering from inflammations of the ovary (oophoritis) sometimes feel. In multiparous women, the ovary corresponds to Claudius' fossa, enclosed between the border of the sacrum behind and the ureter and the uterine artery in front; its area is crossed by the inferior gluteal vasculonervous bundle.

The medial surface is covered by the uterine tube and the mesosalpinx and has relations with the loops of the ileum and the sigmoid colon to the left, the caecum and the vermiform appendix to the right, and even with the rectal ampulla when it is full.

The salpingo-ovarian relation explains why a salpingitis is nearly always accompanied by a certain degree of oophoritis (salpingo - oophoritis).

The appendiculo-ovarian relation shows also the frequency of the right appendico-adnexitis and the relation with the rectum explains why the ovary may be often palpated by rectal examination, especially in cases in which the prolapse of the ovary within the rectocolpo-uterine recess occurs

The ovary is fastened by the broad ligament, the proper ligament of the ovary, the vasculonervous pedicle, the suspensory ligament of the ovary, the tubo-ovarian ligament and the mesovarium.

The suspensory ligament of the ovary (ligamentum suspenserium ovarii) is a fibromuscular formation, parallel to the superior vasculo-nervous pedicle of the ovary. Starting from the mesoappendix on the right, below the sigmoid mesocolon on the left, it descends at the level of the external iliac vessels, enters the superolateral angle of the broad ligament and reaches the tubal extremity of the ovary, the mesovarium.

The proper ligament of the ovary (ligamentum ovarii proprium seu chorda utero-ovarica) is sited in the posterior wing of the broad ligament, between the uterine extremity of the ovary and the angle of the uterus.

The tubo-ovarian ligament unites the tubal extremity of the ovary with the infundibulum of the tube.

The mesovarium is a component of the posterior leaflet of tne broad ligament, which connects the ovary to it and through which the vessels and nerves reach the ovary.

The mesovarian (anterior) border (margo mesovaricus), connected through the mesovarium to the broad ligament, is the place where the hilum of the ovary (hilus ovarii), is sited, through which the vasculonervous elements penetrate, anteriorly being the ascending portion of the tube.

The free (posterior) border (margo liber) is related to the loops of the small intestine.

The tubal (superior) extremity (extremitas tubaria) is the portion to which the suspensory ligament of the ovary and the tubovarian ligament are attached.

The uterine (inferior}extremity is the portion to which the proper ligament of the ovary is attached. It is the pole that is the easiest palpated in the multiparous women by vaginal or rectal examination, as it is very close to the pelvic floor (sometimes even in contact with it).

The histologic structure of the ovaries. The ovary is covered with a simple, cubical or pavement epithelium, the germinal epithelium, corresponding to the peritoneal epithelium; beneath the germinal epithelium lies a tunica albuginea, made up of two areas: a centra! one, the medullar area, abundantly supplied with blood vessels, and a peripheral one, the cortical area, rich in cells, dense and scattered with ovarian follicles.

257

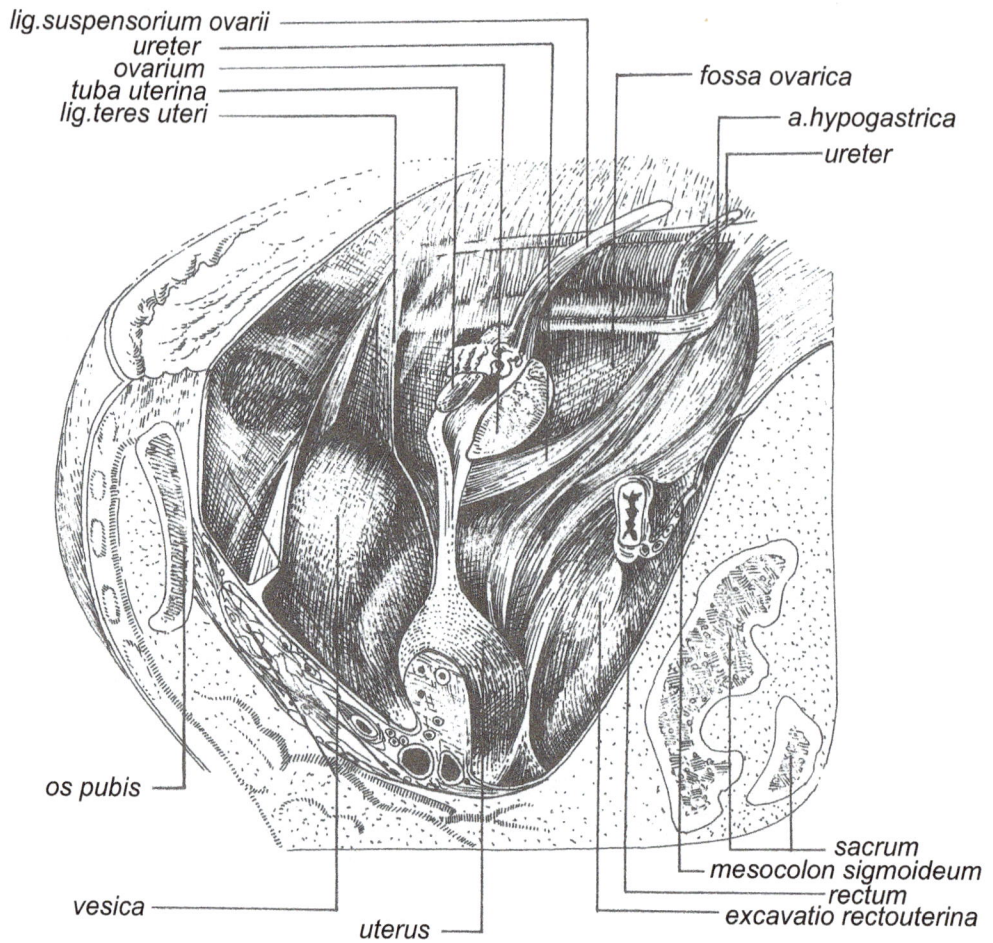

lig.suspensorium ovarii
ureter
ovarium
tuba uterina
lig.teres uteri

fossa ovarica

a.hypogastrica

ureter

os pubis

sacrum
mesocolon sigmoideum
rectum
excavatio rectouterina

vesica

uterus

Fig. 254. The ovary - relations (pelvis sagittal section).

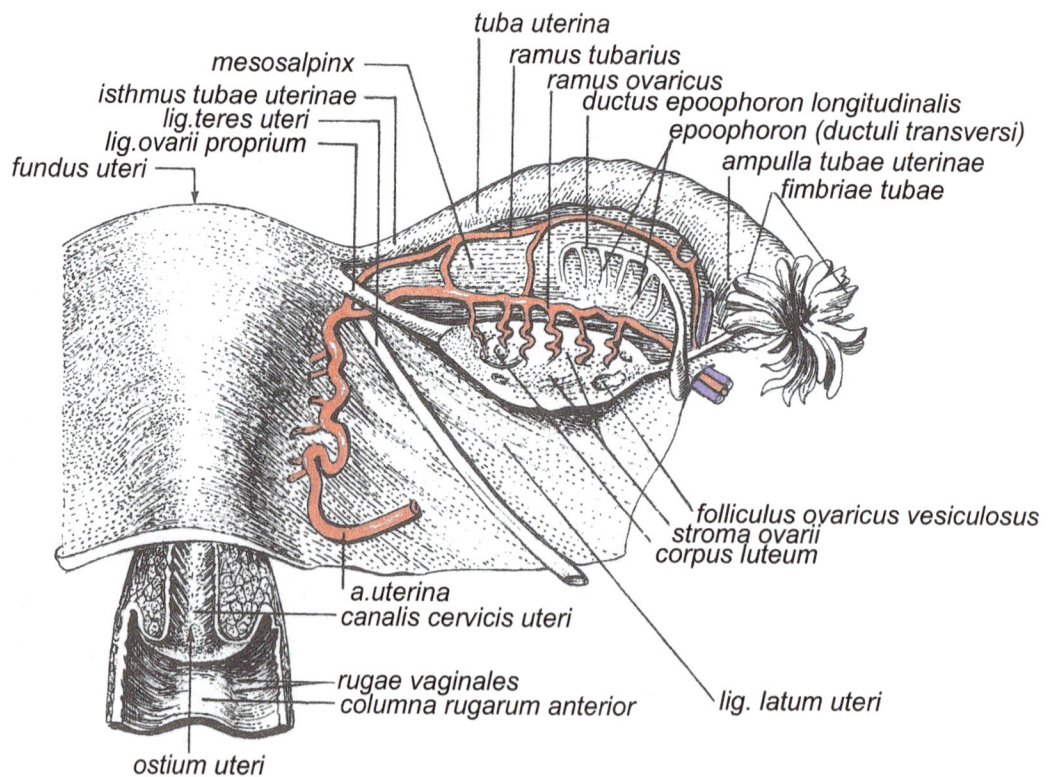

tuba uterina
ramus tubarius
ramus ovaricus

mesosalpinx

isthmus tubae uterinae
lig.teres uteri
lig.ovarii proprium

fundus uteri

ductus epoophoron longitudinalis
epoophoron (ductuli transversi)
ampulla tubae uterinae
fimbriae tubae

folliculus ovaricus vesiculosus
stroma ovarii
corpus luteum

a.uterina
canalis cervicis uteri

lig. latum uteri

rugae vaginales
columna rugarum anterior

ostium uteri

258

Fig. 255. Blood supply of the ovary

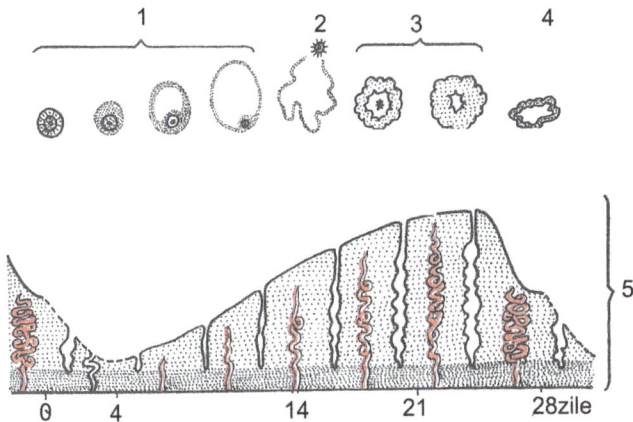

Fig. 256. Structure of the uterine mucosa. Changes of the uterine mucosa under the conditions of occurrence of fertilization 1-maturation of the ovarian follicle; 2-ovulation;3-the formed yellow body (corpus luteum);4-the yellow body of pregnancy;5-the implanted blastocyst. Phases of the menstrual cycle (duration 28 days): days 0 to 4-menstrual phase (shedding of the uterine mucosa); days 4 to 14-follliculinic phase (glandular proliferation); days 14 to 28 progesteronic phase (ovum implantation)

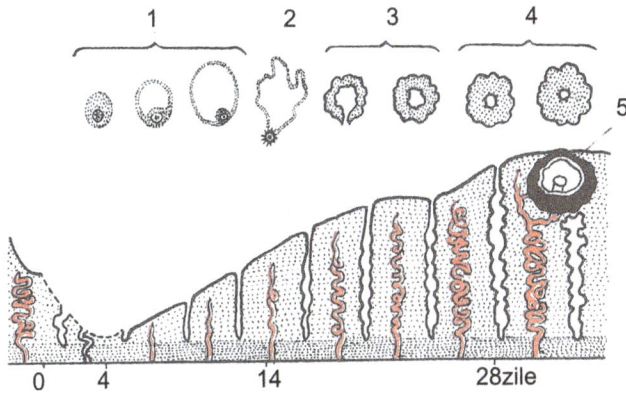

Fig. 257. Structure of the uterine mucosa. Correlation of the ovarian cycle with the uterine cycle under the conditions in which fertilization has not taken place (Longman)
1 - maturation of the ovarian follicle,
2 - ovulation (the non - fertilized ovule lives 48 hours at the most);
3 - the formed yellow body;
4 - diminishing of the yellow body;
5 - the uterine mucosa. Phases of the menstrual cycle (duration 28 days); days 0 to 4 menstrual phase (shedding of the mucosa); days 4 to 14 - folliculinic phase (glandular proliferation); days 14 to 28 - secretory progesteronic phase.

The ovarian cortex. The cortex of the ovary is characterized by cell density and by the presence of ovarian follicles in

different evolutive or involutive stages, as well as of corpora lutea, concomitantly with the existence of corpora albicantia (cicatriceal foci), all these elements being included in the ovarian stroma, which is connective-fibrous.

The ovarian follicles, resulting from the fragmentation of the cortical cords (short, but thick, arising from the second proliferation of the germinal epithelium), appear initally under the form of full cell corpuscles (primordial and primary follicles) and afterwards as cavitary structures (secondary follicles) and mature (tertiary) follicles. Of the enormous number of ovarian follicles present at birth (200,000-400,000 in the two ovaries), in the course of the active sexual life of the female (from puberty, which occurs at 12-14 years of age, to menopause, at 45-52 years), only 300-400 follicles reach the maturation stage, whereas the remaining undergo involution and cicatrize.

The primary follicles appear as spheroidal corpuscles, of 30-50 microns in diameter.

The primordial follicle is characterized by a primary oocyte (rarely ovogenic), inconstantly and incompletely surrounded by a flattened epithelium. These primary follicles hold a particular position among the follicular population of the ovary, they may be both "inactive" and "activated", representing in the latter case the initial stage of the evolutive processes. The actually evolutive structures are represented by secondary cavitary follicles and by Graafian tertiary mature follicles *(folliculi ovarici vesiculosi).*

The cavitary ovarian follicle, a histologically and cytologically well individualized structure, represents the formation stage of all the structures characteristic of the mature follicle and corresponds to an oocyte of about 100 microns in diameter. The follicle grows in size (up to 10 mm in diameter), becomes ovoid and shifts to the superficial zones of the cortex, which it elevates.

The lot of the cavitary follicles is varied: the greatest part undergo involution and only one or at most two reach the final stage of mature follicle. The mature or Graafian follicle (gametogenic or tertiary follicle) represents the stage of full development of the cavitary (secondary) follicle. The mature follicle, 20-25 mm in diameter, bulges at the surface of the ovary under the form of a large round vesicle.

259

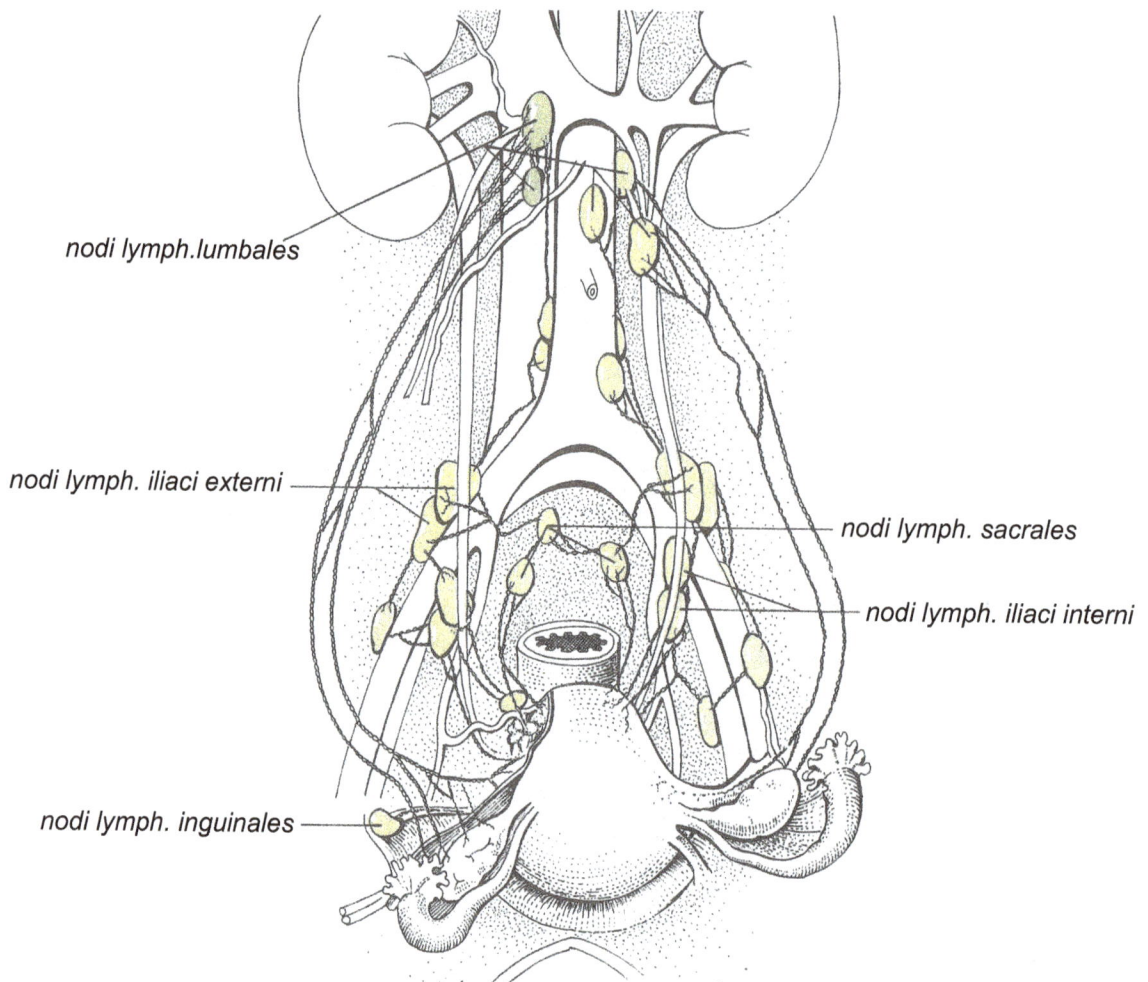

nodi lymph.lumbales

nodi lymph. iliaci externi

nodi lymph. sacrales

nodi lymph. iliaci interni

nodi lymph. inguinales

Fig. 258. Lymph vessels of the female genital apparatus.

It consists of a cavity, the antrum, which contains the follicular fluid, and of the oocyte situated eccentrically and attached by a pedicle to the membrana granulosa.

The oocyte of the mature follicle, initially of the first order, i.e. diploid, undergoes before ovulation the first maturation division and becomes a secondary oocyte, a haploid cell, which is expelled from the surface of the ovary during ovulation (fig.256, 257).

The cortex contains also the yellow body *(corpus luteus)* or the progesteronic body, which is a temporary endocrine gland that develops into a mature ovarian follicle after ovulation. The evolution of this yellow body varies according to whether the ovum has been or not fecundated. If the ovum has been fecundated, the corpus luteum develops considerably, forming the corpus luteum of pregnancy or gestation; if fecundation does not occur, the development of the corpus luteum is interrupted.

The medulla or *vascular area (zona vasculosa)* is made up of a loose connective tissue in which, besides reticulinic fibres, lie also numerous elastic structures and, towards the hilum, smooth muscle fibres, too. The medulla contains a great number of blood and lymph vessels, nerve fibres and even sympathetic cells, located at the level of the hilum. Characteristically, the medulla is a richly vascular structure, the vessels, especially the veins, are very wide, often dilated and full with blood, so that they may be quite easily confounded with haemorrhagic foci. From these vessels arise the arteries that assure the nutrition of the cortex.

The ovary is supplied by the spiral arteries (derived from the ovarian and uterine arteries) which, at the boundary between the medulla and the cortex, ramify, forming plexuses. These plexuses give off radial branches which enter the cortex, forming a capillary network of the sheaths of the cavitary and mature follicles. The veins, which arise from the capillaries of the cortex, make up an abundant plexus in the medulla and then leave the ovary at the level of the hilum.

The capillaries emerge at the external sheath of the ovarian follicles and, after penetrating the medulla in the form of larger lymph vessels, leave the organ at the level of the hilum; the lymph vessels of the ovary empty into the lumbar lymph nodes.

260

The nerve fibres, most of them non-myelinated, enter the ovary with the vessels. Many of these fibres end in the walls of the vessels and those which penetrate the cortex form slender plexuses in the follicular sheaths, but without penetrating the membrana granulosa.

Vessel and nerve supply of the ovary and the uterine tube (fig.253,255).

The main arterial vessels are the ovarian artery (a. ovarica) and the uterine artery *(a. uterina).*

The ovarian artery is a branch of the abdominal aorta, from which it descends into the retroperitoneal space, crosses anteriorly the ureter, passes laterally, reaches the pelvis, in which it enters through the suspensory ligament of the ovary, in the thickness of which it crosses anteriorly, for the second time, the ureter and runs towards the tubal extremity of the ovary. In the suspensory ligament, it divides into an ovarian and a tubal branch.

The uterine artery is a ramus of the internal iliac artery and, at the level of the uterine horn, it breaks up, like the ovarian artery, into two terminal branches; tubal and ovarian.

The homonymous branches of these two arteries anastomose with each other and form two arches: paraovarian, in the mesovarium, and subtubal, in the mesosalpinx. The two arches give off branches of a terminal character to the ovary, which penetrate through the hilum of the ovary, and to the tube. *The veins* are satellites of the arteries. They make up a network in the medulla of the ovary, reach the hilum of the ovary, anastomose with veins coming from the uterus, receive the subtubal venous network, form a pampiniform plexus and in the iliac fossa make up a single trunk - the ovarian vein *(v. ovarica),* which opens on the left into the renal vein and on the right, directly into the inferior vena cava. We mention that the veins of the tube form a subtubal network which anstomoses medially with the veins of the uterus and laterally, with those of the ovary.

The lymphatics of the ovary accompany the ovarian vessels, ascending along them towards the lumbar lymph nodes, and empty, on the right and the left, into the latero-aortic nodes around the renal pedicle. In addition, Marcille has described collecting lymph vessels which empty into the external iliac lymph nodes.

The tubal lymphatics anastomose wiht the ovarian lymph vessels and follow the course of the latter. The lymph is partially collected by the external iliac lymph nodes and by a hypogastric lymph node (Pelle) (fig. 258).

The nerve supply of the ovary is represented by fibres which emerge from the ovarian plexus, made up of fibres derived from the renal, superior mesenteric and abdominal aortic (Kuntz) plexuses. The ovarian plexus gives off nerve fibres to the ovary and to the tube and has connections with the uterovaginal plexus. Some preganglionic sympathetic fibres of the above described plexuses and the afferent fibres that traverse the ovarian plexus are components of the splanchnic nerves. The tube, besides the ovarian source of nerve supply, receives branches also from the uterine plexus and from the mesenteric nerves.

The Female Urethra

The female urethra is situated in the pelvis and extends from the base of the urinary bladder to the vulva. It is directed obliquely downwards and forwards and runs parallel to the vagina, situated behind it; the urethra describes a slight curve with the anterior concavity and is not quite rectilinear. It is about 3-5 cm long and approximately 7 mm in diameter. The narrowest and least dilatable area is that of the inferior orifice (the urinary meatus), the remainig urethra being very extensible.

The inner configuration is similar to that in the male. There are two longitudinal folds, of which the postero-median is more evident, called the urethral crest *(crista urethralis).* The inner surface of the urethra is provided with numerous orifices info which open diverticula similar to Morgagni's lacunae, which we find in the male urethra.

We distinguish two portions of the female urethra: a superior, intrapelvic portion, situated above the middle perineal aponeurosis, and an inferior, infraperineal portion.

The pelvic portion of the urethra is surrounded by the striated shpincter. It is related: in front, to the dorsal vein of the clitoris, Santorini's venous plexus, the pubovesical ligaments and the pubic symphysis; laterally, to the sacrorecto-genitopubic connective laminae and the levatores ani; behind, it is in close relations with the vagina through a dense connective tissue, the paracolpium.

The perineal urethra is situated anteriorly to the vagina and is separated from it by a tissue made up of connective and smooth muscle fibres, which form the urethra-vaginal septum. In front and laterally it is related to the middle perineal fascia, the transverse muscle and the sphincter of the urethra.

Below the middle perineal aponeurosis, the urethra is crossed by the cavernous bodies *(corpora cavernosa)* of the clitoris and the anterior part of the bulb.

The inferior orifice of the urethra *(orificium seu ostium urethrae externum)* is situated 20-25 mm behind the clitoris and immediately in front of the vaginal tubercle; the borders of the orifice present folds of an indented aspect, which often bulge, forming an elevation variable in shape and size, called the urethral papilla.

As regards the structure, we mention that the initial portion is lined by an urothelium, the middle portion by a stratified or pseudostratified columnar epithelium, whereas the epithelium of the terminal segment is of stratified columnar type.

The corium contains numerous elastic fibres, as well as a rich venous plexus. The disposition of the muscle fibres that form the middle tunic is similar to that in the male - an inner layer of longitudinal fibres and an outer layer of circular fibres which, in the initial portion of the urethra, form a smooth sphincter (distal to the external orifice lies the external sphincter, made up of striated fibres, which strengthen the circular smooth fibres).

The urethra contains also, in the thickness of the muscular wall, two paraurethral glands (Skene's glands), placed on each side of the urethral canal; their excretory ducts open at the level of the mucous membrane, on either side of the meatus. Owing to its shortness, the female urethra is often the portal of entry of germs which produce the inflammation of the urinary bladder (cystitis).

Vessel and nerve supply

The arterial supply is assured, for the pelvic portion, by the inferior vesical arteries and the vaginal arteries, branches of the internal iliac artery, as well as by the anterior vesical artery, a branch of the internal pudendal artery, and for the perineal portion, by the bulbar and bulbo-urethral arteries, branches of the internal pudendal artery. *The veins* of the urethra drain the blood, above, into Santorini's plexus and below, into the bulbar veins. *The lymphatics* convey the lymph into the external and internal iliac nodes. *The nerves* are derived from the inferior pelvic (hypogastric) plexus and from the internal pudendal nerve (fig. 259, a and b).

The Perineum

The perineal region contains in its structure a number of anatomical formations, which bound, below, the pelvic cavity. It is bounded superficially by the two genitofemoral folds (laterally), by the arcuate ligament of the pubic symphysis *(iigamentum arcuatum pubis)* (in front) and by the tip of the coccyx (behind); these elements form the boundaries of a rhombic space. In the depth, the boundaries are the following: above, the pelvic diaphragm, constituted of the levatores ani and the coccygeus muscle with the fasciae that invest them; in front, the pubic arch and the subpubic ligament *(ligamentum arcuatum pubis);* behind, the tip of the coccyx; laterally, the inferior branches of the pubis *(rami inferior ossis pubis),* the inferior branches of the ischium *(rami inferior ossis ischii),* the ischial tuberosity and the sacrotuberous ligament.

The fascia of the pelvic diaphragm *(fascia diaphragmatis pelvis)* covers the surfaces of the levatores ani. That of the inferior surface of the muscle *(fascia diaphragmatis pelvis inferior)* is very thin. It forms the medial wall of the ischiorectal fossa and is continuous upwards with the obturator fascia, along the line of origin of the levator ani muscle; downwards, it is continuous with the fascia of the sphincter urethrae.

The leaflet that covers the superior surface of the levator ani muscle follows the line of origin of the muscle and is therefore variable.

Inside, the fascia that invests the superior surface of the pelvic diaphragm *(fascia diaphragmatis pelvis superior)* blends with the fibrous capsules of the pelvic viscera; the fascia that invests that part of the internal obturator muscle which lies above the origin of the levator ani is, thus, a complex formation, comprising: a) the obturator fascia, b) the fascia of the levator ani muscle and c) the atrophied fibres of origin of the levator ani.

The space between these boundaries is approximately rhombic in shape. A transverse line, which unites the ischial tuberosities, divides the perineal region superficially into two triangular portions: the posterior portion comprises the anal orifice and is termed the anal region; the anterior portion contains the external urogenital organs and is called the urogenital region.

The perineum is made up of a number of muscles, with their fasciae, nerve and vessel supply, and of formations belonging to the urogenital and digestive systems (fig. 260-270).

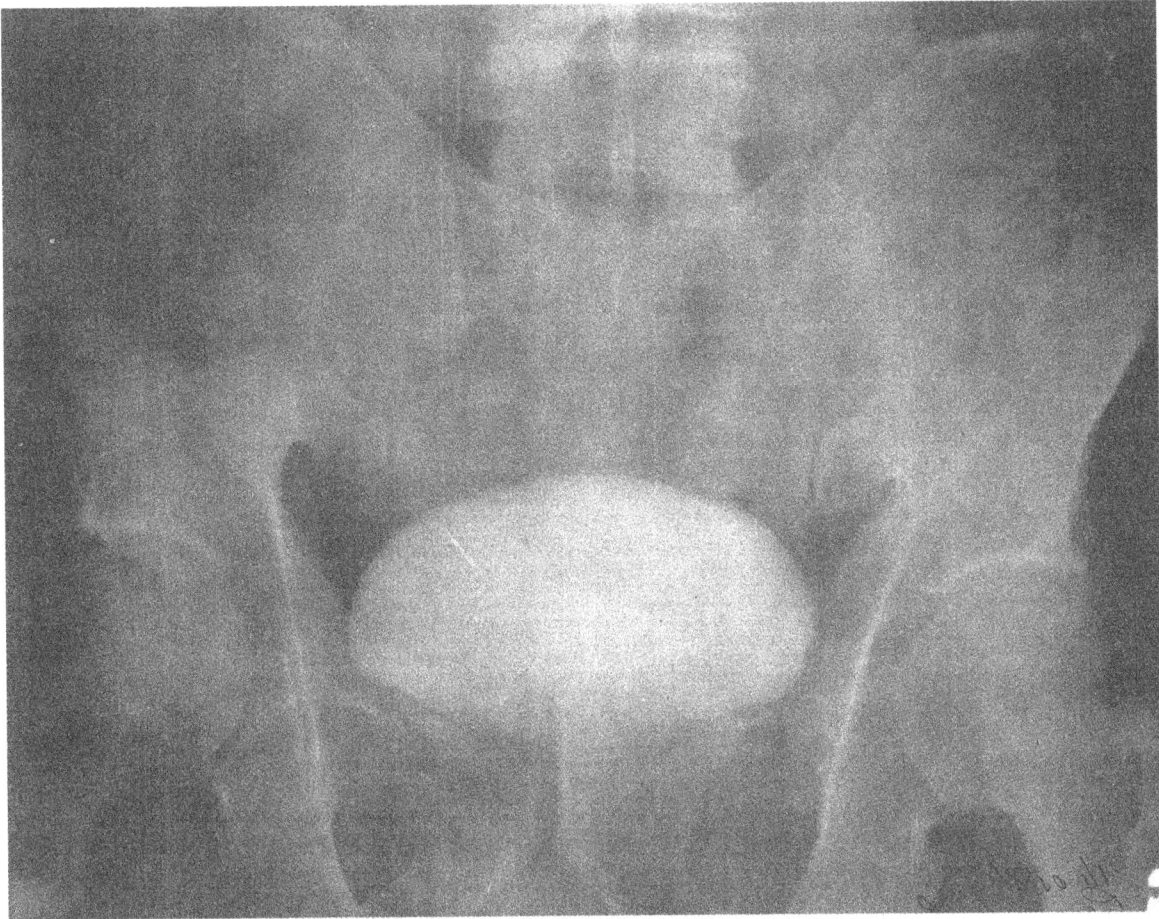

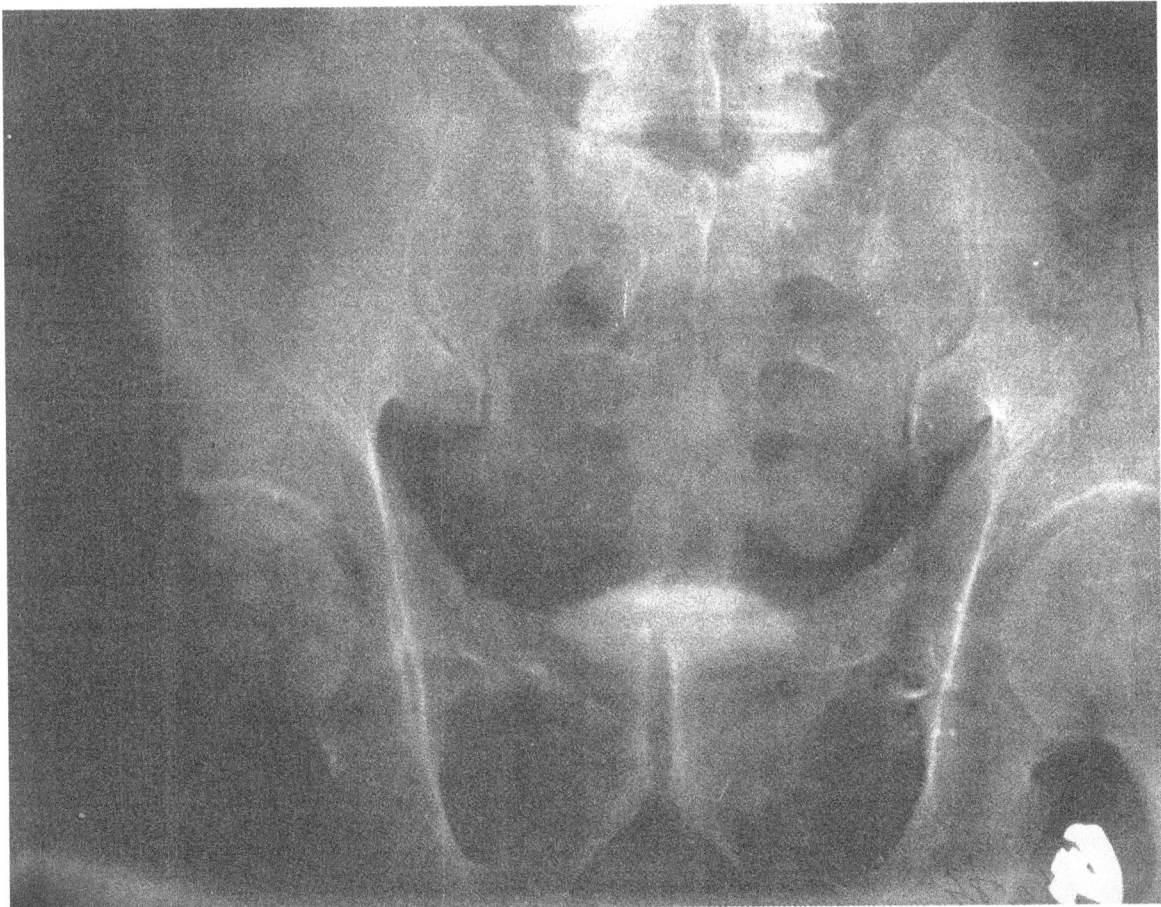

Fig. 259. Premiction and postmiction cystography.

263

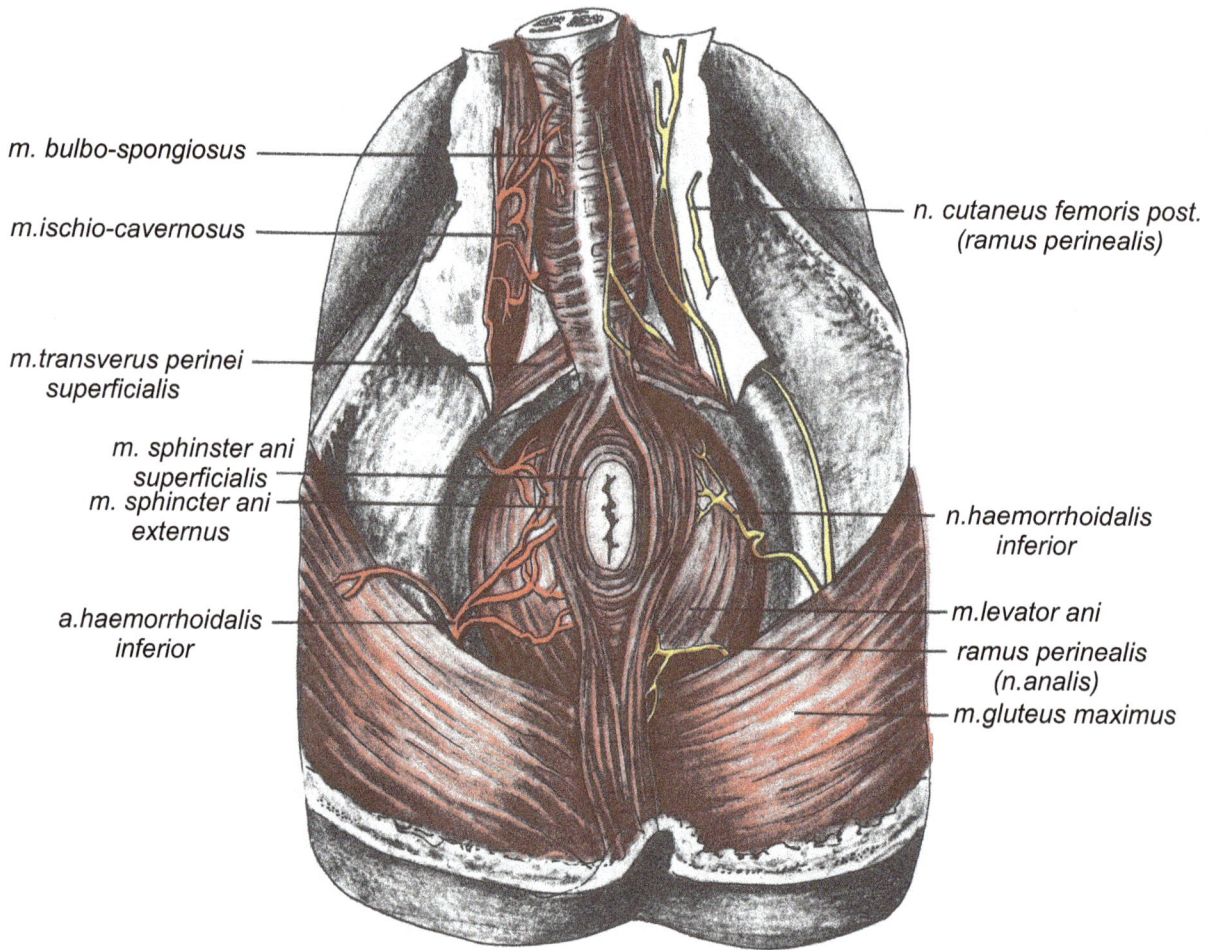

m. bulbo-spongiosus

m.ischio-cavernosus

m.transverus perinei
superficialis

m. sphinster ani
superficialis
m. sphincter ani
externus

a.haemorrhoidalis
inferior

n. cutaneus femoris post.
(ramus perinealis)

n.haemorrhoidalis
inferior

m.levator ani
ramus perinealis
(n.analis)
m.gluteus maximus

Fig. 260. The perineum in the male.

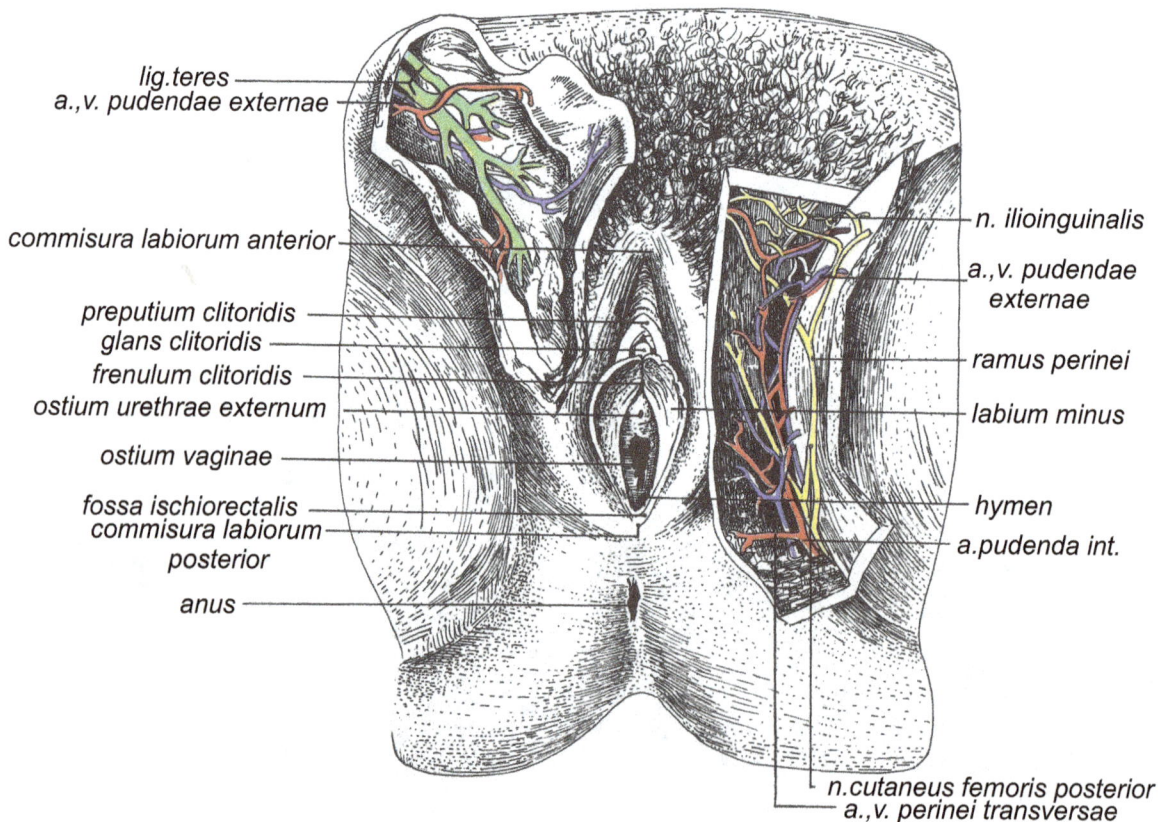

lig.teres
a.,v. pudendae externae

commisura labiorum anterior

preputium clitoridis
glans clitoridis
frenulum clitoridis
ostium urethrae externum
ostium vaginae
fossa ischiorectalis
commisura labiorum
posterior
anus

n. ilioinguinalis

a.,v. pudendae
externae

ramus perinei

labium minus

hymen

a.pudenda int.

n.cutaneus femoris posterior
a.,v. perinei transversae

264   Fig. 261. The perineum in the female (stratigraphic dissection of planes I and II).

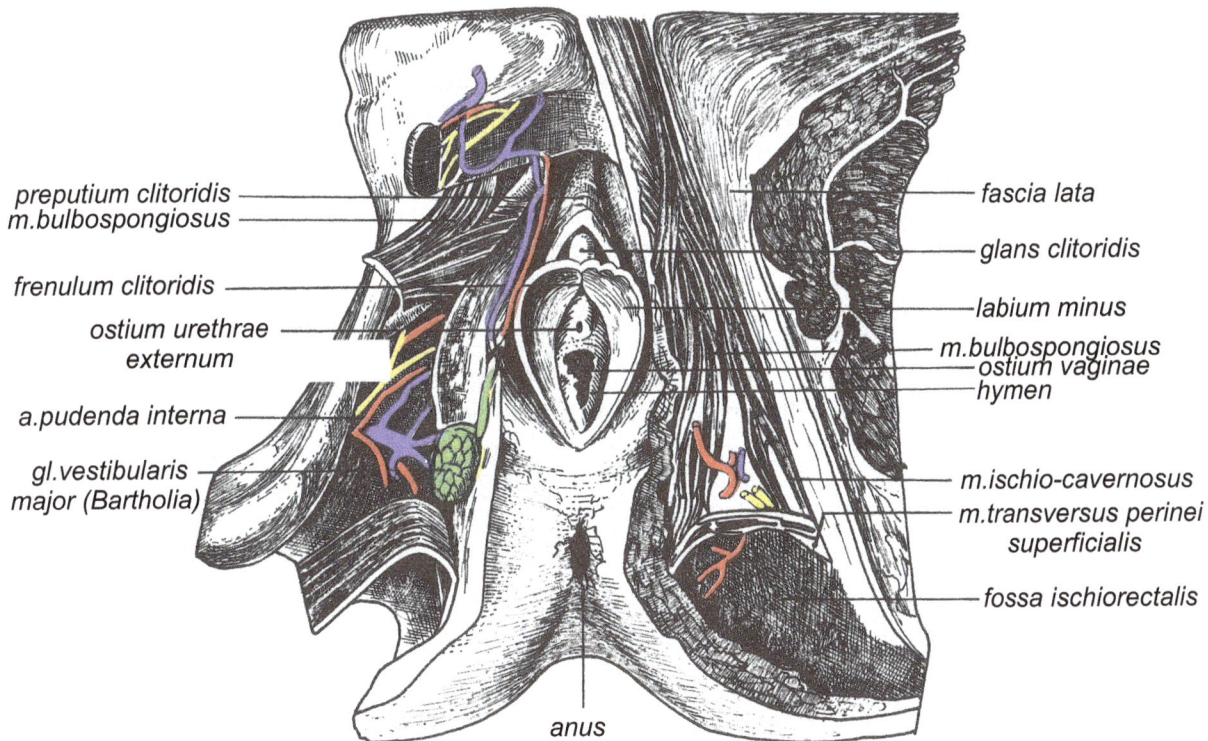

Labels on figure:
- preputium clitoridis
- m.bulbospongiosus
- frenulum clitoridis
- ostium urethrae externum
- a.pudenda interna
- gl.vestibularis major (Bartholia)
- anus
- fascia lata
- glans clitoridis
- labium minus
- m.bulbospongiosus
- ostium vaginae
- hymen
- m.ischio-cavernosus
- m.transversus perinei superficialis
- fossa ischiorectalis

Fig. 262. The perineum in the female (stratigraphic dissection of planes III and IV).

The Muscles of the Anal Region

The levator ani muscle (m. levator ani) has its origin: in front, on the pelvic surface of the body of the pubis; laterally on the symphysis; behind, on the internal surface of the spine of the ischium and between these two points, on the fascia of the obturator muscle. The fibres run towards the median plane, with varying degrees of obliquity.

a) The most anterior fibres are directed backwards and downwards on the lateral surface of the prostate and attach on the perineal node or central fibrous nucleus of the perineum. They constitute, in the male, the levator muscle of the prostate *(m. levator prostatae),* whereas in the female they cross the sides of the vagina, and form thus an important additional sphincter for this organ.

b) The next fibres run backwards and downwards over the lateral surface of the prostate and over the superior extremity of the anal canal and turn medially towards the anorectal flexure, to intermingle with the corresponding fibres on the opposite side; some of them become lost in the wall of the anal canal; this part of the muscle is termed puborectal *(m. puborectalis).*

c) The remaining fibres attach on the side of the two last segments of the coccyx and on a median fibrous raphe, which stretches from the coccyx to the wall of the anus *(flexura perinealis recti)* (fig.260,264).

Morphologically, the levator ani may be divided into an iliococcygeal muscle and a pubococcygeal muscle.

The iliococcygeal muscle *(m. iliococcygeus)* has its origin on the ischial spine and on the posterior portion of the tendinous arch of the pelvic fascia, then attaches to the coccyx and the median raphe; usually thin, it may be sometimes completely absent or it may be replaced to a great extent by fibrous tissue. An additional bundle, at its posterior pari, is sometimes called the iliosacral muscle *(m. iliosacralis)* (see also fig. 228, 237).

The pubococcygeal muscle *(m. pubococcygeus)* arises from the posterior surface of the pubis and from the anterior part of the fascia of the obturator muscle and is directed backwards, almost horizontally, along the side of the anal canal. Between the coccyx and the anal canal, the pubococcygeal muscles come close to each other and form a thick fibromuscular layer, situated on the raphe formed by the iliococcygeal muscles.

265

The levator ani muscle is supplied by a branch of the third and fourth sacral nerves, from the sacrococcygeal plexus.

The levatores ani induce the constriction of the lower end of the rectum, consequently of the anal orifice and of the vagina, and strengthen the perineal node (the fibrous central nucleus of the perineum), forming a diaphragm which supports the pelvic organs and opposes itself to their downwards thrust, produced by the increase of the intra-abdominal pressure.

The coccygeal muscle (m. coccygeus) lies behind the levator ani muscle, has its origin on the spine of the ischium and on the least sacrosciatic ligament *(lig. sacrospinale)* and attaches to the margin of the coccyx and to the lateral border of the last sacral vertebra. It contributes, with the levator ani muscle and the pyramidal muscle, to the closure of the posterior part of the pelvic outlet, completing posteriorly the pelvic diaphragm.

The coccygeal muscle is supplied by a branch of the fourth and the third sacral nerves.

During parturition, ruptures of the muscles forming the pelvic floor may sometimes occur. If the tendinous centre *(centrum tendineum perinei),* resulted from the intrication of the muscular and tendinous fasciae and fibres of most muscles of the perineum, has been torn, the contraction of the anterior fibres of the levator ani increases instead of diminishing the normal gap in the pelvic floor and, as a consequence, a uterine prolapse may occur and, in severe cases, the ovary, the urinary bladder and the rectum may also prolapse.

The external sphincter of the anus (m. sphincter ani externus) is a broad band of muscle fibres, elliptical in shape. It is made up of two portions: a superficial and a deep portion. The superficial portion constitutes the main part of the muscle and has its origin, through a narrow tendinous band, on the tip of the coccyx and on the anococcygeal raphe, so that the muscle consists of two widened bands, passing each on either side of the anus, to meet in front and attach to the perineal fibrous node (the fibrous centre of the perineum), where they unite with the superficial transverse muscle of the perineum *(m. transversus perinei superficialis),* the levator ani muscle and the bulbo-spongiose or bulbocavernous muscle *(m. bulbocavernosus).* The deep portion forms a complete sphincter for the anal canal. Its fibres surround the canal, closely applied on the puborectal muscle *(m. puborectalis)* and on the internal sphincter of the anus *(m. sphincter ani internus),* and interlace anteriorly with the other muscles, on the fibrous centre of the perineum.

The external sphincter of the anus is supplied by a branch arising from the fourth sacral nerve and by fibres derived form the inferior haemorrhoidal or anal branch (S_2-S_3) of the pudendal nerve.

Normally, the external sphincter of the anus is in a state of tonic contraction and, as it has no antagonistic muscle, it maintains the anal orifice and canal closed. It may be voluntarily brought in a state of stronger contraction and in this way it occludes more firmly the anal orifice.

The Muscles off the Urogenital Region in the Male

The superficial fascia of this region is made up of a superficial fatty and a deep, membranous portion (fig.265, 266). The fatty portion is continuous, in front, whit the dartos muscle of the scrotum and, behind, with the subcutaneous areolar tissue, that surrounds the anus.

The membranous portion of the fascia is aponeurotic. It is continuous, anteriorly, with the dartos muscle; laterally, it attaches to the borders of the branches of the pubis and ischium; posteriorly, it unites with the posterior border of the perineal membrane or the inferior fascia of the urogenital diaphragm *(fascia diaphragmatis urogenitalis inferior)* and with the "fibrous centre of the perineum".

The fibrous centre of the perineum is a fibromuscular node situated in the median plane, about 1.25 cm in front of the anus and very close to the urethral bulb of the penis *(bulbus urethrae).* Towards this node converge six muscles, which attach to it: the external sphincter of the anus; the m. bulbospongiosus, the two superficial transverse muscles of the perineum and the anterior fibres of the two levatores ani. The superficial transverse muscle of the perineum *(m. transversus perinei superficialis)* arises, by tendinous fibres, from the medial and anterior portion of the tuberosity of the ischium and inserts into the tendinous centre. The simultaneous contraction of the two superficial transverse muscles of the perineum contributes to fix the fibrous centre of the perineum.

The bulbocavernous muscle (m. bulbospongiosus sive bulbocavernosus) is sited in the median part of the perineum, in front of the anus, and consists of two symmetrical parts, connected by a median tendinous raphe; if has its origin on the median raphe and the perineal fibrous centre. Its fibres diverge: the most posterior are lost on the perineal membrane (or on the inferior fascia of the urogenital diaphragm);

The middle fibres encircle the bulb and the spongy body of the penis *(corpus cavernosum urethrae);* the anterior fibres spread out on the side of the cavernous body *(corpus cavernosum penis),* to attach partly to the latter (in front of the ischiocavernous muscle).

The bulbospongiosus muscle pushes out the contents of the urethra! canal The middle fibres contribute to the erection of the spongy body of the penis *(corpus spongiosum penis).* The anterior fibres contribute to the erection of the penis, by compressing the dorsal vein of the penis.

The ischiocavernous muscle *(m, ischiocavernosus)* arises,by tendinous and fleshy fibres, from the inner surface of the ischial tuberosity, behind the crus of the cavernous body of the penis *(crus corporis cavernosi penis).* The muscle fibres end on an aponeurosis,which attaches to the sides and the inferior surface of the penis.

The ischiocavernous muscle compresses the crus of the cavernous body of the penis and delays the backward flow of blood through the veins, maintaining thus the penis in erection.

Between the muscles described above, a triangular space is enclosed, bounded medially by the bulbospongiosus (bulbocavernous) muscle, laterally by the ischiocavernous muscle and behind, by the superficial transverse muscle of the perineum.

The scrotal vessels and nerves, as well as the perineal branch of the posterior femoral cutaneous nerve *(n. cutaneus femoris posterior),* traverse this space from behind forwards; the transverse perineal artery *(arteria perinei)* runs along its posterior boundary to the superficial transverse muscle of the perineum.

The deep fascia of the urogenital region forms a coat for the deep transverse muscle of the perineum; if contains also the deep vessels and nerves of the anterior perineum. It is made up of two membranous layers.

The strongest and most superficial of these layers is the perineal membrane or inferior fascia of the urogenital diaphragm *(fascia diaphragmatis pelvis inferior).* Its tip, directed forwards, is thickened to form the transverse ligament of the perineum *(tig. transversum pelvis);* between this ligament and the subpubic ligament *(lig. arcuatum pubis)* the deep dorsal vein of the penis (or of the clitoris) enters the pelvis.

The fascia is traversed by: the urethra, the arteries and nerves of the bulb and, near the urethra, by both excretory ducts of Cowper's bulbo-urethral glands *(glandulae bulbo-urethrales);* by the deep arteries of the penis, one along each side, just near the pubic arch and about at midway from the attachment border of the membrane; by the dorsal arteries and veins of the penis, near the tip of the membrane; at the base, by the scrotal vessels and nerves.

Beneath the fascia there are; the membranous portion of the urethra, the deep transverse muscle of the perineum and the sphincter of the urethra, the bulbo-urethral glands with their excretory ducts, the internal pudendal vessels and the dorsal nerves of the penis, the arteries and nerves of the bulb of the penis and a venous plexus.

These formations are separated from the pelvis by a second layer of the fascia, less evident.

The deep transverse muscle of the perineum *(m. transversusperinei profundus)* arises from the branches of the ischium and courses towards the median plane, where if interlaces, within a tendinous raphe, with the homonymous muscle of the opposite side. It is a tensor of the perineal centre.

The sphincter muscle of the urethra *(m. sphincter uretrael)* invests the membranous portion of the urethra and is sited between the two layers of the deep fascia of the urogenital region. The muscles of the two parts act together as a sphincter, compressing the membranous portion of the urethra.

All the muscles of the urogenital region are supplied by the perineal branch of the pudendal nerve *(rami perineales nervi pudendalis).*

The Muscles of the Urogenital Region in the Female

The superficial transverse muscle of the perineum *(m. transversus perinei superficialis)* differs very slightly from the corresponding muscle in the male, it is only more developed (fig. 261-263).

The bulbospongiosus or bulbocavernous muscle *(m. bulbospongiosus sive bulbocavernosus)* surrounds the orifice of the vagina. It covers the lateral parts of the vestibular bulbs and is attached, behind, to the fibrous centre of the perineum, where its fibres blend with the external sphincter of the anus. Its fibres pass forwards, on each part of the vagina, to be attached to the cavernous bodies of the clitoris; a fascicle crosses the body of the clitoris, so that it compresses the dorsal part of the clitoris, respectively the dorsal vein.

The bulbocavernous muscle narrows the orifice of the vagina. The anterior fibres contribute to the erection of the clitoris, by the compression of its deep dorsal vein.

267

The ischiocavernous muscle *(m. ischiocavernosus),* smaller than the corresponding muscle in the male, covers the unattached portion of the crus of the clitoris *(crus ciitoridis).* If arises, by tendinous and muscle fibres, from the inner surface of the tuberosity of the ischium, behind the crus of the clitoris, and from the adjacent surface of the branch of the ischium.

The muscle fibres end in an aponeurosis, which is attached on the sides and the inferior surface of the crus of the clitoris.

The ischiocavernous muscle compresses the crus of the clitoris and delays the return of blood through the veins, thus contributing to maintain the clitoris in erection.

The perineal membrane or inferior fascia of the urogenital diaphragm *(fascia diaphragmatis pelvis inferior)* in the female is weaker than in the male and is traversed by the aperture of the vagina, with the external tunic of which if interlaces.

It covers the following anatomical structures: portions of the urethra and the vagina, the deep transverse muscle of the perineum, the sphincter of the urethra, the greater vestibular glands *(glandula vestibularis major or Bartholin's gland)* and their excretory ducts, the internal pudendal vessels, the dorsal nerves of the clitoris, the arteries and nerves of the vestibular bulbs and a venous plexus. The deep transverse muscle of the perineum *(m. transversus perinei profundus)* arises from the ramus of the ischium and courses transversally, behind the vagina, to meet the homonymous muscle on the opposite side; it contributes to fix the fibrous centre of the perineum.

The sphincter muscle of the urethra *(m. sphincter urethrae)* is constituted of external and infernal fibres. The external fibres are directed across the pubic arch, in front of the urethra. The interna! fibres encircle the lower end of the urethra.

Both muscles contract simultaneously and act as a constrictor of the urethra.

Vessel and nerve supply

The perineum is supplied by the infernal pudendal artery (see fig. 243-245).

The internal pudendal artery *(a. pudendalis interna)* is one of the most important branches of the hypogastric artery. If supplies the perineum and the external genital organs.

Course and relations. The internal pudendal artery emerges from the pelvis through the inferior part of the incisure of the infrapiriform orifice *(foramen infrapiriforme),* below the piriform muscle, turns round the ischial spine *(spina ossi ischii)* and enters, through the lesser sciatic foramen *(foramen ischiadicum minus),* the ischiorectal fossa, where if divides info two terminal branches: the perineal artery *(a. perinealis)* and the artery of the penis *(a. penis seu clitoridis).* Before its bifurcation, the artery gives also off 2-3 inferior haemorrhoidal or anal branches.

In the gluteal region, the infernal pudendal artery overlaps the ischial spine. At this level if is situated outside the sciatic artery; the pudendal nerve is behind the artery.

In the ischiorectal fossa, the artery is situated on the external wall of the latter, together with the nerve, in a splitting of the aponeurosis of the infernal obturator.

The infernal pudendal artery gives rise to the haemorrhoidal branches and then breaks off into the perineal artery and the penile artery:

a. The inferior haemorrhoidal arteries *(aa. anales)* are two or three arterial branches which separate below the ischial spine and run through the ischiorectal fossa towards the external sphincter of the anus and the levator ani.

b. The perineal artery *(a. perinealis),* usually termed the superficial perineal artery, arises from the ischiorectal fossa and passes info the anterior perineum, traversing or crossing the superficial transverse muscle. The artery runs into the subcutaneous fibroadipose tissue, between the skin and the superficial perineal fascia, up to the level of the bursae *(aa. scrotales posteriores)* or of the labia majora *(aa. labiates posteriores).*

c. The artery of the penis continues the course of the internal pudendal artery. If runs into the urogenital diaphragm, passes between the two leaflets of the middle perineal fascia, near the ischiopubic rami. Reaching the inferior border of the pubic symphysis (below the arcuate ligament), it becomes continuous with the dorsal artery of the penis.

The branches of this artery are:

- the bulbo-urethral artery *(a. bulbi urethrae),* called also the deep perineal artery, runs to the ischiobulbar triangle, made up of the ischiocavernous, bulbocavernous and superficial transverse muscles, and sends branches to these muscles and to the corpus cavernosum of the urethra;

- the urethral artery *(a. urethra/is)* is a small artery which arises in front of the preceding and runs towards the corpus spongiosum of the urethra, which if penetrates up to the place where the latter unites with the corpus cavernosum in the glans penis;

268

- the cavernous or deep artery of the penis *(a. profunda penis)* arises at the level of the pubic symphysis and enters the corpus cavernosum of the penis, within which if ramifies, but on the apex of the penis if anastomoses with the homonymous artery and with the dorsal artery of the penis the dorsal artery of the penis *(a. dorsalis penis)* is continuous with the main trunk, runs through the space between the deep transverse muscle and the preurethral ligament and reaches the dorsal aspect of the penis, along which it courses below the fascia penis up to the level of the glans, in company with the subfascial dorsal vein. In its course, the artery gives off branches to the corpora cavernosa and the spongy portion of the urethra, as far as to the glans, where it anastomoses with the other arteries of the penis;

- the artery of the clitoris *(a. clitoridis)* corresponds to the penile artery and has the same course. It gives off the artery of the vaginal bulb, the urethral artery, the cavernous artery and the deep dorsal artery of the clitoris.

The nerve supply of the perineum is derived from the pudendal plexus *(plexus pudendalis),*

This plexus consists of branches which arise from the anterior branches of the second to the fifth sacral nerves. The pudendal plexus is closely related to the sacral plexus, the component of which if is considered, because if is united with the sacrococcygeal plexus. From this plexus the following branches run towards the genital organs and the intrapelvic viscera:

The pudendal *nerve(n. pudendalis)* arises from the pudendal plexus, leaves the pelvis through the infrapiriform foramen, turns round the ischial spine and enters the ischiorectal fossa, where if divides info two branches: (1) the perineal nerve and (2) the penile nerve.

The perineal nerve runs forwards and divides into two branches:

a)   *the cutaneous branch,* which passes between the skin and the superficial perineal fascia and ends within the external genitalia by scrotal, respectively labial branches;

b)   *the muscular (musculo-urethral) branch,* which supplies the three muscles that form the ischiobulbar triangle: transverse, ischiocavernous and bulbocavernous; from if emerges a nerve filament to the bulb and another to the urethra.

The dorsal nerve of the penis *(n. dorsalis penis)* continues the course of the pudendal nerve, passes below the pubic symphysis in company with the homonymous artery and reaches the dorsum of the penis in the groove formed by the corpora cavernosa;if gives off branches to the corpora cavernosa, the skin of the penisand the glans.

The nerve of the levator ani and of the ischiococcygeal muscle derives from the pudendal plexus and supplies the two muscles.

The haemorrhoidal or anal nerve derives from the pudendal plexus or nerve, turns in company with it round the ischial spine and enters the ischiorectal fossa, running towards the external sphincter.

The visceral nerves are branches variable in number and size; they run towards the rectum, the urinary bladder and the vagina, either directly or through the sympathetic hypogastric plexus.

Structures Belonging to the Digestive end Urogenital Apparatuses

In the perineum, these structures, pertaining both to the digestive and to the urogenital apparatus, belong to the topographical anal and urogenital regions.

The anal region (or posterior perineum) is situated behind the ischiadic line.

Its constitution comprises the rectal ampulla, which is continuous with the anal canal, surrounded by the external sphincter of the anus *(m. sphincter ani externus).*

Topographically, on each side of the anal canal lie the two ischiorectal fossae *(fossae ischiorectales),* prismatic triangular in shape.

The ischiorectal fossa has the base directed to the surface of the perineum constituted of integument, and its thin edge at the line of meeting of the obturator muscle with the anal fasciae (of the levator ani). It is bounded medially by the external sphincter of the anus *(m. sphincter ani externus)* and the anal fasciae; laterally; by the tuberosity of the ischium and the fascia of the obturator muscle; in front, by the perineal membrane or the inferior fascia of the urogenital diaphragm *(fascia diaphragmatis urogenitalis inferior);* behind, by the greatest gluteal muscle *(m. glutaeus maximus)* and the great sacrosciatic ligament *(lig. sacrotuberosum).*

The inferior rectal or haemorrhoidal vessels *(vasa haemorrhoidaies inferiora seu anales)* and the inferior haermorrhoidal nerve *(n. haemorrhoidalis inferior)* cross the space transversally, from the lateral to the medial surface; the perineal and perforating cutaneous branches of the sacral plexus are in the posterior part of the fossa, whereas from the anterior part arise the scrotal (labial) vessels and the homonymous nerves.

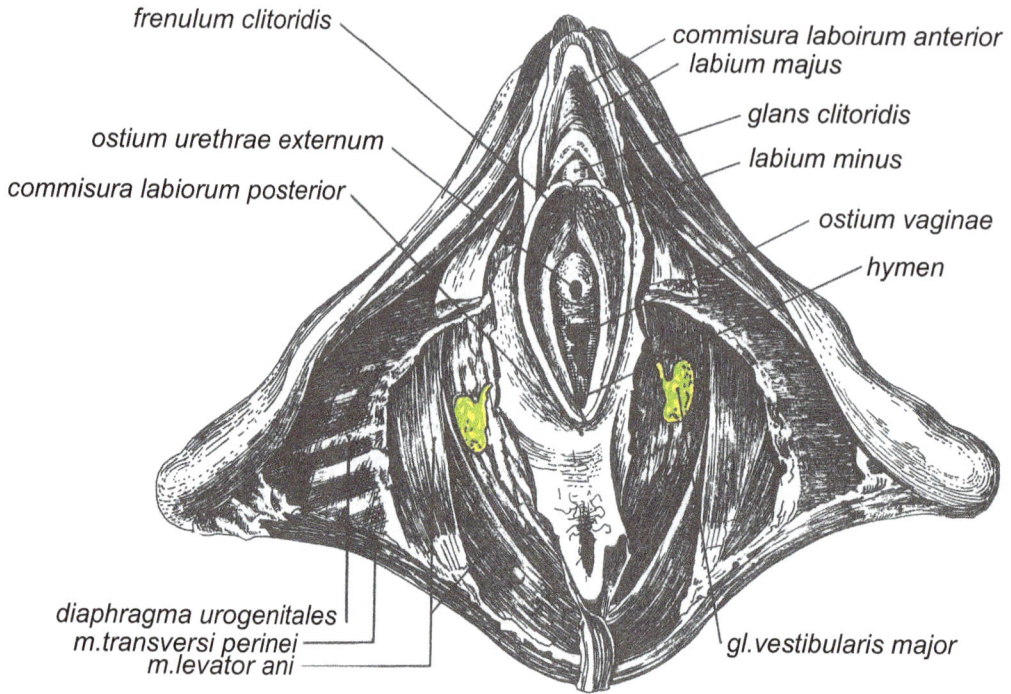

frenulum clitoridis

commisura laboirum anterior
labium majus

ostium urethrae externum

glans clitoridis

labium minus

commisura labiorum posterior

ostium vaginae

hymen

diaphragma urogenitales
m.transversi perinei
m.levator ani

gl.vestibularis major

Fig. 263. The urogenital diaphragm in the female (fifth stratigraphic plane).

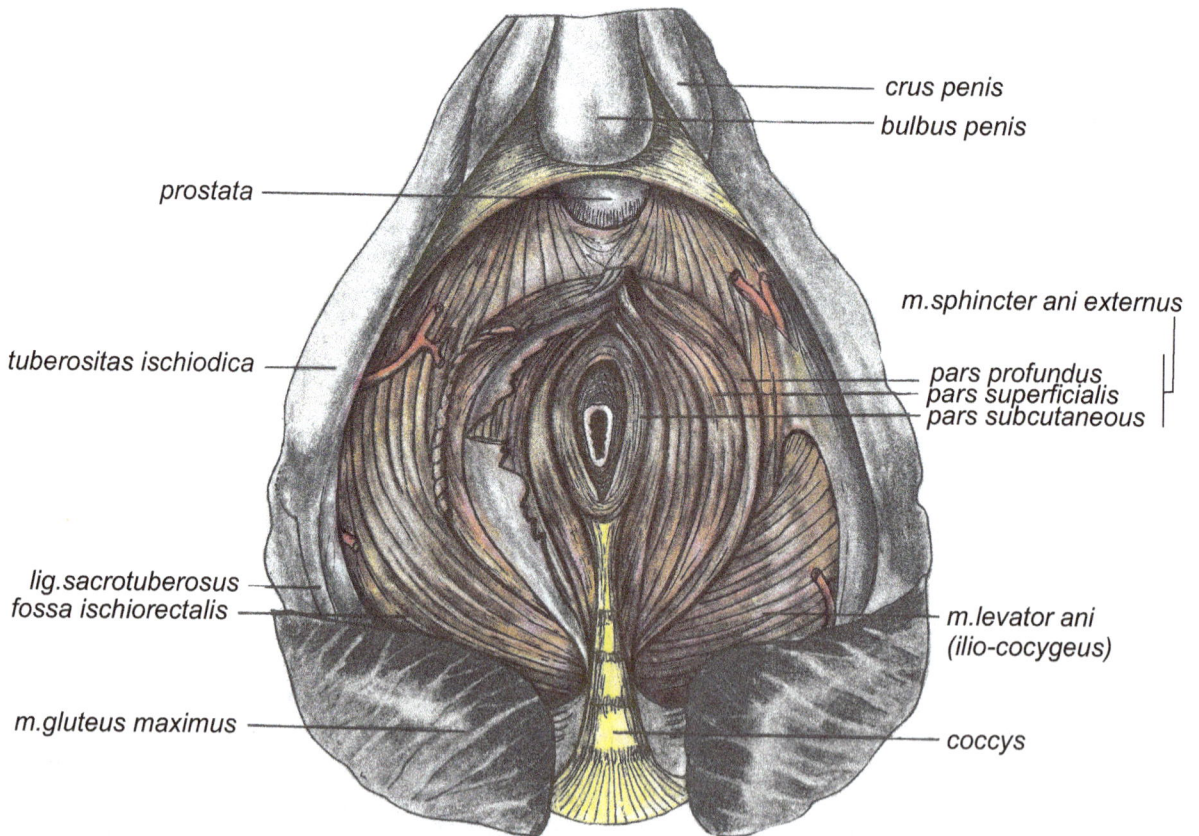

crus penis
bulbus penis

prostata

m.sphincter ani externus

tuberositas ischiodica

pars profundus
pars superficialis
pars subcutaneous

lig.sacrotuberosus
fossa ischiorectalis

m.levator ani
(ilio-cocygeus)

m.gluteus maximus

coccys

Fig. 264. The external anal sphincter muscle.

270

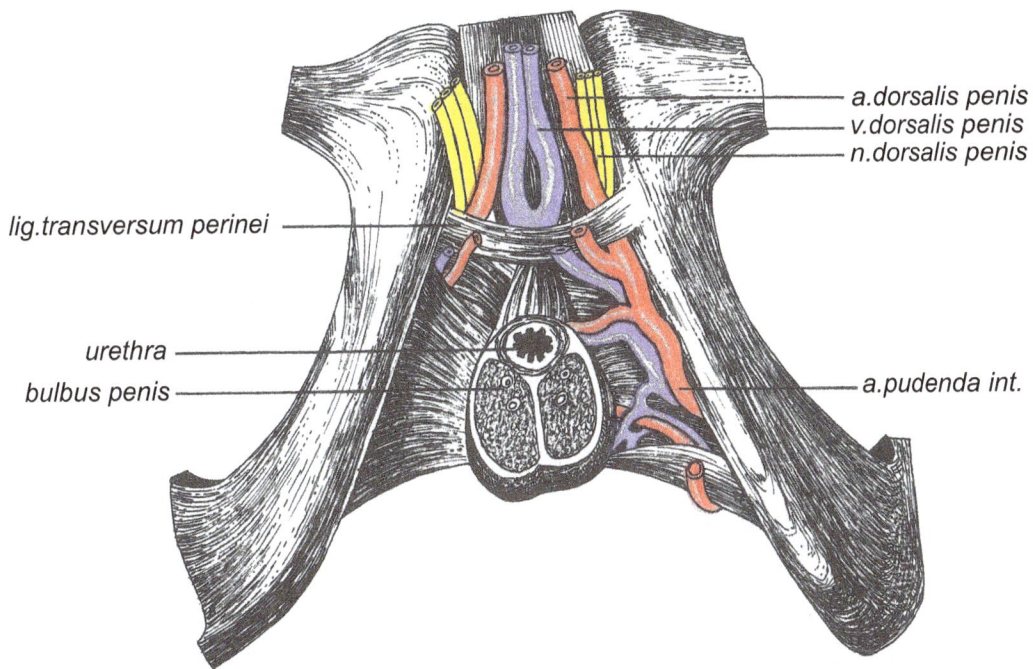

a.dorsalis penis
v.dorsalis penis
n.dorsalis penis

lig.transversum perinei

urethra
bulbus penis

a.pudenda int.

Fig. 265. The deep perineal space in the male (viewed from behind)

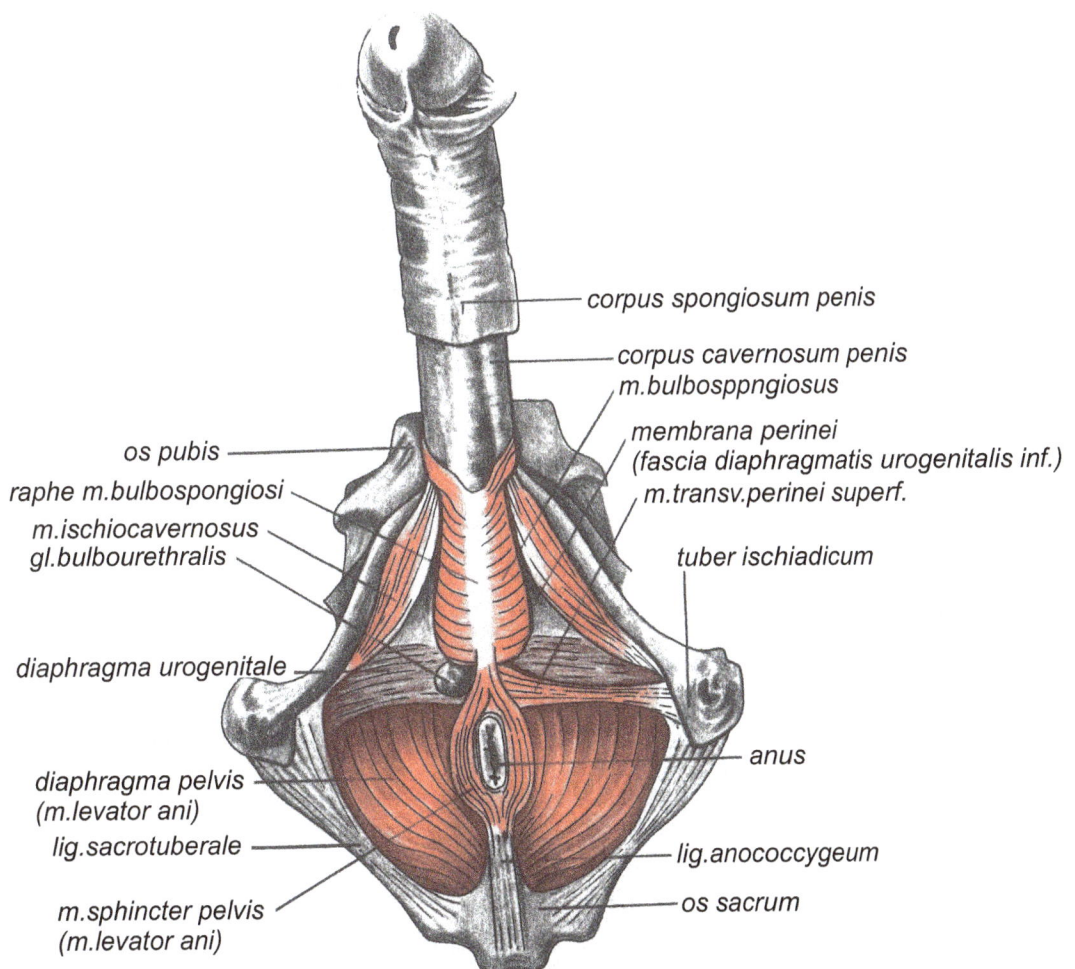

corpus spongiosum penis

corpus cavernosum penis
m.bulbosppngiosus

membrana perinei
(fascia diaphragmatis urogenitalis inf.)
m.transv.perinei superf.

os pubis

raphe m.bulbospongiosi

tuber ischiadicum

m.ischiocavernosus
gl.bulbourethralis

diaphragma urogenitale

anus

diaphragma pelvis
(m.levator ani)

lig.sacrotuberale

lig.anococcygeum

os sacrum

m.sphincter pelvis
(m.levator ani)

Fig. 266. Root of the penis (radix penis).

271

The internal pudendal vessels and the pudendal nerve *(vasa pudendalae internae et n. pudendalis)* are on the lateral wall of the fossa, in the pudendal canal (Alcocq's canal). The fossa is filled with adipose tissue, through which numerous fibrous bands extend.

The External Genital Organs of the Female

The external genitalia of the female in their totality are designated by the term vulva, which includes the greater and lesser lips, the interlabial space, the clitoris, the erectile apparatus and the *adnexal glands* (fig.261-263).

The labial structure

The labial structures consist of the greater lips and the lesser lips *(labia minora pudendi)* (see also fig. 251).

The labia majora are cutaneous folds, ovoid elongated in shape, containing fat, connective tissue, smooth muscles, vessels, nerves and the ramified endings of the round ligaments. They have an external, cutaneous surface, covered with hair and pigmented, and an internal, cutaneous, thin, pink surface, showing the aspect of a mucous membrane; the internal surface presents numerous **seba**ceous glands, which secrete the vulvar smegma. The external surface of the labia majora is in relation with the respective thigh, while the internal surface is related to the corresponding labium minus and to the labium majus on the opposite side, with which it bounds the interlabial space. The labia majora join at the two extremities, forming an anterior commissure *(commissura labiorum anterior)* and a posterior commisure *(commissura labiorum posterior)*. The anterior commissure, poorly visible, is situated below the mons pubis *(mons pubis Veneri),* an anatomical region which lies in front of the pubis and which some authors describe among the external genital structures. The posterior commissure, designated also by the term fork, is more apparent than the anterior; the distance which separates if from the anus is of about 25 mm.

The labia majora receive *branches* from the internal and external *pudendal arteries. The veins* reach the trunks that correspond to the two arteries. *The lymphatics* are tributaries of the external inguinal lymph nodes.

The nerve supply is derived from branches arising from the iliohypoogastrsc and ilio-inguinal nerves, as well as from the internal pudendal nerve.

The labia minora are cutaneous folds of the aspect of a mucous membrane. They are about 3-4 cm long and 1 cm wide.

They have: an external surface, which is in relation with the corresponding labium majus, and an internal surface, related to the labium minus on the opposite side (through the interlabial space, as normally they are covered by the labia majora); an anterior or free border and a posterior or adherent border, in contact with the vaginal bulbs; an anterior end, where the clitoris, the female erectile organ, is situated.

This anterior end divides into two portions: one which is directed towards the posterior surface of the clitoris and forms with that on the opposite side the frenulum of the clitoris *'frenulum clitoridis);* the other, which unites with that on the opposite side in front of the clitoris and forms the cap of the latter, named the prepuce of the clitoris *(preputium clitoridis)*. Behind the clitoris lies the urinary meatus *(orificium urethrae externum)*.

The posterior end unites with that on the opposite side and forms the navicular fossa *(fossa navicularis),* evident in nulliparous women *(fossa vestibuli vaginae)*. Inside the vestibule is the introitus of the vagina or vaginal orifice *(ostium vaginae)*.

In virgins, a semilunar cutaneomucous structure, the hymen *(hymen vaginae),* is here present, which narrows the vaginal orifice. It may assume also other shapes: cribriform, annular or even imperforate.

The labia minora consist of a double cutaneous leaflet of the aspect of a mucuous membrane, containing inside connective tissue rich in elastic fibres, but devoid of fat. In addition, the labia minora contain numerous sebaceous glands, that secrete the vulvar smegma.

The vessel supply is the same as to the labia majora and the nerve supply is derived from the perineal branch of the internal Pudendal nerve.

The role of the labia minora, as well as that of the labia majora, is to protect the vagina.

272

The Interlabial Space
(Rima pudendi)

The interlabial space is the cleft, more or less open according to whether the respective woman is nulliparous or multiparous, which becomes funnel-like if we separate the labia majora. At the base of this space are: (1) the vestibule *(vestibulum vaginae)*; (2) the urinary meatus *(orificium urethrae externum)* and (3) the inferior orifice of the vagina *(ostium vaginae)*.

(1) *The vestibule (vestibulum vaginae)* is a small anatomical region, triangular in shape, bounded: laterally, by the labia minora, in front by the clitoris and behind, by the horizontal line drawn through the urinary meatus.

(2) *The urinary meatus (orificium urethrae externum),* a circular orifice of 3-4 mm in diameter, represents the opening of the urethra to the outside. This orifice is situated in the middle of the base of the vestibular triangle.

(3) *The inferior orifice of the vagina (ostium vaginae)* opens into the vulvar canal and varies in its aspect according to whether we refer to a virgin, a nulliparous or a multiparous woman.

In the virgin, the inferior orifice of the vagina is closed partly by a membranous diaphragm, the hymen, perforated by one or several holes.

The shape, thickness and resistance of the hymen vary.

As regards the shape, the following types are described: the semilunar type, with fringes, bilabiate, biperforate, and the cribriform type.

According to the thickness and resistance, we distinguish: the fleshy, resistant, thin, flexible and elastic hymen.

In the nulliparous woman, the hymen presents under the shape of four or eight fringes (caruncles) resulted from its rupture. Usually, the rupture spares the adherent border of the membrane, so that the vaginal orifice is bounded by an easily discernible fold, called the hymenal ring.

In the parous woman, the ruptures pass beyond the adherent border of the hymen, extending up to the elements of the vulva.

The vessel and nerve supply of the hymen are assured by the *vessels* and *nerves* of the inferior portion of the vagina; structurally, the hymen is a mucous diaphragm, containing between its leaflets connective tissue with numerous elastic fibres.

The Erectile Apparatus

The erectile apparatus consists of a median organ, the clitoris *(clitoris),* and two lateral organs, the vestibular bulbs *(bulbi vestibuli)* (see also fig. 251).

The clitoris (clitoris) is in the female the homologue, but of much lesser proportions, of the penis with the corpora cavernosa in the male.

Cylindrical in shape, it has two tapered roots, the cavernous bodies *(corpus cavernosum dexter et sinister)* which, starting from the middle part of the internal surface of the ischiopubic rami, are directed obliquely upwards and inwards, in front of the pubic symphysis, constituting the body of the clitoris *(corpus clitoridis)*.

The latter is directed upwards and ends by a bulging extremity, named glans *(glans clitoridis),* about 5-6 mm long. The clitoris is made up of a hidden, deep sited portion, and of a free portion, covered by a prepuce *(preputium clitoridis);* between the prepuce and the glans there is a preputial cavity. From the posterior surface of the glans arises the frenulum of the clitoris *(frenulum clitoridis)*.

The body of the clitoris consists of two corpora cavernosa, separated by an incomplete septum. The glans is formed of a central connective nucleus, covered by a dermopapillary mucous membrane; it contains also numerous nerve endings and especially some genital sensitive corpuscles.

Vessel and nerve supply

The arteries derive mostly from the internal pudendal artery (the deep and the dorsal arteries of the clitoris).

The veins run towards the femoral vein and the internal pudendal vein.

The lymph vessels reach the superficial and deep inguinal and the external iliac lymph nodes.

The nerve supply is derived from branches arising from the internal pudendal nerve. There are also an autonomic sympathetic component, from the hypogastric plexus, and a parasympathetic one, from the sacra level.

The vaginal bulbs of the vestibule (bulbi vestibuli) are two erectile formations and represent the homologue of the male urethral bulb, which in the female has been separated into a right and a left half, by the interposition of the vulvar orifice. They are situated on the sides of the vaginal orifice, in the vulvar region, not in the vagina, so that it would be correct to term them vulvar bulbs instead of vaginal bulbs. They have the shape of an ovoid, with an external convex surface, that is in contact with the corpora cavernosa, from which if is separated by a small space, containing an amount of connective tissue and the superficial perineal vessels. More outwards, the external surface is in contact with the ischiopubic branch. This surface is covered by the bulbocavernous muscle which, united behind with that on the opposite side, forms a muscular ring, that surrounds the lower end of the vulvovaginal canal, fact which led some authors to call if the "constrictor vaginae". The internal surface of the bulb is concave and related to the terminal segment of the urethra, to the inner surface of the labia minora and, behind, to Bartholin's glands. The posterior end or the base of the bulbs is round and extends close to the navicular fossa. The anterior end is very slender and unites, in the vestibular region, between the meatus and the clitoris, with that on the opposite side and with the clitoris.

The cavernous bulbs are rather spongy than erectile, and are made up of an albuginea, that contains an erectile tissue with large areolar spaces.

Vessel and nerve supply

The arteries of the bulbs arise from the internal pudendal artery.

The veins run towards the internal pudendal vein.

The nerves derive from the hypogastric plexus.

The Adnexal Glands of the Vulva

Bartholin's glands (glandulae vestibulares majores), two in number, ore situated on the right and on the left of the inferior end of the vagina. They are ovoid in shape and have an elastic consistency.

The external surface of the gland, convex, covered by the posterior end of the bulb and by the bulbocavernous muscle, is situated at the base of the labium majus. The infernal surface, concave, is related to the vagina (this is the surface on which the gland is incised in the case of purulent collections). The excretory duct of the gland, measuring 15-20 mm in length, is directed obliquely forwards, downwards and inwards, runs along the base of the labia minora and opens into the groove that separates the labia minora from the hymen or from its remnants.

Bartholin's glands are ramified tubular glands.

Vessel and nerve supply

The arteries arise from the internal pudendal artery.

The veins run towards the homonymous vein.

As regards the lymph vessels, some of them course to the Iliac lymph nodes and others to the inguinal lymph nodes.

The nerves derive from the perineovulvar branch of the internal pudendal nerve.

The role of Bartholin's glands, the homologues of Cowper's glands in the male, is to moisten the vagina during coitus.

In the anterior portion of the vestibule, on the medial surface of the labia minora, are the lesser vestibular glands (glandulae vestibulares minores), the secretion of which enters info the constitution of the smegma.

The External Genital Organs in the Male
(The scrotum and the penis)

The scrotum
(Scrotum)

The scrotum is a prominent, unpaired and median sac, which hangs in the anterior peritoneum between the thighs and is suspended to the pubic region. Its size and shape vary according to the age. It presents on the antero-inferior surface a median suture (the raphe of the scrotum| (raphe scroti), which gives to it a bilobate aspect It is divided by a wall (septum scroti}into two cavities, one for each testis. The scrotal septum is made up of all the tunics of the scrotum, except the skin (fig. 267-269).

274

As if is a structure arisen from the ventral abdominal wall, the scrotum consists of five superposed funics:

- the skin is pigmented and provided with pores. It is sometimes smooth, sometimes corrugated owing to the dartos tunic, contained in its thickness. Being extensible, if constitutes also a reserve for the penis when if enters in a state of erection;

- the dartos funic *(tunica dartos)* is formed of a cutaneous muscle with many smooth fibres, which adheres closely to the skin, but which lacks in the skin of the abdominal wall; it contains, however, the subcutaneous tissue, which contributes also to the formation of the scrotal septum;

- the cremasteric fascia *(fascia cremasterica Cowperi)* is continuous with the fascia of the external oblique muscle. The cremasteric muscle arises from the internal oblique and transverse muscles of the abdomen. At the level of the bottom of the scrotus, in which the testis is suspended, the cremasteric muscle spreads out and forms thus the raising apparatus of the male genital gland. The cremasteric reflex of pulling up the testis is brought about by excitation of the integument of the superomedial portion of the thigh;

- the common vaginal funic *(tunica vaginalis communis)* is derived from the fascia transversalis and invests directly the spermatic cord and, through the vagina! funic, the testis; inferiorly, this tunic connects the skin and the dartos, on the one hand, with the testis and the epididymus, on the other hand, forming the scrotal ligament, a remnant of the embryonic formation called gubernaculum testis.

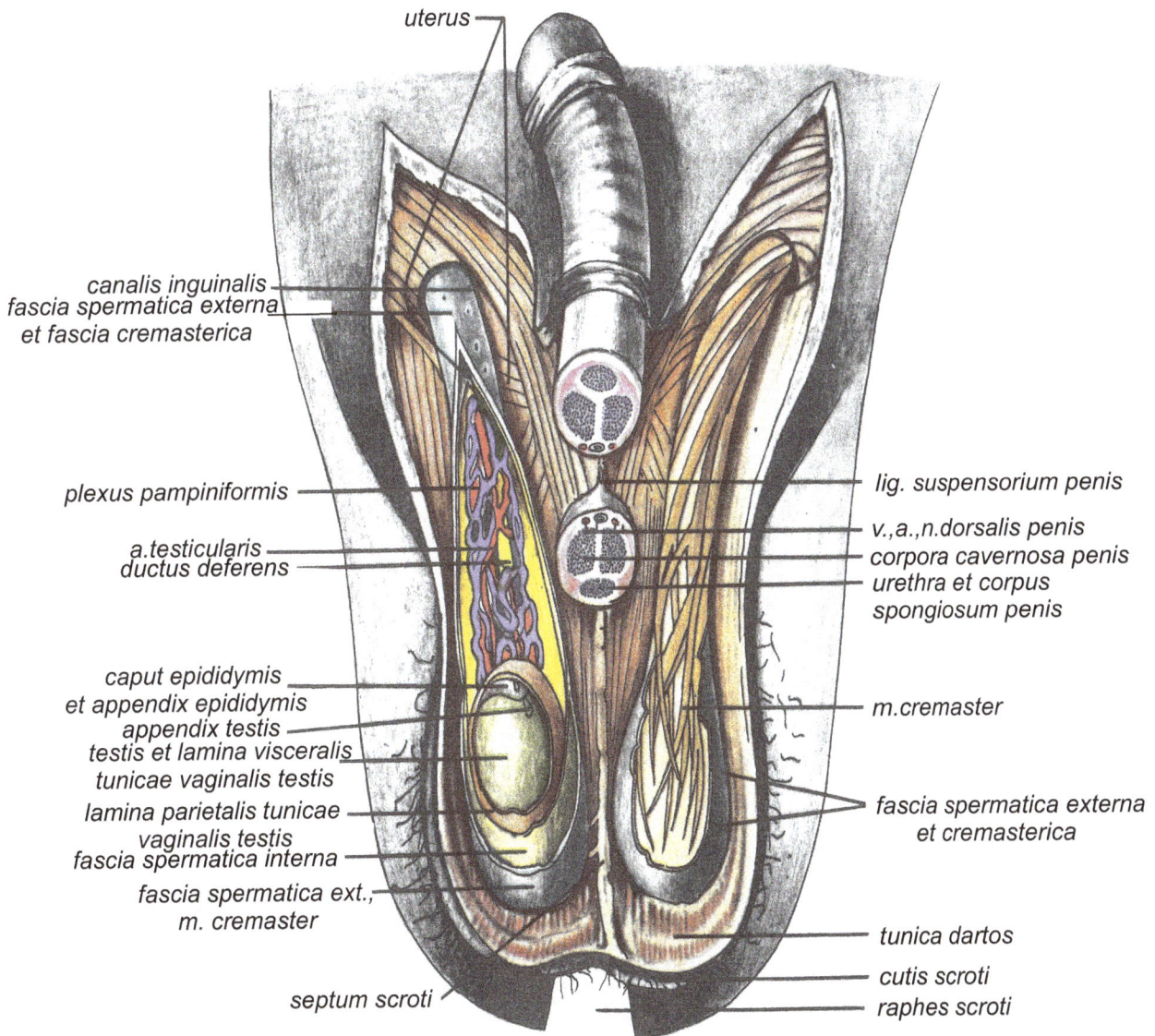

Fig. 267. Scrotum and penis (anterior view).

- the vaginal tunic *(tunica vaginalis propria),* a serous membrane, a derivative of the peritoneum, invaginates before the descent of the testis and of the epididymus within the scrotum; it consists of two layers: parietal *(lamina parietalis seu periorchium)* and visceral *(lamina visceralis seu epiorchium),* which enclose between them, at the beginning of the descent of the testis, the vaginal process and after its descent within the scrotum, the vaginal or scrota! cavity *(cavum scroti).*

The fluid collection in the vaginal cavity is named *hydrocele.*

In the embryonic life, between the peritoneal cavity and the tunica vaginalis lies a connecting canal *(processus vaginalis peritonealis),* which normally is obliterated at birth; in case of persistence, if constitutes a path of descent of the intestinal loops into the scrotum, giving rise to an indirect congenital hernia. When the vaginal process is closed at the infernal orifice of the inguinal ring and, following an effort or owing to another cause, an indirect acquired hernia occurs also, through the internal (peritoneal or deep) ring, this hernia has a proper peritoneal hernial sac, formed by the intestinal loops that push it outside the peritoneal cavity. The scrotal cavity may be inflamed, when an exudate appears within if, constituting the hydrocele.

Vessel and nerve supply

The arteries of the scrotum are divided into superficial and deep arteries. The superficial arteries are the external pudendal artery, which is a ramus of the femoral artery, and the superficial femoral artery, which is a branch of the internal pudendal artery. The deep arteries arise from the funicular artery, which is a ramus of the epigastric artery.

The superficial veins drain into the external pudendal vein (which opens into the internal saphenous vein) and into the internal pudendal vein. The deep veins drain into the venous plexus of the **spermatic** cord. The lymph vessels of the scrotum are tributaries of the superficial inguinal lymph nodes (the superomedial and infero-medial groups).

The nerves are sensory, motor and autonomic. They derive from the internal pudendal nerve through the scrotal nerves and from the genitofemoral and abdominogenital nerves.

The Penis
(Penis sive membrum virile)

The penis is the male organ of copulation; if contains also tne spongy part of the urethra and, consequently, is also the organ of miction (fig. 268-270).

It is constituted of two portions: a posterior perineal portion *(radix penis),* situated in the penile base of the perineum, and an anterior of free one *(corpus penis).* The latter represents the actual penis *(pars peniles).* In the state of flaccidity, the ventral portion descends vertically, forming with the perineal portion the penile angle, which disappears on erection. It is 10-11 cm long and has a circumference of 8-9 cm. In the erect condition, these sizes increase and the length may attain 14-18 cm.

The root of the penis (radix penis) is formed of the two roots of the cavernous bodies *(corpora cavernosa penis),* which are continuous below the pubic symphysis with the cavernous body of the urethra and with the bulb of the penis situated medially *(corpus spongiosum penis et bulbum penis seu corpus cavernosum urethrae).*

The body of the penis (corpus penis), of the shape of a slightly from before backwards flattened cylinder, has a superior surface *(dorsum penis)* and an inferior surface *(facies urethralis),* on which bulges the corpus cavernosum urethrae, especially during erection, and on which lies the raphe of the penis *(raphe penis),* continuous with that of the scrotum and that of the perineum.

The anterior extremity of the penis, the glans (glans penis), is a conoid prominence, belonging to the spongy body of the male urethra, which increases considerably in volume at this level. It has an apex, a base and a hollow posterior surface, into which penetrate the united tips of the corpora cavernosa penis.

The prepuce or foreskin (praeputium) is a portion of the spare skin that covers the glans like a muff. It has: an external surface, continuous - without a clear-cut boundary - with the skin of the penis; an internal surface, of the aspect of a mucous membrane, moulding itself on the glans, to the inferior surface of which if adheres through the frenulum *(frenulum preputii),* that contains numerous rudimentary sebaceous glands (glands of Tyson), which secrete the smegma with the desquamated epithelial cells; a dorsal circumference at the balanopreputial groove and an inferior, free circumference (the preputial orifice), at the level of which the skin and the mucous membrane unite.

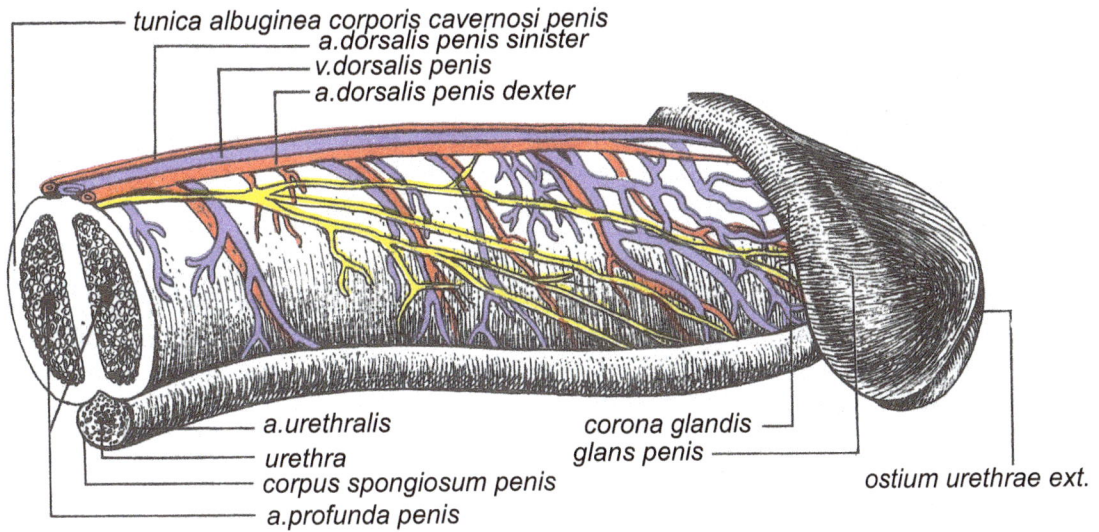

Fig. 268. The penis (vessel supply).

tunica albuginea corporis cavernosi penis
a.dorsalis penis sinister
v.dorsalis penis
a.dorsalis penis dexter

a.urethralis
urethra
corpus spongiosum penis
a.profunda penis

corona glandis
glans penis

ostium urethrae ext.

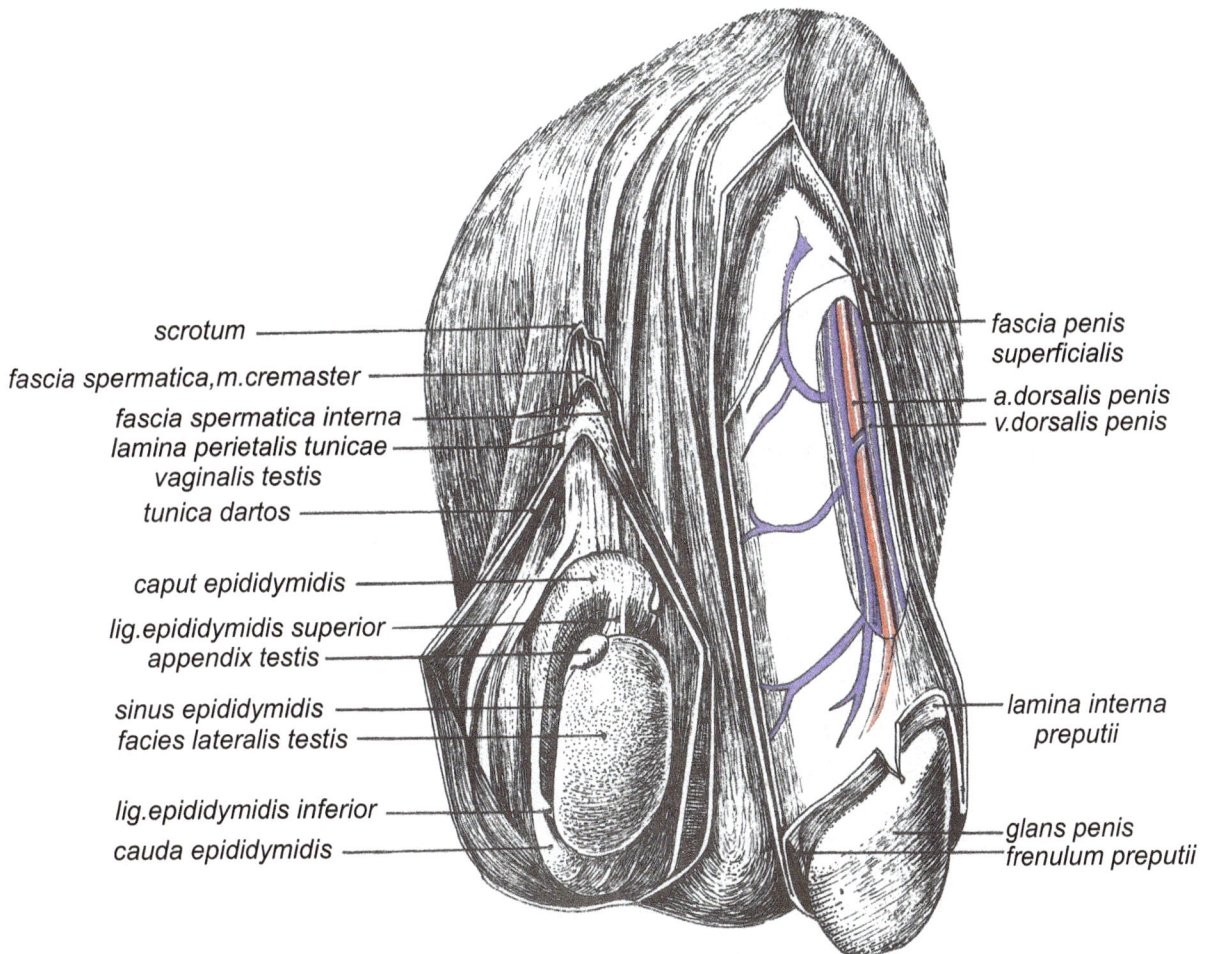

scrotum
fascia spermatica,m.cremaster
fascia spermatica interna
lamina perietalis tunicae
vaginalis testis
tunica dartos

caput epididymidis

lig.epididymidis superior
appendix testis

sinus epididymidis
facies lateralis testis

lig.epididymidis inferior
cauda epididymidis

fascia penis
superficialis

a.dorsalis penis
v.dorsalis penis

lamina interna
preputii

glans penis
frenulum preputii

Fig. 269. Scrotum and penis (lateral view)

277

At the apex of the glans lies the urinary meatus, a sagittal slit of 5-6 mm.

The base of the glans is very oblique from above downwards and from behind forwards. It exceeds in circumference the body of the penis, forming the corona glandis, which is much more marked on the dorsal surface. The glans is bounded behind by the balanopreputial groove, inside the preputial cavity, between the corona glandis and the body of the penis. The portion of the penis which corresponds to this groove is called neck *(collum glandis)*. When the prepuce is too narrow and cannot be refracted over the corona glandis, the condition called phimosis occurs. When the smegma cannot be removed, if may become infected, condition termed balanitis.. The forced passage of the prepuce over the glans results in a paraphymosis.

The penis presents a number of means of attachment. The roof is fixed within the perineum through the perineal anatomical formations. The fundiform ligament *(lig, fundiforme penis)*, of a connective-elastic structure, arises from the abdominal linea alba, descends and passes around the penis, suspending it at the boundary between the roof and the body. Behind this ligament lies the suspensory ligament *(lig. suspensorium penis)*, which has its origin on the pubic symphysis and has the same disposition as the fundiform ligament, being attached on the fascia of the penis.

Structure of the penis. The penis is made up of the erectile organs and their envelopes (fig. 270).

The erectile organs are the cavernous bodies *(corpora cavernosa penis)* and the spongy body *(corpus spongiosum penis)*.

The corpora cavernosa (right and left) are situated on the dorsal surface of the penis. They are 15-16 cm long in the condition of flaccidity and 20-21 cm in the condition of erection. They are cylindrical in shape, united on the median line within the body of the penis, but separated posteriorly, each being firmly attached to the periosteum of the inferior branches of the two pubes, from where they run forwards and join in front of the pubic symphysis.

The two posterior portions are called branches of the corpora cavernosa *(crura penis)*.

The septum of the penis *(septum penis) (septum pectiniforme)* lies between the two corpora cavernosa within the body of the penis and is here and there perforated by acunae, which communicate with each other.

On their dorsal surface lies the dorsal sulcus *(sulcus dorsalis)*, median, longitudinal, in which run the deep dorsal vein *(vena dorsalis penis profunda)*, a very important vein, the dorsal arteries and the nerve with the same name.

On the inferior surface there is also a median, deeper and longitudinal sulcus, the subcavernous sulcus *(sulcus urethralis)*, which lodges the corpus cavernosum urethrae.

The corpora cavernosa are formed of a very thick (1 mm), resistent , fibrous envelope, the albuginea *(tunica albuginea corporum cavernosorum)*, and of a system of trabeculae, consisting of connective, elastic and smooth muscle fibres, which start from the deep surface of the albuginea and from the septum of the penis *(trabeculae corporum cavemosorum)*.

The trabeculae bound a system of areolae *(cavernae coporum cavernosorum)* variable in size, lined with an endothelium, which communicate with each other and are filled with blood. They have the significance of arteriovenous anastomoses and play a part in erection.

The spongy body of the penis (corpus spongiosum penis) is situated on the inferior surface of the penis. It has 12-16 cm in length and is made up of three portions: a median portion, the actual corpus spongiosum; an enlarged posterior extremity, the bulb of the penis *(bulbus penis)*, covered by the m. bulbospongiosus, in which penetrates the urethra and which is traversed by the canals of the bulbo-urethral glands; an anterior extremity, also enlarged, the glans *(glans penis)*, that presents posteriorly a depression in which penetrate the pointed anterior extermities of the corpora cavernosa of the penis. On the inferior surface of the glans, at the level of the corona glandis, there is the frenulum of the prepuce *(frenulum praeputii)*.

The corpus spongiosum has the same structure as the corpora cavernosa, with the difference that the tunica albuginea is thinner, the trabeculae are finer, the caverns smaller and the septum is present only between the two hemispheres of the bulb, suggesting its paired genesis, and lacks in the body of the penis. The envelope of the penis is formed of: skin, continuous with the skin of the pubic region and with that of the scrotum; the muscular *(dartos)* tunic, continuous with the dartos of the scrotum; the connective tunic, made up of a very loose connective tissue, rich in elastic fibres *(fascia penis superficialis)*, which makes possible the mobility of the skin of the penis, and the albuginea *(tunica albuginea penis)*, sited directly on the erectile organs and rich in connective fibres, that impart to it a 1mm thickness and that are less elastic and play a part in erection; however, it does not take part in the formation of the prepuce.

278

Vessel and nerve supply

The arteries of the penis may be classified into two groups: those of the envelopes and those of the erectile organs.

The arteries of the envelopes derive from the pudendal arteries, the femoral artery and the dorsal arteries of the penis.

The arteries of the spongiose corpora are: for the bulb, the bulbar artery; for the proper spongiose body, the bulbar artery and the urethral artery; for the glans, the terminal branches of the dorsal artery of the penis. The arteries of the corpora cavernosa are represented by the two deep cavernous arteries and the two dorsal arteries of the penis. All these arteries destined to supply the erectile formations are branches of the internal pudendal artery, which is a branch of the hypogastric artery (internal iliac artery). In the non-erectile condition of the penis, they follow in the corpora cavernosa a helicoidal course and are therefore called helicine arteries *(aa. helicoidae)*.

The veins are grouped into two sytems: the superficial system, which arises from the envelopes of the penis and forms the subfascial dorsal vein, that opens into the external pudendal veins and then into the right or left great saphenous vein, and the deep system, that empties into the internal pudendal veins.

The lymph vessels form a superficial and a deep network. Their collecting trunks course to the deep and superficial inguinal lymph nodes, the medial group, that may become inflamed, giving rise to an inguinal adenitis (boil), and to the external iliac lymph nodes.

The nerve supply is of two kinds: the envelopes of the penis are supplied by spinal nerves, i.e. branches of the pudendal nerve, of the lumbosacral plexus - through the scrotal, perineal and dorsal nerves of the penis, that receive the excitations which give rise to the voluptuousness sensation; the erectile bodies are supplied by branches of the inferior hypogastric plexus and of the prostatic plexus, branches of the I-III sacral nerves, from these plexuses arising the cavernous nerves, that bring about vasodilation through their parasympathetic fibres, respectively vasoconstriction through their sympathetic fibres, a mechanism which induces the erection of the penis.

The centre which controls the sexual activity lies in the inferior spinal cord (the genitospinal centre). It is subordinate to cerebral centres which, through excitations of visual, olfactive or other origin, may give rise to the erection, followed by the ejaculation brought about by the genitospinal centre.

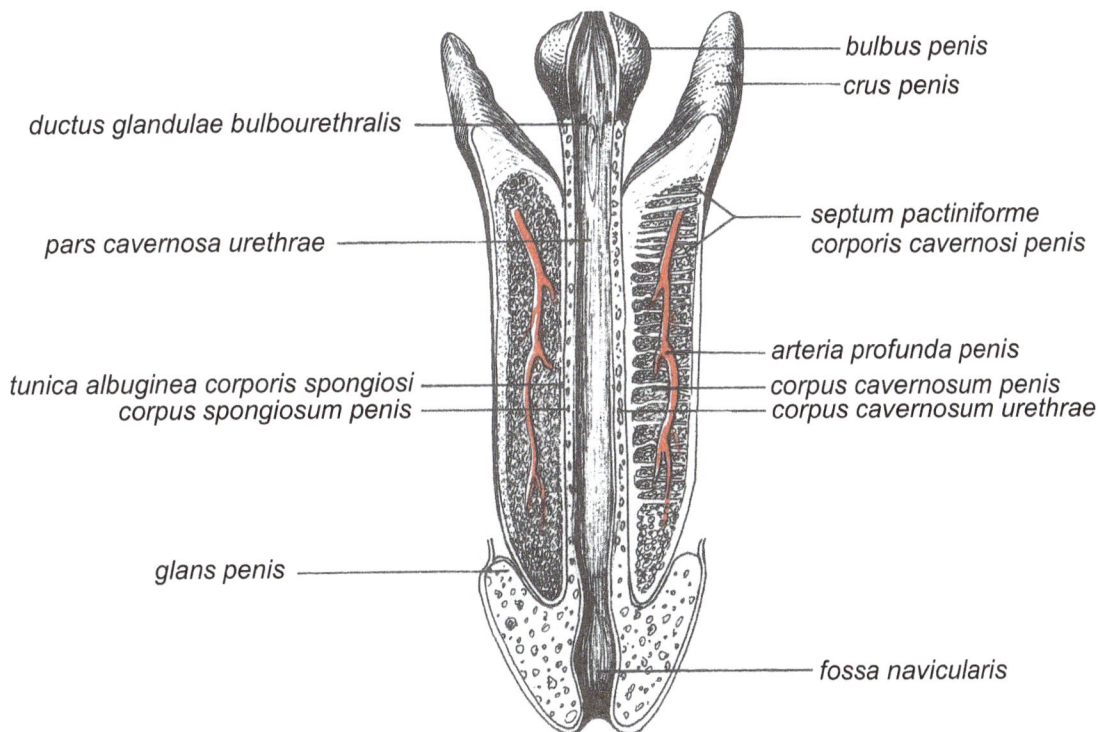

Fig. 270. The penis - structure.

Some Medicosurgical Applications

Chapter 5

Some Medicosurgical Applications

✍ **The tracheotomy** is an operation of extreme emergency, which consists in performing a new opening of the airway, below the level of the glottis; the operation is indicated in all preasphyctic conditions.

According to the level at which it is performed, the procedure may be a cricotracheotomy, a superior tracheotomy or an inferior tracheotomy (fig. 271).

The cricotracheotomy or coniotomy is the operation of extreme emergency which is carried out when the instruments required for the maintenance of the opening of the superior airway created, are not available (fig. 272).

The tracheotomy consists in a longitudinal incision, measuring 5-6 cm, through the skin and the thin thyrostemal muscular planes, below which appear immediately the first four tracheal rings. Sometimes, the isthmus of the thyroid is detached and ligated.

When a tracheotomy is performed as an emergency procedure, the first tracheal rings are sectioned as rapidly as possible and a Krishaber's cannula is inserted into the created orifice (fig. 273). Under conditions of extreme emergency, if no tracheotomy kit is at our disposal, the operation is performed with any cutting instrument available (knife, blade, penknife, glass splinter etc.), the tracheal orifice maintaining itself open by means of the patient's movements of inspiration and expiration.

When a tracheotomy is carried out, avoidance of a damage to the surrounding vascular formations should be kept in mind (fig, 274).

In addition, throughout the period of maintenance of the tracheotomy, the steps required to prevent the infection should be taken.

✍ **Thorax injuries** have continuously increased in frequency as a result of the mechanization and development of the road traffic and in war-time, owing to the use of highly sophisticated weapons. In addition, their severity has increased, too, and approximately 45% of these injured subjects die on the spot of the accident.

Usually, the lesions are polymorphous and occur either as a result of the impaction of the different regions of the trunk between or below hard bodies or as a consequence of gunshot wounds (fig. 275).

✍ **The haemothorax** is a collection of blood in the pleural cavity, which may have various origins and requires a correct diagnosis and treatment. Its severity depends on the accumulated amount of blood (fig.276,277).

✍ **The pneumothorax** is an accumulation of air in the pleural cavity and may be either closed or open with an internal or external valve.The closed pneumothorax may be produced by a parietal or pleuropulmonary solution of continuity, its severity depending on the existent amount of air.

✍ **The thoracic shutter** is a complex polyfracture of the chest wall, characterized by a double fracture on vertical lines of at least two ribs, associated with fractures of the surrounding bones (clavicle, scapula), which give rise to severe cardiorespiratory disturbances as a result of the great instability of the thorax (fig. 278). As a consequence, at the moment of inspiration, the shutter is drawn inwards, whereas during expiration it is pushed outwards, bringing about the "paradoxical respiration", which is often lethal (fig. 279).

✍ **The wounds of the thorax** are particularly severe when they are penetrating and especially transfixing. More frequent among these wounds are the pneumothorax through aspiration and the open pneumothorax, which cause the so-called "traumatopneic syndrome".

In the pneumothorax through aspiration, the communication with the outside is intermittent, especially in forced inspiration or during the cough effort. In the open pneumothorax, the traumatopneic syndrome is more severe when the parietal solution of continuity exceeds in width the diameter of the glottis.

The thoraco-abdominal wounds are much more severe and may be situated at various levels, in accordance with the points of entrance and exit of the vulnerant agent or with the area of action of the contusive agent (fig. 280). The treatment of chest injuries requires the urgent carrying out of measures, sometimes minimal and simple, but which may save the life of the injured.

In case of rib fractures, good results are obtained by the introduction of novocain or xylene with alcohol below the rib, at the neck of the fractured rib or/and at the site of fracture, allowing the patient to breathe easily and precluding the possibility of pulmonary complications (congestion, pneumonia) due to blood stasis in the lobar parenchyma (fig. 281).

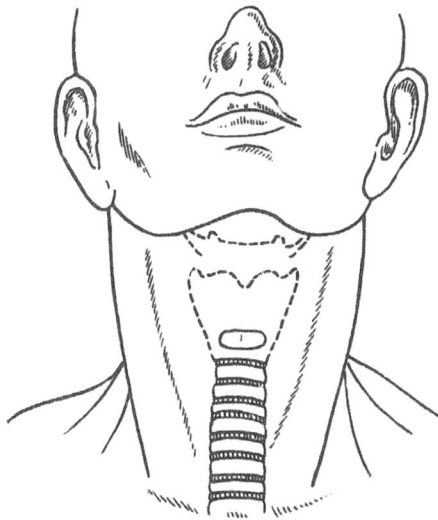

Fig. 271. Schematic representation of the region in which tracheotomy is performed.

a

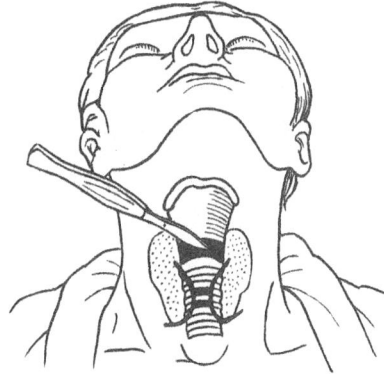

b

Fig. 272. Coniotomy. a) - patient's position;
b - surgical technique.

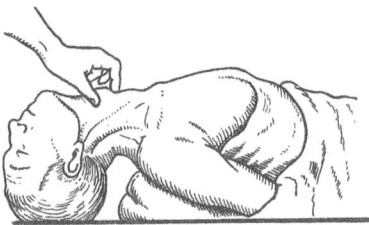

a

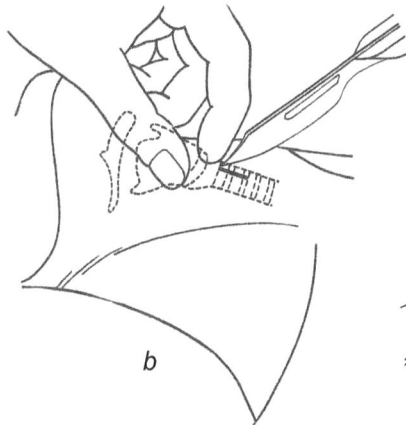

b

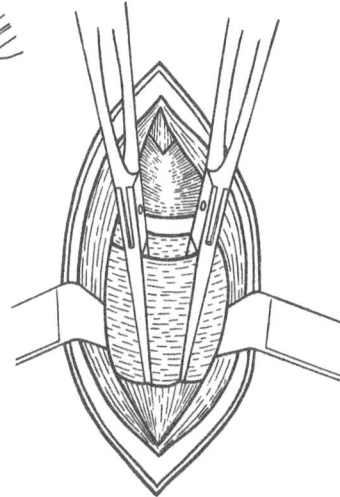

c

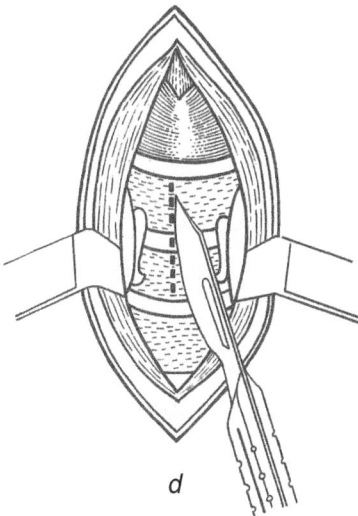

d

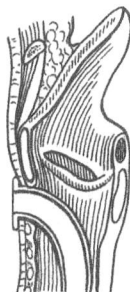

e

Fig. 273. Tracheotomy.
a- patient's position for the purpose of
undergoing operation; b - median cutaneous
incision; the left hand fastens the larynx, the
index of the left hand locates the middle of the
lower border of the cricoid cartilage;
c- the detached isthmus of the thyroid is divided
between two clamps; d -opening of the trachea;
e - position of the cannula in the trachea.

283

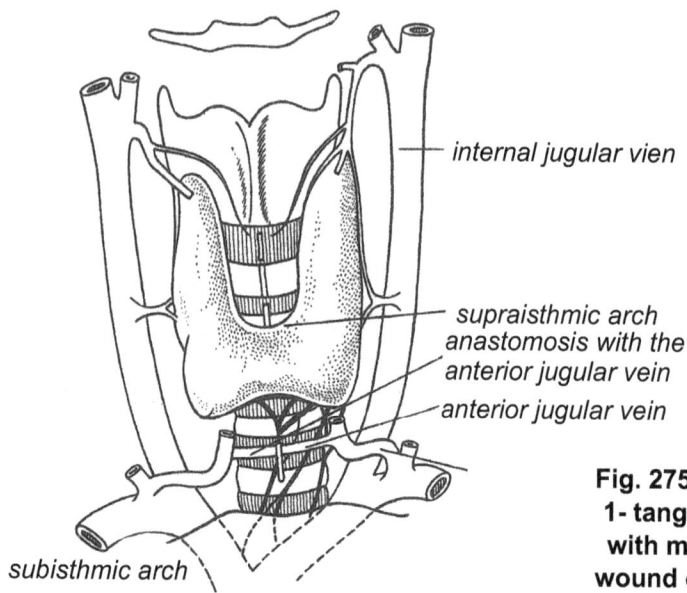

internal jugular vien

supraisthmic arch
anastomosis with the
anterior jugular vein

anterior jugular vein

subisthmic arch

**Fig. 274. Venous risk in carrying
out a tracheotomy**

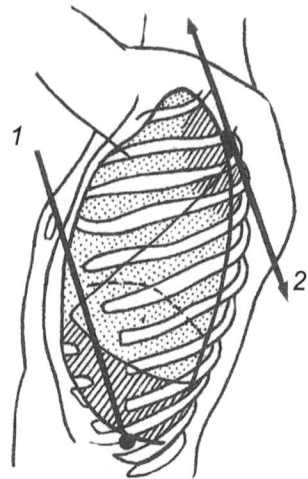

**Fig. 275. Scheme of tangential thoracic wounds.
1- tangential wound of the lateral thoracic wall,
with multiple rib fractures; 2 - deep, tangential
wound of the thoracic wall, with contusion - foci
per contiguitatem at the level of the lung,**

**Fig. 277. Severity of the
haemothorax a - small
haemothorax (less than 30 ml of
blood); b - moderate haemothorax
(less than 150 ml of blood); c -
bulky haemothorax (over 3,000 ml
of blood)**

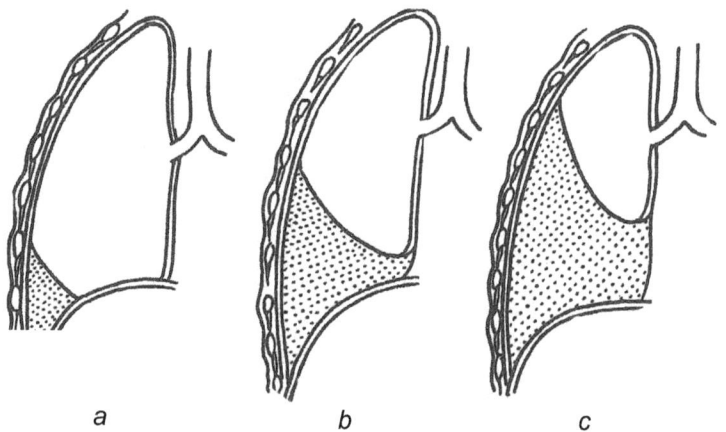

a   b   c

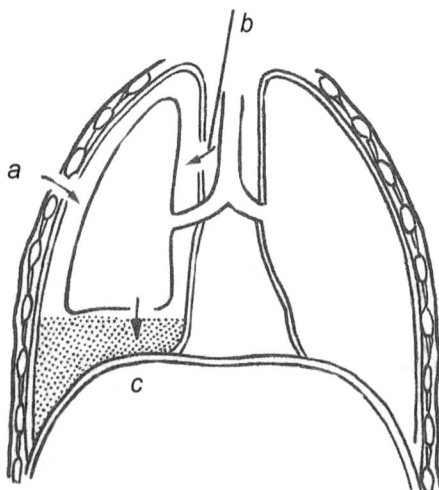

**Fig. 276. Origin of the haemothorax a - peristal
injury (damaging of the intercostal, internal
mammary, pleural arteries etc.); b - pulmonary
wound (damaging of the pulmonary
parenchyma); c - mediastinal injury (vascular
visceral injury)**

284

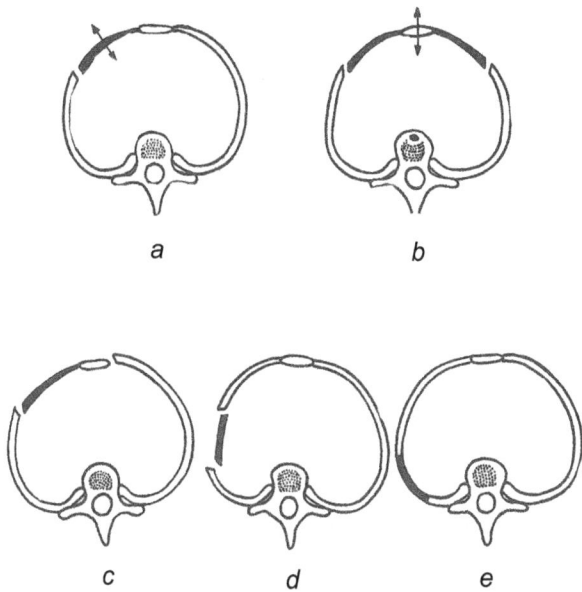

Fig. 278. Anatomoclinical variants of the thoracic shutter: a - anterior; b - anterolateral; c - lateral; d - posterior; e - half-shutter

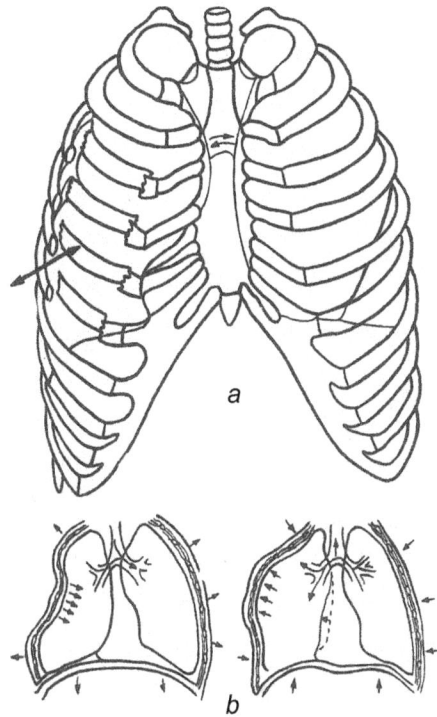

Fig. 279. Costal shutter paradoxical respiration (b) during inspiration, on the left, and during expiration, on the right

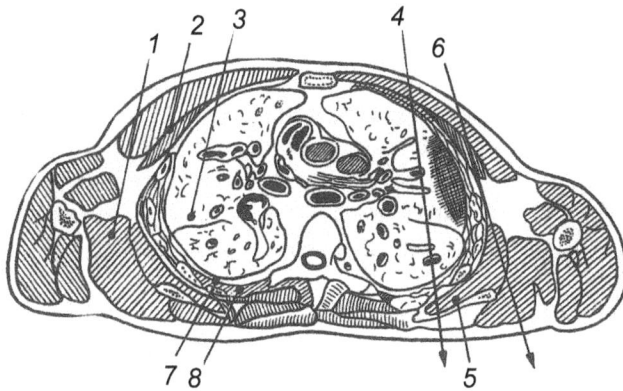

Fig. 280. Horizontal section through the thorax, representing schematically the various types of injuries of the two cavities:] - penetrating wound of the clavipectoral region; 2 - penetrating pulmonary wound; 3 - penetrating wound of the thoracic wall; 4 - perforating thoracopulmonary wound; 5 - perforating pulmonary wound and penetrating thoracic wound with arrest of the projectile in the subcapsular region; 6 - tangential costal wound, accompanied by pulmonary contusion; 7 - penetrating thoracopulmonary wound without damaging the lung; 8 - penetrating thoracic wound without involving the pleura.

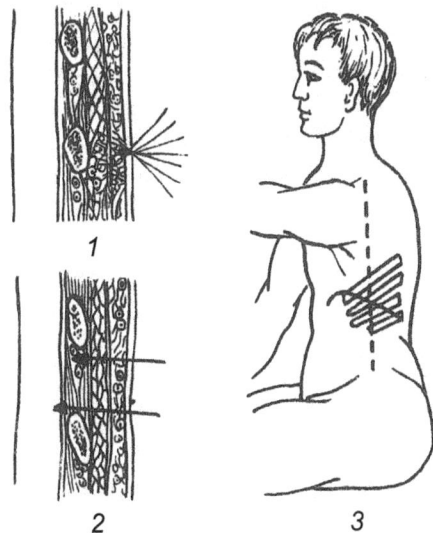

Fig. 281. Performing of local anaesthesia 1 - anaesthesia of superficial planes; 2 - anaesthesia of planes; 3 - anaesthesia area

285

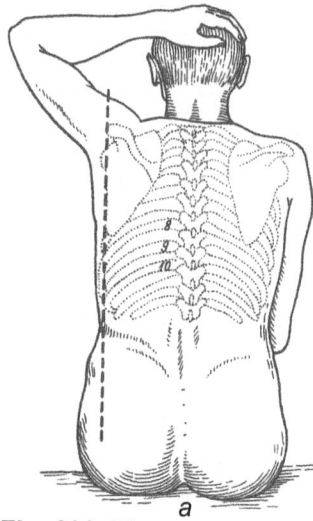

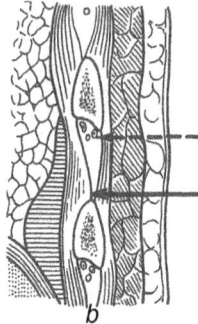

Fig. 282. Thoracocentesis a -patient's position and landmarks for performing the thoracocentesis; b - anatomical landmarks of the intercostal space; 1 – faulty puncture (injury to the vasculonervous costal bundle; 2 ▪ correct puncture (on the upper costal edge).

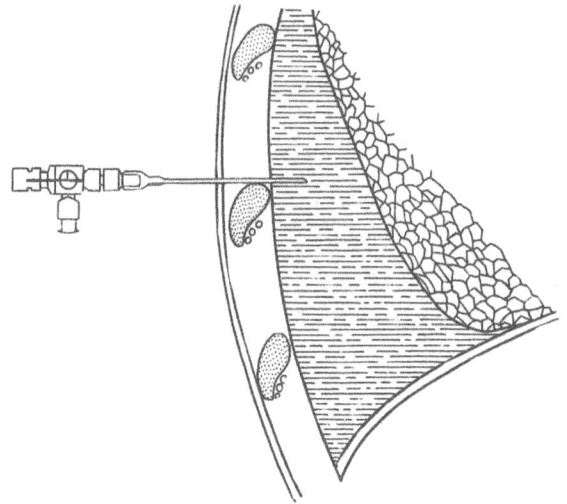

Fig. 283.Introduction of the trocar

-*The thoracocentesis* represents the method of puncture of the pleural cavity with the aim of exploration, evacuation (blood, air, pus) or therapy (introduction of antibiotics after previous evacuation). The patient will be placed in lateral decubitus (on the healthy side) or on.the buttocks the most suitable position. After a previous disinfection with tincture of iodine of the respective region (the posterior axillary line) *at* the level of the corresponding number of ribs, local anaesthesia is induced. The trocar for puncture is introduced on the superior border of the inferior rib, beyond the parietal pleura, the pleural space is penetrated and the pathological product (blood, pus, air) contained in it is extracted (fig. 282, 233). The loss of blood may be successfully corrected by means of the blood extracted from the thoracic cavify (haemothorax).

-*The minimal pleurotomy* is used in case of pyothorax (pus in the pleural cavity) or in some forms of pneumothorax. After a preliminary anaesthesia, under perfectly aseptic conditions, the usual trocar or the Monod's trocar is introduced. If the usual trocar is utilized, the mandrin is removed and a catheter of India rubber is passed through the cannula. The cannula of the trocar is removed and the catheter is fastened to the skin (fig. 284). The Monod's trocar is specially adapted for the minimal pleurotomy, the catheter being inserted into the lumen of the trocar itself and, after the removal of the latter, the catheter is left in place (fig. 285).

In operations on the thorax, drainage with Pezzer's catheter is performed postoperatively, the intercostal musculature forming a sphincter around the catheter, which it fastens. The tube of India rubber is fitted to a system of pleural aspiration.

Fig. 284. Minimal pleuroomy by means of the common trocar. a- introduction of the trocar; b- introduction Pezzer's chateter ; c- removal of the trocal;

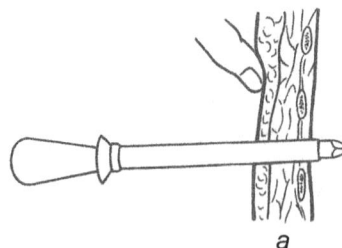

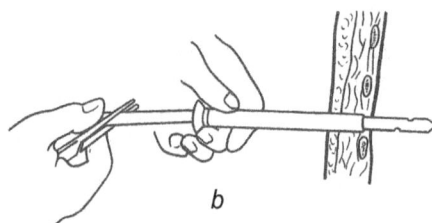

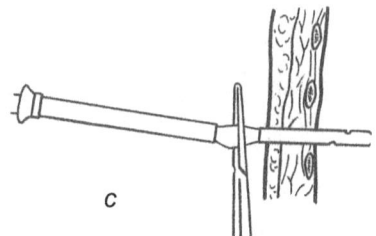

286

Since the occurrence of the thoracic shutter is accompanied by the formation of chest wall areas in which an normal movements and also pleuropulmonary lesions are present, it is necessary to perform emergency surgery for the treatment of the lesions and the fastening of the movable shutter. For this purpose, the metallic blade fixed at the site of the lesion, the retrosternal metallic blade, improvised external tractions or the device of O. Constantinescu may be utilized.

In the case of major operations on the lung, various ways of surgical approach may be used.

A troublesome accident, which may have severe repercussions, is the aspiration of various foreign bodies in the airways. They stop at different levels, in accordance with their volume and with the level up to which they have penetrated the tracheobronchial tree (fig. 286). For their removal, it is necessary to resort to the cough reflex and/or Heimlich's manoeuvre, a sudden, powerful pressure applied at the base of the xiphoid appendix.

✍ **The congenital cardiac malformations** may be isolated or associated and can be present in one of the following eventualities: vascular positional anomalies, abnormal intercavity communications and vulvar or vascular stenoses or atresiae. In the clinic, we distinguish "white" or non-cyanotic heart diseases and cyanotic heart diseases. Non-cyanotic heart diseases are the: interatrial communication and the interventricular communication, and cyanotic heart diseases are: Fallot's tetralogy, Fallot's pentalogy, Fallot's trilogy, obstructive congenital angiopathies, complex congenital angiopathies or patent arterial duct (Botallo's duct) between the pulmonary artery and the aorta (fig. 287-295).

✍ **The pericardial puncture** is performed only when indubitable signs of pericardial tamponade are present (fig. 296), in order to establish the presence and nature of the fluid, with the purpose of introduction of antibiotics, and especially in the imminence of cardiac arrest (introduction of epinephrine, calcium).

Various points have been described where the pericardial puncture may be performed (fig. 297), but the best procedure is that proposed by Marfan, in which the puncture is made beneath the tip of the xiphoid process (fig. 298).

In case of cardiac arrest, the ribs are counted obliquely down up to the fifth rib and the needle is inserted into the fifth intercostal space to the left of the border of the sternum (fig. 299).

Mastitis or abscess of the mammary gland occurs almost exclusively during the period of lactation and appears under various clinical forms, in dependence on the site and severity of the infection (fig. 300). The only treatment is surgical, which in association with an intense and complex antibiotherapy may cure these rather severe lesion (fig. 301).

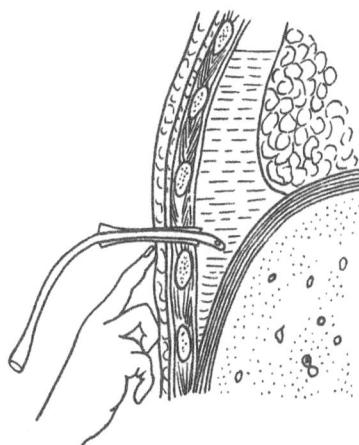

Fig. 285. Minimal pleurotomy by means of Monad's trocar.

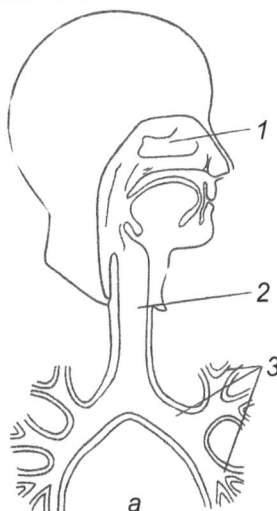

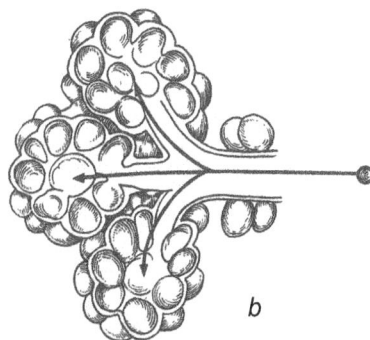

Fig. 286. Particles aspirated in the airways and mode of their expulsion: a - site of the particles in the bronchial tree: 1 - nasopharynx (10 microns in size); 2 – trachea (5-10 microns in size); 3 - bronchi and bronchioles (2-3 microns in size); b - site of the particles in the alveoli (1-2 microns); c - discharge of the particles through the cough reflex

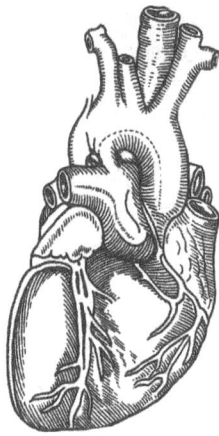

Fig. 293. Botallos duct
(ductus arteriosus).

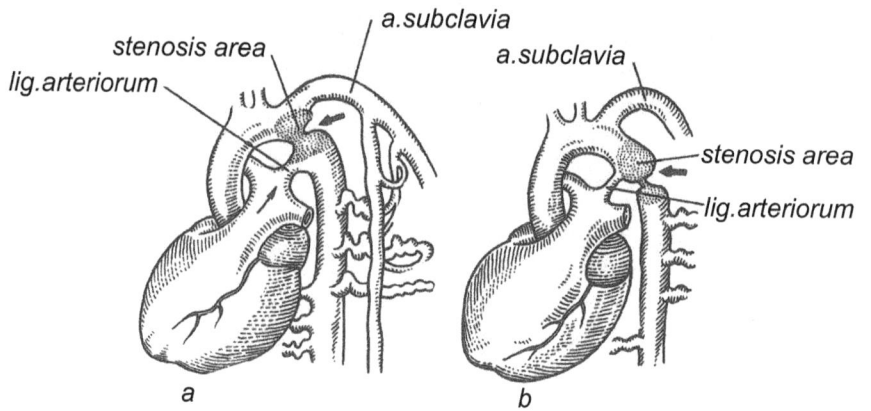

stenosis area
lig.arteriorum
a.subclavia
a.subclavia
stenosis area
lig.arteriorum

a

b

Fig. 294. Obstructive congenital angiopathy a) coarctation of
the aorta, infantile type; b) coarctation of the aorta, adult type

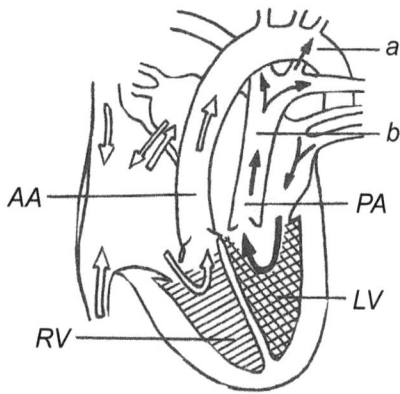

AA
PA
RV
LV
a
b

Fig. 295. Complex congenital angiopathy-
transposition of the great vessels AA -
anterior atrium; PA - posterior atrium; RV=
right ventricle; LV- left ventricle; a = the
aorta arises from the right ventricle; b - the
pulmonary artery arises from the left
ventricle.

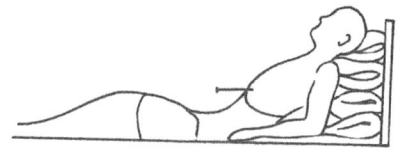

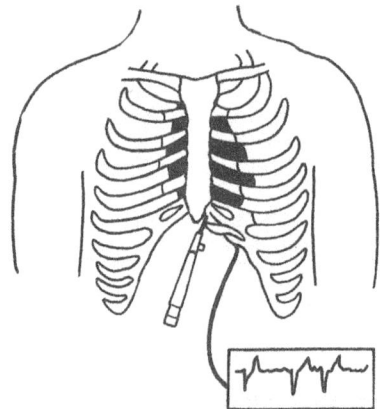

Fig. 296. The pericardiocentesis should be
performed most rapidly in case of cardiac
tamponade, the injured subject being in a
half – sitting posture.

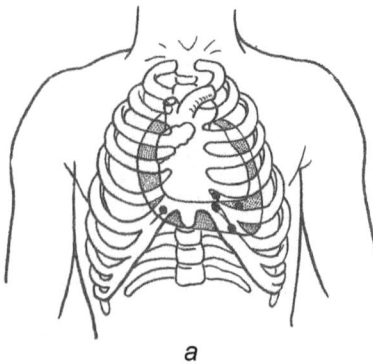

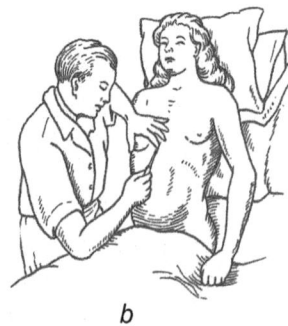

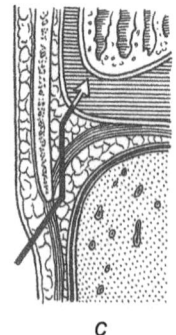

a

b

c

Fig. 297. Pericardial puncture a - points where the pericardial puncture may be
performed; b - patient's position; c - direction of the needle.

288

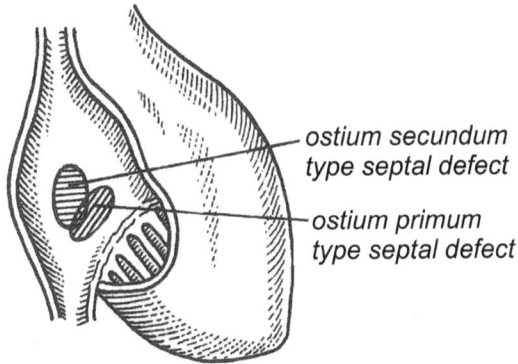

Fig. 287. Non-cyanotic barf disease through interatrial communication.

ostium secundum type septal defect

ostium primum type septal defect

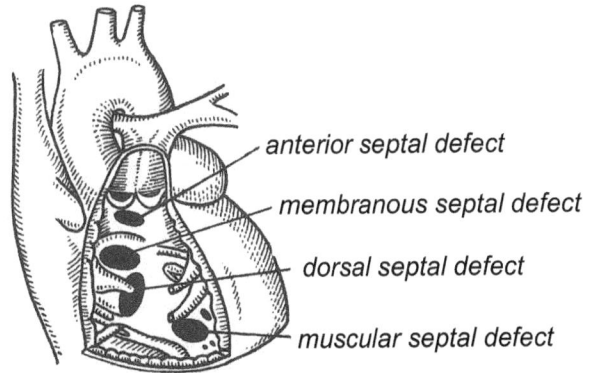

anterior septal defect

membranous septal defect

dorsal septal defect

muscular septal defect

Fig. 288. Non-cyanotic heart disease through interventricular communication (Roger's disease). A large communication occurs in Eisenmenger's syndrome

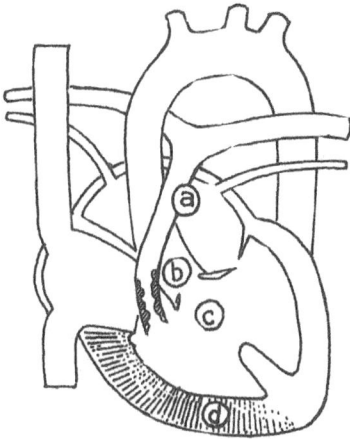

Fig. 289. Cyanotic heart disease due to Fallot's tetralogy: a - atresia of the pulmonary artery; b - dextroposition of the aorta; c - interventricular communication; d - hypertrophy of the right ventricle

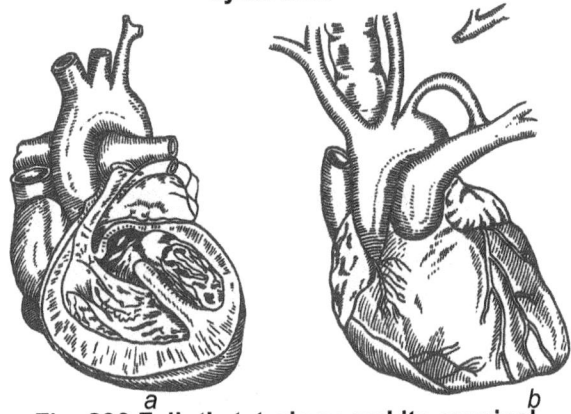

a   b

Fig. 290 Fallot's tetralogy and its surgical correction: a - anatomopathological appearance; b - anastomosis between the subclavian artery and the left branch of the pulmonary artery

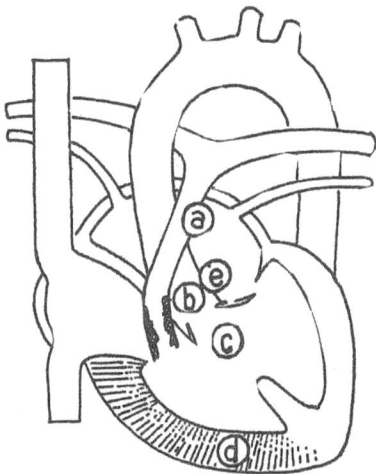

Fig. 291. Cyanotic heart disease due to Fallot's pentalogy: a - atresia of the pulmonary artery; b - dextroposition of the aorta; c - interventricular communication; d - hypertrophy of the right ventricle; e - interatrial communication.

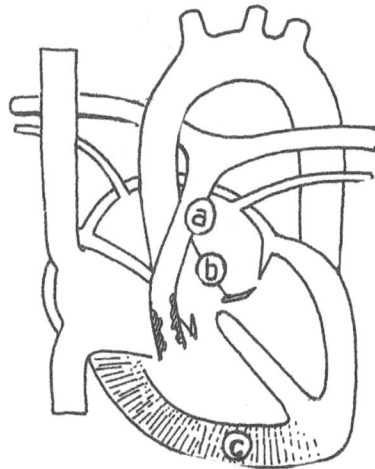

Fig. 292. Cyanotic heart disease due to Fallot's trilogy: a- atresia of the pulmonary artery; b - interatrial communication; c - hypertrophy of the right ventricle

289

Hernias are the most frequent diseases of the abdominal wall and they consist in the partial o total protrusion of an organ from the abdominal cavity, concomitantly with its normal envelopes (fig. 302). Post-traumatic or postoperative hernias are termed eviscerations or eventrations. Infernal hernias occur through abnormal orifices of the mesenteries and manifest themselves under the form of intestinal obstructions.

The tumefactions that bulge under the skin and are produced through the weak spots of the abdominal wall are called external hernias. They may occur either along the physiological orifices (inguinal canal, umbilicus), or at the level of muscular crossings (linea alba, Spiegel's line, Petit's triangle, Grynfeltt tetragon etc.).

The crossed arrangement of the muscle fibres of the abdominal walls assures a high resistance to the mechanical factors which act whenever the intra-abdominal pressure increases, as in coughing, sneezing, vomiting, prolonged straining for evacuation of the faeces in case of constipation etc.

In other regions of the abdominal wall, too, owing to the above mentioned causes, to which also other factors may join, a lowering of the resistance capacity of the connective tissues may occur, leading to the formation of hernias.

-*The inguinal hernia* is the most frequent of the external hernias and occurs along the inguinal canal.

The inguinal canal is a passage through which in the male, in the intra-uterine life, the testis with The spermatic vessels descends from the abdomen into the scrotum.

At birth, it contains in the male the spermatic cord and in the female, the round ligament of the uterus. This canal runs between the muscles of the abdominal wall, which it perforates. Although it is protected by a number of fibrous formations, if may be forced and dilated by repeated abdominal hypertensions, leading to the occurrence of inguinal hernias.

The inguinal hernias may be external oblique, direct and internal oblique, in dependence on the fossa they pass through. The external oblique inguinal hernias penetrate through the internal ring of the inguinal canal and lie in the lateral inguinal fossa *(fovea inguinalis lateralis),* which is situated outside the epigastric vessels *(a.,v. epigastrica inferior).*

Direct hernias develop in the medial inguinal fossa *(fovea inguinalis medialis,* which is situated between the epigastric vessels and the lateral vesico-umbilical fold *(plica vesico-umbilicalis lateralis),* due to the obstructed vesico-umbilical arteries. Internal oblique hernias have their site in the supravesical fossa   *(fovea supravesicalis),* situated between the lateral vesico-umbilical fold and the median vesico-umbilical fold, which is constituted by the obstructed urachus (fig. 303).

As regards the evolution, there are four varieties of the external oblique hernia: hernial point, bubonocele, funicular hernia and inguinoscrotal hernia, form in which the herniated organs have penetrated within the scrotum (fig. 304).

In the female, inguinal hernias occur rarely, because the round ligament of the uterus traverses easily the muscles of the abdominal walls, owing to its very small size.

-*The femoral hernia* is brought about by dilatation and penetration of the abdominal contents (omentum and/or intestine) through one of the orifices situated in the femoral region.

The femoral region is separated above from the inguinal region by the femoral arch, a fibrous bridge extending between the anterosuperior iliac spine and the spine of the pubis.

Projected on the inferior abdominal wall, it is called Malgaigne's line and separates inguinal from femoral hernias.

Laterally are the psoas muscle and the femoral nerve and dorsally, Cowper's ligament From within outwards lie Gimbernat's ligament, the femoral vessels and the inguinopectineal ligament, sites at which femoral hernias may occur In dependence on these anatomical landmarks, the following anatomical variants of the femoral hernia are described: transligamental (Gimbernat's), paravascular and intervascular (Laugier's) hernias, the most frequent forms met (fig. 305).

The femoral hernia occurs much more often in the female because, owing to the conformation of the pelvis, the space created below the inguinal ligament, through which the vessels, the nerves, the lymphatics and the iliopsoas muscle pass, is much larger than in the male.

-*The lumbar hernia* may occur either through the Petit's triangle or through Krause's tetragon (fig. 306).

The J.L Petit's triangle is formed by the superior border of the iliac crest, the posterior border of the external oblique muscle and the anterior border of the latissimus dorsi muscle.

Krause's tetragon is constituted of the inferior border of the twelfth rib, the quadratus lumborum muscle, the latissimus dorsi muscle and the external oblique muscle.

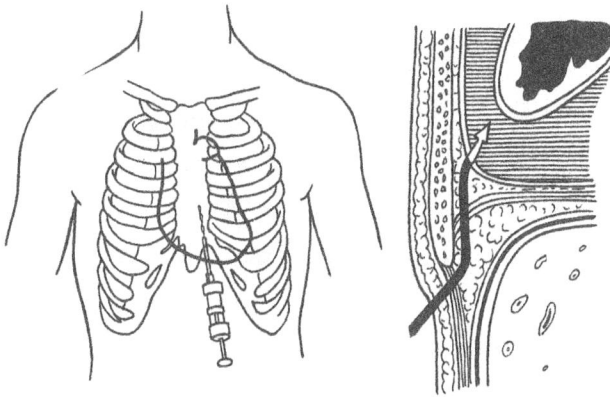

Fig. 298. Marfan's method.

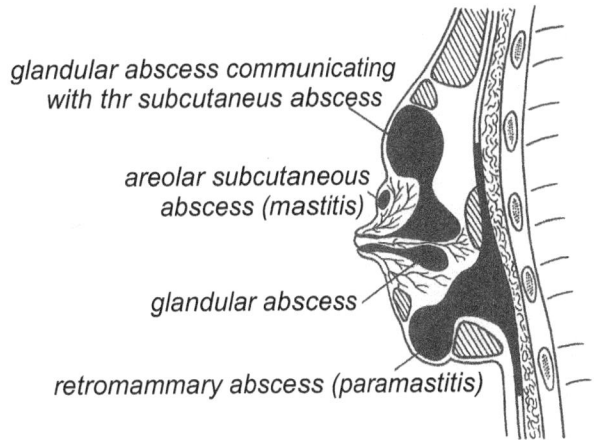

glandular abscess communicating with thr subcutaneus abscess

areolar subcutaneous abscess (mastitis)

glandular abscess

retromammary abscess (paramastitis)

Fig. 300 Inflammatory lesions of the mammary gland (mastitis).

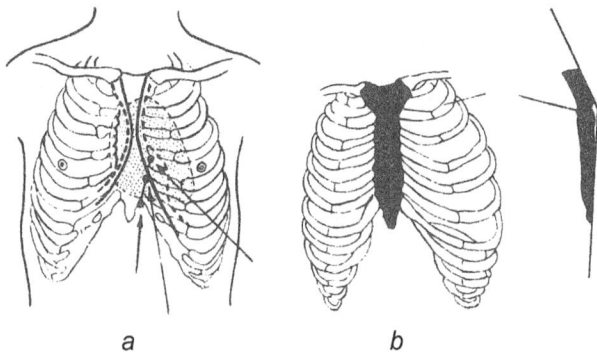

a   b

Fig. 299. Pericardial puncture in cose of cardiac arrest a - elective points; b - location of Louis's angle.

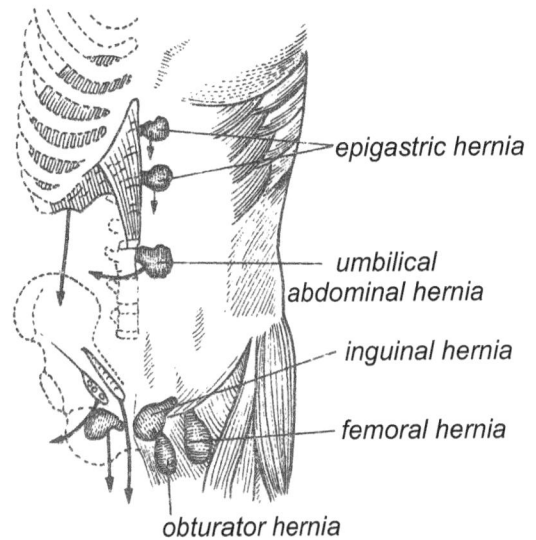

epigastric hernia

umbilical abdominal hernia

inguinal hernia

femoral hernia

obturator hernia

Fig. 302. Types of hernias of the anterior abdominal wall.

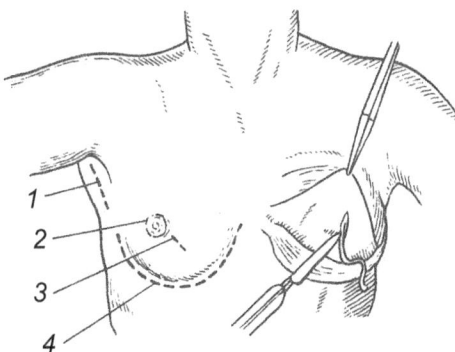

Fig. 301. Various incisions for the drainage of mammary abscesses. 1 - retropectoral incision; 2 - pariareolar incision; 3 - radial incision; 4 - submammary incision

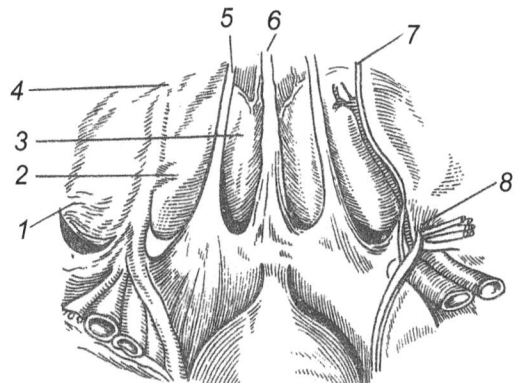

Fig. 303. The inguinal fossae 1. lateral inguinal fossa; 2. medial inguinal fossa; 3. supravesical fossa; 4. epigastric fold; 5. lateral vesico-umbilical fold; 6. median vesicoumbilical fold; 7. inferior epigastric a.,v.; 8. internal ring of the Inguinal canal.

291

-**The umbilical scar**, when it is weakened owing to infections or as a consequence of whooping cough in children, may represent a site where the occurrence of umbilical hernias is possible.

-At the linea alba, between the umbilicus and the xiphoid process, on the very midline, through the small orifices existing at this level may occur *hernias of the linea alba*, which appear under different clinical variants, according to their contents: subperitoneal lipoma, subperitoneal lipoma with a serous diverticulum, epiplocele, hernia of the small intestine (Richter's hernia) (fig. 307).

-**The hernia of Spigelius' line** *(spigelian hernia)* is much less frequent (fig. 308) and occurs in the supero-external part of the junction of two projection lines (Douglas'arch and Spigelius' line) on the abdominal wall

-Exceptionally, *ischiatic hernia* and **perineal hernia** may occur.

When hernias are bulky, there is danger of strangulation of the intestinal loops, with their possible necrosis, that may lead to exitus if the operation is not performed as an extreme emergency. To prevent these accidents, the early operation, in the period of formation of hernias, is advisable. The operation consists in the anatomical restoration of the abdominal walls.

-At the **level of the diaphragm**, herniations towards the thorax may occur, either following the lowering of resistance of the tissue at the site of passage of the oesophagus *(oesophageal hiatus)* or at the insertion of muscle bundles, or following trauma (operative). Between the oesophagus and the borders of the oesophageal opening in the diaphragm lies a fibro-elastic connective tissue - the Bertelli-Leimer's membrane-, an elastic connexion between the diaphragm and the oesophagus, which assures the mobility and sliding of the oesophagus during deglutition and respiration (fig. 309). In formation of the hiatus takes part predominantly the right pillar *(crus dexter)*, the His's angle being also formed between the stomach and the oesophagus. In some situations, the oesophageal hiatus is abnormally enlarged, which makes possible the herniation of the abdominal oesophagus and of the stomach in the thorax, leading to the formation of hernias of the oesophageal hiatus or *hiatal hernias*. In dependence on the mechanism and on the anatomicoclinical form, hiatal hernias may be: sliding or axial, para-oesophageal or lateral (through uncoiling, the cardia remaining subdiaphragmatic) or hernias through brachyoesophagus, i.e. hernias of the stomach in the case of a short oesophagus (fig. 310). Likewise, at the level of the sternal part *(pars sternalis)* of the diaphragm, between the two bundles, lies a hiatus (Morgagni's sinus) through which the mediastinal cell tissue may communicate with the subperitoneal tissue. These bundles are separated from the costal portion through Larrey's hiatus or Larrey's cleft *(trigonum sternocostale)*. Between the costal and the lumbar portion is enclosed the Bochdalek's hiatus *(trigonum lumbocostale)*. Through all these openings, congenital or acquired diaphragmatic hernias may protrude in some cases, or infections may spread from the abdominal to the thoracic cavity and vice versa (fig. 311).

-**Diaphragmatic eventrations** occur as a consequence of malformations during the embryonic development, following severe injuries or after operations. In some cases, merely loosenings of the dipahragm may be observed.

-**Internal hernias** occur following the protrusion of a segment of the small intestine through a peritoneal fossa, through Winslow's hiatus or through a cleft at the level of the mesenteries or of the broad ligament of the uterus in the female. There are situations in which, owing to an error of surgical technique, intestinal loops may protrude through these orifices, like at the level of the transverse mesocolon during operations of gastric resections and anastomosis between the gastric stump and the duodenum.

✍ **The occlusive syndrome** is an abdominal drama, caused by the presence of a dynamic or mechanical hindrance along the intestine. At the first stage, the organism endeavours to overcome the hindrance, as show the distension and the projection of peristaltic waves on the anterior abdominal wall. If the removal of the hindrance cannot be obtained by medical means, an emergency operation is imperative.

✍ **Abdominal injuries** are also recorded with an increasing frequency in hospital practice. According to the depth of the lesion, nonpenetrating wounds (the peritoneum is spared), penetrating wounds (the peritoneum is damaged) and perforating wounds (the intra-abdominal organs are injured, too) may be distinguished (fig. 312). As a consequence, in the peritoneal cavity may be present blood *(haemoperifoneum)* or pus *(pyoperitoneum)*. Some diseases of the abdominal organs are associated with the presence of ascitic fluid etc.

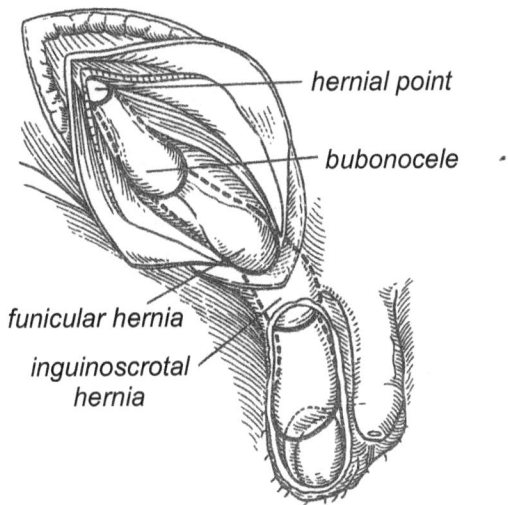

Fig. 304. Evolutive varieties of the external oblique inguinal hernia

hernial point

bubonocele

funicular hernia

inguinoscrotal hernia

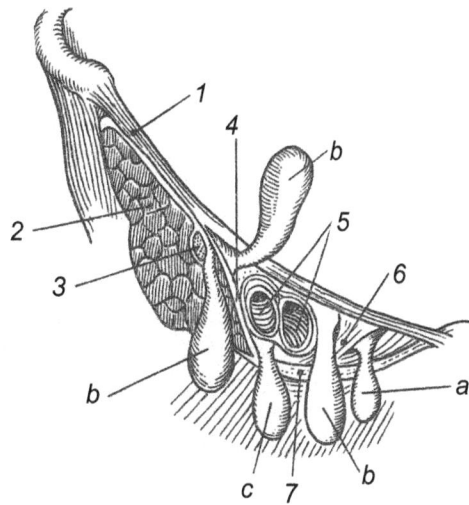

Fig. 305. Anatomical variants of the femoral hernia 1. femoral arch; 2. psoas muscle; 3. femoral nerve; 4. inguinopectineal ligament; 5. femoral vessels; 6. Gimbernat's ligament; 7. Coles ligament: a - transligamental hernia (Gimbernat); b - paravascular hernia and c - intervascular hernia (Laugier), the most frequent forms.

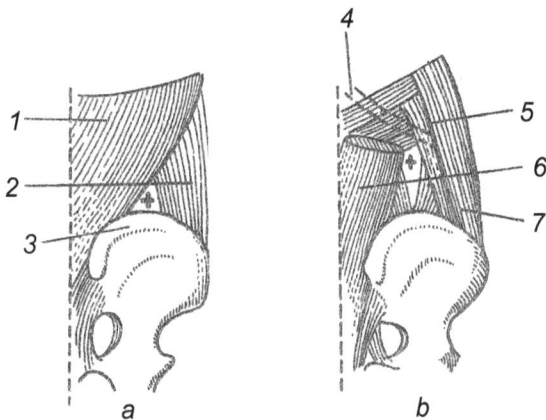

Fig. 306. Lumbar hernia ▪ occurrence sites a - triangle of J.L.Petit; b ▪ Krause's tetragon 1. m, latissimus dorsi; 2. external oblique muscle; 3. iliac crest; 4. twelfth rib; 5. m. quadratus lumborum; 6. m. latissimus dorsi; 7. internal oblique muscle

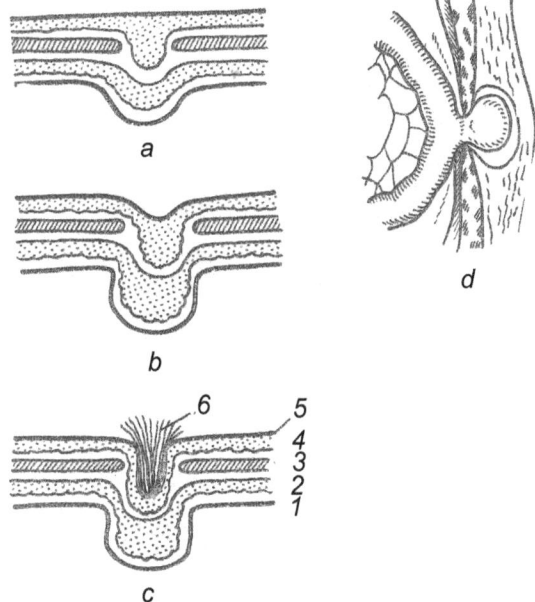

Fig. 307. Hernia of the linea alba - clinical variants a - subperitoneal lipoma; b - subperitoneal lipoma with a serous diverticulum; c - epiplocele: 1 - skin; 2 - subcutaneous cell tissue; 3 – linea alba; 4 - subperitoneal fat; 5 - peritoneum; 6 - omentum; d – Richters hernia (contents, small intestine)

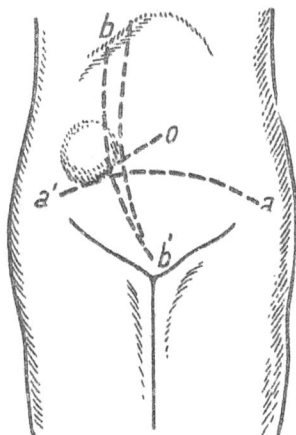

Fig. 308. Spiegel's hernia a'o - spino-umbilical line; aa' - projection line of Douglas' arch; bb' ▪ projection of Spiegel's line

293

The haemoperitoneum occurs as a result of the rupture of a parenchymal organ (liver, spleen) or of the vessels of the mesentery, as well as following the rupture of an extra-uterine or ectopic pregnancy (pregnancy which develops outside the uterine cavity). According to the damaged organ, the amount of blood effused and the speed of bleeding appears the condition of traumatic shock (accelerated pulse, exceeding 120/minute, arterial pressure below 10 cm Hg etc). The treatment is only surgical, under an adequate resuscitation.

The pyoperitoneum occurs following the infection of the peritoneal serosa, mostly after perforating lesions of the digestive tract, and is characterized by spontaneous pain, abdominal immobility, cutaneous hyperesthesia (Dieulafoy's sign), muscular defence (Blumberg's sign) (fig. 313) and muscle contracture (pathognomonic sign) (fig. 314). Peritoneal infections may be diffuse or localized. The most severe are the diffuse infections, since they produce generalized peritonitis, requiring an extremely urgent and highly competent medicosurgical treatment. The intraperitoneal pathological fluids diffuse either in the supramesocolic or in the submesocolic space, as well as in the whole peritoneal cavity (fig. 315).

Besides the clinical signs (earth-coloured facies, fever, leucocytosis, vomiting etc.), the rectal examination renders possible the detection of the infection in the cul-de-sac of Douglas (fig. 316). Both the exploratory and the therapeutic peritoneal punctures are performed at this level, since this is the place where the pathological fluid usually accumulates (fig. 317). In case of incertitude regarding the site of the generalized infection of the peritoneum, the abdominal puncture in the four quadrants is performed (fig. 318).

Renal injuries are quite frequent and severe, although the kidney is protected by a strong muscular mass, since it is situated in the retroperitoneal region. The abundant vagosympathetic nerve supply at this level causes and maintains the condition of post traumatic shock and the rich vascular network produces extensive and bulky haematomas, which may simulate an intestinal obstruction (fig. 319). In dependence on the integrity of the surrounding thissues and on the communication with the outside, these

Injuries are closed or open. The most frequent are the closed injuries, that may appear under various anatomicoclinical forms: intrarenal rupture with a subcapsular haematoma, extrarenal rupture with a perirenal haematoma, cortical open fissure in cavities with haematonephrosis and total rupture with perirenal haematoma and haematonephrosis (fig. 320). The two last clinical varieties are very severe as there is danger of infection, owing to the fact that they manifest themselves through total haematuria, which is often very intense and particularly severe. Renal wounds are observed especially in war-time and involve usually also the surrounding organs.

As regards the mechanism of production, it may be direct or indirect (contusion of the lumbar transverse processes, of the ribs or of the bones of the pelvis).
The lesions require, on account of their severity, a prompt diagnosis and an adequate therapy.

The injuries of the urinary bladder are perhaps more severe than those of the kidney, since they are followed by overflowing of urine which is toxicoseptic within the peritoneum or the surrounding tissues. It is a well-known fact that 100 ml of human urine may kill a rabbit weighing 2 kg (fig. 321, 322). Doubtless, the state of fullness of the urinary bladder exerts an influence on the severity of the respective wounds. If the bladder is empty, lesions of medium severity occur, without damage to the bladder, but with the possible involvement of the peritoneum. If the bladder is full, the injury is very severe, with a septic effusion into the surrounding organs, although the wound is extraperitoneal. These injuries can have a favourable course only if the diagnosis has been established accurately and in due time (fig. 323-325) and if all therapeutic medicosurgical measures have been taken: resuscitation, emergency operation with drawing up of an inventory of the lesions and their repair, followed by cystotomy with Pezzer's drainage and massive antibiotherapy, adapted to the antibiogram. There are cases in which the urinary bladder cannot be emptied by urethral catheterization, owing to strictures of the urethra, ruptures of the urethra, adenoma of the prostate etc. In these situations the acute retention of urine occurs, which requires urgently *the puncturing of the urinary bladder*. After shaving the hair in the pubic region and local asepsis of the integument, the puncture is performed exactly on the midline, immediately above the symphysis. A small-sized trocar or a thicker syringe needle, 12-14 cm long, is used. A perpendicular or slightly oblique direction towards the symphysis is imparted to the needle. Urine is discharged slowly, without aspiration. Quick aspiration, as well as rapid discharge through the catheter should be avoided, since "ex vacuo" haematurias may occur (fig. 326).

294

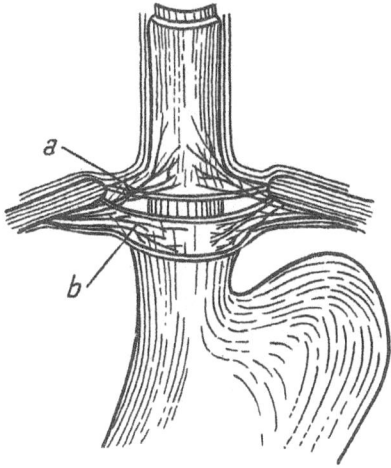

Fig. 309. Bertelli - Leimer's phreno - oesophageal membrane: a - Juvara's fibres; b - Rouget's fibres.

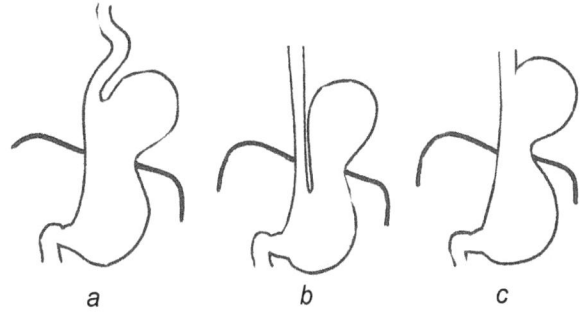

Fig. 310. Anatomoclinical variants of hiatal hernias: a - through sliding (paraaxial); b - through unraveling (paraoesophageal); c - through brachyoesophagus.

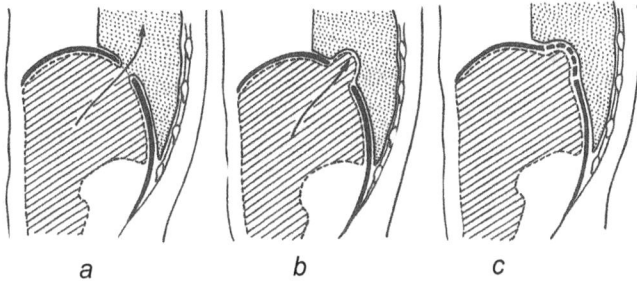

Fig. 311. Lesional variants at the level of the diaphragm: a - Bochdalek's hernia (embryonic); b - hernia following a congenital malformation; c - diaphragmatic loosening.

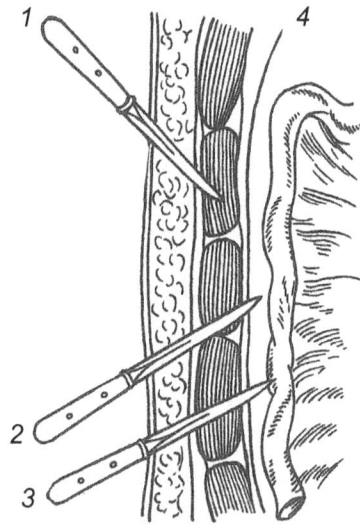

Fig. 312. Classification of abdominal injuries according to the depth of the lesion: 1 - non-penetrating wound; 2 - penetrating wound; 3 -perforating wound; 4 - peritoneal serosa.

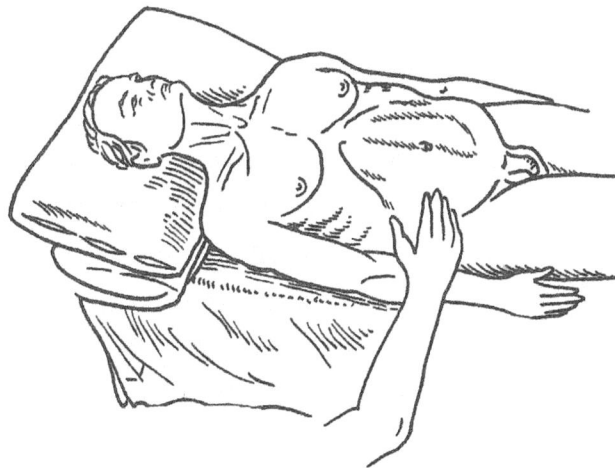

Fig. 313. Muscular defence.

Haematuria may be often caused by *the presence of a vesical calculus* or a *vesical tumour*. In these cases, after a cystoscopy performed to establish the diagnosis, one may proceed - before deciding the operation which suppresses radically the disease - to lithotrity by means of the lithotrite cystoscope (the calculus is crushed and then eliminated by natural ways) (fig. 327) or to biopsy followed by cauterization (fig. 328).

The injuries of the urethra are not very frequent and may be closed or open. It is important, clinically and therapeutically, to establish the site of the lesion, because any injury inflicted to the anterior urethra - from the meatus to the median perineal aponeurosis - is followed by the formation of a sclerotic tissue at the site of impact, as this portion of the urethra is invested by a spongy sheath. Owing to the absence of this sheath at the posterior urethra,the scar is at this level less elastic, but the vicious cicatrizations lead to direction and calibre deformations, fact which explains the severity of the urethra injuries. The ruptures of the anterior urethra may be brought about by several mechanisms: pressure on the pubic symphysis, pressure on the symphysis and/or the arcuate ligament and pressure on the ischiopubic branch.

Among other diseases of the urethra we mention: retention of foreign bodies, presence of urethral calculi, perilithiasic infections localized at the urethra, urethral strictures and fistulae.

Foreign bodies are retained only by the male urethra and, if they are not eliminated by miction, they should be removed, in order to prevent severe complications: inflammations, fistulae etc.

The urethral calculi, too, are quite infrequent and occur only in the male, either originating in the superior urinary tract or in the bladdder, or being autochthonous, formed in the urethra, as a result of the local pathological conditions, that favour the urinary stasis and the appearance of infection in the posterior urethra. According to their site, they may be penile, scrotal or perineal (fig. 329).

The perilithiasic infections localized at the urethra appear at the same sites as the calculi, if an emergency operation was not performed for their removal The most severe are the perineal abscesses (fig. 330).

The urethral strictures are permanent lesions which diminish the calibre of the urethra and may be: congenital, traumatic or inflammatory. More frequent are the traumatic strictures, following urethral damages or ruptures; they occur shortly after the accident.

The treatment of strictures is indicated only after the stabilization of the lesion and consists in dilatations of the urethra by means of rubber catheters (Nelaton's catheter, filiform-tipped catheters etc.) or of metallic instruments (Benique).

In cases in which the dilatation of the urethra to assure a physiological miction could not be obtained, the operation is indicated, consisting in the resection of the stricture with terminoterminal urethrorrhaphy.

The urethral fistulae are communications, through pathological routes, of the urethra with the outside. They occur follwing infections or injuries and appear under two clinical variants: urethrocutaneous (urethrobulbar, urethroprostatic or urethropenilobalanic) fistulae and urethrorectal (urethrorectal or urethroprostaticorectal) fistulae (fig. 331).

The urethrorectal fistulae are the most severe and occur as a result of the opening of a prostatic abscess, both into the rectum and the urethra.

Among the diseases of the prostate we mention the inflammations of the prostate and the adenoma of prostate.

The inflammations of the prostate are very severe and, if not treated in due time and correctly, they give rise to complications which require therapeutic measures of an extreme complexity.

The abscesses of the prostate may open spontaneously into the rectum, the urethra, the ischiorectal space and the prerectal space, the last two being particularly severe (fig. 332).

The adenoma of the prostate, periurethral adenoma (Marion) or hypertrophy of the prostate (Mercier) represents a benign tumour of the prostate, which occurs in elderly subjects and is clinically expressed as an enlargement of the gland, which hinders the emptying of the urinary bladder, causing simultaneously miction disturbances.

The only curative treatment is surgical and consists in the extirpation of the adenoma by enucleation from its bed, from which it is separated by a good cleavage plane.

Phimosis is the narrowing of the orifice of the prepuce and represents the most frequent of the regional anomalies. When the conservative treatment through progressive dilatations, is inefficacious, the best results are obtained by means of the circumcision. The operation consists in the cutaneous resection, the sectioning of the mucosa and then the circular suture between skin and mucosa.

296

Among the complications of phimosis, the most severe is the paraphimosis, manifested by strangulation of the glans in the orifice of the prepuce, narrowed and inextensible. In case of failure of the manual reduction, the strangulation ring should be sectioned surgically.

The puncture of the tunica vaginalis testis is indicated in the serous, purulent or haemorrhagic effusions, with the purpose of exploration, evacuation or therapy.

The routes of surgical approach to the urinary apparatus should take into account the anatomical data: they should penetrate among the muscular interstices, be as wide as possible without sectioning the important anatomical formations (vessels, nerves) and render possible a successful restoration (fig. 333-334).

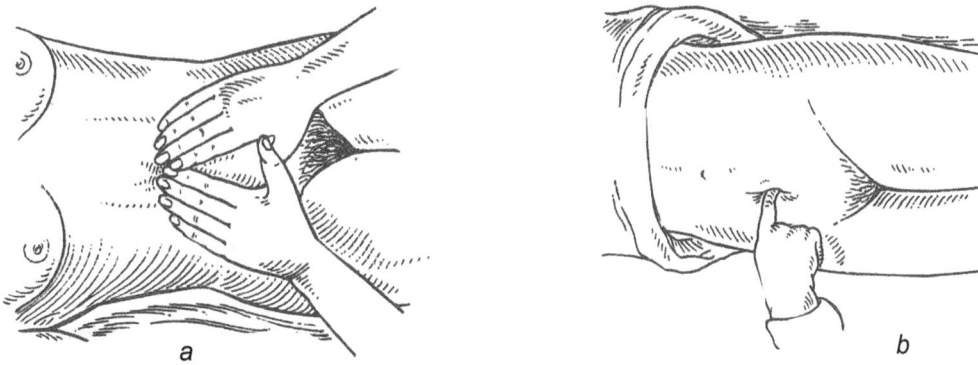

Fig. 314. Exploration of the abdominal contracture in case of peritonitis; a - correct method: the region is explored with both hands and with the whole palm; b - faulty method: exploration with a finger deep implanted.

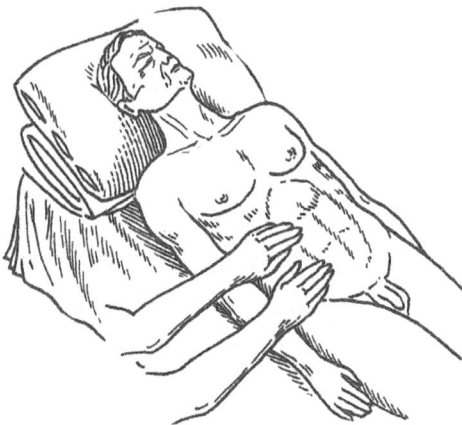

Fig. 315. Correct technique of deep exploration of the abdomen - the abdominal contracture („wooden belly") is detected.

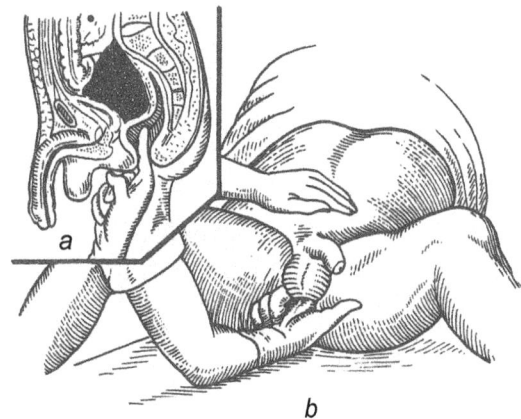

Fig. 316. Rectal touch in the male, by which the infection sited in Douglas' cul-de-sac is detected a - schematic sagittal section; b - technique of rectal touch.

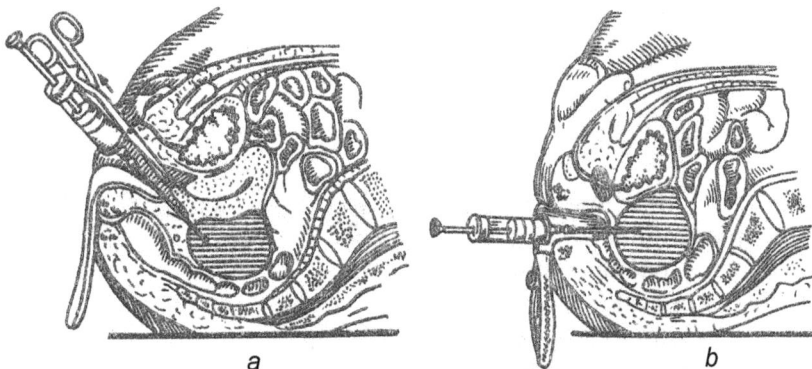

Fig. 317. Puncture of Douglas cul-de-sac a - puncture in the female (transvaginal) b - puncture in the male (transrectal)

297

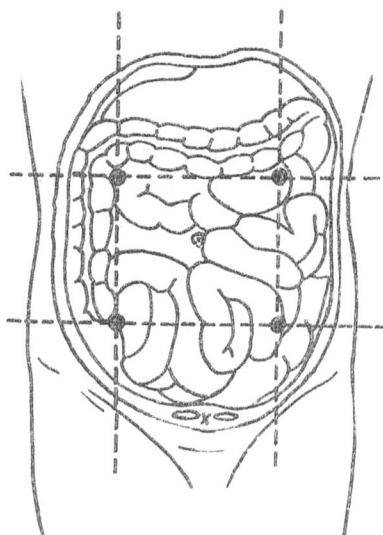

Fig. 318. Puncture of the peritoneum in its four points, in case of intraperitoneal collections (blood, infections, ascites)

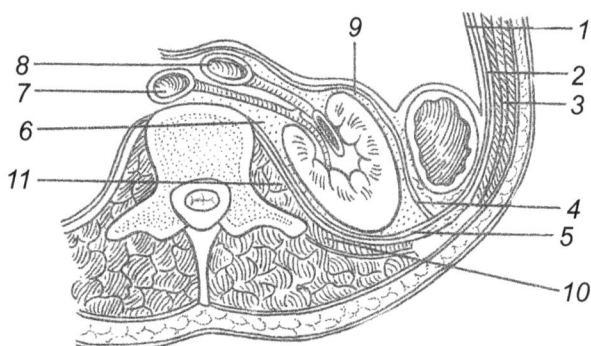

Fig. 319. Relations of the kidney and formation of the renal bed (transversa section): 1 - peritoneum; 2 - transverse fascia; 3 - muscles of the abdominal wall; 4 - prerenal leaflet; 5 - Zuckerkandl's rectorenal leaflet; 6 - pararenal space; 7 - aorta; 8 - vena cava; 9 – pararenal space; 10 - m. quadratus lumborum; 11 - iliopsoas muscle

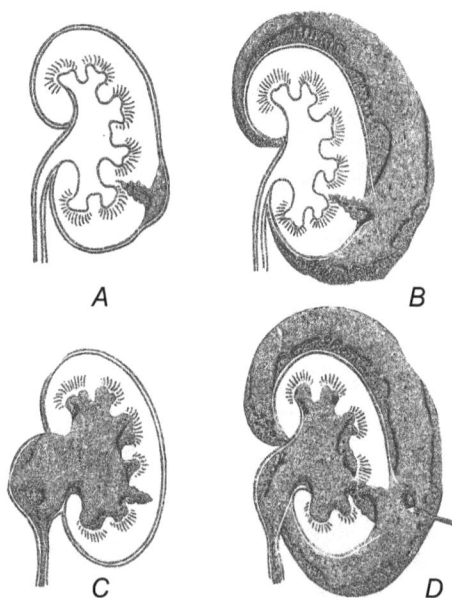

Fig. 320. Types of renal injuries; A - intrarenal rupture with subcapsular haematoma; B extrarenal rupture with perirenal haematoma; C - cortical open fissure in cavities, with haematonephrosis; D - total rupture with perirenal haematoma and haematonephrosis.

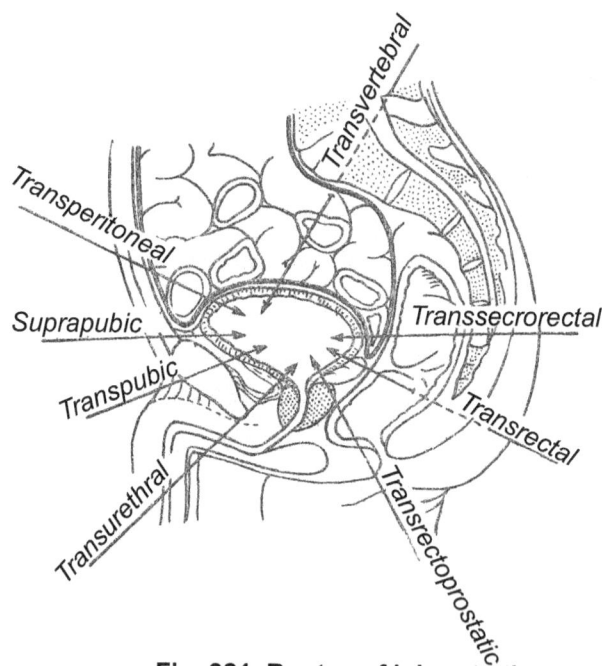

Fig. 321. Routes of injury to the urinary bladder

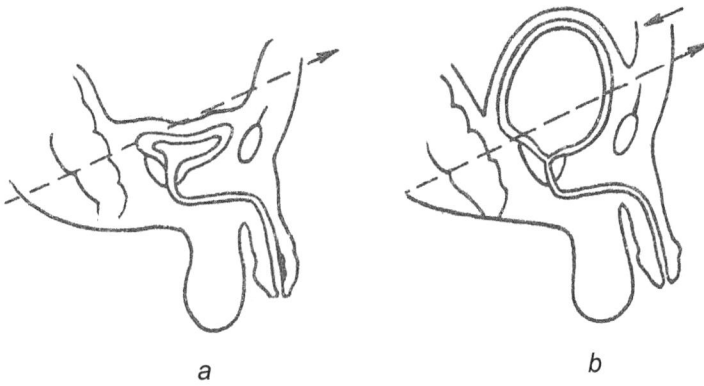

Fig. 322. Influence exerted by the state of fullness of the urinary bladder on the severity of injuries: a - empty bladder - injury of medium severity, without involvement of the bladder, but with the possible damage of peritoneum; b - full bladder - very severe injury, with septic effusion into the surrounding organs, although the wound is extraperitoneal

a   b

Fig. 323. Anterior topographic relations of the urinary bladder 1- Transverse fascia; 2. Umbilico-prevesical aponeurosis; 3. Peritoneum; 4. Suprapubic space; 5. Prevesical space

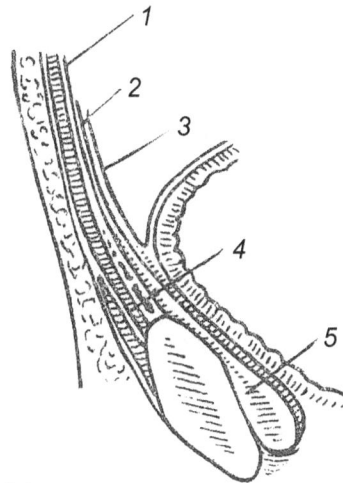

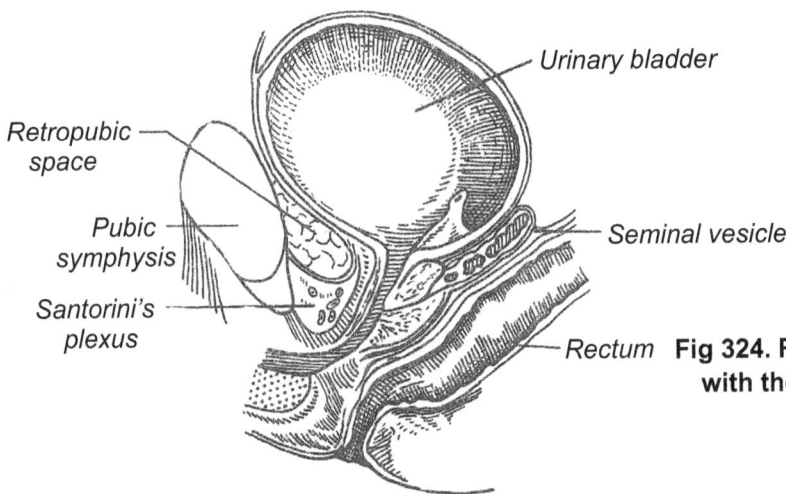

Urinary bladder

Retropubic space

Pubic symphysis

Santorini's plexus

Seminal vesicle

Rectum

Fig 324. Relations of the urinary bladder with the neighbouring organs in the female

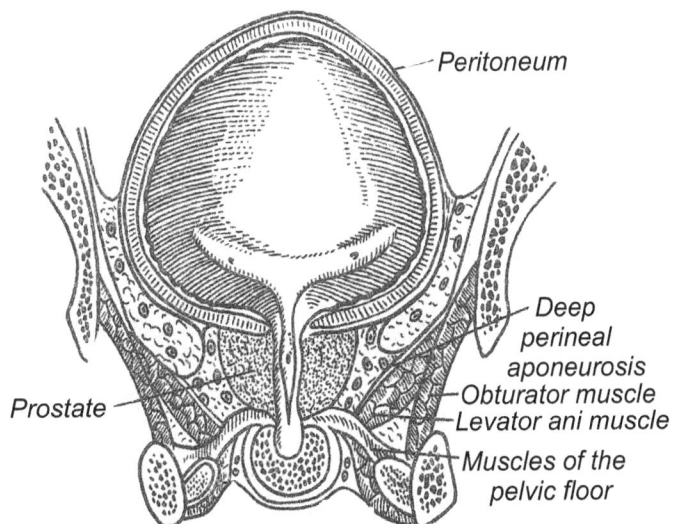

Peritoneum

Fig. 325. Relations of the urinary bladder with the neighbouring organs in the male

Prostate

Deep perineal aponeurosis
Obturator muscle
Levator ani muscle
Muscles of the pelvic floor

299

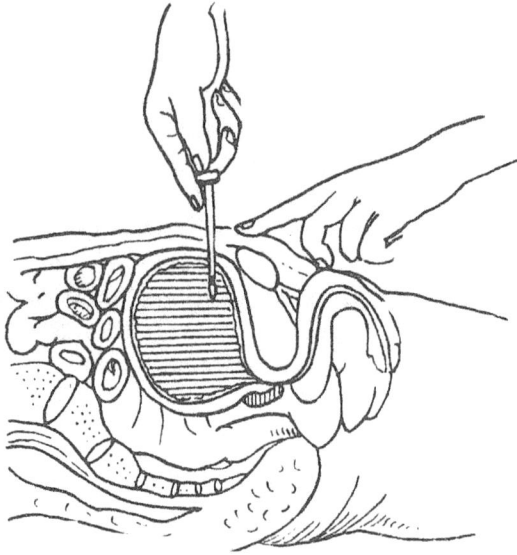

Fig. 326. Puncture of the urinary bladder

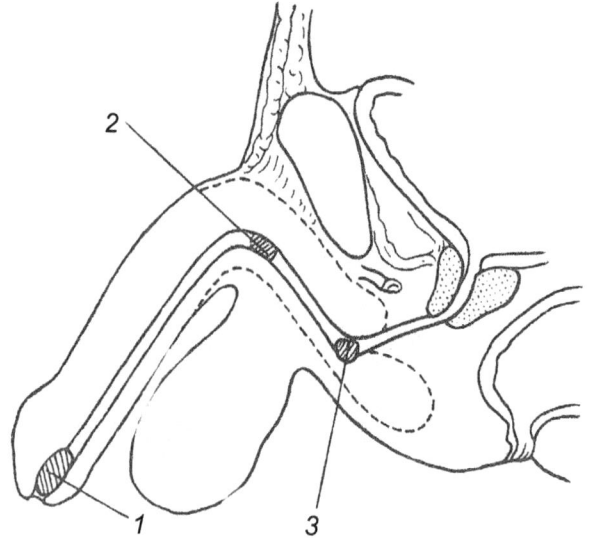

Fig. 329. Site of the urethral calculi;
1. penile; 2. scrotal; 3. Perineal.

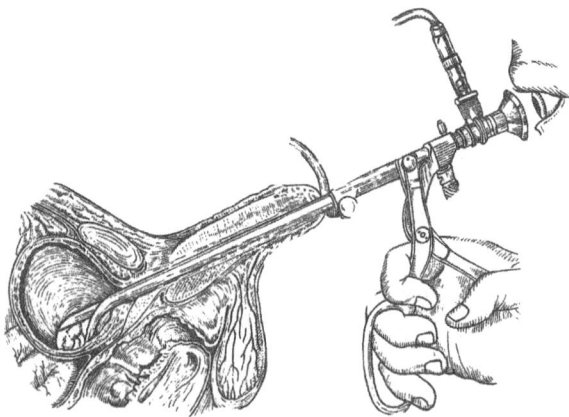

Fig. 327. Lithotripsy by means of the lithocystoscope

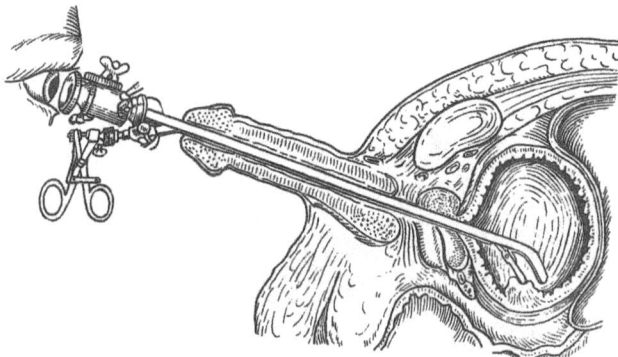

Fig. 328. Endoscopic biopsy from a vesical tumour.

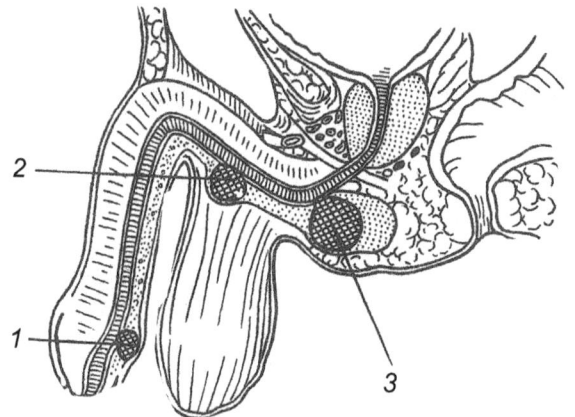

Fig. 330. Development of infections
around urethral calculi; 1. calculus +
penile abscess; 2. calculus + scrotal
abscess; 3. calculus + perineal abscess

300

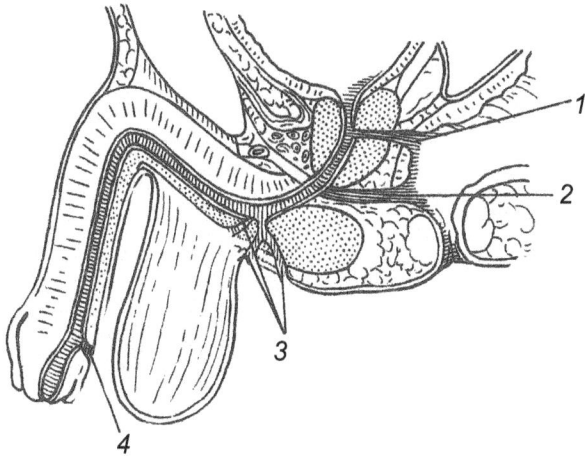

Fig. 331. Urethral fistula; 1. urethroprostato-rectal;
2. urethrorectal; 3. perineoscrotal; 4. Balanic.

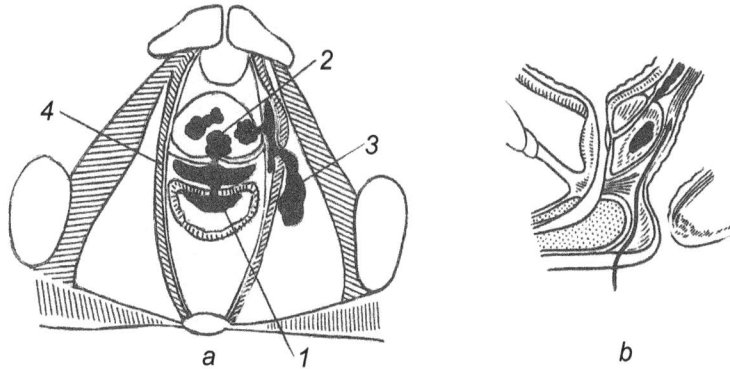

Fig. 332. Course of the prostatic abscess: a - transverse section; b - mediosagittal section.
Spontaneous opening of the abscess into: 1 - rectum; 2 - urethra; 3 - ischiorectal space; 4 -
prerectal space.

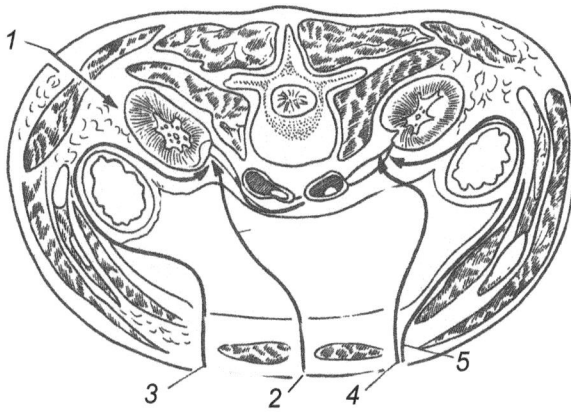

Fig. 333. Routes of access to the kidney
1. Lombotomy; 2. transperitoneal route;
3. extracolic intraperitoneal median route;
4. intraperitoneal median route; 5. extraperitoneal
paramedian route.

Fig. 334. Route of access to the urinary
apparatus. 1. subcostal incision; 2. prolonged
subcostal incision (Sabadini); 3. transverse
incision (Bazy); 4. hypogastric incision
completed with median laparotomy; 5.
paramedian incision; 6. lumbo - iliac incision;
7. iliac incision (for exploration of the ureter).

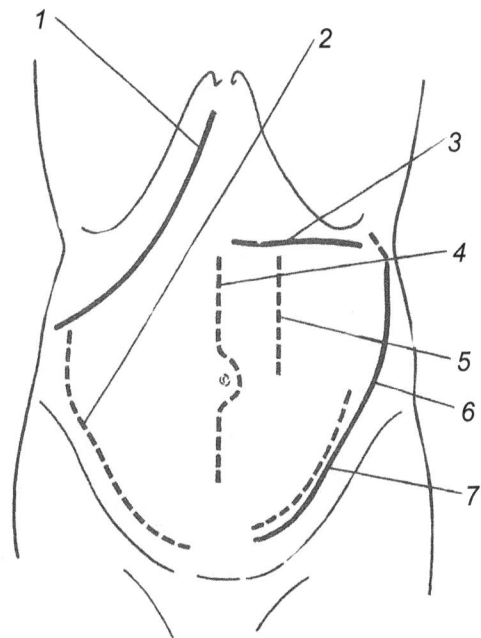

301

In Vivo Anatomy and Endoscopy Notions

Chapter 6

In Vivo Anatomy and Endoscopy Notions

The knowledge of the in vivo exploration of the internal **organs** has a particular practical significance, forming the basic, indispensable element for clinical investigation.

These examinations include the clinical exploration, as well as the instrumental exploration, the latter completing the data obtained by means of the physical examination of the human organism. These parameters have an utmost practical value, since they permit the early detection of various diseases and the carrying out of the adequate treatment.

Clinical Exploration of the Organs of the Trunk

In the framework of the clinical exploration, the obtention of the projection of internal organs on the surface of the integument, in dependence on the known osteomuscular landmarks, is aimed at.

The main methods which are used in the objective examination and familiarize the student with the exploration of the internal organs are the inspection, the palpation and the percussion. The auscultation integrates these objective data.

The inspection is the objective examination by the naked eye of the topographical location of an organ. The image obtained is nothing else than the precise projection of the latter on the wails of the trunk. *The palpation* is the method which allows, by means of various manoeuvres, to feel the shape and position of the examined viscus. It is a particularly valuable technique, indispensable for the investigation of the studied organ.

The percussion consists in the obtention of variable tonalities by tapping directly or indirectly on the different areas of the wails of the trunk. If may furnish, in dependence on the sounds perceived, valuable data regarding both the shape and the size of the explored viscus, as well as its consistency, in comparison with the normal state, by detecting the deviations from the normal aspect of morphofunctional organization.

Projection of the Intrathoracic Organs on the Surface of the Walls off the Thorax

The thoracic cavity *(cavum thoracis)* is the superior cavity of the trunk, comprised between the two thoracic openings, separated from the abdomen by the diaphragm and communicating with **the** cervical region through the superior opening of the thorax *(apertura thoracis cranialis)*. Its walls are made up in front by the sternum, anterolaterally by the twelve pairs of ribs and the intercostal muscles and posteriorly by the twelve vertebrae of the thoracic vertebral column, covered by the erector muscles of the trunk. The thoracic cavity lodges the organs of the respiratory apparatus, the heart and the large vessels, the thymus and the oesophagus, arranged in three topographical regions: the right and left pleuropulmonary regions *(cavum pleurae)* and the mediastinal region, situated between them. The mediastinum, in its turn, is subdivided by a conventional frontal plane, which passes through the bifurcation of the trachea, into an anterior and a posterior mediastinum. The anterior mediastinum is made up of two parts: a superior or thymic and an inferior or cardiopericardial part. The posterior mediastinum, too, is formed of two parts: a superior or tracheal and an inferior or infratracheal part.

Projection of the pericardium. The pericardium is projected on the surface of the sternocostal wail, using the following landmarks: on each side of the sternum, at a distance of about 1 cm from its borders, a point is marked in the first intercostal space (above and laterally from the Louis's angle). Then, the sternal end of the fifth right intercostal space and the heart apex beats, visible in the fifth left intercostal space, are marked on the midclavicular line, at 10 cm from the mediosternal line. The points marked on the right and on the left are united by two outwards convex lines and the ends of the superior lines and the above mentioned inferior points, situated in the fifth right and left intercostal spaces, are united by a curved line, with the convexity in proximity of the jugular notch of the sternum (jugular incisure of the sternum), an irregular quadrilateral area, with curved sides, being obtained.

Projection of the heart Examining obliquely, against the light, the anterior aspect of the thorax, beneath the left nipple, the apex pulsations may be observed, which are transmitted to the thoracic wall during the systole (apex shock). The apex shock projects in the fifth left intercostal space, on the midclavicular (or mamillary) line. The projection varies according to the age, the constitutional type, the various functional changes or in some pathological conditions.

304

Thus, in children, the apex shock is perceptible in the fourth left intercostal space, slightly outwards the midclavicular line. In the aged, as a consequence of the general ptosis of viscera and implicitly of the heart, the apex shock is perceived in the sixth left intercostal space. The cardiac area projects on the anterior wall of the thorax, in the surface enclosed by the curved lines which unite, to the right, the second intercostal space (its sternal end) and the fifth chondrosternal articulation and, to the left, the sternal end of the second left intercostal space and the point which marks the apex shock.

Projection of the cardiac orifices. The right atrioventricular (tricuspid) orifice, which is the most anterior, projects at the level of the articulation of the fifth right costal cartilage with the sternum, reaching the midsternal line. The left atrioventricular (mitral) orifice projects at the level of the sternal extremity of the third left intercostal space, on the line which unites this space with the articulation between the fifth right costal cartilage and the sternum. The aortic orifice projects behind the left border of the sternum, at the level of the third left costal cartilage, whereas the orifice of the pulmonary artery projects behind the left border of the sternum, on an oblique line which prolongs anteriorly the upper border of the third left rib. However, the projection of the orifices does not correspond to the site of auscultation, with maximum efficiency, of the cardiac sounds.

These auscultation foci are situated: for the aortic orifice, in the second right intercostal space, at the border of the sternum (in the new-born, on the midsternal line, at the level of the third intercostal space); for the pulmonary orifice, on the left border of the sternum, at the level of the second left intercostal space (in the new-born, at the level of the fourth left chondrosternal articulation); for the mitral orifice, at the apex of the heart, in the fifth left intercostal space, at 8-10 cm lateral from the midsternal line (in the new-born, at the level of the fourth left chondrosternal articulation) and for the tricuspid orifice, at the base of the xiphoid process (in the new-born, at the level of the fourth intercostal space, on the midsternal line). The anterior projections of the heart undergo numerous physiological and pathological variations, whereas much more constant are the posterior relations of the heart with the vertebral column, corresponding to the thoracic vertebrae IV-VIII (Giacomini's cardiac vertebrae): T_4 = supracardiac vertebra; T5= infundibular vertebra; T_6 = atrial vertebra; T_7 = ventricular vertebra; T_8 - apexian vertebra (fig. 335).

Projection of the pleural sinuses. At the sites where the parietal pleura, which lines the chest wall, reflects from one wall on the other, changing consequently its direction, appear the pleural recesses (pleural sinuses): anterior costomediastinal, costodiaphragmatic, pleural dome (the superior recess) or superior costomediastinal and vertebropleural (posterior costomediastinal). In addition to these sinuses, the mediastinal pleura forms also, in relation with the organs of the posterior mediastinum, other recesses: interaortico -oesophageal (on the left) and interazygo - oesophageal (on the right), united with each other by an interpleural ligament (described by Morozov), formed of a plate of connective tissue arranged in a frontal plane.

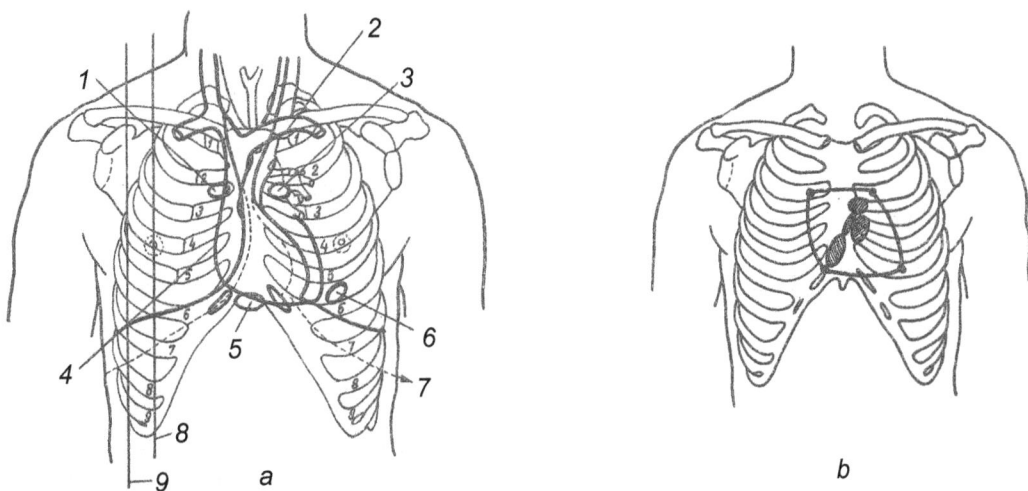

Fig. 335, Projection of the heart and its orifices on the chest wall (a,b) 1. Auscultation focus of the aorta; 2. Sternoclavicular joint; 3. Auscultation: focus of the pulmonary artery; 4. Projection of the pericardium; 5. Auscultation focus of the tricuspid orifice; 6. Auscultation focus of the mitral orifice; 7. Projection of the anterior costomediastinal sinus; 8. Mamillary (medioclavicular) line; 9. Anterior axillary line.

305

The projection lines of the anterior costomediastinal sinus on the chest wall begin, both on the right and on the left, at the level of the respective sternoclavicular articulation.

They descend obliquely-medially nearly up to the midsternal line, which they reach behind Louis's sternomanubrial angle, descending then closely united obliquely to the left up to the level of the fourth costal cartilage, without passing beyond the left border of the sternum, where they diverge from each other laterally, following a different direction. To the right, the line of the pleura! reflection descends obliquely towards the junction of the sixth costal cartilage with the sternum.

From here begins the projection of the right costodiaphragmatic sinus, which descends obliquely along the sixth costal cartilage, crosses the seventh rib on the midclavicular line, the eighth rib on the anterior axillary line, the ninth rib on the axillarx midline, the tenth rib on the posterior axillary line, the eleventh rib on the scapular line, then the upper border of the twelfth rib towards its vertebral end up to the transverse process of the twelfth thoracic vertebra, from where the posterior costomediastinal sinus ascends vertically up to the base of the neck, at the level of the seventh cervical vertebra. From here, it reflects anteriorly up to the sternoclavicular articulation, forming the medial limit of Kronig's isthmus. The lateral limit of the latter is detected by percussion, which gives rise to a tympanic sound, owing to the aerial resonance of the apex of the lung.

Normally, the Kronig's isthmus on the right side exceeds by a fingerbreadth that on the left side, which is two fingerbreadths wide.

When this isthmus is narrowed, a pathological change should be present at the apex of the right lung or at the dome (cupula) of the pleura. On the left, the line of projection diverges more from the sternum, crosses the fifth chondrosternal articulation and bends outwards, crossing the sixth costal cartilage at the junction of the median 1/3 with the lateral 1/3, after which it reaches the seventh rib on the midclavicular line, becoming the left costophrenic sinus and following the same direction and the same landmarks as on the right. In this way, the ventral end of the fifth and sixth intercostal spaces and the sixth and seventh spaces remain non-invested by the pleura of the left haemithorax.

Projection of the lungs. On the left, the anterior border of the left lung presents a more marked curvature ,which crosses the sixth costal cartilage at its middle (the cardiac notch or incisure). The projection of the lung continues along a line which crosses the seventh rib on the axillary line and the ninth rib on the scapular line, running dorsally along a horizontal line up to the eleventh rib, in the proximity of the transverse process of the eleventh thoracic vertebra.

The right and left lines of projection describe thus, behind the sternum, a clepsydra-shaped area, non-invested by pleura: two triangles united at the level of their apices by adjoining of the two pleurae on a distance of 4-5 cm, with the bases directed respectively upwards and downwards.

The superior, thymic triangle corresponds to the thymus (in the adult to the fibre-adipose tissue that replaces the most part of the gland) and the inferior, pericardial triangle, to the area in wich the pericardium comes into direct contact with the sternocostal wall, area of choice for pericardial taps. The dome of the pleura passes beyond the superior thoracic aperture and rises at 2 cm above the clavicle and at 4 cm above the first rib, reaching nearly by the plane of the sixth cervical vertebra (higher on the right then on the left) (fig. 336-338).

Clinically, the apex of the lung may be explored ventrally, in the tetragon enclosed by the anterior border of the clavicle, the superior border of the second costal cartilage and the second rib; laterally, on the bottom of the axilla (the lifting of the arm uncovers the first rib and the first intercostal space); posteriorly, in the first two intercostal spaces, projected within the interscapulovertebral space, which is enlarged by extension of the arm also into the surapspinous fossa, bounded below by the superior lip of the spine of scapula.

These exploration areas are united into a percussion area, normally resonant, which passes like a brace over the shoulder, the above-mentioned space described by Kronig or Chauvet's alarm area (fig. 339- 341),

Clinical exploration of the mammary gland. Among the most frequent diseases of the chest wall in the female, the diseases of the mammary gland rank on the first place.

Therefore, an attentive, systematic and periodic examination by palpation of the mammae is necessary for the early detection of possible diseases of the mammary gland (fig. 342-345),

By its frequency, severity and high death rate, the cancer of the mammary gland has drawn the physicians attention on the necessity of knowledge of the mammary lymph supply and of the modality of remote metastase formation.

The investigation of lymph node groups makes possible the detection of initial forms with the view of early surgery, leading to the recovery of these patients (fig. 346-349).

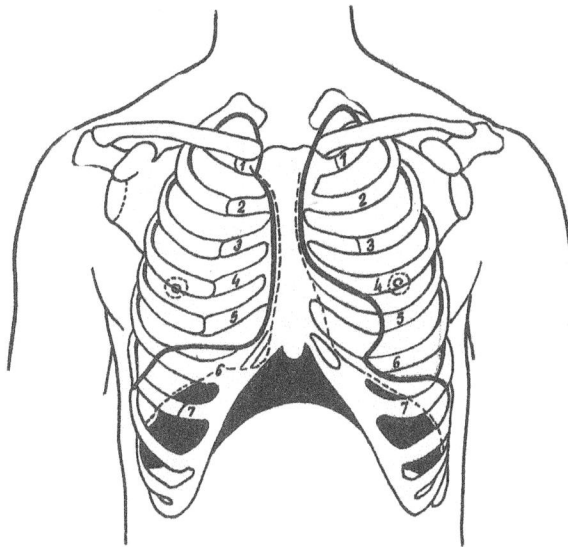

Fig. 336, Anterior fhoraco-pleuropulmonary topography.

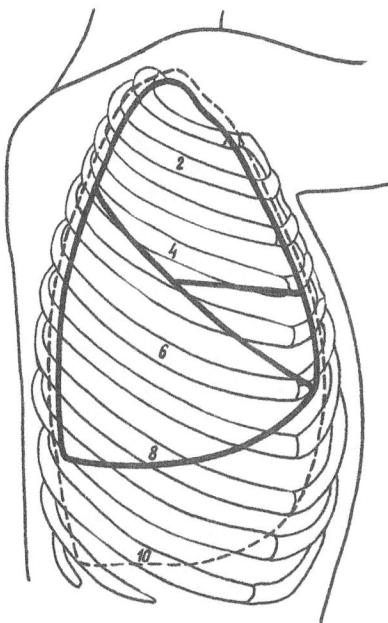

Fig. 337. Pleurocostal topography (right lung, lateral view).

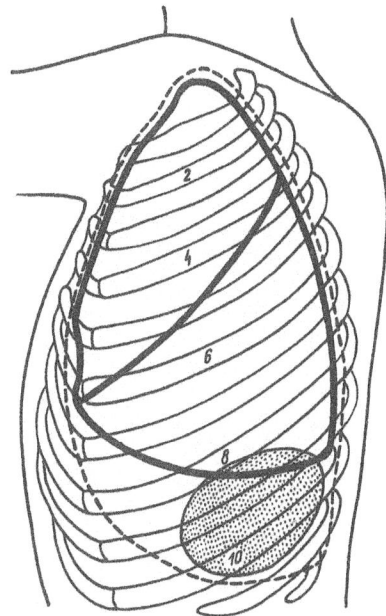

Fig. 338. Topography of the left pleura and lung (lateral view).

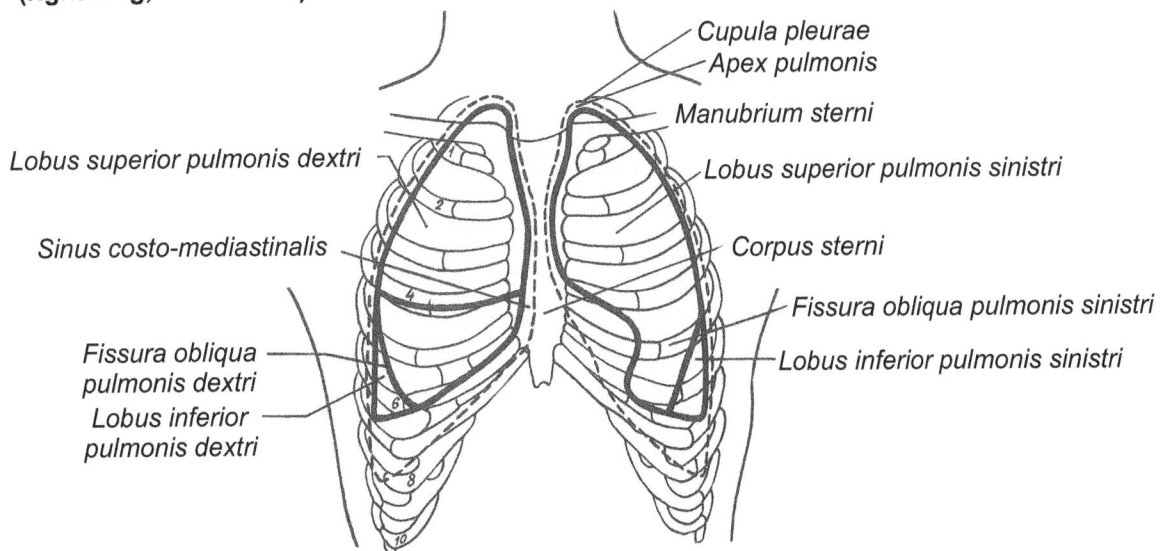

Cupula pleurae
Apex pulmonis
Manubrium sterni
Lobus superior pulmonis dextri
Lobus superior pulmonis sinistri
Sinus costo-mediastinalis
Corpus sterni
Fissura obliqua pulmonis sinistri
Fissura obliqua pulmonis dextri
Lobus inferior pulmonis sinistri
Lobus inferior pulmonis dextri

Fig. 339. Pleuropulmonary topography (anterior view).

307

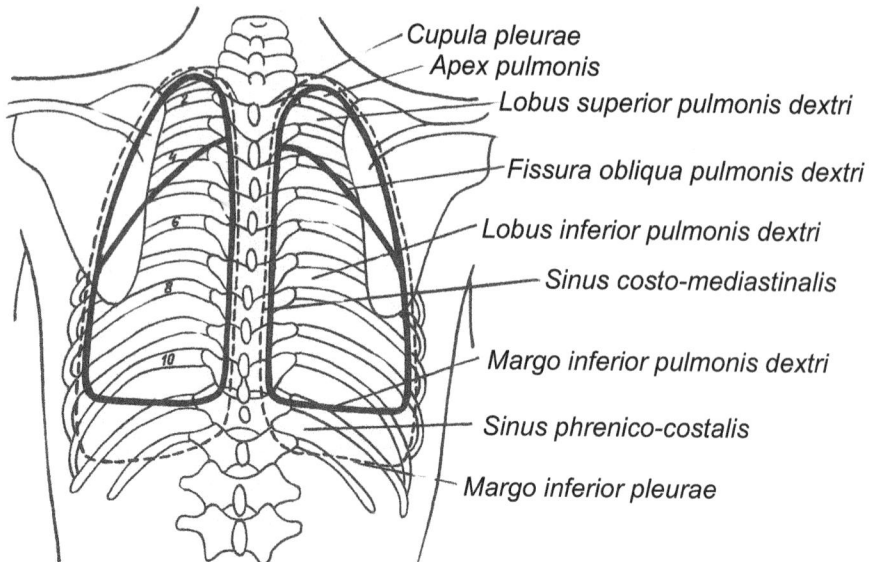

Cupula pleurae
Apex pulmonis
Lobus superior pulmonis dextri
Fissura obliqua pulmonis dextri
Lobus inferior pulmonis dextri
Sinus costo-mediastinalis
Margo inferior pulmonis dextri
Sinus phrenico-costalis
Margo inferior pleurae

Fig. 340. Pleuropulmonary topography (posterior view).

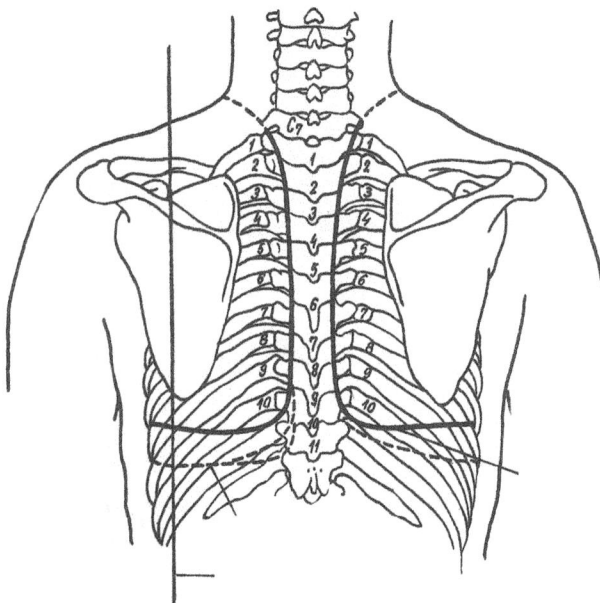

Fig. 341. Thoracopulmonary topography (posterior view).1) Projection of the posterior costomediastinal sinus; 2) Projection of the lungs; 3) Scapular line

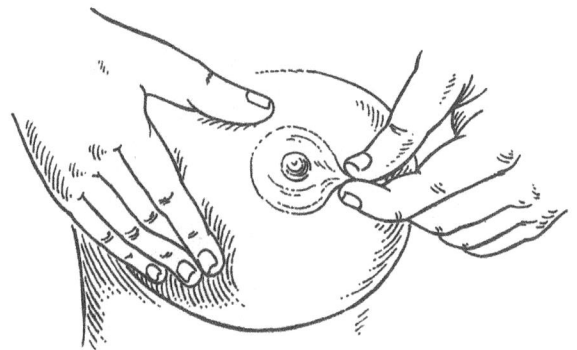

Fig. 343. Examination of a normal mamma: integument of normal aspect, possible folding of the integument, absence of adherence of the gland to the deep planes.

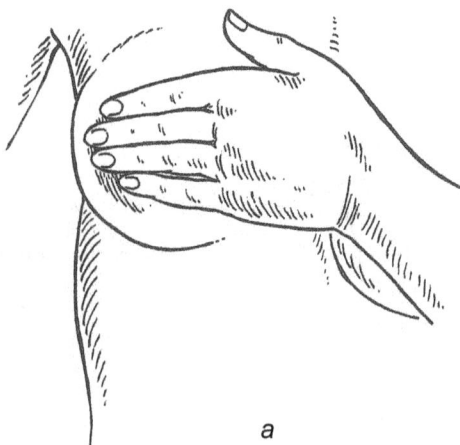

a

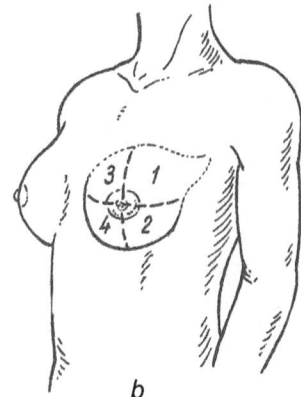

b

Fig. 342. Palpation technique of the mamma: a) correct palpation; b) palpation sequence of the quadrants of the mamma

308

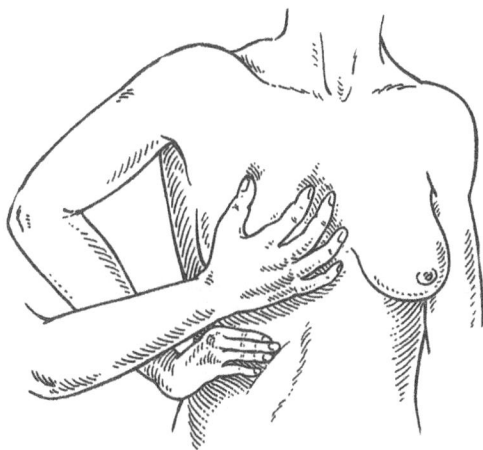

Fig 344. Investigation technique of the fixity degree of the mammary gland on the deep planes.

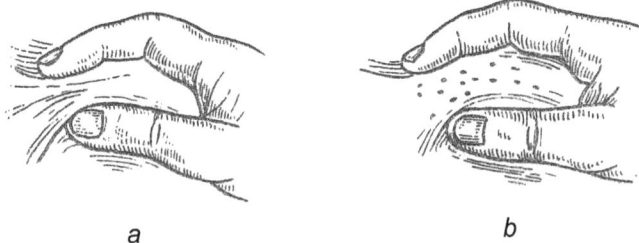

Fig. 347. Palpation of the mamma - sign of the fold: a - normal aspect: the skin of the mamma gripped between the fingers wrinkles; b - pathological aspect: the skin is no more wrinkling (it is rigid)

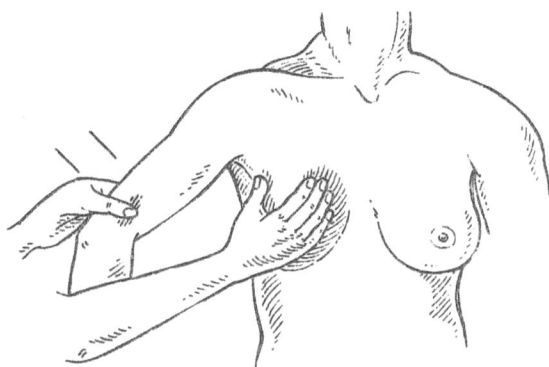

Fig. 345. Tillaux's manoeuvre. The mobility of the mammary gland on the plane of the pectoralis major muscle is displayed.

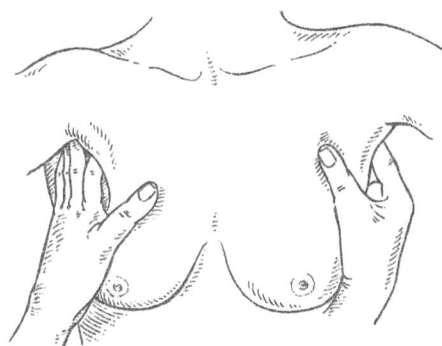

Fig. 348. Examination of the axillary lymph node groups in the erect posture.

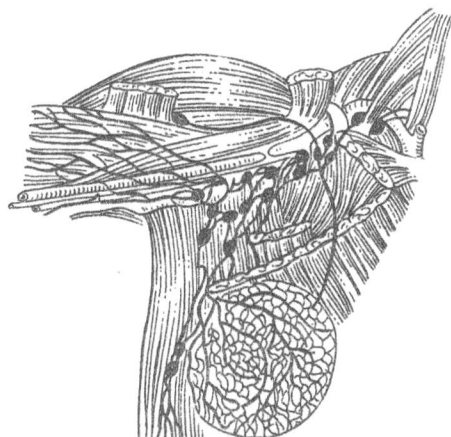

Fig. 346. Lymph vessels of the mammary gland.The lymphatic network of the mamma is directly related to the axillary lymph node groups (external mammary, central mammary, interpectoral etc.)

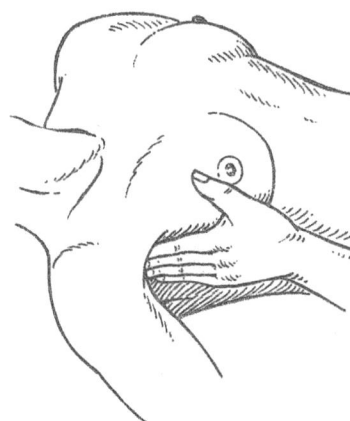

Fig. 349. Examination of the axillary lymph node groups in dorsal decubitus

309

Projection of the Intra-Abdominal Organs on the Surface of the Abdominal Walls

The abdomen is the segment comprised between the thorax, from which it is separated by the diaphragm, and the pelvis, with which it largely communicates through the pelvic excavation. The external boundaries are, above, the costal edge and the xiphoid process and below, the line which unites the iliac crests. These boundaries do not correspond to the actual volume of the abdominal cavity.

Topographic delineation of the ventral abdominal wall. The ventral wall of the abdomen is subdivided into nine areas. The anatomical areas are delineated by means of two horizontal lines and two vertical lines.

Of the two horizontal lines, the superior joins the anterior ends of the ribs of the tenth pair (bicostal line) and the inferior unites the cranial points of the iliac crests (bicristal line). By drawing these lines, three parts of the abdominal wall are obtained: the epigastrium. situated above the bicostal line, extending up to the dome of the diaphragm in the fifth intercostal space; the mesogastrium, situated between the two lines; the hypopgastrium, below the bicristal line and above the inguinal fold and the pubic symphysis.

The two vertical lines, which prolong the midclavicular lines up to the meeting point with the inguinal fold (or the lines which run along the lateral border of the recti abdominis muscles), divide each part into three regions: the epigastric region proper, situated in the median plane, and the right and left hypochondrium on each side of if; the right flank, the umbilical region and the left flank; the right iliac fossa, the suprapubic or supravesical region and the left iliac fossa. On the surface of these abdominal regions project the intraabdominal organs (fig. 350).

Projection of the stomach. Into the epigastric region and into the left hypocondrium projects the stomach, delineated as follows:

-above, by a horizontal line which passes through the fifth left intercostal space, percussable owing to the cardiac dullness;

-below, by a horizontal line which passes through the inferior border of the ribs of the tenth pair;

-laterally, by the plane tangent to the left lateral chest wall, delineated by the dullness of the spleen, obtained by percussion of the left hypochondriac region.

Corresponding to the portion of the stomach hidden below the left costal arch, on the anterior thoracoabdominal wall, is situated Traube's semilunar space, bounded above by the fifth costal cartilage and the cardiac dullness and laterally (at the level of the ninth and tenth rib), by the dullness of the spleen. This space is hyperresonant, tympanic in percussion, owing to the air chamber in the fornix of the stomach. Medially, on the costal edge, in the epigastric region, the anterior surface of the stomach comes into direct contact with the anterior wall of the abdomen, in the area of a triangular zone (the trigone described by Labbe), bounded laterally and to the left by the left costal arch, below by the greater curvature of the stomach, normally projected on the line that unites the anterior ends of the tenth ribs on the right and the left, and above by the bisector of the angle formed by the above-mentioned bicostal line of the tenth pair and the right costal arch.This landmark corresponds to the anterior border of the liver (fig. 351).

Projection of the liver. The liver projects into the right hypochondrium and the epigastric region. On percussion a dullness is obtained, which corresponds to the hepatic parenchyma and reaches, above, the fifth right inferior intercostal space, usually without passing beyond the right costal arch. In the epigastrium, the inferior border of the liver projects on the above-mentioned line, which represents the bisector of the angle formed by the right costal arch and the bicostal line (the superior side of the triangle described by Labbe). According to Murchinson, the superior line of the hepatic dullness is arcuate and begins posteriorly at the level of the tenth or eleventh thoracic vertebra, passes on the right axillary line in the seventh intercostal space and reaches the right midclavicular line in the fifth intercostal space and medially, the base of the xiphoid process (fig. 352).

The hepatic dullness attains normally 9-10 cm in height on the right midclavicular line, the inferior border descending by approximately 1 cm in the upright posture (fig. 353); at the site where the line that unites the umbilicus with the anterior fold of the right axilla crosses the costal arch, lies the cystic point corresponding to the fundus of the gall bladder. Dejerne situates the cystic point at the meeting place of the lateral border of the rectus abdominis with the right costal arch, below the anterior end of the cartilages of the ninth and tenth ribs, at 9-10 cm from the midline, but in depth (fig. 354).

310

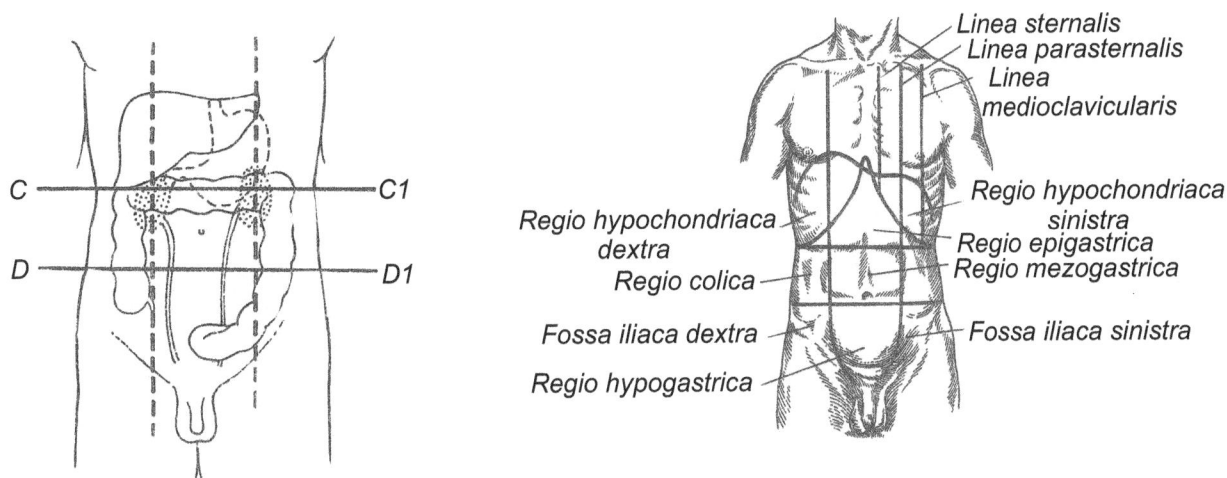

Fig. 350. Topographical division of the abdominal wall and visceral correspondence.

Projection of the pancreas. In the epigastrium and left hypochondrium lies also the pancreas, enclosed by the following planes: above, a horizontal line which passes through the anterior end of the eighth rib; below, a line which passes horizontally at 3-4 cm above the umbilical cicatrix; on the right, a vertical line at 3-4 cm from the midline; on the left, a vertical line situated at about 2 cm medially from the left midclavicular sagittal plane (fig. 355).

Round the head of the pancreas, the duodenal horseshoe (the right para-umbilical duodenal point) projects on the anterior abdominal wall (fig. 356).

Projection of the spleen. The spleen projects into the left hypochondrium on an oval, dull area, the long axis of which is represented by the tenth rib; the superior boundary of the dullness reaches the ninth rib and the inferior boundary, the eleventh rib. The superior and posterior poles are situated at about 4 cm from the midline, in proximity of the transverse processes of the ninth, tenth and eleventh vertebrae, and the inferior and anterior poles do not pass, normally, beyond the left costal arch (fig. 357). Splenomegalies are seen in chronic infectious diseases like malaria, in acute diseases with a prolonged course such as typhoid fever, leukoses etc.; splenomegalies associated with hepatomegalies occur in the syndrome described by Banti, in which the spleen passes beyond the costal groove, reaching the umbilicus and even the left iliac fossa.

Projection of the colon and small intestine. The middle part of the anterior abdominal wall comprises the right and left flanks and the umbilical region (fig. 358).

The ascending colon projects into the right flank, up to the hepatic angle of the colon. The ascending colon is continuous with the transverse colon, forming an angle of 70-80°, oblique medially and anteriorly, angle which projects into the right hypochondrium, below the projection of the liver. The transverse colon, projected between two horizontal lines, passes through the anterior extremity of the costal cartilage of the tenth pair, respectively through the umbilical cicatrix or slightly below it. It may assume various shapes (arcuate, U-, V-, W-, italic S-shaped etc.), as well as different sizes, its length varying between 50 and 90 cm. The transverse colon is continuous at its left extremity with the descending colon, after the splenic angle of the colon, always acute and higher situated than the hepatic angle (by approximately 4-5 cm), has been formed. Its lumbar portion projects into the depth of the left flank. At the level of the iliac crest it is continuous with the iliac segment of the sigmoid colon up to the left iliac fossa. Here begins the pelvic segment of the descending colon, which extends up to the third sacral vertebra, where the colon is continuous with the rectum.

In the frame formed by the segments of the colon lie the loops of the jejuno-ileal intestine, which project on the anterior abdominal wall in the region of the mesogastrium.

Projection of the kidney. The kidneys project into the flanks, into their upper half, and dorsally, in a region bounded above and below by two horizontal planes, which pass through the eleventh thoracic vertebra and the iliac crest respectively, and medially and laterally by two sagittal planes, which pass through the paravertebral plane, on either side, and approximately at 2-3 cm outside the sacrolumbar muscular mass (fig.359).

The renal pelvis projects on the anterior abdominal wall at the crossing point of the horizontal line which passes through the umbilicus with the lateral border of the rectus abdominis (point described by Bazy) (fig. 360).

311

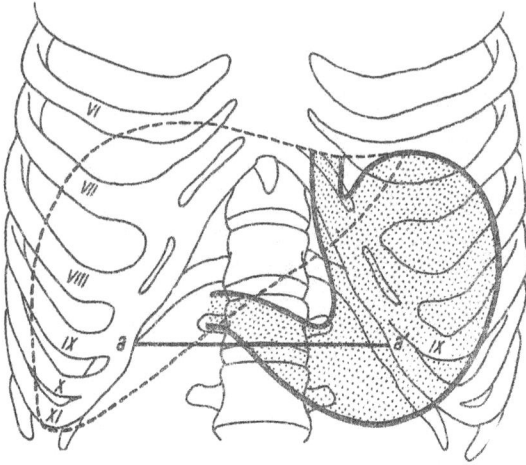

Fig. 351. Ventral projection of the stomach-
Labbe's triangle: aa' - line which connects
the costal cartilages

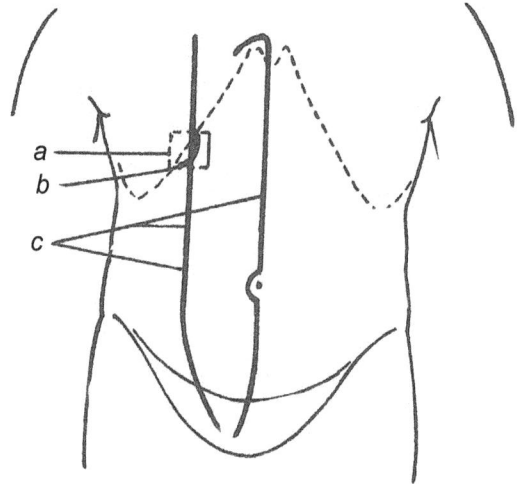

Fig. 352. Projection of the liver on the
anterior thoraco-abdominal wall

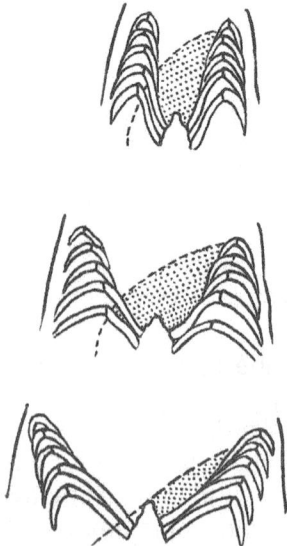

Fig. 353. Parietal projection surface of
the liver, in dependence on the
thoracic conformation.

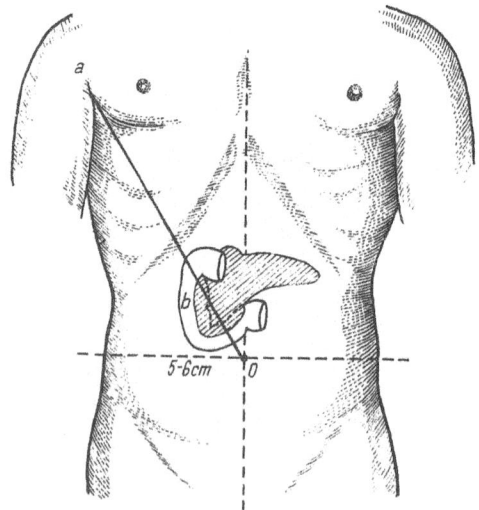

Fig. 354. Projection of the gal bladder on the
anterior abdominal wall a - area of the fundus of the
gal bladder; b - cystic point; c - limits of the sheath
of the rectus abdominis muscle.

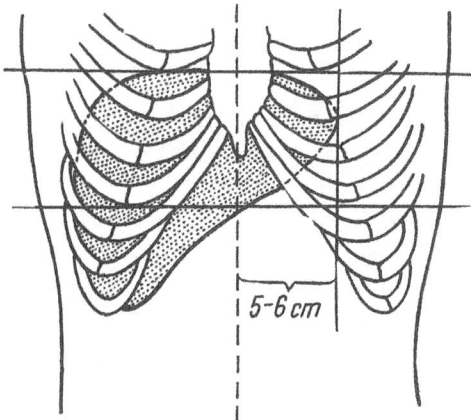

Fig. 355. Projection of the pancreas on the
anterior abdominal wall, the "pancreatico-
choledochal area a - axilla; Ao - Desjardins'
line; b - pancreatic point
(choledochoduodenal confluence)

312

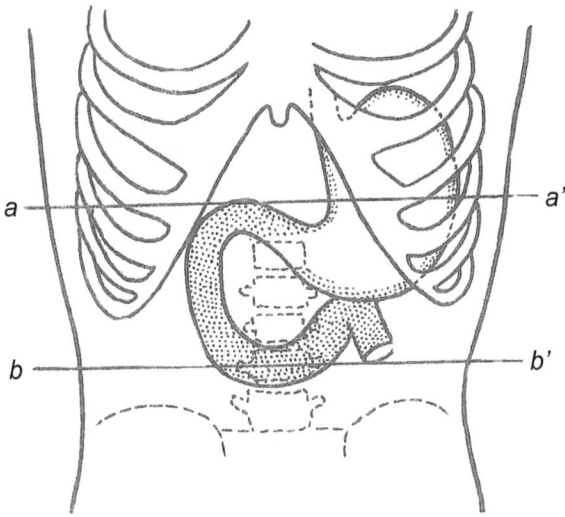

Fig. 356. Projection of the duodenum on the anterior abdominal wall: aa' - line which connects the anterior extremities of the eighth ribs; bb' - line of the umbilical plane

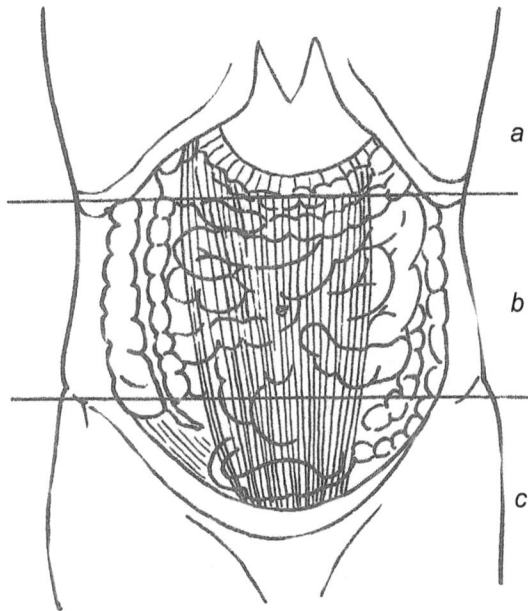

Fig. 358. Projection of the loops of the small intestine on the anterior abdominal wall (a), into the middle part (b) and into the lower part (c).

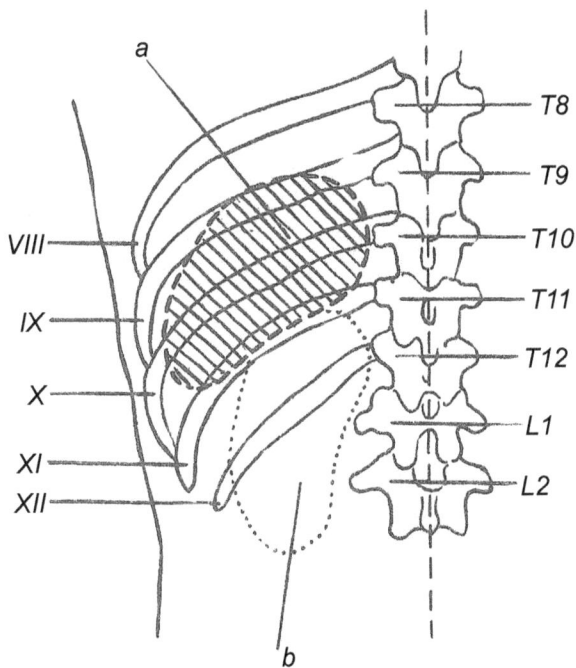

Fig. 357. Projection of the spleen on the dorsal wall of the left haemithorax and its relation with the left kidney: a - spleen; b - left kidney.

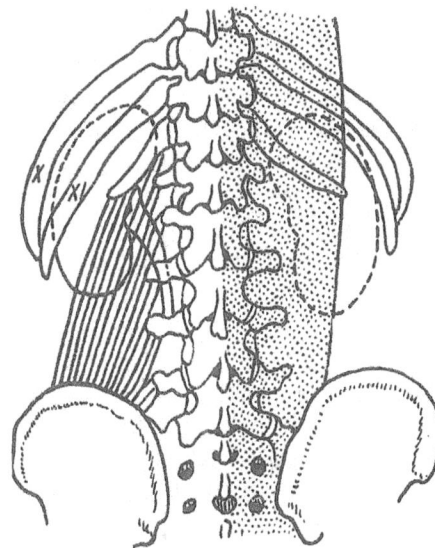

Fig. 359. Anatomical position of the kidneys and their relations.

313

Projection of the ureter. The course of the abdominal ureter is marked on the anterior abdominal wall by a number of points,the ureteral points. The superior ureteral point corresponds to the paraumbilical point; the iliac point is situated at the crossing of the bricristal line with the lateral border of the rectus abdominis or, according to Halle, at the intersection of the bi-iliac line with the vertical drawn through the spine of the pubis (site where the abdominal ureter becomes pelvic, crossing the iliac vessels). On the bicristal line, but at its half, lies also the projection place of the aortic bifurcation. The inferior ureteral (ureterovesical) point may be palpated by rectal or vaginal touch (fig. 361).

Projection of the vermiform appendix. The inferior part of the anterior abdominal, hypogastric wall is divided into the two iliac fossae, right and left, comprising between them the suprapubic or vesical region. Into the right iliac fossa project the caecum and the vermiform appendix. The appendicular point represents the projection of the implantation site of the appendix into the caecum (visible on the organ at the place where the three taeniae join, at 2 cm below the ileocaecal valve). According to Monro, this implantation site is midway between the umbilicus and the ventrocranial iliac spine, McBurney places the apppendicular point at two fingerbreadths from the anterosuperior iliac spine, on the spino-umbilical line.

Lanzmann states that this point is situated at the junction of the right lateral third with the middle third of the line which unites the two anterosuperior iliac spines; according to Sonnenburg, it is sited at the intersection of the bi-iliac line with the lateral border of the rectus abdominis on the right side, lacobovici describes an appendicular triangle, formed by the umbilicospinous line, the bispinous line and the vertical which descends through the lateral border of the rectus abdominis (fig. 362). Into the left iliac fossa projects the iliac segment of the sigmoid colon.

Projection of the urinary bladder. In the hypogastrium is palpated the urinary bladder which, when full, passes beyond the pubic symphysis (fig. 363).

Projection of the pregnant uterus. From the second month, the pregnant uterus appears above the pubic symphysis, increasing then gradually by approximately 4 cm every month of pregnancy (fig. 364).

The clinical exploration of the abdomen is performed by palpation, which may be superficial, deep and with pressure, according to the type of the abdominal disorder.

The superficial palpation is used in acute abdominal lesions, the deep palpation in chronic lesions and that with pressure, for the detection of deep lesions (fig, 365, 366). In addition, it is necessary to bow also the areas of referred pain in the various abdominal diseases (fig 367).

Clinical exploration of the kidney. The deep, retroperitoneal site of the kidney allows its detection only in case of bulky tumours and hydro-or pyonephroses. Usually, it is detected by palpation, by means of one of the following procedures:

Guyon's procedure: the patient is in supine position, with the head slightly raised on a cushion and the thighs flected on the pelvis and in slight abduction. The examiner applies the left hand on the lumbar region, with the tip of fingers introduced within the costolumbar angle (intersection of the twelfth rib with the lateral border of the lumbar mass), and the right hand on the flank to be investigated. During inspiration, the kidney is gripped betwen the examiners hands, under the form of a deep tumefaction. By means of this manoeuvre, the sign of "renal ballottement", described by Guyon, is perceived (fig. 368).

Glenard's procedure: the patient is in the same position.The palmar surface of the left hand is applied on the

lumbar region with the tip of fingers within the costolumbar angle and the thumb on the anterior wall of the abdomen, below the false ribs. During inspiration, the fingers are approached to each other and the inferior pole of the kidney is felt (fig. 369).

Israel's procedure: the patient is in lateral decubitus, opposite the part to be examined. The examiner uses the same manoeuvre as in Guyon's procedure, the renal mass being much better felt (fig. 370).

Clinical exploration of the abdomen in the pregnant woman. The palpation of the abdomen in the obstetrical practice is reliable.The age of the foetus, as well as some therapeutic conducts are established by means of this method. D. Savulescu has described five successive moments (fig. 371).

The anorectal digital examination. It is particularly valuable in the diagnosis of some diseases localized at this level or at the surrounding organs. It is performed with the index gloved and lubrified with vaseline, the whole anorectal surface being examined circularly, to a depth of 10 cm at the most (fig. 372, 373).

314

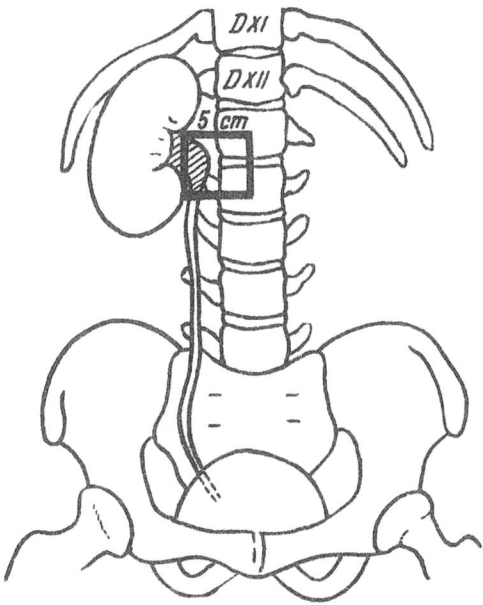

Fig. 360. Projection of the renal pelvis into the Bazy-Moyrand's tetragon

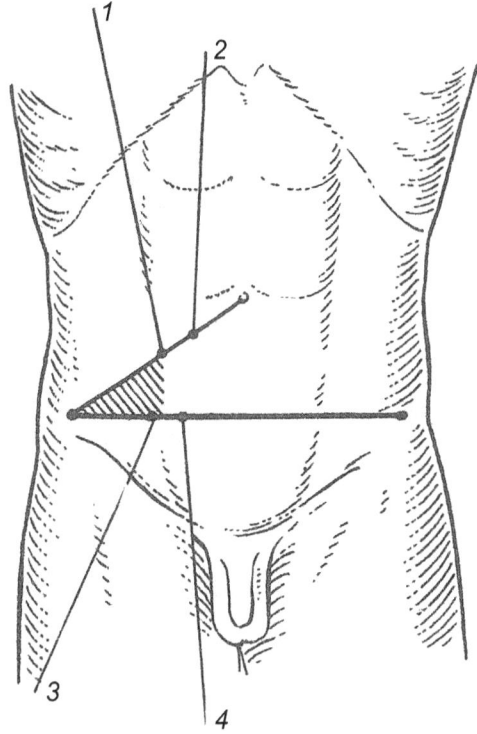

Fig. 362. Abdominal painful points in case of appendicits. Hatched area- triangle of Iacobovici; 1) McBurney's point; 2) Morris point;3) Lanz's point; 4) Sonnenburg's point.

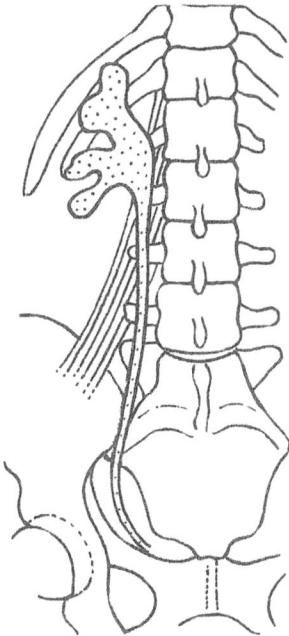

Fig. 361. Relation of the ureter to the lumbar spine

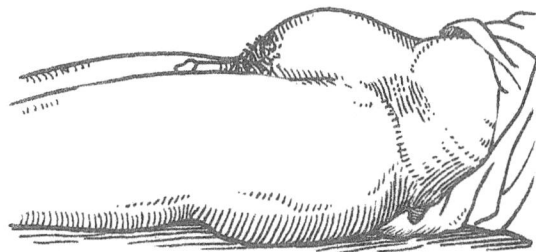

Fig. 363. Projection of the urinary bladder on the anterior abdominal wall

315

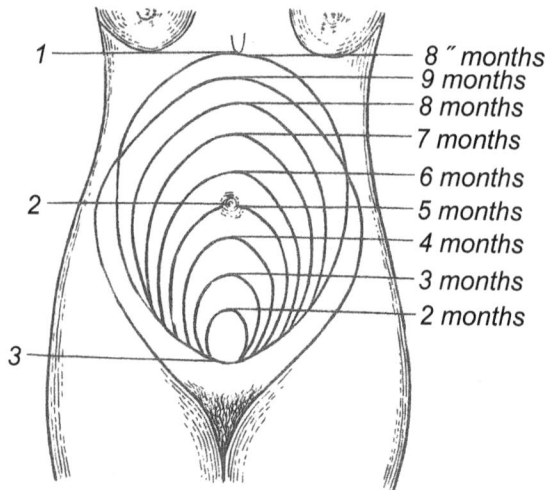

Fig. 364. Height of the fundus of uterus at different gestation ages, in relation to the symphysis, the umbilicus and the xiphoid process 1) xiphoid process; 2) umbilicus; 3) pubic symphysis

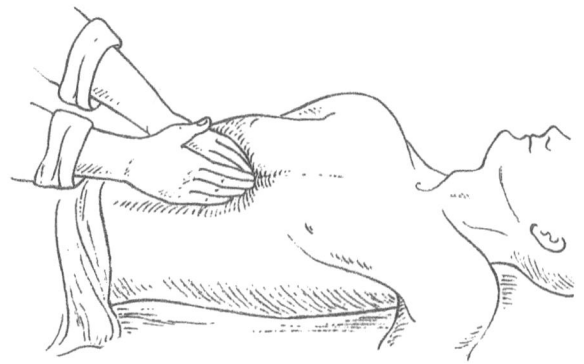

8 ″ months
9 months
8 months
7 months
6 months
5 months
4 months
3 months
2 months

Fig. 365. Muscle contracture.

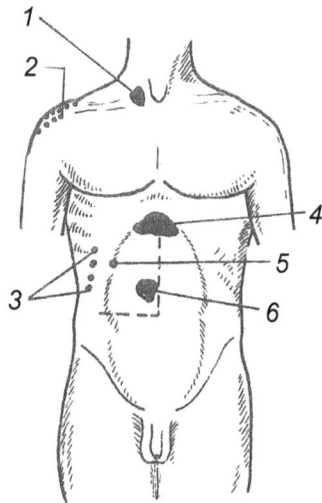

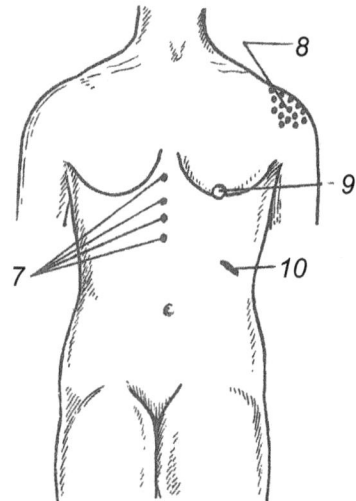

Fig. 367. Anterior and posterior pain referral areas in the hepatic colic.1)superior right parasternal point;2)deltoid point;3)scapular point; 4)vertebral points;5)costal point (eleventh rib);6)inferior right parasternal point;7)vertebral points;8)deltoid point;9)scapular point;10)costal point (eventh rib).

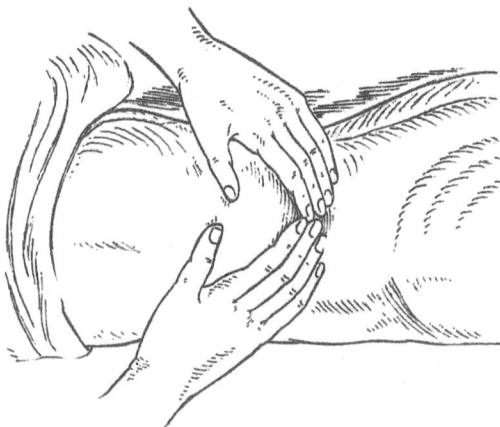

Fig. 366. Exploration of the lumbar region in case of acute retrocaecal appendicitis.

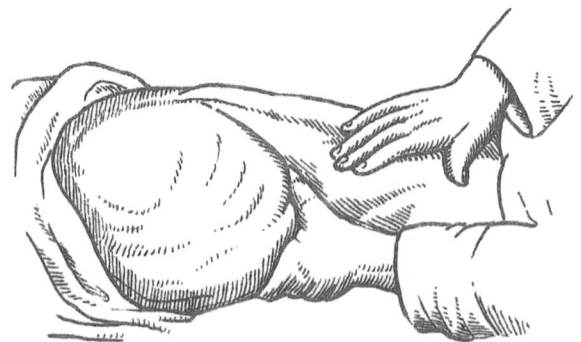

Fig. 368. Guyon's procedure

316

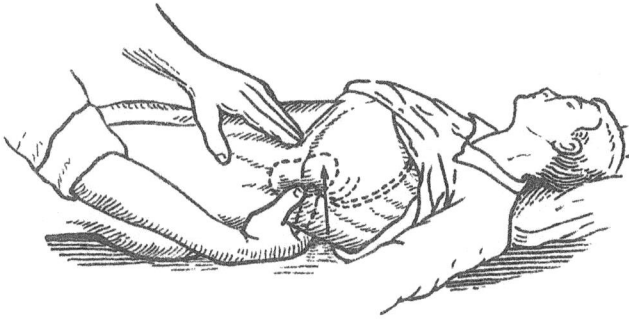

Fig. 369. Glenard's procedure.

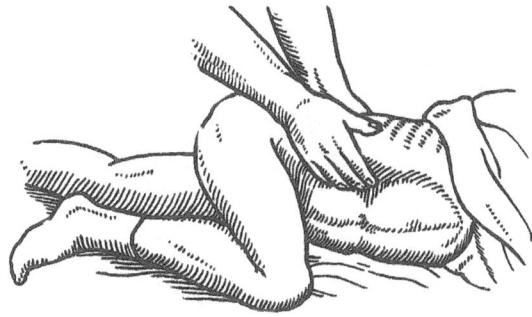

Fig. 370. Israel's procedure

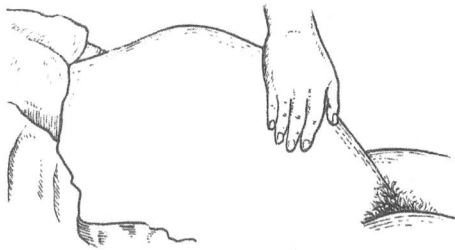

I

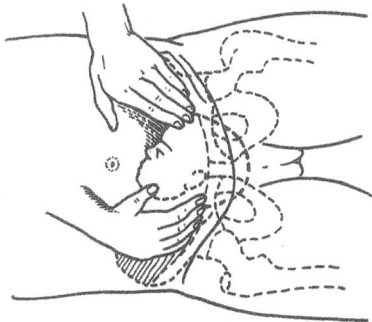

II

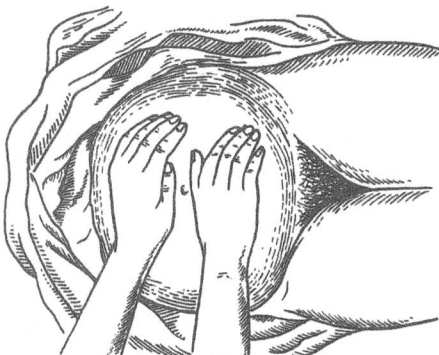

III

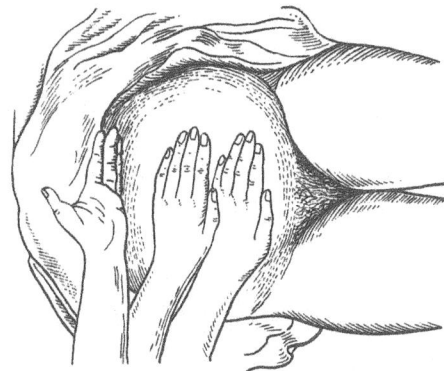

IV

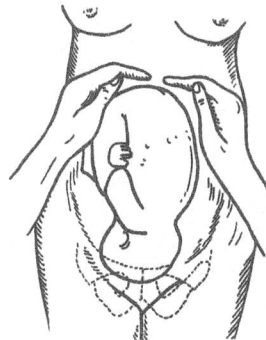

V

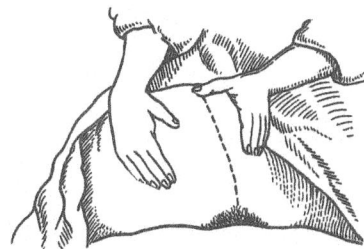

VI

Fig. 371. Palpation of the abdomen in obstetrical practice: > first stage - application of the examiners hand to the abdomen of the pregnant woman and superficial exploration; second stage -palpation in order to delineate the fundus of the uterus; third stage - palpation of the lower segment; fourth stage - bimanual palpation of the fundus of uterus; fifth stage - bimanual palpation of a side of the uterus; sixth stage - palpation of the opposite side of the uterus

317

In case of distension, the urinary bladder is perceived by combined rectal touch as a median, round, regular in shape and renitent. *The exploration of the prostate* is also performed by rectal touch, after emptying the urinary bladder, the patient being in one of the two positions (fig. 375). At about 5 cm from the anal orifice, the volar surface of the index palps the prostate, which is triangular in shape, with the base upwards and the apex downwards (fig. 375).

The vaginal touch is the compulsory method of exploration both in gynecology and in obstetrics, with the concomitant palpation of the abdomen with the free hand. It this way, the vaginoabdominal examination becomes a bimanual palpation (fig. 376). The internal pelvimetry is also carried out by vaginal touch. If is necessary to retain that the useful diameter, called also *conjugata vera,* should measure 10.8 cm (fig. 377).

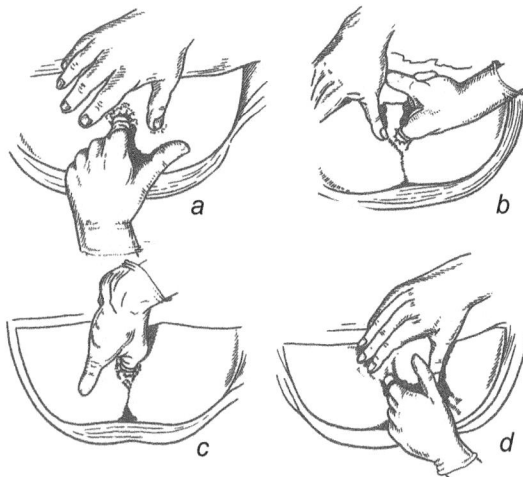

Fig. 372. Digital anorectal examination. a,b,c,d, - various positions imparted to the finger for the purpose of clinical exploration

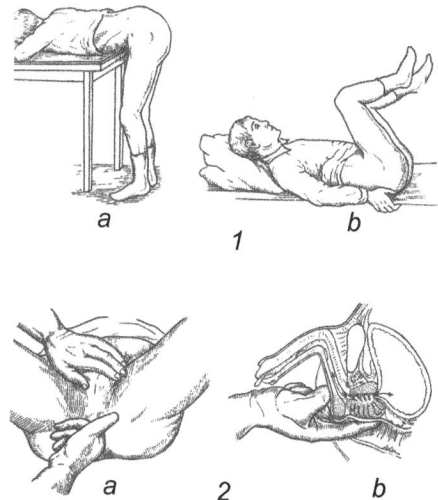

Fig. 375. Prostate exploration technique.
1. patient's posture; a - erect, the trunk flexed on the lower limbs and propped against the examination table; b - in dorsal decubitus, the thighs flexed on the pelvis, in moderate abduction, and the legs flexed on the thighs;
2. exploration technique: a - rectal touch; b - prostate palpation

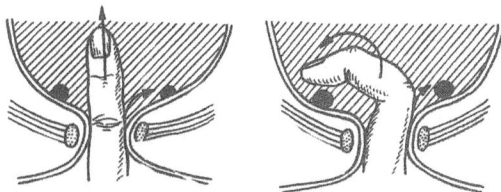

Fig. 373. Position of the finger during the anorectal examination above the strap of the levatores ani.

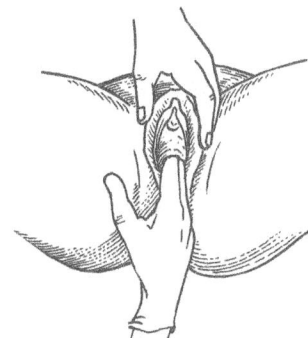

Fig. 376. Vaginal touch technique

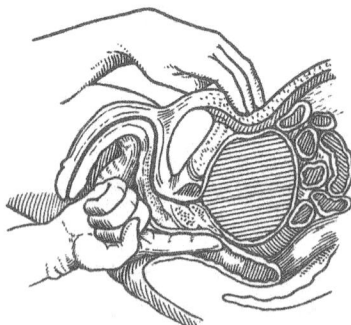

Fig. 374. Examination of the urinary bladder by combined rectal touch.

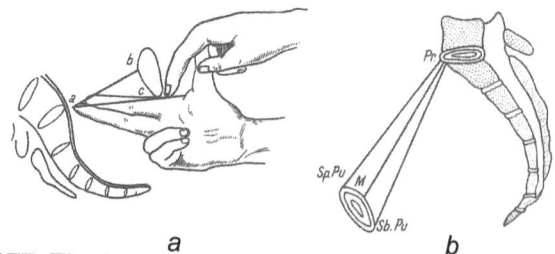

Fig. 377. The internal pelvimetry is practised by vaginal touch (a); the useful diameter, called also conjugata vera, measures normally 10.8 cm (b).

318

Instrumental Exploration of the Organs of the Trunk

Often, the diagnosis cannot be established with certitude on the basis of the clinical examination and therefore various instrumental explorations have been introduced in the routine practice as compulsory methods of investigation, of which we shall mention the most frequently used.

The electrocardiogram is the method of exploration by means of which lesions of the myocardium or of the coronary arteries are detected and the nature of heart rhythm disturbances is established. It presents as a tracing arbitrarily marked by the succesive letters of the alphabet P, Q, R, S, T and is based on the recording of voltage variations at the level of the myocardium at the different stages of the cardiac cycle. These voltage variations are produced by the depolarization and repolarization of myocardial cells. The P wave is generated by atrial depolarization. It has a duration of 0.11 seconds and measures 3 mm. The P-R interval represents the time between atrial depolarization and ventricular depolarization. The QRS deflection occupies the QRS interval of ventricular depolarization. The ST segment represents the end of ventricular depolarization and the beginning of repolarization. In this way, three phases are distinguished: the atrial phase, the phase of impulsion invasion within the ventricles and the withdrawal phase.

In pathological cases, the electrocardiographic aspect assumes various, specific forms, allowing to establish the diagnosis of heart disease (fig. 378, 379).

The angiography detects the presence of congenital or acquired heart diseases. It may be non-selective or selective, according to whether the contrast solution is introduced either into a peripheral vein, or through a catheter inserted into the superior or inferior cava system. The iodine solution at the temperature of the body and in the concentration of 70 percent is introduced with an automatic syringe under a pressure of 5-7 kg/cm^2. After 2-3 seconds is obtained the dextrogram, after 4-5 seconds the angiopneumography and after 7-8 seconds, the image of the pulmonary veins and of the left heart.

The cardiac catheterization permits the direct exploration of the chambers of the heart, the measuring of intracardiac (atrial and ventricular) and intravascular (pulmonary artery and pulmonary capillaries) pressures, the introduction of radiopaque and radioactive material, as well as blood sampling for the determination of gases in the blood of the right heart or of the pulmonary artery.

The opaque catheter is introduced by denudation of a vein at the bend of the elbow and is passed under roentgenologic control along the subclavian vein and the superior vena cava within the right heart. From here it may be pushed towards the right pulmonary artery. After its arrival at the desired point, small amounts of blood necessary for the determination of oxygen and carbon dioxide values are withdrawn and the intracavitary pressure is measured. The intracardiac pressures obtained furnish significant data for establishing the diagnostis, both in clinics of cardiology and of pneumology.

The method of cardiac catheterization, introduced by Forssman (1929), Cournard and Ranges (1941), was improved by Bradley and Grandjean, who recoursed to microcatheterization, using to this purpose a somewhat thicker needle for venous puncture, through the lumen of which is passed a very flexible, slender, catheter, with an inner diameter of 1 mm and measuring 130 cm in length. The catheter is borne along by the blood flow and pushed towards the chambers of the heart and the pulmonary artery. The procedure is well tolerated by the patient, it does not produce significant cardiac disturbances and may be repeated, if necessary (fig. 380, 381).

The bronchography is obtained by introducing a contrast medium (lipiodol) into the bronchial tree; it elucidates the nature of lessons localized in areas exceeding the possibilities of the bronchoscopy and permits thus the detection of lesions which are overlooked at the radiological examinations. It is indicated in all cases suspect of tumours, in bronchial dilatation, bronchial cavities and stenoses. It has the disadvantage of favouring infection and atelectasis (fig. 382, 383).

Before performing the bronchoscopy, a laryngoscopy examination is compulsory, to detect a possible lesion of the recurrent nerve (bitonal voice) and to exclude in this way an accusation which could be raised against the examining physician.

The bronchoscopy is the best method of exploration of the trachea and of the main and lobar bronchi. It is frequently used in the clinics of penumology and thoracic surgery for the elucidation of the nature of atypical pleuropulmonary opacities. To this purpose is used the fibrobronchoscope (fig. 384), which permits to photo graph these visual images, to extract foreign bodies included in the tracheobronchial tree, to remove pathological tissue for biopsy and even to electrocauterize certain benign bronchial tumours (fig. 385). For performing a bronchoscopy it is necessary to know the topography of the larynx, the distribution of the bronchial tree and the size of the orifice of passage of the bronchoscope.

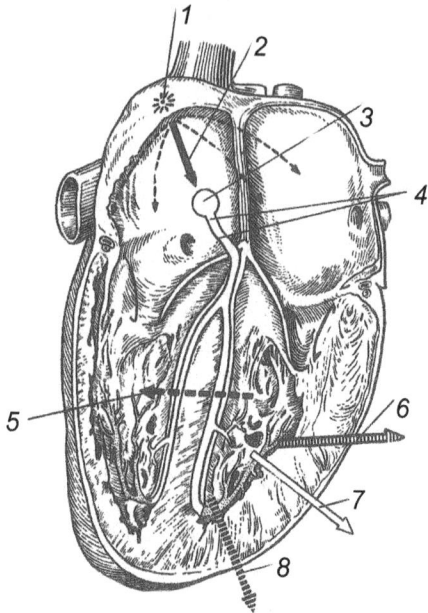

Fig. 378. Anatomical bases of the
electrocardiogram
1. Sinoatrial node; 2. Atrial depolarization
vector; 3. Atrioventricular node; 4. Bundle of
His;5 - Septal depolarization vector; 6. Vector
of the end of ventricular depolarization;
7. Repolarization vector; 8. Vector of the
beginning of apical and left ventricular
depolarization

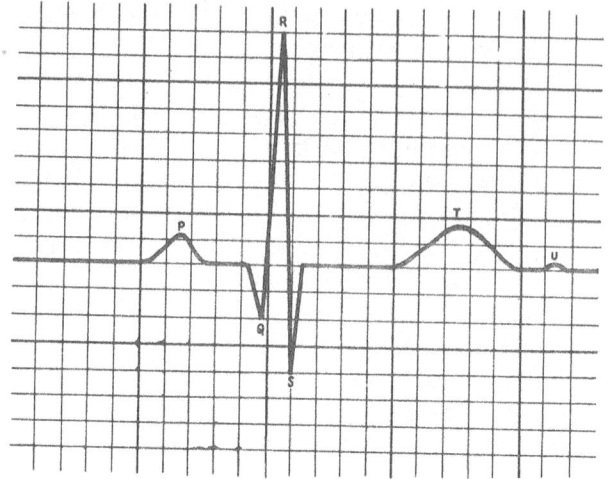

Fig. 379. Aspect of a normal electrocardiogram

The oesogastro-intestinal endoscopy has become a routine procedure in practice through introduction into usage of an instrument of high technical quality, which permits the detection of the finest lesions. The first employed in practice was Schindler's semiflexible gastroscope (1932), followed by Hirschowitz's fiberoptic gastroscope (1958), then by Overholt's colonoscope (1960) and nowadays by the fiberoptic instruments with "cold" lighting.

The gastro-intestinal endoscopy not only confirms the radiologically detected lesions, but permits also to make the diagnosis even in lesions affecting small areas and to establish the condition of the perilesional tissues. In addition, it allows to perform biopsies and even treatments (foreign body extractions, electrocauterizations of benign tumours eta) (fig. 386-390).

According to the site of the lesion, the procedure performed is an oesophagoscopy, a gastroscopy or a colonoscopy.

The oesophagoscopy permits the direct inspection of the lumen of the oesophagus. By means of the oesophagoscopy, the mucous membrane, normally of a pale colour, with longitudinal folds, that are smoothed out on the passage of the oesophagoscope, is explored. Moreover, the site, nature and extension of the lesion are established and, at the same time, biopsies or foreign body extractions, ulceration cauterizations and incisions of oesophageal abscesses are performed. Of a great utility is this method in establishing the diagnosis of hiatal hernia. Barrett's oesophagitis, oesophageal varices, oesophageal strictures and oesophageal carcinomas (fig. 396, 387).

The gastroscopy and *the gastrophotography* are extremely useful methods, often very necessary in the study of the gastric mucosa with the purpose of establishing the diagnosis, of performing the biopsy or of extracting foreign bodies. The site, form, extension and condition of the surrounding mucosa in gastric ulcer, the nature, form and number of gastric polyps, the severity of gastric cancer etc. are established with certitude (fig. 386, 387).

The duodenal endoscopy permits to make the diagnosis of duodenal ulcer and duodenitis, which may be concealed at the routine radiological examination. If detects also the duodenal polyps and diverticula, the carcinoma of Voter's papilla, the pancreatic carcinoma, etc (fig. 388).

The colonoscopy makes possible the direct exploration of the whole colic frame, the terminal ileon inclusive.

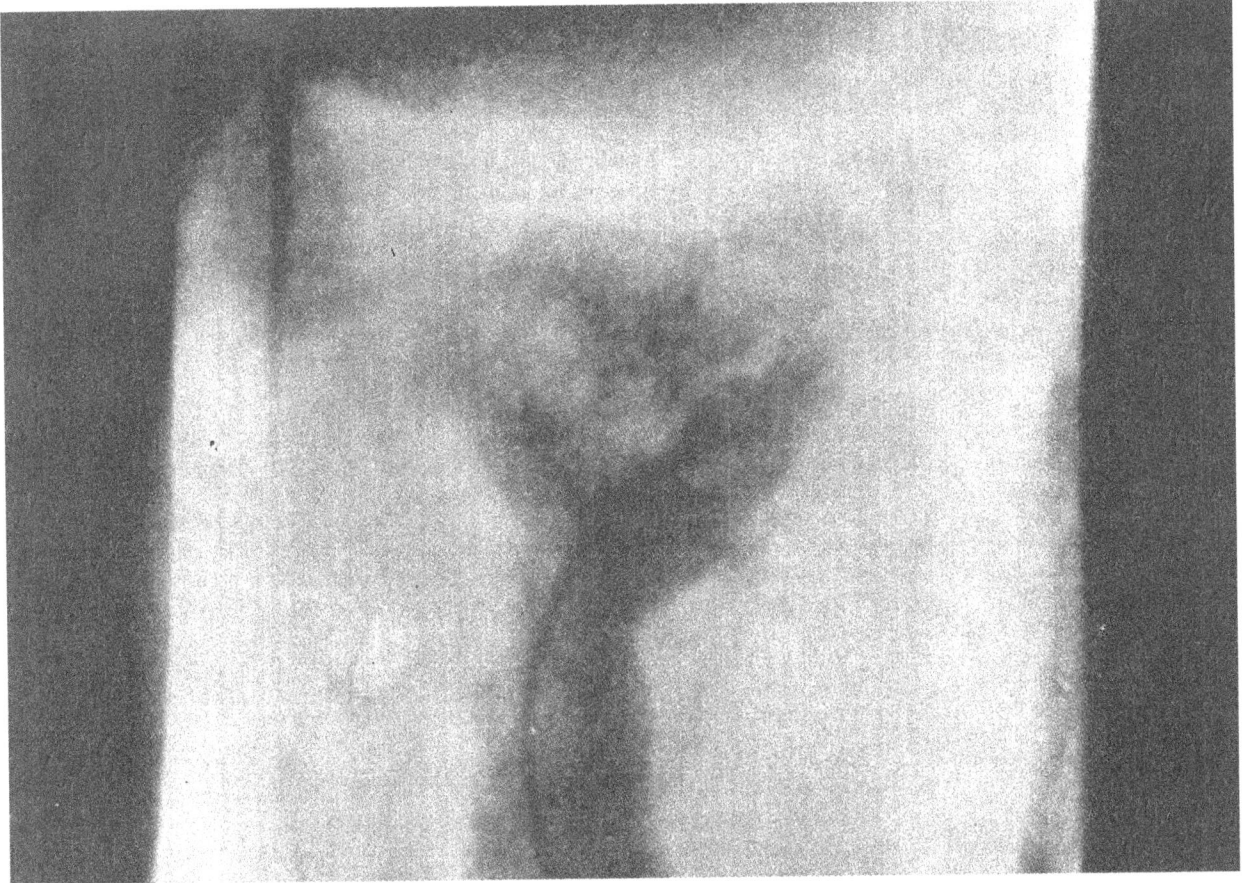

Fig. 380. Cardiac catheterization: the contrast medium has penetrated into the ventricle (ventriculography).

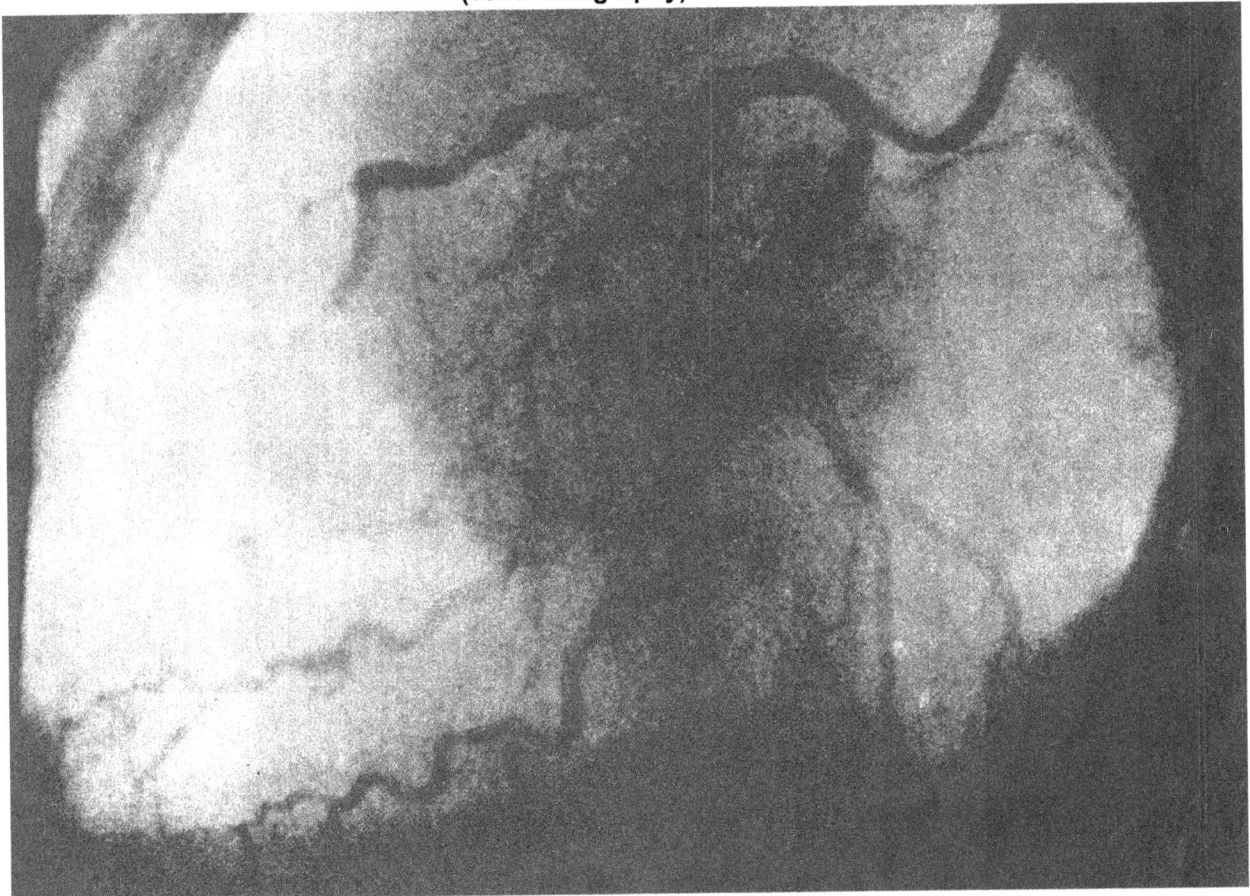

Fig. 381. Cardiac catheterization: the contrast medium has penetrated into the coronaries (coronorography).

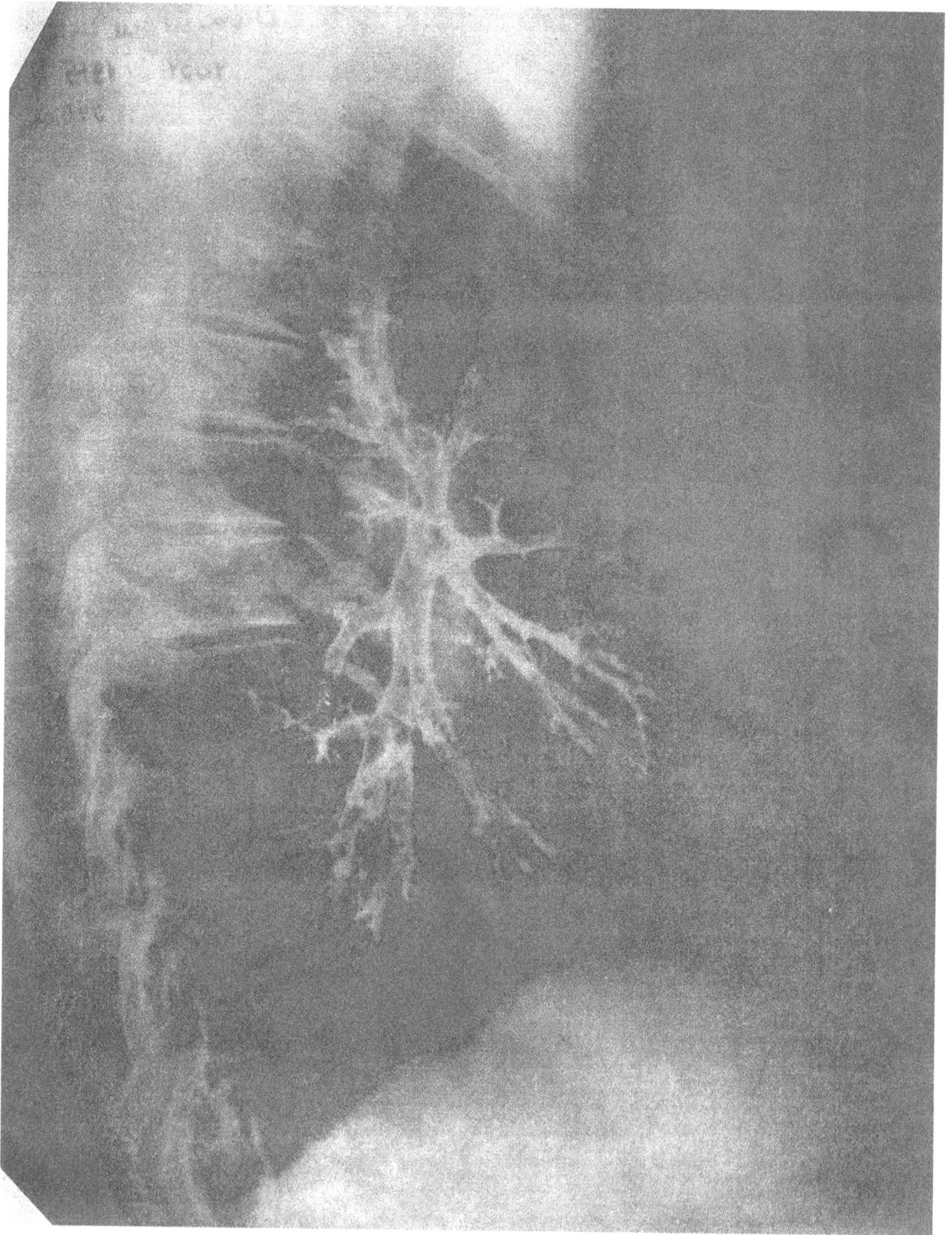

Fig. 382. Bronchography: left lung and branching of the bronchial tree (lateral view).

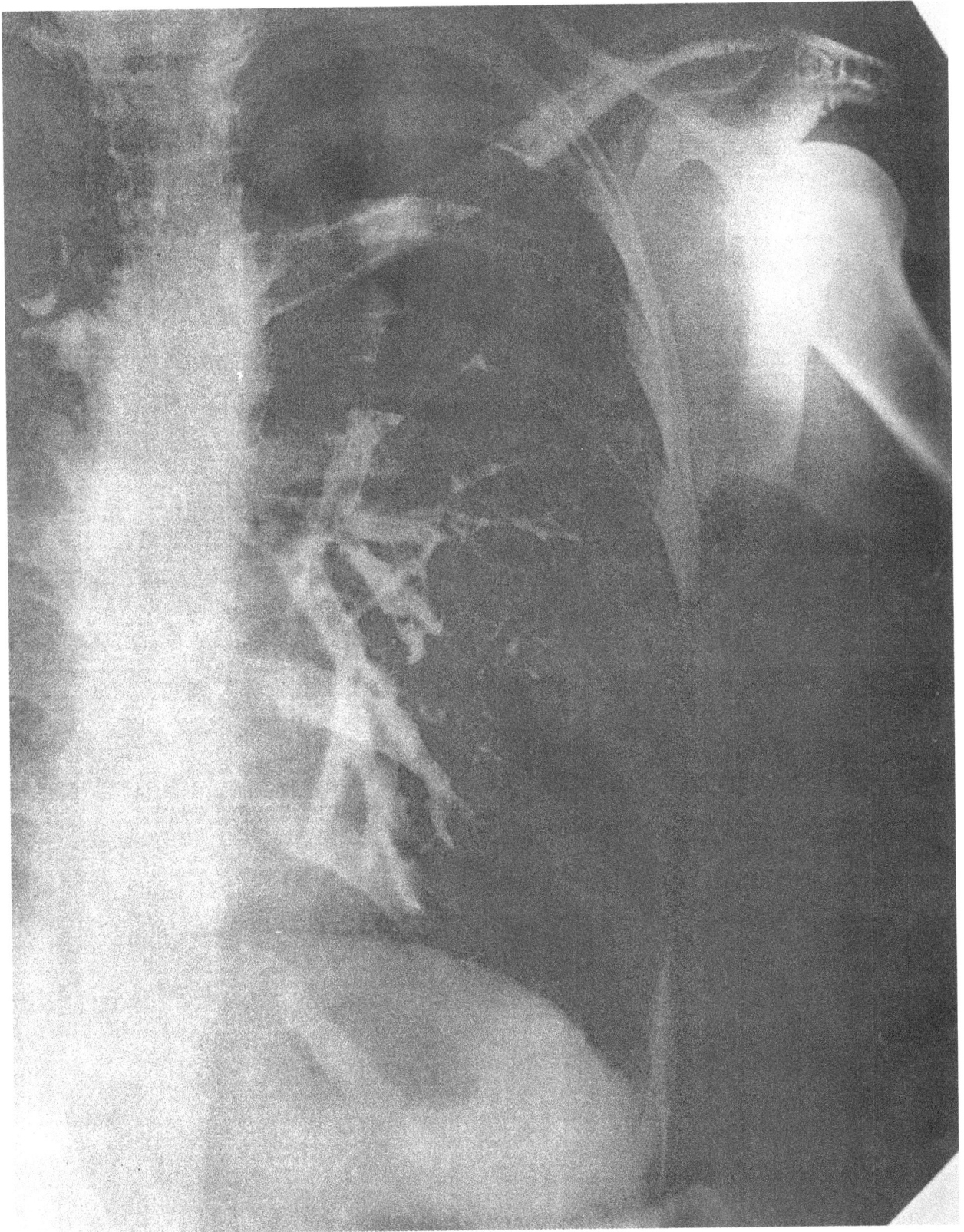

Fig. 383. Bronchography: left lung and branching of the bronchial free (anterior view).

The colonoscope is available under two sizes: the average model, 130 cm in length, and the long model, measuring 180 cm. By its help are established the diagnosis of inflammatory diseases of the colon, the nature, number and site of diverticula, the presence of non - neoplastic, hyperplastic or neoplastic polyps, the presence of a carcinoma of the colon, of angiodysplasiae etc. (fig. 389, 390)

The rectosigmoidoscopy allows a detailed exploration of the rectum and the sigmoid colon. The position for exploration is genupectoral, the chest resting on the examination table. The examination instrument should be abundantly smeared with vaseline. First, a direction is given as if we want to reach the umbilicus; after having penetrated sufficiently deep, the mandrin is removed and the examination begins. At the level of the striated sphincter, the lumen appears round, with radiating striae. At a depth of 2.5-3 cm., the striae become less frequent, but deeper, the rectal ampulla appears open and much longer than the rectoscope, being furrowed by transversal folds. At 12-14 cm, in front of the rectoscope appears the rectosigmoid valve. Immediately after penetration into the sigmoid, the aspect changes: small folds appear, the rectoscope advances with more difficulty and there may be observed elongation, congestion and oedema, ulcerations of various forms, bleeding or covered with false membranes, strictures and tumours (fig. 389, 390).

ENDOSCOPIC ASPECT

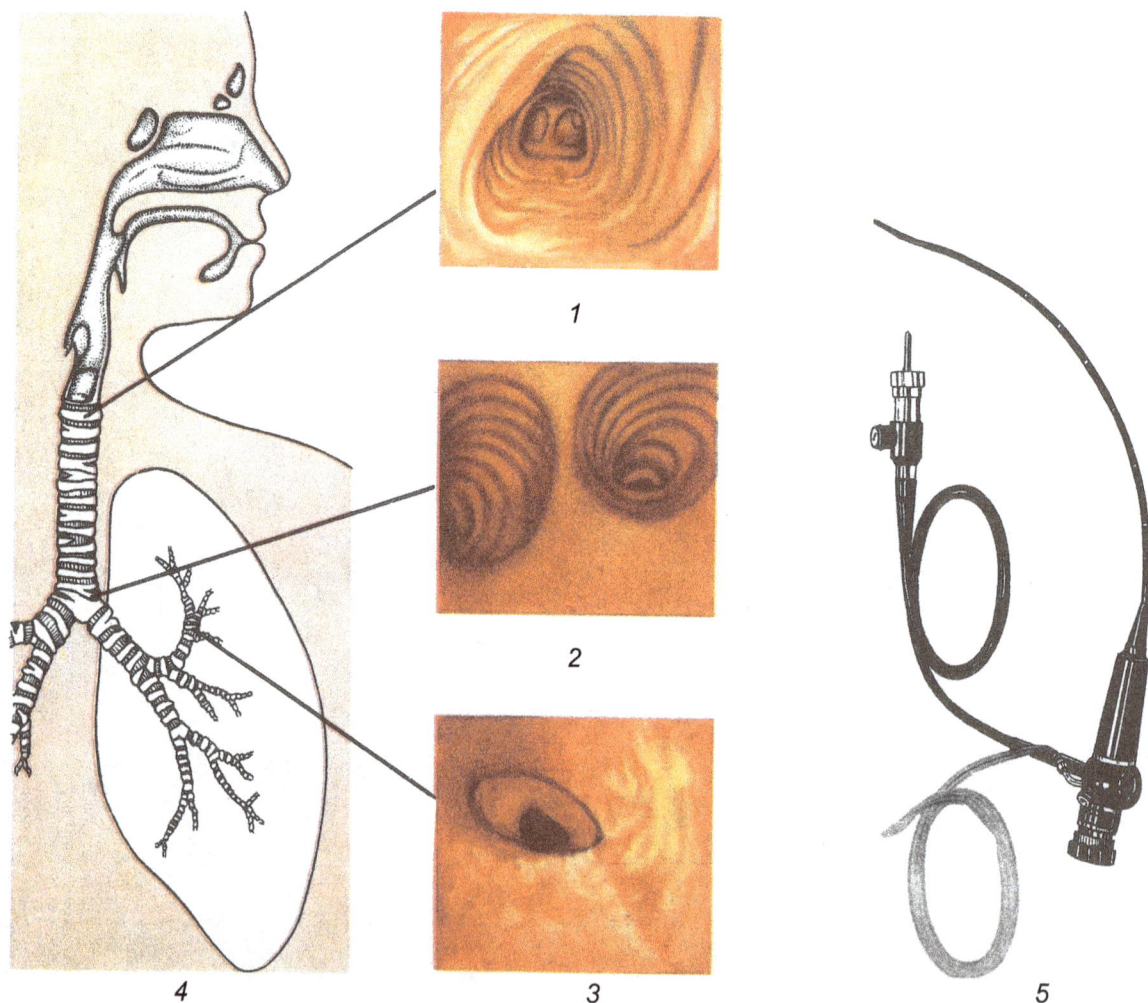

Fig. 384. Tracheobronchial endoscopy: 1) ephglottis; 2) tracheal carena with the emergence orifices of the main bronchi; 4) upper airways and distribution of the tracheobronchial tree; 3) division site of the upper lobar bronchus; 5) fibroscope with cold lathing.

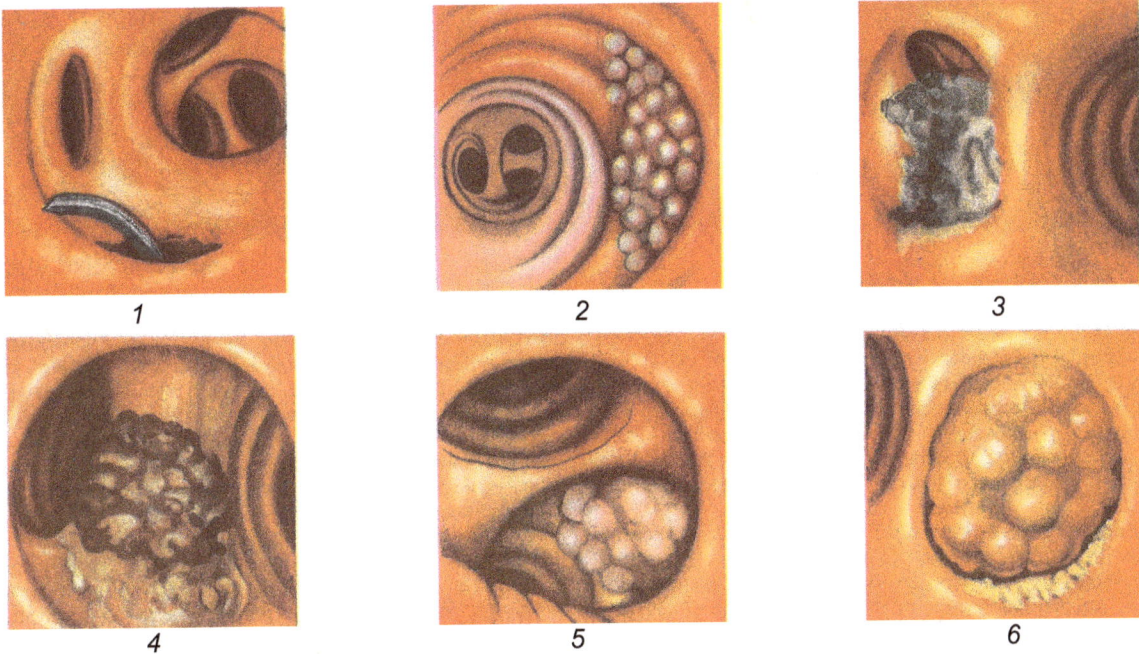

Fig. 385. Bronchoscopic images (1-6) 1. Foreign body in the apical bronchus of the right lower lobe; 2) Inflammatory bronchial tumour similar to tuberculosis; 3) Tuberculous ulcerative image of the left bronchus; 4) Malignant tumour at the level of the left bronchus - epithelioma, vegetative form; 5) Benign tumour of the right lower lobe; 6) Endoscopic image in Hodgkin's disease with bronchial location.

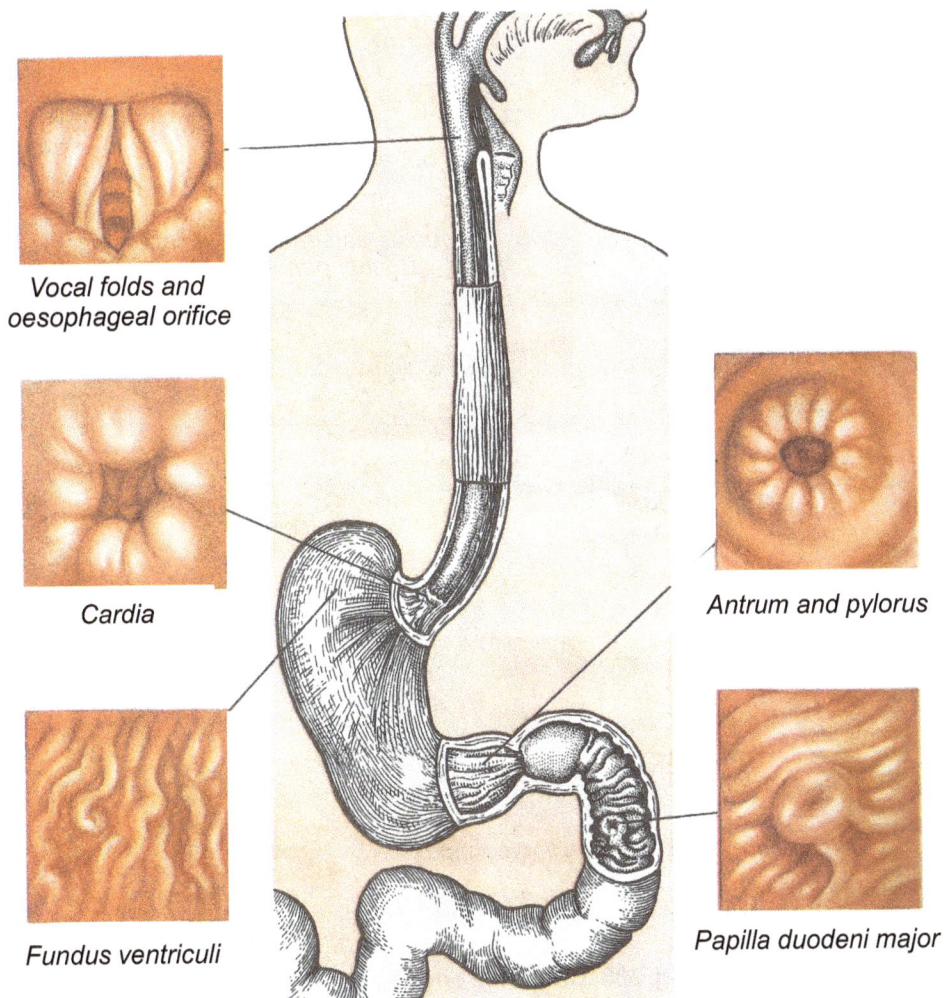

Vocal folds and oesophageal orifice

Cardia

Fundus ventriculi

Antrum and pylorus

Papilla duodeni major

Fig. 386. Oesogastro-intestinal endoscopy

325

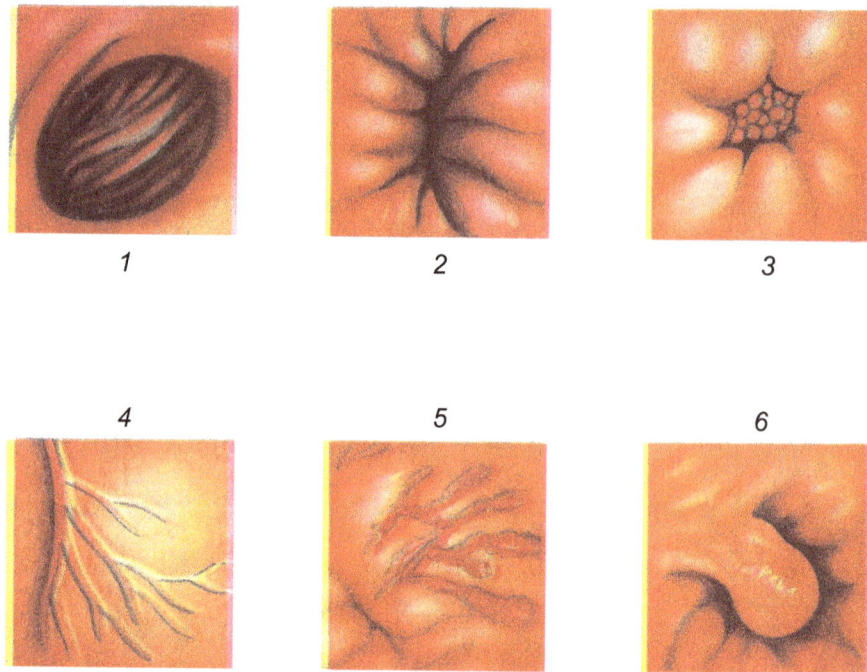

Fig. 387. Oesogastroscopic images (1-6) 1) Foreign body (kernel of almond) wedged in the oesophagus; **2)** Esophageal stricture due to the presence of a malignant tumour; **3)** cinoma at the level of the cardia; **4)** Atrophic gastritis with display of the vessels in case of hyperchromic anaemia; **5)** Lesional aspect in Malory- Weiss disease; **6)** Gastric polyp

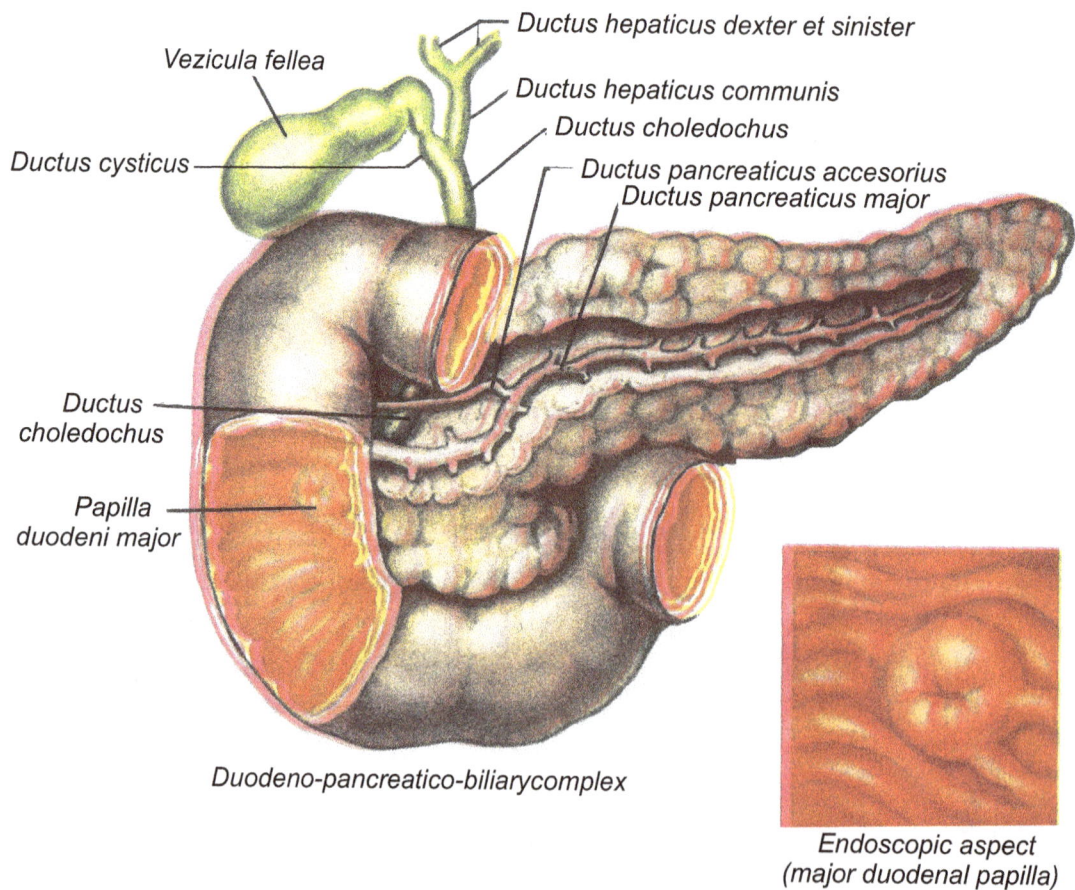

Ductus hepaticus dexter et sinister

Vezicula fellea

Ductus hepaticus communis

Ductus choledochus

Ductus cysticus

Ductus pancreaticus accesorius
Ductus pancreaticus major

Ductus choledochus

Papilla duodeni major

Duodeno-pancreatico-biliarycomplex

Endoscopic aspect (major duodenal papilla)

Fig. 388. Duodenal endoscopy

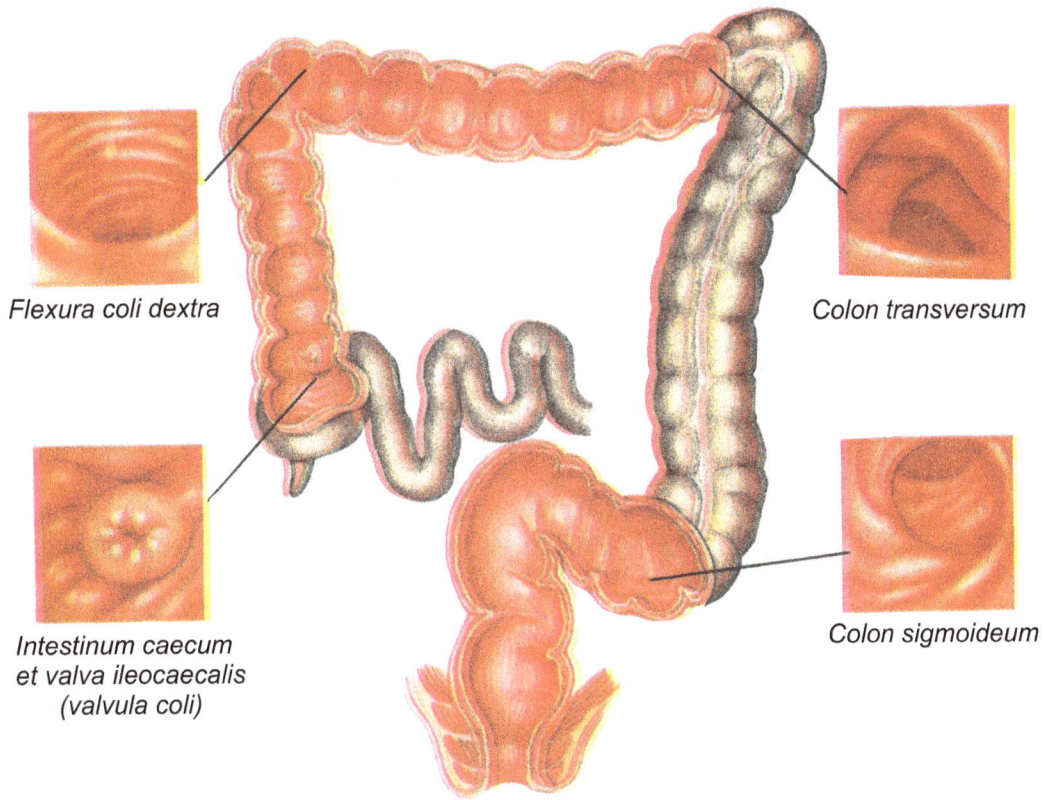

Flexura coli dextra

Colon transversum

*Intestinum caecum
et valva ileocaecalis
(valvula coli)*

Colon sigmoideum

Fig. 389. Colonoscopy

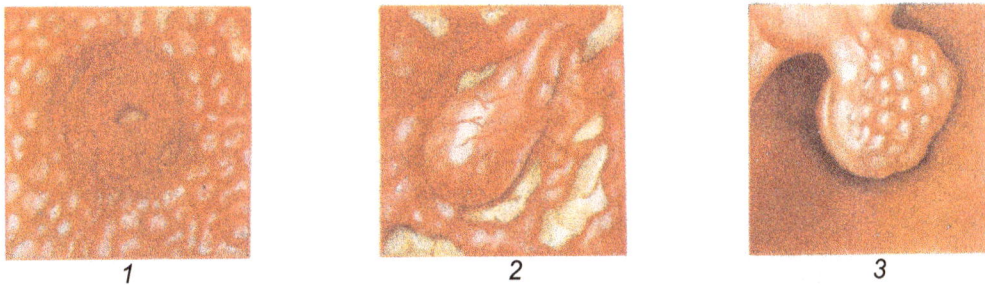

1

2

3

**Fig. 390. Colo-endoscopic images 1. Acute colitis; 2) Pseudopolyp; 3) Neoplastic lesion
on a colitic background.**

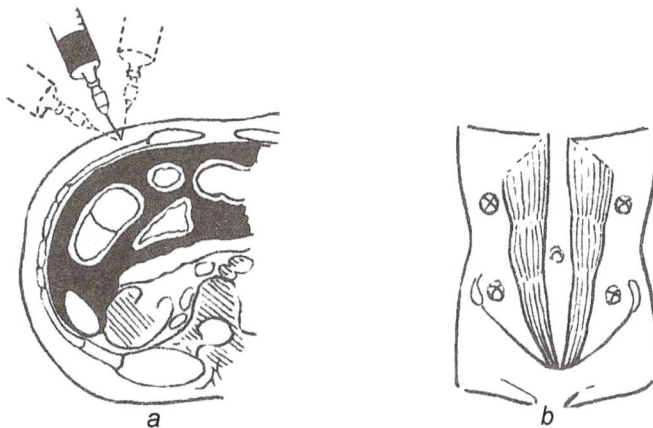

a

b

**Fig. 391. Spots used for the abdominal puncture a - anterior abdominal wall; b -
transverse section through the abdominal region with display of the puncture site
and technique.**

327

The anoscopy, in addition to the rectal touch, permits the exploration of the terminal sigmoid and of the anal canal.

The abdominal puncture is often used in current practice, in emergencies, in order to establish the presence and nature of the intra-abdominal fluid (blood, pus), as well as in chronic cases for the extraction of the ascitic fluid (fig. 391).

The laparoscopy is used in Europe by surgeons and gastroeoferologists and in USA by gynecologists; it permits to establish the diagnosis of various intraabdominal diseases, especially of the liver, often sparing the patient the exploratory laparotomy(fig. 392, 393).

The exploration of the urethra will always be preceded by an antiseptic lavage of the urethral canal. Then, under absolutely aseptic conditions, the olive-tip catheter No. 16 or 18 (fig. 394J or a Benique sound No. 40 is introduced. By means of this manoeuvre the calibre and the elasticity of the urethra, as well as the presenceof possible obstacles (polyps, calculi, strictures etc.) are assessed. In women, either the soft catheter (especially in pregnant women) or the metallic sound will be used.

In special cases, the indwelling balloon catheter (Foley) will be inserted for a variable lapse of time, talcing into account the possibilities of later infection (fig. 393).

The urethroscopy is performed with the urethroscope with direct or indirect viewing (Mac Carty). Most frequently the Mac Carty urethroscope is used, since it permits to inspect concomitantly the posterior urethra, the neck of the bladder and the trigone (urethrocystoscopy).

The urethroscopy allows to establish a reliable urologic diagnosis (fig. 396) *The cystoscopy* is the method by means of which the exploration of the urinary bladder is performed. To this purpose there may be used: the simple cystoscope, which permits only the investigation of the bladder; the cystoscope for unilateral or bilateral catheterization when simultaneously the ureter is explored (for a diagnostic purpose or for pyelography); the cystolithotriptor, with the aim of crushing an eventual vesical calculus.

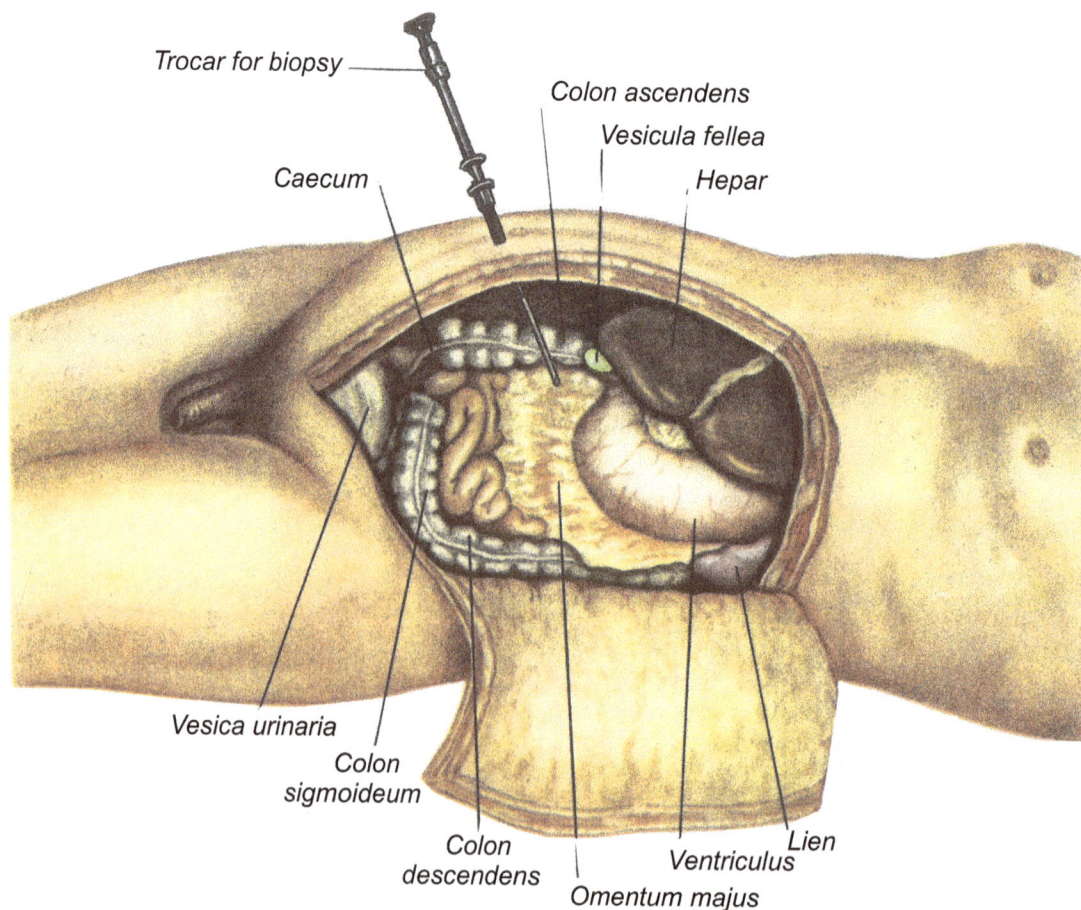

Trocar for biopsy

Colon ascendens

Vesicula fellea

Caecum

Hepar

Vesica urinaria

Colon sigmoideum

Colon descendens

Ventriculus

Omentum majus

Lien

Fig. 392. Laparoscopy: technique of performing a puncture; normal appearance of the abdominal viscera.

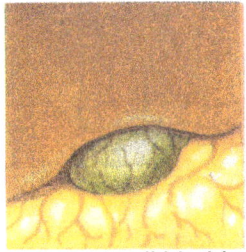

Liver and gall bladder   Liver cirrhosis   Heptaic metastases   Adhesions

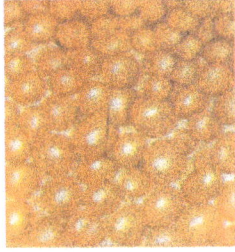

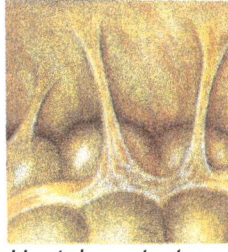

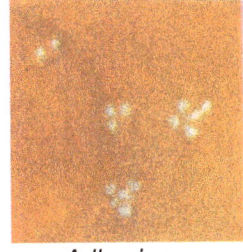

Fig. 393. Pathological images obtained by laparoscopy.

The requirements to perform a cystoscopy are a good permeability of the urethra, a certain degree of vesical capacity and good clearness of the medium introduced within the bladder.

After introduction of the cystoscope, the examination should be as complete and methodical as possible. For orientation in the urinary bladder, it is necessary to recognize three guiding marks: the air bubble, the ureteral orifices and the neck of the bladder. The air bubble results from the air introduced with the distension fluid within the bladder and is situated at the apex of the bladder. Through the rotation movement of the cystoscope around its axis, at "4 hours" and "8 hours" (like the hands of a clock), the left and right ureteral orifices are displayed (fig. 397). By slightly withdrawing the cystoscope, the neck of the urinary bidder (the trigone of the bladder) is visualized, too.Normally, the vesical mucous membrane is glossy, yellow pink,darker at the level of the trigone and crossed by slender arborizations. By means of cystoscopy, congenital or acquired deformities, inflamatory lesions, the presence of eventual calculi, foreign bodies, tumours etc. may be studied (fig. 398). *The chromocystoscopy* is performed after a previous cystoscopy and intravenous introduction of 4 ml of a 4 percent indigo-carmine solution. In the case of a normal renal function, the elimination of the dye through the ureteral orifice begins 6-8 minutes after the injection of the substance (fig. 399). *The pyelography* is performed after an ureterocystoscopy, by introduction of the contrast medium along the ureteral catheter (fig. 400).To the same extent are used also the roentgenographic examinations of these viscera. In clinical investigation for establishing the diagnosis, the oesophagography (fig. 401), the gastroduodenography (fig. 402), the colonography (fig. 403) or the pyelography (fig. 404) are of a great utility

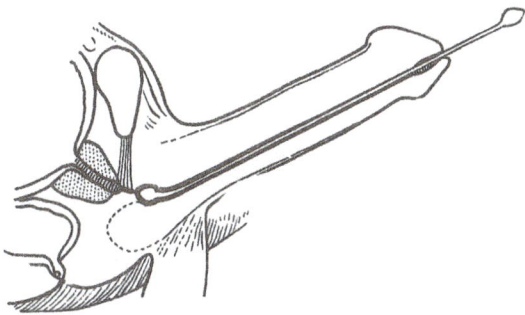

Fig. 394. Exploration of the urethra by means of the olive-tip catheter

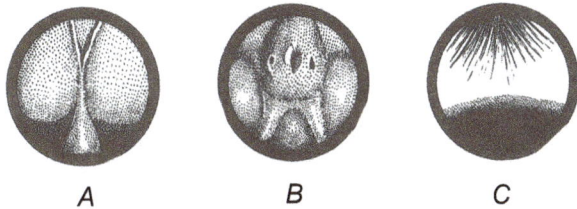

A   B   C

Fig . 396. Urethroscope images (normal urethra) A - supracollicular fossa; B - verumontanum; C - urethral crest.

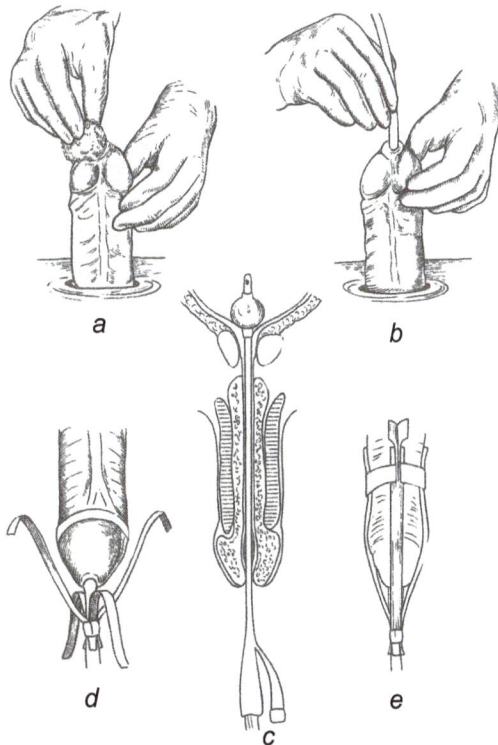

a   b

d   e

c

Fig. 395. Urethral catheterization technique in man a - lavage of the urethral orifice with a sterile solution; b - introduction of the catheter under aseptic conditions; c - the Foley balloon is inflated;d - preparation of the catheter fastening; e - the catheter is fastened

1

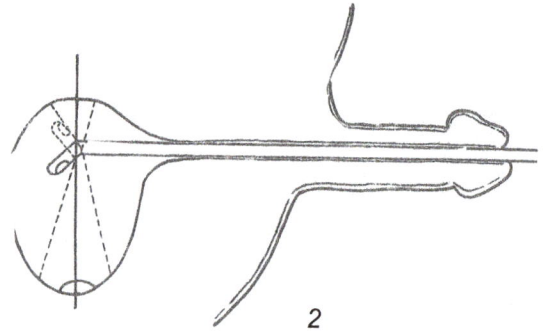

2

Fig. 397. Cystoscopy. 1 - direction of the site of the urethral orifices; 2 - detection of the air bubble at the level of the bladder; 3 - cystoscopic aspect: A - right urethral orifice; B - interureteral ridge; C - left urethral orifice

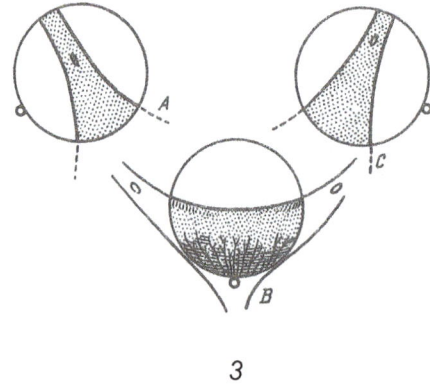

3

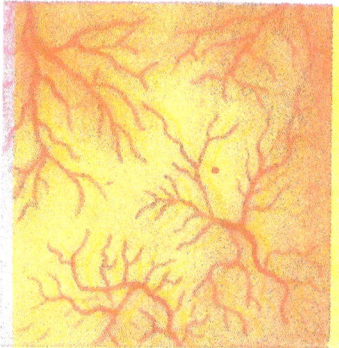

More strongly marked vascular outlone

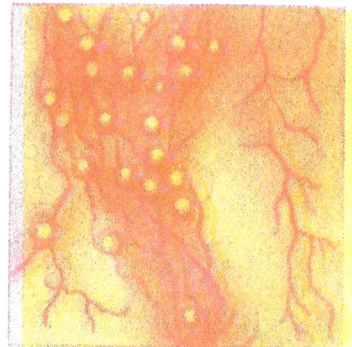

Presence of tuberculous granulations

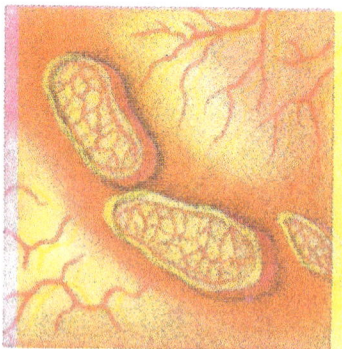

Presence of tuberculous ulcerations

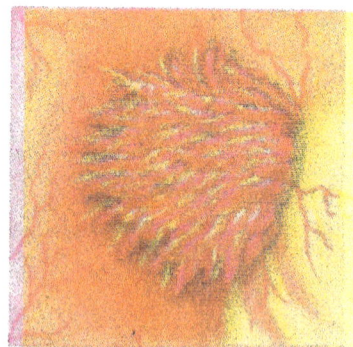

Vesical epithelioma

Fig. 398. Cystoscopic aspects

The dye is eliminated through the ureteral orifice.

Fig. 399. Chromocystoscopy

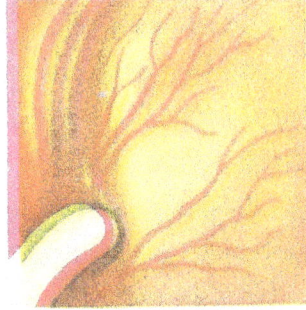

The ureteral cather penetrates through the ureteral orifice.
Normal appearance of the mucosa

Fig. 400. Ureteral catheterization

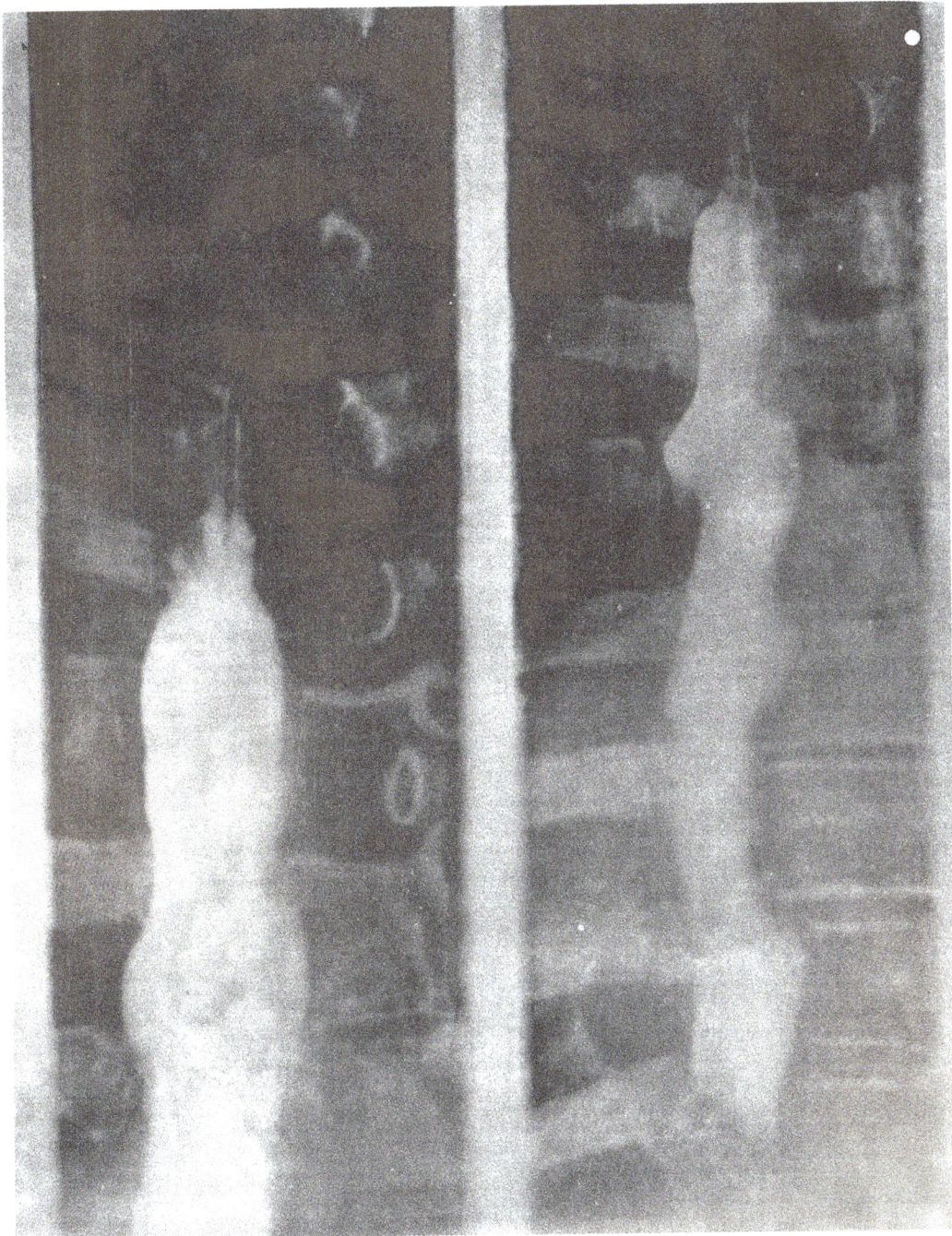

Fig. 401. Oesophagography (front and profile)

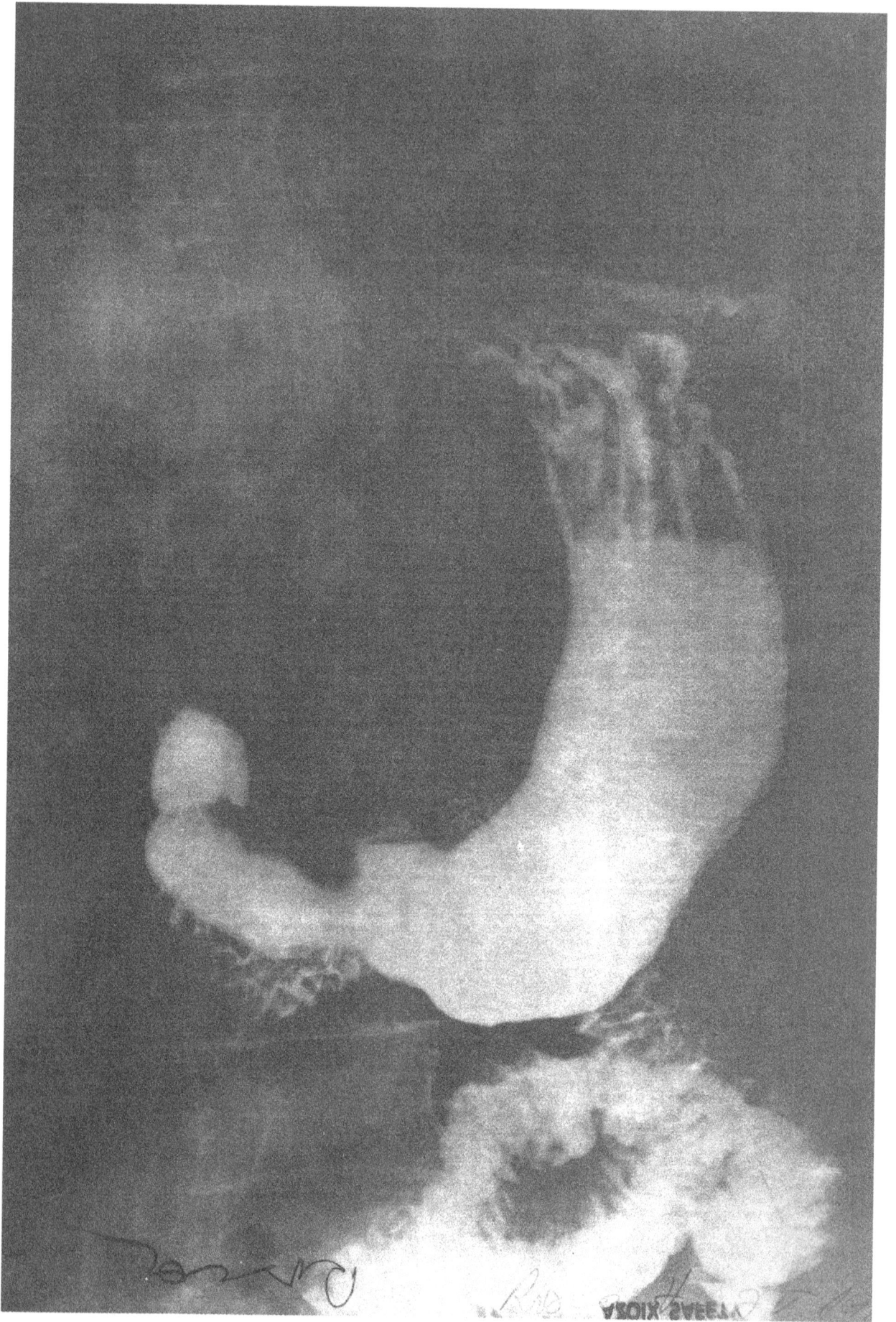

Fig. 402. Gastroduodenal radiography.

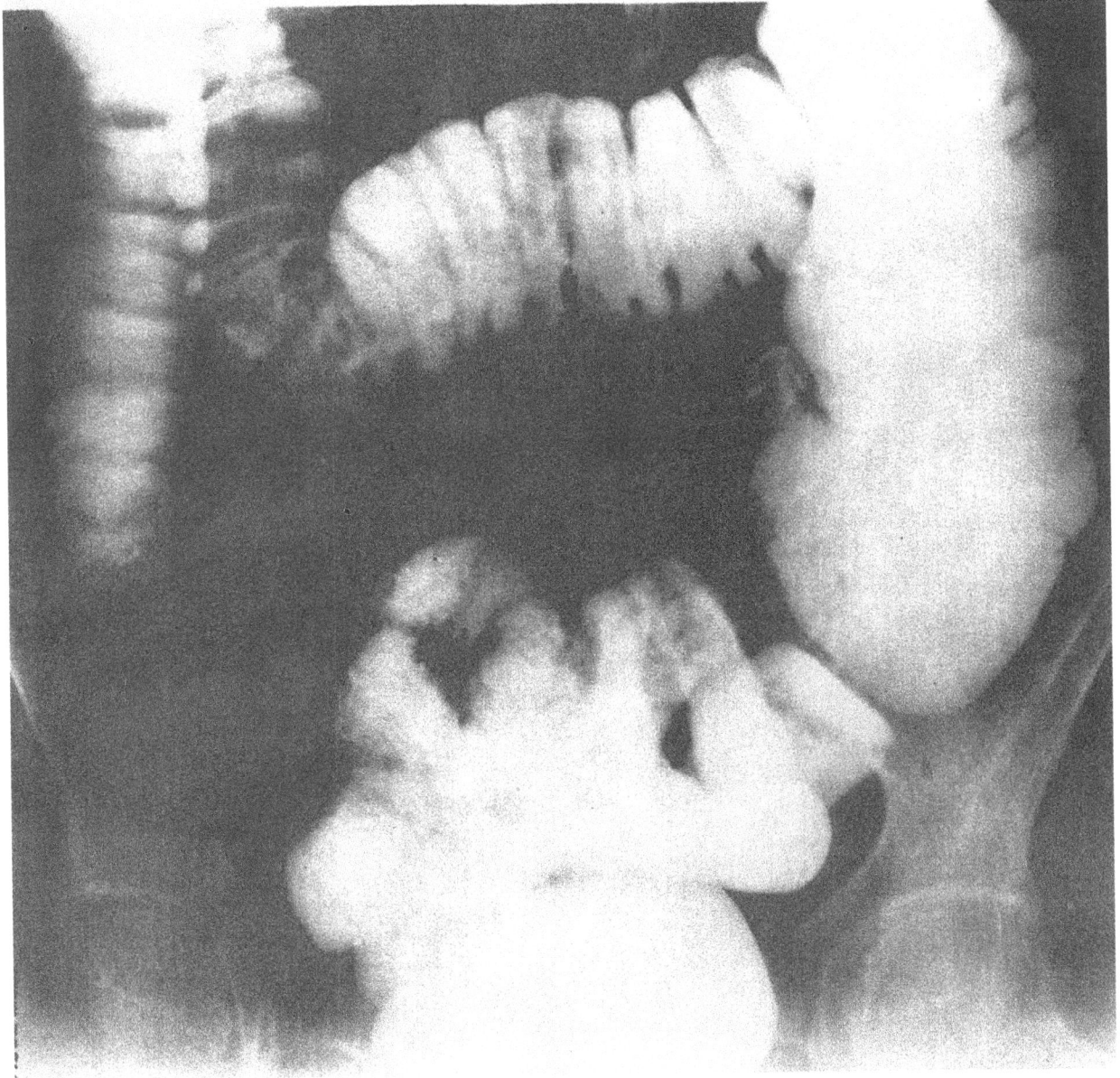

Fig. 403. Radiography of the colon

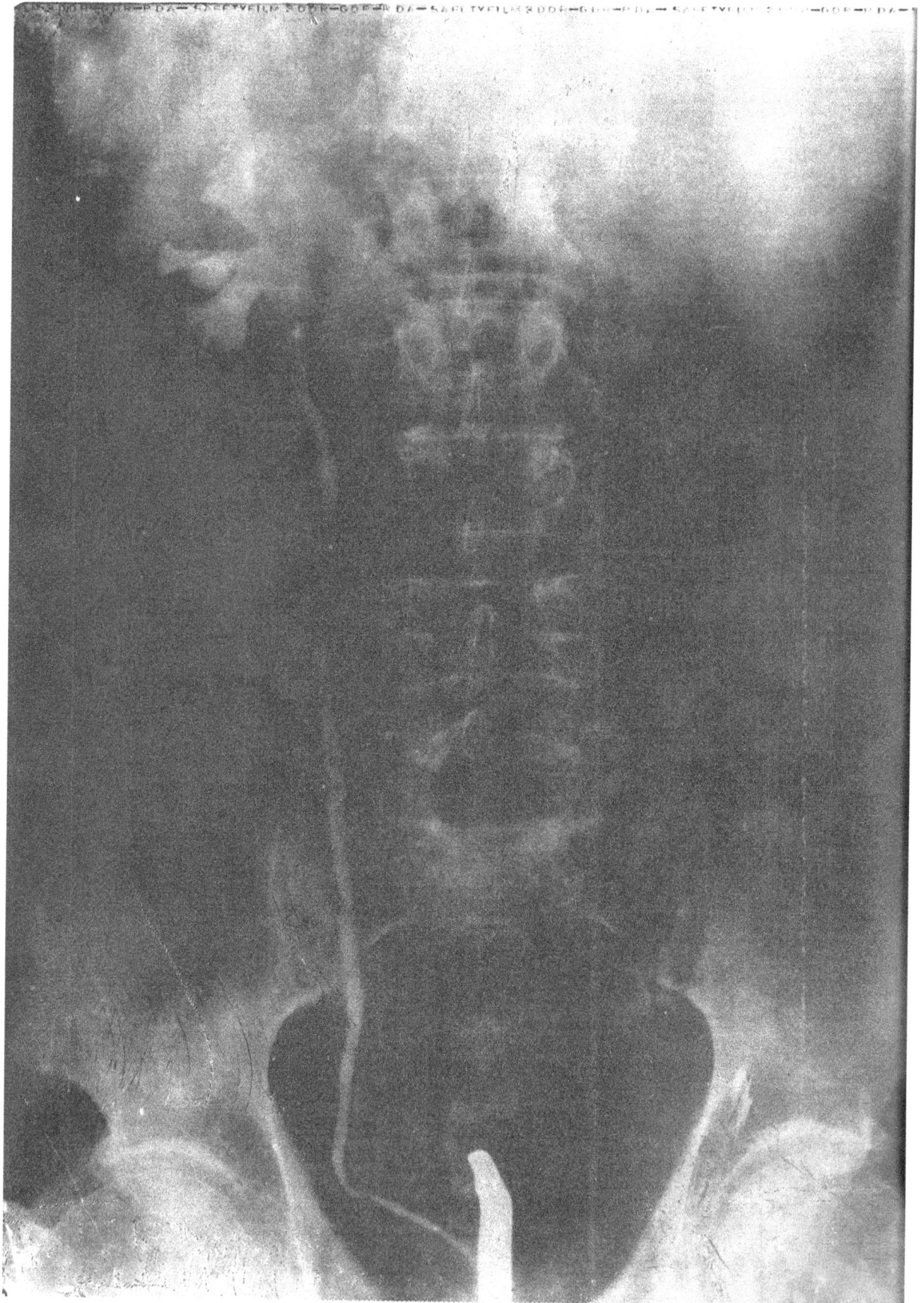

Fig. 404. Pyelography. The cystoscope is introduced within the urinary bladder and the contrast medium which opacifies the ureter and the upper urinary tract is injected through the ureteral catheter. The presence of a right pyelic calculus may be observed.

www.
Lightn
Chamb com/pod-product-compliance
CBHW C
41598C
42